CONTRIBUTING AUTHORS

LYNN S. BEEDLE, *Lehigh University*

JACK H. BLACKMON, *C. F. Braun and Co.*

PETER B. COOPER, *Kansas State University*

GEORGE C. DRISCOLL, JR., *Lehigh University*

WILLIAM J. ENEY, *Lehigh University*

SAMUEL J. ERRERA, *Bethlehem Steel Corporation*

FIORELLO R. ESTUAR, *Estuar and Associates (Philippines)*

JOHN W. FISHER, *Lehigh University*

THEODORE V. GALAMBOS, *Washington University*

WILLIAM C. HANSELL, *Bethlehem Steel Corporation*

MAXWELL G. LAY, *Broken Hill Proprietary, Ltd. (Australia)*

VICTOR LEVI, *University of Panama*

LE-WU LU, *Lehigh University*

A. OSTAPENKO, *Lehigh University*

HAROLD S. REEMSNYDER, *Bethlehem Steel Corporation*

JOHN L. RUMPF, *Temple University*

ROGER G. SLUTTER, *Lehigh University*

LAMBERT TALL, *Lehigh University*

B. T. YEN, *Lehigh University*

STRUCTURAL STEEL DESIGN

SECOND EDITION

LAMBERT TALL

Lehigh University

EDITOR

JOHN WILEY & SONS
New York • Chichester • Brisbane • Toronto

ISBN 0 471 06674-5

Library of Congress Catalog Card Number: 72-96969

PRINTED IN THE UNITED STATES OF AMERICA

10 9 8 7 6 5 4 3 2

PREFACE TO THE SECOND EDITION

Since the first edition of this book was published a decade ago, many changes have occurred, not in the basic structural principles on which the book is based, but in the application of these principles. A new awareness has developed that no part of structural engineering is an entity unto itself. To this new awareness this book is dedicated.

The purpose remains the same—the design of structural steel members and frames from the standpoint of an understanding of the basic behavior of structures. The impetus for the first edition, which received wide acceptance and use, was provided by the major revision in 1961 of the specifications of the American Institute of Steel Construction. These specifications have received a number of changes over the years, with a major revision in 1969 and minor changes since then. In the past few years, there have been major revisions also in the specifications of the American Association of State Highway Officials, the American Railroad Engineering Association, and the American Welding Society. The preparation of the second edition of this book has been prompted not only by these changes, but by advances in the understanding of structural behavior.

The authorship remains the same, but now reflects an even broader background of engineering practice, research, and education, as the outlook and responsibility of the contributors have diversified.

The titles of chapters and the overall arrangement of the book also remain the same. The contents have been revised to reflect recent developments in research, design, and specifications.

The excellent assistance given by many individuals and organizations during the preparation of the first edition has continued. In addition to those who helped at that time, the following have assisted with this edition: Göran Alpsten (Swedish Institute of Steel Construction), Omer Blodgett (Lincoln Electric Company), A. C. Davenport (University of Western Ontario), T. C. Kavanagh (Praeger-Kavanagh-Waterbury), W. A. Litle (Massachusetts Institute of Technology), and C. K. Yu (Praeger-Kavanagh-Waterbury).

The manuscript of the second edition was typed by Mrs. Linda Welsch, Mrs. Grace Mann Roberts, and Mrs. Jean Neiser; their assistance with many phases of the preparation of the manuscript is appreciated. Many organizations have given permission to reproduce quotations, graphs, tables, and photographs; this is appreciated and credit is given in the text at the appropriate places.

LAMBERT TALL
EDITOR

Bethlehem, Pennsylvania
January, 1974

PREFACE TO THE FIRST EDITION

This book presents the design of structural steel members and frames from the standpoint of an understanding of the basic behavior of structures. Such a basis includes the properties of steel, the behavior of the component parts of a structure, and the behavior of the complete structure as a unit. The attempt has been made to relate design specifications to the basic behavior of structures, and to show how such specifications may be used in the solution of practical design problems.

The book is intended to provide a comprehensive source of information to explain fundamental behavior. As such, the book can be used on the one hand by students in colleges and universities, and on the other hand by researchers and by engineers in practice. Numerous references have been appended for each chapter, especially in those cases where it has not been possible to give a fairly complete outline of theory and confirming experiments.

Both allowable stress design and plastic design concepts are treated. The former, which is based on specified working loads and allowable stresses, is contrasted in many cases with the latter which is based on the ultimate load of the element or structure. Inherently, allowable stress design procedures utilize many of the concepts of plastic design, and the adoption of plastic design methods into specifications has emphasized this fact. Throughout, an attempt has been made to give the best available technique for describing the behavior of a steel member or frame, to show the extent to which this behavior is confirmed by experiment, to present design approximations (specification provisions), and to illustrate the design procedures with examples. Both bridge and building specifications are considered, but with somewhat greater emphasis on the latter.

The book is based on a set of notes originally prepared for a weekly seminar conducted by the staff of the Civil Engineering Department of Lehigh University in the spring semester 1962. (At the time, all of the authors were associated with Lehigh University.) The purpose of the seminar was to present to students and to practicing engineers the subject of steel design with particular emphasis on the behavior of structural members and frames. The impetus for both the seminar and the notes was provided by the publication of a revised American Institute of Steel Construction Specification in November, 1961.

The book is arranged in six parts. The first part is on structural behavior and materials: Chapter 1 provides information on the concepts of design; Chapter 2 presents the properties of the materials used in steel structures. The second part of the book deals with structures and presents the initial steps in design: Chapters 3 and 4 present various aspects of the design of buildings and bridges from the initial stages of selecting the type, through preliminary designs and exact analysis to the final design procedure; Chapter 5 outlines the use of models in the design of frames.

The third part of the book is on the design of individual structural members. Chapters 6 through 13 consider tension members, beams, plate girders, compression members (subjected to axial compression or to combined bending and compression), light-gage cold-formed members, and composite steel-concrete members. Consideration is given to the effect of torsion, lateral buckling, instability, and residual stresses. The factor of safety is discussed. The elastic, buckling, and post-buckling behavior of members is explained.

The fourth part of the book covers Chapters 14 through 17, which are concerned with the process of welding, the phenomenon of brittle fracture, design for fatigue, and the limitations required to prevent local buckling. The fifth part, Chapters 18 and 19, deals with riveted, bolted, and welded connections.

The sixth part is on frames. Both allowable stress and plastic design concepts are presented. Chapter 20 considers single-story rigid frames, and Chapters 21 and 22 consider the design of multi-story frames.

Many chapters contain brief historical notes showing the development of design concepts. The material of the book has been arranged in such a way that the basic behavior of structures and the basis for design rules may be studied or used first without reference to any specific code or specification. Hence, although the major United States specifications are referred to and used in design examples, any other specification or code from the United States or elsewhere may be used to illustrate local conditions where necessary.

A book of this magnitude would not have been possible without the assistance of many persons and organizations. Acknowledgment is first due to those organizations that have sponsored research pertinent to this topic at Lehigh University—research which is reflected throughout and which is the basis for the content of this book. These include the American Institute of Steel Construction, the American Iron and Steel Institute, the American Railway Engineering Association and the Association of American Railroads, the Association of Iron and Steel Engineers, Bethlehem Steel Company, the Column Research Council, Engineering Foundation, Fort Pitt Bridge Works, Institute of Research of Lehigh University, Modjeski and Masters, National Science Foundation, the Pennsylvania Department of Highways, Research Council on Riveted and Bolted Joints, the United States Department of Commerce Bureau of Public Roads, the United States Department of Navy (Bureau of Ships, Bureau of Yards and Docks, Office of Naval Research), the United States Steel Corporation, and the Welding Research Council.

Appreciation is due to Bruce G. Johnston (University of Michigan) who reviewed the seminar notes and the complete manuscript, to Theodore R. Higgins (American Institute of Steel Construction) and Konrad Basler (Consulting Engineer, Switzerland) who reviewed the seminar notes and portions of the manuscript, and to Omer Blodgett (Lincoln Electric Company), Henry J. Degenkolb (Consulting Engineer, California), Frederick H. Dill (United States Steel Corporation), Frank E. Fahy (Bethlehem Steel Company), Gerard F. Fox (Howard, Needles, Tammen and Bergendoff, New York), William G. Kirkland (American Iron and Steel Institute), Robert D. Stout (Lehigh University), and George Winter (Cornell University) for their reviews of portions of the manuscript. Their encouragement and contributions have been of great help. Jackson L. Durkee and E. T. Moffett (both Bethlehem Steel Company) were of great assistance in obtaining photographs.

The assistance of the authors' colleagues who helped with the design examples and drawings is appreciated. Special thanks are due to William B. Seaman and Ching-Kuo Yu for their assistance with the preparation of the nomenclature, references, and index, and for their careful check of the manuscript.

The manuscript was typed by Miss Grace Mann and Mrs. Rita Tall; their assistance with many phases of the preparation of the manuscript is appreciated. Special acknowledgment is due Joseph A. Yura for his assistance with the design examples and illustrations in the seminar notes. Many organizations have given permission to reproduce quotations, graphs, tables, and photographs; this is appreciated and credit is given in the text at the appropriate place.

The authors are sincerely indebted to all of these.

THE AUTHORS

Bethlehem, Pennsylvania
February, 1964

CONTENTS

Structural Behavior and Materials

Structures

Members

Welding, Material and Structural Considerations

Connections

Frames

STRUCTURAL STEEL DESIGN

INTRODUCTION

1.1 STRUCTURAL DESIGN

The process of designing a structure is often thought of as arranging elements or forms in such a way that the expected loads are carried safely. Whereas the ability to support load is perhaps the most important requirement in design, there are three overall objectives to be met if a structural design is to be satisfactory. These objectives are:

1. To provide a structure to meet *functional requirements*
2. To select structural elements and frames to *support loads*
3. To satisfy *economical requirements*

These three concepts were stated in the following manner when the American Society of Civil Engineers prepared its *Commentary on Plastic Design*: "An engineering structure is satisfactorily designed if it can be built with needed economy and if, throughout its useful life, it carries its intended loads and otherwise performs its intended function." [1.1]*

In this regard the word "functional" must be viewed in its larger context, more like that envisioned by the "old-time" civil engineers who considered not only the technical aspects of the problem but the broader sociological and economic ones as well. Indeed these men frequently carried out the financing, construction, and operation of the project. It is fashionable now to call their approach the "systems approach," and civil engineers now are using in the more formalized "systems engineering" that which these engineering pioneers accomplished through their judgment and experience. [1.2]

The components of the systems approach in the case of engineered construction have been given in fairly detailed form in Ref. 1.3 and are shown in condensed and slightly modified layout in Fig. 1.1. Since systems engineering requires consideration of alternative approaches, interacting components,

* The superscript numbers denote listings in the References (pp. 822–49).

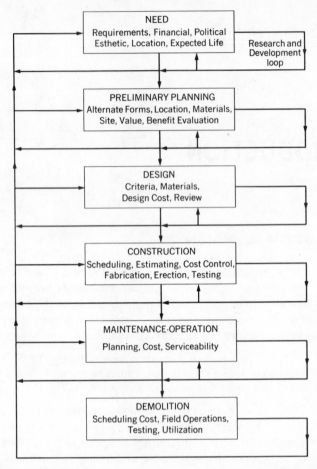

Fig. 1.1 Components of the systems approach in the building process.

and re-evaluation—all with the objective of reaching the best result in terms of cost, time, and other criteria—each successive step can involve a "Research and Development" loop. These loops, in Fig. 1.1, provide for the essential "feedback" into the system so that the final design best meets the objectives. It is especially important to note that the procedure includes maintenance, operation, and even demolition at the end of the useful life of the structure.

1. Aids in Meeting Design Objectives

In the process of meeting design objectives, there are three principal technological aids to assist in arriving at a satisfactory solution. These "tools" are the properties of the material, methods of analysis, and codes, specifications, and standards.

Knowledge of Mechanical and Chemical Properties. The yield point and modulus of elasticity are probably the most important of the mechanical properties. These and others are described in Art. 2.2.

Analysis of Structural Strength. The two basic design techniques in current use are allowable stress design and plastic design.* As seen later when these methods are defined, described, and illustrated for indeterminate structures, the former leads to a trial-and-error process of selecting member sizes, followed by an analysis of the stresses in the structure to see if the allowable values are exceeded. In the plastic design method, the process of selecting member sizes is more direct.

Codes, Specifications, Standards. These classify the types of structures, establish design values for loading, provide legal limits of loading, define acceptable materials, and govern zoning for such things as height, area, and use restrictions. Another of their functions is to assure safety. Specifications and standards often serve as a guide to checking local details; they facilitate design through tables and charts and are frequently the source of standard forms and details that are accepted by the profession as representing "good practice," thus contributing to economy of design and construction. In this group might also be listed design aids in the form of computer programs.

2. Steps in Design Process

There are certain logical steps into which the design process falls. Although there are no hard-and-fast rules in this regard, the following factors are considered, and usually in the order indicated:

Functional Requirements. It is essential first to establish the purpose for which the structure is to be used. This is especially true if its size, form, or use departs from the usual pattern that has seen service in the past. An example would be in the design of a large structure to house launching rockets where the requirement to operate doors under all circumstances places unusual deformation limits on the structure. Another aspect of their performance requirement is esthetics. Civil engineering works by their very nature are always in the public view and frequently tend towards the monumental; they always contribute to and sometimes dominate their surroundings. It is thus essential that the appearance of the structure be carefully considered; indeed, there are times when esthetics may constitute a major design criterion, as in the case of a bridge at a point of scenic

* A Glossary is included at the end of the book as part of the Nomenclature.

interest.* An essential part of the functional requirement is the consideration of how the structure relates to the larger system of which it is a part.

Structure and Loading. Hand in hand with the functional requirement
goes the development of the structural form to be used. The major forms of
structures and structural elements are noted elsewhere in this chapter and are
examined in detail in Chapters 3 and 4. The types of loads acting on the
structure also are considered at this stage since the load and the structure are
often mutually dependent. An example is the amplification of wind effects
due to aerodynamic response of the structure. The loads that are to be
considered for the different types of structures are given in Chapters 3 and 4.
The fire loading and the method of protecting steel from the effects of fire
would be examined in this phase. Along with the functional requirement
and the development of structural form to support the loads goes the
necessary consideration of the cost of the structure. It is an essential aspect
of structural design.

Loading Conditions. The selection of loading conditions follows the
determination of the type and magnitude of the loads that act on the structure. It involves, in the main, a decision as to which loads act in combination
with the dead load. Some codes specify the minimum loading conditions
that must be considered, but it is part of the designer's basic responsibility to
determine the extent and magnitude of the actual loading conditions.
Occasionally, the structural form has a very significant influence on the
loading conditions that must be considered. Again, where the form is
unusual, more attention must be given to this feature. Details concerning
loading conditions are given in Arts. 3.3 and 4.5, and are illustrated in
designs in Chapters 20, 21, and 22.

Preliminary Design. The initial selection of member sizes is called
"preliminary design." As indicated earlier, it is a necessary step in allowable stress design, since the initial estimate of section size must be checked
to see if the allowable stress is exceeded. In the plastic design of continuous
beams and industrial building frames, this step frequently involves no more
than a determination of the ratios of required member sizes.[1.5] Preliminary
design techniques for bridges are discussed in Art. 4.7.

Analysis. With the information obtained in the previous step, an analysis
(theoretical, model, or full-scale experimental) is made to determine points
of critical moment, force, or stress, from which are calculated the required
design properties of each member.

Selection of Section. This step consists of the final or revised selection
of size and shape of members to be used in the design. The selection is
based on the analysis of structural strength which leads to the required

---

* Waddell's injunction is too often ignored: "If a bridge is to be located where it will be
seen constantly by many people, it is well to spend extra money to make it sightly, beautiful,
and in keeping with its surroundings."[1.4]

Fig. 1.2 Empire State Building.

section property—which could be the section area, the section modulus, or the plastic modulus.

Secondary Design Items.   Prior to completing a design, a most important final step is to check on such factors as shear, local buckling, bracing, column buckling; to design connections; and, where required, to examine deflections, stiffness requirements, and any special material and design provisions to obviate brittle fracture under adverse conditions.

Some of the steps listed and described above may be bypassed, and frequently the step-by-step sequence will need to be modified.   For example, in the interests of meeting an immediate objective which involves the national interest it may be necessary to give less attention to finding the most economical design and more attention to some other aspect.   It is in a circumstance such as this that reliance must be placed upon judgment—a quality that is gained through experience, but which will have a better basis when founded on an understanding of structural behavior and systems.   It is in part for this reason that such emphasis is laid in this book on the behavior of structures throughout the entire range of structural response: elastic, inelastic, and failure.

1.2 TYPES OF STRUCTURAL MEMBERS AND FRAMES

The various types of steel structures are listed and classified in the tabulation below.   In some ways the table reflects a consideration of the function of the structure since the form frequently is controlled or influenced by the intended use.

Frames
  Industrial buildings
  Multi-story buildings
  Bridges
Trusses
  Bridges
    Through
    Deck
    Pony
  Buildings
    Roof, floor, and bracing
Plate Girders
  Bridges
    Through
    Deck
  Building girders

Plate-type (Surface) Structures
  Orthotropic bridge decks
  Shells
  Folded plates
  Deck panels
  Wall panels
Arches, Rings
Towers
Space Frames
Suspension Structures
  Cable roofs
  Suspension and cable-stayed bridges
  Cable-supported structures

The above structures are normally found in a land environment.   For other environments one could add: marine and harbor structures, underwater structures, buried structures, and structures in space.   One of the unusual design features of the latter is the influence of a gravity force different from that on earth.

Frames are usually thought of as the skeleton of beams and girders and the columns to which they are connected; the former carry load primarily by bending actions, the latter primarily by compression.   Trusses are pinned or rigidly jointed structures whose members carry primarily axial forces. Steel frames make up the principal load-carrying elements of many buildings and bridges; this is also true of trusses, although fewer of these are used as the main structural form in buildings.   A steel frame is the load-carrying structure for the Empire State Building, shown in Fig. 1.2.   Figures 1.3 and 1.4 also depict monumental buildings with steel frames, but their structural action is radically different from that of the Empire State Building. Typical of modern design, both of the new designs involve a more complete interaction of the frame—resisting load as a complete structural system, rather than as an assortment of beams resting on columns.

Fig. 1.3 John Hancock Building.   (Courtesy of Hedrich-Blessing.)

Fig. 1.4 World Trade Center, in model form.   (Courtesy of Port Authority of New York and New Jersey.)

Although some plate girders are used in buildings, this form is found most frequently in bridges.   Marine and harbor structures can include a number of the other types of structural members and frames, but their principal characteristic in this regard is the type of loading involved.

A plate-type structure that is being used more and more frequently in this country and abroad is the orthotropic plate bridge, a form of construction in which floor beams, stringers, and stiffened-plate traffic deck all form a single structural load-carrying unit.   The term "orthotropic" is a contraction of "orthogonal" and "anisotropic," and thus conveys the concept of different properties in the longitudinal transverse directions.   Stiffened and unstiffened plates are often the main load-carrying elements in marine structures, ships, aircraft, tanks, tunnels, and casings.   Shells and stiffened rings are essential features of these same forms at times, but shells are also commonly used in special building forms.   Towers are usually vertical trusses.   The term "space frame" covers those rigid and trussed structures whose analysis and design involve a consideration of the form as a three-dimensional structure. Suspended structures are found both in buildings and bridges; two examples

Fig. 1.5 Golden Gate Bridge.   (Courtesy of AISC.)

of suspension bridges are shown in Figs. 1.5 and 1.6.   Cable-suspended roofs are frequently used, a recent example being shown in Fig. 1.7.

The scope of this book is restricted to but a few of the types of structural members and frames tabulated above, namely, to buildings and bridges of the first three categories: Frames, trusses, and plate girders.   Chapters 3 and 4 give further discussion of the structures listed in the tabulation as well as details concerning the types of members and frames that make up buildings and bridges.

Structural Elements

In the light of the foregoing restriction, it is of value to examine in a general way the principal types of structural elements which are used in the

Fig. 1.6 Verrazano Narrows Bridge. (Courtesy of Triborough Bridge and Tunnel Authority.)

most common buildings and bridges. Table 1.1 designates opposite the different principal structural elements (tension members, beams, compression members, connections) the type of structure in which the element would be found. To these items the main listing adds certain governing failure modes (frame instability, local buckling, fracture, and fatigue) which enter into a

Fig. 1.7 Madison Square Garden under construction. (Courtesy of Bethlehem Steel Corporation.)

consideration of various structures. Corresponding forms and common loading conditions are shown to the right in the table.

1.3 HISTORICAL NOTES

Some of the chapters of this book include a brief historical account of developments of structural form and methods of design. In order to relate these to one another and to provide a broad perspective of the historical record of steel structures, Table 1.2 is presented, the material being based in part on Refs. 1.4, 1.6–1.11. This tabulation contains information in four principal categories: (1) the major fundamental concepts upon which design theory rests; (2) major developments in connection with materials and manufacturing or fabricating process; (3) the development of specification information; and (4) highlights in structural applications.

Table 1.1 Structural Elements

	Building and bridge frames	Trusses	Plate girders	Forms and loadings
Tension members	X	X		
Beams	X	X		(shapes as above)
Compression members	X	X		(shapes as above)
Bending and compression	X	X		(shapes as above)
Frame stability	X			
Local buckling	Flanges, webs, stiffeners			
Lateral buckling	Flanges			
Column buckling	Element or as part of a frame			
Fatigue	Elements under severe cyclic load			Alternating Repeating Fluctuating or pulsating
Brittle fracture	Elements under adverse combination of temperature, material and design conditions.			Conditions in which tensile stresses predominate; tri-axial tension.
Bolted, riveted joints	X	X	X	
Welded connections	X	X	X	
Composite beams	X		X	

Table 1.2 Historical Outline of Developments in Steel Structures

1676	Hooke's law developed
1744	Buckling of bars (L. Euler)
1779	First iron bridge (Coalbrookdale, England)
1786	Tests by Paine of cast-iron arch bridge models[1.6]
1820	Cast iron columns used in Philadelphia building
1823	Navier formulated differential equation for buckled plate
1828	First use of steel in a bridge (Vienna, Austria)
1840	First iron truss in U.S.A. (Baltimore and Ohio Railroad)
1843	Wrought iron lighthouse built on Block Island
1847	Squire Whipple presented stress analysis of truss systems (iron truss in 1853)
1853–58	First building with wrought iron frame (Cooper Union six-story frame)
1856	Steel first made in U.S.A.
1862	Bessemer steel bridge, Holland
1869–74	First major bridge built with steel shapes (double-deck Eads Bridge, St. Louis)
1873	First tabulated values of properties of rolled shapes[1.9]
1876	Eiffel Tower
1877–1900	First specifications in U.S.A. (individual consulting engineers and railroad companies)
1879	All-steel (Bessemer) railroad bridge, Glasgow, Missouri
1881	Electric arc welding introduced
1883	Roebling's suspension bridge (Brooklyn)
1884	First building with steel frame (Home Insurance Co. building, Chicago, designed by W. Jenney)
1888	Riveted connections used in Tacoma Building in Chicago
1890	Firth of Forth Bridge (Scotland)
1893	Formation of Office of Road Inquiry, Dept. of Agriculture (forerunner of Bureau of Public Roads)
1905	First AREA Specification
1907	Grey mill installed at Bethlehem, Pa.; first wide-flange shapes rolled in 1908
1909	First building erected with wide-flange shapes (American Optical Co., Worcester, Mass.)
1914	First tests to demonstrate "plastic hinges" conducted by Kazinczy in Hungary
1914	AASHO first organized
1921	AISC first organized
1923	First AISC Specification for buildings issued (written by five practicing engineers; chairman, Prof. G. Swain of Harvard)
1926	First AASHO Specification issued
1936	Studies of plastic design initiated at Bristol University by J. F. Baker, later (1944) at Cambridge University
1936	First revision in AISC Specification (minimum required yield stress for ASTM A7 steel raised from 30 to 33 ksi)
1944	Column Research Council organized. First edition of "Guide," 1960
1945	Provision made in AISC Specification for welded connections, refinements for rivets and bolts, lateral buckling formula, "20% increase" provision
1946	Research on ultimate strength of structures and components commenced at Lehigh University. Mathematical theories of plastic behavior of materials studied at Brown University
1946	First AISI Specification
1949	First specification for high-strength bolts
1957	First plastic design in North America (D. T. Wright)
	First plastic design in U.S.A. (W. A. Milek)
1961	Complete revision to AISC Specification. Publication of first edition of ASCE Manual No. 41, "Plastic Design in Steel" (second edition, 1971)
1963	Revision to AISC, AREA, AWS Specifications
1965	Design methods issued for plastic design of multi-story frames.
	A7 steel dropped from Specification. Replaced by A36
1969	Major revisions in AISC and AREA Specifications
1970	Tower 1 of World Trade Center completed: world's tallest building (1350 ft)
1973	Sears Tower (Chicago) became tallest building in the world (1450 ft)

Hooke's law and Euler's formula provided the basic concepts upon which much of the early work was based. It was not until many years later (and only about fifty years ago) that any conscious attempt was made to utilize the ductile characteristics of steel. As has been true throughout history, many buildings and bridges were built before formal specifications were available, and this has been true even in recent years for some types of structures and materials. Engineers have been prompt to use new materials and techniques as soon as they have been convinced of their virtue. Various consulting engineers had their own building specifications and the individual railroads had their own bridge specifications. Examples of the latter were the specifications for metallic bridges of the Atcheson, Topeka, and Santa Fe Railroad (1895) and the Canadian Pacific Railroad specifications (1901).[1.7] In 1877 there was issued a general specification for a combined railway and highway bridge at Killbourn City, Wisconsin.[1.8] Naturally, rapid acceptance and use is possible only as specifications and codes incorporate new provisions.

Perhaps the most striking feature of the events shown in Table 1.2 is the remarkable advance of steel construction over a relatively short period. It was only about one hundred years ago that steel was first made. Wide-flange shapes became available sixty years ago. Specification-writing groups have been in existence but a short while, the first of the major groups being the American Railway Engineering Association (AREA, 1905) followed by the American Association of State Highway Officials (AASHO, 1914) and the American Institute of Steel Construction (AISC, 1921). The first American Iron and Steel Institute (AISI) Specification was issued in 1946. Intensive research into the maximum strength of steel structures began in this country only recently and yet the results of much of that work now find application in even the most routine of designs. With such a background of recent rapid advance, it must be expected that there will be even greater developments in the future.

Not the least significant development is the relative frequency with which specifications are being changed to keep pace with new advances in theory and technology. The AISC Specification was first issued in 1921 and significant changes were made in 1936, 1945, 1961, 1963, and again in 1969. Provision has now been made for annual revision to that Specification which will mean a more rapid incorporation of research results into design practice. The AASHO Specification, first available in 1926, was printed and revised in 1935, 1941, 1944, 1949, 1953, 1957, 1961, and in 1965.

Because specific provisions change rapidly, primary attention is given in this book to the behavior of structures (and especially to the theory predicting that behavior) rather than to the detailed provisions of specifications that are susceptible to revision.

1.4 BEHAVIOR OF STEEL STRUCTURES

A brief, general review of the behavior of steel structures is given below. It will be seen that, except for such modes of failure as fatigue and fracture, there is remarkable similarity in the behavior of different elements and structural forms.

1. Material

The two main characteristics describing the behavior of a structural material are strength and ductility. Figure 1.8 shows the engineering stress-strain curve that is characteristic of most steels with structural applications. Such a curve is obtained from a tension test. The figure notes the four typical ranges of behavior: the elastic range, the plastic range (during which the material flows at near-constant stress), the strain-hardening range, and the range of strain at and beyond the ultimate stress during which necking occurs in a tensile bar, this range terminating in fracture.

Figure 1.9 shows the initial portion of Fig. 1.8 to an expanded scale and in somewhat idealized form. The curve is drawn for ASTM A36 steel with a yield stress level F_y of 36.0 ksi. (Curves for other structural carbon steels are similar and are described in further detail in Chapter 2.) The following points are to be noted in Fig. 1.9:

1. After the initiation of yield there is a flat "plateau." The stress at this level is termed the "yield stress level." It is one of the characteristic features of the structural steels, especially structural carbon steel.

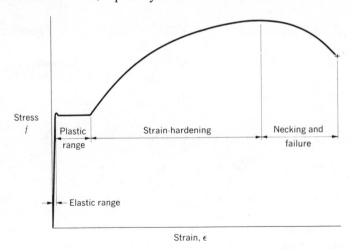

Fig. 1.8 Stress-strain curve for structural carbon steel.

2. The extent of the yield zone (or "plastic range") is considerable. The value ϵ_{st} is about 10 times ϵ_y for structural carbon steel, varying from a low of about 6 to a maximum of about 16.

3. At the end of the plateau strain-hardening begins, with consequent increase in strength. The magnitude of the strain-hardening modulus varies from 250 ksi to 1000 ksi, averaging about 1/50 of Young's modulus.

4. Some tension test curves do not show an upper yield point. The result is a gradual transition from the elastic to the plastic region, as shown by the dashed curve. This is also the condition most usually encountered in full-size members.

The region designated as the plastic range in Fig. 1.8 is thus made up, in fact, of two regions. One of these is called "contained plastic flow";

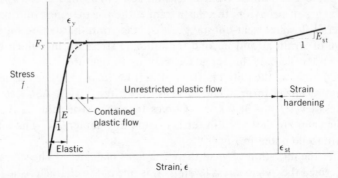

Fig. 1.9 Initial portion of stress-strain curve (A36 steel).

although there is plastic flow in part of the cross section, the deformation is restricted or "contained" by the remaining elastic part. The other region is characterized by the term "unrestricted plastic flow"; as shown in Fig. 1.9, it is a region in which the strains increase markedly at a constant yield value up to the point of strain-hardening.

2. Behavior of Structural Elements

The characteristic behavior of structural elements is somewhat similar to that of the idealized tension specimen as sketched in Fig. 1.9. Such elements also exhibit an initial elastic region followed by one of contained plastic flow and one of unrestricted plastic flow.

Tension. The behavior of a tension member is shown in Fig. 1.10 on a load-elongation basis. Two curves are shown: one is the complete curve to rupture, and the other is the initial part to expanded scale. The plate specimen has a reduced section length of 18 in., is 1 in. thick, and is loaded in tension.   The average stress at which local yielding first started was about half the load at which unrestricted plastic flow developed.

Local plastic flow can and usually does commence at an average stress which is less than the yield value.   The reason for this usually lies in the fact that perfect alignment of load seldom exists in practice, and fabrication operations may introduce a variation in cross-sectional characteristics.   Also, unless a member is stress relieved, it will contain residual stresses—internal stresses that are introduced in the member during manufacture or fabrication; upon application of external load their combined effect will cause local yielding.   On further loading, the entire cross section eventually yields at a load corresponding to the tension specimen yield value.   The point of local yielding corresponds to the onset of "contained plastic flow" sketched in

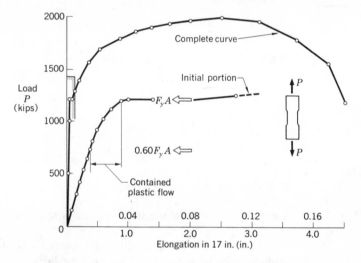

Fig. 1.10 Typical behavior of tension member.

Fig. 1.10.   It ends with the plastification of the entire cross section and is followed by unrestricted flow.

Similarly, a tension member that is slightly crooked will yield locally at one edge before the average yield stress is reached.   Eventually, however, the entire section will yield at $P/A = F_y$.

A plate with a hole exhibits this same phenomenon and illustrates especially the region of contained flow.   Due to stress concentrations, local yielding will commence at the edges of the hole (stage 1 of Fig. 1.11) and initiate flow (strain-hardening is neglected in this discussion).   The corresponding point on the load-elongation curve is shown in Fig. 1.11.   On further loading, an intermediate condition is reached (stage 2) with additional fibers brought to yield; however, the elastic part of the member "contains" the deformation until further load is applied (see sketch of yielded cross section). Finally, at stage 3, each fiber has reached the yield value (the term for this phenomenon is "plastification"), and unrestricted plastic flow commences.

The point at which contained plastic flow commences could be termed the "proportional limit" of the member, and in light of the discussion above it is evident that there will be wide variation in its value. However, the load which terminates this region and marks the beginning of unrestricted plastic flow is one that is calculable and corresponds to a real limit of usefulness of the member, namely, a significant and relatively uncontrolled elongation. Thus, the allowable stress for a tension member (in the instance of the AISC Specification: $F_t = 0.6F_y$, as sketched in Fig. 1.10) is based not on the proportional limit (where contained plastic flow commences) but instead it is

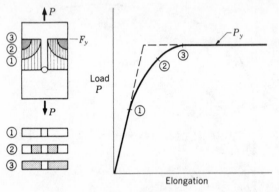

Fig. 1.11 Tensile stresses and deformation in plate with a hole.

based upon $P/A = F_y$, which marks the point at which unrestricted plastic flow begins. (The difference between gross and net area has been ignored in this discussion.)

Bending. Figure 1.12 shows a typical moment-curvature relationship for a shape under pure bending.[1.12] Here again is a region of contained plastic flow. In part it is the result of the early yielding caused by residual stresses and stress concentrations, and in part it is the result of the gradual plastification of the cross section as deformation is continued. M_y is the moment at first yield, M_p is the plastic moment (after plastification is complete) and ϕ_{st} is the curvature at which strain-hardening commences.

Following the plastification of the cross section, a region of unrestricted plastic flow ensues at the moment-value of M_p; this rotation at near-constant moment characterizes the "plastic hinge," which is one of the fundamental concepts of plastic design.

As in the case of tension members, the "proportional limit" or the beginning of the region of contained plastic flow is subject to wide variation and

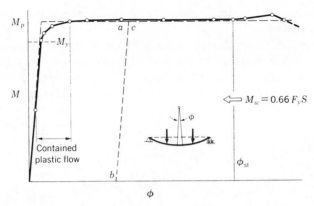

Fig. 1.12 Typical behavior of bending member under pure moment.[1.12]

cannot, in fact, be defined sufficiently to constitute a design criterion. The plastic limit M_p, however, is not subject to such variation and does constitute a real limit of usefulness upon which design specifications can be based.

The broken lines *a-b-c* in Fig. 1.12 show the behavior to be expected if a member is unloaded after reaching the plastic region and then is reloaded. The relationship follows the original elastic slope, and the effect of the prior plastic deformation is obliterated insofar as subsequent reloading is concerned. The post-yielding behavior is thus elastic for subsequent loads less than the maximum value previously attained.

Beam-Columns. Figure 1.13 shows the typical behavior of a beam-column.[1.13] Again the region of contained plastic flow is evident; in the case of an isolated member, its upper limit represents a suitable maximum to which a factor of safety may be applied for design. In this case the limit is

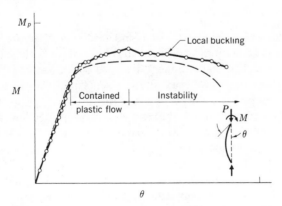

Fig. 1.13 Typical behavior of beam-column.[1.13]

defined by stability. A similar situation would exist in columns subjected to transverse loading.

Connections. Of critical importance in structures are the regions making up the connections between beams and columns. Figure 1.14 shows the behavior of a connection fabricated by welding. Due to the loading configuration, the effect of strain-hardening is evident: the assembly carried

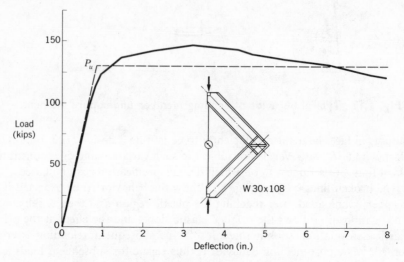

Fig. 1.14 Typical behavior of connection[1.5]

considerably more load than the value P_p as predicted by the plastic theory. It is an effect customarily ignored in design. The shape of the curve is similar to that of other structural members.

3. Behavior of Rigid Structural Frames

If the load-deformation behavior of an indeterminate structure is observed, it is found that the resulting curve is remarkably similar to the curves shown above for beams and beam-columns. This has been done in Fig. 1.15 for a single-span portal frame; the dashed line represents the theoretical calculation, and the connected points show the result of the test.[1.14] It is a gabled frame of 40-ft span, loaded as shown in the inset with concentrated loads simulating uniformly distributed vertical and horizontal loading. The abscissa is the vertical deflection at the peak of the gable.

Figure 1.15 illustrates the typical behavior of a rigid frame. There is an initial elastic region, but the inelastic region of contained plastic flow is the dominant one both on the load scale and on the deflection scale. Actual yielding begins at a load very much lower than the hypothetical yield load

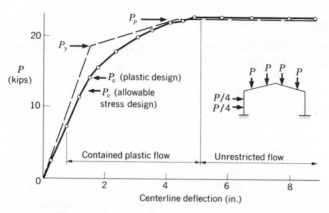

Fig. 1.15 Load-deflection curve of a rigid frame.[1.14]

P_y; on the other hand, the load at which unrestricted plastic flow commences is in remarkable agreement with theory. The service loads that would be permitted on this test structure according to allowable stress design and to plastic design are also shown in this figure, the latter being based on the attainment of maximum plastic strength.

4. Instability

The "maximum plastic strength," referred to above, neglects instability effects. As seen later in this text, the extension of plastic design to multi-story frames, for example, requires that the instability effect be considered.[1.15] The overall instability of the entire frame becomes in certain instances the appropriate limit of usefulness of the structure.[1.16]

The example given in Fig. 1.12 is one that is "typical" so long as the beam does not fail due to local or lateral buckling—in which case the limit of usefulness is the stability limit load unless provision is otherwise made to prevent such failure. If the beam were a plate girder then other instability effects also would have to be considered.

The buckling of the centrally loaded column is the classic case of instability failure. No structural member has been tested more extensively nor is there more variation in design approach than there is for this element, all of which simply reflects the complication of the problem and the accounting for it in design.

Thus, instability effects constitute a significant design problem and are the subject of much on-going research. The Column Research Council was organized to provide a forum for discussion and evaluation in this field and to provide designers with comprehensive and consistent design concepts. Its Guide summarizes the latest information available.[1.17]

5. Fatigue and Fracture

Two modes of failure of steel structures are of quite a different nature from those discussed thus far. One is failure by fatigue as the result of application of many cycles of stress; the other is failure due to brittle behavior of the material as a result of adverse combinations of temperature, material, and design conditions. Although instances of such failures are rare, special measures are taken to avoid them because of the serious consequences that may be involved. In the case of fatigue, an adjustment in allowable stresses frequently is made; but when fatigue is known to be the dominant design criterion, the greatest attention is given to design details and material properties, especially in the loading region where failure would be expected. Similar attention is given when brittle behavior is a possibility.

6. Summary

In summary, Fig. 1.16 shows the typical behavior of a steel structure under load. The heavy solid line that reaches "maximum plastic strength" is the

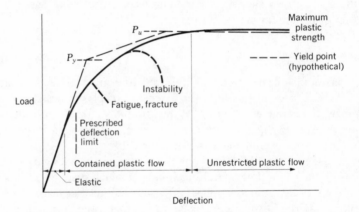

Fig. 1.16   Behavior of a steel structure under load and limits of structural usefulness.

typical behavior of a structure not limited by fatigue, fracture, or instability, or subject to a prescribed maximum deflection limit (these limitations being shown by dashed excursions from the heavy line). Shown also are the three regions that are observed in the behavior of a structure under load. The elastic region is frequently very limited; it is followed by the region of contained plastic flow whose onset is subject to wide variation; and this latter region merges into the third one, which is that of unrestricted plastic flow. The transition is gradual from one to the other; the exact point of departure has no real physical meaning; and it is especially important to note that the "yield point" of a structural member or frame has only hypothetical significance.

1.5 DESIGN CONCEPTS

The review of the behavior of steel structures in the previous article shows that the design load may be controlled by one or more of several criteria. These may be termed "limits of structural usefulness." They are suggested diagrammatically in Fig. 1.16 and are listed as follows:

1. Hypothetical attainment of yield point
2. Attainment of maximum plastic strength (onset of unrestricted plastic flow)
3. Excessive deflections at service load, drift limits
4. Instability
5. Fatigue
6. Fracture

One or more of these must form the basis of any rational design, and their consideration enters into much of the material presented in Chapters 6 through 22.   Strictly speaking, the term "limit design" could apply to any one of these six criteria, and as a consequence a certain confusion has resulted from its use in engineering literature.   Therefore, the term will not be used in this book, and the particular limit will be defined in each case.

As a result of the various "limits " two major design methods have evolved in practice: "allowable stress design" and "plastic design."   Allowable stress design embraces items 1, 4, 5, and frequently item 2.   Plastic design is based mainly on item 2 (maximum plastic strength), but also embraces item 4 (instability), especially in unbraced multi-story frames.   Design for stiffness is particularly concerned with deflection or flexibility limits (item 3) and occasionally with item 6; it can be either the major design criterion or it can enter as a secondary check in both allowable stress and plastic design. Most frequently, whenever items 3, 5, and 6 are involved at all, they are involved as secondary checks.

The allowable stress design criteria and the plastic design criteria are now discussed briefly.

1. Allowable Stress Design

In allowable stress design, a member is so selected that, under expected loads (called "service" or "working" loads), the stress will not exceed a certain permitted or allowable value.   This stress incorporates a factor of safety against one of the previously described limits of usefulness.   Allowable stress design is thus performed by specifying expected design (working) loads and allowable stresses. The factor of safety is inherent, but usually is not stated. Also, the limit of usefulness is usually undesignated.

Insofar as methods of analysis are concerned, allowable stress design is based on an elastic analysis to obtain the moments, shears, and axial forces which the member must be designed to carry. For indeterminate structures, methods such as moment distribution and matrix methods are used for this analysis.

Figure 1.17 shows on a diagrammatic basis the load-deflection curve for a simply supported beam with a central concentrated load, the loading and

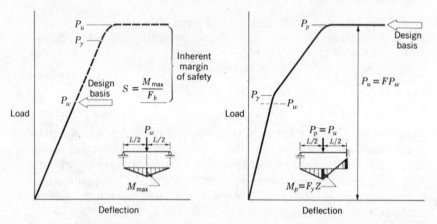

Fig. 1.17 Allowable stress design and the inherent margin of safety.

Fig. 1.18 The load factor and basis for design in the plastic method.

moment diagrams being shown in the inset. As with the beam in bending that was shown in Fig. 1.12, the member is elastic up to P_y. The gradual plastification of the cross section begins there, a process completed when the plastic limit load P_p is reached. This is the theoretical maximum load-carrying capacity. For purposes of illustration, the design ultimate and plastic limit loads are assumed to coincide ($P_u = P_p$ and $P_w = P_a$). In allowable stress design conscious attention is given only to the point at which the working load P_w is reached, which is at the end of the solid portion of the load-deflection curve and designated by the arrow. The margin of safety (in this case taken against ultimate) is inherent, but the selection of section modulus S is made on the basis of working loads (or moments) and allowable stresses.

The section modulus is computed in the given example from $S = M_{max}/F_b$ where F_b is the allowable stress permitted by the specification and $M_{max} = P_w L/4$. It is thus evident that consideration in the design process is focused entirely on allowable loads, moments, and stresses. In the background, however, is consideration of the margin of safety. To anticipate Art. 7.8 (and as touched on previously), the allowable value is actually obtained by dividing the maximum load (which would be exhibited by the selected beam) by a safety factor. Thus, even though the allowable value appears as the end product of a stress design, it is in fact a "limit" design. Column design has always been, in effect, "limit design." Column formulas have always been based on maximum column strength; the fact that these formulas have been altered in terms of allowable stress does not change the fact that it is a design based on ultimate. Many other examples may be given of the tacit assumption of ductile behavior in allowable stress design.[1.5] Thus, much of allowable stress design may be interpreted as maximum load design in which the allowable values incorporate a load factor.

2. Load-Factor Design

The fact that all design rests on either stated or inherent limits, plus possible design economies, has led to recent active consideration of "load-factor design," a method of designing structures for multiples of service loads. In this method the design ultimate load is determined by multiplying the service loads by load factors (corresponding to the increase in allowable stress that is permitted when the dead and live loads act in combination with wind or earthquake forces). But in addition it permits the use of different load factors for dead load, live load, and unusual or infrequent live loads including special factors to express the lower probability of simultaneous application of multiple loadings, and this can lead to a saving of material. Load-factor design is gaining increasing attention in the United States and is used extensively in Europe, where it is usually referred to as "limit design." The term does not imply any particular strength or deformation limit, and in application the loading function can be equated to any of the previously listed limits of structural usefulness expressed in terms of load-carrying capacity.[1.16]

3. Maximum Load Design—Plastic Design

In maximum load design (variously termed "limit," "ultimate," "collapse," or "plastic" design), the members are so selected that the structure will reach its maximum strength at the factored load—a load which is determined by multiplying the working load by a load factor. Maximum load design is used in plastic design for steel building construction. It is a method which is based on the concept of ultimate applied loads (which are obtained by applying the appropriate load factor), and ultimate (or maximum) resisting moments, which are the so-called "plastic hinge" moments. From the prior discussion, it is evident that plastic design is a load-factor design in which the limit of usefulness is the maximum plastic strength.

In considering design concepts, there are some advantages in maximum load design as compared with other methods. First, the load factor is known because the expected loads are multiplied by it. Second, the limit of usefulness is known by the nature of the design. This is illustrated in Fig. 1.18 on a basis similar to that of Fig. 1.17 but for a beam fixed at one end. In plastic design, one first calculates the maximum load which the required member should support if the entire margin of safety were absorbed. It is the maximum load to which attention is fixed in plastic design, a maximum determined by multiplying the expected loads P_w by the load factor to obtain $P_u = FP_w$. Attention is given essentially to maximum loads, although the entire range occasionally may be of interest.

As a simple illustration of the plastic design approach, for the problem in Fig. 1.18 after the factored maximum load is determined, the required plastic moment would be computed from

$$\frac{P_u L}{4} = M_p + \frac{M_p}{2} = \tfrac{3}{2} F_y Z \tag{1.1}$$

from which Z is obtained as $P_u L/6F_y$. Thus, the entire solution to this problem, in contrast to the allowable stress solution, revolves about a consideration of the conditions at ultimate; concern with the behavior at working load would arise only if deflections were critical.

Although presently the only example of maximum load design is plastic design in steel, in the future it is reasonable to expect that more and more of design will be on this basis. Insofar as buildings are concerned, the trend began with the 1946 revision to the AISC Specification to incorporate a 20-per cent increase of permitted stress at regions of negative moment in continuous beams. "For the first time, the Specification contained a provision that recognized the real strength of steel, insofar as main structural elements were concerned." [1.18]

Recent specification revisions continue this trend, and more and more frequent reference to the factor of safety is found. In commenting on the 1961 edition of the AISC Specification, T. R. Higgins, then AISC Director of Engineering and Research, stated:

Plastic Design uses a long-ignored, but all-important property of structural steels—its ductility. Many of the other provisions of the new Specification are also based on a frank recognition that proven strength is the soundest engineering approach. Working stress provisions now contain more refinements based on improved knowledge of the actual behavior of structures.[1.18]

The term "elastic design" is sometimes used to refer to an allowable stress design—probably because the method is usually based on an elastic analysis.

However in view of the discussion thus far, it is apparent that the term "elastic design" is technically a misnomer. The load-deformation curve shown as the heavy solid line in Fig. 1.16 is typical of the behavior of most structures and structural members. The elastic region exists, but its upper limit cannot be specified because of such things as residual stresses, misalignment, support settlement, and stress concentrations.

It is true that elastic analysis is often used to calculate stresses in a structure, but the term "allowable stress design" is more appropriate than "elastic design." The designation "allowable stress" may incorporate the selection of a permitted stress by applying a factor of safety to any applicable limit of structural usefulness. This, in fact, is the way allowable stresses are determined, recognizing in turn a number of the limits listed earlier.

1.6 FACTOR OF SAFETY

Although the factor of safety is always inherent in allowable stress provisions, the term was scarcely mentioned in early editions of specifications except (in the case of buildings) for the load factor of plastic design. It is only in most recent years that reference to it is made in specifications. There is an increasing tendency among technical groups and specification writers to provide supplementary documents giving an indication of the background of the provisions. References 1.1, 1.17, 1.19 and the commentaries of Refs. 1.20 and 1.21 are such examples.

The factor of safety is not concerned alone with the possibility of overloading. Factors which influence the selection of an appropriate margin of safety are:[1.1]

1. Approximations and uncertainties in the method of analysis
2. Quality of workmanship
3. Presence of residual stresses and stress concentrations
4. Underrun in physical properties of material
5. Underrun of cross-sectional dimensions of members
6. Location and intended use of structure*
7. Loading

It is clear then that the safety factor does not imply safety against overload alone; instead there are many other factors involved, largely related to the strength of the structure.

A precise method of arriving at the proper value of the safety factor would require a statistical analysis of each item in the list, since the possibility of

---

* This can also include the importance of the structural element and the consequences of failure.

variation must be considered. Progress is being made on this matter.[1.22] Another possibility in arriving at the design ultimate load is to apply different factors to the different loads, depending on their nature. This has been touched upon earlier in this chapter. Thus, consideration might be given to a lower load factor for dead load than for live load because the former is subject to fewer uncertainties. Such a consideration requires further study, and it is not part of present specifications for steel structures in the United States, although it is rather widely used abroad.[1.16] This book considers a single load factor for a given loading condition.

Past and present practice with respect to safety factors is based largely on experience and judgment. In allowable stress design, the stress (or load) at failure is reduced by a factor of safety. In plastic design, service loads are multiplied by a load factor. In a design for stiffness, the performance is evaluated at service loads and the margin of safety may be uncertain.

In general, the factor of safety F.S. is given by

$$\text{F.S.} = \frac{P_L}{P_a} \qquad (1.2)$$

where P_L = limit load (which requires that the design criteria be specified)
$\quad\;\; P_a$ = allowable load (which is obtained from a specification or is otherwise prescribed).

Using Eq. 1.2, Table 1.3 shows values obtained for the factor of safety based on the designated design criterion and in accordance with the AISC Specification.[1.21] The table shows the major structural elements, the limit load P_L according to the selected limit of usefulness, the allowable load (or moment) according to the specification, and finally the computed factor of safety. F_u is the tensile strength.

The comparison of the factors of safety for four of the structural elements listed shows a relatively consistent development. A factor of 1.67 for tension members, 1.67 for "short" columns, 1.92 for "long" columns, and 2.0 or more for rivets and bolts are values that reflect a reasonable pattern. Tension members and short compression members logically may have the same load factor, because their limit of usefulness is essentially the same (unrestricted plastic flow). A long column should have a greater margin of safety because of the consequences of a buckling failure. It is reasonable to require a higher factor of safety for groups of fasteners since this assures that the connections will not fail before the main members have reached their limit of usefulness.

It is evident that there is a logical background to the factor of safety that is "designed into" present-day structures—a logic that is increasing with increasing knowledge concerning the behavior of structures.

Table 1.3 Factor of Safety for Selected Structural Elements*

Structural Element	Design Criterion	Limit Load, P_L	Allowable Load, P_a	Factor of Safety (Eq. 1.2)
Tension members	Unrestricted plastic flow	$F_y A$	$0.6F_y A$	$\dfrac{F_y}{0.6F_y} = 1.67$
	Ultimate† strength (tensile strength)	$F_u A$	$0.6F_y A$	For A36 steel: $\dfrac{F_u}{0.6F_y} = \dfrac{58}{20} = 2.6$†
Simple beams	Hypothetical first yield (slender-flanged shapes)	$M_y = F_y S$	$M_a = 0.6F_y S$	$\dfrac{F_y}{0.6F_y} = 1.67$
	Unrestricted plastic flow (compact shapes)	$M_p = F_y Z$	$M_a = 0.66F_y S$	$\dfrac{F_y Z}{0.66F_y S} = \dfrac{1.12}{0.66} = 1.70$
Columns	Maximum load (instability)	CRC column formula	Depends on $\dfrac{L}{r}$	$\dfrac{L}{r} = 0,\ \text{F.S.} = 1.67$ $\dfrac{L}{r} = 130,\ \text{F.S.} = 1.92$
High-strength bolts (A325)	Shear failure on fastener (bearing-type joint)	Depends on joint length	$22A_b n$	Maximum $= 3.3$ Minimum $= 2.1$
Rivets	Shear failure	Depends on joint length	$15A_b n$	Maximum $= 3.0$ Minimum $= 2.2$
Continuous beams Industrial frames (plastic design)	Unrestricted plastic flow	$1.70P_w$	P_w	1.70
Multi-story frames (plastic design)	Plastic limit load or stability limit load	$1.70P_w$	P_w	1.70
Plastic design (frames with wind)	Unrestricted plastic flow	$1.30P_w$	P_w	1.30

* Based on AISC Specification.

† The deformations at ultimate load of a tension member are such that a factor of safety computed on this basis may not be significant. Also there is considerable variation, depending on the steel. Unrestricted plastic flow is the logical criterion for such members.

1.7 SPECIFICATIONS AND CODES

In words chosen from a statement by a specification writer:

A specification is the backbone of construction. To the architect and engineer, it is a guide to safe and accepted design procedures, a convenience in selecting structural members and outlining construction methods. To the contractor and building code official, it is a document setting forth rules of safe construction that must be strictly followed. And to the owner, it is a guarantee that the resulting structure will comply with basic standards to ensure safety, utility, and economy.

The following are some important specifications for steel structures in the United States and Canada:

AASHO American Association of State Highway Officials[1.23]
AISC American Institute of Steel Construction[1.21]
AISI American Iron and Steel Institute[1.24]
AREA American Railway Engineering Association[1.25]
AWS American Welding Society[1.26]
AISE Association of Iron and Steel Engineers[1.27,1.28]
CISC Canadian Institute of Steel Construction[1.29]
CSA Canadian Standards Association[1.30]

There are also the specifications of certain U.S. federal agencies, including the Corps of Engineers, and Naval Facilities Engineering Command of the Department of Defense; the General Services Administration; and the U.S. Department of Transportation, Bureau of Public Roads.

In addition to specifications, designs are frequently controlled by local, regional, or national codes. These codes sometimes incorporate specifications either by direct reference, by incorporation of all or part of the specification, or by rewriting and revision to suit their particular needs. Local and regional codes include city, county, and state codes. The major national codes concerned with building construction are those of the Building Officials Conference of America, the International Building Officials Congress, the American Insurance Association (successors to the National Board of Fire Underwriters), and the Southern Building Congress.

As already noted, the most recent specification changes mark, in many respects, a significant departure from past practice. It is a trend that will continue with new developments in materials, in material processes, in fabrication, and in new design concepts and approaches.

The effect of these revisions is manifold. They permit more attention to matters of economy through selection of materials with the most appropriate strength and other characteristics; there is choice among rivets, among bolts, and among welding materials to suit the conditions. A mixture of steels may be contemplated in design, as in the case of the second Carquinez Bridge crossing in California (which was the first example of the use of constructional alloy steel in bridges), and as in the case of the World Trade Center in New York City.   Although approximations for rapid design use remain, specifications are permitting more and more design refinements.   In some instances these refinements necessarily involve more complex procedures, but developments in the use of computers have made such rapid strides that the matter of complexity is not the problem it once was.   In the long run, the improved procedures lead to more logical designs.   As formulas appear for needed cases, revised concepts will stimulate use of other shapes, such as box-type members for frames.   There is a narrowing of the supposed gap between allowable stress design and plastic design, primarily because of the recognition of plastic strength and post-buckling strength in the allowable stress method.[1.31]

These changes, and the recognition that a specification is not static but dynamic and subject to change, are the result of what has been termed a "revolution in structural thought" that has been going on for a number of years.   It represents the effort to take advantage of the results of research work and to be consistent with both theoretical and practical knowledge of steel structures.   The following chapters set forth structural steel design in this same spirit.

MATERIALS

2.1 INTRODUCTION

The wide use of steel in our civilization today can be attributed to the remarkable properties of this metal, the abundance of the raw materials required for its manufacture, and its competitive market price. Steel can be produced with widely varying characteristics which can be controlled to satisfy the intended use. The final product may be anything from a stainless steel surgical instrument to a city skyscraper, a giant bridge or ocean liner, a nuclear reactor vessel, or innumerable other objects.

Several types of steel have been given standard designations by the American Society for Testing and Materials[2.1] (hereafter referred to as ASTM). The purposes of this organization are the promotion of knowledge of the materials of engineering and the standardization of specifications and methods of testing.

This chapter discusses those mechanical properties of steel which are most important to the structural designer, together with the factors affecting these properties, and the methods by which they are determined. A comparison is made of some of the structural steels presently available.

2.2 PROPERTIES OF IMPORTANCE TO DESIGNER

As a basis for discussing mechanical behavior, complete tensile stress-strain diagrams for three ASTM-designation steels are shown in Fig. 2.1, and a small portion of the diagram for A36 steel is drawn to a different scale in Fig. 2.2. Although the curves shown are from the results of tension tests

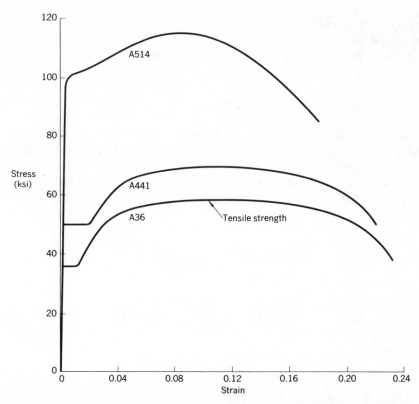

Fig. 2.1 Tensile stress-strain curves for three ASTM-designation steels.[2.2]

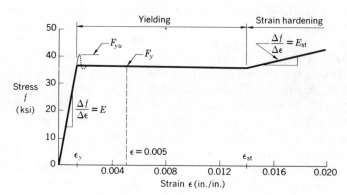

Fig. 2.2 Portion of stress-strain diagram for A36 steel.

for each steel, the early part of the curve, as in Fig. 2.2, would be much the same for compression as it is for tension. The properties of steel of most importance in structural design are:

1. Yield Point. The yield point is defined as the first stress in a material less than the maximum attainable stress, at which there is a marked increase in strain without increase in stress. This phenomenon is indicated by the horizontal portion of the solid curve in Fig. 2.2. Some test pieces of structural steel exhibit behavior indicated by the dotted portion of the curve, producing an upper yield point F_{yu} followed by a lower yield point and a plateau. The appearance of an upper yield is affected by the test technique (speed of testing, shape of the specimen, accuracy of alignment) and the condition of the test piece (especially the presence of residual stresses in a test on the full cross section). The lower yield point is as transient as the

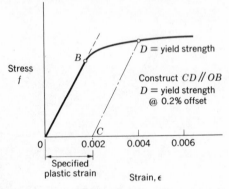

Fig. 2.3 Stress-strain diagram for material without yield point.

upper yield point. The consequence of the sudden flow at the upper value is a drop to this lower value, somewhat below the plateau. The plateau itself, on the other hand, is a more stable characteristic of the material.

2. Yield Strength. Not all materials exhibit a yield point; for those which do not, the yield strength is a useful measurement of behavior and is defined as the stress at which the material shows a specified limiting plastic strain, usually 0.1 or 0.2 per cent (0.001 or 0.002 in./in.). It is determined as indicated in Fig. 2.3. When quoting the value of the yield strength, the numerical value of the offset used should be stated.

3. Yield Stress Level. The yield stress level is the stress corresponding to a strain of 0.5 per cent (see Fig. 2.2). This stress will usually correspond to the constant stress at yielding when the stress-strain relationship exhibits such yielding.

4. Yield Stress. This is a generic term covering the yield point, yield strength, and yield stress level, as defined above.

5. Proportional Limit. The highest stress a material can withstand without deviation from a straight-line proportionality between stress and strain is called the proportional limit. Measured values of the proportional limit depend considerably on the sensitivity of the strain-measuring equipment used.

6. Tensile Strength. The tensile strength is the maximum axial load observed in a tension test (Fig. 2.1) divided by the original area.

7. Ductility. Ductility is the ability of a material to undergo large plastic deformations without fracture. It is measured by reduction of area and elongation in a tension test, usually expressed as percentages. In the latter case the gage length should be stated. As will be seen in other chapters, it is the ductility of steel which "justifies" many incorrect assumptions made knowingly in allowable stress design; it is the basis for plastic design. Under certain conditions, the ductility of steel can be inhibited, as is the case in brittle fracture and fatigue failures. (See Chapters 15 and 16.)

8. Modulus of Elasticity. The ratio of the normal stress to the normal strain in the direction of the applied load (the slope of the stress-strain curve) in the elastic range is called the modulus of elasticity or Young's modulus E. It defines the stiffness of the material, governs deflections, and influences buckling behavior. The modulus of elasticity is practically constant for all structural carbon steels at room temperature; 29,600 ksi is the average value, and values of 29,000 to 30,000 ksi are used in design calculations. The modulus decreases with increasing temperature at a rate of about 650 ksi per 100°F up to a temperature of 800°F, and more rapidly thereafter.[2.2] This stiffness, which is much higher than that of any other common structural material, is an important asset of steel.

9. Strain-Hardening Modulus. The strain-hardening modulus is the slope of the stress-strain curve in the strain-hardening range E_{st}, as shown in Fig. 2.2. Its value varies over a much greater range than the modulus of elasticity, and is usually about 700 to 900 ksi for structural carbon steels in the early part of the strain-hardening range.

10. Poisson's Ratio. The absolute value of the ratio of transverse strain to longitudinal strain under axial load is called Poisson's ratio, μ. It is approximately 0.30 in the elastic range and 0.50 in the plastic range.

11. Shearing Modulus of Elasticity. The shearing modulus of elasticity is the ratio of the shearing stress to the shearing strain within the elastic range and is designated as G. For structural steel, G is usually 11,500 to 12,000 ksi. From the theory of elasticity,

$$G = \frac{E}{2(1 + \mu)}$$

Hence, experimental determination of any two of the quantities E, μ, and G enables calculation of the third one.

12. Weldability. Weldability is the capability of a material to be welded without seriously impairing its mechanical properties. Weldability varies considerably for different types of steels and for different welding processes. (See Chapter 14.)

13. Machinability. The ease with which a material can be sawed, drilled, or otherwise shaped without seriously impairing its mechanical properties defines its machinability.

14. Formability. Formability is the ease with which a material can be bent or pressed to shape without fracture or other damage.

15. Durability and Corrosion Resistance. The ability to resist deterioration in a given environment defines the durability and corrosion resistance.

16. Fatigue Strength. Fatigue strength is the ability to withstand repeated applications of load or stress, and is usually expressed either as a fatigue limit or as the stress causing failure at a given number of cycles under a prescribed loading condition. (See Chapter 16.)

17. Toughness. In traditional usage, toughness is defined as the capacity to absorb large amounts of energy. It is related to the area under the

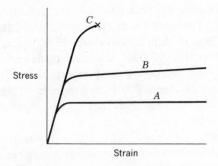

Fig. 2.4 Comparison of toughness (area under curves) for three different materials.

stress-strain curve, and hence is dependent on both strength and ductility. Figure 2.4 shows stress-strain curves for three materials: curves *A* and *C* exhibit about the same toughness with different strength and ductility, while *A* and *B* have equal ductility with different strength and toughness. Curve *B* has greater toughness than either *A* or *C*. In recent years the term toughness has been used in referring to notch toughness or fracture toughness defined below.

18. Notch Toughness. Notch toughness is the resistance to brittle fracture in the presence of notches or other stress concentrations.

19. Fracture Toughness. Fracture toughness of a material is a quantitative measure of notch toughness which relates critical stress and flaw size. (See Chapter 15.)

20. Impact Strength.   The ability to absorb energy at high rates of loading defines the impact strength.

21. Creep.   Creep is a gradual flow or change in dimension under a sustained constant stress.   It is usually not important in steel structures except at elevated temperatures, relatively high stresses, or a combination of the two.

22. Relaxation.   Relaxation is a decrease in load or stress under a sustained constant deformation.   It is important under the same conditions noted above for creep.   (Steel wire strand and wire rope exhibit a form of relaxation even at room-temperature.)

Of the foregoing properties, those of most general interest to the designer are strength, modulus of elasticity, and ductility.   In Figs. 2.1 and 2.2, note the linear behavior of the first part of the curve and the sudden change from this behavior to yielding.   For most carbon steels the extension at strain-hardening, ϵ_{st}, varies from about 6 to 16 times the elastic strain at yielding, ϵ_y. The plateau following initial yielding is often a desirable characteristic, for it permits structures to absorb high local strains without harmful effects.   For structural carbon steels, the elongation at fracture is about 20 per cent in 8 in. or approximately 15 times the elongation at the beginning of strain hardening, and 160 to 200 times the elongation at the beginning of yielding. There is no generally accepted criterion for the minimum ductility required for structural use; it obviously varies with the application.   Until further research results are available, past history of structural behavior is a useful guide.

Because the initial part of the stress-strain curve of steel is much the same in tension and compression, steel can be equally effective in resisting normal stresses of both types.   Except for wire rope and wire strand, tensile strength is usually not directly useful as a basis for design, due to the large deformations associated with it.   However, the margin between yield and tensile strength acts as reserve strength under some loading conditions.

The yield stress of steel in shear is about 0.57 times the yield stress in tension, and the ultimate strength in shear is $\frac{2}{3}$ to $\frac{3}{4}$ times the tensile strength.[2.2]

2.3 FACTORS AFFECTING MECHANICAL PROPERTIES

Factors affecting the measured values of the mechanical properties listed above include:

Chemical composition	Geometry
Heat treatment	Temperature
Prior strain history	Strain rate
	State of stress

The factors in the left column are largely dependent on the steel manufacturing process and fabrication method; those in the right column depend on the application, functional design, and detail design of the structure or structural element.

1. Chemical Composition

The most important single factor in determining the properties of a given heat of steel is the chemical composition. In carbon steels the elements carbon and manganese have a controlling influence on strength, ductility, and weldability. Most structural carbon steels are over 98 per cent iron, roughly $\frac{1}{4}$ of 1 per cent carbon, and about 1 per cent manganese by weight. Carbon increases the hardness or tensile strength, but has adverse effects on ductility and weldability. Therefore, small quantities of various alloying elements are sometimes used to increase the "hardenability" of a steel to get the maximum effectiveness from a given low percentage of carbon content.

Phosphorus and sulfur have a harmful effect on steel, especially on its impact strength; hence the fractional percentages of these elements must be kept low. Small amounts of copper increase corrosion resistance, and fractional percentages of silicon are used mainly to eliminate unwanted gases from the molten metal. Steels are classified as rimmed, killed, semikilled, or capped, according to the deoxidation process used in their manufacture, which controls the amount of gas evolved during solidification. Nickel, columbium, and vanadium also have a generally beneficial effect on steel behavior.

In general, some ductility must be sacrificed to obtain increased strength. This is tolerable as long as "surplus" ductility is available in the material. *What is of prime importance is that adequate ductility be exhibited by the final structure as fabricated.* This is a function of the material, the design (including design of details), the fabrication procedure, and the service conditions.

2. Heat Treatment and Strain History

Heat treatment and strain history can be interpreted broadly as including the rate of cooling, the finishing temperature of the hot-rolling operation, and the reduction of cross section that takes place in the normal rolling process. Faster cooling rates, lower finishing temperature, and greater reduction of cross section increase the final yield and tensile strength of hot-rolled steel. The effect of the amount of rolling is indicated in Fig. 2.5. Even though the chemical composition and original thickness are the same, the bar which is rolled more ($\frac{1}{4}$ in.) is stronger than the thicker bar.

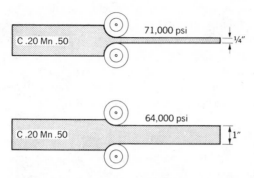

Fig. 2.5 Effect of amount of rolling on tensile strength.

Some steels are given prescribed heat treatments such as quenching and tempering. The rapid cooling caused by quenching increases strength and reduces ductility. Tempering then restores part of the ductility although giving up some of the strength gained by quenching. This process permits attainment of higher strengths while retaining good ductility.

Other steels are purposely cold rolled to obtain higher strength levels. The cold-working strain hardens the material and may be thought of, in effect, as utilizing or exhausting the initial part of the stress-strain curve, as indicated in Fig. 2.6. The true situation is more complex than this, however. For example, work-hardening in tension can reduce the yield strength in compression. Also, the phenomenon of strain aging (cold working followed by aging) sometimes increases the yield strength beyond that indicated in Fig. 2.6, and, in addition, increases the tensile strength and further reduces the ductility.[2.2] Cold-formed sections may have their yield strength raised considerably in the forming process (see Art. 12.1).

Differential cooling rates and cold-working both produce residual stresses whose effect is discussed in Art. 9.7.

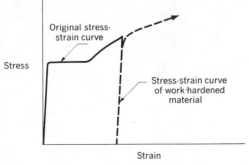

Fig. 2.6 Effect of work-hardening (both loadings in same direction).

The grain size of steel, which is affected by its composition and heat treatment, is frequently important. Steels of smaller grain size are usually less notch-sensitive and more effective in resisting impact and brittle fracture. (See Chapter 15.)

"Accidental" heat treatments such as those which occur in welding or burning also affect mechanical properties and residual stress patterns, and must be considered as part of the total design problem.

3. Geometry, Temperature, Strain Rate

The size and shape of a structural element influence its stress distribution and structural behavior. In general, smaller elements tend to give higher failure stresses than large ones, especially in fatigue and brittle-fracture situations. Notches or other changes in section can greatly affect the stress distribution, stress gradient, and structural behavior. Surface finish is also sometimes important, as in the case of fatigue, with smoother surfaces giving higher strengths.

Low temperature and high strain rate tend to increase the observed yield strength and tensile strength of steel, but they may also reduce the ductility. Furthermore, both low temperature and high strain rate have a greater influence on the yield strength than on tensile strength, and so the usual margin between yield and tensile strength is reduced. The effect of very high strain rate (100 in./in./sec) in comparison with a more common "slow" strain rate is indicated in Fig. 2.7. Figure 2.8 shows the influence of strain rate on yield stress at much lower test speeds, including a "zero" strain rate. For instance, at a strain rate of 313 micro-in./in./sec, the yield stress is 35.4 ksi, whereas it is 36.4 ksi at 549 micro-in./in./sec.

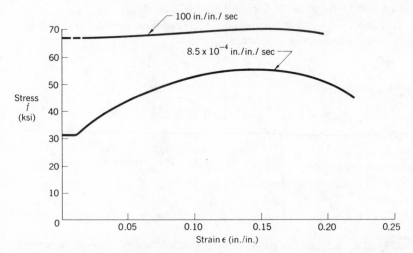

Fig. 2.7 Effect of high strain rate on shape of stress-strain diagram.[2·3]

4. State of Stress

The combination of stresses applied to a structural member greatly affects some measured properties. A normally brittle material can be induced to fail in a ductile manner if hydrostatic compressive stress is applied to the exterior of a tensile test bar. On the other hand, the application of equal triaxial tensile stresses will cause a completely brittle fracture in a ductile material. Notches and stress gradients which tend to promote triaxial stress conditions also tend to promote brittle fracture. (See Reference 2.5 and Chapter 15.)

Apart from the problem of brittle fracture, several theories have been developed to relate failure (by either yielding or fracture) under combined stresses to strength under uniaxial stress conditions. Problems 2.5 and 2.6 and Ref. 2.3 present a discussion of this important subject of failure theories or strength theories.

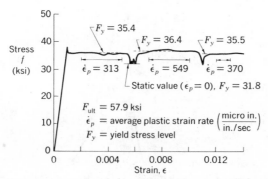

Fig. 2.8 Influence of strain rate on yield stress level.[2.4]

2.4 DETERMINATION OF MECHANICAL PROPERTIES

Standard test specimens and test procedures have been devised by ASTM to determine the various properties of steel. Because of the effects noted in Art. 2.3, these or other appropriate standards should be adhered to in order that results may be correlated with similar tests conducted by others. For example, the ratio of diameter to gage length should be fixed if comparable elongations are to be obtained in tension tests of round bars, and prescribed strain rates should be used in all tests. Even under normally controlled manufacturing procedures and carefully controlled test conditions, there is still some variation in observed mechanical properties of any material. Figure 2.9 shows the distribution curve of yield point values obtained in over 3900 mill tests on structural carbon steel.[2.6]   To ensure that practically all their steels meet the specified minimum yield point, steel mills must arrange their production so that the average yield stress is well above the specified minimum.   Considerable research has been done on the statistical behavior of steel and other structural materials, and the relationship of this statistical behavior to structural safety.[1.22]

As indicated above, the ordinary stress-strain curve is of major importance in determining several important properties, including the modulus of elasticity, yield point or yield strength, tensile strength, and ductility. Another diagram, a true stress-strain diagram, is obtained by plotting the true stress (axial load divided by the actual instantaneous area) as the ordinate, versus the true strain (change in length divided by instantaneous

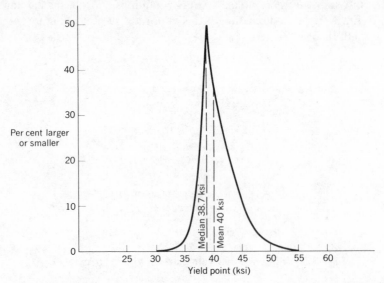

Fig. 2.9 Distribution of yield-point values, mill tests of structural carbon steel.[2.6]

gage length) as the abscissa.   It can be shown that the true strain can be expressed by $\log_e A_o/A_i$ where A_o and A_i are the original and instantaneous area respectively.[2.3]   Using this expression, a true stress-strain diagram for the minimum section at the neck of a tensile test specimen can be obtained as shown in Fig. 2.10.   While of limited value in structural design, the true stress-strain curve tends to give a better picture of local behavior and has certain advantage in correlating tension data with results from torsion tests, notched bar impact tests, and combined stress tests.   The data also correlate better with observations of metal forming behavior.

Fatigue tests are discussed in Chapter 16.   Impact tests, in addition to indicating ability to absorb energy under high rates of loading, are also used

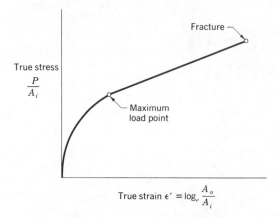

Fig. 2.10 True stress-strain diagram.

with notched specimens at low temperature to indicate the brittle-fracture resistance of a material. The standard Charpy impact test uses a small, simply supported, notched beam subjected to an impact load by a swinging pendulum. A series of tests conducted at different temperatures produces the relationship of absorbed energy versus temperature indicated in Fig. 15.3, which shows the marked reduction in energy-absorbing capacity at the "transition temperature" for the particular material, specimen configuration, striking velocity, and other prevailing test conditions. (See Art. 15.3.) Standard procedures for these tests, as well as for creep and relaxation tests, are covered in appropriate ASTM publications.

The discussion above relates to small coupon tests to determine basic material properties under prescribed test conditions. Because geometry, state of stress, and initial residual stresses play an important part in mechanical behavior, it is often desirable or necessary to test rather large pieces to determine behavior under simulated service conditions. As examples, compression tests of full cross sections of short columns (called "stub column tests"; [9.30] see Chapter 9) are made in order that the effects of residual stresses may be included directly in the results, and tests of large eye bars or cables of suspension bridges are often made on full cross sections of the material so that the effect of size and geometry of the cross section are taken into account in appraising the material behavior.

2.5 STRUCTURAL STEELS AVAILABLE

Prior to 1960 the AISC Specification for the Design, Fabrication, and Erection of Structural Steel for Buildings included only one grade of structural steel, ASTM A7, Steel for Bridges and Buildings, a carbon steel with a specified minimum yield point of 33 ksi. The 1961 Edition of the AISC

Table 2.1   Structural Carbon Steel ASTM A36[2.1]

Product	Thickness or ASTM Group	Minimum Yield Point (ksi)	Tensile Strength (ksi)	Minimum Elongation in 8 in. (%)
Plates	To 8 in. incl.	36.0	58.0 to 80.0	20
Shapes	Groups 1, 2, 3, and 4(a)	36.0	58.0 to 80.0	20
	Groups 4(b) and 5	36.0	58.0 min.	19*

* In 2-in. gage length.

Specification included six grades of structural steel, adding two carbon steels (ASTM A36 and A373) and three "high strength" steels (ASTM A242, A440, and A441). The 1969 Edition of this Specification deleted A7 and A373, but added nine others to the listing to include a total of thirteen different structural steels with ASTM designations, several of which have subgrades within the designation. Other recognized specifications include additional grades, and there are still other steels which are not listed in specifications largely because they are proprietary. All the available structural steels cannot be considered individually in this chapter, and for ease of discussion it is convenient to group them in the following general classifications.

Structural Carbon Steels. These are the least expensive and most widely used of all structural steels. They depend on the carbon content to develop their strength, and usually have yield points of 25 to 40 ksi. ASTM A36 Structural Steel and ASTM A570 Hot Rolled Carbon Steel Sheets and Strip of Structural Quality are examples of this group. The mechanical properties of ASTM A36 are listed in Table 2.1. A36 steel is suitable for riveted, bolted, or welded construction, and is available in plates, bars, and all structural shapes.

As noted earlier, the mechanical properties of a given steel may vary with the thickness. Table 2.2 divides the various rolled sections into five groups for tensile property classification, according to the dimension of the thickest element of the section.

Throughout this article, the strengths quoted apply to mill tests at ASTM allowable strain rates. As shown in Fig. 2.9, mill test results usually exceed these values considerably, but they are based on coupons taken from the web of rolled shapes and record the upper yield point if one exists. Laboratory tests from the thicker flanges of similar shapes conducted at "zero" strain rate to determine "static" strength may give yield points from 5 to 30 per cent below the values indicated by mill tests.[2.4,2.6]

A partial listing of the chemical requirements for ASTM A36 structural carbon plates and shapes appears in Table 2.3. The amount of phosphorus

Table 2.2   Structural-Shape Size Groupings for Tensile Property Classification

Structural Shape	Group 1	Group 2	Group 3	Group 4	Group 5
W Shapes	W 24 × 55, 61 W 21 × 44, 49 W 18 × 45 to 60 incl W 18 × 35, 40 W 16 × 26 to 50 incl W 14 × 22 to 53 incl W 12 × 14 to 58 incl W 10 × 11.5 45 incl W 8 × 10 to 48 incl W 6 × 8.5 to 25 incl W 5 × 16 to 18.5 incl W 4 × 13	W 36 × 135 to 194 incl W 33 × 118 to 152 incl W 30 × 99 to 210 incl W 27 × 84 to 177 incl W 24 × 68 to 160 incl W 21 × 55 to 142 incl W 18 × 64 to 114 incl W 16 × 58 to 96 incl W 14 × 61 to 136 incl W 12 × 65 to 106 incl W 10 × 49 to 112 incl W 8 × 58 to 67 incl	W 36 × 230 to 300 incl W 33 × 200 to 240 incl W 14 × 142 to 211 incl W 12 × 120 to 190 incl	W 14 × 219 to 550 incl (See Note 2)	W 14 × 605 to 730 incl
M Shapes	to 35 lb/ft incl	over 35 lb/ft			
S Shapes	to 35 lb/ft incl	over 35 lb/ft			
HP Shapes		to 102 lb/ft incl	over 102 lb/ft		
American Standard channels (C)	to 20 lb/ft incl	over 20 lb/ft			
Miscellaneous channels (MC)	to 28.5 lb/ft incl	over 28.5 lb/ft			
Angles (L), structural & bar-size	to ½ in. incl	over ½ to ¾ in. incl	over ¾ in.		

NOTES: (1) Structural tees from W, M, and S shapes fall in the same group as the structural
shape from which they are cut.
  (2) Group 4 is sometimes subdivided into 4(a), up to 426 lb/ft, and 4(b), over 426 lb/ft.
SOURCE: AISC Manual.

and sulphur is limited, while the carbon, manganese, and silicon content are
controlled to insure weldability.   Higher percentages of these latter elements
are required in the heavier sections to develop the required minimum yield
point, which is the same for all thicknesses.

Another carbon steel designation is ASTM A529, which covers steel
plates and bars ½ inch and under in thickness, and Group 1 shapes, with a

Table 2.3 Chemical Requirements, ASTM A36 Steel[2.1]

Product	Thickness or ASTM Group	C_{max}	Mn	P_{max}	S_{max}	Si
Plates	To $\frac{3}{4}$ in.	0.25	—	0.04	0.05	—
	Over $\frac{3}{4}$ in to $1\frac{1}{2}$ in.	0.25	0.30/1.20	0.04	0.05	—
	Over $1\frac{1}{2}$ in. to $2\frac{1}{2}$ in.	0.26	0.30/1.20	0.04	0.05	1.15/1.30
	Over $2\frac{1}{2}$ in. to 4 in.	0.27	0.30/1.20	0.04	0.05	1.15/1.30
	Over 4 in. to 8 in.	0.29	0.30/1.20	0.04	0.05	1.15/1.30
	Over 8 in. to 15 in.	0.29	0.30/1.20	0.04	0.05	1.15/1.30
Shapes	Groups 1, 2, 3, and 4(a)	0.26	—	0.04	0.05	—
	Groups 4(b) and 5	0.26	—	0.04	0.05	0.15/0.30

minimum yield point of 42 ksi and 60-to-85-ksi tensile strength, for use in buildings and in similar riveted, bolted, and welded construction.

High-Strength Low-Alloy Steels. This group of steels uses various alloying elements in addition to carbon to attain higher yield strengths, which are usually between 42 and 65 ksi. Examples are ASTM A242 High-Strength Low-Alloy Structural Steel, ASTM A441 High-Strength Low-Alloy Structural Manganese Vanadium Steel, ASTM A572 High-Strength Low-Alloy Columbium-Vanadium Steels of Structural Quality, and ASTM A440 High-Strength Structural Steel. The specified minimum mechanical properties of these steels are listed in Table 2.4. Because of its chemical composition, the A440 grade steel technically cannot be defined as "high-strength low-alloy steel," but it is listed in this group because of its mechanical

Table 2.4 Mechanical Properties of High-Strength Low-Alloy Steels[2.1]

ASTM Grade	Thickness or ASTM Group	Minimum Yield Point (ksi)	Tensile Strength (ksi)	Minimum Elongation in 8 in. (%)
A242, A440, and A441	$\frac{3}{4}$ in. and under Groups 1 and 2	50.0	70.0	18
	Over $\frac{3}{4}$ in. to $1\frac{1}{2}$ in. Group 3	46.0	67.0	19
	Over $1\frac{1}{2}$ in. to 4 in. Groups 4 and 5	42.0	65.0	16
A572-42	To 4 in. inclusive Groups 1, 2, 3, 4(a)	42.0	60.0	20
-45	To $1\frac{1}{2}$ in. inclusive Groups 1, 2, 3, 4(a)	45.0	60.0	19
-50	To $1\frac{1}{2}$ in. inclusive Groups 1, 2, 3, 4(a)	50.0	65.0	18
-60	To 1 in. inclusive Groups 1 and 2	60.0	75.0	16
-65	To $\frac{1}{2}$ in. inclusive Group 1	65.0	80.0	15

properties. It usually costs slightly less than A441, and is intended primarily for riveted and bolted structures. A242, A440 and A572-42, -45, and -50 steels are suitable for riveted, bolted or welded construction, while ASTM A572-55, -60, and -65 steels are intended for riveted or bolted construction. These steels can be furnished in plates, bars, sheets, and a wide variety of structural shapes, but suppliers should be consulted regarding availability of specific shapes.

Several of the steels in this classification have atmospheric corrosion resistance two to eight times that of structural carbon steel; but if this property is desired, the designer should specify it. A group of steels known as "weathering steels" are used exposed and unpainted in buildings and bridges; many of these are included in ASTM designation A588. In a short length of time such steels develop their own protective oxide coating.

Table 2.4 shows that the minimum yield points for A242, A440, and A441 steels are the same, but that the yield point decreases for thicker material, from 50 ksi down to 42 ksi. This approach differs from that used in A36 steel, where the yield point is held constant for all thicknesses by altering the chemistry. The required minimum elongation in 8 inches for these higher strength steels is slightly less than that for A36.

Heat-Treated High-Strength Carbon Steels. This group includes carbon steels that have been heat-treated to obtain certain desirable mechanical properties. They provide yield strength levels of about 50 to 80 ksi. A number of proprietary steels are available in this category; they are rolled mostly in plates, and are usually weldable.

Quenched and Tempered Alloy Steels. This group of steels uses various alloying elements, in addition to carbon, plus a quench and temper heat treatment to attain minimum yield strength levels of about 90 to 100 ksi. These steels, sometimes called constructional alloys, are usually suitable for welding, and because of their alloy content, they usually provide better corrosion resistance than carbon steel. They are available in plates and bars, and in some shapes. An example of this group is ASTM A514, High Yield Strength Quenched and Tempered Alloy Steel Plate, Suitable for Welding. Another example is ASTM A517, a pressure-vessel-quality, quenched and tempered steel plate with a minimum yield strength of 100 ksi. Specified minimum mechanical properties for A514 steel are given in Table 2.5.

Table 2.5 Mechanical Properties of Quenched and Tempered Alloy Steel Plates, ASTM A514[2.1]

Thickness (in.)	Minimum Yield Point (ksi)	Tensile Strength (ksi)	Minimum Elongation in 2 in. (%)
To $2\frac{1}{2}$ in. inclusive	100	115–135	18
Over $2\frac{1}{2}$ in. to 4 in. inclusive	90	105–135	17

Stainless Steels.   These steels are available for structural use primarily in light gages, and contain sizeable amounts of chromium, nickel, and other alloys to give very high corrosion resistance, ease of maintenance, and pleasing appearance.   Stainless steels are finding increased structural application, particularly in the transportation, chemical, and processing industries, where their inherent advantages are important.   Yield strengths range widely, depending largely on the composition and amount of cold working.   Stainless steels usually exhibit gradual yielding with no distinct yield plateau, and their proportional limit often is a lower percentage of the yield strength than is the case for carbon and low alloy steels.   Also, these

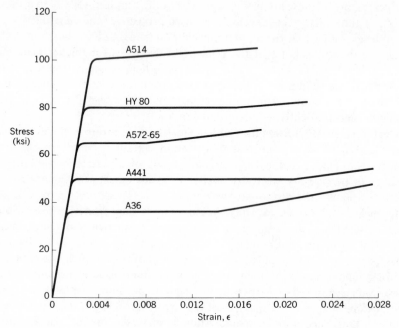

Fig. 2.11   Initial portion of stress–strain curves for various steels.

steels may have somewhat different stress-strain curves in tension and com-pression, and in directions parallel and transverse to their rolling direction. Because of their influence on buckling and strength, such factors require consideration in structural design.

Other Steels.   Many other special-purpose steels are available, including for example, HY80, an 80-ksi-yield-strength steel which has been widely used in ship construction, and "Maraging steels" which develop yield strengths in the 200-ksi to 300-ksi range using a high nickel alloy content and a heat treatment to age the iron-nickel martensite.   (The theoretical binding force between iron atoms is over 4000 ksi.) Of course, as strength is increased, price is usually increased, and ductility is usually decreased. Furthermore, the modulus of elasticity, which governs deflections and elastic buckling, is relatively constant for all structural steel. Hence, because of cost, buckling considerations, and greater deflections at higher stresses, these very-high-strength materials probably will have limited, but important, structural design applications.

Initial portions of the stress-strain curves for several of the materials mentioned above appear in Fig. 2.11.

2.6 MATERIALS COVERED BY DESIGN SPECIFICATIONS

All the steels listed in Tables 2.1, 2.4, and 2.5, as well as several others, are covered by the AISC Specification for the Design, Fabrication, and Erection of Structural Steel for Buildings,[1.21] and plastic design is permitted for the included steels with yield points up to 65 ksi. The availability of plates and shapes in some of the ASTM-designation steels is summarized in Table 2.6.

The AISC Specification covers design using hot-rolled steel plates and shapes. The design of cold-formed carbon steel structural members is covered by the AISI Specification for the Design of Cold-Formed Members.[1.24] Some of the sheet and strip steels listed in the 1972 printing of that Specification are given in Table 2.7, along with the minimum mechanical property requirements. ASTM A570 Grades D and E are hot-rolled carbon steel grades which are covered also by the AISC Specification. ASTM A611 is a cold-rolled carbon steel sheet, ASTM A606 and ASTM A607 are high-strength low alloys, and ASTM A446 covers galvanized sheet steels. A611 Grade E and A446 Grade E steels are severely cold-rolled full-hard products with a minimum yield and tensile strength of 80 ksi and 82 ksi, respectively, and no specified minimum elongation. These products are used mainly for roofing material and, pending additional research, are not intended for primary structural applications.

The AISI Specification also permits the use of other steels whose "properties and suitability" are appropriately established and controlled. The various buckling provisions in the AISI Specification are written for gradually yielding steels whose proportional limit is not lower than about 75 per cent of the specified minimum yield point or yield strength. This Specification permits utilization of the increased strength of a material due to cold-forming (see Chapter 12). Another AISI publication, "Design of Light Gage Cold-Formed Stainless Steel Structural Members," [2.7] gives design provisions for stainless steel material with anisotropic behavior and relatively low proportional limit compared with its yield strength.

Table 2.6   Availability of Shapes, Plates, and Bars According to ASTM Structural Steel Specifications

Steel Type	ASTM Designation	F_y Minimum Yield Stress (ksi)	1†	2	3	4	5	To ½″ Incl.	Over ½″ to ¾″ Incl.	Over ¾″ to 1″ Incl.	Over 1″ to 1½″ Incl.	Over 1½″ to 2½″ Incl.	Over 2½″ to 4″ Incl.	Over 4″ to 5″ Incl.	Over 5″ to 8″ Incl.	Over 8″
Carbon	A36	32														▨
	A36	36	▨	▨	▨	▨	▨	▨	▨	▨	▨	▨	▨	▨	▨	
	A529	42	▨					▨								
High-strength	A440	42				▨	▨					▨	▨			
		46			▨					▨	▨					
		50	▨	▨				▨	▨							
High-strength, low-alloy	A441	40												▨	▨	
		42				▨	▨					▨	▨			
		46			▨						▨	▨				
		50	▨	▨				▨	▨	▨						
	‡A572 Gr. 42	42	▨	▨	▨	▨	▨	▨	▨	▨	▨	▨	▨			
	‡A572 Gr. 45	45	▨	▨	▨	▨		▨	▨	▨	▨	▨				
	‡A572 Gr. 50	50	▨	▨	▨			▨	▨	▨	▨					
	‡A572 Gr. 55	55	▨	▨				▨	▨	▨						
	‡A572 Gr. 60	60	▨					▨	▨							
	‡A572 Gr. 65	65	▨					▨								
Corrosion-resistant, high-strength, low-alloy	A242	42				▨	▨					▨	▨			
		46			▨					▨	▨					
		50	▨	▨				▨	▨							
	A588	42													▨	▨	
		46				▨								▨		
		50	▨	▨	▨			▨	▨	▨	▨	▨				
Quenched & tempered alloy	A514	90											▨			
		100						▨	▨	▨	▨	▨				

* Grouped per ASTM A6.
† Includes bar size shapes.
‡ To W 14 × 426 only.
▨ Available.
□ Not available.
SOURCE: AISC Manual.

Table 2.7 Mechanical Properties of Structural Sheet and Strip Steels[2.1]

Description	ASTM Designation	Grade	Thickness (in.)	Minimum Yield Point or Yield Strength (ksi)	Tensile Strength (ksi)	Minimum Elongation in 2″ in. (%)
Hot-rolled carbon steel sheets and strip, structural quality	A570	A	0.0255	25	45	18–27
		B	to	30	49	17–25
		C	0.2299	33	52	16–23
		D		40	55	14–21
		E		42	58	12–19
Cold-rolled carbon steel sheet, structural quality	A611	A		25	42	26
		B		30	45	24
		C		33	48	22
		D		40	52	20
		E*		80	82	—
High-strength, low-alloy steel sheet and strip with improved corrosion resistance, hot rolled and cold rolled	A606	Hot rolled		45–50	65–70	22
		Cold rolled		45	65	20–22
High-strength, low-alloy steel sheet and strip with columbium and/or vanadium, hot rolled and cold rolled	A607	45		45	60	22–25
		50		50	65	20–22
		55		55	70	18–20
		60		60	75	16–18
		65		65	80	15–16
		70		70	85	14
Zinc-coated (galvanized) steel sheet, structural quality	A446	A		33	45	20
		B		37	52	18
		C		40	55	16
		D		50	65	12
		E*		80	82	—
		F		50	70	12

* Not intended for general structural use.

NOTE: ASTM designations A570 and A611 recently replaced older designation A245. Also, A606 and A607 replaced A374 and A375.

The AASHO Standard Specifications for Highway Bridges provides for use of structural carbon steel A36, and high-strength low-alloy steels A242, A440, A441, A588, A572 as well as the quenched and tempered steels, A514/A517.

2.7 SIGNIFICANCE OF MATERIAL SELECTION

To meet the functional, economic, and safety requirements of a structure, there are a broad variety of steels from which to choose, with appropriate design guides available to ensure their proper use.   One task of the structural designer, then, is to select the material most suitable for the particular needs.

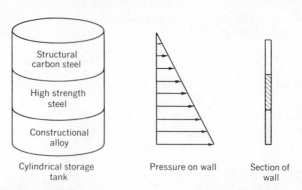

Fig. 2.12   Hybrid construction of cylindrical storage tank.[2.8]

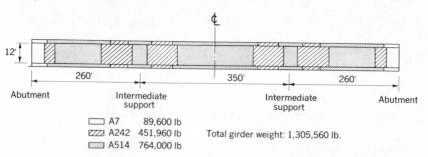

	A7	89,600 lb
	A242	451,960 lb
	A514	764,000 lb

Total girder weight: 1,305,560 lb.

Fig. 2.13   Hybrid steel girder, Whiskey Creek Bridge, Shasta County, California.   (Courtesy of California Division of Highways.)

The selection will be based on the proposed service conditions, the mechanical properties of the materials, and economics. The question of economics includes the base price of the material, fabrication costs, freight, effect of dead weight of the structure on foundation costs, optimum space utilization, and other factors. Even though their base price is higher, the higher-strength steels may still have considerable cost advantage if their full allowable stress is utilized, or if the other economic factors mentioned above are important.

A frequent use of the higher-strength steels is in hybrid construction, that is, using higher-strength steels for the more severely stressed elements of a structure, and lower-strength, lower-price steels elsewhere, permitting greatest over-all economy. This concept is illustrated in Fig. 2.12, where the wall thickness of a storage tank is maintained constant by the use of higher-strength steel at the bottom, where the pressure is the greatest. Figure 2.13 shows a bridge in which the girder is fabricated from three different steels. Provisions covering hybrid flexural members are included in the AISC Specification.

Proper selection of material is one of the essential steps in the design to ensure that the structure will meet its functional requirements with adequate safety and minimum cost.

PROBLEMS

2.1. An ASTM standard 1.50-by-0.375-in. tension coupon of structural carbon steel is to be tested. To obtain additional information, the standard 8-in. gage length is subdivided into 1-in. lengths. After testing, the gage lengths measure in inches as follows: 1.17; 1.19; 1.24; 1.32; 1.60 (the fracture occurred within this original 1-in. length); 1.26; 1.20; 1.17. The maximum load was 36,400 lb, the load at fracture was 31,000 lb, and the reduction of area was measured as 41.4 per cent.

    (a) Compute the tensile strength.
    (b) Compute the true stress at fracture.
    (c) Compute the per cent elongation in 8 in.
    (d) Compute the maximum per cent elongation in 2 in. (Note, however, that the specimen does not have dimensions meeting ASTM requirements for standard 2 in. gage length specimens.)

2.2. A material is to be selected for the propeller shafts of an ocean liner. What properties of the material would be important in the selection, and why?

2.3. A new material has been suggested for use in a small steel bridge for a mining railroad in northern Canada.

    (a) In addition to cost, what considerations would you take into account in deciding whether the material is suitable?
    (b) What considerations would you make in deciding whether the bridge should be of welded, riveted, or bolted construction?

2.4. A full-size test bar, 6 in. wide, $1\frac{1}{2}$ in. thick, and 20 ft long, is made from the same ASTM A36 material which gave the stress-strain curve shown in Fig. 2.1 for a standard coupon with an 8-in. gage length. Give a rough estimate of the per cent elongation that would be measured over a 10-ft gage length of the full-size test bar.

2.5. Yielding in steel occurs along planes of maximum shear stress, and the shear stress theory of failure under combined stresses predicts yielding when the maximum shearing stress becomes equal to the maximum shearing stress at the yield point in a simple tension test.   Draw Mohr's circle[2.5] for:

    (a) a simple tension test, when $F_1 = 36$ ksi;
    (b) a biaxial stress condition when $F_1 = 45$ ksi, $F_2 = 15$ ksi;
    (c) a biaxial stress condition when $F_1 = 20$ ksi, $F_2 = -20$ ksi;
    (d) $F_x = 20$ ksi, $F_{xy} = 15$ ksi;

where F_1 and F_2 are principal stresses, F_x is a normal stress and F_{xy} is a shear stress on the same plane.

    Comment on the results, assuming the material is ASTM A36 steel.

2.6. One of the theories of failure under combined stresses most applicable to the failure of steel by yielding is the maximum distortion energy theory, which states that yield failure will occur under combined stresses when

$$(F_1 - F_2)^2 + (F_2 - F_3)^2 + (F_3 - F_1)^2 = 2F_y^2$$

where F_1, F_2, and F_3 are the principal stresses, and F_y is the yield stress in simple tension. For $F_y = 36$ ksi, compute the predicted value of tensile stress F_1 to cause yielding if:

    (a) $F_2 = F_3 = 6$ ksi
    (b) $F_2 = \frac{1}{2}F_1$, and $F_3 = 0$
    (c) $F_2 = F_3 = \frac{1}{2}F_1$
    (d) $F_2 = \frac{1}{2}F_1$, and $F_3 = -\frac{1}{2}F_1$
    (e) $F_2 = -\frac{1}{2}F_1$, and $F_3 = -\frac{1}{2}F_1$
    (f) $F_1 = F_2 = F_3$

    Comment on the results.

3

BUILDINGS

3.1 INTRODUCTION

A guide to building design is presented in this chapter by reducing the complex building structure into its basic systems: the roof system, the floor system, the walls and partitions, the steel frame and its bracing, and special provisions for operational facilities. Each system is presented by giving the various schemes commonly used for a particular type of building.

The design of a building involves both the satisfactory completion of its planning and its structural design. Planning includes both functional planning and the selection of a framing scheme. The steps in meeting the design objectives are outlined in general in Art. 1.1.

Functional planning is the development of the layout of the building to meet the service requirements in the most efficient manner. This involves the consideration of the purpose for which the building is designed, its general size and shape, its location, and other related features. The choice of a framing scheme is very much dependent on the layout. For this reason, the selection and arrangement of the structural members that make up a building frame is done usually in conjunction with the planning of the layout.

With the completion of the preliminary plans, the detailed design of the structure follows. This involves load analysis, structural analysis, and the structural design of the members. Load analysis is discussed in this chapter, and structural member design is presented in succeeding chapters.

3.2 HISTORICAL REVIEW

The art of building construction probably dates back to the Greek civilization of the seventh century B.C. However, of all the ancient civilizations, the Egyptians had an architecture which was the most unified, the most artistic, and which endured the longest.[3.1]

For centuries, masonry was the main structural material. Concrete and steel are relative newcomers to the field. The first rolled metallic beams were made of wrought iron, which was made available in 1853.

The development of the skyscrapers started with the introduction of rolled steel shapes in 1884. In the same year, the first building with a steel skeleton frame was built in Chicago by William LeBaron Jenny. This was a 10-story office building for the Home Insurance Company.[3.2] The wide acceptance of tall steel-framed buildings which then followed the construction of this structure is traced in Art. 21.1. For years, the most notable of skyscrapers has been the Empire State Building (Fig. 1.2). Its construction in 1930 drew the attention of both engineers and laymen to its 102 stories, extending through a record building height of 1250 feet.[3.3] In 1950, a TV tower was added extending its height to 1472 feet. However, the twin towers of the World Trade Center in New York (Fig. 1.4), completed in 1972, each have 110 stories and a building height of 1350 feet. The Sears Tower in Chicago, currently (1973) in construction, will soar to a height of 1450 feet.

The future outlook in steel structures shows a tendency to higher office buildings designed as space structures acting as a unit, and to the increasing use of unique framing schemes in exhibition buildings.

The introduction of weldable high-strength steels in the early sixties brought in a new dimension to structural steel design, that of the ratio of material strength to weight. The use of model studies and high-speed computers, and the increased emphasis on research, have led to considerable advancement of the knowledge of structural analysis and design. New design techniques combined with the greater ability of fabricators to shape steel into any desired configuration promise more and more imaginative structures that are not only esthetic but functional and economical as well.

3.3 LOADS

The loads for which a building must be designed may be classified into dead loads, vertical live loads, and lateral live loads. Dead loads include the weight of permanent equipment and the weight of the fixed components of the building, such as floors, beams, girders, roofs, columns, fixed walls and partitions, and the like. All vertical loads other than dead loads may be included under vertical live loads. Lateral live loads are external forces due to the action of wind, earth, and hydrostatic pressure. The probable effects of seismic disturbance are also lateral loads. The dynamic effects of crane movement may also be considered, in part, as lateral loads.

1. Dead Loads

The weight of the structural framing and all other building components fixed to and permanently supported by it are considered as dead loads. A reasonable estimate of the total weight of a building can usually be obtained from a preliminary sketch showing the structural and architectural layouts. This information is necessary for such steps in the analysis as the evaluation of preliminary values of earthquake loads and the determination of the loads on the foundation. A correction to this approximation should not be neglected in the final design if the actual calculation of the total weight shows a significant difference from the original assumptions.

For the design of the individual members, a more detailed dead load analysis usually is required. This involves a step-by-step procedure starting with the components directly supporting the live loads (floor and roof systems) and proceeding along the path of stresses (beams, girders, and columns) to the foundations. The dead loads on a member are determined only after the member itself, and the portion of the structure it supports, have been designed. Thus, the actual dead load for each member should be checked and corrected; the design proceeds only after the necessary adjustments have been made.

Information on the weights of building materials is given in local building codes. A more complete list of weights is given in Ref. 3.4.

2. Vertical Live Loads

The loads on a building due to its occupancy, as well as snow loads on roof surfaces, are regarded as vertical live loads. Occupancy loads include personnel, furniture, machinery, stored materials, and other similar items. In buildings, live loads are often regarded as a uniformly distributed loading.

Table 3.1   Minimum Snow Loads*

Roof Slope	3 in 12	6 in 12	9 in 12	12 in 12 or more
Southern States	20	15	12	10
Central States	25	20	15	10
Northern States	30	25	17	10
Great Lakes, New England, and Mountain Areas†	40	30	20	10

* Pounds per square foot on horizontal projection.

† These areas include the northern portions of Minnesota, Wisconsin, Michigan, New York, and Massachusetts; the states of Vermont, New Hampshire, and Maine; and also the Appalachians above 2000-ft elevation, the Pacific Coast range above 1000 ft, and the Rocky Mountains above 4000 ft.

The amount of live load due to snow is dependent upon the location of the structure, the slope of the roof, and the orientation of the building with respect to wind direction. Snow load provisions are usually included in local building codes. The recommendations of the Housing and Home Finance Agency based on a snow study conducted by the U.S. Weather Bureau are given in Table 3.1 for the United States.[3.5]

Several attempts have been made to standardize live load requirements due to occupancy, but building codes and specifications still have very diverse provisions on this subject. Table 3.2 gives a comparison of live load

Table 3.2   Typical Minimum Live Load Requirements*

Type of Occupancy	New York City (1942)	Philadelphia (1941)	Chicago (1941)	ANSI (1955)	Natl. Board of Fire Und. (1955)
Residential					
Halls, dining					
rooms, etc.	100	100	100	100	100
Rooms	40	40	40	40	40
Institutional					
Halls, dining					
rooms, etc.	100	100	100	100	100
Rooms	40	50	50	40	40
Assembly Halls					
With fixed seats	75	60	75	60	60
Assemblage	100	100	100	100	100
Stage floor	—	—	—	150	150
Business					
Office space	50	60	50	80	80
Corridors	100	100	100	100	100
Mercantile	75	110	100	125	125
Industrial	120	200	100	125	125
Storage†	120	150	100–250	125–250	125–250

* Pounds per square foot of floor area.
† The building should carry whatever is stored, with minimum values as shown.

recommendations taken from several building codes.[3.6]   The requirements are classified according to occupancy as follows:[3.7]

1. Residential (including hotels)
2. Institutional (hospitals, sanitariums, jails)
3. Assembly (theatres, auditoriums, churches, schools)
4. Business (office-type buildings)
5. Mercantile (stores, shops, salesrooms)
6. Industrial (manufacturing, fabrication, assembly)
7. Storage (warehouses)

Except in structures for storage and certain types of manufacturing and warehousing, the maximum loading on each floor is not likely to occur at any one time. In recognition of this fact, most building codes allow a reduction in live loading. The recommendations of ANSI A58.1-1969[3.4] have been adopted in the National Building Code, the Uniform Building Code, and the Basic Building Code. The recommendations are given below:

a) No reduction shall be applied to the roof live load.
b) For live loads of 100 pounds or less per square foot, the design live load on any member supporting 150 sq ft or more may be reduced at the rate of 0.08% per sq ft of area supported by the member, except, that no reduction shall be made for areas to be occupied as places of public assembly. The reduction shall exceed neither R as determined by the following, nor 60%:

$$R = 100 \times \frac{D + L}{4.33L} \tag{3.1}$$

in which R = reduction in per cent
$\qquad D$ = dead load per sq ft of the area supported by the member
$\qquad L$ = design live load per sq ft of the area supported by the member.
For live loads exceeding 100 pounds per sq ft, no reduction shall be made except that the design live loads on columns may be reduced 20%.

3. Lateral Live Loads

Lateral live loads include earth and hydrostatic pressure on the building, and the effects of wind and earthquake. Although the evaluation of lateral forces due to earth and water pressure may or may not be covered by codes or specifications, it is the responsibility of the designer to adhere to "good practice," that is, the proper consideration of horizontal pressures as well as uplift forces due to the hydrostatic head.[3.8,3.9]

The requirements of wind pressure and earthquake forces are usually specified in local building codes. These requirements normally represent values for average conditions. There is a wide variation that exists among codes of different areas. In the following sections, an attempt is made to present data and information which will serve as a basis for a rational approach to the problem of determining appropriate values of wind and earthquake forces acting on building structures.

Wind Loads. The behavior of a structure in wind is analogous to the behavior of a stationary object submerged in flowing water. Wind is essentially air in motion. A structure constitutes an obstacle placed in the path of the wind such that the moving air is stopped or deflected from its original path. The kinetic energy of motion is transformed into the potential energy of pressure. The intensity of the pressure depends on the velocity of the wind, the mass density of the air, and the orientation, shape, and area of the structure.[3.10]

The determination of wind forces on a structure is a complex problem in aerodynamics. In structural design applications, the usual practice is to convert the dynamic pressure arising from wind action to the equivalent static forces.[3.10]   The dynamic pressure is a function of the mass density of air and the velocity of the wind, that is,

$$q = \tfrac{1}{2}\rho v^2 \tag{3.2}$$

where q = dynamic pressure in psf; ρ = mass density of air in slugs/cu ft; v = wind velocity in fps.

The conversion of the dynamic pressure into an equivalent static force involves the determination of the "drag force" or the reactive force exerted by the structure in opposing the forward motion of the wind.   The drag force (F_d, in pounds) can be expressed in terms of the dynamic pressure q by the expression

$$F_d = C_d q A \tag{3.3}$$

where A, in sq ft, is the exposed area normal to the wind direction, and C_d is a drag coefficient which depends on the shape of the structure and its orientation with respect to the wind stream.

The average static pressure on the structure is then

$$p = \frac{F_d}{A} = C_d q \tag{3.4}$$

For standard air ($\rho = 0.765$ pcf at 15°C at sea level), $q = 0.00256 V^2$ (V in mph) and thus

$$p = 0.00256 C_d V^2 \tag{3.5}$$

Therefore, the evaluation of the wind load is reduced to the determination of the drag or shape coefficient C_d and a design wind velocity V.   A summary of the Final Report (1961) of the ASCE Task Committee on Wind Forces is presented in the following sections.

Wind Velocity.   The determination of a design wind velocity involves the statistical study of wind recordings for the vicinity under consideration.   The United States Weather Bureau has voluminous records of observations taken at different locations in the United States.   The results of the statistical analysis of the data are presented in Fig. 3.1.   The general considerations involved in arriving at the final results are given in Ref. 3.10.

Shape Coefficient.   The recommended values for the drag or shape coefficient are given as follows.   More detailed information on the effect of building form, wind direction, and air-tightness on the external and internal pressure is available in Ref. 3.10.

For rectangular buildings with the wind direction perpendicular to one of the sides, the average coefficient for the front face is about 0.9 regardless of building proportions.   The rear suction depends on the ratio of various

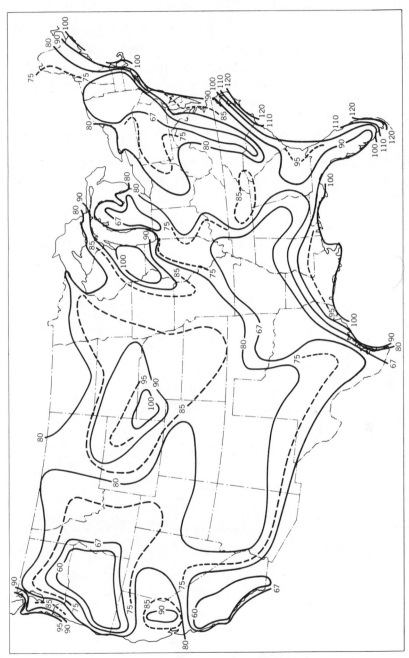

Fig. 3.1 Contours showing extreme basic winds with 50-year recurrence. (Courtesy of ASCE.)

dimensions. For example, for a building square in plan and with a height-to-width ratio of unity, the coefficient is -0.5; for a ratio of 2.5 or greater, a value of -0.6 is recommended.

For gabled roofs the values recommended by the ASCE Subcommittee 31, the Danish Building Code, and the Swiss Building Code are given in Fig. 3.2. The ASCE curve represents only average values, as no consideration is given to the proportions of the building. The negative pressures recommended by the Danish Building Code are relatively small. However, an additional provision of this code requires that all walls and roofs be anchored to resist a suction of $0.8q$. No reference is made to the proportions of the building. In the Swiss Building Code, the effect of the dimensions of the building is taken into account by limiting the application of the rule to buildings with ratio of wall height to depth (parallel to the wind direction) of 0.67 to 1.5, and ratio of width (perpendicular to the wind direction) to depth of 0.4 to 2.5.

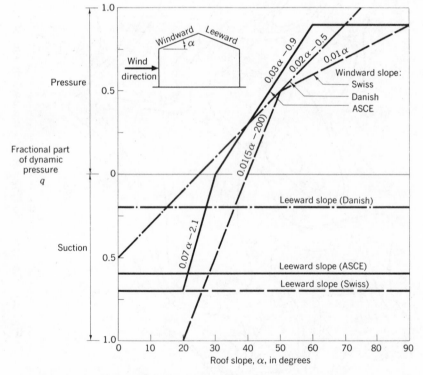

Fig. 3.2 External coefficients for wind pressures and suctions on symmetrical gable roofs. (Courtesy of ASCE.)

Effect of Terrain. The wind records which form the basis of the reported data are usually taken in weather stations located in open fields or airports and not in city areas where the presence of buildings will affect the basic wind velocities. Therefore, for complex and major buildings, it is necessary to conduct wind tunnel tests to study the effects of surrounding buildings on the forces generated by different levels of basic wind velocity values.[3.11]

Wind Forces. For direct design applications in areas of the United States where average wind conditions occur, the recommendations of Subcommittee 31 on external and internal wind forces on buildings are summarized in Figs. 3.3, 3.4, and 3.5.[3.12,3.13] Figure 3.3 shows the wind pressure to be

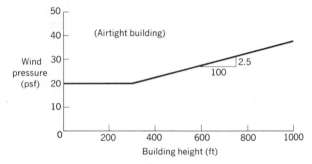

Fig. 3.3 Wind pressure on tall buildings (airtight). (Courtesy of ASCE.)

used on the vertical projection of a multi-story building. For inclined roof surfaces of heights less than 300 ft above the ground, the values are as given in Fig. 3.4. For buildings with wall openings, the internal pressure or suction acting normal to the walls and the roof are given in Fig. 3.5.

Earthquake Loads. The forces developed during an earthquake are not physical loads applied to the structure but are inertial forces resulting from the resistance of the mass of the system to the imparted motion. Therefore, the inertial forces generated by the dynamic disturbance are dependent on (a) the nature of the earthquake motion which may be described in terms of acceleration, velocity, time, and direction, and (b) the response of the structure which is defined by its elastic and mass properties, as well as its stiffness and damping properties.

A rigid structure set in motion (such as the one in Fig. 3.6), is subjected to an inertial force determined from Newton's First Law as

$$F_i = \frac{a}{g} W \qquad (3.6)$$

where a is the acceleration of the ground at any distance, g is the acceleration due to gravity, and W is the weight of the structure.

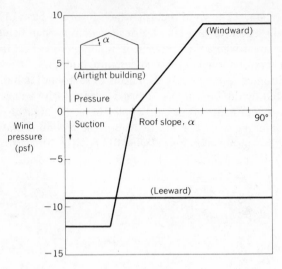

Fig. 3.4 Wind pressure on gabled roofs (airtight).   (Courtesy of ASCE.)

From statics, the base shear is given by

$$V = \frac{a}{g}W = CW \qquad (3.7)$$

where C is the ratio of the ground acceleration to gravitational acceleration (seismic coefficient).   This expression is the basis of some of the earliest earthquake codes.   By assuming a value for the maximum acceleration of the ground, the base shear can be expressed as a certain percentage (say 10 to 12 per cent) of the weight of the structure.

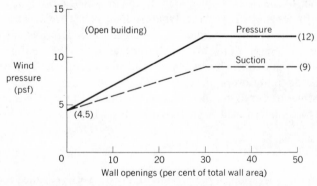

Fig. 3.5 Internal and external pressure for open buildings.   (Courtesy of ASCE.)

However, a building structure possesses some degree of flexibility. During an earthquake, the structure vibrates. The vibration can be defined in terms of relatively simple deflected shapes of the sinusoidal type, called the natural modes of vibration. Each natural mode of vibration results in a lateral force distribution related to the shape of the mode. The effect of an earthquake can be expressed in terms of lateral forces varying along the height of the structure which represents the cumulative influence of all the modes of vibration. The action of the lateral forces results in an envelope of maximum dynamic shear at each point along the height of the structure. The general shape of the maximum shear envelope is approximately parabolic.

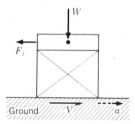

Fig. 3.6 Dynamic response of rigid structure to earthquake.

For practical applications in a design office, seismic building codes attempt to translate the dynamic solution into relatively simpler static design procedures.

To define the base shear, V, several factors are considered, namely: Z, a coefficient dependent on the earthquake zone in which the structure is located; K, a coefficient dependent on the type of structure; C, a coefficient dependent upon the vibration characteristics or fundamental period of vibration of the structure; and W, the mass factor which is expressed in terms of the dead load of the structure and portions of storage live load; thus:

$$V = ZKCW \tag{3.8}$$

The values of the coefficients are defined in detail in Ref. 3.14 and the Uniform Building Code.

Recognizing the approximately parabolic shape of the maximum shear envelope and the "whiplash" effect in tall flexible buildings,[3.14,3.15] the lateral load is assumed distributed as follows:

$$F_x = \frac{(V - F_t)w_x h_x}{\sum\limits_{i=1}^{n} w_i h_i} \tag{3.9}$$

where V = base shear

    F_t = portion of the base shear considered concentrated at the top of the structure to account for "whiplash"

w_i, w_x = portion of W located on or assigned to level i or x

h_i, h_x = height in feet, above the base to level i, or x.

For a building with approximately uniform weight distribution Eq. 3.9 results in a triangular lateral load distribution along the building height with the maximum value at the top (with or without a concentrated load depending on the flexibility of the structure). The triangular load distribution results in a parabolic shear distribution closely approximating the dynamic action.

Use of the provisions of seismic codes, such as the SEAOC recommendations,[3.14] results in design loads calculated from coefficients based on observation and past experience. However, there is a certain "good design practice" which cannot be expressed in terms of code values but nevertheless should be considered by the design engineer:[3.14,3.16]

1. The structure must be able to transmit the seismic forces, including all their effects such as shear, torsion, and overturning, all the way from the point of origin to the foundation material.

2. Adjacent sections of a structure with different dynamic properties should either be tied together to force them to oscillate together or be provided with adequate separation to allow for independent oscillation without hammering against each other.

3. Provide structural continuity or a "second line of seismic resistance" such that the failure of one element results only in the redistribution of the forces to the remaining elements which still form a stable system.

4. Failures should occur first in the member rather than in the connection.

5. Structural members must possess adequate strength, ductility, and resistance to reversals of large deformation.

4. Load Combinations

In the design of roof trusses and single-story frames the loads carried by the structure include the dead load and the live load of snow and wind. The design usually is made on the basis of the stresses from the following combinations of loading:

    I   Dead Load + Snow
    II   Dead Load + Wind
   III Dead Load + Wind + Snow
   IV Dead Load + Wind + Partial Snow Load

Partial snow load sometimes is considered due to drifting of snow. This may happen in buildings with uneven bay heights. Due to wind, part of the snow on the higher roof level may be blown away but snow drift may accumulate on the lower roof level. It might even be that the snow load on the lower side of the building might be greater than normal.

Other load combinations may also be considered. A certain degree of engineering judgment is always required in selecting the proper load combinations.

In multi-story buildings, the stress is determined from a combination of:

    I   Dead Load + Live Load, *and*
   II Dead Load + Live Load + Wind or Earthquake

Live loading is a random type of load and should be imposed on the structure such that maximum or critical moments and forces are induced on each individual member. From an analysis of the qualitative influence line for moment on the members of the multi-story frame, it can be shown that the combination of full dead load plus full live load may not always be the critical loading. The loading pattern shown in Fig. 3.7, usually called

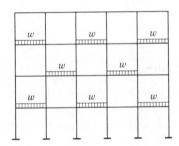

Fig. 3.7 "Checkerboard" pattern of loading.

"checkerboard loading," may produce the critical conditions particularly for the maximum positive moment in the loaded spans.[3.17]

Sections [1.5] and [2.1] of the AISC Specification[1.21] allow an increase of $33\frac{1}{3}$ per cent in the allowable stresses, and a reduction of the load factor for plastic design from 1.7 to 1.3 when the stresses are induced by wind or earthquake forces. These provisions apply to load combinations II, III, and IV for roofs, and load combination II for multi-story frames. The increase in allowable stress is also equivalent to considering three quarters of the value of the above load combinations. Thus, the choice of the critical stress should be based on a comparison of the values produced by the load combinations given in Table 3.3.

Table 3.3 Load Combinations

Building Component	Load Combination	Allowable Stress Design	Plastic Design
Roof	I	D + S	1.70 (D + S)
	II	$\frac{3}{4}$ (D + W)	1.30 (D + W)
	III	$\frac{3}{4}$ (D + W + S)	1.30 (D + W + S)
	IV	$\frac{3}{4}$ (D + W + partial S)	1.30 (D + W + partial S)
Floor	I	D + L	1.70 (D + L)
Multi-story frame	I	D + L	1.70 (D + L)
	II	$\frac{3}{4}$ (D + L + W, E)	1.30 (D + L + W, E)

3.4 TYPES OF STEEL BUILDINGS

Modern buildings may be classified according to their layout and framing into (1) industrial-type buildings, (2) multi-story buildings, and (3) special buildings with unusual framing. Some buildings may combine the characteristic framing of each type to serve multipurpose objectives.

1. Industrial-Type Buildings

One- or two-story buildings used for industrial, institutional, and residential purposes may be included under the classification of industrial-type buildings. The steel framing may be a rigid frame, a two-hinged or three-hinged arch, a truss system supported on columns, or a tier-type frame, as shown in Fig. 3.8. These types of framing are more commonly used in industrial buildings.

For an industrial building where an assembly-line type of production is housed, a structure is needed which has one or more relatively wide aisles. The building is usually of one story and must be unobstructed by columns for a considerable length. Few, if any, partitions should exist to allow for a smooth flow of production. The "mill-type" steel building with a truss-on-column framing scheme has been used extensively for this purpose.

More recently, rigid frame construction is gaining popularity, due to its simplicity and pleasing appearance. The rigid frame has found use in smaller structures, such as light commercial buildings, schools, churches, and residences, and in wide span structures such as airplane hangars, gymnasiums, and supermarkets. The arched frame has been used also for structures

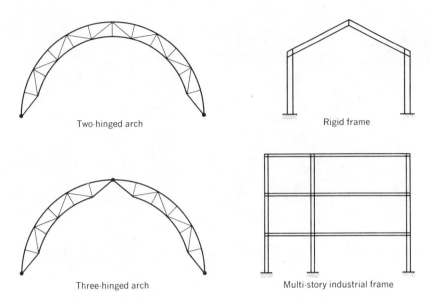

Two-hinged arch   Rigid frame

Three-hinged arch   Multi-story industrial frame

Fig. 3.8 Industrial-type steel framing.

where a relatively wide area free of columns is required.   Although industrial buildings are usually single-storied, a tier-type frame is sometimes used where office space is required and for a "gravity-flow" type of production where the assembly line is laid out on a vertical system.

2. Multi-Story Buildings

The multi-story building is used for apartments, hotels, office buildings, public buildings, and other structures similar to those shown in Fig. 3.9. These buildings are characterized by a regular floor pattern and arrangement of columns, all extending through several levels or stories.   Such an arrangement is usually referred to as a "tier-type" construction, wherein a system of beams and girders on columns forms a multi-story steel skeleton frame which supports the roof, the floors, and the walls.

Buildings with steel skeleton frames combined with an efficient lateral bracing system can be constructed to great heights.   When the cost of real estate is high, an increase in the number of stories may be economically necessary.   However, as the structure attains greater heights, limitations are imposed by the strength of the foundation, legal heights allowed by the local code, higher initial cost of the building, and the greater influence of wind and earthquake loads.

Warren Petroleum Company
Tulsa

Latino-American Tower
Mexico City

Wilshire Comstock Apartments
Los Angeles

International Building
San Francisco

Fig. 3.9 Typical multi-story buildings. (Courtesy of Bethlehem Steel Corporation.)

3. Special Buildings

The versatility of steel as a structural material is becoming increasingly evident with the appearance of novel and unique designs in steel buildings. With the advancement of knowledge in the design and analysis of structures, more imaginative building forms, which are functional as well as esthetic, have been made possible.

Steel buildings with unique framing schemes have been designed for such structures as exhibition buildings, churches, houses, restaurants, gymnasiums, auditoriums, and transportation terminals. In most cases, the special framing is required to provide a large area unobstructed by columns, such as for the Las Vegas Convention Center, which is framed with a dome structure (Fig. 3.10), the "cable-supported" roof for the Madison Square Garden

Fig. 3.10 Las Vegas Convention Center. (Courtesy of AISC.)

Complex (Fig. 1.7) in New York, and the "cable-suspended" system for the Villita Assembly Hall in San Antonio, Texas (Fig. 3.11).

In smaller structures such as churches and exposition buildings, unique framing schemes are used to achieve an architectural effect. Some examples of this type of building are the Air Force Academy Chapel in Colorado, the Beth Torah Temple in North Miami Beach, and the Space Needle of the Seattle World's Fair (Figs. 3.12 to 3.14).

These types of buildings, which combine beauty in form with functional arrangement as well as economy, are possible with such modern framing schemes as arches, domes, space frames, folded plate structures, and suspended beams and frames.

3.5 BUILDING COMPONENTS

The structural components of a building may be grouped into basic systems which contribute to the serviceability of the structure by satisfying some functional requirement. These are: the structural framing, which supports

Fig. 3.11 Villita Assembly Hall. (Courtesy of Bethlehem Steel Corporation.)

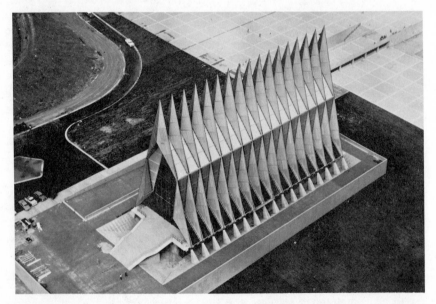

Fig. 3.12 Air Force Academy Chapel. (Courtesy of U.S. Air Force Academy.)

Fig. 3.13 Beth Torah Temple, North Miami Beach.

the loads coming from the roof and floors; the roof system, which provides an enclosure from the top; the floor system, which provides usable space; the walls and partitions, which provide side enclosures and separation of working space; and special provisions for operational facilities. The scheme or arrangement of a system is dependent on the building type, although some components may be used for both industrial-type and multi-story buildings.

More recent developments, introduced in the later part of the 60's, attempt to achieve economy of construction through the application of systems design concepts. Examples of this type of approach include the modular construction method, component systems, stacked modules based on mobile home units, and the staggered truss system.

1. Industrial-Type Buildings

Structural Framing. The trend in modern industrial building design is toward the use of the rigid frame. Although the rigid frame may require more material than a truss, the savings in the cost of fabrication and erection due to its simplicity and in wall heights due to increased headroom make it highly competitive with other framing schemes.

A rigid frame may be fabricated from rolled shapes or from built-up sections with riveted, bolted, or welded connections. Some examples of

Fig. 3.14 The Space Needle, Seattle World's Fair, 1962. (Courtesy of General Public Relations.)

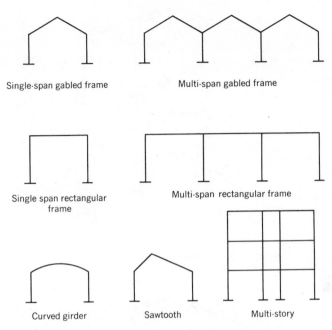

Single-span gabled frame

Multi-span gabled frame

Single span rectangular frame

Multi-span rectangular frame

Curved girder

Sawtooth

Multi-story

Fig. 3.15 Types of rigid frames.

rigid frames are shown in Fig. 3.15. The inclined or horizontal member which directly supports the purlins is called a "rafter." It is rigidly connected to the column to form a continuous structural member which supports the loads in bending, shear, and thrust. The analysis and design of rigid frames is treated in Chapter 20.

The more common forms of single span rigid frames are shown in Fig. 3.16. The rigid frame may be made up of rolled or built-up shapes which are welded together. In some cases a haunched knee is used where the moments are most critical. This results in a lighter section for both the rafter and the column. The haunched knee may be fabricated from plates. In both cases, stiffeners are usually provided at the joints. Another type of rigid frame construction makes use of tapered members which may also be fabricated from plates by welding.

For many years the most commonly used structural framing was the truss-on-columns scheme shown in Fig. 3.17. The line of trusses and its supporting columns constitute a "bent" and the distance between bents is called a "bay." The roof deck is supported directly by the purlins which span the distance between trusses. When the walls are made up of light sheets, a grid system using "girts" is provided for attaching the sheets. A diagonal bracing

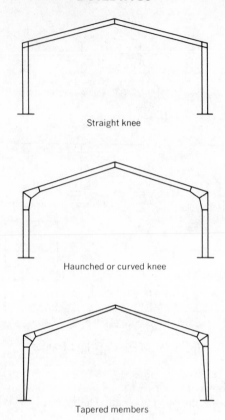

Straight knee

Haunched or curved knee

Tapered members

Fig. 3.16 Types of rigid frame construction.

system also is provided to resist instability in the lateral and longitudinal directions.

Bracing. Under the action of lateral forces, a building frame such as that shown in Fig. 3.17 may fail due to sway in the plane of the bents or by tilting like a series of cards. A bracing system is designed to provide the necessary rigidity in the building to resist this instability. In some cases, the bracing is used to aid in aligning the frame in the erection, to minimize vibration, and to reduce the unsupported length of main members. A discussion of a bracing system is presented in Art. 20.4.

A rigid frame can be designed to carry its load without causing excessive sidesway (see Art. 10.3). However, in a truss-on-column scheme, there is a remote possibility that the corner joints and the base become virtually hinged and that the structure becomes unstable. Thus, it is common practice to stiffen the connection between the truss and the column with knee braces.

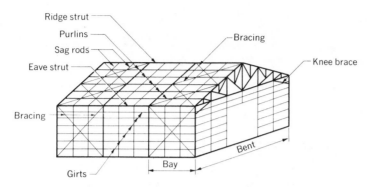

Fig. 3.17 Truss-on-column scheme.

The choice of a scheme for the longitudinal bracing system is a matter of design. The structure is required to resist lateral forces and to transmit them to the foundation. A bent can resist vertical as well as lateral forces in its plane. When the lateral forces act normal to its plane, the bent offers little, if any, resistance. However, two bents may be braced together to form a stable structure. First, diagonal braces are introduced in the plane of the top chord and in the plane of the columns to form a braced bay or tower, as shown in Fig. 3.17. A braced bay with the purlins and the eave struts is able to resist lateral forces in the longitudinal direction. In addition to the bracing in the top chord, vertical sway frames as well as intermediate struts should also be provided at the bottom chord to reduce its unsupported length and to anticipate the possible application of lateral forces from unforeseen loads such as heavy material temporarily hung from below.

Roof System. The roof system includes the roof framing, the roof deck, and the roof covering. The truss or the rafter of the rigid frame makes up the main roof framing. The purlins are provided as a secondary framing system to span the distance between the trusses. Typical roof framing arrangements are shown in Fig. 3.18. Purlins may be spaced from 2 to 5 ft or more, depending on the strength of the roof deck supported. The sag rods are provided to support the purlins in the weak direction and to resist the tangential component of the roof load (Fig. 3.19). Depending on the bay distance and the type of roofing, one or more sag rods may be used. The sag rod is a tension rod designed to resist the total tangential load from the purlins below the peak of the truss and on one side of the roof. Anchorage for the sag rod must be provided at the ridge.

When used on a pitched roof, the purlins present one of the more complex of beam problems. Due to the inclination of the roof, vertical loads subject the purlins to biaxial bending as well as twisting moments (Fig. 3.19).

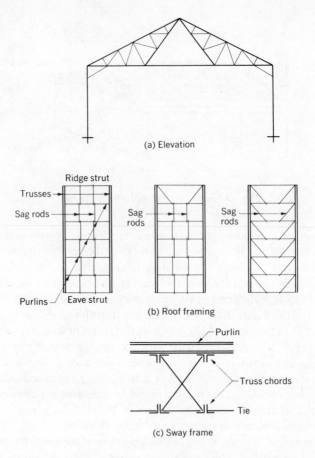

Fig. 3.18 Typical roof framing.

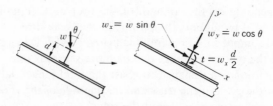

Fig. 3.19 Forces acting on a purlin.

Purlin sections are usually made up of channels, junior beams, or cold-formed sections. Prefabricated open-web joists are widely used also.

The term "roof deck" denotes the structural assembly directly resting on the purlins and providing the enclosing element to the top surface of the building. It may also be the surface to which the roof covering is applied. Several types of roof decks are available with varying degrees of strength, weight, weatherproofness, fire resistance, and cost. These are (1) precast slab planks of concrete, asbestos, or gypsum, (2) interlocking steel plates, (3) concrete, (4) wood planks, (5) corrugated metal or asbestos sheets, and (6) poured-in-place gypsum.

A roof covering may be of the single-unit type or made up of factory-processed, multiple units. The single-unit type, which is specially adaptable to a flat roof deck, is discussed later in this article. The multiple-unit type is designed for pitched roof decks where water can flow directly over the roof surface into gutters and leaders for drainage to the ground. Shingles, slate, tile, and metal panels are included under this type. The imperviousness of this type of roofing depends on its effectiveness to allow the water to flow down without seeping through the joints. A minimum roof slope of about 4 in. vertical to 12 in. horizontal is usually required. In addition, special care is taken in making the joints waterproof.

Floor Systems. The floor for a one-story industrial building is usually a concrete slab laid directly on the ground. Although it appears to be a simple type of construction, special attention should be given to a number of details to avoid the cracking of the slab. A concrete ground slab may crack due to freezing of trapped water, shrinkage in curing, expansion due to a rise in temperature, or excessive loading. Construction techniques to prevent the cracking of the slab are discussed in Ref. 3.18.

Walls and Partitions. The exterior walls of an industrial building may be made of corrugated metal or asbestos sheets, sandwich-type cold-formed sheets and interposed insulation, hollow tile, concrete masonry units, or poured concrete. The selection of one type over the other depends on several factors such as appearance, cost, serviceability, weight, and fire resistance.

Corrugated sheets are lightweight wall covering. These types of covering require a secondary framing system for attachment and support. Masonry and concrete walls may be of the load-bearing or nonload-bearing type. For nonload-bearing walls, the minimum thickness required by building codes is usually of sufficient strength.

Partitions may be of gypsum, masonry, concrete, glass, or steel materials. Some of the factors which affect their construction are fire resistance, lighting characteristics, acoustical properties, ease of erection, and resistance to wear, impact, and abrasion.

Special Facilities: Crane Systems. In the design of industrial buildings, provisions for an overhead crane system to operate within the structure often are required. The crane system consists of the crane trolley, the rail, the crane runway girder, and the supporting columns. Other types of crane systems include bracket, hanging, portal, and monorail cranes.

The design of the crane runway girder is dependent on the capacity of the crane, the weight of the crane and trolley, and certain clearance requirements above the floor as well as below the roof system. Crane wheel loads dependent on the type of the crane are usually available from crane manufacturers' catalogs.

The AISC Specification requires that crane girders be designed for a lateral force equal to 20 per cent of the sum of the lifted load (maximum crane capacity) and the weight of the crane trolley. The load is considered to be applied at the top of the rail, one-half on each side of the runway, and assumed to act in either direction normal to the girder. In addition to this force, an

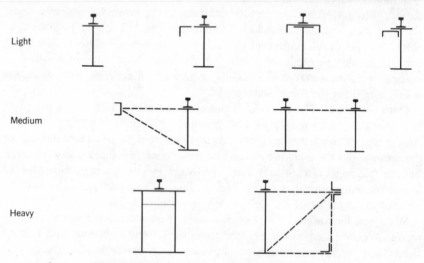

Fig. 3.20 Typical crane girder sections.

impact value of 25 per cent is used on the vertical load. These additional loads take into account the dynamic effect of the moving trolley. Common sections in use are shown in Fig. 3.20. The girder may be designed for the combined effect of biaxial bending and torsion, or the action of the vertical loading and the horizontal thrust may be treated separately. The lateral action of the horizontal thrust may be assumed to be resisted by the top flange area only or by a supplementary horizontal truss system.

The column which supports the crane girder is subjected to forces and moments due to the crane load, the roof load, and lateral wind load or earthquake. The column is designed to resist the maximum effect of the combined loading although such a combination will rarely occur. Three common types of arrangements are used in crane columns. These are shown in Fig. 3.21. In the first scheme, a bracket rigidly connected to the column section is used to provide the support to the crane girder. This system is used for light cranes. For heavier cranes, the column section may be increased at the girder level as shown in the second example. The third arrangement is used for heavy cranes as well as for light cranes in a high bay. Two column sections, connected with intermittent diaphragms, can be designed to act independently. By providing sufficient clearance between the two columns,

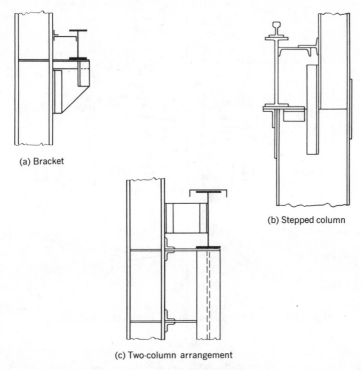

(a) Bracket

(b) Stepped column

(c) Two-column arrangement

Fig. 3.21 Typical crane columns.

the crane girder may be centered on the secondary column. Thus, no bending moment is induced by the vertical crane loads. The main building column can be designed to resist the lateral thrust from the crane in addition to its building loads. The secondary column can be designed to carry the gravity loads as well as the impact loads coming from the crane girder.

More detailed recommendations on structural design and fabrication of crane facilities and their material requirements are given in Ref. 1.27.

2. Multi-Story Buildings

Structural Framing. The steel framing of a multi-story building is a system of beams, girders or trusses, and columns, designed to carry all the gravity loads of the structure and to resist wind and earthquake forces. The roof and the floor systems are directly supported by horizontal beams and girders spanning the distance between the vertical columns. A typical arrangement is shown in Fig. 3.22. The details of a typical tier or floor

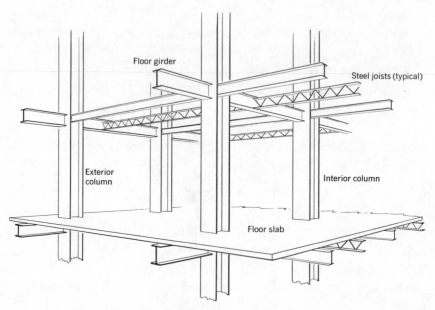

Fig. 3.22 Structural framing of multi-story building.

level usually can be duplicated and extended both horizontally to cover a wide area and vertically to skyscraper heights, except when high wind loads or earthquake loads require a modification in the framing scheme.

In planning the structural framing, early attention should be given to provisions for special facilities such as elevator shafts and stairwells. These components extend through several stories, suggesting the fixed position of some of the columns. The rest of the columns may be arranged in a regular pattern from a consideration of the architectural layout and the typical construction used for the roof, floor, walls, and partitions.

The AISC Specification allows three types of construction in steel frames based on the type and behavior of the connections. Type 1, commonly designated as "rigid-frame" (continuous frame), assumes that beam-to-column connections have sufficient rigidity to hold virtually unchanged the original angles between intersecting members. Type 2, commonly designated as "simple" framing (unrestrained, free-ended), assumes that, in so far as gravity loading is concerned, the ends of beams and girders are connected for shear only, and are free to rotate under gravity load. Type 3, commonly designated as "semi-rigid" framing (partially restrained), assumes that the connections of beams and girders possess a dependent and known moment capacity intermediate in degree between the complete rigidity of Type 1 and the complete flexibility of Type 2. The design of connections appropriate for each type of construction is presented in Chapters 18 and 19. In this section, special attention is given to the additional requirements of wind bracings in Type 2 construction ("simple" frames).

To provide lateral stiffness to the building, especially to those with frames of Type 2 construction, a system of bracing may be designed to resist lateral forces due to wind or earthquake. Wind loads acting on the exterior walls are transmitted by the floor system to braced bents or the shear walls in the steel framing system. If the floor is not sufficiently rigid to transform the loads, a horizontal bracing system must be provided between bents.

Braced bents may be placed in the outside walls, in permanent interior walls, around elevator and stairway shafts, and other service areas where the bracing will not be an obstruction. If possible, the braced bent should be placed symmetrically across the building to avoid the twisting of the structure about its vertical axis. The distribution of the wind loads to the braced bents should take into account the eccentricity between the line of action of the resultant wind force and the center of resistance of the wind bracing system.

The two approaches used to obtain stiffness in regular-shaped buildings are shown in Fig. 3.23. In the core-wall type, the lateral load resisting system is integrated with a central utility core. In the bearing-wall type, the exterior walls are braced and stiffened such that the building acts like a huge canti-levered box. The use of the bearing-wall type of bents was a major factor in reducing the cost of the structural framing for both the World Trade Center in New York and the John Hancock Building in Chicago.

Bents may be braced in a number of ways, as shown in Fig. 3.24. The full diagonal bracing is the most economical type of wind bracing. Its use is somewhat restricted by frequent requirements for window and door openings. Portal bracing is usually more costly and its use is limited to special cases. The Vierendeel type of braced bents has the advantage of being easily treated architecturally and can be used on the outside walls as part of the building facade.

Floor System. The floor system of the multi-story building is a more elaborate construction than the type used in industrial buildings. It

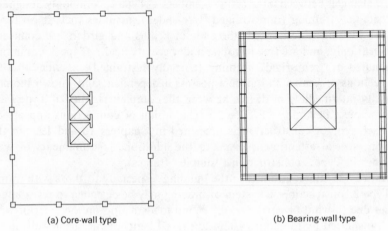

(a) Core-wall type   (b) Bearing-wall type

Fig. 3.23 Arrangement of walls for stiffness.

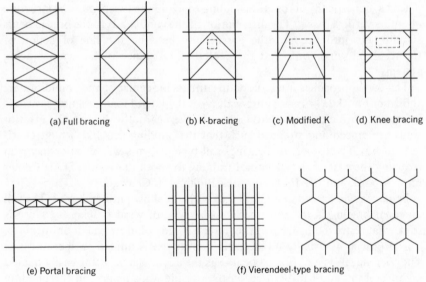

(a) Full bracing   (b) K-bracing   (c) Modified K   (d) Knee bracing

(e) Portal bracing   (f) Vierendeel-type bracing

Fig. 3.24 Bracing systems.

includes a framing of girders and beams which supports the floor deck. Several types of floor decks are available with varying durability, fire resistance, weight, and adaptability to the application of finished floors and ceilings and to the installation of utilities.   The selection of a suitable type of floor deck depends on the occupancy requirements, structural adequacy, and cost considerations.

The more common types of floor construction in use are: (1) reinforced concrete slabs, (2) concrete-pan system, (3) open-web joist system, (4) cellular steel type, and (5) other adaptations of these four basic systems.

A reinforced concrete slab is appropriate where a high degree of rigidity is desired.   This system is one of the heaviest types used for floor construction in steel framed buildings, although the use of lightweight concrete has reduced the dead weight considerably.

In the concrete-pan system shown in Fig. 3.25, the metal pan and the supporting wood planks make up the forms on which the concrete is poured. This results in a ribbed concrete floor which is a lighter floor system than the conventional reinforced concrete slab.

One of the lightest types of floor construction is the open-web joist system (Fig. 3.26).   The system consists of a 2 to $2\frac{1}{2}$ in. concrete slab poured in place over ribbed-wire fabric backed by heavy paper laid directly on the

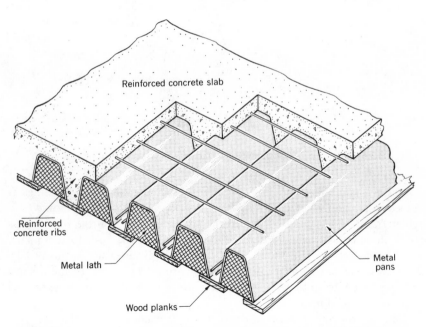

Fig. 3.25 Concrete-pan system.   (Courtesy of John Wiley & Sons, Inc.)

joists. The satisfactory design and construction of the joist is assured by adhering to the Specifications of the Steel Joist Institute.[3.19] Several types of prefabricated joists are availablè.

Cellular steel floors have weights comparable to the open-web joist system. A typical arrangement is shown in Fig. 3.27. The concrete on top of the decking usually serves merely as a finishing material, and a low-strength, lightweight concrete is adequate for the purpose. The load-carrying system consists of light gage, cold-formed cellular units. This system facilitates the installation of wiring for utilities because each cell serves as a conduit. This acts to offset the high initial cost of the cellular decking.

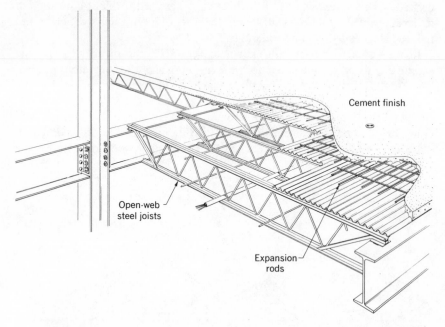

Cement finish

Open-web steel joists

Expansion rods

Fig. 3.26 Steel joist construction. (Courtesy of Bethlehem Steel Corp.)

Roof System. The flat roof commonly used in multi-story buildings is framed in a manner similar to the floor system. In some cases, however, a lighter framing may be used due to the smaller loading on the roof. This can be done by using smaller members or by spacing the joists further apart.

The roof deck may be any of the several types given earlier in this article. Ordinarily, the better-quality types of decks are used in roofs for multi-story buildings to conform to performance requirements higher than those needed in industrial buildings.

The roof covering in flat roof construction is usually of a single-unit type. It may be built up from four or five layers of felt cemented together with

asphalt or coal tar pitch and surfaced with gravel or slag (Fig. 3.28).   Due to
the tendency of the flat roof to retain water, a very high degree of imperviousness must be attained.   Great care should be exercised in the preparation of
the roof deck surface to avoid the possibility of leakage.

Walls and Partitions.   The exterior walls of multi-story buildings may be
made of masonry, concrete, metal and glass facings, or prefabricated panels.
The primary purpose of the wall is to serve as a weather screen.   Factors
such as weight, fire resistance, insulating properties, durability, initial cost,
and maintenance cost should also be considered.

Masonry and concrete walls have been used extensively in the past.   The

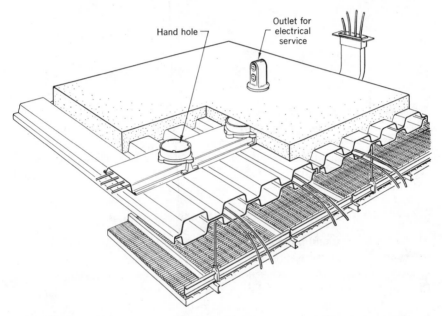

Fig. 3.27 Cellular floor construction. (Courtesy of John Wiley & Sons,
        Inc.)

use of metal and glass facings backed up with insulation, a fire-resistant
material, and an interior finishing is gaining more favor from architects and
engineers.   This is due to its light weight and attractive color and texture.
When used in a regular pattern on the walls of a tall building, a building
facade with a very pleasing appearance results.

Prefabricated panels are made up of an insulation core, sandwiched between
a thin lightweight backing and facing.   They can be designed to satisfy
performance requirements comparable to concrete walls, and have the added
advantage of being lighter and more convenient to erect.

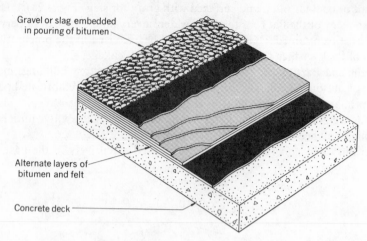

Gravel or slag embedded
in pouring of bitumen

Alternate layers of
bitumen and felt

Concrete deck

Fig. 3.28 Flat roof construction.

Partitions are used to provide a dividing wall which separates the interior
of a building into spaces and rooms.   They may be fixed or movable.   Their
design is based on requirements of fire resistance, visual and acoustical privacy,
and the facility of concealing pipes, conduits, and ducts.   The more common
types used in multi-story buildings are made of masonry, concrete, or metal.

3. Modern Framing Scheme for Special Buildings

The arch frame and the dome, which have been used in a number of structures, are probably the most familiar of modern schemes used in framing
special buildings.   More uncommon types of unusual steel framing include
the trussed space frame, folded plate structures, and suspended roofs.

A circular building has an advantage over other building forms in that it
provides more usable space when used in sports arenas and buildings of
similar functions.   Better spectator participation is achieved through the use
of peripheral seating.   A maximum efficiency of access and egress is also
attained with the radial aisles.

In framing a circular building, one of several types of metallic dome
structures may be used.   There are eight existing types of metallic dome
structures: (1) braced rib domes, (2) solid-rib domes, (3) lamella domes,
(4) curvilinear domes, (5) lattice domes, (6) geodesic domes, (7) stressed-
skin domes, and (8) inverted domes.[3.20]   Reference 3.20 considers basic
dome geometries and the different types of domes and their comparative
characteristics and gives some criteria for the analysis of each type.   Several
examples of existing dome structures are also cited.

For the roof framing of the Air Force Academy Chapel shown in Fig. 3.12, a modification to the standard A-frame was introduced.   Instead of a plane frame, the structure was built up from tetrahedron space frames as shown in Fig. 3.29.   The tetrahedrons were fabricated in three sections: top, intermediate, and bottom.   The erection of the structure was started with ten bottom sections, eight intermediate sections, and six top sections to

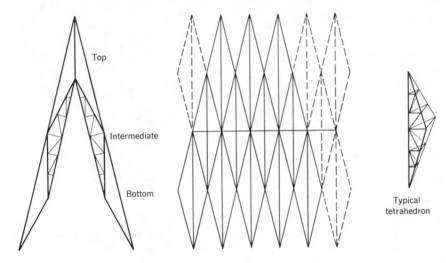

Fig. 3.29   Air Force Chapel framing.

form a self-supporting assembly.   The rest of the frames were then erected and connected to the stable basic system.

The roof for the Beth Torah Temple, shown in Fig. 3.30, makes use of a frame which is designed as a folded space structure.   The roof over each half of the building is formed as a skewed square in plan and is folded into eight triangular plates in the structure.   Each plane is framed with three edge beams: a ridge beam acting in tension and bending, a perimeter beam acting in compression and bending, and a valley beam acting in compression only. Joists span at right angles from the ridge beam to the valley and perimeter beams.[3.21]

The Villita Assembly Hall, San Antonio, Texas, is a classic example of a radial-type cable-suspended system.   See Fig. 3.11.   The cables act as the main supporting system for the roof construction.   The truss-work shown in the diagram is not a primary structural component of the roof system, but simply a series of ceiling frames hung from the cables.[3.22]   The central portion of the roof structure is made up of radial trusses with a peripheral tension ring and an inner ring which also serves as the base for the ventilation frame.   Radial trusses reinforce the tension ring by resisting direct tension

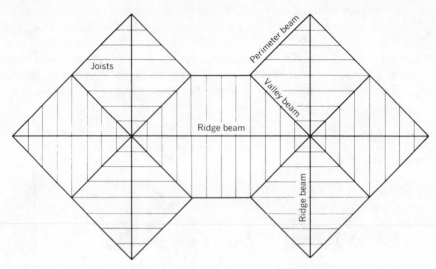

Fig. 3.30 Roof framing—Beth Torah Temple.

from the cables. Bending moments on the tension ring are thus minimized.

At the outer perimeter, the radial cables are attached to a built-up compression ring, fabricated from two I-members which provide stiffness in both vertical and horizontal axes.

Damping of the cables to prevent flutter is achieved through the additional mass of the ceiling frames as well as the continuous mass of the roofing materials. Roof loading is resisted in tension by the cables, which in turn transmit their reactions to the inner tension ring (reinforced with radial trusses), and the compression ring at the perimeter.

Cables may also be used with roof trusses and rigid frames (Fig. 3.31) to contribute to the load-carrying capacity of the basic structure and to minimize

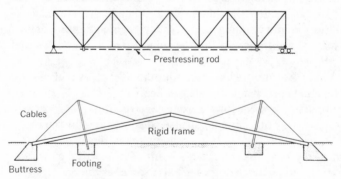

Fig. 3.31 Cables in roof framing.

secondary stresses and deflections. These types of roof framing were used in the United Airlines hangar at O'Hare International Airport, Chicago, and in the main arena at Squaw Valley, California.[3.23,3.24]

3.6   ADDITIONAL DESIGN CONSIDERATIONS

Aside from the architectural planning and the over-all structural design of the building members, a number of details require the full attention of the building engineer. For example, in certain types of construction and occupancy, the provision for fire protection of the steel frame is necessary.

Other operations, such as fabrication and erection, the installation of conduits for electrical and mechanical utilities, plans for future expansion and similar considerations, affect the final design. In many cases a large structural member has to be provided with field splicing at several points to facilitate its handling in the fabrication plant. The manner of erection also suggests certain types of connections and in many cases influences the design of the members. The installation of conduits may also influence the structural design. This is one of the phases of the work where close co-ordination is required between the structural engineer and the mechanical and electrical engineer. It is not unusual to use cutouts in deep floor girders to provide a passageway for the conduits. The proper size and location of such cutouts depends on the utility requirements and the structural adequacy of the section.

Structural steel is an incombustible material and may be used without fireproofing in industrial type buildings and similar structures which contain little or no combustible materials, such as hangars, stadiums, and warehouses. However, sustained extreme heat applied to steel affects its properties, in particular its strength. At 1300°F the yield point of mild steel is about 8 ksi.[3.25] Thus, where sustained fire and high temperatures are possible, it becomes necessary to protect the steel frame with a fire-resistant insulating material. Fire-resistance requirements depend on the type of construction, building height, floor area, type of occupancy, fire-protection system available, and the location of the building.

Fire protection may be obtained in the form of coverings such as brick, stone, concrete, tile gypsum, and various fire-resistant plasters. An AISI booklet[3.26] on fire-resistant construction gives a valuable guide to the selection of suitable fireproofing schemes which will conform with architectural, economic, and building code requirements.

BRIDGES

4.1 INTRODUCTION

It is the purpose of this chapter to present an introduction to the design of steel bridges. The chapter is not intended to be a complete study of the science and art of bridge design, but rather to provide a consistent design approach and guidelines to the solution of the major problems encountered.

4.2 HISTORICAL REVIEW

From tree trunks and vines over small streams, the timber bridge evolved through such forms as Caesar's famous timber pile bridge over the Rhine to the timber arches of the eighteenth century.[4.1] These elaborate timber bridges were able to span crossings in excess of 200 ft. Meanwhile, the Romans had developed the masonry arch to a stage of perfection which was not re-attained until many centuries later.[4.2] The greatest span achieved was probably 160 ft and many of the aqueducts and bridges are still standing, for example the Pont du Gard, Nimes, France.

The first European truss bridges were of timber and, by their arch action, they exerted considerable horizontal thrusts on the abutments of the bridge. The nineteenth century saw major steps taken to simplify truss design (for example, by Howe and Pratt in the United States) and iron became a more frequently used bridge material.[4.3] Some relevant historical dates will be found in Table 1.2. The first iron bridge was built of cast iron at Coalbrookdale, England, by A. Darby in 1779. Its main span was about 100 ft in length and it was of arch design.   Early trusses were of a highly redundant lattice type of design; probably the first iron bridge of modern truss design was built by Squire Whipple in the United States in 1853.   This cast iron bridge spanned 146 ft.

Later developments continued to improve both the materials (cast iron, wrought iron, steel) and the techniques (arch-truss, truss-suspension, cantilever).   The latter half of the nineteenth century saw the construction of such structures as the Roeblings' Brooklyn suspension bridge (1595 ft) in 1883, Eads' St. Louis arch bridge of 1874, which spanned 520 ft and was the first bridge to use structural carbon steel (and also the first to use high-strength steel), and Fowler and Baker's 1710-ft-span cantilever-truss Firth of Forth bridge (1890).[4.1,4.2]   It was many years before some of these spans were exceeded.   In fact, the cantilever bridge record of 1800 ft set by the Quebec railway bridge in 1917 has yet to be surpassed.

Today there are larger bridges, but the forms of arch, cantilever, and suspension remain basic in the design of great bridges.   Currently, the Verrazano-Narrows suspension bridge (Fig. 1.6) has an unsurpassed span of 4260 ft, exceeding that of the world-renowned Golden Gate bridge (Fig. 1.5) by only 60 ft.   The longest cable-stayed span is (1973) the 990-ft Severin bridge in Cologne, and this form of bridge appears to have taken over for spans up to 1000 ft.   In small bridges the trend is now towards the use of plate girders in many instances in which trusses would previously have been used.   Composite action of the various components of the bridge is now considered and utilized, and orthotropic (single-unit) systems are becoming more popular.

4.3 BRIDGE TYPES

Bridges can be classified by the way in which their members transmit load (axial force, flexure, or torsion), by the type of members used (truss, plate girder, and so on), or by their method of transferring load to the foundations (for example, arch or tied arch, cantilever or beam).   Later chapters of the book discuss the action of the individual members and connections which form the bridge.   This article illustrates how the various types of bridges result from different combinations of their components.

The bridges shown in Figs. 4.1a, b, and c are by far the most common bridge types in present use.   When trusses are used as the main flexural members, these bridges can span many hundreds of feet.   The arch bridges (Figs. 4.1d to 4.1i) are a versatile form which can be used to meet special requirements over short spans or to provide clear crossings over intermediate and long spans.   The Kill Van Kull arch bridge (New York) spans 1652 ft. The suspension and cable-stayed bridges shown in Figs. 4.1j to 4.1l are

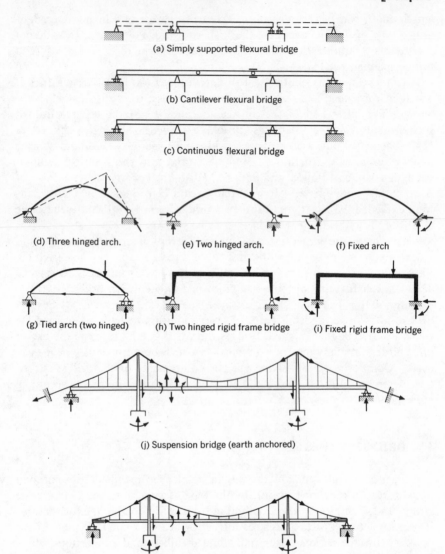

(a) Simply supported flexural bridge

(b) Cantilever flexural bridge

(c) Continuous flexural bridge

(d) Three hinged arch.   (e) Two hinged arch.   (f) Fixed arch

(g) Tied arch (two hinged)   (h) Two hinged rigid frame bridge   (i) Fixed rigid frame bridge

(j) Suspension bridge (earth anchored)

(k) Suspension bridge (self anchored)

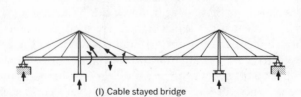

(l) Cable stayed bridge

Fig. 4.1 Main bridge types.

commonly used for long spans, although they have also become popular for carrying light loads over small spans.

Figure 4.1 shows a classification of bridges according to their method of structural action.   Bridges also can be differentiated as deck- or through-type depending on the position of the traffic deck.   Figure 4.2 illustrates this distinction; Fig. 4.2a shows a through truss in the upper picture and a deck girder bridge in the lower picture; Fig. 4.2b is an arch-type through bridge, and Fig. 4.2c is an arch-type deck bridge.   The deck bridge has become increasingly popular in recent years, mainly due to its pleasing appearance and the absence of obstructions to the user's line of sight.

Housatonic River, Connecticut.

Twin, Florida.

Fig. 4.2a Flexural bridges: *upper*, truss type; *lower*, girder type. (Courtesy of Bethlehem Steel Corp.)

Fig. 4.2b Through-arch bridge, Mohawk River, New York. (Courtesy of AISC.)

Fig. 4.2c Deck-arch bridge, Lewiston-Queenstown. (Courtesy of Bethlehem Steel Corp.)

The flexural type of bridge may be simply supported or continuous over its supports (Figs. 4.1a and 4.1c). Continuous bridges of up to seven spans have been constructed in recent years, whereas causeway-type bridges may contain literally hundreds of simply supported spans connected in a long series. Under the same loading, a continuous bridge will be stronger than a succession of simply supported spans; however, the continuous bridge is liable to additional forces due to foundation movements and temperature changes. To some extent, however, this problem can be overcome by the use of jacking devices built into the bridge abutments.

The truss form is a versatile one and may be used as a flexural member (Figs. 4.1a, b, and c) or an axial member in arch bridges. The truss may take a variety of internal arrangements, some of which are illustrated in Fig. 4.3. For the Warren truss in Fig. 4.3a, the truss shears are carried only by diagonal members and these are in alternate compression and tension; the Pratt truss (Fig. 4.3b) is arranged to keep most of the diagonals in tension whereas the Howe truss (Fig. 4.3c) has its diagonals in compression; the Vierendeel truss (Fig. 4.3d) requires moment-resisting connections and the use of a more involved analytical procedure; the Wichert truss (Fig. 4.3e) is a quasi-continuous beam which is not subjected to settlement loads; curved chord trusses (Fig. 4.3f) are more economical over long spans, as the distribution of material can be made to approximate the moment diagram; subdivided trusses allow the designer to increase the truss depth (Figs. 4.3g, 4.3h, and

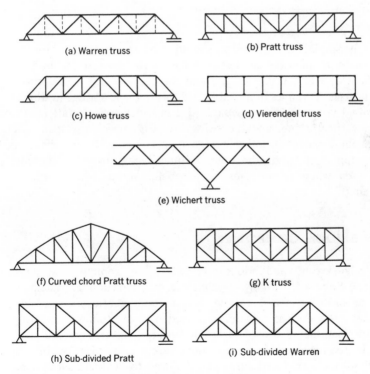

(a) Warren truss   (b) Pratt truss

(c) Howe truss   (d) Vierendeel truss

(e) Wichert truss

(f) Curved chord Pratt truss   (g) K truss

(h) Sub-divided Pratt   (i) Sub-divided Warren

Fig. 4.3 Common truss types.

4.3i) while maintaining the most economical spacings of floor beams and panel points.

The members used in a flexural type of bridge may be rolled shapes for small spans, with plate girders and trusses being used as the spans increase in size. Table 4.1 shows guidelines which are sometimes used. The values shown in Table 4.1 are for simple spans and are not inclusive. Long-span, continuous plate girders are replacing trusses in many applications; 450-ft-span plate girders have been built.[4.4] In addition, the use of composite design (Chapter 13) has increased the span lengths possible with both rolled shapes and plate girders, and orthotropic design has resulted in plate girders with spans up to 856 ft (Art. 4.9).

Table 4.1   Bridge Spans and Types

Member	Railroad Bridges	Highway Bridges
Rolled sections	Span < 50 ft	Span < 80 ft
Plate girders	50 ft < Span < 150 ft	80 ft < Span < 300 ft
Trusses	Span > 150 ft	Span > 300 ft

Some special types of bridges should also be mentioned. Movable bridges are used to provide the required clearance between the bridge and the traffic-way which it crosses, without the necessity for high approaches and supports. Examples of movable bridges are the lift bridge in which the deck is raised vertically, the bascule bridge in which spans rotate about one end in a vertical plane, and the swing bridge in which the spans rotate in a horizontal plane. Another bridge type is the pontoon bridge, in which the supports float on the water channel being crossed. Pontoon bridges are used where foundations are unsatisfactory or where rapid erection is required.

4.4 BRIDGE COMPONENTS

Bridge layouts and bridge nomenclature in common use are shown in Figs. 4.4 and 4.5. Figure 4.4 illustrates two typical floor systems.

The bridge traffic is carried directly on the bridge deck. In highway bridges, this deck is sometimes topped with a bituminous wearing surface about 2 in. thick. Raised curbs and steel railings are provided on either side of the road surface to increase traffic safety. In colder regions, open-type railings and steel-faced curbs may be used to facilitate snow removal. In railroad bridges, the rail track is carried on cross ties which may bear directly

on the girders or on a layer of ballast.   The ballast helps to reduce noise and
vibration and allows continuous track maintenance.

The concrete slab is a very common deck type,[4.5] and its usefulness is
further increased when composite methods of construction are used.   Other
deck systems are the battledeck,[4.6] steel grid, laminated or plain timber,[4.7]
masonry and, for railroad bridges, wrought iron plate decks.   Orthotropic
bridges employ a deck slab very similar to that used in the battledeck systems.
(See Art. 4.9.)

Fig. 4.4a   Stringer and floor beam system, Walt Whitman Bridge, Pennsylvania–New Jersey.   (Courtesy of Bethlehem Steel Corp.)

Fig. 4.4b Stringer floor systems with vertical crossed bracing, Delaware Water Gap, Pa.–N. J. (Courtesy of Bethlehem Steel Corp.)

The deck and its loading are carried by the floor system, which distributes these loads to the main members of the bridge. A basic system is illustrated in Fig. 4.4a; the members running parallel to the bridge centerline are called "stringers" and are supported on transverse members called "floor beams." In Fig. 4.4a the floor beams span between the main vertical trusses and cantilever out from either side of them. This type of floor system is used in most truss bridges and in other bridges with only two main members. Occasionally the stringers are omitted by using a closer spacing of floor beams. Another form of deck-type bridge (Fig. 4.4b) utilizes the stringers spanning between the main supports as the prime load-carrying members. The stringers are then connected by light cross-beams, diaphragms, or braced panels as shown in Fig. 4.4b. These cross-systems usually are placed at approximately 25 ft centers. In addition to helping to equalize the distribution of load amongst the stringers, they also facilitate erection and improve the transverse stability of the bridge.

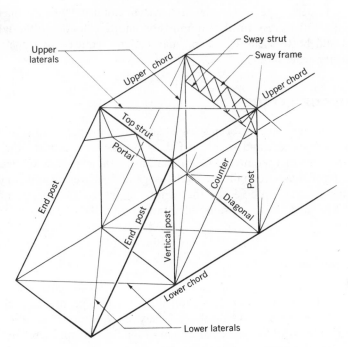

Fig. 4.5 Nomenclature for truss bridges.

Lateral loads on the bridge are resisted primarily by systems of lateral braces and vertical portal frames. In truss bridges, these also serve to stabilize the compression chords. Figure 4.5 shows the general nomenclature used in describing such systems. K bracing (Figs. 4.3g and 4.2b) is frequently used for lateral bracing. Crossed diagonals may also be used and usually are designed under the assumption that diagonal members in compression are not capable of carrying any load. When this assumption is used in main member design the crossed diagonals are called counters. Cross-bracing has some disadvantages when used in lateral systems. The most serious of these is that the chords of the system are also the chords or flanges of the main members (Fig. 4.5) and so, under vertical live loads, the two chords will compress uniformly. This compression may be of sufficient magnitude to cause simultaneous buckling of both crossed diagonals. The problem does not exist for K bracing because the shortening can be absorbed by bending of the transverse brace (the vertical leg of the letter K), thus relaxing any forces in the two sloping members. The pony truss is a through truss which is an exception to the top chord bracing layouts. In this type of bridge there is no compression chord bracing and the only lateral support comes from the lateral stiffness of the truss cantilevering from the floor system. In deck-type bridges the deck system can usually be assumed to be sufficiently stiff to restrain the compression chords laterally, although some vertical and bottom chord bracing may still be required to transfer wind loads to the reactions.

4.5 BRIDGE LOADS

After a bridge type has been selected according to spans, foundation conditions, appearance, and so on, it is next necessary to analyze the chosen structure; and this requires a determination of the expected applied loads. The designer must ascertain the likely usage and environmental conditions; the loads are then specified by the various design codes—for example, the AREA Specification[1.25] for railroads, and the AASHO Specification[1.23] for highways.

1. Dead Loads

The dead load acting on a bridge is the sum of the weights of all its components. If the bridge cross section does not change radically along the length of the bridge, the dead load may be considered as a uniformly distributed load. Only the dead weight of the deck is considered in the deck design, but the dead weight of the entire structure is carried by the main members.

Some typical values useful in dead weight calculation are:

Steel	490 lb/cu ft
Timber	50 lb/cu ft
Concrete (plain or reinforced)	150 lb/cu ft
Ballast, etc.	120 lb/cu ft
Loose sand, etc.	100 lb/cu ft
Macadam	140 lb/cu ft
Railway rails, guard rails, fastening	200 lb/lin ft of track

The designer must include all other loads carried by the bridge, such as those applied by sidewalks and utilities. Snow load is usually disregarded in bridge design, due to the accompanying reduction of traffic and traffic impact effects. An allowance of 10 to 15 per cent is sometimes applied for such details as bolts, rivets, appurtenances, gussets, paintwork, and so on.

2. Live Loads

Live loads depend on the bridge usage and the relevant bridge code. The following sections will outline only the loading provisions of the two main bridge codes in the United States.

Railroad Bridges (AREA[1.25]). The standard loading recommended is known as the Cooper E80 and is shown in Fig. 4.6. This loading represents two locomotives with maximum axial loads of 80 kips, followed by a uniformly distributed load of 8 kip/ft of track. If E80 is not a satisfactory representation of the bridge usage, as for example a bridge on a minor branch

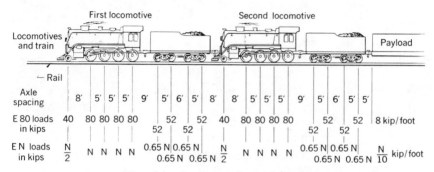

Fig. 4.6 Cooper railroad loadings.

line, it is necessary to use another loading with the same axle spacing but with all the loads altered by a constant ratio. For example, EN would have loads N/80 times those shown in Fig. 4.6 for E80. Figure 4.7 illustrates the case of a multi-track bridge in which it is necessary to decide on track loadings. The code does not require all tracks to be fully loaded simultaneously for bridges with three or more tracks because of the improbability of this occurring in critical conditions. Consequently, the reduced loads of N/2 and N/4 shown in Fig. 4.7 may be used.

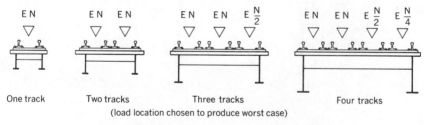

Fig. 4.7 Loads on multi-track rail bridges.

It will be seen in Art. 4.7 that a series of concentrated loads is inconvenient to handle analytically. Equivalent uniform loads have been calculated for EN train loadings by Steinman, and his work is reported in Ref. 3.2. Although these equivalent loads remain functions of the span and of the location in the span at which the action values are required, they nevertheless reduce the time and effort required to determine maximum loads and stresses.

Highway Bridges (AASHO[1.23]). The American Association of State Highway Officials uses five standard truck loadings which are as follows:

Two axle trucks   H20, H15, and H10
Two axle trucks with semi-trailers   H20-S16 (or HS20) and H15-S12
                                     (or HS15).

Figure 4.8 illustrates this loading system, and it can be seen that the number after the H gives the truck load in tons, and after the S in the hyphenated designation, the trailer load in tons. The trailer axle is placed within the limits shown in Fig. 4.8 in such fashion that it produces the maximum loads at the point under consideration. In special cases where the standard truck loading is not suitable, it is permissible to vary the loads in the same ratio as the standards, although the standard truck axle spacing must be maintained.

The uniformly distributed lane loading shown in Fig. 4.8 is used when it produces the more severe loading condition. For simple spans it will be found for HS loadings that truck loads are critical for moment for spans under 140 ft and for shear for spans under 120 ft.

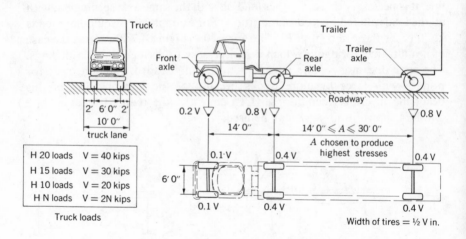

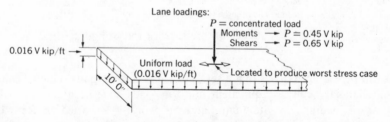

Fig. 4.8 AASHO highway loads.

If three lanes are loaded simultaneously, the resultant combined live load stresses can be reduced by 10 per cent; the reduction for four or more lanes is 25 per cent.

3. Impact Loads

The member loads which are calculated for a static applied load do not represent the loads which would exist if the same load were to be applied dynamically. It is obvious that the momentum of the load itself must produce loads above the static values. Bridge live loads are dynamic in nature, and it is necessary to account for their dynamic or impact effects. The formulas used in bridge codes simply give an increased static live load which is used in the calculation of the member stresses due to impact loads.

The impact formulas are all simple functions of the length of the member being designed and their theoretical foundations are somewhat vague. It appears[4.8] that the early impact formulas were developed primarily to eliminate the possibility of member stresses changing sign, which is part of the fatigue problem discussed in Chapter 16. The use of higher live loads achieved this goal by requiring larger members than the static analysis would predict. The increase in member size also decreases dynamic resistance. Experimental investigations[4.9,4.10] have indicated that the provisions are usually conservative.

Railroad Impact Loads. If L is the loaded length of the member in feet, then the vertical live load P is increased by the ratio p to pP, where p is given by the following simple functions for open decks, and is taken as 90 per cent of these values for ballasted decks.

Case	Equation	Equation No.

Steam Locomotives with Hammer Blow*

$L < 100$ ft	$p = 1.6 - \dfrac{1}{5}\left(\dfrac{L}{100}\right)^2 + \dfrac{1}{S}$	(4.1)
$L > 100$ ft	$p = 1.1 + \dfrac{18}{L - 40} + \dfrac{1}{S}$	(4.2)
Trusses	$p = 1.15 + \dfrac{40}{L + 25} + \dfrac{1}{S}$	(4.3)

* Although steam locomotives are in little use, the data shown are relevant for heavy locomotives.

Diesel and Electric Locomotives, and Others Without Hammer Blow

$$L < 80\,\text{ft} \qquad\qquad p = 1.4 - \frac{3}{16}\left(\frac{L}{100}\right)^2 + \frac{1}{S} \qquad\qquad (4.4)$$

$$L > 80\,\text{ft} \qquad\qquad p = 1.16 + \frac{6}{L-30} + \frac{1}{S} \qquad\qquad (4.5)$$

Spacing in feet of the member being designed.

If two tracks are loaded, then the following modifications are allowed.

Case	Two Tracks	More Than Two Tracks
$L < 175\,\text{ft}$	Full impact on both	
$175\,\text{ft} < L < 225\,\text{ft}$	Full impact on one track, p^* times full impact on the other	Full impact on any two tracks
$225\,\text{ft} < L$	Full impact on one, none on the other	

where

$$p^* = 4.5 - \frac{L}{50} \qquad\qquad (4.6)$$

For more than two tracks, full impact is taken on any two tracks.

Highway Impact Loads.[1.23]   If L is the loaded length of the member in feet, then the vertical live load P, to be used, is increased by the ratio p to pP, where

$$p = 1 + \frac{50}{L + 125}, \qquad p \leqslant 1.30 \qquad\qquad (4.7)$$

Longitudinal Forces.   The braking and traction of vehicles will cause longitudinal forces to be exerted on the bridge deck.   These forces act in the direction of the bridge centerline and AASHO assumes them to be 6 ft above the bridge deck.   For AREA only one track is considered, and the force is 15 per cent of the live load for braking, although a modification is made for continuous rails (AREA Art. [1.3.12]).   AASHO requires the braking forces to be taken as 5 per cent of the live load in lanes carrying traffic in the same direction.   Much of this longitudinal load will actually be carried directly by the floor system without participation of the main members, and for this reason AREA allows it to be distributed on the basis of relative stiffnesses.

4. Lateral Loads

Lateral loads result predominately from the wind forces.   Model studies have indicated that the code provisions are not unreasonable for 100-mph winds.[3.10]   Wind forces appear to be dependent upon the type of bridge used, and a complete summary of the effect of wind on plate girders and trusses will be found in Ref. 3.10.   For large structures, particularly suspension bridges, a frequently adopted precaution is to carry out preliminary wind tunnel tests on models of the bridge.   A discussion of the geographical distribution of wind pressures has been given in Chapter 3.

Wind Loads on Railroad Bridges.   It is necessary to consider wind as a moving load acting on the bridge and train in any horizontal direction.   The specifications further require that the loads on the train act 8 ft above the track.   The loads to be applied are:

0.3 kip/ft   on the train (one track only)
0.045 kip/sq ft   on the vertical projection of the area of girders
0.030 kip/sq ft   on the vertical projection of the area of trusses.

A second case occurs if the bridge is carrying no traffic; in this situation the wind loads on the structure are increased by 5/3, as the traffic is no longer shielding the leeward side.   A third case to be considered is a load of 0.2 kip/ft on a loaded chord or flange and 0.15 kip/ft on an unloaded chord or flange.   The design is based on whichever of the three cases produces the most severe stresses.

Wind Loads on Highway Bridges.   AASHO specifies that the wind forces should be:

Trusses and arches   0.075 kip/sq ft
Girders and beams   0.050 kip/sq ft

These forces are assumed to act on the vertical projection of the structure exposed to the wind.   However, the total effect shall not be less than 0.3 kip/ft on the loaded chord and 0.15 kip/ft on the unloaded chord for trusses, nor less than a load of 0.3 kip/ft for girders.

The wind loads quoted are for 100-mph winds and, as wind pressure varies with the square of the speed, the loads for other expected wind speeds may easily be estimated.   For instance, a 150-mph wind would cause forces $(150/100)^2 = 2.25$ times greater than a 100-mph wind.

Other Lateral Loads.   If the bridge is curved, the moving loads will create centrifugal forces on the bridge deck.   These are proportional to the square of the vehicle's speed and are assumed to act 6 ft above the deck surface.   AASHO gives the force as C times the live load, where

$$C = 6.68 \frac{s^2}{R} \tag{4.8}$$

and s = speed in mph and R = radius of the curve in feet.

Another lateral load to be considered is that due to the train nosing or snaking laterally as it travels across the bridge. This load is assumed to be $(5/18)N$ for an EN train (22 kips for the standard E80 train) and acts at the top of the rail.

Earthquake effects must be considered in regions subjected to seismic disturbances. The normal provisions for earthquake design must be fulfilled; however, the lateral bracing systems employed for wind resistance frequently will be capable of resisting these loads. The actual loads applied to a bridge during a disturbance will be approximately proportional to the weight of the bridge. AASHO specifies the following loads to be applied horizontally in any direction through the center of gravity of the structure:

$$0.02D \quad \text{for structures on spread footings with a base material}$$
$$\text{resistance of 8 kip/sq ft or more}$$
$$0.04D \quad \text{for structures on spread footings with a base material} \quad (4.9)$$
$$\text{resistance of 8 kip/sq ft or less.}$$
$$0.60D \quad \text{for structures on piles}$$

where D is the dead weight of the structure.

As is the case in building design (Chapter 3), the allowable stresses used in bridge design may be increased for loads resulting from a combination of dead load, live load, and impact and centrifugal forces with the loads resulting from lateral and longitudinal forces. However, the resulting member sections must not be less than those required when the lateral and longitudinal forces are not considered. For highway bridges, the allowable stresses used may be increased in the ratio 1.25 if (1) the wind is considered without the structure being loaded, or (2) the structure is loaded and the bridge is considered loaded by 0.30 times the above wind loads and by a load of 0.1 kip/ft to represent the wind load on the vehicles. For railroad bridges, the allowable stresses under combined loadings may be increased in the ratio of 1.25.

5. Miscellaneous Loads

The loading conditions discussed above are intended only as a guide. They should not be regarded as embracing all possible load situations or as supplanting the real usage to which the bridge is to be subjected. Some additional loads which must be accounted for are now discussed.

Aerodynamic Effects. The effect of wind on certain objects placed in its path is not always predictable from the simple wind analyses discussed above. For instance, an airplane wing and the side of a building will behave quite differently in the same strong wind. There is the possibility that a bridge placed in a wind current may also be subjected to aerodynamic effects[4.11] and, among other things, may develop periodic movements at right angles to the wind direction. The presence of gusts in the wind will accentuate the problem. Aerodynamic behavior contributed to the failure of the Tacoma Narrows suspension bridge.[4.12]

Overloads. Many local authorities specify additional loads which bridges under their jurisdiction must carry. Military loadings and earth-moving equipment are common instances of such loads. AASHO specifies that, if a bridge is designed for loads lighter than those of H20, provision shall be made for infrequent heavy loads. The ultimate load capacity of members and connections is discussed in the various other chapters of this book, and the capacity of bridges is further discussed in Art. 4.9.

Temperature Effects. With temperature effects, the first problem is the definition of likely temperature ranges. AASHO suggests the use of a temperature range from 0 to 120°F in moderate climates and −30 to 120°F in cold climates. Temperature changes will produce internal forces in any redundant structure, even if this redundancy is only the result of malfunction-ing bearings (see Art. 4.8). A uniform temperature change will produce axial deformations and, hence, axial forces. A temperature change across a member will cause curvature changes and, hence, internal moments. Longitudinal forces in a bridge can be eliminated by the use of expansion bearings (see Art. 4.8). AREA recommends the use of an expansion allowance of 1 in., and AASHO, $1\frac{1}{4}$ in. per 100 ft of span.

If the bridge is to be continuous over a number of spans, considerable attention must be paid to the design of expansion bearings. In addition, moments will be induced by temperature differences between the upper and lower faces of the bridge.

The deflections due to temperature may be significant. For instance, the closing of an arch during erection can be significantly affected by temperature changes. Temperature changes in the completed bridge can also be large: the 3800-ft center span of the Mackinac suspension bridge has a 7-ft deflection change due to extreme changes in temperature.[4.13]

Settlement Loads. Support settlement is important in externally redundant structures, because the deflection of one reaction will induce internal forces in the structure. For this reason, simply supported structures are sometimes preferred to continuous structures, although the problem of support settle-ment has probably been over-emphasized in the past. It can be overcome, for instance, by allowing for jacking at supports and abutments.

Erection Loads. Erection loads can be critical and should be considered in the final design.[4.14] The following are typical examples: cantilever methods of erection will change the sign of the stress in some members; falsework will cause unusual load distributions; erection cranes will impose large concentrated loads; member misfits may make it necessary to jack final members into place; concrete decks in suspension bridges must be poured in a controlled sequence to avoid overstress. The failures which have occurred during the erection process are evidence that it is a critical stage in the life of a bridge.[4.15]

4.6 FLOOR SYSTEM ANALYSIS AND DESIGN

The live loads discussed in Section 4.5.2 are applied to the bridge deck as concentrated loads or as loads distributed over a limited lane width. If the loads are transferred to the slab through ballast or fill, some distribution can be assumed to apply through the ballast. For an earth fill of more than 2 ft, AASHO[1.23] allows the load to be spread at a slope of 41° from the vertical.

When the live loads have been obtained, the deck system can then be designed. The selection of the deck slab is beyond the scope of this chapter, and the reader is referred to the references quoted in Art. 4.4 for further details. Concrete highway deck slabs are usually between 6 and 8 in. thick, and AASHO has a number of specific requirements for slab design.

Once the deck has been chosen, it is necessary to evaluate the loads on the floor system. Floor system design is basically a process of distributing the traffic load transversely across the bridge, whereas the design of the main members (Art. 4.7) deals primarily with the longitudinal distribution of these loads. The transverse assignment of loads is a quite complex analytical problem as the system of deck slab, stringers, transverse diaphragms, and floor beams is highly redundant. The two United States bridge specifications circumvent the problem by the use of certain simplifying provisions and assumptions.

The AASHO requirements are given in the table partially reproduced below as Table 4.2 (from Art. [1.3.1] of Ref. 1.23, courtesy of AASHO). The factors shown in the table give the proportion of a load carried by the adjacent stringers when the load is applied between the stringers. If stringers are omitted, there are similar provisions for floor beams. A separate tabulation is also used for the exterior stringers. Shear calculations assume that the immediate loads are not distributed.

The AREA[1.25] rules depend on the type of deck (open or ballasted) and on the arrangement and type of cross-ties, deck, stringers, and transverse (floor) beams used. Article [1.3.4] of AREA[1.25] presents a detailed discussion of these factors.

Floor system beams are normally assumed to be simply supported, although this assumption can be quite conservative. Methods have been proposed[4.16,4.17] which allow the floor system to be treated as a type of steel mesh or grillage. In this method the torsional strength of one set of beams provides the end moments for the set at right angles to it. Figure 4.9 illustrates this interdependence and shows the real behavior of a loaded

Table 4.2 Lateral Distribution of Wheel Loads on Concrete Slabs and Steel Grids

	Bridge Design	
	One Traffic Lane	Two Traffic Lanes
Concrete		
On steel I-beam stringers*	$S/7.0$	$S/5.5$
	See† if $S > 10$ ft	See† if $S > 14$ ft
Steel grid		
Less than 4 in. thick	$S/4.5$	$S/4.0$
4 in. or more	$S/6.0$	$S/5.0$
	See† if $S > 6.0$ ft	See† if $S > 10.5$ ft

S = average stringer spacing in feet.

* "Design of I-Beam Bridges" by N. M. Newmark, *Proc. Am. Soc. Civil Engrs.* (March 1948).

† In this case the load on each stringer shall be the reaction of the wheel loads, assuming the flooring between the stringers to act as a simple beam.

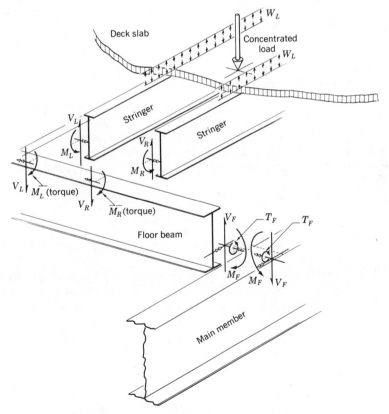

Fig. 4.9 Behavior of a floor grid under load.

floor system. The torsional stiffness of the floor beams provides end moments for the stringers and the main girders similarly restrain the ends of the floor beams. It is common practice to utilize at least some of this reserve of strength by making the stringers continuous over the floor beams.

Finally, the floor beam reactions transmit the floor loads to the main members. If these main members are trusses, it is necessary for the floor beams to connect to the truss at panel points (Fig. 4.4a); otherwise, flexural stresses will develop in the truss chord members.

Example 4.1 illustrates a typical design procedure for a system of floor beams and stringers. References, such as 3.6, 4.18, and 4.19 contain more details of practical design layouts.

4.7 MAIN MEMBER LOAD ANALYSIS

The previous article considered the prediction of panel point or girder loads for a given set of vehicle loads. This required the lateral distribution of the vehicle loads. These loads are now to be distributed along the length of the bridge in order to determine the moments in the main longitudinal members.

1. Preliminary Design

In order to analyze redundant structures and to find dead weights of determinant structures, it is first necessary to produce a preliminary design. The preliminary and the final designs must conform to the route layout required and must satisfy the requirements of clearance, economics, and esthetics. Of these, only economics requires further discussion here, and if certain simplifying assumptions are made, the minimum cost of a bridge is indicated by the criterion:

$$\text{Cost/ft of main members} = \text{Cost of substructure/ft of span} \quad (4.10)$$

Although approximate, Eq. 4.10 affords an estimate of the economic efficiency of a bridge.

Another factor to consider is the possibility of using higher-strength steels for more highly loaded sections of the bridge. This allows members to be kept to a more uniform size. The economy of such a bridge will require a more specific study (see Fig. 2.13).

Once the span lengths L are known, the depths d of the members may be estimated from experience or judgment. The United States specifications have suggested limiting L/d to the following maximum values:

Rolled beams: $L/d = 15$ (railway), 25 (highway)
Plate girders: $L/d = 12$ (railway), 25 (highway)
Trusses: $L/d = 10$ (railway and highway)

EXAMPLE 4.1

PROBLEM:
    Design the stringer and floor beams for the deck system shown below. Assume the stringers to be simply supported at the floor beams. The bridge is to be designed for a live loading of H20−S16. (HS20) The steel is A36.
      Cross−section of bridge:

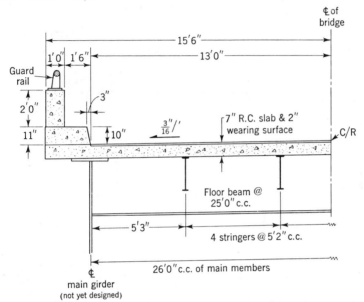

SOLUTION:
    As the main girder is much stiffer than the stringers it will be assumed that the safety walk and parapet dead loads are carried directly by the main girders.

DEAD LOADS:

Slab (7")	$\frac{7}{12}$ x 150	= 87.5 psf
Surface	$\frac{2}{12}$ x 140	= 23.3 psf
		111 psf

Estimated stringer weight	= 62 lb/ft
Estimated floor beam weight	= 160 lb/ft
Estimated weight of connections, etc.	= 15 lb/ft
	237 lb/ft

LIVE LOADS:

      H20−S16

    Impact: (eq 4.7)

      Floor beam: $p = 1 + \dfrac{50}{26 + 125} = 1 + 0.33 = 1.33$

              > 1.30 ∴ Use $p = 1.30$

      Stringers: $p = 1 + \dfrac{50}{25 + 125} = 1 + 0.33 = 1.33$

              > 1.30 ∴ Use $p = 1.30$

EXAMPLE 4.1 (Continued)

DESIGN LOADS:
In all cases therefore

$$V_{DES} = V_D + 1.3\, V_L$$
$$M_{DES} = M_D + 1.3\, M_L$$

LOAD DISTRIBUTION TO STRINGERS:

Table 4.2
$$S = \frac{5.17}{5.5} = 0.94$$

Shear: Adjacent loads are not distributed (S=1)

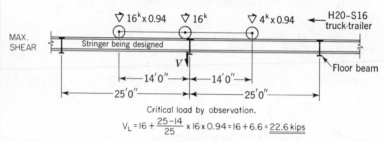

MAX. SHEAR

Critical load by observation.

$$V_L = 16 + \frac{25-14}{25} \times 16 \times 0.94 = 16 + 6.6 = \underline{22.6 \text{ kips}}$$

Moments: All loads distributed (S=0.94)
Two loading cases to be checked.

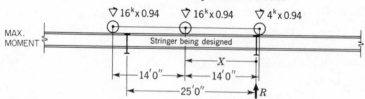

MAX. MOMENT

One axle on span:

Case I, X = 12'6"
$$\therefore M = \frac{1}{2} \times 16 \times 0.94 \times 12.5$$
$$= 94 \text{ kip-ft}$$

Two axles on span:

Case II, 11'0" ≥ X ≥ 5'6"
$$\therefore R = \left[\frac{25-X}{25} + \frac{11-X}{25} \right] \times 16 \times 0.94$$
$$R = \frac{2 \times 16 \times 0.94}{25} (18-X)$$
$$\therefore M = R \times X$$
$$= \frac{32 \times 0.94}{25} (18X - X^2)$$
$$\frac{dM}{dX} = 0 \text{ when } 18 - 2X = 0$$
$$X = 9'0"$$
$$\therefore M = 1.28 \times 0.94 \times 9 \times 9$$
$$= 97.4 \text{ kip-ft}$$
$$\therefore \text{Case II critical}$$

$$M_L = \underline{97.4 \text{ kip-ft}}$$

STRINGER DESIGN:

$$V_{DESIGN} = V_D + 1.3\, V_L$$
$$V_{DESIGN} = [0.111 \times 5.17 + 0.062] \times \frac{25}{2} + 22.6 \times 1.30$$
$$= 0.636 \times \frac{25}{2} + 29.4 = 37.3 \text{ kip}$$

EXAMPLE 4.1 (Continued)

$$M_{DESIGN} = M_D + 1.3\,M_L$$
$$= [0.111 \times 5.17 + 0.062] \times \frac{25^2}{8} + 97.4 \times 1.30$$
$$= 49.7 + 126.8 = 177 \text{ kip-ft}$$

Allowable bending stress, $F_b = 20$ ksi

$$Z = M_{DES}/F_b = 177 \times 12/20 = 106 \text{ in.}^3$$

<u>USE W 21 × 55</u>

Check shear, $F_v = 12$ ksi

$$f_v = \frac{V_{DES}}{A_w} = \frac{37.3}{19.76 \times 0.375} = 5.0 \text{ ksi} < 12 \text{ ksi}$$

$$\therefore O.K.$$

LOAD DISTRIBUTION TO FLOOR BEAMS:

Shear

Critical longitudinal location, center axle on floor beam.

∴ Distribution of front and rear axles to floor beam is V_e,

$$V_e = \left[\frac{11}{25} \times 16 + \frac{11}{25} \times 4\right] = 8.8 \text{ kip}$$

Critical lateral location, as far to one side as possible.

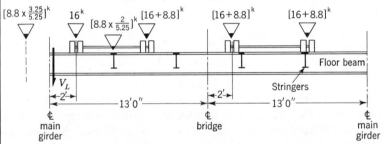

$$V_L = 16\left[\frac{24}{26} + \frac{18}{26} + \frac{11}{26} + \frac{5}{26}\right] + 8.8\left[\frac{18}{26} + \frac{11}{26} + \frac{5}{26}\right] + 3.4 \times \frac{20.75}{26}$$
$$= 35.7 + 11.5 + 2.7 = \underline{49.9 \text{ kip}}$$

Moments: No longitudinal or lateral distribution. Critical longitudinal location as before. Critical lateral location as close to center as possible (2'0").

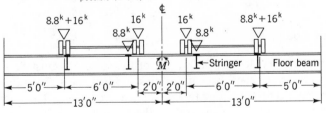

$$M = 16(5 + 11) + 8.8(5 + 10.4) = 256 + 136$$
$$\therefore M_L = 392 \text{ kip-ft}$$

FLOOR BEAM DESIGN:

Floor beam is assumed to carry 5'0" of slab as a u.d.l.
∴ u.d. load = $5 \times 0.111 + 0.185 = 0.740$ kip/ft
Load from stringers = $20 \times 0.111 \times 5.17 + 0.062 \times 25 = 13.1$ kip

EXAMPLE 4.1 (Continued)

$V_{DES} = V_D + 1.3\,V_L$

$\qquad = 0.740 \times 13 + 2 \times 13.1 + 2 \times 49.9 \times 1.3\,\text{kip}$

$\qquad = 9.6 + 26.7 + 64.8 = \underline{101\ \text{kip}}$

$M_{DES} = M_D + 1.3\,M_L$

$\qquad = 0.740 \times 26^2/8 + 13.1 (5.25 + 10.42) + 392 \times 1.3$

$\qquad = 62 + 204 + 510 = \underline{775\ \text{kip-ft}}$

$\quad Z = M_{DES}/F_b = 775 \times 12/20 = 465\ \text{in.}^3$

USE W 36 x 150
floor beam

Check shear:

$$f_v = \frac{V}{A_w} = \frac{101}{34 \times 0.625} = 4.8\ \text{ksi} < 12\ \text{ksi}$$

O.K.

CONNECTIONS: Using the calculated shears, web angle connections may be designed as discussed in Chapters 18 & 19. The stringer floor beam connection is designed in Example 18.2.

|← 5'3" →|← 5'2" →|← 5'2" →|← 5'2" →|← 5'3" →|

W 21 x 55   W 36 x 150

Floor
system
cross-
section

|← 26'0" →|

These values are based mainly on deflection requirements. The optimum inclination of the diagonals of a truss is approximately 45°.

The structure is analyzed first as a beam of constant moment of inertia, and standard tabulated solutions may be used. The values of moment and shear so obtained can be used to make a rapid preliminary design of the members. If the moment diagram is such as to suggest a structure with varying moment capacity, then this may be estimated and a new preliminary design undertaken using the new section properties. This process will in turn allow a more accurate beam analysis to be made, and so the procedure is repeated and refined until a sufficient degree of accuracy is obtained.

2. Dead Loads

For a preliminary design, the weight of the structure must be estimated. This is an important step, because in most bridges the stresses due to dead load are greater than those due to any applied live loads. The weight of the floor system is largely independent of the main member used and can be estimated with some accuracy. A number of approximate formulas have been proposed[3.2,4.20] to find the weight of steel in the remainder of the bridge. Care should be taken in using the formulas of earlier decades as the use of higher stresses and more efficient design processes has made their predictions conservative. Design offices usually rely on past experience with similar bridges, whereas the novice must be prepared to make some unsuccessful trials.

Dead loads may be considered as distributed along flexural members or as acting at the panel points of trusses. The consideration of the further distribution of these loads is a problem in structural mechanics. For continuous flexural members, such as beams and plate girders, the standard beam techniques may be used. The method of superposition of redundants is useful for the type of problem which can be solved in tabular form. An alternative method is moment distribution, which can be applied to members of non-uniform cross section by using published tables of stiffness factors, carryover factors, and fixed-end moments,[4.21] cardboard analogs,[4.22] or direct analytical computation. The method is illustrated in Example 4.2.

Truss redundancies may be internal or external. An externally redundant truss can be treated in a manner similar to that described above for continuous beams. If the truss is internally redundant, it is more convenient to use the method of internal superposition.[1.6] It may be seen that the method is suitable for tabulation and by writing in matrix form;[4.23] it is also convenient for electronic computation.

3. Live Load Analysis

The problem of main member design is complicated by the moving location of the load. For member stresses, it is necessary to find that position of the live load which will cause the highest stress in the member under consideration. There are two main methods for locating this critical position: the first is by inspection, and the second is by the use of influence lines.

In addition, the dynamic nature of the loads will cause shock, vibration, and fatigue problems. These are dealt with by increasing the applied loads (for shock and vibration) and by decreasing the allowable stresses (for fatigue).

Inspection Methods. It may be immediately obvious where to place the live load in order to produce the maximum stress in a member. This would be the case for a single concentrated load moving across a simply supported span. For a series of loads, it is possible to calculate the forces for each position of the live load and thus to find the critical position by trial and error. This process can be laborious and can be circumvented by the use of influence lines. However, the trial-and-error method may be expedited by the use of moment tables.[3.2,4.24] Such a presentation is given in Table 4.3 for Cooper E 80 loading.

EXAMPLE 4.2

PROBLEM:
  Find the moments and shears due to dead weight in the three span continuous bridge shown below. The floor system is as designed in Example 4.I. The loading is H20-SI6.

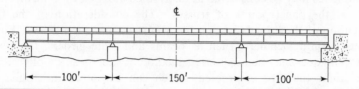

SOLUTION:
  For purposes of analysis and comparison two distributions will be assumed for the flexural stiffness, $S = EI$;

UNIFORM STIFFNESS-

STIFFNESS VARIATION DUE TO COVER PLATES, ETC-

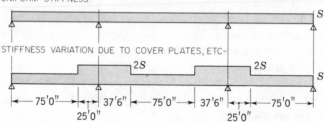

It is very likely that the final bridge chosen for this crossing will have a stiffness distribution lying between these two cases.

DEAD LOADS:

	Hand Rail	2×0.023	$= 0.046$ kip/ft
Concrete	Parapet	$2 \times 2 \times 1 \times 150$	$= 0.600$
	Safety walk	$4.75 \times \frac{10.5}{12} \times 150$	$= 0.624$
	Slab	0.0875×31	$= 2.713$
	Surface	26×0.0233	$= 0.606$
	Stringer	4×0.062	$= 0.248$
	Floor beam	$\frac{26}{25} \times 0.185$	$= \underline{0.193}$
			5.03 kip/ft = deck weight

  From past experience (see Art 4.7) the dead weight of the girders is estimated at 0.60 kip/ft each. This estimate will probably be conservative but is good for a first design.

Total Bridge Dead Load = $5.03 + 1.20 = \underline{\underline{6.23}}$ kip/ft

(note that deck weight is the critical quantity)

Dead weight/girder (or truss) = $\underline{\underline{3.12}}$ kip/ft

DISTRIBUTION OF DEAD LOADS:

  UNIFORM STIFFNESS
                    Using Tables such as Table 3.5 of Reference 4.25, the dead load produces moments at the interior supports of
$$M = -0.1682 \times 3.12 \times 100 = -\underline{5250} \text{ kip-ft}$$
                    from Table W   L
The moments at other locations are found by simple statics.

  VARIABLE STIFFNESS
                    In this case it is first necessary to derive certain beam constants. These are stiffness, carry-over and fixed-end moments, and will also be used later in the calculations of Influence Lines.

EXAMPLE 4.2 (Continued)

Unit systems :–

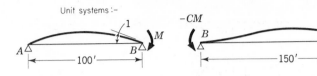

Stiffness diagrams :–

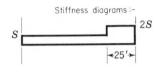
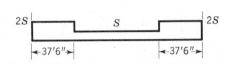

Bending moments :–

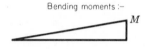

Conjugate beams :–

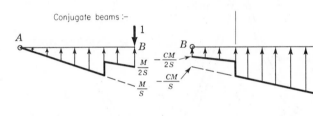

Equilibrium of conjugate beam:–

Moments about A

    I.L = moment of load diagram

$$\therefore M = \frac{384}{91} \times \frac{S}{L}$$

$$\therefore K' = M = \underline{4.22} \frac{S}{L}$$

    (cf. $K' = 3\frac{S}{L}$ for uniform case)

Moments about C

    moment of load diagram = 0

$$\therefore C = \frac{3}{5} = \underline{0.6}$$

    (cf. C = 0.5 for uniform case)

Vertical equilibrium

    area of load diagram = I

$$\therefore K = M = \frac{20}{3} \cdot \frac{S}{L} = \underline{6.67} \frac{S}{L}$$

    (cf. $K = 4\frac{S}{L}$ for uniform case)

$$K_{sym} = (I - C)K = (I - 0.6)\frac{20}{3} \cdot \frac{S}{L}$$

$$= \frac{8}{3} \cdot \frac{S}{L} = \underline{2.67} \frac{S}{L}$$

Proceeding similarly the fixed–end moments are found to be:–

    Exterior span $M'_F = \dfrac{wL^2}{6.55}$

    (cf. $\dfrac{wL^2}{8}$ for uniform case)

    Center span $M_F = \dfrac{wL^2}{10.7}$

    (cf. $\dfrac{wL^2}{12}$ for uniform case)

MOMENT DISTRIBUTION:

    Making use of symmetry and pin – ends the one – cycle moment distribution is:–

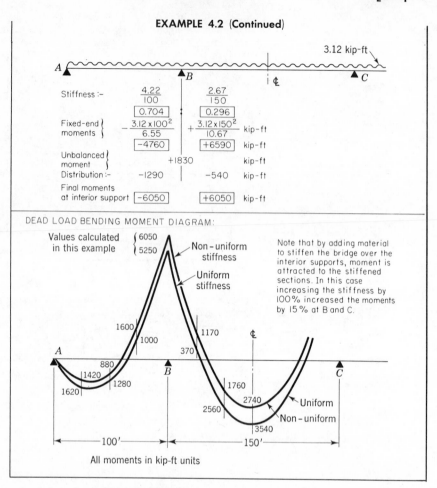

EXAMPLE 4.2 (Continued)

DEAD LOAD BENDING MOMENT DIAGRAM:

Values calculated in this example
$\begin{cases} 6050 \\ 5250 \end{cases}$ Non-uniform stiffness

Uniform stiffness

Note that by adding material to stiffen the bridge over the interior supports, moment is attracted to the stiffened sections. In this case increasing the stiffness by 100% increased the moments by 15% at B and C.

All moments in kip-ft units

In using the table it is convenient to draw the bridge to scale, carefully marking the position of the load points on a paper cutout drawn to the same scale. The cutout is then moved across the bridge, and the moments and shears are calculated for each position, using the summations given in the moment table. This eliminates considerable calculation; but the shears and moments must still be maximized. The use of more extensive moment tables allows calculations to be reduced still further. In addition, loading tables are available[4.25] which give directly the maximum moments and shears in a large variety of two-, three-, and four-span uniform continuous beams for H20-S16 loadings. Such tables circumvent the processes described above and those given in the following section. Their use is recommended in any practical design situation.

Table 4.3 Simple Moment Table for Cooper E 80 Loading*

Wheel Number	Axle Load (kips)	Spacing (ft)	Position (ft)	Shear (kips)	Moment (kip-ft)
1	40	8	8	40	0
2	80	5	13	120	320
3	80	5	18	200	920
4	80	5	23	280	1,920
5	80	9	32	360	3,320
6	52	5	37	412	6,560
7	52	6	43	464	8,620
8	52	5	48	516	11,404
9	52	8	56	568	13,984
10	40	8	64	608	18,528
11	80	5	69	688	23,392
12	80	5	74	768	26,832
13	80	5	79	848	30,672
14	80	9	88	928	34,912
15	52	5	93	980	43,264
16	52	6	99	1032	48,164
17	52	5	104	1084	54,356
18	52	5	109	1136	59,776
Continuous load	8 kips/ft				

* For other loadings, EN, the above shears and moments are multiplied by N/80.

Influence Lines. If the value of a structural action is M_{ab}, where M is the action, point a its location, and point b the location of its cause, then M_{ab} will vary as point b varies. For a given point a, this variation will produce a graph of M_{ab} against the location of point b, and such a graph is called an "influence line" for M_{ab}. Figure 4.10 illustrates the above concept for the moment caused at point a in an outer span of a three-span beam. In this case the cause of the moment is a vertical load P. The influence line is seen to be a graph drawn with the loaded chord of the structure as its abscissa.

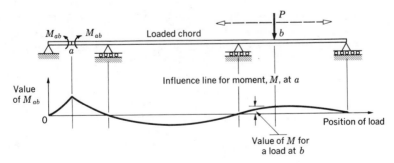

Fig. 4.10 Influence-line illustration.

If influence lines are available, they allow both the visualization of critical load locations and also the calculation of the actions due to loads applied at any point or series of points on the structure. The following paragraphs will illustrate how influence lines may be calculated with relative rapidity. Use is made of the principle of virtual work, which states that, for a body in static equilibrium, zero total work will be done when the body moves through a virtual displacement.[1.6] Consider the example shown in Fig. 4.11. The structure and loading condition for which the influence line is required (Fig. 4.11a) is a system in static equilibrium. To obtain a set of virtual

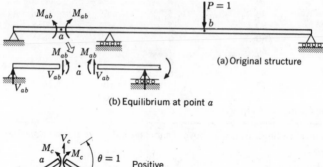

(a) Original structure

(b) Equilibrium at point a

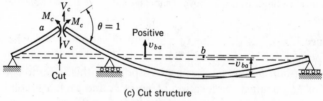

(c) Cut structure

Fig. 4.11   Influence-line theory.

displacements, the unloaded structure is cut at the point a for which the influence line is required, and a unit deformation is introduced at point a in the direction and sense of the required action. For the case of the action M_{ab} in Fig. 4.11a, this deformation is an angle change as shown in Fig. 4.11c. The principle of virtual work then gives (external work positive)

$$-(P \times v_{ba}) + M_{ab} \times \theta = 0 \tag{4.11}$$

where v_{ba} is the displacement upwards of the point b due to the virtual cut introduced at point a. Then with P and θ both unity

$$M_{ab} = v_{ba} \tag{4.12}$$

Equation 4.12 is very useful as it shows that the deflected shape of the cut system is the influence line for the action corresponding to the cut in direction, sense, and location.   In other words, the structure is made to draw its own influence line by imposing upon it a unit deformation; this may be done by constructing the resultant deformed shape by using models (see Chapter 5) or by direct calculation.

Two procedures may be followed to calculate influence lines for redundant structures.   The first adopts the general technique of introducing a unit deformation at the point under consideration and thus calculating the influence line for this point.   However, in a bridge the influence line will be required for many points and this will involve many repetitions of similar procedures.   It is often convenient to calculate the influence lines for a number of loads and moments (usually reactions) equal to the redundancy of the structure.   This allows the influence line for any other point to be calculated by simple statics.   The method lends itself to tabulation and is very suitable for design purposes.   Example 4.3 illustrates the use of the latter procedure for calculating influence lines.

The actual calculation process will require a redundant analysis if the structure is more than once indeterminate.   Any of the standard techniques may be used; usually either moment distribution or slope deflection will be found to be the most efficient.   For non-uniform beams, it is necessary to determine the stiffness and carryover factors and the fixed-end moments for each non-uniform beam.   These quantities can be found by conjugate beam methods, the three-moment equation, numerical integration,[4.26] or by semi-graphical techniques such as the cardboard analog.[4.22]   Constants for the more common types of non-uniform beams can also be found in standard tabulations such as Ref. 4.21.   Whichever method is used, it is best to try to impose the unit deformation directly on the structure rather than to work in terms of unknown forces, and for this reason conjugate beam processes are useful whenever they are applicable.

Influence lines can also be obtained by the use of models (Chapter 5) or by the step-by-step calculation of the action M_{ab} for each load position.   The step-by-step procedure is illustrated by a very complete example in Ref. 4.18.   Model analysis emphasizes the importance of allowing the structure to draw its own influence lines.   Influence lines have also been tabulated for a number of cases: Reference 4.25 gives all the relevant design data for two-, three-, or four-span structures of constant moment of inertia and with varying span ratios, together with further useful design information.   Such tabulations are extremely convenient for members of uniform section.   It is stated that they may also be used where cover plates do not increase the second moment of inertia by more than 50 per cent.[4.25]

Application of Influence Lines.   Influence lines are associated with unit loads.   For a load P, the ordinates are multiplied by P.   If the concentrated load P is replaced by a load $w \cdot dl$ acting over an infinitesimal length dl, then from Eq. 4.12 the influence line for $w\,dl$ is $w\,dl \cdot v_{ba}$.   If the load $w\,dl$ is

EXAMPLE 4.3

PROBLEM:
   Find the influence lines which will allow a live load analysis of the bridge treated in Examples 4.I and 4.2.

SOLUTION:
UNIFORM STIFFNESS
   These values are completely tabulated in sources such as Reference 4.25, and thus are not calculated here.

NON-UNIFORM STIFFNESS
   As the structure is twice redundant and symmetrical it is only necessary to calculate one influence line; the most convenient is the influence line for R_A, although it might be more direct to construct the influence line for a midspan moment.

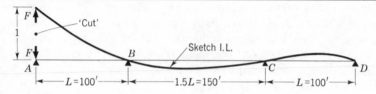

The bending moment diagram resulting from this deformation is :—

The value of M is found by moment distribution, using the stiffness and carryover factors found in Design Example 4.2.

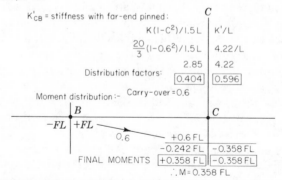

K'_{CB} = stiffness with far-end pinned:

	K(I−C²)/I.5L	K'/L
	$\frac{20}{3}(I-0.6^2)/I.5L$	4.22/L
	2.85	4.22
Distribution factors:	0.404	0.596

Moment distribution :— Carry-over = 0.6

	B		C
−FL	+FL	0.6	+0.6 FL
			−0.242 FL	−0.358 FL
FINAL MOMENTS	+0.358 FL		−0.358 FL

∴ M = 0.358 FL

The conjugate beam method is now used to evaluate F (F must cause unit deflection at A)

Conjugate beam

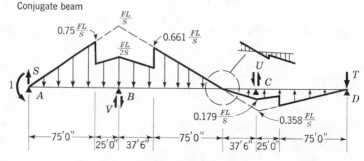

EXAMPLE 4.3 (Continued)

Eqm of CD:
     Moments about C gives $T = 0.055\, FL^2/S$
     Vertical equilibrium gives $U = 0.085\, FL^2/S$

Eqm of AB:
     Moments about B gives $R = \dfrac{1}{L} + 0.153\, FL^2/S$
     Vertical equilibrum gives $V = \dfrac{1}{L} - 0.237\, FL^2/S$

Eqm of BC:
     Vertical equilibrum gives $\underline{F = 1.95\, S/L^3}$
     Moments about C gives $O = 0\ \therefore$ calcs O.K.

The value of F found allows evaluation of the conjugate beam ordinates.

The influence line for R_A is then the bending moment diagram for the above conjugate beam.

Simple statics calculations give the influence line as:—

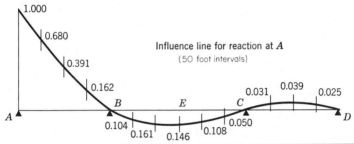

Influence line for reaction at A
(50 foot intervals)

INFLUENCE LINES FOR OTHER LOCATIONS.
   1) The influence line for R_D is obtained from symmetry.

   2) The influence line for R_B is found by taking moments about C.

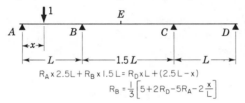

$$R_A \times 2.5L + R_B \times 1.5L = R_D \times L + (2.5L - x)$$

$$R_B = \frac{1}{3}\left[5 + 2R_D - 5R_A - 2\frac{x}{L}\right]$$

For example R_B for midspan (point E) is:—

$$\therefore R_B = \frac{1}{3}\left[5 + 2(-0.146) - 5(-0.146) - 1.75 \times 2\right]$$

$$R_B = 0.645 \text{ at } E.$$

   3) The influence line for M_E (midspan) is:—

$$M_E = R_A \times 1.75L + R_B \times 0.75L - 1(1.75L - x) \quad x \leqslant 1.75L$$
$$ = R_A \times 1.75L + R_B \times 0.75L \qquad\qquad x \geqslant 1.75L$$

   4) Using these and similar equations the results tabulated below
      were obtained.

STAᴺ	A	25'	50'	75'	B	125'	150'	E	200'	225'	C	275'	300'	325'	D
R_A	1	.680	.391	.162	0	−.104	−.161	−.146	−.108	−.050	0	.031	.039	.025	0
R_B	0	.383	.709	.916	1	.972	.864	.645	.406	.180	0	−.111	−.138	−.088	0
V_B	0	−.320	−.609	−.838	±1	.870	.702	.500	.298	.130	0	−.079	−.099	−.063	0
M_F	0	12.2	15.6	6.5	0	−4.2	−6.4	−5.8	−4.3	−2.0	0	1.2	1.5	1.0	0
M_B	0	−7.0	−10.9	−8.7	0	−10.4	−16.1	−146	−10.8	−5.0	0	3.1	3.9	2.5	0
M_E	0	−2.2	−3.5	−2.8	0	4.8	11.6	22.8	11.6	+4.8	0	−2.8	−3.5	−2.2	0
V_E	0	.063	.099	.079	0	−.130	−.298	∓.500	.298	.130	0	−.079	−.099	−.063	0

EXAMPLE 4:3 (Continued)

M_F is for a point 40'0 from A. Moments are in foot units

COMPARISON WITH UNIFORM STIFFNESS INFLUENCE LINES:

In the comparison below the effect of increased stiffness on moments should again be noted. It would not have been very incorrect to use the tables for uniform moment.

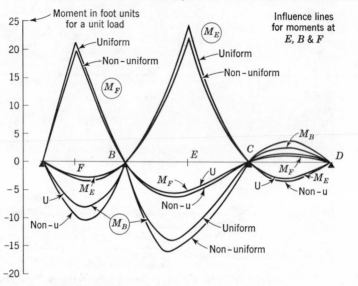

placed at a series of adjacent points b (Fig. 4.10) along the structure, then the total effect of the distributed load $w \, dl/dl = w$ is the sum of all the values of $w \, dl \cdot v_{ba}$. This can be written as

$$A_w = \int_c^d w \, dl \cdot v_{ba} \tag{4.13}$$

where A_w is the action resulting from w, and c and d are the limits within which w acts. If w is constant, then Eq. 4.13 becomes

$$A_w = w \int_c^d v_{ba} \, dl \tag{4.14}$$

$$= w \sum_{cd} \tag{4.15}$$

where Σ_{cd} is the algebraic area of the influence lines between c and d. Thus, the maximum value of A_w is found by maximizing Σ_{cd}, that is, by placing w on all and only those parts of the influence line which are of the same algebraic sign.

The maximization procedure for an action M is rapid for uniform loads or for the "equivalent uniform loads" discussed in Art. 4.5. With a series of concentrated loads, such as truck loads or Cooper E 80, the maximization is a more tedious process. It may be satisfactory to find the critical position by trial and error, locating the train at successive points on the structure. As a first approximation, the heaviest concentrated load should be placed over the maximum influence line ordinate and the next heaviest loads in the adjoining portion with the flatter influence line gradient. The procedure is analogous to, but more rapid than, that described in the preceding paragraphs. Example 4.4 illustrates this technique for locating the critical load, and for most practical cases the method will be found to be adequate. General criteria for locating the maximum load positions have been well discussed elsewhere.[4.27]

4. Maximum Action Envelopes

Once the influence line is known for a particular location and is used to find the maximum action at that location, it is possible to plot these maximum values along the bridge. The dead load stresses are added to the plot and a maximum action envelope is obtained. This is particularly useful in beam or girder type bridges, and the calculation of a typical envelope is illustrated in Example 4.4. In simple cases it is possible to draw a sufficiently accurate maximum action envelope without resort to influence lines. The maximum action envelope is also useful for determining the location of erection splices. These should be located at low moment regions if this is feasible from an erection point of view. The splices are designed as ordinary connections (Art. 18.4) carrying the moments and shears given by the envelopes.

5. Truss Analysis

The influence line procedures adopted above are general and may be applied to beams, trusses, suspension bridge cables, and so forth. Their application to trusses assumes a knowledge of basic truss analysis procedures.[1.3,4.24] Indeterminate systems are handled most conveniently by tabulation procedures[4.23] which allow relatively rapid analysis of a variety of loading conditions.

4.8 DESIGN CONSIDERATIONS

1. Design of Components

The design of individual members and connections is carried out in the manner described in the following chapters of this book. Layout and

EXAMPLE 4.4

PROBLEM:
   Find the maximum action envelopes for the bridge considered in Examples 4.1, 4.2 and 4.3.
For convenience it will be assumed that the load is applied directly to the main members and not distributed
longitudinally by the stringers. Thus the case analysed would be directly applicable to the type of bridge
having no floor beams (e.g. Figures 4.4 b and c). However the error resulting from this assumption is small.

SOLUTION:

INFLUENCE LINE AREAS:
   These are needed for continuous load analyses. They can be found from the influence lines by the
trapezoidal or by Simpson's rule.

NON-UNIFORM CASE

I.L.	AB	BC	CD
R_A	42.9.	−14.4	2.4
R_B	62.8	89.4	−8.6
V_B	−56.8	75.1	−6.2
M_F	929	−579	96
M_B	−680	−1440	240
M_E	−220	1380	−220

) FOOT UNIT

) FOOT² UNITS

Note:
   The dead load values found in Example 4.2
may be checked by using these influence
line areas, together with eq 4.15.

LOADING:
   As shown in Example 4.1 trucks may be displaced to one side of the bridge so that the reaction is
increased.

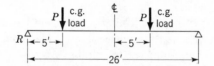

$$\therefore R = P \times \frac{21}{26} + P \times \frac{8}{26} = \frac{29}{26} P$$
$$= 1.11 P$$
$\therefore$ Increase all loads by 11%

POSITIVE SHEAR AT A:
   In this case the critical load location is obvious from influence line.

   1) Truck loads

1.11×32^k 1.11×32^k 1.11×8^k

From I.L. for V_A,
$$V_A = 35.7 + 0.818 \times 35.7 + 0.644 \times 8.9 = \underline{\underline{70.5}} \text{ kip}$$

   2) Lane loads

1.11×26^k ─1.11 x 0.64 kip/ft─

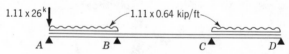

From I.L. areas,
$$V_A = 29 + 45.3 \times 0.71 = 61.5 \text{ kips}$$
$\therefore$ Truck load critical, max pos $V_A = 70.5$ kip

TRIAL AND ERROR METHOD:
   The maximum negative moment at B due to truck loading will be found to illustrate the trial and
error process. As a first guess the center axle is place on the maximum ordinate of the influence
line, the truck is faced so that the rear axle ordinate is greater than the front axle ordinate.

EXAMPLE 4.4 (Continued)

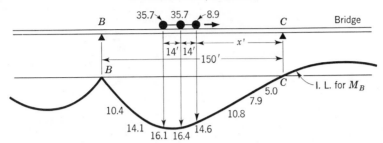

$x = 76'0$ $M_B = 35.7 \times 15.5 + 35.7 \times 16.4 + 8.9 \times 14.8 = 1270$ kip-ft
$x = 74'0$ $\Delta M_B = 35.7 \times 0.3 + 35.7 \times 0 + 8.9 \times (-0.3) = +8$ (first guess)
$x = 72'0$ $\Delta M_B = 35.7 \times 0.3 + 35.7 \times (-0.1) + 8.9 \times (-0.2) = +6$ ⟵ MAX as
                                                                increments
                                                                change sigr.

$x = 70'0$ $\Delta M_B = 35.7 \times 0.1 + 35.7 \times (-0.1) + 8.9 (-0.2) = -3$
             $\therefore M$ (max) = $\underline{1284}$ kip-ft

(The small ΔM_B values indicate that the exact location is often not very critical for spans considerably
longer than the truck)

SUMMARY:
   Repeating the above process for all the influence lines gives:—

QUANTITY	MAX			MIN		
	UNIFORM	NON-U	LOAD	UNIFORM	NON-U	LOAD
V_A	71.1	70.7	T	13.3	15.3	L
V_B	85.8	79.5	L	82.2	86.5	L
M_F	1410	1330	T	482	540	L
M_B	1770	2040	L	192	290	T
M_E	1680	1520	T	333	390	L
V_E	34	33	T	34	33	T

L = lane loading ⎫ critical
T = truck loading ⎭

units: kip-ft

IMPACT:
        AASHO 1.2.12 gives spans to use (function of loading)
                $M_F, V_A, L = 100'$ $p = 1.222$
                $M_B, L = 125'$ $p = 1.200$
            $V_B, M_E, V_E, L = 150'$ $p = 1.182$

MAXIMUM ACTION ENVELOPES:
   The envelopes may now be prepared for each location. As an example the calculations for point B are
shown.

VALUES FOR B

QUANTITY	MOMENTS (UNIFORM)	MOMENTS NON-U	SHEAR (UNIFORM)	SHEAR NON-U
Dead Load	−5250	−6050	−208/+234	−216/+234
p x +Live Load	−2130	−2450	+101	+94
+ Design Load	−7380	−8500	+335	+328
p x − Live Load	+230	+350	−97	−102
− Design Load	−5020	−5700	−305	−318

When this has been repeated for for each location the following envelope
may be drawn:

EXAMPLE 4.4 (Continued)

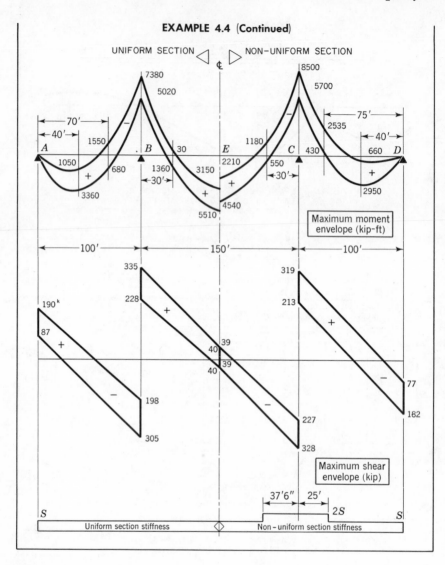

detailing practice may be found in many design texts.[3.6,4.19] In Example 8.1 the maximum stress envelopes obtained in Example 4.4 are used to design a plate girder bridge. Chapter 13 contains a composite design of a similar type of bridge. The design of compression members is illustrated in Chapter 9, tension members in Chapter 6, and connections in Chapters 18 and 19.

2. Fatigue and Brittle Fracture

As the loads on a bridge will vary with time, the problem of fatigue is a relevant one.   This is particularly so with bridges in which the ratio of dead to live load stresses is small.   Chapter 16 discusses the fatigue problem and methods for overcoming its effects.   Brittle fracture is also a problem in bridge design due to the possibility of low temperatures and the dependence of a bridge on the adequate performance of each individual member.   A number of large bridge failures in recent times have been attributed to brittle fracture, and particular attention should be given to such items as design details and steel quality.   A general discussion of these problems is presented in Chapter 15.

3. Secondary Stresses

In a structure, the primary load system will produce deflections which may create loads due to their presence.   For instance, there is the common example of column action (Art. 9.4) in which the lateral deflections of the column produce significant moments due to axial load acting through this lateral deflection.   In truss design the transverse deflections Δ of truss joints relative to each other produce sway moments in the members equal to $6EI\,\Delta/L^2$, where EI is the flexural rigidity and L the length of the member. The procedure for calculating these stresses is dealt with in texts on structural analysis.[1.6]   AREA and AASHO suggest that such stresses are important only when the member depth is more than $\frac{1}{10}$ of its length.   Secondary stresses may also arise from the eccentricity of a joint detail, the torsional behavior of the floor system beams (that is, the stringer end-moments are carried by the torsional resistance of the floor beam), and the self-weight of a member.

4. Camber

Initial deflections are built into beams and trusses to eliminate unsightly geometrical effects, dips and water traps on the bridge surface, and to allow for vertical curves or cross slopes in the road surface.   Beams may be bent into the required shape or camber continuously or at discrete points.   For trusses, the camber is generally introduced by altering the fabrication lengths of various members.   AREA[1.25] recommends a camber equal to the dead load deflection plus a live load of 3 kips/ft of track.   Williot-Mohr graphical techniques are useful for truss calculations.

The total deflection of a bridge under live load and impact is restricted by the codes. AASHO[1.23] requires deflection to be less than $\frac{1}{800}$ of the span; for AREA[1.25] the corresponding value is $\frac{1}{640}$. If the L/d ratios of Art. 4.7 are used, deflections will not be a problem, as these ratios can be derived by assuming the use of mild steel, maximum permissible stresses, uniform moments, and the appropriate deflection limitations.

5. Bearings and Abutments

Bridge bearings are of two types, fixed and expansion. The term "fixed" applies only to the longitudinal displacement of the bearing and not to its rotational behavior (Fig. 4.12a). The simplest fixed bearing (Fig. 4.12b) is a bearing plate; this should be used only for short spans as it tends to develop excessive contact pressures. A typical pin-fixed bearing is shown in Fig. 4.12c. It consists of a vertical tongue fixed to the pier. The beam rests on the rounded top of this tongue and is held in place by a keeper plate. This bearing type is useful in short spans. Further details may be found in Refs. 3.6 and 4.18. It is important to note that bearings may need to be designed for the possibility of uplift and to provide the anchorage necessary during flood conditions.

Expansion bearings are necessary (1) to allow free thermal expansion, (2) to reduce the redundancy of the structure, (3) to reduce foundation settlement stresses, and (4) to keep longitudinal loads from arising in certain parts of the structure. For spans under 100 ft, sliding bearings may be used (Fig. 4.12d); however, it is usually more desirable to use rocker, roller, or self-lubricating bearings. Figure 4.12e shows a rocker bearing, and Fig. 4.12f a roller bearing. For short spans, only one segmental roller bearing may be used, in which case it is held in place by a keeper plate at the top and a pintle (similar to Fig. 4.12e) at the base. Care must be taken that these bearings do not seize with age due to an accumulation of rust and debris. This problem is overcome by the use of proprietary types of bearings using rubber or neoprene elastomeric pads (Fig. 4.12b), low-friction polymers such as Teflon, or self-lubricating bronze alloy materials.[4.28] The use of such simple but effective bearings has increased in recent years and AASHO now defines the requirements for elastomers. Roller bearings can be used as fixed bearings by curving the bottom roller surface. Normally this surface is horizontal, as shown in Fig. 4.12f.

The maximum bearing stress between a roller and a flat surface is given in the various specifications. It may be calculated elastically[4.29] as f_b, where

$$f_b = \sqrt{\frac{PE}{2\pi R(1 - \mu^2)}} \qquad (4.16)$$

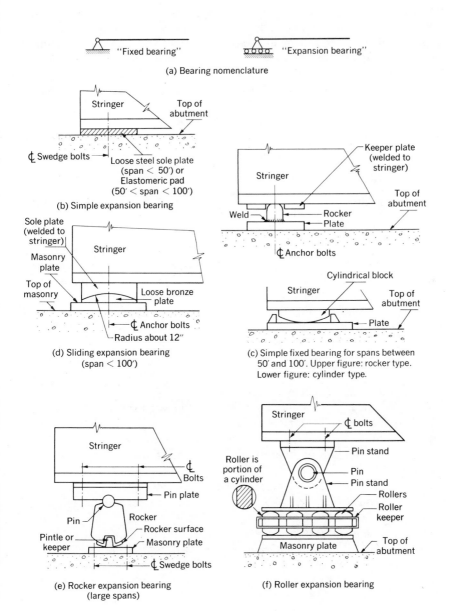

(a) Bearing nomenclature

(b) Simple expansion bearing

(d) Sliding expansion bearing
(span < 100′)

(c) Simple fixed bearing for spans between
50′ and 100′. Upper figure: rocker type.
Lower figure: cylinder type.

(e) Rocker expansion bearing
(large spans)

(f) Roller expansion bearing

Fig. 4.12 Bridge bearings.

where P = reaction per length of bearing, E = modulus of elasticity, μ = Poisson's ratio, R = radius (and f_b, R, P, E are in consistent units).

Internal reactions are usually pins or linkages. The advantage of a linkage is that it restricts the reaction to a force in the direction of the linkage. Design details are given in Refs. 3.6 and 4.19.

Piers and abutments are designed for two sets of loadings. The first are from superstructure and comprise the vertical reactions calculated under Arts. 4.6 and 4.7, and the lateral loads found in Art. 4.7. The second set of loads are from the environment of the pier and may comprise such loads as wind, water pressure, wave action, ice drifts, earthquake, and the like. Actual design details are given in references such as Ref. 3.6.

6. Lateral Load Design

Wind loads have been discussed in Art. 4.5. These loads are carried to the piers and abutments by relatively light truss systems placed in a horizontal plane (Fig. 4.5). In addition, most deck systems are sufficiently stiff to be capable of transferring their proportion of the wind loads to the supports without any additional assistance. The design of the truss systems follows standard procedures, although the diagonals are frequently crossed and only those diagonals acting in tension are assumed to be effective (that is, a Pratt truss [Fig. 4.3b]). However, in this case the transverse members will always be in compression. The chords of the wind truss will frequently be part of the main flexural members of the bridge.

At least one of the horizontal systems resisting wind will not be at the same level as the tops of the piers or abutments. Some means must then be provided to bring the reactions from such systems down to the bridge supports. If there is no intervening traffic space (as in most deck bridges), systems of vertical crossed bracing (similar to those in Fig. 4.4b) may be used. Where clearance is a problem, as in most through bridges, it is necessary to use end portals which transfer the reactions by the flexural strength of the portal legs (Fig. 4.13).

The lateral loads also produce a tendency to turn the bridge over about a longitudinal axis. This rotation may take place about the bridge bearings or about the base of the piers. AASHO specifies that an additional over-turning force of 0.02 kip/sq ft times the deck area must be applied at the windward quarterline of the unloaded bridge. AREA requires the applied live load to be limited to 1.2 kip/ft on the leeward track. Wind tunnel tests have shown[3.10] that overturning can be a significant factor in the performance of a bridge. In the models tested, horizontal wind forces produced vertical overturning forces on the bridge.

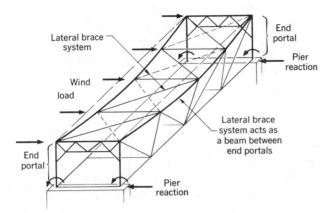

Fig. 4.13 Diagrammatic illustration of wind brace and portal action.

Example 4.5 illustrates the design of lateral systems for the bridge analyzed in earlier examples.

4.9 RECENT BRIDGE DEVELOPMENTS

1. Load Capacity Considerations

The load capacity of a bridge may be governed by either its behavior in the plane of loading or by various instability modes occurring out of the plane of loading. If the first case applies, the load capacity of girders can be predicted by ultimate load methods (plastic design for beams [Chap. 7] and tension field action for plate girders [Chap. 8]). The AISC Specification allows these methods to be used in building design.[1.21] The plastic hinge method has also been proposed for trusses and the method allows realistic predictions to be made of the load capacity of practical trusses.[4.30,4.31]

All the above methods were developed for static loads, and thus they predict only the load capacity under static loading. The behavior under dynamic loading is more complicated analytically and also accentuates the fatigue and brittle-fracture problems. However, the static ultimate load of a bridge is an important structural parameter and should be known for each design. The designer should not assume that a bridge designed by allowable stress methods will behave in a purely elastic manner. Studies have indicated (see Chapter 7) that the results of an allowable stress design are not directly related to either the advent of yielding or to the load capacity of a structure.

EXAMPLE 4.5

PROBLEM:
　Design wind bracing for the bridge considered in Examples 4.1 to 4.4. In Example 8.1 it is found this bridge would need plate girders 10'-4" deep.

SOLUTION:
　The section of the bridge exposed to the wind is:

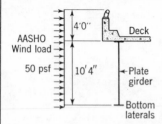

Distribution of wind to deck & bottom laterals:

Deck　　$50 \times (4'\text{-}0'' + 5'\text{-}2'') = 50 \times 9\frac{1}{6} = 0.458$ kip/ft

B. Laterals　$50 \times 5'\text{-}2''$　　　$= 50 \times 5\frac{1}{6} = 0.258$ kip/ft

[Check: $50 \times 14'\text{-}4'' > 300$ min. allowed, ∴ O.K.]
Group III loading, $(0.100 + 0.3 \times 0.258) = 0.177 < 0.258$ ∴ O.K.

Taking the bracing as continuous, the influence tables for a uniform section (Ref: 4.25) give the following support reactions: (max. values)

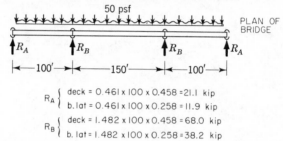

PLAN OF BRIDGE

$R_A \begin{cases} \text{deck} = 0.461 \times 100 \times 0.458 = 21.1 \text{ kip} \\ \text{b. lat} = 0.461 \times 100 \times 0.258 = 11.9 \text{ kip} \end{cases}$

$R_B \begin{cases} \text{deck} = 1.482 \times 100 \times 0.458 = 68.0 \text{ kip} \\ \text{b. lat} = 1.482 \times 100 \times 0.258 = 38.2 \text{ kip} \end{cases}$

DECK LOAD:
　The deck wind loads must be carried to the piers by vertical bracing.

SECTION @ PIER:

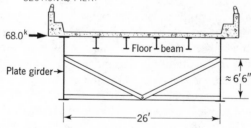

Bracing system chosen to avoid shallow angle between diagonal and horizontal.

Force in diagonals $= \dfrac{68.0}{2} \times \dfrac{\sqrt{13^2 + 6.5^2}}{13} = \dfrac{68.0}{2} \times \dfrac{14.5}{13} = 38$ kip (tension or compression)

Min L/r for compression $= 140$　∴ min $r = \dfrac{14.5 \times 12}{140} = 1.24''$

Try WT 5 x 19.5,　$r_{min} = 1.25'' > 1.24''$

Appendix C of Ref 1.23 $\Big\}$ gives, $\dfrac{e_g c}{r^2} = \dfrac{0.88 \times 0.88}{1.25^2} = 0.49$ ∴ $F_a = 5.5$ ksi

∴ Compression capacity $= 1.25 \times 5.5 \times 5.74 = 39.4$ kip > 38 kip (includes 25% wind allowance) $\Big\}$ O.K.

Tension capacity $= 20$ (5.74 − holes, etc.) ≈ 80 kip

As L/r controls, there is no need to check allowable stress increases.

∴ Use WT 5 x 19.5 for vertical bracing.

EXAMPLE 4.5 (Continued)

BOTTOM LATERAL SYSTEM (horizontal)
Using I.L. tables gives:
  Max. moment = $0.185 \times 100^2 \times 0.258 = 478$ kip-ft @ B
  Max. shear = $0.798 \times 100 \times 0.258 = 20.6$ kip   @ B

∴ Force in chords = moment / truss depth = $478/26 = 18.4$ kip   (note: wind truss chord is
    bottom flange of plate girder)

Force in braces = $20.6 \times \dfrac{\sqrt{26^2+25^2}}{26} = 28.6$ kip $\sqrt{26^2+25^2} = 36.1$ (active in tension only)

Brace area needed = $\dfrac{28.6}{1.25 \times 20} = 1.15$ in.^{2}

  $L/r \big|_{max} = 240$ ∴ $r_{min} = 36.1 \times 12/240 = 1.81$ in.

∴ Use WT 7 x 21.5 for braces

Check stress: $f_a = \dfrac{28.6}{6.32} = 4.5$ ksi   O.K.

PLAN OF BOTTOM LATERAL SYSTEM

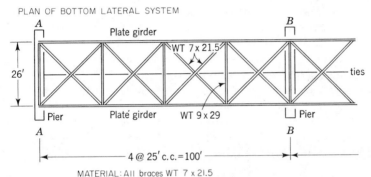

MATERIAL: All braces WT 7 x 21.5

Check tranverse members of bracing system:

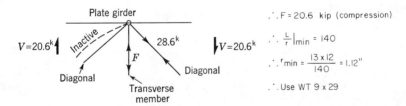

∴ F = 20.6 kip (compression)

∴ $\dfrac{L}{r}\big|_{min} = 140$

∴ $r_{min} = \dfrac{13 \times 12}{140} = 1.12''$

∴ Use WT 9 x 29

CONCLUSION
  It is left to the reader to check the systems under the various combinations of live loads and
  wind load. It will be found that the above case is critical. The reader should also check the
  resistance of the bridge to overturning.

It was noted at the beginning of this section that a load-carrying member may buckle out of its plane of loading, and the design provisions for beams (Art. 7.6) and plate girders (Art. 8.2) have been set to prevent this occurrence. However, for trusses the compression chord is frequently separated from the stiff deck system by the depth of the truss. The lateral wind braces serve to tie both chords together (Fig. 4.5) but are not completely effective, as the entire system is self-contained. In the pony truss the only lateral support comes from the lateral stiffness of the vertical members of the truss. These are usually designed as cantilevers and fixed to the floor system by strong moment connections. The most effective systems for providing lateral restraint in ordinary trusses are the sway frames and portal frames (Fig. 4.5). Sway frames reduce effective lengths and portal frames provide lateral stiffness; both increase the torsional stiffness of the entire system. The dimensions are usually such that there is no possibility of the bridge itself buckling laterally as a beam.

Under certain conditions, trusses may also buckle in the plane of loading.[4.32] If the loads in the members are purely axial, then a pin-jointed truss will fail by buckling of the first of its members to reach its buckling load (Art. 9.3). Figure 4.14a shows this process. If the truss is rigidly jointed, there will be end-moments applied to all members due to the axial contractions and extensions of the members, even if the joints are designed to eliminate eccentric loads. Hence only a hypothetical truss could buckle as an observable whole (Fig. 4.14b). As most of the main compression members of a bridge are relatively stocky, they will fail in the inelastic zone and thus localize failure effects;[4.30] hence the study of in-plane buckling has little relevance to bridge design.

Two series of tests on large-scale trusses are described in Refs. 4.33 and 4.34. Both series indicated that the floor systems participated in the overall flexural behavior of the bridge.

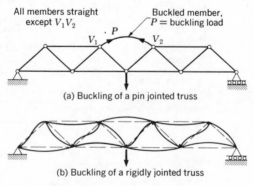

All members straight
except $V_1 V_2$

Buckled member,
P = buckling load

V_1 · P V_2

(a) Buckling of a pin jointed truss

(b) Buckling of a rigidly jointed truss

Fig. 4.14 Truss buckling.

2. Orthotropic Systems and Box Girders

The studies of load distribution mentioned above illustrate the effectiveness of the floor system in carrying longitudinal flexural loads, and this leads to one of the important recent developments in bridge design. Conventional design methods consider the bridge to act as a series of separate units, whereas obvious economies would result if the bridge could be made to act as a single-unit system, and, in addition, could be designed as such. Such developments have already been discussed under the topics of composite design and grillage design of floor beams and stringers. The battle-deck floor system could also be placed in this category.[4.6] If a steel flange is to be used as a traffic deck it needs either frequent stiffening or frequent supporting. Recent bridges which have adopted the integral deck-and-flange system have stiffened the deck by longitudinal stiffeners (analogous to the stringers in Art. 4.4) comprised of either simple vertical members (Fig. 4.15a) or torsionally strong U sections (Fig. 4.15b). These elements are then carried by transverse floor beams and so the general arrangement follows the "traditional" pattern of Art. 4.4. It is found that this type of bridge results in significant dead-load reductions and is thus particularly appropriate for long span bridges where the dead weight is such a large proportion of the actual weight.

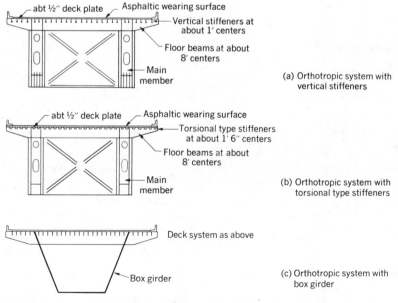

(a) Orthotropic system with vertical stiffeners

(b) Orthotropic system with torsional type stiffeners

(c) Orthotropic system with box girder

Fig. 4.15 Orthotropic bridge systems.

The longitudinal stiffeners function partly as stringers and partly as plate stiffeners. It will be appreciated that such a stiffened top flange presents a difficult analytical problem. It is essentially a plate but has different (anisotropic) properties in the longitudinal and transverse (orthogonal) directions. The design assumption usually made is that the plate is uniformly and orthogonally anisotropic, and the name "orthotropic" which is used for this type of bridge, is a contraction of "orthogonal-anisotropic." For actual design details the reader is referred to the very comprehensive design data in Ref. 4.35.

The advent of welding has been a significant factor in the development of such bridges,[4.18] and in Europe welded orthotropic bridges have been constructed[4.36] with spans of up to 856 ft. To some extent, the increasing use of box girders has been coupled with the development of orthotropic systems. The closed box section has a torsional strength which is far superior to that of the open I section. When using orthotropic systems (Fig. 4.15) the economy is improved by cantilevering the deck beyond the line of the main girders. With such a layout any unsymmetrical lane loading will produce torsional effects. These can be conveniently resisted by a main girder of closed section. Torsional loads also arise with curved girders. Such bridges are frequently required to meet highway alignment conditions but create structural problems as any load on the bridge will generally be eccentric to a line joining the centers of abutments. Thus, even symmetrical loads on curved bridges will lead to torsional effects and curved box girders offer a ready way of handling the stresses involved.

PROBLEMS

4.1. It is decided to replace both the Brooklyn suspension bridge (New York) and the Sydney Harbor through-arch bridge (Australia). In the light of present techniques and developments, discuss the various types of bridges which could be used as replacements. Make a final choice in each case and present preliminary sketches. (See Ref. 4.1 if further information on these two bridges is required.)

4.2. Preliminary designs are required for a road bridge to cross the Mississippi River. Foundation conditions require three spans of 500 ft, 600 ft, and 500 ft, but all other choices are up to the designer. There are at least eight bridge types which could be used; sketch feasible possibilities and comment on each. Comment on your final choice.

4.3. Prepare a first design for the bridge of Problem 4.2. Indicate the possible changes which could be made to the first design.

4.4. Assume that the floor system in Example 4.1 is built with full moment and torsion connections (the floor beams may be assumed to be fixed-ended torsionally). Calculate the maximum concentrated load which could be placed on the deck.

4.5. Design a pair of simply supported pony trusses required to carry a total of 600 lb/ft each over a 50 ft span. Use the allowable stresses in AASHO and consider the 600 lb/ft as the only load acting.

4.6. Calculate the maximum overload which the pony truss can carry (distributed load). Allowable stresses may be increased 50 per cent if this appears reasonable.

4.7. Check the deflections which would be produced in the bridge designed in Example 8.1 if there were a temperature differential of 50°F between top and bottom flanges.

4.8. Assuming a uniform section, redesign the bridge designed in the text using the stringers as the main flexural members and omitting floor beams (see, for example, Fig. 4.4b). Use tables wherever possible; it is only necessary to find required section moduli and shear areas.

4.9. Select, sketch, and dimension suitable bearings for the bridge designed in the text (see also Example 8.1).

4.10. Calculate the influence line for the moment at an interior support of a three-span continuous bridge (75 ft, 150 ft, 75 ft) whose center span has twice the flexural stiffness of the outer two.

4.11. Using Ref. 4.35, prepare preliminary orthotropic designs for a bridge to replace the one designed in Examples 4.1 through 4.5, with the two midstream piers removed.

USE OF MODELS IN DESIGN

5.1 INTRODUCTION

Simple structural models of small scale which can be manipulated on a drawing board, either with or without special instruments, can be a valuable design tool to the structural engineer. There are many reasons for using a model to solve a structural problem. A model can be used (1) when the mathematical analysis of the problem is either impractical or impossible, (2) when an independent check is needed on the mathematical analysis, such as when a designer attempts to solve a new problem or must check his own analysis, and (3) when the model analysis provides a quicker solution for those cases in which the mathematical analysis of a complex structure is tedious or is based on uncertain assumptions. The solutions with a model can be as accurate as desired. Models approximating the structure are often sufficiently accurate for design purposes. Generally, an engineer checking a mathematical analysis is more concerned with eliminating gross errors than in disclosing minor errors of computations. Even when digital computers are used, an independent check with a model may be desirable.

The model provides a physical interpretation of the structure being studied. A designer should have a feeling for the relative stiffness of the component parts of a structure as well as the ability to visualize the deformation of the loaded structure. The model aids the designer in developing an intuitive sense of structural behavior. Experienced designers using simple models can supplement a preliminary "trial design" procedure and further shorten the design time.

When the deflections of a structure are desired, frequently models may be used to advantage. A method is presented in Ref. 5.1 for obtaining the deflections with a model, without need of an elastic analysis. This elastic model method could be applied as an adjunct to plastic design to determine the deflections at working load when the structure is in the so-called elastic range. The model method is especially advantageous when the cross sections of the structure vary considerably within their length. The model is easily shaped to simulate tapered and varying or curved sections such as at the haunches of welded rigid frames. Furthermore, if the cross section of the structure is modified following a preliminary design, the model is quickly altered and the deflection of the final frame is then obtained by repeating the observations.

When one keeps in mind that no model or mathematical analysis can take into account many factors such as any initial slip of bolted or riveted joints, and the stiffening effect of partial composite action, then the futility of a precise determination of the deflection of a structure is apparent. Frequently the limiting-deflections are set by empirical procedures or experience, and a good estimate of the deflection is all that is needed. Models can give a better determination of the deflection than any calculation, give it more economically, and give both upper and lower bounds which consider possible influence of interaction of floor slabs, and other factors.

It is not possible to present in a single chapter a full treatment of the use of models. The discussion is limited to the methods employed to obtain influence lines for elastic analysis, and to determine the deflection of and stresses in rigid frames. Models have also been used effectively to determine forces and deflections caused by wind, temperature changes, and foundation displacements,[5.2,5.3] the stiffness, carryover factors, and fixed-end moments of structural members,[5.4] the analysis of trusses and suspension bridges,[4.12,5.5,5.6] and in many other situations. Many structures today are being analyzed by observing the strains and deflections of three-dimensional plastic models.

5.2　DIRECT AND INDIRECT MODELS

There are in general two methods of model analysis. The first, called the "direct method," employs a model which is a scaled replica of its prototype. This model is loaded in a manner simulating that of the actual structure. This method is often used in the analysis of stresses in large and complex structures when an analytical solution requires many assumptions and costs considerable time and money. The three-dimensional models used in the direct method are fabricated by joining sheets and shapes of suitable material, or are formed by drawing in a vacuum a plastic sheet over a mold. Strains are observed with various types of strain gages, and deflections by dial gages or micrometers.

In the second method, called the "indirect method," the model is only an elastic replica and is loaded in a manner bearing no relation to the actual loads, but such that the influence lines for stress or deflections of the structure may be determined. This procedure can be used easily in the solution of statically indeterminate beams, frames, and other types of structures as discussed hereafter. In this chapter, emphasis is on the indirect method.

5.3 STRUCTURAL SIMILITUDE FOR INDIRECT MODELS

The model must satisfy the laws of structural similitude to permit prediction of the forces and deflections of its prototype. For all applications described in this chapter, these laws may be derived from elementary principles of mechanics. In more complex structures the prediction equations require a more rigorous application of structural similitude through dimensional analysis.

In all the problems hereafter considered, strain energy accompanying direct and shearing stresses is so small relative to the bending strain energy that it may be neglected in proportioning the model. Consequently, the shearing modulus and Poisson's ratio of the prototype need not be considered. Also, the cross-sectional areas of the structure need not be simulated exactly.

Numerous authors have presented the application of the Maxwell-Betti reciprocal theorems and the Müller-Breslau principle to establish similitude requirements. Referring to Fig. 5.1, the fundamental equation is

$$X_a = nP_c \frac{\Delta_{ca}}{\Delta_{aa}} \tag{5.1}$$

in which, as shown in Figs. 5.1a and 5.1b,

$X_a =$ the unknown force component acting at point A of the structure;

$P_c =$ the force acting on the structure at point C;

$n =$ a constant dependent upon the dimensions of the model. If X_a is a thrust or shear force component, n has a value of unity. If X_a is the moment, n is equal to the number of units of length on the structure represented by one unit on the model;

$\Delta_{aa} =$ the displacement introduced into the model at point A in the direction of the force component X_a, and

$\Delta_{ca} =$ the deflection of the load point C parallel to the line of action of force P caused by the induced displacement Δ_{aa}.

Each unknown force component acting either on or within the structure can be found by determining on its model the ratio of the deflection of the point corresponding to the point of load on the structure, and measured

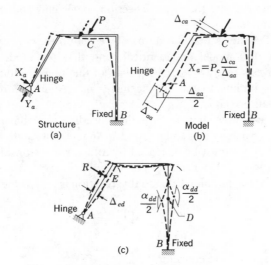

Fig. 5.1 Fundamental deformations.

parallel to the line of action of that load, to the displacement induced in the model along the line of action of the unknown force component. The proof of this fundamental equation now follows.

In Fig. 5.1b, if Δ_{aa} is the displacement induced at some point in the structure such as point A, and Δ_{ca} is the deflection of a second point such as point C due to this induced displacement, one may write

$$\frac{\Delta_{paa}}{\Delta_{pca}} = \frac{C''(L_p^3/E_pI_p)}{C'(L_p^3/E_pI_p)} \tag{5.2}$$

in which C' and C'' are deflection coefficients and the subscripts p identify the structure, hereafter referred to as the prototype.

If a model of this prototype is now proportioned such that:

1. It is geometrically similar, with n units on the prototype represented by one unit on the model;
2. The moment of inertia of the prototype is k times that of the model so that there is a fixed ratio between corresponding sections;
3. The modulus of elasticity of the prototype is p times that of the model;

then if a similar displacement Δ_{maa} is imposed on the model, the corresponding ratio of this imposed displacement Δ_{maa} and the resulting deflection Δ_{mca} is

$$\frac{\Delta_{maa}}{\Delta_{mca}} = \frac{C''(L_m^3/E_mI_m)}{C'(L_m^3/E_mI_m)} = \frac{(pk/n^3)C''(L_p^3/E_pI_p)}{(pk/n^3)C'(L_p^3/E_pI_p)} = \frac{\Delta_{paa}}{\Delta_{pea}} \tag{5.3}$$

in which C' and C'' are the same deflection coefficients and the subscripts m identify the model.

Equation 5.3 shows that the ratios of deflection for corresponding points in prototype and model are identical, and that the model without scale correction gives directly the correct values of the influence line for forces and shears corresponding to the displacement induced in accordance with the Müller-Breslau principle.

However, as is now shown, if the influence line for moment in the prototype at a point such as D is to be found by inducing a rotation to the model, the model moment must be increased by a factor n. Consider Fig. 5.1c, in which a rotation α_{dd} has been applied to the prototype at point D, and the deflection of a second point E due to this rotation is Δ_{ed}. One may then write

$$\frac{\alpha_{pdd}}{\Delta_{ped}} = \frac{Q''(L_p/E_pI_p)}{Q'(L_p^2/E_pI_p)} \tag{5.4}$$

in which Q' and Q'' are constants.

If an identical rotation, α_{mdd}, is now applied to the corresponding point in the model, the corresponding ratio for the model is

$$\frac{\alpha_{mdd}}{\Delta_{med}} = \frac{Q''(L_m/E_mI_m)}{Q'(L_m^2/E_mI_m)} = \frac{(pk/n)Q''(L_p/E_pI_p)}{(pk/n^2)Q'(L_p^2/E_pI_p)} = n\frac{\alpha_{pdd}}{\Delta_{ped}} \tag{5.5}$$

Therefore, for equal values of α_{pdd} and α_{mdd}, $\Delta_{ped} = n\,\Delta_{med}$.

Equations similar to 5.2 to 5.5 were first presented by L. F. Stephens in Ref. 5.7. More detailed discussions on structural similitude can be found in Refs. 5.8 and 5.9.

5.4 PROPORTIONING AND FABRICATION OF INDIRECT FRAME MODELS

As is demonstrated by Eq. 5.3, the ratio $\Delta_{maa}/\Delta_{mca}$ will equal $\Delta_{paa}/\Delta_{pca}$; and by Eq. 5.5, $\alpha_{mdd}/\Delta_{med}$ will equal $n(\alpha_{pdd}/\Delta_{ped})$ if the three conditions stated in Art. 5.3 are fulfilled, thereby satisfying the conditions of similitude. It should be noted that the scales selected for the geometric axis and moments of inertia are chosen arbitrarily and are independent of each other.

The selection of the material for the model depends primarily upon the shape of the structure. If the members are of constant moment of inertia, the brass wire model is excellent. The individual brass wires can be soldered together and bent to the desired shape and the model is ready for service. If the moments of inertia vary, this can be approximated by joining together short lengths of wire of appropriate diameter. Selection of the wire sizes should be based on the product of the modulus of elasticity and the moment of inertia obtained by observing the relative deflections of the wires loaded as cantilever beams, and not by assuming the modulus to be constant for all rods and computing the moment of inertia from the measured diameter. The soldered joints must be carefully made, since minor rotations in the joint will invalidate the results.

A model will duplicate satisfactorily the elastic action of its prototype if it is geometrically similar and proportioned so that the moment of inertia of the model bears a fixed ratio to that of the structure at corresponding sections. When the model is fabricated from a homogeneous isotropic sheet material of constant thickness, this is accomplished by making the width of the model proportional to the cube root of the moments of inertia of the structure. Strict adherence to the conditions for similitude dictates that the cross-sectional areas of corresponding sections of the model and structure be proportional, but no serious error results if they are not proportional, for generally the deformations due to shear and direct stress are relatively small.

If the structure involves curved members and variable moments of inertia, a model cut from seasoned sheets of cellulose acetate is best. This plastic is readily available in sizes 20 by 50 in. of uniform thickness up to $\frac{1}{8}$ in. The cellulose acetate model is cut preferably as a unit from one sheet; however, individual members can be joined together with either acetone and similar solvents, or with a "body cement" made by dissolving shavings of the plastic in acetone. These joints must age 48 hours before they are sufficiently elastic. Both the pure acetone and the body cement must be used sparingly as the solvent is absorbed rapidly by the plastic and diffused throughout, thereby lowering the modulus of elasticity in that area. For the same reason, when the thickness of the model is increased by cementing together strips of plastic, the body cement should be used very sparingly.

Methyl-methacrylate plastics sold as Lucite and Plexiglas are also suitable for models. The thickness of the sheet is not as uniform and cementing is more difficult. However, this material is used for three-dimensional models, the military grade being preferred.

The care with which the plastic model is finished to exact dimensions depends upon the degree of accuracy required of the analysis. For preliminary studies the model could be carefully sawed on a band saw, eliminating all precise dimensional finishing except where the dimensions are small relative to the rest of the model. If the greatest possible precision is desired, such as for a comparison with a mathematical analysis, the model must be brought to an exact size.

Cardboard of even the finest grade, although formerly extensively used, has not always proven satisfactory. The paper fibers are so oriented that the material has a grain and therefore is not isotropic. This can be detected by flexing a square piece in the hands. The errors resulting from this variation in the cardboard may not be large. However, cardboard can be used if only an approximate solution is desired.

5.5 ELIMINATION OF GEOMETRIC ERRORS CAUSED BY LARGE DISPLACEMENTS

An examination of Eq. 5.1 discloses that only when the induced displacement Δ_{aa} is so small as not to change the geometric position of the model, will the ratio Δ_{ca}/Δ_{aa} be identical to that of the structure. When the deflected position of the model and its prototype are not similar, geometric errors enter into the measured deflection Δ_{ca}. These errors can, however, be eliminated for all practical purposes so that relatively large displacements can be employed to good advantage.

The principal source of geometric error is illustrated in its simplest form in Fig. 5.2. Here is shown, in exaggerated fashion, the elastic axis of a simple rigid frame in its normal position and with equal but opposite displacements induced at point A. When point A is displaced to the position A', the side member assumes the curved sloping position $A'B'$ and point B displaces a distance e to point B'. The effect of this is to increase the apparent elastic deflection of point C by some portion of the displacement e at point B. If A is now displaced to the right to point A'' such that AA' and AA'' are equal, the side member AB again shortens, taking the curved sloping position $A''B''$. The effect of this is to decrease the apparent elastic deflection of point C by an identical portion of the displacement e at point B. If now the deflection of point C be taken as $d_1 - d_2$ the effect of the displacement e essentially is

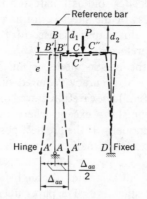

Fig. 5.2 Geometric errors.

eliminated and the deflection measured is that due to the elastic deformations alone. Here the deflection of point C is taken as the difference in two readings taken on a scale sliding along a reference bar and oriented parallel to the load which acts at point C on the prototype.

Obviously, this hypothesis is sound only when the proportional limit of the model material has not been exceeded. When there is uncertainty, it is quite simple to vary the induced displacement Δ_{aa} and determine if the deflection Δ_{ca} varies proportionately.

Proof of the accuracy of this hypothesis has been obtained by both model experiments[5.10] and by mathematical analysis.[5.11]

5.6 DISPLACEMENTS INTRODUCED INTO MODEL

The displacements which must be introduced to determine the force components, and moments acting at the supports A and B and at an intermediate point D on the frame of Fig. 5.3a are shown in Figs. 5.3b to 5.3i. The frame is hinged at A and fixed at B.

It should be noted that in Fig. 5.3b the displacement Δ_1 must be introduced in equal amounts from the normal position. The deflection Δ_2 at the point of load must be the component parallel to the direction of load. The fact that the horizontal reaction H_A acts to the right (Fig. 5.3b) is established by observing that in Fig. 5.3b, as A is displaced to the right, the displacement of point C opposes the direction of the load acting at this point on the prototype.

Note that in Fig. 5.3c the model must be allowed to rotate at A as the displacement Δ_3 is introduced, but in Fig. 5.3f no rotation can be permitted at the displaced end B. The deflection of the model (Fig. 5.3f) must be entirely the effect produced when V_B, the force being sought, alone displaces the model; but all other forces are simultaneously acting so that point B neither rotates nor displaces horizontally.

The moment M_B (Fig. 5.3a) is found by rotating point B as shown in Fig. 5.3d. It is observed that if this rotation is introduced in a clockwise direction, point C moves against the load, establishing the clockwise direction of the moment M_B shown in Fig. 5.3a. It should be further observed that if identical rotations α_5 were introduced into both the model and its prototype (which is n times larger), the deflection of point C on the prototype would be $n\Delta_6$, defined in Fig. 5.3d.

The moment, thrust, and shear at an internal point in the frame such as point D in Fig. 5.3a are best determined by introducing the displacements at that point and measuring the subsequent deflection rather than by computing these values from the experimentally determined reactions. Figures 5.3g to 5.3i show the half displacements which must be introduced in each direction

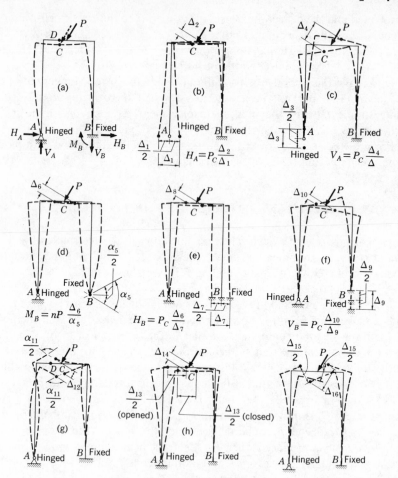

Fig. 5.3 Model displacements.

from the normal position and the deflections which must be measured. The model must be cut to introduce these displacements.

The application of these principles is presented in Example 5.1, with the determination of the horizontal reactions of a welded hinged building frame.

5.7 DEFORMETER GAGES

Simple procedures for obtaining influence lines for continuous beams and girders without special gages are given in Refs. 5.12 and 5.13. Although the displacements to be introduced at the support points of a model such as in

EXAMPLE 5.1

PROBLEM:
 Determine the support reaction by model analysis for a hinged welded building frame — Grace Hall, Lehigh University.

SOLUTION:
 Figure A shows the details of the steel prototype frame. The plate girder serves as a tie for the hinged frame made of two rolled sections. The cellulose acetate model of the structure is shown in Figure B. At six points in the model along the roof member, holes with a #50 drill were made to record with a prick point the deflections when the hinged support was displaced an amount $\delta_{aa} = 3.00$ in. This displacement and the frame deflections are shown in Figure C. By dividing each vertical deflection by the induced displacement the influence line values for the horizontal reaction at the hinge were obtained as shown in Figure D. For a uniform load on the roof the area of the influence diagram may be obtained by graphical integration as shown in Figure E or by using Simpson's Rule. The horizontal reaction obtained with the model and by mathematical analysis is given in Figure F.

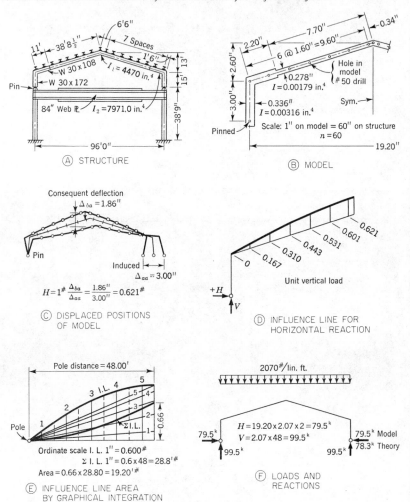

Ⓐ STRUCTURE

Ⓑ MODEL

$$H = 1^\# \frac{\Delta_{ba}}{\Delta_{aa}} = \frac{1.86''}{3.00''} = 0.621^\#$$

Ⓒ DISPLACED POSITIONS OF MODEL

Ⓓ INFLUENCE LINE FOR HORIZONTAL REACTION

Ordinate scale I. L. $1'' = 0.600^\#$
Σ I. L. $1'' = 0.6 \times 48 = 28.8'^\#$
Area $= 0.66 \times 28.80 = 19.20'^\#$

Ⓔ INFLUENCE LINE AREA BY GRAPHICAL INTEGRATION

$$H = 19.20 \times 2.07 \times 2 = 79.5^k$$
$$V = 2.07 \times 48 = 99.5^k$$

Ⓕ LOADS AND REACTIONS

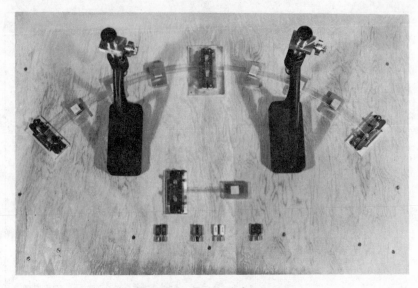

Fig. 5.4 Beggs deformeter.

Fig. 5.5 Eney deformeter.

Fig. 5.3 can be directed by carefully positioned holes in a drawing board, better results are obtained if special gages are employed. When the moment, thrust, or shear is to be determined by cutting the model at an interior point, some form of a special gage is necessary. In 1922 G. E. Beggs invented special deformeters to introduce very small displacements into the model and used micrometer microscopes to measure model deflections.[5·14,5·15,5·16] Figure 5.4 is a typical application of the Beggs deformeter.

A simplified method for introducing the displacements and measuring the deflections was developed by W. J. Eney in 1932.[5·8,5·10,5·11,5·17] "It is less costly than the Beggs deformeter, is easier to use, and gives excellent results."[5·8] A typical assembly is shown in Fig. 5.5.

Otto Gottschalk,[5·18,5·19] J. S. Kinney,[5·8] A. J. S. Pippard,[5·20] and others contributed other instruments for model analysis of structures. Special reaction gages were developed for continuous beams, and for obtaining elastic constants and fixed-end moments for structural members needed for the moment-distribution method of analysis.[5·4,5·12,5·17] All these developments subsequent to the introduction of the Beggs gage have made the model method more generally applicable in a design office.

5.8 DETERMINATION OF DEFLECTIONS BY MODELS

If the model of a frame structure is geometrically similar and so pro-portioned that the moment of inertia of the model bears a fixed ratio to that of the structure at corresponding sections, Eq. 5.6 expresses the relation between the deflection of the structure and that of its model.

$$\Delta_p = \Delta_m n^3 (I_m/I_p)(E_m/E_p) \tag{5.6}$$

When the deflection of a structure is caused primarily by flexure and the deflection due to direct stress or shear is negligible, then Eq. 5.6 is exact. This is the case with most building or bridge structures.

The deflection of a structure therefore can be obtained by observing the deflection of its model under a suitable load, determining the modulus of elasticity, and then using Eq. 5.6. The modulus of the model material can be determined by hanging a known weight on the end of a model cantilever beam, observing the deflections, and then calculating the modulus from the formula for the deflection. However, cellulose acetate and other plastic models creep under constant load; that is, the observed deflection is usually time-dependent. This creep may influence the deflection of the model by as much as 5 per cent. When a more accurate value of the deflection is needed, two alternate methods may be employed. From Eq. 5.6 it can be seen that, if identical forces are applied to (1) the model of a frame structure, and (2)

EXAMPLE 5.2

PROBLEM:

Construct the influence line by model analysis for the deflection of the ridge of the gable frame shown in Figure A of Example 5.1.

SOLUTION:

The cellulose acetate model used in Example 5.1 was also used in this investigation. In order to compensate for the creep effect of the model material, a cantilever model made of the same material was used as an aid for stabilizing the deflections. The general arrangement of the model frame and the cantilever is shown in Figure A. If an arbitrary displacement is introduced between point D on the model frame and point A on the cantilever beam, the deflection of both structures usually would not change with time. As the cantilever and the model frame tend to creep to a certain extent toward each other (or away from each other, depending on the type of displacement introduced), each offsets the effect of creep on the other. Thus, the effect of creep is completely eliminated. Figure B shows the resulting deflection δ_{aa} of the model cantilever to be 0.87 in. at point A and the deflection δ_{dd} of the model frame to be 0.51 in. at point D.

The force which is required to produce a displacement of $\delta_{dd} = 0.51$ in. in the prototype frame, and 0.81 in. on the prototype cantilever, can be determined by considering the bending of the prototype cantilever shown in Figure C. This prototype cantilever is a steel beam 15.2 in. x 60 ÷ 12 = 76 ft long, with a moment of inertia of $(0.50^3/0.336^3) \times 7910 = 26 \times 10^3$ in.^{4}. From Eq. 5.6 it can be seen that if identical forces are applied to (1) the model of a structure and (2) the model of a cantilever beam, the ratio of the deflection of the frame model to the deflection of the cantilever model is equal to the ratio of the deflection of the frame prototype to that of the cantilever prototype. Consequently, if a force sufficient to deflect the cantilever model 0.87 in. is then applied to the frame model, the deflection of the frame model is identical to the deflection of its prototype when acted upon by the same structural force that would deflect the cantilever prototype 0.87 in. This structural force is computed by the moment area principle as follow (see Figure C):

$$\Delta_{A_s} = \frac{800P \times 26.67 \times 12^3}{30 \times 10^3 \times 26 \times 10^3} = 0.047P \text{ (in inches)}$$

for $\Delta_{A_s} = 0.87$ in., $P = 18.4$ kip. Therefore, a force of 18.4 kip on the prototype frame would produce a deflection of 0.51 in. at the ridge. Then, by proportion the structure would deflect 0.028 in. under a 1-kip load acting at point D.

The influence line for the deflection of the ridge can be found from the measured deflections at the six points along the roof member by the following relationship:

$$\text{Influence line ordinate} = 0.028 \times \frac{\text{measured deflection}}{0.51 \text{ in.}}$$

The resulting influence line is given in Figure D.

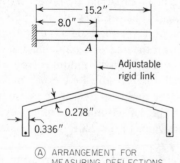

(A) ARRANGEMENT FOR
MEASURING DEFLECTIONS

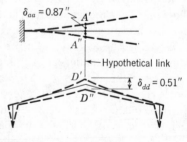

(B) DISPLACED POSITIONS
OF THE MODELS

EXAMPLE 5.2 (Continued)

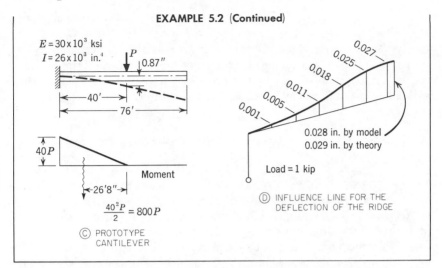

© PROTOTYPE CANTILEVER

Ⓓ INFLUENCE LINE FOR THE DEFLECTION OF THE RIDGE

the model of a cantilever beam of constant moment of inertia, the ratio of the deflection of the frame model to the deflection of the cantilever beam model is equal to the ratio of the deflection of the frame prototype to the cantilever prototype. This is true only if both models have the same modulus of elasticity and are proportioned to the same similitude scales and the forces are independent of time.

The first method uses a creep-compensating spring to apply identical forces first to a calibration cantilever beam and then to the model of the prototype frame. Many different designs of compensating springs to overcome creep can be devised; some of these are described in Refs. 5.1 and 5.21.

The second method uses a rigid link to apply simultaneously equal and opposite forces to a calibration cantilever beam and the model of the prototype frame, thereby overcoming creep. This link is adjustable for lengthening or shortening. Example 5.2 illustrates the second method.

Also, since the deflected model is the influence line for the deflection of the loaded point, the influence line is obtained by observing the deflection of the entire structure. This is especially useful when the deflection is desired for uniform or various concentrated loads.

5.9 OTHER APPLICATIONS OF STRUCTURAL MODELS

1. Beams on Elastic Foundations

The continuous beam on an elastic foundation is encountered in many engineering problems. Foundation beams on subgrades, grillages distributing column loads, culverts, subway structures, and beams supported by elastic tie bars, are but a few of the problems.   Numerous theoretical methods for solving these types of problems are available.[5.22,5.23]   However, except for a few simple cases, the mathematics involved in the theoretical analyses is often very lengthy.   A simple model method employing coil springs which permits consideration of varying soil properties is given in Ref. 5.24.   The method has not been widely publicized, but is worth the reader's attention.

2. Structural Model Analysis by Means of Moiré Fringes

When two grids of closely spaced lines are superimposed on each other with one grid displaced with respect to the other, an optical phenomenon known as the moiré effect is observed.   A pattern of alternating dark and light fringes due to the mechanical interference or shadow effect results.   If one array of lines is on a model of a structure and a second array is on a base board, when the model is displaced the moiré fringes resulting can be directly interpreted to give the displacement of the model.   This convenient method of recording the deflections of the model by moiré fringes has been developed in Ref. 5.25.

Fig. 5.6 Testing of a model three-story, two-bay frame. (Courtesy of Dr. William A. Litle.)

3. Model Steel Frames for Studying Inelastic Structural Response

Small-scale models have been used successfully to study the inelastic response of steel frames subjected to static and dynamic loads. In fact, models played a very important role in the early development of the plastic theory of structural analysis.[5.26]   Figure 5.6 shows the testing of a model frame under combined vertical and horizontal loads. The vertical loads on the beams were applied through the M-shaped linkages which could sway with the structure as it deflected laterally; thus the line of action of the loads always remained vertical. The model frame was a 1/8 scale reproduction of the three-story, two-bay frame shown in Fig. 22.7. The prototype frame was tested to study the behavior and load-carrying capacity of unbraced frames with particular emphasis on the effect of frame instability (see Art. 22.2). The details of the model and the prototype frame tests are given in Refs. 5.27 and 5.28.

Fig. 5.7 Wind-tunnel testing of an aeroelastic model of the U.S. Steel Building in Pittsburgh, Pa. (Courtesy of Dr. Alan G. Davenport.)

4. Wind Tunnel Testing of Building Models

An important recent application of structural models is in the study of the dynamic response of tall buildings to fluctuating wind. In this application, reduced-scale aeroelastic models are tested in a specially designed wind tunnel. Figure 5.7 shows an upstream view of the boundary layer wind tunnel of the University of Western Ontario with a 1/400 scale model of the 844 ft U.S. Steel Building (Pittsburgh, Pennsylvania) in the foreground. Dominant aerodynamic features in the immediate surrounding were reproduced by block outline models of the neighboring buildings. Results of wind tunnel testing have helped greatly in the design of tall buildings, stadiums, chimney stacks, and suspension bridges.

PROBLEMS

5.1. The frame shown in the accompanying figure sways and deflects under the 500-lb force. Draw the pressure line and approximate the reactions.

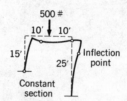

Problem 5.1

5.2. Using a steel spline $\frac{3}{32}$ by $\frac{3}{32}$ by 36 in. long, construct the influence line for reaction at support C in the accompanying figure and determine the reaction corresponding to the given loads.

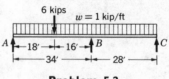

Problem 5.2

5.3. Using the model from Example 5.1, determine the horizontal deflection of the ridge due to a horizontal load of one kip acting at the ridge.

5.4. Using the model from Example 5.1, determine the deflection of a point on the sloping member 24 ft 0 in. from the knee caused by a unit load acting perpendicular to the sloping member.

5.5. Using a brass wire model, determine the horizontal and vertical reaction at supports A, E, and H.   Draw the curve of bending moment.

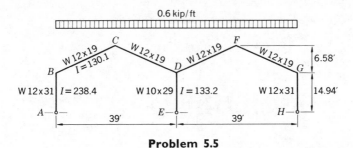

Problem 5.5

6

TENSION MEMBERS

6.1 INTRODUCTION

The selection of sections for tension members is one of the simplest and most straight-forward problems encountered in design. Since stability is of minor concern with tension members, the problem is reduced to selecting a section with sufficient area to carry the design load without exceeding the allowable tensile stress as determined by the factor of safety. Reduced to equation form,

$$A = \frac{P}{F_t} \tag{6.1}$$

where A is the required area, P is the working load, and F_t is the allowable tensile stress. The rigidity of a tension member must be considered. The slenderness ratio is limited by the specifications to prevent "whipping."

The allowable stress F_t is given as $0.6F_y$ in Section [1.5.1.1] of the AISC Specification. Referring to Art. 1.4 and Fig. 1.10, it is seen that the allowable stress is based on "unrestricted plastic flow" at the yield stress level. The resulting factor of safety for structural carbon steel is F.S. = $36/22 \approx$ 1.65. Retaining this value for use on steels with other yield points,

$$F_t = \frac{F_y}{(\text{F.S.})} = \frac{F_y}{1.65} = 0.6F_y$$

In 1969, a further provision was added that is applicable only to the net section of axially loaded members. This limits the allowable stress on the net section to $\frac{1}{2}$ the minimum tensile strength of the steel. This provision recognizes the fact that it may not be possible to develop unrestricted plastic flow in the gross area of high-strength steel tension members.[6.1]

The AREA and AASHO Specifications traditionally have provided allowable stresses that were about 10 per cent less than stresses used in building construction to provide for factors such as exposure to weather and corrosion of structures. Dynamic loading may require further reduction since the fatigue strength of the tension member may be reduced because of some detail. (See Chapter 16.)

Table 6.1 gives values of allowable stresses for the three specifications.

Table 6.1   Allowable Stresses for Axial Tension

Specified Yield Point (ksi)	Thickness (in.)	Allowable Stress (ksi)		
		AISC	AREA	AASHO
36.0	All	22.0	—	—
42.0	$1\frac{1}{2}$–4	25.0	—	22.0
46.0	over 1	—	25.0	—
46.0	$\frac{3}{4}$–$1\frac{1}{2}$	27.5	—	24.0
50.0	up to 1	—	27.0	—
50.0	up to $\frac{3}{4}$	30.0	—	27.0

6.2   TYPES OF MEMBERS

In early practice the most commonly used tension members were rods, cables, or eyebars with pin-connected ends. In more recent times, these members have given way for the most part to angles used singly or in pairs, or members built up of plates and rolled sections. Some of the more common configurations are shown in Fig. 6.1. Sections which extend the

Fig. 6.1   Typical sections used as tension members.

full length of the member are shown solid. Intermittent connections which are used to maintain alignment of the sections or perforated members are shown dotted.

Eyebars are used in current practice as anchorages for suspension bridge cables and, of course, the cables and the suspenders are themselves tension members. Eyebars are also used in large continuous truss bridges where large forces may be developed over the supports in the top chords of the trusses.[6.2]

6.3 EYEBARS AND PIN-CONNECTED LINKS

The design of eyebars and pin-connected members is covered by AISC Section [1.14.6], AREA Section [1.12], and AASHO Section [1.6.93]. The provisions of these sections concern the general geometry of eyebars and pin-connected links. The relationships between width of body, diameter of pin, and diameter of head or net section adjacent to the pin hole are such that "dishing" of the head or region adjacent to the pin hole will be avoided. If "dishing" of the head occurs, there will be an associated loss of strength of the member. The restriction on width-to-thickness ratio of the body is included to provide a compact section and to avoid large-diameter pins. Since the diameter of the pin is a function of the width of the body and the allowable bearing stress, an extremely wide or thin body will require an excessively large pin diameter. Reference 6.2 contains the results of an extensive research program on pin-connected links. The design of an eyebar is illustrated in Example 6.1.

6.4 TENSION RODS

Threaded rods used as tension members have a net area based on the diameter at the root of the thread. "Upset" rods have enlarged ends such that the area at the root of the thread is greater than or equal to the area of the main portion of the rod. The design of the rod may be based on the "stress area" or on the gross unthreaded area of the rod depending on which is the governing condition. The tensile "stress area" is the assumed area of an externally threaded part. It is taken as

$$A_s = 0.7854 \left(D - \frac{0.9743}{n} \right)^2$$

where D is the nominal size in inches and $n =$ threads per inch.

6.5 USE OF ROLLED SHAPES AS TENSION MEMBERS

In current practice the structural shape most commonly employed as a tension member is the angle, used either singly or in pairs. If the end connections of the member are welded, the gross area of the section can be used. The selection of the section reduces to the application of Eq. 6.1 and a check of the slenderness ratio to assure adequate rigidity. If the end connections are riveted or bolted, the required gross area will be somewhat

EXAMPLE 6.1

PROBLEM:
Eyebars are to be used as cable anchorages for a suspended roof over a sports arena. The design load for a typical eyebar is 150 kip

Using flame-cut A440 steel plate, design the eyebar in conformance with AISC Specification.

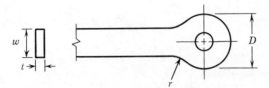

SOLUTION:

From Eq. 6.1

$$A = \frac{P}{F_t}$$

From Sec $\left[1.5.1.1\right]$ or Table 6.1

$$F_t = 0.60F_y$$

From the Appendix of the Specification

For $t \le \dfrac{3}{4}$ in. $F_y = 50{,}000$ psi
                                  $F_t = 30{,}000$ psi

For $\dfrac{3}{4} < t \le 1.5$ in. $F_y = 46{,}000$ psi
                                      $F_t = 27{,}500$ psi

To take advantage of the greater allowable unit stress for the thinner plate, try a thickness of 0.75 in

$$w = \frac{P}{F_t t} = \frac{150{,}000}{30{,}000\,(0.75)} = 6.67 \text{ in.}$$

$$\text{USE: } w = 6\,\frac{11}{16} = 6.6875 \text{ in.}$$

Check width to thickness ratio, Sec $\left[1.14.6\right]$:

$$\frac{w}{t} \le 8$$

$$\frac{6.6875}{0.75} = 8.9 > 8 \qquad\qquad \underline{N.G.}$$

Try: $t = 1.0$ in ; $F_t = 27{,}500$ psi

$$w = \frac{P}{F_t t} = \frac{150{,}000}{27{,}500\,(1.0)} = 5.45 \text{ in.}$$

$$\text{USE: } w = 5\,\frac{1}{2} \text{ in.} = 5.50 \text{ in.}$$

$$\frac{w}{t} = \frac{5.50}{1.00} = 5.5 < 8 \qquad\qquad \underline{OK}$$

From Sec $\left[1.14.6\right]$

"The diameter of the pin shall not be less than 7/8 the width of the body."

$$\underline{\text{USE: 5 in. diameter pin}}$$

Since the design of the pin falls in the category of connections, assume that the pin is satisfactory.

EXAMPLE 6.1 (Continued)

From Sec [1.5.1.1] the allowable stress on the net section at the pin is $0.45\,F_y = 20.5\,\text{ksi}$. Therefore the required net width w_n is

$$w_n = \frac{150,000}{20,500 \ \ (1.0)} = 7.33$$

$$\text{USE}: w_n = 7\frac{3}{8} = 7.375 \text{ in.}$$

This gives the diameter of the "circular" head as

$$D = 5.0 + 7.375 = 12.375 \text{ in.}$$

$$\text{USE}: D = 12\frac{3}{8} \text{ in.}$$

If the diameter of the hole is $5\frac{1}{32}$ in. Sec [1.14.6], the net width is

$$w_n = 12.375 - 5.031 = 7.344 \text{ in.}$$
$$\text{Net Area} = 7.344 \ (1.0) = 7.344 \text{ in.}^2$$
$$\text{Req'd Area} = 7.33 \text{ in.}^2 \qquad\qquad \underline{\underline{\text{OK}}}$$

From Sec [1.14.6]

"The net section of the head through the pin hole transverse to the axis of the eyebar, shall not be less than 1.33 nor more than 1.50 times the cross-sectional area of the body of the eyebar."

$$\frac{7.344}{5.50} = 1.335 \qquad\qquad \underline{\underline{\text{OK}}}$$

"The radius of transition between the circular head and the body of the eyebar shall be equal to or greater than the diameter of the head."

$$\text{USE}: r = D = 12\frac{3}{8} \text{ in.}$$

FINAL DESIGN:

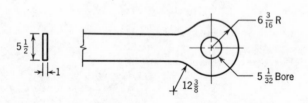

larger than that given by Eq. 6.1. The additional area is necessary to compensate for the reduction in area due to the presence of the fastener holes.

Generally, rivet and bolt holes are fabricated $\frac{1}{16}$ in. larger than the nominal diameter of the fastener to facilitate its installation. The punching operation will cause damage to the metal adjacent to the hole. To compensate for these conditions specifications provide that, for the purposes of computing net area, the diameter of a rivet or bolt hole shall be taken as $\frac{1}{8}$ in. greater than the nominal diameter of the fastener. When plates are greater than $\frac{3}{4}$ in. thickness, punched holes are not usually permitted in bridge construction. Holes are usually subpunched or subdrilled and then reamed to size.

The gross section or gross area of a member is defined as the sum of the products of the thickness and the gross width of each element as measured normal to the axis of the member. The gross width of an angle is defined by the AISC Specification as the sum of the widths of the legs less the thickness. This definition of the gross width of an angle is actually the tabulated area of the angle divided by the thickness. Consider Fig. 6.2. The net

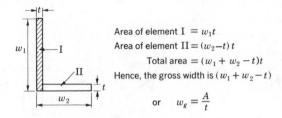

Area of element I $= w_1 t$
Area of element II $= (w_2 - t)\, t$
Total area $= (w_1 + w_2 - t)t$
Hence, the gross width is $(w_1 + w_2 - t)$

or $w_g = \dfrac{A}{t}$

Fig. 6.2   Gross width of an angle.

section or net area of a member is determined by summing the products of the thickness and the net width of each element as measured normal to the axis of the member. The simplest fastener pattern is one in which each row of holes lies on a line perpendicular to the axis of the member, and each row contains the same number of holes. In this case, the net width is obtained by deducting the sum of the diameters of the holes from the gross width. The evaluation of net width for more complex fastener patterns will be considered in Art. 6.6.

In order to prevent excessive sag, vibration, or lateral movement of a tension member, restrictions are placed on the flexibility of the members. The restriction is made by limiting the slenderness ratio, defined as the ratio of the laterally unbraced length of the member to the least radius of gyration. The limiting slenderness ratios for tension members are:

	AISC	AREA	AASHO
Main members	240	200	200
Bracing and secondary members	300	—	240

As was the case with allowable stresses in Art. 6.1, the character of the loading under AREA and AASHO Specifications results in lower limits for the slenderness ratio.

An examination of the results of extensive testing of riveted joints by many investigators over a period of fifty years prior to 1952 has been conducted.[6.3] This study has shown that the ratio of the ultimate strength of a

EXAMPLE 6.2

PROBLEM:

The roof truss for the steel mill building shown below is to be shop fabricated of A36 steel with 3/4 in. diameter rivets. The design forces for the bars are shown below.

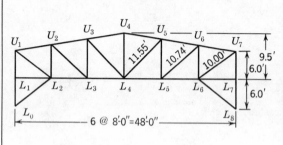

Member	Tension	Comp.
$U_1 - U_2$	0	-28.03^K
$U_2 - U_3$	0	-38.54^K
$U_3 - U_4$	0	-38.09^K
$L_1 - L_2$	0	0
$L_2 - L_3$	27.73^K	-6.60^K
$L_3 - L_4$	38.17^K	0
$L_1 - U_1$	0	-29.52^K
$L_2 - U_2$	0	-20.79^K
$L_3 - U_3$	0	-11.36^K
$L_4 - U_4$	1.01^K	-0.60^K
$U_1 - L_2$	31.61^K	0
$U_2 - L_3$	18.79^K	0
$U_3 - L_4$	7.80^K	-7.16^K
$L_0 - L_1$	0	-29.52^K

Select a member for $U_1 - L_2$. Use single or double angles. Assume rivets will be located on a single gage line. Use AISC Specification.

SOLUTION:

Design Load, 31.61 kip Tension ; Length, 10 ft = 120 in.

        0 kip Compression

Required net area:

$$A_n = \frac{31.61 \text{ kip}}{22 \text{ kip/in.}^2} = 1.437 \text{ in.}^2$$

Minimum gross area Sec [1.14.3] $= \dfrac{1.437}{0.85} = 1.690$ in.^{2}

The selection lends itself quite readily to a tabular solution in the following form:

    Column 1 is the standard thickness of angles.

    Column 2 is the reduction in area for one rivet hole. These are tabulated in the AISC manual

    Column 3 is the required gross area, that is, net area plus reduction in area.

    Columns 4 and 5 are the smallest angles of a given thickness which have the required area.

These rows are terminated when the required gross area exceeds the area of an angle already selected. The following table uses these headings.

Single Angle Selection:

t	Reduction in Area One Hole	Req'd Gross Area	Single Equal − Leg Angle	Single Unequal − Leg Angle
3/16	0.164	1.690	NONE	NONE
1/4	0.219	1.690	$3\frac{1}{2} \times 3\frac{1}{2} \times \frac{1}{4}$ A = 1.69	$4 \times 3 \times \frac{1}{4}$ A = 1.69
5/16	0.273	1.710	$3 \times 3 \times \frac{5}{16}$ A = 1.78	$3 \times 2\frac{1}{2} \times \frac{5}{16}$ A = 1.78
3/8	0.328	1.765	> 1.69	> 1.69

EXAMPLE 6.2 (Continued)

Most economical section: A. IL $3\frac{1}{2}$ x $3\frac{1}{2}$ x $\frac{1}{4}$;   A = 1.69 in.^{2} ; Min. r = 0.69 in.

B. IL 4 x 3 x $\frac{1}{4}$;   A = 1.69 in.^{2} ; Min. r = 0.65 in.

Check slenderness ratio Sec $\left[1.8.4\right]$:

A. $\frac{L}{r} = \frac{120}{0.69} = 174 < 240$ OK

B. $\frac{L}{r} = \frac{120}{0.65} = 185 < 240$ OK

Double Angle Selection:

Required area per angle = $\frac{1.437}{2} = 0.719$ in.^{2}

Minimum gross area $= \frac{0.719}{0.85} = 0.846$ in.^{2}

t	Reduction in Area One Hole	Gross Area	Double Equal – Leg Angle		Double Unequal – Leg Angle	
1/8	0.110	0.846	NONE		NONE	
3/16	0.164	0.883	$2\frac{1}{2}$ x $2\frac{1}{2}$ x $\frac{3}{16}$	A = 0.90	3 x 2 x $\frac{3}{16}$	A = 0.90
1/4	0.219	0.938	> 0.90		> 0.90	

Most economical section: A. 2LS ⌐L $2\frac{1}{2}$ x $2\frac{1}{2}$ x $\frac{3}{16}$;   A = 1.80 in.^{2} ; Min. r = 0.78 in.

B. 2LS ⌐L 3 x 2 x $\frac{3}{16}$;   A = 1.80 in.^{2} ; Min. r = 0.97 in.

with long legs back to back

A. $\frac{L}{r} = \frac{120}{0.78} = 154 < 240$ OK

B. $\frac{L}{r} = \frac{120}{0.97} = 124 < 240$ OK

POSSIBLE CHOICES FOR $U_1 - U_2$:

IL $3\frac{1}{2}$ x $3\frac{1}{2}$ x $\frac{1}{4}$ (A = 1.69)

or IL 4 x 3 x $\frac{1}{4}$ (A = 1.69)

or 2LS ⌐L $2\frac{1}{2}$ x $2\frac{1}{2}$ x $\frac{3}{16}$ (A = 1.80)

or 2LS ⌐L 3 x 2 x $\frac{3}{16}$ (A = 1.80)

conventional riveted structural joint to the ultimate strength of the uninterrupted section seldom exceeds 85 per cent. As a consequence, AISC Section [1.14.3] requires that the net section taken through a hole cannot be considered as more than 85 per cent of the gross section. The behavior of joints under load is considered in detail in Chapter 18.

6.6 PROBLEMS IN DESIGN

Example 6.2 illustrates the selection of a tension member for the diagonal of a roof truss.

Eccentricity of Load. The single angle selected in Example 6.2 is the most economical section.   However, this section warrants a closer examination. The selection was made using Eq. 6.1.   The use of this equation assumes that the load is applied through the centroid of the cross section.   This assumption is in accord with AISC Section [1.15.3], which states that the eccentricity between the gravity axis and the gage line may be neglected for the single angle, the double angle, and similar types of members.

The equation for biaxial bending is

$$f = \frac{P}{A} \pm \frac{M_1 y'}{I_1} \pm \frac{M_2 x'}{I_2} \tag{6.2}$$

where I_1 and I_2 are the principal moments of inertia, f is the stress at a given point, x' and y' are the principal coordinates of the point, and M_1 and M_2 are the moments about the respective principal axes 1 and 2, due to the eccentrically applied load, P.

Using Eq. 6.2 for the given loading condition, a stress of 45.3 ksi would be obtained.   This stress would be the result of using the gross area and the gross moment of inertia.   A more refined calculation considering the net area and net moment of inertia would produce an even higher stress.   At first glance, it would appear that the member is overstressed.   Actually this stress is fictitious.   The stress in the member cannot exceed the yield stress. As the load is applied to the member as shown in Fig. 6.3, localized yielding

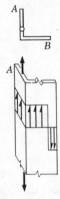

Fig. 6.3 Stress distribution in an eccentrically loaded single angle as unrestricted plastic flow begins.

will begin at *A*. The outer end of the leg will remain at the yield stress as the yielding penetrates the section. Under these conditions, end *B* is in compression. As the yielding progresses across the section, there is a redistribution of the elastic stresses until the entire section is plastified.

In addition, the member will deflect in a direction such that the centroidal axis of the member will tend to approach the loading axis. This in turn will reduce the bending stress. An extensive investigation[6.4] has shown that, as the ultimate load is approached, the centroidal axis actually coincides with the loading axis over most of the length of the member. The same investigation indicates that the ultimate load for the eccentrically loaded member is substantially the same as the ultimate load for the axially loaded member. In the plane of the gusset plate, the connection can always be considered as concentrically loaded. In the first place, the centroidal axis is placed in the line of the system as shown in Fig. 6.4. Secondly, it is standard practice to

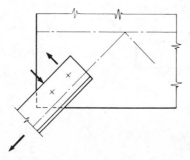

Fig. 6.4   Restraining couple at the end of an eccentrically loaded tension member.

use a minimum of two connectors (rivet or bolt) for even nominal loads. This gives rise to a couple equal to the moment due to the eccentricity (see arrows).

AREA Section [1.6.5] and AASHO Section [1.6.13] require that, if a single angle connected in one leg only is used as a tension member, the effective section of the member shall be the net area of the connected leg plus one half of the area of the unconnected leg.

Although local yielding will occur at end *A* of the section shown in Fig. 6.3 under working load, a large portion of the section will remain elastic. The elastic portion will restrict the elongation of the member, thus eliminating the problem of excessive deflection.

Repeated loading into the plastic range would be undesirable in case of a large number of repetitions. If the member is to be subjected to cyclic

EXAMPLE 6.3

PROBLEM:
It is desired to use a single angle $4 \times 3 \times \frac{1}{2}$ as a secondary tension member in a structure. Determine the maximum allowable load by the interaction method.

SOLUTION:
From the AISC Manual:

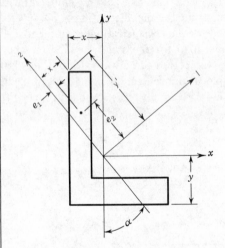

$A = 3.25$	$x = 0.83$
$I_x = 5.1$	$r_x = 0.64$
$I_y = 2.4$	$\operatorname{Tan} \alpha = 0.543$
$y = 1.33$	$g = 2\frac{1}{2}$

The simplest procedure for determining e_1, e_2, w, z is graphically.

$e_1 = 0.05$ in.
$e_2 = 1.30$ in.
$y' = 2.745$ in.
$x' = 0.53$ in.

From Eq. 6.2 :

$$f_b = \frac{Pe_2 y'}{I_1} + \frac{Pe_1 x'}{I_2}$$

$$I_2 = r_2^2 A = 1.33 \text{ in.}^4$$

$$I_x + I_y = I_1 + I_2$$

$$I_1 = 6.17 \text{ in.}^4$$

Then:

$$f_b = P \left[\frac{(1.3)(2.745)}{6.17} + \frac{(0.05)(0.53)}{1.33} \right] = 0.598 P$$

$$f_t = \frac{P}{A} = 0.308 P$$

Using the interaction eq. 6.3 $= \dfrac{0.308 P}{0.6 F_y} + \dfrac{0.598 P}{F_b} \leq 1.0$

Usually $0.6 F_y = F_b$

Hence $\qquad P = 1.1(0.6) F_y$

loading, the allowable stress could be reduced or the size of the section could be increased to eliminate the possibility of partial yielding of the member. The problem of cyclic loading is considered in detail in Chapter 16.

If local yielding under working load is undesirable, a rational approach to the problem can be made using the interaction method of AISC Section

[1.6.2]. Essentially, the method consists of limiting the sum of the ratios of the computed axial and bending stresses to the respective allowable stresses to unity. In equation form,

$$\frac{f_t}{0.6F_y} + \frac{f_b}{F_b} \leqslant 1.0 \tag{6.3}$$

where $f_t = P/A_n$, the computed tensile stress; $0.6F_y =$ maximum allowable tensile stress; $f_b = (M_1 x'/I_1) + (M_2 y'/I_2)$, the computed bending stress; and $F_b =$ maximum allowable tensile stress in bending.

The interaction method is by no means a one-step solution to the problem. Several sections must be examined until one is found which satisfies the interaction equation. Example 6.3 illustrates the procedure.

Symmetry about the gusset plate (double angles) is usually sought as a means of eliminating the problem of the eccentrically loaded single angle. For this reason, the double angle in Example 6.2 would probably be used even though the area is slightly greater than required. In general, the use of single angles for any purpose other than bracing and secondary members is not recommended.

Net Section. In Example 6.2, the reduction in area due to the presence of a fastener in one leg was considered. In many cases it is necessary that fasteners be used on more than one gage line.

Inspection of Fig. 6.5 indicates that failure may occur in one of three ways, depending on the fastener pattern. In case a the failure will occur through

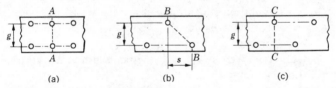

(a)   (b)   (c)

Fig. 6.5 Possible failure paths for a section traversed by a chain of holes.

section A-A. The reduction in area for this configuration will be that corresponding to two holes. For case c with failure across section C-C, the reduction in area will be that of one hole. For case b the reduction will be greater than one but less than two. It is apparent that the area to be deducted is a function of the stagger s and the gage g.

Many investigators have presented equations relating reduction in area to hole diameter, stagger, and gage distance. Among the earliest was Cochrane,[6.5] who presented a rather complicated equation in 1908. In 1915 Smith[6.6] presented a simplification of Cochrane's formula. Other contemporary contributors were Steinman[6.7] in 1917 and Young[6.8] in 1921. In

1922 Cochrane[6.9] presented a simplification of the Cochrane-Smith equation in the following form:

$$w = d + \sum \frac{s^2}{4g} \tag{6.4}$$

where w is the width in inches of a strip to be deducted from a transverse section through one hole because of the weakening effect of a staggered hole, d is the diameter of the hole, s is the stagger, and g is the gage distance. This is the reduction in width given in AISC Section [1.14.3]. The same provision is made by AREA Section [1.5.8] and AASHO Section [1.6.38].

Reference 6.10 has shown that Eq. 6.4 predicts the behavior of the joint adequately with respect to the variables which it considers. However, it was also shown that this approach does not take into consideration the ductility of the material, the method of fabricating the hole, or the ratio of the bearing stress to the tensile stress. Further consideration will be given to these additional variables in Chapter 15. An expression was proposed in 1940 which is slightly more complicated than Eq. 6.4.[6.11] However, an investigation has shown that the two expressions give substantially the same results.[6.3] The same investigation shows that the maximum deviation of net area as determined by Eq. 6.4 from net area as determined by tests is approximately 13 per cent.

In 1955 an analysis of the problem based on plastic strength was presented.[6.12] The solutions indicate that Eq. 6.4 is essentially an upper bound. Significant contributions to the solution of the problem by plastic analysis have also been presented in Ref. 6.13.

Although it has limitations in extreme cases, it would appear that Eq. 6.4 as presented by Cochrane provides an adequate and simple solution to an extremely complex problem.

Using the deduction given by Eq. 6.4, the net width of a section traversed by a chain of holes can be expressed as

$$w_n = w_g - \sum d + \sum_{n=1}^{n} \frac{s_n^2}{4g_n} \tag{6.5}$$

where w_n = net width, w_g = gross width, Σd = sum of the diameters of all the holes in the chain, s = stagger, g = gage distance, and n = the number of gage distances.

Obviously, if more than one combination of gage distance and stagger exists, then each must be evaluated individually and then the sum taken. Alternate failure paths may also have to be considered in order to determine which yields the least net width. Examples 6.4 and 6.5 illustrate the procedure for applying Eq. 6.5. The odd pattern of holes in Example 6.4 was used to make ABD the critical chain.

EXAMPLE 6.4

PROBLEM:
Determine the net width in the plate shown below.
All holes are 1 in. diameter.

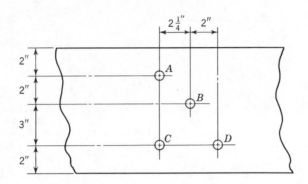

SOLUTION:

Using Eq. 6.5

$$w_n = w_g - \Sigma d + \sum_{n=1}^{n} \frac{S_n^2}{4g_n}$$

For Chain AC

$$w_n = 9 - 2(1) + 0$$

$$w_n = 7 \text{ in.}$$

For Chain ABC

$$w_n = 9 - 3(1) + \frac{(2.25)^2}{4(2)} + \frac{(2.25)^2}{4(3)}$$

$$w_n = 7.06 \text{ in.}$$

For Chain ABD

$$w_n = 9 - 3(1) + \frac{(2.25)^2}{4(2)} + \frac{(2)^2}{4(3)}$$

$$w_n = 6.97 \text{ in.}$$

Critical Chain is ABD

$$w_n = 6.97 \text{ in.}$$

Axial Stiffness. Tension members are often comprised of two or more elements of different stiffness along their length. This is particularly true of members with mechanically fastened joints. Normally bolt holes have no effect on the member stiffness and elongations can be calculated on the basis of the gross area. At splices, the lap plates increase the area of the member and result in an increase in stiffness. Numerous experiments have illustrated that the total gross area including the splice plates contribute to

EXAMPLE 6.5

PROBLEM:
  A heavy truss is to be used to carry the loads from the upper floors of a building across an arcade at street level. The lower chord of the truss is to be angles arranged as shown. Width requirements for the upper chord dictate the arrangement shown. A splice in the chord at the centerline of the truss will require fasteners in both legs of the angles. The load to be carried by the member is 143 kip. The panel length is 19'–0". Use A36 steel and 3/4 in. fasteners, with AISC Specification.

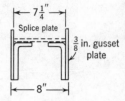

Select an economical angle

SOLUTION:

Required area: $A = \dfrac{P}{F_t} = \dfrac{143}{22} = 6.5 \text{ in.}^2$

Area per angle: $A = \dfrac{6.5}{2} = 3.25 \text{ in.}^2$

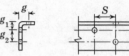

  A check of the AISC manual indicates that the long leg will probably be at least 5 in. This will require at least two gage lines in the long leg or a total of three gage lines for the angle. Using the fastener arrangement shown in the figure, the reduction in width will be between 2 and 3 holes. Assume the reduction in width is $2\frac{1}{2}$ holes.

t	Reduction in Area	Required Gross Area	Angle	Actual Area
7/16	0.958	4.208	$7 \times 4 \times \frac{7}{16}$	4.62
1/2	1.095	4.345	$6 \times 3\frac{1}{2} \times \frac{1}{2}$	4.50
9/16	1.230	4.480	NONE	NONE
5/8	1.368	4.890	$5 \times 3\frac{1}{2} \times \frac{5}{8}$	4.92

The first angle in the table will not meet the space restriction.

TRY 2 Ls $6 \times 3\frac{1}{2} \times \frac{1}{2}$:

$\dfrac{A_N}{A_G} = \dfrac{3.25}{4.5} = 0.72 < 0.85$ <u>OK</u>

Maximum allowable reduction in area = $4.50 - 3.25 = 1.25 \text{ in.}^2$

Maximum allowable reduction in width = $\dfrac{1.25}{t} = \dfrac{1.25}{0.5} = 2.5 \text{ in.}$

Rearranging Eq. 6.5

$w_g - w_n = \Sigma d - \Sigma \dfrac{s^2}{4g} =$ Maximum allowable reduction in width

  The recommended gage distances from the AISC manual are shown in the figure. The average distance between holes 2 and 3 is the sum of the gages less the thickness, Sec. [1.14.4].

$g_a = 2.25 + 2 - 0.5 = 3.75 \text{ in.}$

$g_2 = 2.5 \text{ in.}$

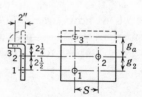

EXAMPLE 6.5 (Continued)

From the rearranged Eq. 6.5 above,

$$\frac{s^2}{4g_a} + \frac{s^2}{4g_2} = 3d - (w_g - w_n)$$

$$s^2\left[\frac{1}{3.75} + \frac{1}{2.5}\right] = 4\left[3(0.875) - 2.5\right]$$

$$s = 0.87 \text{ in.}$$

This represents a minimum value of s.

USE $s = 1\frac{1}{2}$ in. to conform to the standard pitch of 3 in.

Check $\frac{L}{r}$:

$$\frac{L}{r_x} = \frac{19(12)}{1.92} = 119 < 240 \qquad \underline{\underline{OK}}$$

$$I_y = 2\left[4.3 + 4.5(2.795)^2\right] = 79.0 \text{ in.}^4$$

$$r_y = \sqrt{\frac{79.0}{9.00}} = 2.96 \text{ in.}$$

$$\frac{L}{r_y} = \frac{19(12)}{2.96} = 77 < 240 \qquad \underline{\underline{OK}}$$

To assure that the angles will behave as a unit, tie plates must be provided at a spacing such that the slenderness ratio of each angle between tie plates is less than 240 Sec. [1.18.3.2].

$$\frac{L}{r_z} \le 240$$

Maximum allowable unbraced length, L = 240 (0.76) = 182 in. max.

USE: Tie plates at midpanel points

FINAL DESIGN:

$$2 \text{ L}^\text{S} 6 \times 3\frac{1}{2} \times \frac{1}{2}$$

$$S = 1\frac{1}{2} \text{ in.}$$

Tie plates at midpanel points

the stiffness.[18.13,18.18] When large perforations exist, the net area of the perforation is used to determine the member stiffness.

6.7 TENSION MEMBERS IN PLASTIC DESIGN

The maximum usable strength of a tension member is simply the product of the yield stress and the area. The factor of safety for a tension member is the ratio of the yield stress to the working stress:

$$\frac{F_y}{0.6F_y} = 1.67$$

Occasionally, tension members are included in a structure which would otherwise be designed on a plastic analysis or maximum load basis. For example, a tied arch or gable frame might fall in this category. If a load factor of 1.85 is used for the frame analysis, there is no reason why the tension member cannot be selected on a maximum load basis. In fact, a comparison of the factor of safety for allowable stress design and the load factor for maximum load design indicates that the tension member selected in this manner will be conservative.

PROBLEMS

6.1. Using the AISC Specification, A440 steel, design an eyebar to carry a load of 290 kips.

6.2. Make a double angle selection for members L_3-L_4 of Example 6.2. The angles will be installed with long legs back to back. A splice at joint L_4 requires a rivet in each outstanding leg.

6.3. Select an economical single angle for a bracing member which carries a design load of 7 kips. Length of the member is 12 ft; $F_y = 36$ ksi, rivets are $\frac{3}{4}\phi$.
  (a) Use AASHO Specifications (neglect $\frac{1}{2}$ the area of outstanding leg).
  (b) Use AISC Specification.
  (c) Compare the selections.

6.4. A tension member composed of two angles 6 by 4 by $\frac{3}{8}$ in. is shown. If the rivets are 1 in. diameter, what is the maximum load the member can carry in terms of the allowable stress F_t?

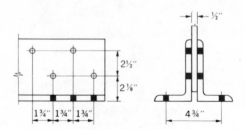

Problem 6.4

6.5. For the member of Problem 6.4, what is the minimum value of s such that the reduction in width will be two holes?

6.6. Determine the maximum load that the single angle selection of Example 6.2 can carry by the interaction method.

6.7. Select an economical single angle to carry a design load of 25 kips. The length of the member is 10 ft. Use $\frac{7}{8}$ in. bolts, $F_y = 33$ ksi.
  (a) By AISC Specification
  (b) By AREA Specifications.

6.8. Compute the "apparent" maximum stress in each selection of Problem 6.7.

6.9. Make double angle selections for the member of Problem 6.7.

6.10. The accompanying figure shows the repeated portion of a rivet pattern used as a joint in an oil storage tank. Determine the net width of the repeated portion of the pattern.

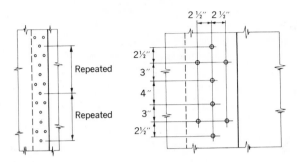

Problem 6.10

6.11. What is the maximum load that the member of Problem 6.4 can carry by the interaction method? Assume each rivet in the long legs carries $P/3$ and each rivet in the short legs carries $P/6$.

(a) Consider the gross moment of inertia (neglecting holes).

(b) Consider the moment of inertia (consider the reduced I).

6.12. Using the interaction method, select an economical single angle for member U_1-L_2 of Example 6.2.

6.13. The two channels shown in the figure are to be used as a tension member to carry a load of 230 kips on a span of 40 ft. At the ends they are connected to the webs by gusset plates and across the flanges by tie plates. Using A36 steel, AISC Specification, and $\frac{3}{4}$ in. rivets, select the channels. Check the slenderness ratio and indicate the location of tie plates. The minimum value of s for driving clearance is $1\frac{1}{2}$ in.

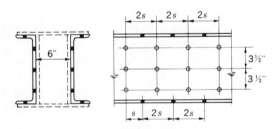

Problem 6.13

6.14. What is the minimum value of s such that the reduction in width is only 3 diameters for the channel selected in Problem 6.13?

7

BEAMS

7.1 INTRODUCTION AND HISTORICAL REVIEW

The object of this chapter is to summarize the strength and stability requirements of beams. Beams are structural members which carry loads transverse to their length. These members resist flexure (bending) and shear, and sometimes torsion, introduced by transverse loads. Purlins, rafters, joists, spandrels, lintels, floor beams, stringers, and other similar structural parts are all beams. Members subjected to bending and axial compression simultaneously are beam-columns which are discussed in Chapter 11. Loads on beams are discussed in Chapters 3 and 4.

Investigation of the strength of beams was undertaken first by Leonardo da Vinci in the fifteenth century.[7.1] In the early seventeenth century, Galileo brought forward his speculative thinking on bending moment and stress distribution in a cantilever beam. It was almost two hundred years later when Navier formulated the flexural formula of beams. Even though more complicated and accurate analyses of stresses have resulted from investigations on torsion and lateral buckling since the time of Navier, his beam theory is still the fundamental theory for the design of bending members.

Torsion of members was investigated as early as the eighteenth century by C. A. Coulomb and by Thomas Young.[7.1] The first rigorous treatment of torsion with constant warping and bending was given by St. Venant in 1853. In 1903, L. Prandtl introduced the membrane analogy.[7.1,7.2] In 1905, S. Timoshenko presented his investigation of non-uniform torsion of I-beams, considering the bending of flanges.[2.3] A general discussion of the torsion of thin-walled open sections was then given,[7.3] and the effect of web deformation on the torsion of I-beams considered.[7.4] Most of the results obtained are not yet incorporated in present design procedures for beams.

The lateral buckling strength of beams is closely related to the torsional strength. L. Prandtl and A. G. M. Michell simultaneously published a theoretical analysis of lateral stability in 1899, followed by S. Timoshenko, H. Wagner, E. Chwalla, and others.[7.5] Contributions to this problem have been numerous,[7.6,7.7] and more investigations are in progress, particularly in the inelastic range.

By considering the plastic deformation of ductile material, St. Venant initiated a new field of mechanics of materials, which led to the analysis of the plastic strength of beams and structures.[7.1] Considerable research on this subject since the Second World War has fostered the growth of the plastic design method.[1.5] Beams are now proportioned either according to the plastic or the allowable stress ("elastic") design procedure, as pointed out in Art. 1.5.

Irrespective of the design method, two aspects must be considered in the design of beams: (1) the strength requirement, that the cross section of a member is adequate to resist the applied bending moment and the accompanying shear force, and (2) the performance consideration, that the deformation of the member is not excessive under load. These considerations lead to requirements for allowable stress design and for plastic design, and are reviewed accordingly.

Depending on their function, beams sometimes are designed on limitations of fatigue loading or deflection. The former are discussed in Chapter 16; the control of deflection is reviewed in Art. 7.9.

7.2 SIMPLE BENDING

When bending of a beam takes place parallel to the plane of loading without twisting, it is called "simple bending." Members are generally chosen with symmetrical cross sections (Fig. 7.1) and with loads in a plane of symmetry, so that simple bending is a frequently encountered condition.

Fig. 7.1 Some symmetrical cross-sectional shapes of beams.

For simple bending in the elastic region, the flexural stresses f_b in a beam vary linearly across a section as shown in Fig. 7.2. The nominal extreme fiber stress is found from the formula (see, for example, Ref. 7.8):

$$f_b = \frac{Mc}{I} = \frac{M}{S} \qquad (7.1)$$

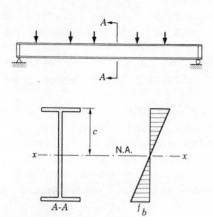

Fig. 7.2 Simple bending.

where f_b = computed flexural stress, M = applied bending moment at the section under consideration, c = distance from the centroidal neutral axis x to the extreme fiber, I = moment of inertia of the section about the same axis x, and S = section modulus, $S = I/c$.

When the magnitude of a computed maximum flexural stress reaches the yield point of the beam material, the beam is considered to have attained its limit of usefulness in the allowable stress design procedure. If F_y is the yield point (and M_y the corresponding moment, $F_y = M_y/S$), the allowable stress F_b as a reference for comparison in design is obtained by incorporating a factor of safety against F_y (see Art. 1.6).

$$F_b = \frac{F_y}{\text{F.S.}} \qquad (7.2)$$

Then for a given bending moment, the minimum section modulus required of a beam is

$$S = \frac{M}{F_b} \qquad (7.3)$$

The attainment of the first yielding in a beam constitutes only an imaginary limit for reference. Gradual plastification of a cross section continues, and moments beyond M_y can be carried by the beam. For beams with stocky components of cross section and with adequate lateral support so that local and over-all instability will not occur, the strength limit is that corresponding to the full yielding of the beam section (Art. 7.8). For beams of "non-compact sections," deformation of the cross section at moments above M_y may be beyond the acceptable limit. These phenomena are recognized by AISC in specifying an allowable flexural stress of $0.66F_y$ for properly braced compact sections and $0.60F_y$ otherwise[1.21] (see Art. 7.8). The AISC,[1.21]

Table 7.1   Allowable Flexural Stresses

| | | Allowable Flexural Stresses, F_b (ksi) | | | | |
| | Yield Point (ksi) | AISC | | AISI* | AASHO | AREA |
Material*		$0.60F_y$	$0.66F_y$	$0.60F_y$	$0.55F_y$	$0.55F_y$
Structural carbon steel	36.0	22.0	24.0	22.0	20.0	20.0
High-strength low-alloy structural steels	42.0	25.2	28.0	25.0	23.0	23.0
	45.0	27.0	29.7	27.0	25.0	25.0
	50.0	30.0	33.0	30.0	27.0	27.0
	55.0	33.0	36.3	33.0	30.0	30.0
	60.0	36.0	39.6	36.0	33.0	33.0
	65.0	39.0	42.9	39.0	36.0	36.0
High-yield-strength quenched and tempered alloy steel	90.0	54.0	—	—	49.0	—
	100.0	60.0	—	—	55.0	—

* For AISI, different material designations with the listed minimum yield points.

AISI,[1.24] AASHO,[1.23] and AREA [1.25] allowable stresses for some steels are listed in Table 7.1.

Section moduli of all rolled shapes are listed in the AISC Manual (Ref. 7.9). Once the minimum required section modulus is known, it is a simple matter to choose the lightest rolled shape with the aid of a section economy table in this manual. For cross sections other than rolled, it is necessary to compute their section moduli.

In Example 7.1 the weight of the wide-flange beam is 84 lb per ft or 4.2 per cent of the uniform load. A 4.2 per cent increase in stress results in $f_b = 21.5$ ksi, which is still less than the allowable; thus, the design need not be revised. For cases where the dead weight of beams is expected to be high, an estimated beam weight is usually added to the applied load for moment computation and the actual weight checked later (Art. 3.3). From Example 7.1, it is seen that the W-shape profile has about the same section modulus as the rectangular section whereas the cross-sectional area is much less for the former. The W-shape is lighter in weight than the rectangle and is therefore more economical. In general, for the same cross-sectional area, profiles with more material distributed to the flanges or with greater depth have greater section moduli and thus are more economical to resist moment. This is why rolled beam or wide-flange shapes are commonly used for bending members.

7.3   BIAXIAL BENDING

It has been pointed out that ordinarily symmetrical sections are used for simple bending. The general case of simple bending is not limited to symmetric sections. Simple bending will result for prismatic beams of any cross-sectional shape, as long as the plane of loading is parallel to a principal plane and produces no torsion.

The principal axes of a cross section are defined as two perpendicular centroidal axes for which the product of inertia is zero. If x and y are two arbitrary, rectangular, centroidal axes of which the moments of inertia (I_x and I_y) and the product of inertia (I_{xy}) are known, then the direction of the principal axes, x_1 and y_1, is given by the angle α (see Fig. 7.3).[7,8]

$$\tan 2\alpha = -\frac{I_{xy}}{\frac{1}{2}(I_x - I_y)} \tag{7.4}$$

The moments of inertia about these principal axes are:

$$I_1 = I_x \cos^2 \alpha + I_y \sin^2 \alpha - 2I_{xy} \sin \alpha \cos \alpha \tag{7.5}$$

$$I_2 = I_x \sin^2 \alpha + I_y \cos^2 \alpha + 2I_{xy} \sin \alpha \cos \alpha \tag{7.6}$$

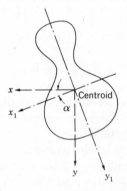

Fig. 7.3 Principal axes.

For example, the principal axes of a zee profile are inclined with respect to the flanges and the web (Fig. 7.4). Hence, a load applied as shown in Fig. 7.4b is not parallel to a principal axis.

When the loading plane of a beam does not coincide with either of the principal planes (as in Fig. 7.4b), but the loading causes no torsion, bending occurs along both principal axes and this then becomes a case of biaxial bending. Biaxial bending is a superposition of simple bending in two perpendicular directions. The applied load can be resolved into components in the direction of the principal axes and the flexural stresses in the beam found by using

$$f_b = \frac{M_1 y'}{I_1} + \frac{M_2 x'}{I_2} = \frac{M_1}{S_1} + \frac{M_2}{S_2} \tag{7.7}$$

where M_1, M_2 and I_1, I_2 are component moments and moments of inertia about the principal axes, S_1 and S_2 are the corresponding section moduli, and x' and y' are the coordinates of the point in question. Example 7.2

EXAMPLE 7.1

PROBLEM:

A simply supported building beam with continuous lateral support for the entire span of 30 ft is subjected to a uniform load of 2 kip per linear ft. Select an adequate cross section for flexural stress, assuming that A36 steel is used throughout the building, that the beam depth is limited to not more than 14 in., and that the AISC Specification is used neglecting the merits of compact sections.

SOLUTION:

Maximum bending moment:

$$M = \frac{wL^2}{8} = \frac{2 \times 30^2 \times 12}{8} = 2700 \text{ kip-in.}$$

Allowable flexural stress:

$$F_b = 22 \text{ ksi}$$

Required section modulus:

$$S = \frac{M}{F_b} = \frac{2700}{22} = 122.8 \text{ in.}^3$$

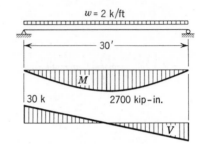

From AISC Manual the lightest rolled section is W 14 x 84 with

$$S = 130.9 \text{ in.}^3, \quad \text{and} \quad A = 24.71 \text{ in.}^2$$

If a rectangular section of 4 x 14 is used

$$S = 130.7 \text{ in.}^3, \quad A = 56.0 \text{ in.}^2$$

The computed flexural stresses are:

$$\text{W 14 x 84} \quad f_b = \frac{2700}{130.9} = 20.6 \text{ ksi} < 22 \text{ ksi}$$

$$\text{4 x 14 } \square \quad f_b = \frac{2700}{130.7} = 20.7 \text{ ksi} < 22 \text{ ksi}$$

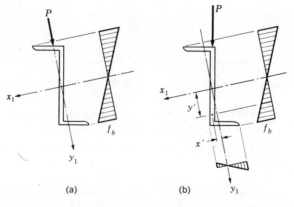

(a)   (b)

Fig. 7.4 Simple and biaxial bending of a zee section.

EXAMPLE 7.2

PROBLEM:

A simple rectangular beam is 24 ft between supports. A transverse load of 12 kips at mid-span is inclined 30° from the vertical axis of symmetry but causes no torsion. Find a section of A36 steel for the beam. Use the AISC Specification.

SOLUTION :

Load components:

$P_x = 12 \sin 30° = 6.0$ kips

$P_y = 12 \cos 30° = 10.4$ kips

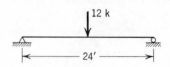

Maximum moment components at mid-span:

$$M_y = \frac{P_y L}{4} = \frac{6 \times 24 \times 12}{4} = 432 \text{ kip-in.}$$

$$M_x = \frac{P_x L}{4} = \frac{10.4 \times 24 \times 12}{4} = 749 \text{ kip-in.}$$

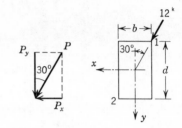

Assuming $\dfrac{S_x}{S_y} = \dfrac{P_y}{P_x} = \dfrac{10.4}{6.0} = 1.73$ then $S_x = 1.73 \, S_y$

and $\dfrac{S_x}{S_y} = \dfrac{bd^2}{6} \div \dfrac{db^2}{6} = \dfrac{d}{b} = 1.73$ $d = 1.73 b$ $S_y = \dfrac{1.73 b^3}{6}$

Maximum bending stress (at points 1 and 2), from Eq. 7.7 :

$$f_b = \frac{M_x}{S_x} + \frac{M_y}{S_y} = \frac{749}{1.73 S_y} + \frac{432}{S_y} = \frac{864}{S_y}$$

$f_b \leq F_b = 22$ ksi, hence required section modulus:

$$S_y = \frac{864}{F_b} = \frac{864}{22} = 39.3 \text{ in.}^3 = \frac{1.73 b^3}{6}$$

Therefore $b = 5.15$ in. and $d = 1.73 b = 8.9$ in.

Use $5\frac{1}{4} \times 9$ section, $S_x = 70.9 \text{ in.}^3$ $S_y = 41.3 \text{ in.}^3$

$$f_b = \frac{749}{70.9} + \frac{432}{41.3} = 21.0 \text{ ksi} < 22 \text{ ksi}$$

illustrates the design of a rectangular member for biaxial bending using Eq. 7.7.

Another equation for the computation of flexural stresses in biaxial bending refers to any set of centroidal, rectangular coordinate axes x and y (Fig. 7.5).[7.8] The product of inertia with respect to the axes then enters into the bending equation,

$$f_b = \frac{M_x I_y - M_y I_{xy}}{I_x I_y - I_{xy}^2} \, y + \frac{M_y I_x - M_x I_{xy}}{I_x I_y - I_{xy}^2} \, x \tag{7.8}$$

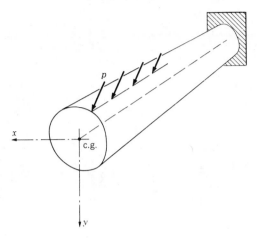

Fig. 7.5 Biaxial bending.

where M_x and M_y are moment components about the x and y axes, respectively, and are positive when tension results in the positive side of x and y. This formula may be simpler to use for stress computation for unsymmetrical cross sections of which the direction of principal axes and the corresponding sectional properties are not known. In Example 7.3, the stress is computed for an angle by Eqs. 7.7 and 7.8; it is evident that the process of using the latter equation is simpler in this case.

Other approaches for computing stresses in unsymmetrical sections under biaxial bending can be found in Refs. 7.2 and 7.8.

7.4 SHEAR STRESSES IN BEAMS

Transverse forces that exist in beams introduce shear stresses f_v in the transverse and the longitudinal direction (Fig. 7.6). Within the elastic limit these stresses can be computed using the formula[7.8]

$$f_v = \frac{VQ}{It} \tag{7.9}$$

for beams of open cross section subjected to bending without torsion. This formula corresponds to the simple bending formula, Eq. 7.1. In Eq. 7.9, the shear force V is parallel to a principal plane; the first moment of area Q and the moment of inertia I are taken about the principal axis perpendicular to V; the area for computing Q is the portion of the cross section beyond the point where f_v is evaluated; and the thickness t is taken at the same point.

EXAMPLE 7.3

PROBLEM:
An unequal leg angle is used as a simple beam and is subjected to 0.2 kip per linear ft in the plane of the long leg. If the span is 8ft with proper lateral support, is a L5 x 3 x 12.8 of A36 steel adequate in flexure by the AISC rules?

SOLUTION:

For L5 x 3 x 12.8

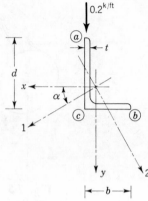

$A = 3.75 \text{ in}^2$ $r_2 = 0.65$ in.

$d = 5$ in. $I_x = 9.5 \text{ in}^4$

$b = 3$ in. $I_y = 2.6 \text{ in}^4$

$t = \dfrac{1}{2}$ in. $\text{Tan } \alpha = 0.357$

Moment at mid-span:

$$M_x = \frac{wL^2}{8} = \frac{0.2 \times 8^2 \times 12}{8} = 19.2 \text{ kip-in.}$$

$$M_y = 0$$

Solve by Eq. 7.7:

$$I_2 = Ar_2^2 = 3.75 (0.65)^2 = 1.6 \text{ in}^4$$

$$I_1 + I_2 = I_x + I_y$$

$$I_1 = I_x + I_y - I_2 = 9.5 + 2.6 - 1.6 = 10.5 \text{ in}^4$$

$x_a = 0.75$ in. $x_b = -2.25$ in.

$y_a = -3.25$ in. $y_b = 1.75$ in.

$$\text{Sin } \alpha = 0.336 \quad \cos \alpha = 0.942$$

$$M_1 = M_x \cos \alpha = 19.2 \times 0.942 = 18.10 \text{ kip-in.}$$

$$M_2 = M_x \sin \alpha = 19.2 \times 0.336 = 6.46 \text{ kip-in.}$$

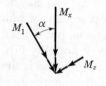

$$x'_a = x_a \cos \alpha + y_a \sin \alpha = 0.75 \times 0.942 - 3.25 \times 0.336 = -0.385 \text{ in.}$$

$$y'_a = y_a \cos \alpha - x_a \sin \alpha = -3.25 \times 0.942 - 0.75 \times 0.336 = -3.312 \text{ in.}$$

$$(f_b)_a = \frac{M_1 y'_a}{I_1} + \frac{M_2 x'_a}{I_2} = \frac{18.10 \times (-3.312)}{10.5} + \frac{6.46 \times (-0.385)}{1.6} = -5.7 - 1.6 = -7.3 \text{ ksi}$$

$$x'_b = x_b \cos \alpha + y_b \sin \alpha = -2.25 \times 0.942 + 1.75 \times 0.336 = -1.530 \text{ in.}$$

$$y'_b = y_b \cos \alpha - x_b \sin \alpha = 1.75 \times 0.942 + 2.25 \times 0.336 = 2.408 \text{ in.}$$

$$(f_b)_b = \frac{M_1 y'_b}{I_1} + \frac{M_2 x'_b}{I_2} = \frac{18.10 \times 2.408}{10.5} + \frac{6.46 \times (-1.530)}{1.6} = 4.1 - 6.2 = -2.1 \text{ ksi}$$

Solve by Eq. 7.8:

$$\tan 2\alpha = 0.819$$

$$I_{xy} = -\tan 2\alpha \left(\frac{I_x - I_y}{2}\right) = -0.819 \times \frac{9.5 - 2.6}{2} = -2.8 \text{ in}^4$$

$$(f_b)_a = \frac{M_x I_y}{I_x I_y - I_{xy}^2} y_a + \frac{-M_x I_{xy}}{I_x I_y - I_{xy}^2} x_a = \frac{19.2 \times 2.6}{9.5 \times 2.6 - 2.8^2} \times (-3.25) + \frac{-19.2 \times (-2.8)}{9.5 \times 2.6 - 2.8^2} \times 0.75$$

$$= -9.8 + 2.4 = -7.6 \text{ ksi}$$

$$(f_b)_b = \frac{M_x I_y}{I_x I_y - I_{xy}^2} y_b + \frac{-M_x I_{xy}}{I_x I_y - I_{xy}^2} x_b = \frac{19.2 \times 2.6}{9.5 \times 2.6 - 2.8^2} \times 1.75 + \frac{-19.2 \times (-2.8)}{9.5 \times 2.6 - 2.8^2} \times (-2.25)$$

$$= 5.3 - 7.3 = -2.0 \text{ ksi}$$

Since $F_b = 22$ ksi $> f_b$, the angle section is adequate. In fact, the flexural stresses are quite low and a smaller section may be used. The neutral axis can be located by first computing the flexural stress at point c, and then plotting the stress diagram.

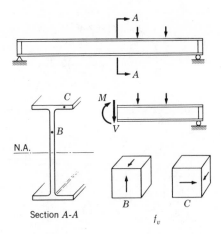

Fig. 7.6 Shear stresses.

From Eq. 7.9 it may be shown that the distribution of shear stresses in a rectangular cross section follows a parabolic curve from the top to the bottom fiber, with the maximum at the neutral axis (Fig. 7.7). The value of the maximum shear stress is

$$f_{v(\text{max})} = \frac{3}{2}\frac{V}{bd} \tag{7.10}$$

which is 50 per cent greater than the average shear stress on the cross-sectional area. However, for W- and S-shapes which are the most common beam profiles, the maximum shear stress at the neutral axis is only slightly greater than the average shear stress obtained by dividing the shear force by the area of the web (see Example 7.4). Therefore, for simplicity in design, the

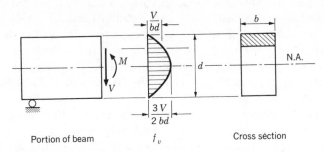

Fig. 7.7 Shear stress in a rectangular cross section.

EXAMPLE 7.4

PROBLEM:
   Compute the maximum and the average shear stress of the profiles selected for flexural stress in Example 7.1.

SOLUTION:

Maximum shear force $V = \frac{1}{2} \times 30 \times 2 = 30$ kip

W 14 x 84

$I = 928.4$ in.^{4}

$w = 0.451$ in. $d = 14.18$ in.

$t = 0.778$ in. $b = 12.023$ in.

$Q_{N.A.} = bt\left(\frac{d-t}{2}\right) + \frac{(d-2t)w}{2} \times \frac{(d-2t)}{4}$

$= 12.023 \times 0.778 \times \dfrac{14.18 - 0.778}{2} + \dfrac{(14.18 - 2 \times 0.778)^2 \times 0.451}{8}$

$= 62.7 + 9.0 = 71.7$ in.^{3}

Maximum shear stress at N.A. by Eq. 7.9:

$f_v = \dfrac{30 \times 71.7}{928.4 \times 0.451} = 5.1$ ksi < 13 ksi $= F_v$

Average shear stress in web by Eq. 7.11:

$f_v = \dfrac{30}{14.18 \times 0.451} = 4.7$ ksi

4 x 14 ▨

At N.A. $f_v = \dfrac{3}{2}\dfrac{V}{bd} = \dfrac{3}{2}\dfrac{30}{4 \times 14} = 0.8$ ksi $<< 13$ ksi

Average $f_v = \dfrac{V}{bd} = \dfrac{30}{4 \times 14} = 0.54$ ksi

N. A. — — — — — 5.1 ksi

f_v

4.7 ksi

average shear stress for these shapes is used instead of that given by Eq. 7.9.

$$f_v = \frac{V}{A_w} \tag{7.11}$$

Analogous to the comparison of flexural stresses, shear stresses are also compared with an allowable value to ensure a sufficient margin of safety. The specified allowable shear stresses are listed in Table 7.2. Unless a beam is very short and is subjected to a high concentrated load, the shear stress usually does not govern the design of beams. It is calculated only as a check after a section is selected on the basis of bending moment. As seen from the results of Example 7.4, the members selected in Example 7.1 are evidently safe against shear.

In the case of biaxial bending without twisting, the shear stresses can be superimposed in a manner similar to the superposition of bending stresses. If the principal axes and the related cross-sectional properties can be found easily, Eq. 7.12 (corresponding to Eq. 7.7) gives the shear stress at any point

Table 7.2 Allowable Shear Stresses

Material*	Yield Point (ksi)	Allowable Shear Stresses, F_v (ksi)			
		AISC $0.40F_y$	AISI* $0.40F_y$	AASHO $0.33F_y$	AREA $0.35F_y$
Structural carbon steel	36	14.5	14.5	12.0	12.5
High-strength low-alloy structural steels	42	17.0	17.0	14.0	14.5
	45	18.0	18.0	15.0	15.5
	50	20.0	20.0	17.0	17.5
	55	22.0	22.0	18.0	19.0
	60	24.0	24.0	20.0	21.0
	65	26.0	26.0	22.0	22.5
High-yield-strength quenched and tempered alloy steel	90	36.0	—	30.0	—
	100	40.0	—	33.0	—

* For AISI, different material designations with the listed minimum yield points.

of an open cross section.

$$f_v = \frac{V_2 Q_1}{I_1 t} + \frac{V_1 Q_2}{I_2 t} \tag{7.12}$$

The quantities I_1, I_2, and t are as defined before. V_1, V_2 and Q_1, Q_2 designate shear forces parallel to, and static moments about, the principal axes x_1 and y_1.   Care must be taken in using this equation because the directions of shear stresses by the shear force components V_1 and V_2 may differ from each other at various points of a cross section.   The computation of shear stresses for the solid member of Example 7.2 is shown in Example 7.5, and that for a thin-walled open cross section is illustrated in Example 7.6.

If an unsymmetrical open cross section is used as a beam (see Fig. 7.4), it is sometimes simpler to use Eq. 7.13 for shear stress computation, similar to the use of Eq. 7.8 instead of Eq. 7.7 for bending.   The direction of shear stresses can be obtained by considering the equilibrium condition, as in Example 7.6. It must be noted that Eq. 7.13 applies only to open cross sections.

$$f_v = \frac{V_y I_y - V_x I_{xy}}{I_x I_y - I_{xy}^2} \frac{Q_x}{t} + \frac{V_x I_x - V_y I_{xy}}{I_x I_y - I_{xy}^2} \frac{Q_y}{t} \tag{7.13}$$

For rolled shapes, shear stresses are usually low in magnitude when compared to the allowable values.   Furthermore, the maximum shear stresses occur at points where the flexural stresses are smallest.   Hence, in general, it is not necessary to check the combined (principal) stress in rolled beams. For beams of slender web and for plate girders, shear stresses and the interaction between shear and bending are discussed in Art. 8.3.

EXAMPLE 7.5

PROBLEM:

What is the magnitude of the maximum shear stress of the section selected for biaxial bending in Example 7.2 ?

SOLUTION:

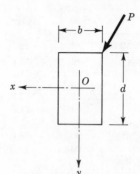

Maximum shear force components:

$$V_x = \frac{P_x}{2} = \frac{6.0}{2} = 3.0 \text{ kips}$$

$$V_y = \frac{P_y}{2} = \frac{10.4}{2} = 5.2 \text{ kips}$$

Sectional properties:

$$\frac{Q_x}{I_x t} = \frac{\frac{bd^2}{8}}{\frac{bd^3}{12} \times b} = \frac{3}{2bd}$$

$$\frac{Q_y}{I_y t} = \frac{\frac{bd^2}{8}}{\frac{b^3 d}{12} \times d} = \frac{3}{2bd}$$

b = 5.25 in., d = 9.0 in.

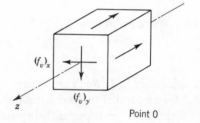

Point O

Maximum shear stresses at centroid:

$$(f_v)_x = \frac{V_x Q_y}{I_y t} = 3.0 \times \frac{3}{2 \times 5.25 \times 9.0} = 0.10 \text{ ksi}$$

$$(f_v)_y = \frac{V_y Q_x}{I_x t} = 5.2 \times \frac{3}{2 \times 5.25 \times 9.0} = 0.16 \text{ ksi}$$

Since the two components are at right angle, the maximum shear stress is the vector sum of the two

$$(f_v)_{max} = \sqrt{(f_v)_x^2 + (f_v)_y^2} = \sqrt{0.10^2 + 0.16^2}$$

$$= 0.19 \text{ ksi} << 14.5 \text{ ksi} = F_v$$

7.5 UNSYMMETRICAL BENDING

Unsymmetrical bending is defined as bending in one or both principal directions and simultaneous torsion in the beam. An example is a beam with an overhead crane rail where both vertical and horizontal loads exist (Fig. 7.8).

1. Shear Center of Cross Sections

The imposed condition that no twisting takes place for simple and biaxial bending of beams cannot always be met. This condition requires that the

EXAMPLE 7.6

PROBLEM:

Compute the shear stresses at the junction of the legs (point c) and at point d of the unequal leg angle of Problem 7.3.

SOLUTION:

Consider the centerline of the section.

For point c:

By principal axes, Eq. 7.12:

$$V_1 = \frac{1}{2} \times 8 \times 0.2 \times \sin \alpha = 0.27 \text{ kip}$$

$$V_2 = \frac{1}{2} \times 8 \times 0.2 \times \cos \alpha = 0.76 \text{ kip}$$

$$Q_1 = (2.75 \times 0.5)\left[1.5 \cos \alpha + \left(\frac{2.75}{2} - 0.5\right) \sin \alpha\right]$$

$$= 2.35 \text{ in.}^3$$

$$Q_2 = (2.75 \times 0.5)\left[-\left(\frac{2.75}{2} - 0.5\right)\cos \alpha + 1.5 \sin \alpha\right]$$

$$= -0.44 \text{ in.}^3$$

$$I_1 = 10.5 \text{ in.}^4 \quad I_2 = 1.6 \text{ in.}^4$$

$$(f_v)_c = \frac{V_2 Q_1}{I_1 t} + \frac{V_1 Q_2}{I_2 t} = \frac{0.76 \times 2.35}{10.5 \times 0.5} + \frac{0.27 \times (-0.44)}{1.6 \times 0.5} = 0.34 - 0.15$$

$$= 0.19 \text{ ksi} \quad (<< 14.5 \text{ ksi} = F_v)$$

For equilibrium, the resistant shear stress at point c in the long leg acts against the applied load.

By Eq. 7.13:

$$V_x = 0, \quad V_y = \frac{1}{2} \times 8 \times 0.2 = 0.8 \text{ kip}$$

$$I_x = 9.5 \text{ in.}^4, \quad I_y = 2.6 \text{ in.}^4, \quad I_{xy} = -2.8 \text{ in.}^4$$

$$Q_x = (2.75 \times 0.5) \times 1.5 = 2.06 \text{ in.}^3$$

$$Q_y = (2.75 \times 0.5)\left[-\left(\frac{2.75}{2} - 0.5\right)\right] = -1.20 \text{ in.}^3$$

$$(f_v)_c = \frac{V_y I_y}{I_x I_y - I_{xy}^2} \frac{Q_x}{t} + \frac{-V_y I_{xy}}{I_x I_y - I_{xy}^2} \frac{Q_y}{t} = \frac{0.8 \times 2.6}{9.5 \times 2.6 - 2.8^2} \times \frac{2.06}{0.5} + \frac{-0.8 \times (-2.8)}{9.5 \times 2.6 - 2.8^2} \times \frac{(-1.20)}{0.5}$$

$$= 0.51 - 0.32 = 0.19 \text{ ksi}$$

For point d:

By Eq. 7.13:

$$Q_x = \left[(4.75 - 1.5) \times 0.5\right]\left[-\left(\frac{4.75 - 1.5}{2}\right)\right] = -2.64 \text{ in.}^3$$

$$Q_y = \left[(4.75 - 1.5) \times 0.5\right] \times 0.5 = 0.81 \text{ in.}^3$$

$$(f_v)_d = \frac{V_y I_y}{I_x I_y - I_{xy}^2} \frac{Q_x}{t} + \frac{-V_y I_{xy}}{I_x I_y - I_{xy}^2} \frac{Q_y}{t} = \frac{0.8 \times 2.6}{9.5 \times 2.6 - 2.8^2} \times \frac{(-2.64)}{0.5} + \frac{-0.8 \times (-2.8)}{9.5 \times 2.6 - 2.8^2} \times \frac{0.81}{0.5}$$

$$= -0.65 + 0.22 = -0.43 \text{ ksi} << 14.5 \text{ ksi} = F_v$$

Again, the resistant shear stress acts against the applied load.

Note: Maximum shear stresses occur at N.A. Since the shear stresses at points c and d are very low, it is not necessary to check the maximum shear stresses against F_v.

Fig. 7.8 Unsymmetrical bending, crane rail.

line of action of an applied load must pass through a specific point of the cross section, the so-called shear center.[7.2]

The shear center of any open cross section can be located by using the equilibrium condition that the resultant moment of external forces and internal stresses in the plane of the section must be zero. If, as an example, an angle supports a transverse load as shown in Fig. 7.9, the resultant shear

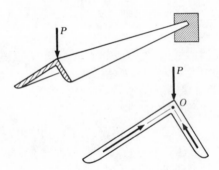

Fig. 7.9 Shear center of an angle.

force of both legs passes through the point O. The line of action of the applied force P must also converge at this point for equilibrium. Point O is therefore the shear center. In fact, for all cross sections composed of plates meeting at a common point, the shear center is this point of intersection (see Fig. 7.10). For doubly symmetric and antisymmetric sections, the shear center coincides with the centroid. For some other sections, the location of this center is available in the literature.[7.2] In Example 7.7 the equilibrium condition is used for a channel to give the location. A more general procedure is given in Art. 12.7.

EXAMPLE 7.7

PROBLEM:
 If a C8 x 11.5 section is used as a beam, where shall a load parallel to the web be applied so that no twisting will occur?

SOLUTION:

$I_x = 32.3$ in^4

d = 8.0 in., w = 0.22 in.

b = 2.26 in., t = 0.39 in.

$A_f = 2.26 \times 0.39 = 0.88$ in^2

$(Q_x)_1 = 3.81\, A_f = 3.36$ in^3

Assuming that V acts downward, the direction of the resistant shear stresses in the section are as shown in the profile.

$(f_v)_1 = \dfrac{VQ}{It} = \dfrac{V \times 3.36}{32.3 \times 0.39} = 0.267\, V$ ksi in flange

$= \dfrac{V \times 3.36}{32.3 \times 0.22} = 0.474\, V$ ksi in web

Resultant force in each flange:

$F = \dfrac{1}{2} \times 2.26 \times 0.267\, V \times 0.39 = 0.118V$ kip

Moment produced by flange forces:

$0.118\, V \times 7.61 = 0.895\, V$ kip-in.

Therefore $m = \dfrac{0.895\, V}{V} = 0.895$ in.

The shear center is the intersection of this line of action and the axis of symmetry.

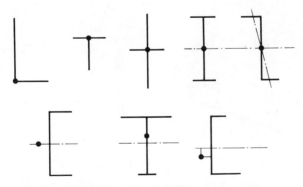

Fig. 7.10 Shear centers of some cross sections.

2. Torsion of Beams

For cases such as the one of Example 7.7 or for an inclined purlin supporting a vertical roof load (Fig. 7.11a), or a floor beam with stringers attached to one side of its web (Fig. 7.11b), the applied loads do not pass through the shear centers. In these cases the member sustains torsion in addition to flexure, and thus is under unsymmetrical bending. Shear stresses or shear and normal stresses are produced by the torsion and must be superimposed on flexural stresses in the members.

In the treatment of torsional problems, the condition of constant warping specified by St. Venant—that a member is free to warp during torsion and that no deformation of the cross section takes place—is often called "uniform," "pure," or the "St. Venant" torsion. The non-uniform torsion considering the bending of flanges and deformation of cross-sectional shape often is termed warping torsion. Because most structural members are not free to warp, but rather are partially restrained against warping under an applied torsion, both St. Venant and warping torsional stresses usually are present.

The St. Venant torsional formula can be derived directly from Hooke's law. If a bar is subjected to a twisting moment M_z^T (Fig. 7.12), the ratio of twisting per unit length is proportional to this torque,

$$\frac{d\phi}{dz} = \frac{1}{K_T} \frac{M_z^T}{G} \tag{7.14}$$

where ϕ is the angle of twisting, G is the shear modulus, and the coefficient K_T is a torsional constant. The magnitude of K_T depends on the cross-sectional shape of the beam. It can be determined by using the membrane analogy method. For a rectangular cross section it is approximated by[7.10,7.11]

$$K_T = (bt^3/3)\left(1 - 0.630\frac{t}{b} + 0.052\frac{t^5}{b^5} - \cdots\right) \tag{7.15}$$

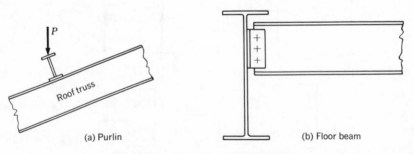

(a) Purlin   (b) Floor beam

Fig. 7.11 Bending with torsion.

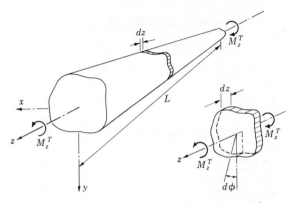

Fig. 7.12   St. Venant torsion of a prismatic bar.

or simply by $K_T = bt^3/3$ when the rectangle is narrow and t/b is small.   The maximum shear stress of the rectangular section occurs along the longer sides of the cross section (Fig. 7.13) and is given by

$$f_{vs} = \frac{M_z^T t}{K_T} \tag{7.16}$$

Under the assumption that cross sections maintain their shape (but are free to warp), it can be shown that the torsional constant for a section composed of components is the sum of constants for the individual components.[7.11] The maximum shear stress in any component part is obtained by substituting in Eq. 7.16 the appropriate value of K_T and the component thickness t (see Example 7.8).

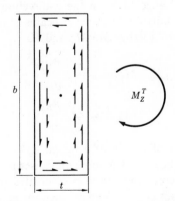

Fig. 7.13   St. Venant torsional stresses, f_{vs}.

EXAMPLE 7.8

PROBLEM:

A shear force of 4 kip is applied in the plane of the web of a C 8 x 11.5 section (Example 7.7). What are the St. Venant torsional stresses in the section?

SOLUTION:

From Example 7.7

$$m = 0.895 \text{ in.}$$

$$M_Z = Vm = 4 \times 0.895 = 3.58 \text{ kip-in.}$$

Taking only the first term of Eq. 7.15.

$$K_T = \Sigma \frac{1}{3} bt^3$$

$$= \frac{1}{3} [2 \times 2.26 \times 0.39^3 + 8 \times 0.22^3]$$

$$= 0.118 \text{ in.}^4$$

Maximum shear stresses by Eq. 7.16:

Flange $\quad f_{vs} = \dfrac{3.58 \times 0.39}{0.118} = 11.8 \text{ ksi}$

Web $\quad f_{vs} = \dfrac{3.58 \times 0.22}{0.118} = 6.7 \text{ ksi}$

Since torsion without warping seldom occurs, normal stresses are introduced to the member. Consider the doubly symmetric wide-flange beam of Fig. 7.14a which is simply supported against both bending and torsional moment and is free to warp at the ends. Due to symmetry, the section at mid-span remains plane without warping deformation. Thus, for the consideration of torsion, half of the beam can be regarded as a cantilever subjected to a twisting moment at the end (Fig. 7.14b). The flange at this end is displaced a relative distance of u_f in the x-direction (Fig. 7.14c). If the flanges are regarded as individual cantilever beams, then the moment and the shear force introduced to a flange are M_w and V_f.

$$M_w = -EI_f \frac{d^2 u_f}{dz^2} \tag{7.17}$$

$$V_f = -EI_f \frac{d^3 u_f}{dz^3} \tag{7.18}$$

where I_f is the moment of inertia of a flange, $I_f = I_y/2$, and direction z is along the length of the beam. The moment causes normal stresses f_{bw} in the flange (Fig. 7.14d) whereas the shear force creates shear stresses f_{vw} (Fig. 7.14e) in addition to those by St. Venant torsion, f_{vs} (Eq. 7.16, Fig. 7.14f). The shear forces in the two flanges form a couple (the warping torsional

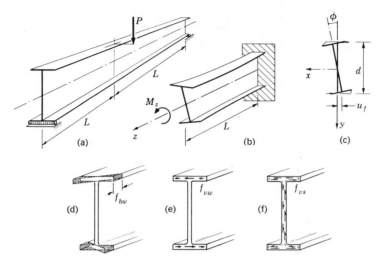

Fig. 7.14 Warping torsion of a wide-flange beam.

moment) which has the magnitude

$$M_z{}^w = V_f d = -EI_f d \frac{d^3 u_f}{dz^3} \tag{7.19}$$

d being depth of the cross section. But $u_f = \phi \, d/2$ (Fig. 7.14c), hence

$$M_z{}^w = -EI_f \frac{d^2}{2} \frac{d^3 \phi}{dz^3} \tag{7.20}$$

Defining $I_f \, d^2/2$ as the warping torsional constant I_w (see Art. 12.7 for cross sections other than doubly symmetric wide-flange shapes) and adding $M_z{}^w$ to the St. Venant torsion $M_z{}^T$ (Eq. 7.14), the total twisting moment is

$$M_z = M_z{}^T + M_z{}^w = GK_T \frac{d\phi}{dz} - EI_w \frac{d^3\phi}{dz^3} \tag{7.21}$$

Equation 7.21 takes into consideration both St. Venant torsion and warping torsion. The solution of this differential equation is

$$\phi = A \sinh \lambda z + B \cosh \lambda z + C + \frac{M_z}{GK_T} z \tag{7.22}$$

where $\lambda = \sqrt{GK_T/EI_w}$ and the coefficients A, B, and C are to be determined from the boundary conditions of the beam. When the function ϕ is known, the warping torsional moment is given by Eq. 7.20 and the corresponding

EXAMPLE 7.9

PROBLEM:

A wide-flange shape is to be used as a member 16 ft long supporting a vertical load of 8 kips which is applied 5 in. from the centroid of the section at mid-span. The ends of the member are simply supported and free to warp. Determine the maximum normal and shear stresses caused by torsion if a W 12 x 50 is selected.

SOLUTION:

$$M_z = \frac{1}{2} \times 8 \times 5 = 20 \text{ kip-in.}$$

$$b = 8.077 \text{ in., } t = 0.641 \text{ in.}$$

$$w = 0.371 \text{ in.}$$

The depth is taken as the distance between flange centroids:

$$d = 12.19 - 0.64 = 11.55 \text{ in.}$$

$$I_y = 56.4 \text{ in.}^4 \quad I_x = 394.5 \text{ in.}^4$$

$$K_T = \frac{1}{3}\left[2 \times 8.077 \times 0.641^3 + 12.19 \times 0.371^3\right] = 1.63 \text{ in.}^4$$

$$I_w = \frac{I_y}{4}d^2 = \frac{56.4(11.55)^2}{4} = 1880 \text{ in.}^6$$

$$\lambda = \sqrt{\frac{GK_T}{EI_w}} = \sqrt{\frac{1.63}{2.6 \times 1880}} = 0.0183 \left(\frac{1}{\text{in.}}\right)$$

(K_T and $1/\lambda$ of W and zee sections can be obtained from the hand book "Torsional Stresses in Structural Beams" by Bethlehem Steel Corporation. For W 12 x 50, the respective values are 1.82 in.^{4} and 51.78 in. = $1/0.0193$ in. The values obtained here are only approximates. See Ref. 7.11)

Referring to sketch above and Fig. 7.14, the boundary conditions are:

Mid-span, $z = 0$, angle of twist $\phi = 0$

slope of twist $\dfrac{d\phi}{dz} = 0$ and

Ends, $z = L$, no flange bending: $\dfrac{d^2\phi}{dz^2} = 0$

By applying these to Eq. 7.22 and solving for the coefficients, the function ϕ for the angle of twist is:

$$\phi = \frac{M_z}{GK_T\lambda}\left[\text{Tanh }\lambda L(\text{Cosh }\lambda z - 1) - \text{Sinh }\lambda z + \lambda z\right]$$

$$\frac{d\phi}{dz} = \frac{M_z}{GK_T}\left[1 - \frac{\text{Cosh }\lambda(L-z)}{\text{Cosh }\lambda L}\right]$$

$$\frac{d^2\phi}{dz^2} = \frac{M_z\lambda}{GK_T}\frac{\text{Sinh }\lambda(L-z)}{\text{Cosh }\lambda L}, \quad \frac{d^3\phi}{dz^3} = -\frac{M_z\lambda^2}{GK_T}\frac{\text{Cosh }\lambda(L-z)}{\text{Cosh }\lambda L}$$

From Eq. 7.23, the flange bending stress is maximum at $z = 0$ where $\dfrac{d^2\phi}{dz^2}$ is maximum:

$$f_{bw} = -E\frac{bd}{4}\frac{d^2\phi}{dz^2}$$

$$= -\frac{bd}{4}\frac{E}{G}\frac{\lambda}{K_T}M_z\text{Tanh }\lambda L$$

$$= -\frac{11.55 \times 8.077}{4} \times 2.6 \times \frac{0.0183}{1.63} \times 20 \times \text{Tanh }(0.0183 \times 8 \times 12)$$

$$= 12.8 \text{ ksi at flange tips (Figure 7.14d)}$$

EXAMPLE 7.9 (Continued)

From Eq.7.24, flange shear stress due to warping is maximum also at $z=0$ where $\dfrac{d^3\phi}{dz^3}$ is maximum:

$$f_{vw} = -E \frac{b^2 d}{16} \frac{d^3\phi}{dz^3} = \frac{b^2 d}{16} \frac{E}{G} \frac{\lambda^2}{K_T} M_z$$

$$= \frac{8.077^2 \times 11.55}{16} \times 2.6 \times \frac{0.0183^2}{1.63} \times 20$$

$$= 0.5 \text{ ksi} \quad \text{at mid-width of flange (Figure 7.14 e)}$$

Other stress at $z=0$ (mid-span) are:

St. Venant torsional shear (Eq. 7.16)

$$f_{vs} = \frac{M_z^T t}{K_T} = Gt \frac{d\phi}{dz} = 0$$

Shear stress at mid-width of flange, due to applied shear force V (Eq. 7.9)

$$f_v = \frac{VQ}{It} = \frac{\frac{8}{2}\left[\left(\frac{8.077}{2} \times 0.641\right) \times \frac{11.55}{2}\right]}{395.4 \times 0.641} = 0.2 \text{ ksi}$$

Bending stresses in the flanges, due to applied bending moment M (Eq. 7.1)

$$f_b = \frac{Mc}{I} = \frac{\frac{8 \times 16 \times 12}{4} \times \frac{12.19}{2}}{395.4} = 5.9 \text{ ksi}$$

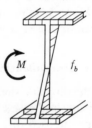

The total normal stress is maximum at flange tips at mid-span,

$$(f_b)_{max} = f_b + f_{bw} = 5.9 + 12.8 = 18.7 \text{ ksi (tension and compression)}$$

The total shear stress at mid-width of flange at mid-span is:

$$f_{vw} + f_{vs} + f_v = 0.5 + 0 + 0.2 = 0.7 \text{ ksi}$$

The maximum total shear stress occurs at the ends of the beam ($z = L$):

$$f_{vw} = -E \frac{b^2 d}{16} \frac{d^3\phi}{dz^3} = \left(\frac{b^2 d}{16} \frac{E}{G} \frac{\lambda^2}{K_T} M_z\right) \frac{1}{\cosh \lambda L}$$

$$= 0.5 \times \frac{1}{\cosh(0.0183 \times 8 \times 12)} = 0.16 \text{ ksi} \quad \text{(Fig. 7.14e)}$$

$$f_{vs} = Gt \frac{d\phi}{dz} = \frac{M_z t}{K_T}\left[1 - \frac{\cosh \lambda(L-L)}{\cosh \lambda L}\right]$$

$$= \frac{20 \times 0.641}{1.63}\left[1 - \frac{1}{\cosh(0.0183 \times 8 \times 12)}\right] = 5.3 \text{ ksi} \quad \text{(Fig. 7.14f)}$$

$$f_v = \frac{VQ}{It} = 0.2 \text{ ksi} \quad \text{(as before)}$$

$$(f_v)_{max} = f_{vw} + f_{vs} + f_v = 0.16 + 5.3 + 0.2 = 5.7 \text{ ksi}$$

maximum flange bending stress and the maximum shear stress are, respectively,

$$f_{bw(max)} = \frac{M_w b}{2I_f} = -EI_f \frac{d}{2} \frac{d^2\phi}{dz^2} \frac{b}{2I_f} = -E \frac{bd}{4} \frac{d^2\phi}{dz^2} \qquad (7.23)$$

$$f_{vw(max)} = \frac{V_f Q_f}{I_f t} = -EI_f \frac{d}{2} \frac{d^3\phi}{dz^3} \frac{b^2 t}{8I_f t} = -E \frac{b^2 d}{16} \frac{d^3\phi}{dz^3} \qquad (7.24)$$

These stresses and the shear stresses of uniform torsion (Eq. 7.16) together make up the torsional stresses in a member (Example 7.9).

The evaluation of torsional stresses as illustrated by Example 7.9 assumes that the angle of twist is small, that the stresses are within the elastic limit, and that the cross section does not change shape. The last assumption is adequate only when the section of a beam is stocky.[7.12] If section components (particularly webs) are slender, the shape of a section is liable to be distorted under a warping torsional moment[7.13] (see Fig. 7.15). The "Goodier-Barton effect" of web deformation[7.4] would have to be considered. For rolled shapes and other commonly used built-up beam sections, the web deformation is usually quite small, and thus can be neglected.

More detailed information on the torsional behavior of members with open or closed cross sections can be found in Refs. 7.2 and 7.11.

3. Simplification for Unsymmetrical Bending

Bending and shear stresses resulting from the biaxial bending and from the torsion in a case of unsymmetrical bending must all be computed and then superimposed to evaluate the total stresses at different points of the beam. However, the computations may be quite lengthy and tedious. Simplifications often are made in practice. For example, in the case of crane rails on relatively deep beams, it may be assumed that the horizontal load is resisted by the upper flange alone (Fig. 7.16). Similarly, for cases such as

Fig. 7.15 Stocky and slender cross sections under torsion.

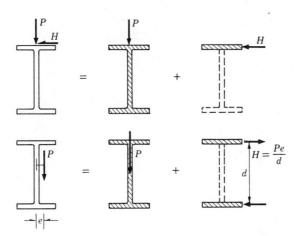

Fig. 7.16 Design simplification for unsymmetrical bending.

Fig. 7.11b and Fig. 7.14a a simplification often is made by replacing the twisting moment by two equal and opposite flange forces, evaluating the stresses in both vertical and horizontal directions, and summing them. This process is demonstrated in Example 7.10. Caution is advisable in the use of these simplifications, for reasons of safety.

7.6 LATERAL BUCKLING

The predicted strength or load-carrying capacity of beams sometimes cannot be attained because failure occurs by instability or excessive deformation at lower loads. When a beam is bent by the action of moments, lateral buckling accompanied by twisting may occur if enough resistance is not provided. This is a phenomenon usually referred to as "lateral-torsional buckling" and is illustrated in Fig. 7.17. For beams, the term "lateral buckling" is used. Some experimental results are shown in Figs. 7.18 and 7.19. Open shapes, such as beam, wide-flange, and channels, are weak against lateral buckling whereas rectangular and box shapes are much stronger.

Because twisting as well as bending take place in lateral buckling, the general theoretical expression for the elastic buckling stress of a beam is very involved (see, for example, Refs. 7.7, 7.14, and 7.15). Since W- and S-shapes are most frequently used for beams (and girders), simple design formulas are derived for them.

For simply supported beams with doubly symmetric cross sections and under uniform bending in the plane of the web (Fig. 7.20), the differential

EXAMPLE 7.10

PROBLEM:
  What are the maximum normal and shear stresses of the member in Example 7.9 if the applied load is two inches from the centroid?

SOLUTION:

  A simplification for unsymmetrical bending may be used here since the applied torsion is not excessive.

$$W \ 12 \times 50 \quad S = 64.7 \ in.^3, \quad b = 8.077 \ in., \quad t = 0.641 \ in., \quad w = 0.371 \ in.,$$

$$I_y = 14.0 \ in.^4 \quad d = 12.19 - 0.641 = 11.55 \ in.$$

Applied Torsion: $\quad M_z = \frac{1}{2} \times 8 \times 2 = 8 \ kip\text{-}in.$

$$H = \frac{M_z}{d} = \frac{8}{11.55} = 0.69 \ kip$$

At flange tip, mid-span:

$$\text{By P} \quad f_b = \frac{M}{S} = \frac{\frac{8 \times 16 \times 12}{4}}{64.7} = 5.93 \ ksi$$

$$\text{By H} \quad f_b = \frac{\frac{HL}{4}}{I_y/2} = \frac{0.69 \times 16 \times 12}{4 \times 14.0 \times 1/2} = 4.73 \ ksi$$

  Maximum normal stress = 5.93 + 4.73 = 10.66 ksi

At points ⓐ and ⓑ, over support:

$$\text{By P} \quad f_v = \frac{VQ}{It} = 0.2 \ ksi \ (\text{see Ex. 7.9})$$

$$\text{By H} \quad f_v = \frac{3}{2} \frac{H}{A_f} = \frac{3}{2} \frac{0.69}{8.077 \times 0.641} = 0.2 \ ksi$$

  Maximum shear stress = 0.2 + 0.2 = 0.4 ksi

For comparison, from Example 7.9, by direct proportioning of torsion:

  Maximum normal stress at mid-span = $\frac{2}{5} f_{bw} + \frac{M}{S} = \frac{2}{5} \times 12.8 + 5.93 = 11.0 \ ksi$

  Maximum shear stress at support = $\frac{2}{5} (f_{vw} + f_{vs}) + \frac{VQ}{It} = \frac{2}{5} \times (0.16 + 5.3) + 0.2 = 2.4 \ ksi$

equations governing lateral buckling are[7.15]

$$EI_y \frac{d^4u}{dz^4} + M \frac{d^2\phi}{dz^2} = 0 \qquad (7.25)$$

$$EI_w \frac{d^4\phi}{dz^4} - GK_T \frac{d^2\phi}{dz^2} + M \frac{d^2u}{dz^2} = 0 \qquad (7.26)$$

where I_y = moment of inertia about the y-axis, u = displacement in the x-direction, ϕ = angle of twist (Fig. 7.14c), I_w = warping torsional constant = $\frac{1}{4}I_y d^2$, G = shear modulus, and K_T = St. Venant torsional

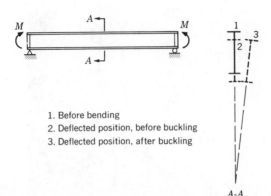

1. Before bending
2. Deflected position, before buckling
3. Deflected position, after buckling

Fig. 7.17 Lateral (lateral-torsional) buckling of beams.

Fig. 7.18 Compression flange of a laterally buckled beam.

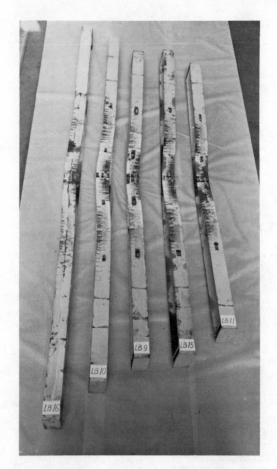

Fig. 7.19 Laterally buckled beams.

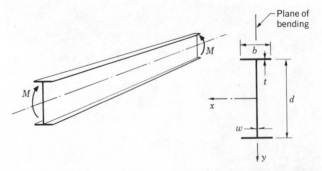

Fig. 7.20 Bending of doubly symmetric wide-flange sections.

constant. The solution of these equations in conjunction with the boundary conditions gives the buckling moment

$$M_{cr} = \sqrt{\frac{\pi^2 EI_y \omega K_T}{L^2} + \frac{\pi^4 E^2 I_y I_w}{L^4}} \qquad (7.27)$$

By taking the distance between the bracing points (the unbraced length) of a member as the length of the simply supported beam in the derivation, the elastic buckling stress of the member is

$$F_{cr} = \frac{M_{cr}}{S} = \frac{1}{S}\sqrt{\frac{\pi^2 EI_y \omega K_T}{L^2} + \frac{\pi^4 E^2 I_y I_w}{L^4}} \qquad (7.28)$$

where S = section modulus about the x-axis and L = unbraced length.

The first term in the radical of Eq. 7.28 corresponds to the resistance to lateral buckling offered by St. Venant torsion and lateral bending,[7.16,7.17] whereas the second term corresponds to the resistance offered by flange bending.[7.14,7.18] It has been demonstrated generally that for rolled shapes and stocky sections, the St. Venant torsion resistance dominates.[7.16] It has also been pointed out that for members for which warping torsion is the main resistance, it may be very much on the conservative side to approximate the buckling stress F_{cr} by the first term of the radical alone.[7.13,7.19] Because of these practical conditions and the complexity in using Eq. 7.28 it becomes desirable to have two equations, each associated with a particularly dominating resistance to torsion, that is, each to a term in Eq. 7.28.

By substituting the values of π and $G = E/2(1 + \mu) = E/2.6$, into the first term of Eq. 7.28, and neglecting the web, $I_y = 2(b^3 t/12) = b^3 t/6$, $K_T = 2(bt^3/3)$, and $S = I_x/(d/2) = 2bt(d/2)^2/(d/2) = btd$, the simplified equation for shallow, stocky-walled beams is

$$F_{cr} = \frac{0.65E}{Ld/bt} \qquad (7.29)$$

The corresponding formula from the second term of Eq. 7.28 and applicable for deep, thin-walled beams is

$$F_{cr} = \frac{\pi^2 E}{(L/r)^2} \qquad (7.30)$$

with

$$r = \sqrt{\frac{I_f}{A_f + A_w/6}} \qquad (7.31)$$

where $I_f = b^3 t/12$, A_f = the area of the compression flange, and A_w = area of the web. Equation 7.30 is obtained by considering the relationship among S, I_y, and I_w of doubly symmetric cross sections; it can also be derived

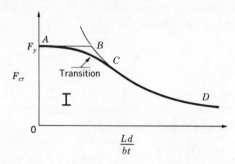

Fig. 7.21 Buckling stresses, stocky sections.

through consideration of equilibrium of the cross section when the compression flange deflects laterally.[7.18]

Equations 7.29 and 7.30 can be applied directly in the elastic region. In the inelastic range, plastic action affects the behavior of beams and transition formulas are often used for these equations (see Figs. 7.21 and 7.22). Because compressive flange in bending is analogous to the condition of a loaded column, a possible transition is the so-called "basic column curve" of the Column Research Council, assuming the proportional limit to be one-half the yield stress of the material[7.14] (see Art. 9.7):

$$F_{cr} = F_y \left(1 - \frac{F_y}{4F_{cr}} \right) \tag{7.32}$$

Equations 7.27 to 7.31 are for simply supported beams under a uniform moment. Actual loading and restraining conditions may differ and modify the buckling stresses. These conditions include end restraints provided for the beam in both the strong and the weak direction, the moment gradient, the point of application of load, and others. One way of incorporating the

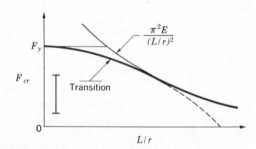

Fig. 7.22 Buckling stresses, slender sections.

influence of moment gradient and end restraint in the direction of the web
is to apply a modifier C_b to Eq. 7.27 to 7.30,[7.14,7.20]

$$C_b = 1.75 - 1.05\left(\frac{M_1}{M_2}\right) + 0.3\left(\frac{M_1}{M_2}\right)^2 \leqslant 2.3 \tag{7.33}$$

$$\left(-1 \leqslant \frac{M_1}{M_2} \leqslant +1\right)$$

where M_1 and M_2 are the smaller and the larger moment at the ends of the
unbraced length, respectively (see Fig. 7.23). The factor C_b equals 1.0

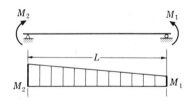

Fig. 7.23   Moments for Equation 7.33.

when the moment within the unbraced length is greater than the end moments.

In the application of the foregoing to design, a safety margin must be
maintained against the estimated maximum stresses. By incorporating a
factor of safety and a transition curve in the form of Eq. 7.32, AASHO
specifies the allowable compressive stress against lateral buckling from Eq.
7.30:

$$F_b = 0.55F_y\left[1 - \frac{(L/r')^2F_y}{4\pi^2E}\right] \tag{7.34}$$

where r' is the radius of gyration of the compression flange about the axis in
the plane of the web.   For flange plates, $r'^2 = b^2/12$ and

$$F_b = 0.55F_y\left[1 - \frac{3(L/b)^2F_y}{\pi^2E}\right] \tag{7.35}$$

AREA similarly specifies a formula:

$$F_b = 0.55F_y\left[1 - \frac{F_y}{1.8 \times 10^9}\left(\frac{L}{r_y}\right)\right] \qquad \left(\text{in psi}; L/r_y \leqslant \frac{29,900}{\sqrt{F_y}}\right) \tag{7.36}$$

with r_y defined as the radius of gyration of the compression portion of the
cross section above the neutral axis, about the axis in the plane of the web.
In addition it also recognizes the St. Venant torsional resistance of Eq. 7.29

by permitting

$$F_b = \frac{10,500,000}{Ld/A_f} < 0.55F_y \qquad \text{(in psi)} \qquad (7.37)$$

Whichever of Eqs. 7.36 and 7.37 gives the higher allowable stress shall be considered the limiting equation for the member in question.

There are three formulas of allowable stress design against lateral buckling in the AISC Specifications, all incorporating the modifier C_b for moment gradient. The first two consider the flange bending resistance of a beam in the elastic and inelastic range (Eqs. 7.30 and 7.32).

$$F_b = \left[\frac{2}{3} - \frac{(L/r_T)^2 F_y}{1530 \times 10^3 C_b} \right] F_y \qquad (7.38)$$

for

$$\sqrt{\frac{102 \times 10^3 C_b}{F_y}} \leqslant \frac{L}{r_T} \leqslant \sqrt{\frac{510 \times 10^3 C_b}{F_y}}$$

and

$$F_b = \frac{170 \times 10^3 C_b}{(L/r_T)^2} \qquad \text{for} \qquad \sqrt{\frac{510 \times 10^3 C_b}{F_y}} \leqslant \frac{L}{r_T} \qquad (7.39)$$

The limit between the two formulas corresponds to an assumed proportional limit of about $F_y/2$. The term r_T is computed from the compression flange and one-third of the compression web area.

The third allowable stress formula of AISC is obtained by dividing Eq. 7.29 by a factor of safety without introducing a transition curve, and is applicable only to solid, approximately rectangular compression flanges.

$$F_b = \frac{12 \times 10^3 C_b}{Ld/A_f} \qquad (7.40)$$

These three formulas are used simultaneously. The higher value of stress from Eq. 7.40, or from Eqs. 7.38 and 7.39 in their appropriate range, is the allowable stress. The maximum permissible stress of $0.60F_y$, derived from the yield condition, defines the lower limit of L/r_T in Eq. 7.38. Further discussion on permissible stresses and unbraced distance is given in Art. 7.8.

Equations 7.37 and 7.40 are based on elastic buckling without correction for plastic action. In such cases as short unbraced lengths where inelastic buckling governs, the equation becomes unconservative (lines *ABC* vs. *AC* in Fig. 7.21). However, considering the beneficial effect of moment gradient, of the disregarded warping strength, and so forth, the unconservative error is not likely to be large.

For comparison of the more exact formula Eq. 7.28 and its component parts, the allowable stresses as given by Eqs. 7.28, 7.38, 7.39, and 7.40 for cross sections of two wide-flange shapes and one girder are plotted against the unbraced length in Figs. 7.24, 7.25, and 7.26. For the shallow, stocky

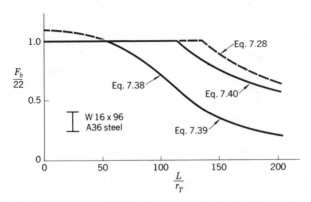

Fig. 7.24 AISC allowable compressive stresses, stocky sections.

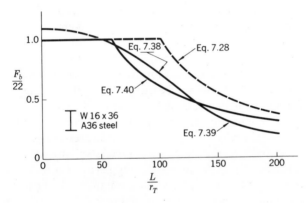

Fig. 7.25 AISC allowable compressive stresses, "intermediate" sections.

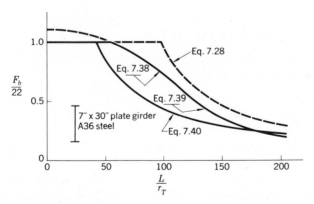

Fig. 7.26 AISC allowable compressive stresses, slender sections.

W 16 × 96 section (Fig. 7.24), the Ld/A_f formula gives values quite close to the more accurate ones which are given by Eq. 7.28. For the 7-by-30-in. girder profile (Fig. 7.26), evidently the column formulas (Eqs. 7.38 and 7.39) give more satisfactory values. (Note that no transition curve is provided for Eq. 7.28 in these figures.) For "intermediate" shapes such as the W 16 × 36, either Eqs. 7.38 and 7.39 or Eq. 7.40 may give a closer approximation depending upon the slenderness ratio, as can be seen from Fig. 7.25.

The application of some of the various provisions against lateral buckling is given in Example 7.11.

It must be pointed out again that all the equations in this article for stress evaluation are for doubly symmetric wide-flange (W) or beam (S) shape cross sections under bending in the plane of the web. For sections symmetrical only about the loading plane, an expression more complicated than Eq. 7.28 is found for estimating F_{cr}.[7.7,7.14,7.15] Nevertheless, Eq. 7.28 and its component terms can still be used as an approximation. In cases where the compressive flange has an area larger than the tension flange for lateral loading, the resistance to lateral buckling is often offered by bending of the compression flange, Eq. 7.34, 7.36, 7.38, and 7.39 can be used to determine the compressive stress (see Example 7.12). If the compression flange of a cross section is smaller then the tension flange, Eqs. 7.37 and 7.40 may be unconservative and thus should not be used. When the loading and supporting conditions are such that lateral buckling of channel or zee sections is possible, Eqs. 7.28 again provides the approximate buckling stresses. Since these sections are generally stocky, the St. Venant resistance accounts for a large part of the strength; the allowable compressive stress against buckling thus can be obtained from the corresponding part of Eq. 7.28, or Eq. 7.29.

Lateral instability of beams affects the strength and behavior of an entire span or a part of a beam between lateral supports. There are instability and deformation problems of a more local nature involving only the components of a beam. These are considered in Art. 7.8.

7.7 PLASTIC DESIGN OF BEAMS

1. Plastic Action of Beams

Thus far, the discussion on beams has been based on the allowable stress design concept. The assumption is that the useful limit of a beam is the hypothetical attainment of the first yielding at the most stressed point, and the allowable stress is obtained by incorporating a factor of safety. However, as has been pointed out in Art. 7.2, gradual plastification of a cross section continues after first yielding and moments beyond M_y can be carried by beams. This strength is recognized in the concept of plastic design which assumes the

EXAMPLE 7.11

PROBLEM:
Select an adequate section for a beam 24 ft in span with 8 ft overhanging on each side. An estimated dead weight of 1.5 kip per linear ft applies throughout the entire length. In addition, concentrated loads of 4 kip each act at the third points of the span and loads of 3 kip at the free ends. Lateral supports are provided at supports and loading points. Use both AISC and AASHO rules but neglect impact.

SOLUTION:

Maximum moment at supports

$M = 864$ kip-in.

Assume that high-strength low-alloy structural steel of 50 ksi yield point is used.

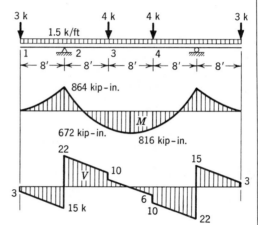

<u>AISC</u>

$F_b = 30$ ksi for tension

Try $F_b = 27$ ksi for compression

$S_{req} = \dfrac{864}{27} = 32$ in.^{3}

Try W 12 x 27 $S_x = 34.1$ in.^{3}, $I_y = 16.6$ in.^{4}

$\qquad d = 11.96$ in., $b = 6.5$ in., $t = 0.40$ in., $w = 0.24$ in.

$\qquad A_f = 6.5 \times 0.40 = 2.60$ in.^{2} $A_w = (11.96 - 2 \times 0.4) \times 0.24 = 2.68$ in.^{2}

$$r_T = \sqrt{\frac{I_y/2}{A_f + \frac{1}{3}\left(\frac{A_w}{2}\right)}} = \sqrt{\frac{16.6/2}{2.60 \times 2.68/6}} = 1.65 \text{ in.}$$

$\qquad L/r_T = \dfrac{8 \times 12}{1.65} = 58.1,$ $(L/r_T)^2 = 3380$

From Eq. 7.33

$\quad$ 1–2 $C_b = 1.75$

$\quad$ 2–3 $C_b = 1.75 - 1.05\left(-\dfrac{672}{864}\right) + 0.3\left(-\dfrac{672}{864}\right)^2 = 2.80 > 2.3$ $C_b = 2.3$

$\quad$ 3–4 $C_b = 1.0$

Limit between Eqs. 7.38 and 7.39: $(L/r_T)_L = \sqrt{\dfrac{510 \times 10^3 C_b}{F_y}} = \sqrt{10200\, C_b}$

$\quad$ 1–2 $(L/r_T)_L = 134$ ⎫
$\quad$ 2–3 $(L/r_T)_L = 153$ ⎬ Larger than $L/r_T = 58.1$. Eq. 7.38
$\quad$ 3–4 $(L/r_T)_L = 101$ ⎭ (or Eq. 7.40) to be used.

Allowable stresses:

From Eq. 7.38:

$\quad$ 1–2 $F_b = \left[\dfrac{2}{3} - \dfrac{3380 \times 50}{1530 \times 10^3 \times 1.75}\right] \times 50 = 30.2$ ksi > 30 ksi

$\quad$ 2–3 $F_b = \left[\dfrac{2}{3} - \dfrac{3380 \times 50}{1530 \times 10^3 \times 2.3}\right] \times 50 = 30.9$ ksi > 30.0 ksi

$\quad$ 3–4 $F_b = \left[\dfrac{2}{3} - \dfrac{3380 \times 50}{1530 \times 10^3 \times 1.0}\right] \times 50 = 27.8$ ksi

EXAMPLE 7.11 (Continued)

From Eq. 7.40, $\quad F_b = \dfrac{12 \times 10^3 C_b}{Ld/A_f} = \dfrac{12 \times 10^3 C_b}{8 \times 12 \times 11.96/2.60} = 27.2\, C_b$

1–2	F_b = 47.6 ksi > 30.0 ksi	30.0 ksi
2–3	F_b = 62.6 ksi > 30.0 ksi	30.0 ksi
3–4	F_b = 27.2 ksi < 27.8 ksi	27.8 ksi

$\left. \right\} = F_b$

Applied stresses:

Maximum bending stress in 3–4,

$$f_b = \frac{M}{S} = \frac{864}{34.1} = 25.4 \text{ ksi} < 27.8 \text{ ksi} = F_b \qquad \text{O.K. all portions}$$

Check shear stress, point 2.

$$f_v = \frac{V}{A_w} = \frac{22}{2.68} = 8.2 \text{ ksi} < 20 \text{ ksi} = F_v \qquad \text{O.K.}$$

Use W 12 x 27.

AASHO

$\qquad\qquad F_b$ = 27 ksi for tension; assume F_b = 25 ksi for compression

$$S_{req} = \frac{864}{25} = 34.6 \text{ in.}^3$$

Try W 14 x 30: $\quad S = 41.8 \text{ in.}^3$

$\qquad\qquad$ d = 13.86 in.,   b = 6.733 in.,   t = 0.383 in.,   w = 0.270 in.

Allowable stress from Eq. 7.35:

$$F_b = 0.55 F_y \left[1 - \frac{3(L/b)^2 F_y}{\pi^2 E} \right] = 0.55 \times 50 \left[1 - \frac{3(8 \times 12/6.733)^2 \times 50}{\pi^2 E} \right]$$

$$= 0.55 \times 50 \times 0.894 = 24.6 \text{ ksi}$$

Applied stresses:

$$f_b = \frac{864}{41.8} = 20.7 \text{ ksi} < 24.6 \text{ ksi}$$

$$f_v = \frac{22}{13.86 \times 0.27} = 5.9 \text{ ksi} < 17 \text{ ksi} = F_v \qquad \text{O.K.}$$

Use W 14 x 30.

the useful limit as the actual failure of a structure. The philosophy of allowable stress design and plastic design has been presented in Chapter 1. In regard to the design of beams, it can be stated that the plastic design procedure generally results in smaller, and thus more economical, cross sections for continuous or rigidly connected beams than those by the allowable stress design method. Further discussion on the merit and details of plastic design is given in Refs. 1.1 and 1.5. The fundamentals for the design of beams are presented here.

Consider a laterally supported simple beam of rectangular cross section as an example. If the material has the idealized stress-strain relationship of Fig. 1.9, the various stages of stress distribution at a cross section are depicted in Fig. 7.27. At a certain magnitude of the applied load, yielding initiates at the extreme fibers (stage 1). This load is the hypothetical limit of

EXAMPLE 7.12

PROBLEM:
A crane runway beam 24 ft long has a cross section consisting of a W 18 x 50 and a C 12 x 20.7 of A36 steel. Lateral supports are provided at the ends and at the mid-span of the beam. The crane has a capacity of 15 kip with a 4 kip auxiliary, weighs 6 kip, and is carried by wheels 6 ft apart. The trolley and hoist weighs 2 kip. Is the section satisfactory according to the AISC Specification?

SOLUTION:

Maximum loads on one wheel:

Vertical:

$\frac{1}{2}(15 + 4 + \frac{6}{2} + 2) = 12.0$ kip

25 % impact $\quad\quad \dfrac{3.0 \text{ kip}}{15.0 \text{ kip}}$

Horizontal:

$\frac{1}{2}(15 + 4 + 2) \times 20\% = 2.1$ kip

Estimated weight of beam: 0.1 kip/ft

Maximum moments (see sketch for wheel positions):

Vertical:

$\left[\frac{10.5}{24} \times 30 \times 10.5 + \frac{1}{2} \times 0.1 \times 10.5 \times 13.5 \right] \times 12$

$= 1740$ kip-in.

Horizontal:

$\frac{10.5}{24} \times (2 \times 2.1) \times 10.5 \times 12 = 232$ kip-in.

Sectional properties:

W 18 x 50

d = 18.0 in., b = 7.50 in., t = 0.57 in., w = 0.358 in.

A = 14.71 in^2, I_x = 800.6 in^4, I_y = 37.2 in^4

C 12 x 20.7

d = 12.0 in., b = 2.94 in., t = 0.501 in., w = 0.28 in.

A = 6.03 in^2, I_x = 128.1 in^4, I_y = 3.9 in^4, $\bar{x}$ = 0.70 in.

N.A. at $\frac{\Sigma Ay}{\Sigma A} = \frac{14.71 \times 9.28 + 6.03 \times 0.70}{14.71 + 6.03} = 6.79$ in.

I_x = 14.71 x 9.28^2 + 6.03 x 0.70^2 + 800.6 + 3.9 − (14.71 + 6.03) x 6.79^2

= 1120 in^4

$I_f = 128.1 + \frac{37.2}{2} = 146.7$ in^4

A_f = 6.03 + 7.50 x 0.57 = 10.31 in^2

$r_T = \sqrt{\dfrac{146.7}{10.31 + (1/3)(6.79 - 0.28 - 0.57) \times 0.358}} = 3.65$ in.

$L/r_T = \frac{12 \times 12}{3.65} = 39.4$

Allowable stresses:

C_b = 1.0 (Eq. 7.33)

$\sqrt{\dfrac{510 \times 10^3 C_b}{F_y}} = 119 > 39.4 = L/r_T$ use Eq. 7.38

$F_b = \left[\frac{2}{3} - \dfrac{(L/r_T)^2 F_y}{1530 \times 10^3 C_b} \right] F_y = 22.7$ ksi > 22 ksi

F_b = 22 ksi, compression and tension

EXAMPLE 7.12 (Continued)

Computed stresses:

Compression flange tip:

$$f_b = \frac{1740 \times 6.79}{1120} + \frac{232 \times 6}{146.7} = 19.8 \text{ ksi} < 22 \text{ ksi}$$

Tension flange:

$$f_b = \frac{1740 \times 11.5}{1120} = 17.9 \text{ ksi} < 22 \text{ ksi} \qquad \underline{\underline{O.K.}}$$

(Note: the compression flange can be considered as a beam–column subjected to axial stresses from vertical loads and bending stresses from the horizontal loads. A longitudinal force, 10 percent of the maximum wheel load, can then be considered.)

Check shear:

Maximum vertical shear: $V = 15 + 15 \times \dfrac{18}{24} + \dfrac{1}{2} \times 0.1 \times 24 = 27.45 \text{ kip}$

$$f_v = \frac{V}{A_w} = \frac{27.45}{18 \times 0.358} = 4.26 \text{ ksi} < 13 \text{ ksi} = F_v$$

Maximum horizontal shear: $H = 2.1 + 2.1 \times \dfrac{18}{24} = 3.7 \text{ kip}$

$$f_v = \frac{H}{A_f} = \frac{3.7}{10.31} = 0.28 \text{ ksi} \qquad \underline{\underline{O.K.}}$$

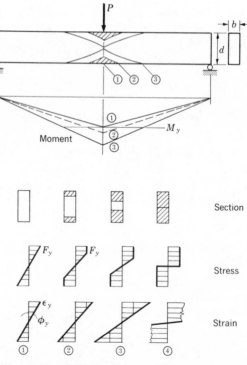

Fig. 7.27 Formation of a plastic hinge.

usefulness for allowable stress design. The moment is $M_y = F_y S$ from elastic analysis. When the load increases, yielding penetrates into the cross section (stage 2). While stress can not be higher than F_y, the strain increases with moment. A further increase of load causes an even higher moment and strain (stage 3) until the flexural stress intensity throughout the whole section equals the yield stress (stage 4). The cross section can then take no additional moment. This is the maximum bending strength of the cross section. The corresponding moment is called the "plastic moment," M_p, and the fully yielded section at mid-span is called a "plastic hinge."

At stage 4 (Fig. 7.27) the equilibrium condition in axial force requires that (Fig. 7.28)

$$\int_0^c F_y b \, dy' - \int_c^d F_y b \, dy' = 0 \tag{7.41}$$

which locates the neutral axis at $c = d/2$. The magnitude of the plastic moment M_p can be evaluated by integration.

$$M_p = 2 \int_0^{d/2} F_y b y \, dy = F_y \frac{bd^2}{4} \tag{7.42}$$

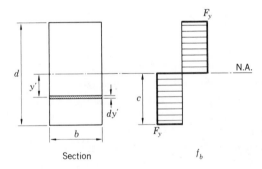

Fig. 7.28 A fully yielded cross section.

If, for rectangular cross section, $bd^2/4 = Z$, then

$$M_p = F_y Z \tag{7.43}$$

where Z is defined as the plastic modulus corresponding to the elastic modulus S. Equation 7.43 is comparable to $M_y = F_y S$. For a simply supported rectangular prismatic beam without the problem of instability, the comparison of the useful limits of the plastic and the allowable stress design procedures is by a ratio of the respective moments.

$$\frac{M_p}{M_y} = \frac{F_y Z}{F_y S} = \frac{Z}{S} = \frac{bd^2}{4} \div \frac{bd^2}{6} = \frac{3}{2} \tag{7.44}$$

In general, the full yielding of a cross section contributes only to a part of the strength above first yielding; the ability of a structural member to redistribute moments after the formation of plastic hinge accounts for more. These points are discussed briefly for design purposes in the following sections.

2. Shape Factor and Redistribution of Moments

The ratio of M_p/M_y indicates, in terms of the yield moment, the magnitude of the moment causing full yielding of the section. As expressed by Eq. 7.44, this ratio is independent of the material properties of the member but depends solely on the cross-sectional properties Z and S. Therefore, the ratio is defined as the shape factor f.

$$f = \frac{Z}{S} \tag{7.45}$$

Shape factors of wide-flange shapes vary between 1.10 and 1.18. The most frequent value is 1.12 for all wide-flange shapes, the average being 1.14. Other shape factors are: 2.00 for a diamond, 1.70 for a round bar, and 1.50 for a rectangle.

It has been pointed out that full yielding of a cross section of a member results in a plastic hinge if no instability occurs. For a simply supported beam, such as the one in Fig. 7.27, the formation of the plastic hinge causes rotation of the two halves and failure of the beam. For statically indeterminate structures, the formation of a plastic hinge reduces the indeterminacy. Further increases in load are possible by moment redistribution until the formation of a sufficient number of plastic hinges so that the structure becomes unstable, as illustrated below.

If a fixed-end beam of length L is subjected to a uniformly distributed load of total weight W, the greatest moment is $WL/12$ at the ends (Fig. 7.29a). The elastic strength is the load magnitude which initiates yielding at these ends.

$$F_y = \left(\frac{W_y L}{12}\right) \div S \tag{7.46}$$

$$W_y = \frac{12 F_y S}{L} \tag{7.47}$$

Upon an increase of load intensity to $W_1 > W_y$, plastic hinges form at the ends of the beam with a moment magnitude of $M_p = M_y f$ (Fig. 7.29b). The beam can now be considered as simply supported with these end moments. The moment at the center of the beam is numerically and proportionally higher than that in Fig. 7.29a in order to maintain equilibrium with the load.

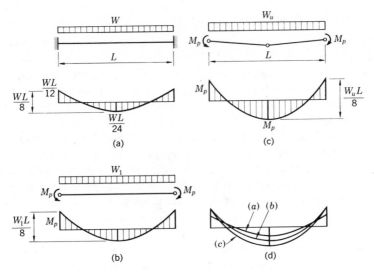

Fig. 7.29 Redistribution of moment.

A further increase of W results in yielding at the center and further redistribution of moment until the formation of the third plastic hinge at W_u (Fig. 7.29c). A "mechanism" is then formed and the beam can take no additional load. This intensity W_u is the load-carrying capacity of the beam. From Fig. 7.29c

$$2M_p = \frac{W_u L}{8} \qquad \text{or} \qquad M_p = \frac{W_u L}{16} = F_y Z \qquad (7.48)$$

$$W_u = \frac{16 F_y Z}{L} \qquad (7.49)$$

Thus

$$\frac{W_u}{W_y} = \frac{16 F_y Z/L}{12 F_y S/L} = \frac{4}{3}\frac{Z}{S} = \frac{4}{3}f \qquad (7.50)$$

which indicates a strength increase of one third, due to the redistribution of moment. For a fixed-end rectangular beam, $f = 1.5$; the total strength is $(\frac{4}{3})(1.5) = 2$, that is, double the yield load W_y. If the section is a wide-flange shape, say W 18 × 60, $f = 122.6/107.8 = 1.12$ and the strength is $W_u = (\frac{4}{3})(1.12)\,W_y = 1.52 W_y$, an increase of 52 per cent. Another illustration is given in Example 7.13. It must be borne in mind that, in order to attain the full plastic moment and the redistribution of moments, proper lateral support and a sturdy cross section are essential (Art. 7.8).

3. Method of Analysis

Because the plastic strength of a beam depends on the formation of plastic hinges and on the redistribution of moments to form a mechanism, it is necessary to find the correct mechanism or its corresponding equilibrium moment diagram for strength evaluation. There are three conditions which must be satisfied: the equilibrium condition that applied loads and reactions must be in equilibrium; the mechanism condition that sufficient plastic hinges form a mechanism; and the plastic moment condition that M_p is reached at each plastic hinge with $M \leqslant M_p$ at all parts of the beam. To fulfill all these requirements, and to find the plastic strength, various methods are available, such as the static (equilibrium) method, the mechanism method, the method of inequalities, the plastic moment distribution method, and others (see Refs. 1.5, 7.21, 7.22, and 7.23). The first two are the most commonly used and are presented below.

The static or equilibrium method is convenient for simple structures such as beams. The method is to find an equilibrium moment diagram in which $M \leqslant M_p$ throughout the structure and which contains sufficient locations of plastic hinges to form a mechanism. Then, with all three conditions fulfilled, the moment diagram must be the correct one and the corresponding capacity of the structure is its plastic strength. This method is employed in Example 7.13.

Occasionally, a structure is so complex that the correct moment diagram is not readily obtainable. An estimate of the plastic strength can be established by using an assumed equilibrium moment diagram in which the moments are not greater than M_p. This estimated strength is less than or at best equal to the true ultimate load according to the lower bound theorem.[7.24] This is evident from the point of view that a load which is not great enough to produce sufficient plastic hinges to form a mechanism is less than the true strength of the structure. Because beams are relatively simple structures, the difficulty of finding the correct moment diagram seldom occurs for beams. Example 7.14 further demonstrates this method of equilibrium.

The mechanism method takes a different approach. First, a mechanism is assumed, then equations are established according to the equilibrium condition and are solved for the load magnitude. If the resultant moment diagram satisfies the plastic moment condition ($M \leqslant M_p$) everywhere, the assumed mechanism is correct and the load is the true ultimate load.

If, however, the plastic moment condition is violated as the result of an assumed mechanism, the corresponding load cannot be regarded as the true ultimate load. A moment diagram with $M > M_p$ indicates that locations other than the assumed hinges are also fully yielded. To produce this situation, the load magnitude must be higher than the true load (the upper bound theorem[7.24]).

EXAMPLE 7.13

PROBLEM:
    What load magnitude can be carried by the continuous beam shown if the section is a W 14 x 34 of A36 steel ?

SOLUTION:
    From elastic analysis, the moment under the load is the highest (a). At first yielding,

$$P_y = \frac{13}{765} M_y = \frac{13}{765} F_y S$$

$$= \frac{13}{765} \times 36 \times 48.5 = 29.7 \text{ kip.}$$

When P is increased, plastic hinges form first at the load point then at the central supports, resulting in a mechanism (b). Assume a determinant structure and redundants as shown (c, d). The corresponding moment diagrams (e, f) are combined to give M_p at locations of the hinges (g).

Hence, from (g),

$$2M_p = 90 P_u$$

$$P_u = \frac{1}{45} M_p = \frac{1}{45} F_y Z$$

$$= \frac{1}{45} \times 36 \times 54.5$$

$$= 43.6 \text{ kip}$$

$$\frac{P_u}{P_y} = \frac{43.6}{29.7} = 1.47$$

For complicated structures which have a high degree of redundancy, the number of possible failure mechanisms is large and each mechanism gives an upper bound estimate. The true mechanism may not be detected without a number of trials. Fortunately, the process of solution for the ultimate load from an assumed mechanism is greatly simplified by using the virtual displacement method (as can be seen from Example 7.14). Also, the moment check of $M \leqslant M_p$ can be made on some fictitious equilibrium moment diagrams. The mechanism method therefore is often used for frames and continuous structures. More details on virtual displacement and moment check are available (for example Ref. 1.5).

4. Load Factor

When the plastic strength of a beam is estimated, a sufficient safety margin must be furnished against the estimate for its use in practice. In the plastic design procedure, the safety margin is provided for by multiplying the expected working load by a designated load factor to compare with the ultimate load. The plastic strength of a beam must be equal to or greater than the factored load.

To maintain the same safety margin against the strength of a simply supported beam as implied in the allowable stress design method, the load factor F is then defined by the ratio of the ultimate load to the allowable load of this beam.

$$F = \frac{W_u}{W_w} = \frac{M_P}{M_w} = \frac{F_y Z}{F_b S} = \frac{F_y}{F_b} f \qquad (7.51)$$

Both AISC and AASHO permit design by plastic strength, but only for hot-rolled or built-up beams (not hybrid members nor beams of A514 steel). Because compact cross sections must be used in the plastic design method so as to prevent local instability (Arts. 7.8 and 17.4), the AISC allowable stress F_b is $0.66F_y$. With a most frequent value of 1.12 for shape factors of wide-flange shapes,

$$F = \frac{F_y}{0.66F_y} \times 1.12 = 1.70 \qquad (7.52)$$

A load factor of 1.70 is applied to the working load in the plastic design of beams according to AISC rules (Example 7.15).

5. Lateral Bracing

The plastic strength of a beam is influenced by various factors. These include material properties and residual stresses, shear force, axial load, unsymmetrical bending, overall and local instability, brittle facture, and repeated loading.[1.1] Some of these are being studied and their effects are considered conservatively in the design procedure. For example, the plastic strength of beams under biaxial bending or under combined bending and torsion are not known completely. Such conditions are not permitted in the AISC Specifications by requiring that the loading plane coincides with the minor plane of symmetry of the beam. The effects of overall or lateral instability are briefly discussed below whereas those of shear, axial load, and local instability are presented later.

In a manner similar to the prevention of elastic lateral buckling prior to yielding in the allowable stress design procedure (Art. 7.6), lateral bracing must be provided so that lateral displacement and twisting do not occur

EXAMPLE 7.14

PROBLEM:

Find the ultimate load which can be supported by the beam. The plastic strength of the cross section of the member is M_p.

SOLUTION:

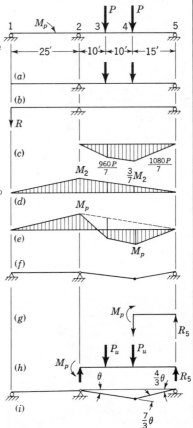

Equilibrium Method

(1) Equilibrum moment diagram: Assume determinate structure (a) and redundant (b). Their moment diagrams are sketched in (c) and (d), and are combined to give the equilibrium diagram (e) with plastic moments at points 2 and 4.

(2) A mechanism is formed (f).

(3) Moment check: At point 4 (c, d, e),

$$M_p + \frac{3}{7} M_p = \frac{1080\,P_u}{7}$$

$$\therefore P_u = \frac{M_p}{108}$$

At point 3 (c, d, e)

$$M_3 = \frac{960\,P_u}{7} - \frac{5}{7} M_p = \frac{960}{7}\,\frac{M_p}{108} - \frac{5}{7} M_p = \frac{5}{9} M_p < M_p$$

$$\therefore M \le M_p \text{ everywhere}$$

Since all three conditions of equilibrium, mechanism, and $M \le M_p$ are satisfied, the ultimate load is

$$P_u = \frac{M_p}{108}$$

Mechanism Method

(1) Mechanism is assumed with plastic hinges at points 2 and 4 (f).

(2) Equilibrium : Take free body as in (g), thus

$$R_5 \times 15 \times 12 = M_p \qquad R_5 = \frac{M_p}{180}$$

Take free body as in (h) and take moment about point 2,

$$P_u \times 10 \times 12 + P_u \times 20 \times 12 - R_5 \times 35 \times 12 = M_p$$

and $P_u = \dfrac{M_p}{108}$

(3) Moment check: Take moment at point 3 (h)

$$M_3 = R_5 \times 25 \times 12 - P_u \times 10 \times 12$$

$$= \frac{M_p}{180} \times 25 \times 12 - \frac{M_p}{108} \times 10 \times 12 = \frac{5}{9} M_p < M_p \qquad \text{OK}$$

Hence $P_u = \dfrac{M_p}{108}$

Alternately, the virtual displacement method may be used to establish equilibrium in step (2). If a virtual displacement (rotation) of θ is allowed for the portion of the beam 2–3 (i), the corresponding displacement at point 5 is $\theta \times \dfrac{20}{15} = \dfrac{4}{3}\theta$, and that at point 4 is $\theta + \dfrac{4}{3}\theta = \dfrac{7}{3}\theta$.

External work done by loads is the sum of the product of force times displacement:

$$P_u\,(\theta \times 10 \times 12) + P_u(\theta \times 20 \times 12) = 360\,P_u\theta$$

Internal work done by rotation is the sum of the product of moment times rotation:

$$M_p\theta + M_p\,\frac{7}{3}\theta = \frac{10}{3} M_p\theta$$

For equilibrium, the work done must be equal, therefore $360\,P_u\theta = \dfrac{10}{3} M_p\theta$

and $P_u = \dfrac{M_p}{108}$

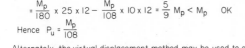

EXAMPLE 7.15

PROBLEM :
A two span continuous beam is subjected to a uniformly distributed load of 3 kip per linear ft throughout its entire length of 60 ft. Lateral bracing is adequately provided. Select a member using the plastic design procedure. A 36 steel is to be used.

SOLUTION :
Assume the required plastic stregth of the beam cross section to be M_p. The design load with a load factor of 1.7 is $3 \times 1.7 = 5.1$ k/ft.

Structure is statically indeterminate to the first degree, requiring two plastic hinges for a mechanism. Three plastic hinges are assumed due to symmetry and according to moment diagram.

Consider left half of structure with virtual displacement:

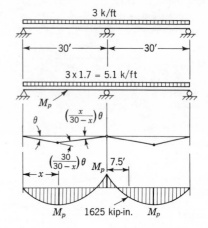

External work done

$$= \text{Force} \times \text{displacement (kip-in.)}$$

$$= (5.1x)(\frac{x}{2}\theta \times 12) + [5.1(30-x)] \times$$

$$\left[(\frac{30-x}{2})(\frac{x}{30-x}\theta) \times 12 \right]$$

$$= (5.1 \times 30)(\frac{x}{2}\theta \times 12) = 918x\theta$$

Internal work done

$$= \text{Moment} \times \text{rotation (kip-in.)}$$

$$= M_p \frac{x}{30-x}\theta + M_p \frac{30}{30-x}\theta = \frac{30+x}{30-x}M_p\theta$$

$$\therefore 918x\theta = \frac{30+x}{30-x}M_p\theta$$

$$M_p = \frac{918(30-x)x}{30+x}$$

Maximum required M_p by $\frac{dM_p}{dx} = 0$, which results in $x = 12.42$ ft and

$$M_p = \frac{918(30-12.42)12.42}{30+12.42} = 4725 \text{ kip-in.}$$

Required plastic modulus:

$$Z = \frac{M_p}{F_y} = \frac{4725}{36} = 131.3 \text{ in}^3$$

Use W 21 x 62, with Z = 144.1 in³ (This is 9% lighter than the section selected in Example 7.18 according to the allowable stress design procedure. See Examples 7.16, and 7.17 for shear and stability checks.)

---

before strain hardening. The AISC limits for the unbraced length are[1.1,1.21]

$$L_{cr} = \frac{76.0b}{\sqrt{F_y}} \tag{7.53}$$

and

$$L_{cr} = \frac{20,000}{F_y}\left(\frac{A_f}{d}\right) \tag{7.54}$$

The first of the above two equations is derived through the consideration of beams subjected to uniform moment and of the average strain hardening modulus of steels employed for plastic design. The second formula corresponds to the St. Venant torsion part of Eq. 7.28 and can be converted from Eq. 7.40 by introducing the factor of safety.

When both requirements of Eqs. 7.53 and 7.54 are met, beams with a compact cross section are insured against overall lateral instability and local buckling or excessive deformation (see Art. 7.8). Further attention must be directed to the bracing requirement at the vicinity of the plastic hinges.

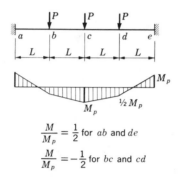

$$\frac{M}{M_p} = \frac{1}{2} \text{ for } ab \text{ and } de$$

$$\frac{M}{M_p} = -\frac{1}{2} \text{ for } bc \text{ and } cd$$

Fig. 7.30 Moment ratio for Eqs. 7.55 and 7.56.

All plastic hinge locations must be adequately braced. Whenever hinge rotation is essential for the redistribution of moment in the development of a failure mechanism, close spacing of lateral support may be necessary. Based on experimental results,[7.25] AISC specifies that the unbraced length from the braced hinge location to the adjacent braced point should not be greater than the length given by Eq. 7.55 and 7.56, as applicable.[1.1,1.12]

$$\frac{L_{cr}}{r_y} = \frac{1375}{F_y} + 25 \quad \text{for} \quad +1.0 > \frac{M}{M_p} > -0.5 \qquad (7.55)$$

$$\frac{L_{cr}}{r_y} = \frac{1375}{F_y} \quad \text{for} \quad -0.5 > \frac{M}{M_p} > -1.0 \qquad (7.56)$$

where L_{cr} is the critical bracing length and M/M_p is the end moment ratio corresponding to the actual distance between bracing points. The ratio is negative when the beam segment is bent in single curvature (Fig. 7.30).

At the last hinge of a failure mechanism, bracing is sufficient if provided according to the requirements of Eqs. 7.53 and 7.54 for points of the beam other than hinge locations. Example 7.16 illustrates the checking of bracing requirements.

EXAMPLE 7.16

PROBLEM:

   Lateral bracing for the beam of Example 7.15 is provided at the supports and at 7.5 ft internals. Check lateral stability.

SOLUTION:

   W 21 x 62   d = 20.99 in., b = 8.24 in., t = 0.615 in., w = 0.40 in.

        r_y = 1.71 in., A_f = 5.07 in.²

For plastic hinge at support: M_p = 4725 kip-in.

   At 7.5 x 12 = 90 in. away, M = 1625 kip-in.

$$\frac{M}{M_p} = + \frac{1625}{4725} = + 0.344 \text{ (positive for double curvature)}$$

Critical bracing length by Eq. 7.55

$$\frac{L_{cr}}{r_y} = \frac{1,375}{F_y} + 25$$

$$L_{cr} = (\frac{1,375}{F_y} + 25) \, r_y = (\frac{1,375}{36} + 25) \, 1.71 = 108 \text{ in.} > 90 \text{ in.} \qquad \text{O.K}$$

The hinges in the spans are the last to form, thus bracing requirements by Eqs. 7.53 and 7.54, as for other portions of the beam.

$$L_{cr} = \frac{76.0 \, b}{\sqrt{F_y}} = \frac{76.0 \times 8.24}{\sqrt{36}} = 104 \text{ in.} > 90 \text{ in.} \qquad \text{O.K.}$$

$$L_{cr} = \frac{20,000}{F_y} (\frac{A_f}{d}) = \frac{20,000}{36} \times \frac{5.07}{20.99} = 134 \text{ in.} > 90 \text{ in.} \qquad \text{O.K}$$

Hence the beam is adequately braced laterally.

6. Shear and Axial Forces

When the cross section of a beam is partially yielded by a bending moment, only the remaining elastic portion is capable of resisting shear (Fig. 7.31a). High shear intensity at this portion may cause yielding in shear and failure of the beam. However, since high shear occurs with high moment gradient, yielding due to combined bending and shear is localized. Experiments show that the actual strength of such a section may even be higher than the full plastic moment M_p.[1.1.1.5]

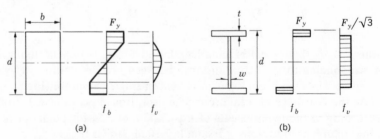

Fig. 7.31 Shear stresses in beam sections.

When the shear-to-moment ratio of a wide-flange beam is high, the moment may be considered to be taken by the flange while shear yielding is produced in the web (Fig. 7.31b). Then, the maximum shear strength is

$$V_u = \frac{F_y}{\sqrt{3}} \, w(d - 2t) \qquad (7.57)$$

where $F_y/\sqrt{3}$ is the shear yield stress.[1.1,7.2] By assuming an effective depth of $0.95d$ to allow for the presence of plastic bending strains in the flanges, this maximum shear force is

$$V_u = \frac{F_y}{\sqrt{3}} \, (0.95d)w = 0.55F_y wd \qquad (7.58)$$

It is necessary that the shear force on the cross section not be greater than this magnitude (Example 7.17).

EXAMPLE 7.17

PROBLEM:
  Is the section selected for Example 7.15 adequate for shear ?

SOLUTION:
  From Example 7.15, for equilibrium:

$$R = \frac{M_p - \frac{5.1}{2}(12.42)^2 \times 12}{12.42 \times 12} = 63.4 \text{ kip}$$

  Shear at central support $= 5.1 \times 30 - 63.4 = 89.6$ kip
W 21 x 62: d = 20.99 in., w = 0.40 in., F_y = 36 ksi
By Eq. 7.58: Shear strength $= 0.55 F_y wd$
                            $= 0.55 \times 36 \times 0.40 \times 20.99$
                            $= 166.1$ kips > 89.6 kip
The section is adequate for shear.

The combination of bending and tensile axial force on a beam can be treated through consideration of stresses and equilibrium. The condition of lateral instability often is improved when a tensile force exists. A load factor of 1.70 (Eq. 7.52) should be applied for design. When compressive axial force is present in a beam, the member becomes a beam-column, which is discussed in Chapter 11. In general, when the axial compression is less than 15 per cent of the yield strength $P_y = F_y A$ (A is the cross-sectional area of the beam), its effect on the moment capacity of the beam can be neglected.

7.8 LOCAL BUCKLING, COMPACT SECTIONS, AND STRESS REDUCTIONS

Lateral instability of beams affects an entire span or a part of a beam between lateral supports.   Local instability involving only components of a beam may cause excessive deformation of the cross-sectional shape and thus affect the behavior and strength of the beam.   Local buckling is discussed in Chapter 17.   Some limits for the allowable stress design and the plastic design procedures are summarized here.

Since the basis of consideration in the allowable stress design concept is the attainment of the first yielding, the limits for the prevention of local buckling of compressive components are derived accordingly.   For a given yield point or a computed compressive flexural stress, the maximum values for outstanding compression flange plates, as expressed by the width-to-thickness ratio of the component plates, are:

$$\text{AISC} \qquad \frac{b}{2t} = \frac{95.0}{\sqrt{F_y}} \tag{7.59}$$

$$\text{AASHO} \qquad \frac{b}{t} = \frac{3250}{\sqrt{f_b}} \leqslant 24 \tag{7.60}$$

$$\text{AREA} \qquad \frac{b}{2t} = \frac{2300}{\sqrt{F_y}} \tag{7.71}$$

The maximum depth-to-thickness ratios of the web are

$$\text{AASHO} \qquad \frac{h}{w} = \frac{23{,}000}{\sqrt{f_b}} \leqslant 170 \tag{7.62}$$

$$\text{AREA} \qquad \frac{h}{w} = 170 \sqrt{\frac{F_b}{f_b}} \tag{7.63}$$

AISC uses a stress reduction formula (see Chapter 8) for values of h/t higher than $760/\sqrt{F_b}$.   All the stresses are in ksi for AISC and in psi for AASHO and AREA.

For plastic design of beams, more stringent rules than those given above must be used to insure that local buckling does not occur prior to the development of full plastic strength.   When strain hardening is used as the reference point for plastic strength development, the AISC width-to-thickness limits as obtained through experimental results are given in numerical values in Table 7.3.   Further studies are necessary for A514 steel and thus it is not used in

Table 7.3   Flange Width-to-Thickness Ratios of Compact Sections

F_y	36	42	45	50	55	60	65
$\dfrac{b}{2t}$	8.5	8.0	7.4	7.0	6.6	6.3	6.0

plastic design.   The maximum depth-to-thickness ratio of webs is

$$\frac{d}{w} = \frac{412}{\sqrt{F_y}}\left(1 - 1.4\frac{P}{P_y}\right) \quad \text{for} \quad \frac{P}{P_y} \leqslant 2.7 \qquad (7.64)$$

where P is the axial force if it exists in the beam.   Cross-sectional shapes satisfying these requirements are compact sections.   Practically all rolled wide-flange and beam shapes are compact sections when they are made of structural steels other than A514.   The non-compactness of a cross section is indicated in handbooks.[7.9]   Only compact sections with proper lateral bracing (Eqs. 7.53 to 7.56) can be used in plastic design.

It is obvious that compact sections with adequate lateral supports are capable of supporting higher stresses in allowable stress design than otherwise permitted for non-compact sections.   Because the full plastic moment of compact shapes is usually more than 10 per cent higher than M_y, a 10-per cent increase in the allowable stress is permitted by AISC.   Thus,

$$F_b = 0.66F_y \qquad (7.65)$$

as is listed in Table 7.1.   The width or depth-to-thickness ratios of Table 7.3 and Eq. 7.64 are expressed in conformity with other allowable stress formulas.

$$\frac{b}{2t_f} = \frac{52.2}{\sqrt{F_y}} \qquad (7.66)$$

$$\frac{d}{w} = \frac{412}{\sqrt{F_y}}\left(1 - 2.33\frac{f_a}{F_y}\right) \quad \text{when} \quad \frac{f_a}{F_y} \leqslant 0.16 \qquad (7.67)$$

For laterally supported compact sections of doubly-symmetric I- and H-shapes and solid bars, subjected to weak axis bending, $F_b = 0.75F_y$ is permitted by AISC in recognition of their favorable shape factors and the condition that these members are not subjected to lateral buckling.

Recognition is also given in the allowable stress design procedure to the capacity of moment redistribution of adequately braced beams with compact sections.   A 10-per cent reduction in maximum negative moment produced by gravity loading is permitted by AISC if a corresponding increase is made in positive moment in the span of a continuous or fixed-end beam.[1.21]   That is, a continuous beam of compact section with lateral bracing placed according to Eqs. 7.53 and 7.54 may be proportioned for $\frac{9}{10}$ of the maximum negative moment at supports, or proportioned for the maximum positive moment in the span plus $\frac{1}{10}$ of the average negative moment at neighboring supports (Fig. 7.32).   Thus, the magnitude of the design moment is

$$M = \tfrac{9}{10}M_1 \qquad (M_1 > M_3) \tag{7.68}$$

or

$$M = M_2 + \frac{1}{10}\left(\frac{M_1 + M_3}{2}\right) \tag{7.69}$$

whichever is greater.   Taken in conjunction with the 10-per cent increase in allowable bending compression (Eq. 7.65), this provision affords a 20-per cent reduction in required bending strength.   The net result of these moment-redistribution provisions is a tendency toward equalization of maximum negative and positive moments.

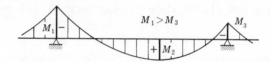

Fig. 7.32   Moments for Equations 7.68 and 7.69.

AASHO specifies that continuous or cantilever beams may be proportioned for negative moment at an interior support for an allowable compressive stress 20-per cent higher than permitted otherwise (Eqs. 7.34 and 7.35). This is a recognition of the fact that partial yielding at such supports has little effect on the strength of the beam; and the 20-per cent increase in allowable stress is comparable to the 20-per cent reduction of bending strength by AISC.

For beams adequately braced and satisfying all the requirements for Eq. 7.65, except that the flange width-to-thickness ratio is greater than $52.2/\sqrt{F_y}$, the allowable stress can not be $0.66F_y$.   So far as the width-to-thickness ratio is less than $95.0/\sqrt{F_y}$, which holds for $0.60F_y$, an allowable stress somewhere in between may be permissible.   The simplest model is by a straightline interpolation.   It is so specified by AISC (Fig. 7.33):

$$F_b = F_y\left[0.733 - 0.0014\,\frac{b}{2t}\,\sqrt{F_y}\right] \qquad \text{for} \quad \frac{52.2}{\sqrt{F_y}} \geqslant \frac{b}{2t} \geqslant \frac{95.0}{\sqrt{F_y}} \tag{7.70}$$

A similar equation is given also by AISC for weak axis bending.

When Eq. 7.70 is employed, the capability of hinge rotation may not exist. Thus, the 10 per cent reduction of moment by Eq. 7.68 is not applicable.[1.1]

It is to be noted that adequate lateral bracing for the above is expressed by Eq. 7.53 in that the unbraced length is not less than $76.0b/\sqrt{F_y}$.   From

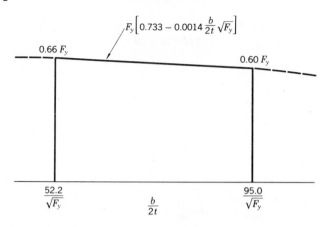

Fig. 7.33 Straightline interpolation of allowable stresses.

Eq. 7.38, the lateral bracing requirement for non-compact sections having an allowable stress of $0.60F_y$ is

$$\frac{L}{r_T} \leqslant \sqrt{\frac{102,000}{F_y}} \qquad (7.71)$$

This results in a limiting unbraced length of $92b/\sqrt{F_y}$ if r_T is taken as $b/\sqrt{12}$ for the compression flange alone. Because no definition is given on the rigidity or restraint at the points of bracing, and linear transition is already allowed for the condition of non-compactness (Eq. 7.70), no interpolation of allowable stresses is permitted at present (1973) for values of L between $76.0b/\sqrt{F_y}$ and that from Eq. 7.71.

Examples 7.18 and 7.19 show the use of the AISC local buckling provisions in the design process.

AASHO Specifications for load-factor design[7.27] contain similar provisions which have been derived on the same rationale.

7.9 DEFLECTIONS

In spite of the differences in the design basis, the working loads for both plastically designed beams and those designed by the allowable stress procedure are usually in the elastic range. Consequently, the calculation of deflection at working load is, in general, identical.

The deflection of beams at working load should not be excessive because of esthetics and discomfort, of possible damage to attached material, and of its effect on the functioning of the whole structure. Limitations are usually

EXAMPLE 7.18

PROBLEM:
Select a member for the beam of Examples 7.15 and 7.16, using the allowable stress design procedure. Consider both AISC and AASHO requirements.

SOLUTION:
Maximum moment:

$$M = 4050 \text{ kip-in. at center support}$$

Maximum shear

$$V = 52.25 \text{ kip at center support}$$

AISC

(1) For A36 steel $F_b = 0.60 \; F_y = 22 \text{ ksi}$

$$S_{req} = \frac{M}{F_b} = \frac{4050}{22} = 184 \text{ in.}^3$$

Try W 24x84 $d = 24.09 \text{ in.}, \; b = 9.015 \text{ in.}, \; t = 0.772 \text{ in.}, \; w = 0.470 \text{ in.}$

$$S = 196.3 \text{ in.}^3, \; f_b = \frac{M}{S} = \frac{4050}{196.3} = 20.6 \text{ ksi}$$

Local stability is assured for all rolled shapes. Checking is made for demonstration.

Flange, by Eq. 7.59 $\dfrac{b}{2t} = \dfrac{95.0}{\sqrt{F_y}} = \dfrac{95.0}{\sqrt{36}} = 15.8$ (See App. A of AISC spec.)

    actual: $\dfrac{9.015}{2 \times 0.772} = 5.85 < 15.8$ O.K.

    web $\dfrac{760}{\sqrt{F_y}} = \dfrac{760}{\sqrt{22}} = 162 > \dfrac{24.09 - 2 \times 0.772}{0.47} = 48.0$ O.K.

Lateral bracing is checked to be adequate. (See Example 7.11 for procedure of checking.) Hence, W 24x84 is suitable for the beam. However, the shape is a compact section. Higher allowable stress can be used.

(2) Assume a compact section. Lateral bracing is adequate by Eqs. 7.53 and 7.54. (See Example 7.16.)

Hence, $F_b = 0.66 \; F_y = 24 \text{ ksi}$

Design moment:

     Eq. 7.68 $M = \dfrac{9}{10} \times 4050 = 3645 \text{ kip-in.}$

     Eq. 7.69 $M = 2280 + \dfrac{1}{10} \dfrac{4050 + 0}{2} = 2483 \text{ kip-in.} < 3645 \text{ kip-in.}$

$$S_{req} = \frac{M}{F_b} = \frac{3645}{24} = 152 \text{ in.}^3$$

Try W 24x68, a compact section (satifies requirements of Eqs. 7.66 and 7.67)

$$S = 153.1 \text{ in.}^3, \; d = 23.71 \text{ in.}, \; w = 0.416 \text{ in.}$$

$$f_b = \frac{M}{S} = \frac{3645}{153.1} = 23.8 \text{ ksi}$$

Check shear

$$f_v = \frac{V}{A_w} = \frac{56.25}{23.71 \times 0.416} = 5.7 \text{ ksi} < 14.5 \text{ ksi} = F_v \qquad \text{O.K.}$$

use W 24x68, which is 19% lighter than W 24x84

AASHO

Assume that local instability will not govern so that a 20% increase in allowable stress is permitted for the negative moment. If $F_b = 20 \text{ ksi}$ after 20% increase, then

$$S_{req} = \frac{M}{F_b} = \frac{4050}{20} = 202.5 \text{ in.}^3$$

Try W 27x84

$$S = 211.7 \text{ in.}^3, \; d = 26.69 \text{ in.}, \; b = 9.963 \text{ in.}, \; t = 0.636 \text{ in.}, \; w = 0.463 \text{ in.}$$

$$f_b = \frac{M}{S} = \frac{4050}{211.7} = 19.2 \text{ ksi}$$

EXAMPLE 7.18 (Continued)

Local stability check

Flange, Eq. 7.60 $\dfrac{b}{t} = \dfrac{3250}{\sqrt{f_b}} = \dfrac{3250}{\sqrt{19,200}} = 23.5 < 24$

actual $\dfrac{9.963}{0.636} = 15.7 < 24$ O.K.

web, Eq. 7.62 $\dfrac{23,000}{\sqrt{f_b}} = \dfrac{23,000}{\sqrt{19,200}} = 166$

actual $\dfrac{26.69 - 2 \times 0.636}{0.463} = 55 < 166$ O.K.

Allowable stress by Eq. 7.35 (lateral instability) and with 20% increase

$$F_b = 120\% \times 0.55\, F_y \left[1 - \dfrac{3(L/b)^2\, F_y}{\pi^2 E} \right]$$

$$= 1.20 \times 0.55 \times 36 \left[1 - \dfrac{3(90/9.963)^2 \times 36}{\pi^2 \times 29,000} \right] = 23.4 \text{ ksi} > 20 \text{ ksi}$$

$$\therefore\ F_b = 20 \text{ ksi as assumed}$$

Check shear:

$$f_v = \dfrac{V}{A_w} = \dfrac{56.25}{0.463 \times 26.69} = 4.6 \text{ ksi} < 12 \text{ ksi} \qquad \text{O.K.}$$

use W 27 x 84.

expressed in beam depth-to-span ratio or deflection-to-span ratio under specific loadings. For instance, AISC specifies that the maximum live load deflection of beams supporting plastered ceilings shall be less than $\frac{1}{360}$ of the span.

For a given uniformly distributed live load w_L, the deflection of a beam is[7.8]

$$\delta = K_1 \frac{w_L L^4}{EI} \tag{7.72}$$

where K_1 is a constant incorporating the end restraint, being 5/384 for simply supported condition and 71/9600 for beams with inflection points at $0.8L$, and L is the span length. Including the effect of live load, the total load w_T creates a bending stress of

$$f_b = \frac{Mc}{I} = K_2 \frac{w_T L^2}{I} \frac{d}{2} \tag{7.73}$$

for doubly symmetrical beams. The coefficient K_2 again depends on end restraint, equal to 1/8 for simple beams and 0.08 for beams with inflection points at $0.8L$. Combining Eqs. 7.72 and 7.73 to eliminate I gives

$$\frac{d}{L} = 2 \frac{K_1}{K_2} \frac{w_L}{w_T} \frac{f_b}{E} \frac{L}{\delta} \tag{7.74}$$

which indicates that for given geometric and loading conditions (K and w) and a given flexural stress, a limitation on depth-to-span ratio is in fact a deflection limit (L/δ).

EXAMPLE 7.19

PROBLEM:

What rolled section can be used for the beam of Example 7.11 if lateral bracing is provided at 4 ft. intervals ? Use allowable stress design method.

SOLUTION:

Maximum moment : M = 864 kip-in.   Maximum shear: V = 22 kips

Lateral bracing at 4 x 12 = 48 in.

F_y = 50 ksi

Assume F_b = 0.66F_y = 33 ksi

$$S_{req} = \frac{M}{F_b} = \frac{864}{33} = 26.1 \ in.^3$$

Try W 14 x 22

S = 28.8 in.³ , I_y = 6.40 in.⁴

d = 13.72 in., b = 5.0 in. , t = 0.335 in., w = 0.230 in.

Check lateral bracing :

Eq. 7.53

$$L_{cr} = \frac{76.0 \ b}{\sqrt{F_y}} = 10.7 \ b = 53.8 \ in. > 48 \ in. = L .$$ O.K.

Check flange width-to-thickness ratio :

Eq. 7.66

$$\frac{b}{2t} = \frac{52.2}{\sqrt{F_y}} = 7.4 < \frac{5.0}{2 \ x \ 0.335} = 7.5$$ non-compact section

Eq. 7.59

$$\frac{b}{2t} = \frac{95.0}{\sqrt{F_y}} = 13.4 > 7.5$$

Allowable stress by Eq. 7.70

$$F_b = F_y \left[0.733 - 0.0014 \ \frac{b}{2t} \sqrt{F_y} \right] = 50 \left[0.733 - 0.0014 \ x \ 7.5 \sqrt{50} \right] = 32.96 \ ksi$$

$$= 33.0 \ ksi$$

Applied stress :

$$f_b = \frac{M}{S} = \frac{864}{28.8} = 30.0 \ ksi < 33 \ ksi$$ O.K.

$$f_v = \frac{V}{Aw} = \frac{22}{13.72 \ x \ 0.23} = 7.0 \ ksi < 20 \ ksi = F_v$$ O.K.

use W 14 x 22.

If $w_L/w_T = 2/3$, $f_b = F_b = 0.60F_y$, and $\delta = L/360$; then $d/L = F_y/1000$ for a simply supported beam. AISC suggests as a guideline that the depth of fully stressed beams in floors shall not be less than $F_y/800$ times their span length, and that of fully stressed purlins $F_y/1000$ times their span. For a yield stress of 36 ksi, this depth-to-span ratio is 1/22 and 1/28, respectively.

The recommended limits for bridges are similar. AASHO suggests a depth-to-span ratio of not less than 1/25 for beams used as girders. The deflection of beams due to live load plus impact is limited to 1/800 of span for AASHO; for AREA, 1/640.

The selection of a rolled section for a beam which is limited by deflection can be made as shown in Example 7.20.   It should be pointed out that depth-to-span ratios are only guides for preliminary selection of members so that the final deflection will not be excessive.   Deflection limits themselves in turn are only approximate guides.

EXAMPLE 7.20

PROBLEM:
  Select an adequate wide-flange section for the simple beam of Example 7.1 if high-strength low-alloy structural steel with $F_y = 50$ ksi is used.

SOLUTION:

  $L = 30$ ft, $w = 2$ kip/ft, $d \leq 14$ in.

AISC suggested guideline : $\dfrac{d}{L} = \dfrac{F_y}{800}$

  $d_{req} = \dfrac{F_y L}{800} = \dfrac{50 \times 30 \times 12}{800} = 22.5$ in. > 14 in.

To keep deflection within guideline:

  $$F_b = \frac{14}{22.5} \times (0.60\, F_y) = \frac{14}{22.5} \times 30 = 18.7 \text{ ksi}$$

  $$S_{req} = \frac{M}{F_b} = \left(\frac{1}{8} \times 2 \times 30^2 \times 12\right) \div 18.7 = \frac{2700}{18.7} = 145 \text{ in}^3$$

Use W 14 x 95, a non-compact section with $S = 151$ in.^{3}

Alternatively, use A36 steel and a compact section :

  $$d_{req} = \frac{F_y L}{800} = \frac{36 \times 30 \times 12}{800} = 16.2 \text{ in.} > 14 \text{ in.}$$

  $$F_b = \frac{14}{16.2} \times (0.66\, F_y) = \frac{14}{16.2} \times 24 = 20.7 \text{ ksi}$$

  $$S_{req} = \frac{M}{F_b} = \frac{2700}{20.7} = 130 \text{ in}^3 \qquad \text{Use W 14 x 84}$$

Shear stress is less than allowable in both cases.

7.10 BUILT-UP, TAPERED, AND CURVED BEAMS

In addition to the most commonly used rolled sections, built-up sections are quite frequently used for beams (Fig. 7.34).   The reason for building up sections is to reinforce rolled beams for strength, to provide strength and

Fig. 7.34 Some reinforced and built-up sections.

rigidity where it is not readily available (for example, with crane girders), to create sectional forms for special purposes (such as eave struts), to utilize different materials, and to satisfy other specific purposes. The use of component parts to make a desirable section has the advantage of effectively utilizing material. However, it also raises the problem of connecting the various parts, which sometimes may offset the advantage of the desirable shape or even create adverse effects of fatigue and brittle fracture.

The design of reinforced or built-up sections with an axis of symmetry in the plane of loading is relatively simple. Both the allowable stress design procedure or the plastic design procedure can be followed (see Example 7.21). The stress computation is more involved for members subjected to unsymmetrical bending. The procedure and necessary precaution against instability have been discussed in Arts. 7.2 through 7.8. When component parts are chosen, the connections between them should be properly designed, following the requirements of Chapters 18 or 19.

Reinforcement for beams may be applied to the entire length of a member, or only to parts of it. A beam with a partial-length cover plate is an example of the latter. Occasionally, different sizes of component parts are used in a beam, as in the case of butt-welded flange plates which differ in width or in thickness. These members have sectional properties constant within a portion, but different from portion to portion. The important thing in design is that the actual capacity of a beam should be higher everywhere than that required by the loading. Figure 7.35 shows some possible arrangements for prismatic beams. An example on the design of cover plates is given in Art. 7.11.

Another arrangement for an economical use of material, for the controlling of stiffness, and for architectural reasons, is the use of tapered beams. Tapered beams are often so constructed that they are symmetrical with respect to their neutral axes and have one flange parallel to the longitudinal direction of the beam (Fig. 7.36). The exact analysis for stress is too complicated for design use.[2.3] One approximation is to resolve the fiber stress f_b into two components normal and parallel to a transverse cross section, as are shown in Fig. 7.36. By using the conventional bending formula, the normal component is

$$f_n = \frac{Mc}{I} \qquad (7.73)$$

with the quantities I and c computed for the section under consideration. The shear component corresponding to this normal stress is $f_n \tan \theta$, which results in a shear force

$$V_n = \int_{\text{area}} f_n \tan \theta \, dA \qquad (7.74)$$

EXAMPLE 7.21

PROBLEM:
The load, including dead weight, of a simply supported beam of a building is 8 kips per linear ft. The beam, 30 ft. long and laterally supported, is limited to 24 inches in over-all depth. Select a section of A36 steel.

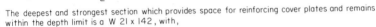

$\frac{3}{4} \times 12$

SOLUTION:

$$S_{req} = \frac{M}{F_b} = \frac{\frac{1}{8} \times 8 \times 30 \times 12}{24} = 450 \text{ in}^3$$

W 21 x 142

No wide-flange section less than 24 in. deep provides such a value. Reinforced wide-flange section is used.

Required $I = 450 \times \frac{24}{2} = 5400 \text{ in}^4$

The deepest and strongest section which provides space for reinforcing cover plates and remains within the depth limit is a W 21 x 142, with,

$$I = 3403 \text{ in}^4, \ d = 21.46 \text{ in.}, \ b = 13.132 \text{ in.},$$
$$A_f = 14.35 \text{ in}^2, \ w = 0.659 \text{ in.}, \ t = 1.095 \text{ in.}$$

Required I of one cover plate : $I_t = \frac{1}{2}(5400 - 3400) = 1000 \text{ in.}^4$

Assume the distance between cover plate centroids to be 22 in.

Required area of one cover plate $= \frac{1000}{(\frac{22}{2})^2} = 8.3 \text{ in}^2$

Try $\frac{3}{4} \times 12$:

$$I_t = \frac{3}{4} \times 12 \left[\frac{1}{2}(21.46 + \frac{3}{4}) \right]^2 = 987 \text{ in}^4$$

Total $I = 2 \times 987 + 3403 = 5377 \text{ in}^4$

$$d = 21.46 + 2 \times 0.75 = 22.96 \text{ in.}$$

$$S = \frac{5377}{\frac{1}{2} \times 22.96} = 467 \text{ in}^3 > 450 \text{ in}^3$$

Check shear

$$f_v = \frac{\frac{1}{2} \times 8 \times 30}{21.40 \times 0.659} = 8.5 \text{ ksi} < 14.5 \text{ ksi} \qquad \underline{O.K.}$$

If plastic design procedure is utilized, the design load is

$$8 \times 1.7 = 13.6 \text{ kip/ft} = W_u$$

and

$$M_p = \frac{W_u L^2}{8} = F_y Z$$

W_u

$$\therefore \quad Z_{req} = \frac{W_u L^2}{8F_y}$$

$$= \frac{13.6 \times 30^2 \times 12}{8 \times 36} = 510 \text{ in}^3$$

Try W 21 x 127, (Z = 317.8 in³, d = 21.27)

$\frac{W_u L^2}{8}$

with $\frac{3}{4} \times 12$ cover plates :

M_p

$$\text{Total } Z = 317.8 + 2(\frac{3}{4} \times 12)(\frac{21.27 + 0.75}{2}) = 317.8 + 198.2$$

$$= 516.0 \text{ in}^3 > 510 \text{ in}^3 \qquad \text{O.K.}$$

The width-to-thickness ratio of the cover plates must be within limit, and the plates must be connected to the flanges by weld or fasteners.

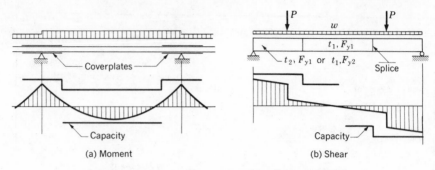

Fig. 7.35 Capacity vs. design requirements.

The remaining shear force $V' = V - V_n$ is assumed to be resisted by the cross section in the conventional way, with shear stresses

$$f_v = \frac{V'Q}{It} \tag{7.75}$$

In the plastic design of tapered beams,[1.5] if the loading and geometry are such that plastic hinges are located at known sections, the design procedure is identical to that for uniform beams. Thus, the relationship for each of the two plastic moments of Figs. 7.37a and 7.37b is

$$M_{p1} + M_{p2} = \frac{wL^2}{8} \tag{7.76}$$

For beams similar to the one in Fig. 7.37c, the location of the central plastic hinge must first be found.

Occasionally, curved beams are used in structures. A crane hook carrying a load sustains both a bending in the plane of the curve and an axial force, as does the end of an eye bar. Because the neutral axis does not coincide with the centroidal axis of the cross section even when there is no axial force, the stress distribution is not linear across the section. Analytical expressions for stresses and deflections can be found in various textbooks (Refs. 7.2, 7.8,

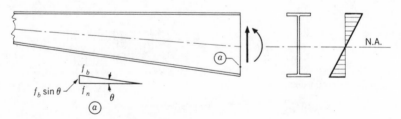

Fig. 7.36 A tapered beam.

and 7.10).   For cases where loads are transverse to the plane of the curve, such as the curved beam of a stair case, or of a bridge, torsion and bending occur simultaneously.   No concise formula is available for design.   References 7.2 and 7.8 furnish additional information.

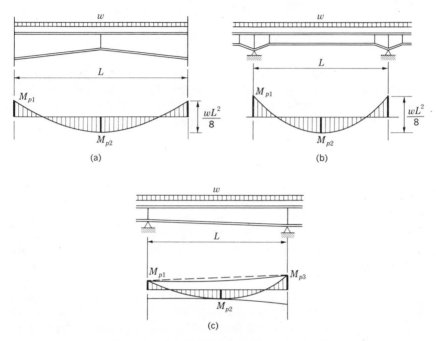

Fig. 7.37   Tapered beams by plastic design.

7.11 DESIGN OF DETAILS

Details that must be considered for beams are connections between components of built-up sections, flange or web splices, rivet or bolt holes, stiffeners, connection between beams and their neighboring members, and bearing.   The details are treated in respective chapters throughout this book. The provisions on the effect of holes on the neutral axis location and on the length of cover plates are treated here.

If the flanges of a beam have holes for rivets or bolts, the net area of a cross section through a group of holes is less than the gross cross-sectional area. When the holes are symmetrical with respect to the centroidal neutral axis of the gross section, no shift of this axis is caused by the holes (Fig. 7.38b). But if the holes are not symmetrical as described above, the neutral axis is

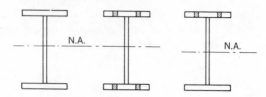

Fig. 7.38 Cross-sectional area and neutral axis.

shifted away from the holes (Fig. 7.38c). Then, hypothetically, the line of zero flexural stress jumps abruptly up and down along the length of a beam while the stresses at the flanges change accordingly. These are not likely to happen in reality; rather, smooth transitions take place. Recognizing this fact as well as that of stress concentration, different provisions are specified.

AASHO and AREA require that rolled beams be proportioned by the moment-of-inertia method. In calculating the net moment of inertia, the gravity axis of the gross section is used and the moment of inertia of all holes on each side of the axis is deducted. Tensile and compressive stresses are computed, using the moment of inertia of net and gross sections respectively. This provision assumes that holes on the compressive side of the gravity axis have less effect on flexure if these holes are filled by rivets. AISC also specifies proportioning by moment of inertia. However, gross sections are used and holes up to 15 per cent of the gross flange area can be neglected in computation. Based on test observations that the stress distribution around holes is the same whether they are filled with a rivet (or bolt) or not, no distinction is made by AISC for holes in the tension or compression flange. These specifications are applied in Example 7.22.

Cover plates for beams may be welded, bolted, or riveted to the flanges. The computation of moment of inertia for the evaluation of stresses should follow the provisions above. In any case, the capacity provided for must be higher than the strength required of the beam. Theoretically, a cover plate can be cut off at points where these two capacities are identical, the "theoretical cut-off points" (Fig. 7.39). However, the stress transfer from a flange to a cover plate is gradual; adequate extension of the cover plate from a theoretical cut-off point into the region of less stressed area (lower moment) is necessary for the utilization of the full strength of the cover plate.

The general method of determination of the necessary extension is to find out the minimum length which is required for a cover plate to develop its full strength.[7.26] This length is sometimes called the terminating length, and is different according to the method of attachment of the cover plates. Based on results of investigations, specifications are formulated. AASHO requires that the length of any cover plate added to a rolled beam shall be not less than twice the beam depth plus 3 ft. With the requirements for rivet hole spacing,

EXAMPLE 7.22

PROBLEM:
The most-stressed sections of the beam in Example 7.11 have two 13/16 in. holes for 3/4 in. bolts in both flanges. Are these sections adequate, considering the holes ?

SOLUTION:

AISC   W 12 x 27:   Gross section $t = 0.40$ in., $b = 6.50$ in., $A_f = 6.5 \times 0.40 = 2.6$ in.^{2},

$$I = 204.1 \text{ in.}^4, \quad d = 11.95 \text{ in.}$$

Area of holes in a flange $= 2 \times \dfrac{13}{16} \times 0.40 = 0.65$ in.^{2}

$A_f \times 15\% = 0.39$ in.$^2 < 0.65$ in.^{2}, deduction of holes must be made.

$0.65 - 0.39 = 0.26$ in.^{2}

Net moment of inertia $= 204.1 - 2 \times 0.26 \left[\dfrac{1}{2} (11.95 - 0.4) \right]^2 = 186.7$ in.^{4}

$$f_b = \frac{Mc}{I} = \frac{864 \times 11.95/2}{186.7} = 27.7 \text{ ksi} < 27.8 \text{ ksi} = F_b \qquad\qquad \text{O.K.}$$

AASHO (AREA)   W 14 x 30: Gross section $I = 289.5$ in.^{4}, $S = 41.8$ in.^{3},

$$d = 13.86 \text{ in.}, \quad t = 0.383 \text{ in.}$$

Area of holes in a flange $= 2 \times \dfrac{13}{16} \times 0.383 = 0.623$ in.^{2}

Net moment of inertia $= 289.5 - 2 \times 0.623 \left[\dfrac{1}{2} (13.86 - 0.383) \right]^2 = 232.8$ in.^{4}

$$f_b = \frac{Mc}{I} = \frac{864 \times 13.86/2}{232.8} = 25.7 \text{ ksi} > 24.6 \text{ ksi} = F_b$$

The section is not adequate.

the actual extension length for riveted or bolted cover plates can be decided. AISC limits the maximum cover plate area of riveted beams or girders to not more than 70 per cent of the total flange area. In addition, according to results of research on welded partial length cover plates,[7.26] special provisions are given for this type of beams and girders. AISC requires that the terminating length shall be 1, 1½, and 2 times the cover plate width if there are continuous welds of not less than ¾ of the plate thickness welded all along the cover plate edges, continuous welds less than ¾ of the plate thickness, and

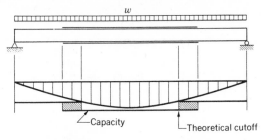

Fig. 7.39 Length of cover plates.

EXAMPLE 7.23

PROBLEM:

At the theoretical cut-off point of the cover plates on a W 18 x 60 beam, the moment is 2370 kip-in. with a moment gradient of 15 kip-in./in., Determine the terminating length of the cover plates, (I) by the AISC Specification for $\frac{1}{2}$ x 6 cover plates with 1/4 in. fillet welds along the edges but not at the ends and (2) by the AASHO Specifications for $\frac{3}{8}$ x 8 cover plates with 7/8 in. rivets at 3 in. pitch near the theoretical cut-off point.

SOLUTION:

AISC

Sectional properties:

without cover plates	with cover plates
I = 984.0 in.^{4}	I = 1511 in.^{4}
S = 107.8 in.^{3}	cover plate area = $6 \times \frac{1}{2}$ = 3 in.^{2}
d = 18.25 in.	N.A. to the centroid of cover
	plates = $\frac{1}{2}\left(18.25 + \frac{1}{2}\right)$ = 9.375 in.

If fully developed, the force in a cover plate at the theoretical cut-off point is

$$\frac{Mc}{I} A_f = \frac{2370 \times 9.375}{1511} \times 3 = 44.2 \text{ kip}$$

This force must be taken by the welds in the terminating length. Since no end welds are used, the minimum terminating length is two times the cover plate width, 2 x 6 = 12 in.
The capacity of the 1/4 in. welds on both sides is

$$2\left[\left(\frac{1}{4} \times 0.707\right) \times 13.6 \times 12\right] = 57.6 \text{ kips} > 44.2 \text{ kip} \quad \text{O.K.}$$

The terminating length is 12 in.

AASHO

Sectional properties with cover plates:

I = 1504 in.^{4}

Cover plate area = $\frac{3}{8} \times 8 = 3$ in.^{2}

N.A. to the centroid of cover plates = $\frac{1}{2}\left(18.25 + \frac{3}{8}\right)$ = 9.31 in.

The force in a cover plate at the cut-off point is:

$$\frac{2370 \times 9.31}{1504} \times 3 = 44.0 \text{ kip}$$

Number of rivets needed for this force is:

$$\frac{44.0}{\frac{\pi}{4}\left(\frac{7}{8}\right)^2 \times 13.5} = 5.41 \text{ rivets}$$

Six rivets are sufficient. With two rivets in a gage line and 3 in. pitch, a 9 in. long terminating length is adequate. (However, the minimum length of a cover plate is $\left(2d + 3\right)$ ft. or

$$\left(2d + 3\right) \times 12 = \left(2 \times 1.5 + 3\right) \times 12 = 72 \text{ in.}$$

If the over-all length of a cover plate, including the terminating length at both ends, is less than 72 in., the ends of the cover plate must be extended so that the over-all length is 72 in.).

continuous welds along the edges only but not at the end of the cover plate, respectively. AASHO specifies the terminating length (terminal distance) be 2 or $1\frac{1}{2}$ times the cover plate width, without or with ends welds. Example 7.23 illustrates the application of these provisions as well as that of the AASHO requirement on riveted cover plates.

It can be seen from the example that the procedure for determining the cover plate length is relatively simple. For welded beams a smooth transition of welds and cover plate configuration are important in the consideration of fatigue strength. For these problems, Chapters 16 and 19 give detailed discussions.

PROBLEMS

7.1. A simply supported building beam of 25-ft span has a profile W 12 × 27 and is made of structural steel ASTM A36. It is fully supported in the lateral direction. What is the allowable load which acts at the mid-span and causes flexure of the beam?

7.2. A small highway bridge has three wide-flange beams as its main member to span a 30-ft crossing. Suppose the load on the central beam is in the plane of its web and is represented by a uniform load w of 2.4 kips per linear foot and a concentrated load of 8 kips; select a proper section for this beam. The compression flange of the beam is restrained from lateral movement by the bridge deck.

7.3. For resisting the wind forces on the side of a building, equal leg angles are used as supports of a 20-ft wall panel. If the equivalent wind load on an angle is 0.12 kip/ft, acting in the plane of a leg, what size angle is required? Stability of an angle is ensured by the wall.

7.4. A beam of 18-ft span overhangs 3 ft at one end to support a heavy load of 54 kips. A uniform load of 1 kip/ft acts on the span. Lateral supports are provided for the entire beam length. A minimum yield point of 50 ksi has been specified for the material. Select an adequate profile.

7.5. If the loading on the side wide-flange beams of the bridge of Problem 7.2 is represented by a uniform load of 1.5 kips per linear foot, parallel to, but 6 in. away from the center line of the web, what section is required? Assume that no instability occurs and the beams are fixed with respect to torsion at their ends. Use the same material as in Problem 7.2.

7.6. What is the allowable load for the beam of Problem 7.1 if lateral supports are provided only at the third points?

7.7. What size of channel is proper to replace the angle of Problem 7.3 if the load acts in the plane of the channel web and the ends of the beam are fully restrained against torsion?

7.8. A wide-flange purlin rests on the top chords of roof trusses which are inclined 30°. It is simply supported with respect to bending, fixed at ends against twisting, and restrained from lateral buckling by the roofing. The distance between trusses is 20 ft. Vertical loads of 0.2 kip/ft are assumed to apply at the mid-width of the top flange of the purlin. Select an adequate section.

7.9. A 20-ft-long crane-runway beam supports two traveling wheel loads 6 ft apart and 12 kips each. Neglecting the longitudinal force of the crane, select a wide-flange section according to the AISC Specification. Assume that no lateral instability occurs.

7.10. Select a section for a two-span continuous beam for a building. Spans are 30 ft long, supporting concentrated loads of 15 kips at third points of each span. Lateral movement of the beam is prevented at mid-spans and supports.

7.11. Select a section for the beam of Problem 7.10 if lateral supports are provided at 6-ft intervals.

7.12. Select a section for the beam of Problem 7.10 by the plastic design procedure. What are the limiting lateral supporting distances?

7.13. If lateral supports are spaced at 6-ft intervals between the supports, what section is needed for a three-span continuous beam subjected to a uniform load of 3 kips/ft? The center span is 36 ft long; the end spans 24 ft.

7.14. If cover plates are permitted and specified for the beam of Problem 7.13, design the member.

7.15. Design the beam of Problem 7.10 if clearance requirements limit the depth of the beam to 22 in.

8

PLATE GIRDERS

8.1 INTRODUCTION

A plate girder is a flexural member fabricated from a collection of plates and employed to carry loads which cannot be carried economically by rolled beams. The use of a plate girder gives the designer the advantage of selecting component parts of convenient size, but this is offset somewhat by the disadvantage (when compared with a rolled beam) of having to connect the flanges to the web with welds or rivets.

Plate girders are commonly used in highway and railroad bridges and in building frames where special loading conditions prevail or long spans are required. A deck-type plate girder bridge is shown in the foreground of Fig. 4.2a while Figs. 8.1, 8.2, and 8.3 illustrate the use of plate girders in buildings. The girders in Fig. 8.1 are needed to provide a clear span of about 150 ft over a turbogenerator unit. The double cantilevered girders in Fig. 8.2 are 365 ft long and provide clear space for the servicing of jet aircraft. The use of plate girders to span over auditoriums or other large areas in a building is illustrated in Fig. 8.3.

Most specifications state that a plate girder is to be proportioned by the moment-of-inertia method, which simply means that the bending stress given by the familiar equation $f_b = Mc/I$ should not exceed the allowable value. In the preliminary design, it is often convenient to use an approximate method. For this purpose, it may be assumed that the applied bending moment is resisted by the flanges alone, so that the required flange area is approximated by

$$A_f = \frac{M}{F_b h} \tag{8.1}$$

Fig. 8.1 Erection of Philadelphia Electric Company's Eddystone Power Plant (Courtesy of Bethlehem Steel Corp.)

Fig. 8.2 United Air Lines jet hangar, San Francisco. (Courtesy of United Air Lines, Inc.)

where M is the externally applied bending moment, h is a specified or assumed depth, and F_b may be taken as the allowable compressive stress without reduction for lateral buckling. In the final design, the girder section selected by this method must be checked by the $F_b = Mc/I$ formula, where F_b is the allowable compressive stress reduced for lateral buckling and I is the moment of inertia of the entire girder section.

The proportions of a plate girder often are limited by a stability consideration, that is, the stress causing web buckling is the criterion for limit of usefulness. Various arrangements of stiffeners are used to reinforce the web and thus raise the buckling stress. The 1969 AASHO Specifications[1.23] and the 1969 AREA Specifications[1.25] are examples of specifications which provide for plate girder design based on buckling strength.

In many cases where plate girder design is based on buckling strength, the existence of post-buckling strength is tacitly recognized by the use of lower factors of safety against web buckling. In 1961 design recommendations were introduced for plate girders used in buildings, based on the maximum load-carrying capacity of a girder.[8.1] Substantiated by extensive theoretical and experimental research,[7.18,8.2,8.3,8.4] these recommendations evaluated satisfactorily the post-buckling strength and are the basis for the provisions pertaining to plate girders in the 1969 AISC Specification.[1.21]

The load-carrying capacity approach to plate girder design is recommended for two reasons. First, this approach will usually result in a more economical use of material than that based on buckling strength. More important, design based on load-carrying capacity is more realistic since it leads to a closer approximation of the actual strength of a girder. However, this

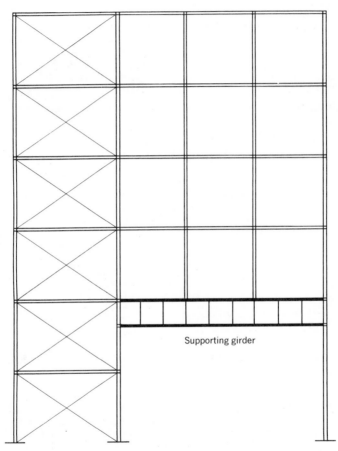

Supporting girder

Fig. 8.3 Use of supporting girders over an auditorium.

approach has not yet been applied satisfactorily to bridge girders and other situations where repeated loads are encountered.

The first portion of this chapter is devoted to the development of design rules for plate girder bridge members, using web buckling as a design criterion. The AASHO Specifications will be referred to throughout this portion since the provisions of the AREA Specifications are based on the same considerations and therefore are very similar. The application of the design rules is illustrated in an example bridge girder design problem. In the second part of the chapter, design rules are formulated from a load-carrying-capacity viewpoint and illustrated in a building girder design problem.

8.2 BUCKLING STRENGTH OF PLATE GIRDERS

A general introduction to the phenomenon of plate buckling is presented in Art. 17.3. The present discussion is limited to buckling of a plate girder web.

As an example, consider a section of a plate girder subjected to pure bending, as shown in Fig. 8.4a. At any section, the stress distribution varies linearly from a maximum compressive stress $-f_b$ at the top fiber to a maximum tensile stress $+f_b$ at the bottom fiber (Fig. 8.4b). Let the lateral deflection of the web or deflection in the z-direction (Fig. 8.4c) at the center of the section

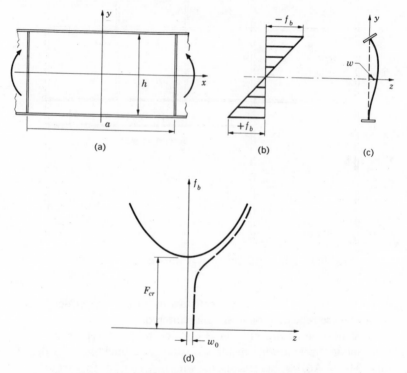

(a)   (b)   (c)

(d)

Fig. 8.4 Plate buckling due to pure bending.

or origin of the *xyz* coordinate system be designated by *w*. The plot in Fig. 8.4d then represents schematically the behavior of the web as the moment on the section is gradually increased. If the web is initially perfectly plane, theoretically no lateral web deflection will occur until the critical buckling stress F_{cr} is reached. At this point, the web is in a condition of unstable equilibrium. Any further increase in the applied moment will cause the web to buckle according to one of the solid curves of Fig. 8.4d, assuming a deflected shape similar to that shown in Fig. 8.4c. However, if the web has some initial deflection w_0 at the center, the behavior will be that shown by the dashed line in Fig. 8.4d; no sudden buckling will occur but deflection will increase gradually as the moment increases. Although in all practical cases some initial out-of-straightness will exist, the theoretical case of an initially perfectly plane web is used as a basis for design rules.

An expression for the critical buckling stress F_{cr} is derived in Art. 17.3 and is rewritten here as

$$F_{cr} = \frac{k\pi^2 E}{12(1 - \mu^2)} \left(\frac{t}{h}\right)^2 \tag{8.2}$$

where k is the buckling coefficient, E is the modulus of elasticity (for steel, $E = 29,000$ ksi), μ is Poisson's ratio (for steel, $\mu = 0.3$), t is the plate thickness, and h is the web depth or clear distance between flanges. It should be noted that F_{cr} can be either a critical compressive stress due to bending or a critical shear stress, depending on the loading condition. The coefficient k is a function of the plate geometry, loading condition, and the edge conditions. The conservative assumption that the web is simply supported at all edges usually is made.

The basic factor of safety used in the AASHO Specifications ranges from 1.80 to 1.85. Although the design rules developed in the following sections are based on the buckling strength of the girder web or a portion of the stiffened web, the existence of post-buckling strength is recognized by the use of appropriately reduced factors of safety for certain types of web buckling.

1. Design for Bending

Limiting Slenderness Ratios. Equation 8.2 can be solved for the slenderness ratio h/t with $\mu = 0.3$, thus:

$$\frac{h}{t} = 0.951 \sqrt{\frac{kE}{F_{cr}}} \tag{8.3}$$

For a plate subjected to pure bending, the buckling coefficient k cannot be less than 23.9.[7.6] Substituting this value in Eq. 8.3, the web slenderness ratio h/t becomes

$$\frac{h}{t} = 4.65 \sqrt{\frac{E}{F_{cr}}} \tag{8.4}$$

Since a plate girder will not fail when web buckling due to bending occurs, the applied loads can be increased beyond the buckling load (see Art. 8.3). This post-buckling strength is utilized by adopting limiting slenderness ratios which inherently provide for a low factor of safety, F.S. = 1.19. Substituting $F_{cr} = 1.19f_b$ and $E = 29,000$ ksi into Eq. 8.4, the AASHO formula for limiting web slenderness ratio is obtained:

$$\frac{h}{t} = \frac{23,000}{\sqrt{f_b}},$$

where f_b is the calculated maximum compressive bending stress in psi. When the calculated compressive bending stress equals the allowable bending stress, the limiting slenderness ratios are those shown in Table 8.1.

Table 8.1   AASHO Limiting Slenderness Ratios—No Longitudinal Stiffener

Minimum Yield Point (ksi)	Maximum h/t
36	165
42	150
45 and 46	145
50	140
55	133
60	127
65	121
90	105
100	100

Longitudinal Stiffeners. From the point of view of buckling, the most effective type of stiffener for web plates subjected to bending is one oriented parallel to the girder flanges and the longitudinal girder axis; this type of stiffener commonly is referred to as a longitudinal or horizontal stiffener (Fig. 8.5a). It has been shown analytically that the best location for a longitudinal stiffener is at $\frac{1}{5}$ of the web depth from the compression flange, if the web is assumed conservatively to be simply supported at all edges.[8.5] Thus, this location is specified in the AASHO Specifications.

A longitudinal stiffener should have a moment of inertia sufficient to force the formation of a nodal line in the stiffened panel (Fig. 8.5b). The determination of this minimum moment of inertia is an involved process. An approximate formula is proposed in Ref. 8.6 for the case of a simply supported web plate subjected to pure bending with the stiffener at the $\frac{1}{5}$-depth position. Using the conservative assumption that the ratio of stiffener area to web area is 1/20,[8.7] the AASHO minimum moment-of-inertia requirement has

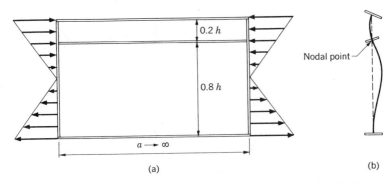

Fig. 8.5 Longitudinally stiffened plate subjected to pure bending.

been derived from the formula of Ref. 8.6.   The AASHO equation is

$$I_s \geqslant t^3 h \left[2.4 \left(\frac{a}{h} \right)^2 - 0.13 \right] \tag{8.5}$$

where a is the distance between transverse stiffeners* (discussed below under "Design for Shear") and I_s is the moment of inertia which is computed as indicated below.   Once the minimum moment of inertia has been supplied, the buckling strength of the reinforced panel cannot be increased further by an increase in stiffener rigidity.

In Fig. 8.6, the ratio of stiffener moment of inertia to web stiffness is plotted against panel aspect ratio a/h for the British,[8.8] German,[8.9] and AASHO Specifications.   From these curves it is clear that for $a/h < 1$, which is the only range allowed by AASHO, both the British and the German Specifications are more conservative than AASHO.

It is of practical interest to compare the economies of two-sided and one-sided stiffeners with reference to their moments of inertia.   From Fig. 8.7a, neglecting the thickness of the web, the moment of inertia of the two-sided arrangement is $I_s = (2b)^3 t/12 = 2b^3 t/3$.   From Fig. 8.7b, the moment of inertia of the one-sided stiffener about the stiffener-web interface is $I_s' = (b')^3 t'/3$.   Letting $t' = t$ and setting $I_s' = I_s$, b' is found to be $1.26b$.   Thus, the area of the two-sided arrangement is $2bt$ and that of the one-sided arrangement is $1.26bt$, showing that the use of a one-sided stiffener requires only 63 per cent of the area required for a two-sided stiffener when only stiffener moment of inertia is required.

---

* Although stiffener spacing is defined as the distance between transverse stiffeners in both the AASHO and AISC specifications, the center-to-center distance is used in this chapter for simplicity and is elsewhere referred to as "stiffener spacing."

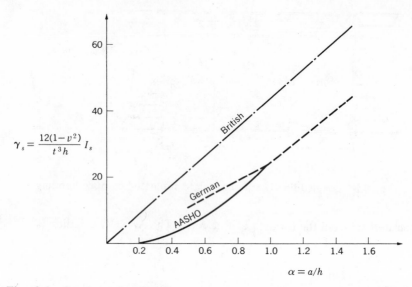

Fig. 8.6 Comparison of stiffness requirements for longitudinal stiffeners.

The effect of flange restraint on the optimum stiffener location and minimum stiffness requirement can be assessed from an analytical study described in Ref. 8.10. If it is assumed that the unloaded edges are restrained from both lateral deflection and rotation by the flanges, the most effective position for a single longitudinal stiffener is 0.22 times the web depth from the compression flange, a value that is not much different than the $\frac{1}{5}$ position obtained assuming simply supported edges. However, the required moment of inertia of the stiffener is reduced appreciably if full fixity at the flanges can be assumed.

Limiting Slenderness Ratios with a Longitudinal Stiffener. The calculation of the maximum allowable web slenderness ratio h/t when one longitudinal stiffener is placed at the $\frac{1}{5}$-depth position is similar to that when no

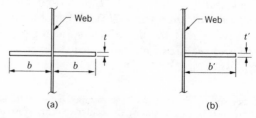

Fig. 8.7 (a) Two-sided and (b) one-sided stiffener arrangement.

longitudinal stiffener is used. The minimum buckling coefficient for the stiffened plate is $k = 129$.[8.10] Using this value in Eq. 8.3 results in

$$\frac{h}{t} = 10.81 \sqrt{\frac{E}{F_{cr}}} \tag{8.6}$$

The smaller panel of the stiffened web (top panel of Fig. 8.5a) is the critical one. Since the longitudinal stiffener forms one of the boundaries of this panel, the selection of an appropriate factor of safety against buckling will depend in part on the behavior of the stiffener after the panel has buckled. The minimum moment of inertia of the stiffener given in the Inequality 8.5 is only sufficient to ensure that a nodal line will form along the stiffener at the panel buckling load; thus, the post-buckling strength has not been defined. However, the stiffened web will develop some post-buckling strength; therefore, web slenderness ratios which provide for a factor of safety of about 1.60 against web buckling are specified in the AASHO Specifications. Substitution of $F_{cr} = 1.60 f_b$ and $E = 29,000$ ksi in Eq. 8.6 yields the AASHO formula for limiting slenderness ratio for plate girders with one longitudinal stiffener,

$$\frac{h}{t} = \frac{46,000}{\sqrt{f_b}} \tag{8.7}$$

When $f_b = F_b$, the limiting ratios are those listed in Table 8.2.

Table 8.2 AASHO Limiting Slenderness Ratios—One Longitudinal Stiffener

Minimum Yield Point (ksi)	Maximum h/t
36	330
42	300
45 and 46	290
50	280
55	266
60	253
65	242
90	210
100	200

Compression Flange Buckling. For a girder section subjected to bending, the stability of the compression flange must be considered as well as the stability of the web. Lateral buckling of the compression flange has been treated in Art. 7.6 (see Fig. 7.18); the AASHO formula, Art. [1.7.1], for allowable compressive stress in psi in the extreme fibers of plate girders having

$F_y = 36$ ksi, is repeated here for completeness:

$$F_b \leqslant 20{,}000 - 7.5 \left(\frac{L_b}{b}\right)^2 \tag{8.8}$$

where b is the flange width in inches, L_b is the length of the flange between points of lateral support in inches and the ratio L_b/b is limited to a value of 36 or less. If the compression flange is laterally supported over its full length, the allowable compressive stress in the extreme fibers is 20,000 psi for A36 steel.

Torsional buckling of the compression flange is essentially a local buckling problem, and as such is treated in Art. 17.4 (see Fig. 17.2). Any possibility of the occurrence of torsional buckling is eliminated by specifying a maximum value of the ratio of the width of the outstanding leg (or $\frac{1}{2}$ flange width) to its thickness.

2. Design for Shear

Web Buckling Due to Shear. Equation 8.2 can be used also to find the average shear stress at which, theoretically, web buckling will occur. The average shear stress is defined simply as the externally applied shear force divided by the web area ht. For simply supported plates subjected to shear (Fig. 8.8), the buckling coefficient is[7.6]

$$\begin{aligned} k &= 4.00 + \frac{5.34}{(a/h)^2} \qquad \text{(for } a/h \leqslant 1) \\[2mm] k &= 5.34 + \frac{4.00}{(a/h)^2} \qquad \text{(for } a/h \geqslant 1) \end{aligned} \tag{8.9}$$

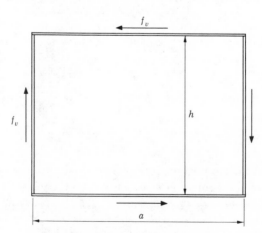

Fig. 8.8 Plate subjected to shear.

where a/h is the aspect ratio or ratio of plate length to depth. In Fig. 8.9, where the shear buckling coefficient has been plotted against the panel aspect ratio a/h, the influence of fixity along the flanges on the k-values can be assessed. For an aspect ratio of 0.5 the k-value differs very little for varying flange restraint. With $a/h = 3.0$, however, the k-value corresponding to a clamped boundary condition is over 60 per cent greater than that corresponding to a simply supported condition. Although some flange restraint will be developed in all girders, usually it is assumed conservatively that the web is simply supported along the flanges.

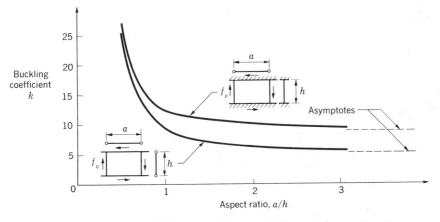

Fig. 8.9 Variation of shear buckling coefficient with aspect ratio.

From Eq. 8.9, $k = 5.34$ for an infinitely long plate ($a/h = \infty$). When this value is used in Eq. 17.22, which is a modified form of Eq. 8.3 and which takes into account imperfections and residual stresses, the maximum slenderness ratio h/t for an unstiffened web can be written as

$$\frac{h}{t} = \frac{8320}{\sqrt{f_v(\text{F.S.})}} \tag{8.10}$$

This equation provides for buckling to occur at a stress of $f_v(\text{F.S.})$, where f_v is the calculated average shear stress in psi. By assuming a value of F.S. = 1.23, the AASHO formula for the maximum web slenderness ratio which can be used for plate girders without transverse stiffeners is obtained from Eq. 8.10,

$$\frac{h}{t} = \frac{7500}{\sqrt{f_v}} \tag{8.11}$$

When the calculated shear stress equals the allowable shear stress, the maximum ratios are those shown in Table 8.3.

Table 8.3 Maximum Web Slenderness Ratios—No Transverse Stiffeners

Minimum Yield Point (ksi)	Maximum h/t
36	68
42	64
45 and 46	60
50	58
55	56
60	53
65	51
90	43
100	41

Transverse Stiffeners. For any given web depth and thickness, the only possible way of increasing the critical buckling stress is by increasing the buckling coefficient k. This is accomplished by subdividing the web into panels through the use of stiffeners oriented perpendicular to the girder flanges, commonly referred to as vertical or transverse stiffeners (Fig. 8.10). The plate length a then equals the stiffener spacing.

The usual design situation is that the web depth and thickness and the externally applied shear force are known; hence the average shear stress $f_v = V/ht$ can be computed readily. The stiffener spacing a then remains to be determined. Rewriting Eq. 8.2,

$$F_{cr} = (\text{F.S.})f_v = \frac{k\pi^2 E}{12(1 - \mu^2)} \left(\frac{a}{h}\right)^2 \left(\frac{t}{a}\right)^2 \tag{8.12}$$

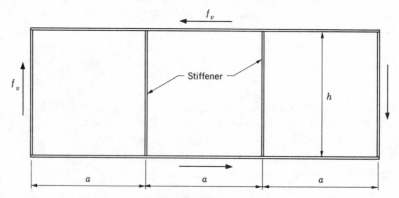

Fig. 8.10 Transversely stiffened plate subjected to shear.

where F.S. is the safety factor. Solving Eq. 8.12 for the required stiffener spacing,

$$a = \left(\frac{a}{h}\right)\sqrt{\frac{k\pi^2 E}{12(1-\mu^2)\text{F.S.}}}\left(\frac{t}{\sqrt{f_v}}\right) = (\text{coefficient})\left(\frac{t}{\sqrt{f_v}}\right) \qquad (8.13)$$

If a safety factor F.S. = 1.5 is assumed, the coefficient of Eq. 8.13 can be evaluated for various values of aspect ratio a/h by utilizing Eq. 8.9. The results of such a calculation are summarized in Table 8.4.

It can be seen from Table 8.4 that the value of the coefficient in the stiffener spacing equation (Eq. 8.13) does not change a great deal for small variations in aspect ratio. In many specifications, this situation is utilized to obtain a simple design formula. In the AASHO Specifications, the stiffener spacing formula is

$$a = \frac{11,000t}{\sqrt{f_v}} \qquad (8.14)$$

The factor of safety in this formula varies from F.S. = 1.37 for $a/h = 0.5$ to F.S. = 2.02 for $a/h = 1.0$. The use of these factors of safety can be justified by the fact that a girder will not fail when the web plate buckles; due to the framing elements of a web panel, the flanges, and transverse stiffeners, a considerable post-buckling strength exists. A method of evaluating this post-buckling strength is discussed in Art. 8.3.

The AASHO web-stiffening requirements are summarized in Fig. 8.11, which is a plot of the average calculated shear stress f_v, against web slenderness ratio h/t for a steel with a yield point of 36 ksi. For h/t between 0 and 68, no stiffeners are required. For the center area where h/t is between 68 and 165, transverse stiffeners are required with a spacing such that the average shear stress is less than the value given by the applicable a/h curve. At the extreme limit of this area (that is, at $h/t = 165$) it is seen that in order to have the allowable shear stress at its maximum level of $F_v = 12$ ksi, stiffeners must be spaced at a distance less than $\frac{2}{3}$ of the web depth. Finally, for the large rectangle at the right in Fig. 8.11, a longitudinal stiffener (at the $\frac{1}{5}$-depth position) as well as transverse stiffeners are required. When $h/t = 330$, the transverse stiffeners would have to be spaced less than $\frac{1}{3}$ of the web depth apart for the maximum allowable shear stress to apply.

A transverse stiffener is similar to a longitudinal stiffener in that a minimum moment of inertia is required to ensure that a nodal line will be formed at the location of the stiffener. The following approximate formula has been proposed for this purpose:[7.6]

$$I = \frac{t^3 a_0 J}{10.92} \qquad (8.15)$$

where $J = 28(h/a)^2 - 20$, a_0 is the actual stiffener spacing, and a is the

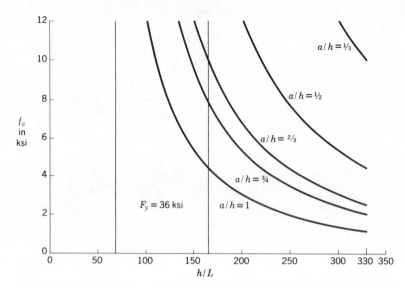

Fig. 8.11 AASHO web-stiffening requirements.

required stiffener spacing. Equation 8.15 has been adopted by AASHO with a slightly modified value of $J = 25(h/a)^2 - 20$, but not less than 5. In addition to this minimum moment of inertia requirement, the AASHO Specifications require that the width of a stiffener plate or outstanding leg of a stiffener angle shall not exceed 16 times its thickness to prevent local buckling (see Art. 17.4) and shall not be less than 2 in. plus $\frac{1}{30}$ of the depth of the girder. The latter provision is an empirical one based on experience.

The ratio of required transverse stiffener moment of inertia to web stiffness has been plotted for various specifications in Fig. 8.12, which indicates that the AASHO requirement is more liberal than that of the British specifications, but more conservative than the German specifications.

Since the size of a transverse stiffener is based on a required moment of inertia, a one-sided stiffener will provide the most economical use of material, as previously discussed above for longitudinal stiffeners. If a one-sided transverse stiffener is used, however, the AASHO Specification requires that it be connected to the outstanding leg of the compression flange. Additional detailing considerations for transverse stiffeners are discussed in Art. 8.3.

The use of the AASHO Specifications for plate girders is illustrated in Example 8.1. The section selected in this example is close to the minimum weight for the depth of 120 in. but is not necessarily the most economical since the close spacing of transverse stiffeners may necessitate too much labor for welding. If this were found to be the case in an actual design situation,

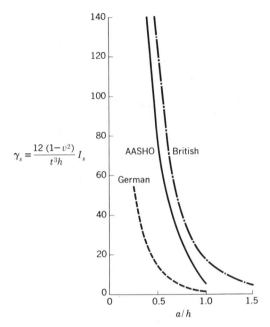

$$\gamma_s = \frac{12\,(1-v^2)}{t^3 h}\,I_s$$

AASHO   British

German

Fig. 8.12 Comparison of stiffness requirements for transverse stiffeners.

the longitudinal stiffener could be eliminated, requiring twice the web thickness but only about 60 per cent the number of transverse stiffeners. Also, other girder depths should be considered in an attempt to arrive at the most economical design.

Table 8.4   Stiffener Spacing Coefficients (Eq. 8.13)

a/h	k (from Eq. 8.9)	Coefficient for Eq. 8.13
0.5	25.36	10,526
0.6	18.83	10,883
0.7	14.90	11,295
0.8	12.34	11,747
0.9	10.59	12,243
1.0	9.34	12,775

8.3 LOAD-CARRYING CAPACITY OF PLATE GIRDERS

In the previous article, different factors of safety have been used for different cases of web buckling in order to take advantage of post-buckling strength.

EXAMPLE 8.1

PROBLEM:

Design the girders for the three span, continuous, deck–type bridge discussed in Ex. 4.4. Use the maximum moment and shear diagrams shown below. A 36 steel is to be used with the AASHO specifications.

SOLUTION:

Select a section to resist the maximum center–span moment of 4540 kip-ft. and use cover plates over the interior supports to resist the 8500 kip·ft. maximum moment. Preliminary trials indicate that a web depth of 10'-0" will provide a suitable section.

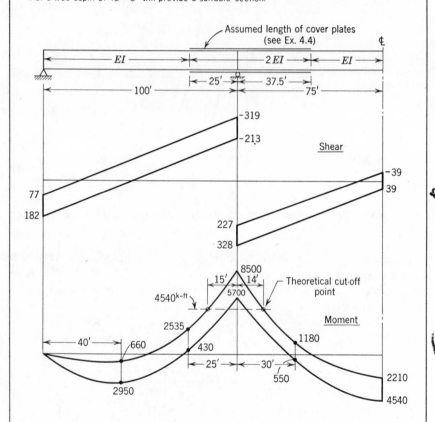

Section: Try 3/8" web (with one longitudinal stiffener)

check $\frac{h}{t} = \frac{120}{3/8} = 320 < 330$ O.K.

using Eq. 8.1, $A_f = \frac{4,540,000 \times 12}{20,000 \times 120} = 22.7$ in.^{2}

keeping in mind that the web will supply some bending resistance, try 1" x 18" flanges which supply A = 18.0 in.^{2}

check $\frac{b}{t}$: allow. $\frac{b}{t} = 23$ if $f_b = 20,000$ psi.   O.K.

Section properties: $I = 2(18)(60.5)^2 + \frac{1}{12}(0.375)(120)^3 = 186,000$ in.^{4}

$S = \frac{186,000}{61} = 3050$ in.^{3}

EXAMPLE 8.1 (Continued)

Bending stress: At the middle of the center span, $f_b = \dfrac{M}{S} = \dfrac{4,540,000 \times 12}{3050} = 17,860 \text{ psi.}$

Lateral support for the compression flange is available at 25 ft intervals,

$L_b = 25 \times 12 = 300$ in. in Eq. 8.6

$L_b/b = \dfrac{300}{18} = 16.7 < 36$ O.K.

$F_b = 20,000 - 7.5 \left(\dfrac{L_b}{b}\right)^2 = 20,000 - 7.5(16.7)^2$

$\qquad = 17,910 \text{ psi} > 17,860$ O.K.

Cover plates: Over interior supports, $M = 8,500$ kip-ft, requiring a moment of inertia

of $I = \dfrac{Mc}{F_b} = \dfrac{8,500,000 \times 12 \times 62}{17,910} = 353,000 \text{ in.}^4$, or 353,000 minus

186,000 = 167,000 in.^{4} to be supplied by the cover plates. Thus the required

area of one cover plate is $A_{cp} = \dfrac{167,000}{(61.5)^2 \times 2} = 22.1 \text{ in.}^2$.

Try 1"x22" cover plates, $I_{total} = 186,000 + 2(22)(61.5)^2 = 352,000 \text{ in.}^4$

$S = \dfrac{352,000}{62} = 5680 \text{ in.}^3$

$f_b = \dfrac{8,500,000 \times 12}{5680} = 17,960 \text{ psi}$ say O.K.

Cut-off points: The theoretical cut-off points (from the moment diagram) are 14 ft from
the support for the interior span and 15 ft from the support for the end
spans. Assuming that 3.5 ft will be required to develop the cover plate
strength, the cut-off points would be 17.5 ft and 18.5 ft from the support
for the interior and exterior spans, respectively. However, it can be shown
that the factor of safety against fatigue at these cut-off points is quite
low. The determination of acceptable locations for the cut-off points
should be based on the AASHO provisions for fatigue stresses and the
considerations presented in Chapter 16 and is left as an exercise for the
reader.

Longitudinal stiffener: min. $I_s = t^3 h \left[2.4 \left(\dfrac{a}{h}\right)^2 - 0.13\right]$

Assuming max. $\dfrac{a}{h} = 1.0$, $I_s = \geq \left(\dfrac{3}{8}\right)^2 (120) \left[2.4(1)^2 - 0.13\right] = 14.36 \text{ in}^3$

Try a one-sided stiffener 3/8"x5"

$I_s = \dfrac{1}{3}\left(\dfrac{3}{8}\right)(5)^3 = 15.62 \text{ in.}^4 > 14.36$ O.K.

Width-thickness ratio $= \dfrac{5}{3/8} = 13.3$

Allow. $\dfrac{b'}{t} = \dfrac{2250}{\sqrt{f_b}} = \dfrac{2250}{\sqrt{17960}} = 16.8 > 13.3$ O.K.

Bearing stiffeners: At interior support, $V_{max.} = 647^k$. Try 4 plates $1"x8\frac{1}{2}"$ as shown below.

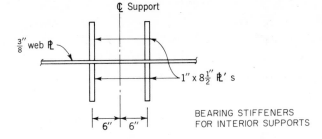

BEARING STIFFENERS
FOR INTERIOR SUPPORTS

Check column strength: $I \approx 2\left[\dfrac{1}{12}(17)^3(1)\right] = 819 \text{ in.}^4$

$A = 4(8\tfrac{1}{2})(1) + \dfrac{3}{8}\left(12 + 18 \times \dfrac{3}{8}\right) = 41.0 \text{ in}^2$

EXAMPLE 8.1 (Continued)

$$r = \sqrt{\frac{I}{A}} = \sqrt{\frac{819}{41.0}} = 4.47 \text{ in., } L = 120 \text{ in., } \frac{L}{r} = \frac{120}{4.47} = 26.8$$

$$F_a = 16,000 - 0.3 \left(\frac{L}{r}\right)^2 = 16,000 - 0.3(26.8)^2 = 15,780 \text{ psi}$$

$$f_a = \frac{647,000}{41.0} = 15,780 \text{ psi} \hspace{3cm} \underline{O.K.}$$

The bearing stress and width-thickness ratio are obviously less than the allowable values.

At end supports, $V_{max.} = 182^k$. Try 2 ℄'s 3/4"x8"

Width-thickness ratio $= \dfrac{8}{3/4} = 10.7$

Allow. $\dfrac{b'}{t} = 12 \sqrt{\dfrac{33,000}{36,000}} = 11.5 > 10.7 \hspace{2cm} \underline{O.K.}$

Check column strength: $I \approx \dfrac{1}{12}(16)^3\left(\dfrac{3}{4}\right) = 256 \text{ in.}^4$

$$A = 2(8)\left(\frac{3}{4}\right) + 18\left(\frac{3}{8}\right)^2 = 14.5 \text{ in.}^2$$

$$r = \sqrt{\frac{256}{14.5}} = 4.20 \text{ in., } L = 120 \text{ in., } \frac{L}{r} = \frac{120}{4.20} = 28.6$$

$$F_a = 16,000 - 0.3(28.6)^2 = 15,760 \text{ psi}$$

$$f_a = \frac{182,000}{14.5} = 12,550 \text{ psi} < 15,760 \hspace{2cm} \underline{O.K.}$$

The bearing stress is obviously less than the allowable value.

Transverse stiffeners: Using the stiffener spacing formula $a = \dfrac{11,000\,t}{\sqrt{f_v}}$, the following table is determined:

a(in.)	$\sqrt{f_v} = \dfrac{11,000\left(\frac{3}{8}\right)}{a}$	f_v (ksi)	$V_{allow.} = f_v A_w$ (k)
48	85.9	7.38	332
54	76.4	5.84	263
60	68.8	4.73	213
66	62.5	3.91	176
72	57.3	3.28	148
78	52.9	2.80	126
84	49.1	2.41	108
90	45.8	2.10	94.5
96	43.0	1.85	83.2
102	40.4	1.63	73.4
108	38.2	1.46	65.7
114	36.2	1.31	59.0
120	34.4	1.18	53.1

Using this table and the maximum shear diagram, the stiffener spacing shown below was selected.

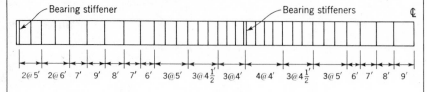

TRANSVERSE STIFFENER SPACING

From Eq. 8.14, required stiffener spacing near interior supports is

$$a = \frac{11,000(3/8)}{\sqrt{328/(3/8 \times 125)}} = 48.3 \text{ in.}$$

EXAMPLE 8.1 (Continued)

$$J = 25\left(\frac{h}{a}\right)^2 - 20 = 25\left(\frac{120}{48.3}\right)^2 - 20 = 134.3$$

Min. $I_s = \dfrac{a \cdot t^3 J}{10.92} = \dfrac{48(3/8)^3(134.3)}{10.92} = 31.1\,\text{in}^4$

Try one-sided stiffeners $1/2'' \times 6''$

$I_s = \dfrac{1}{3}\left(\dfrac{1}{2}\right)(6)^3 = 36.0\,\text{in}^4 > 31.1\,\text{in}^4$ $\underline{O.K.}$

Check width-thickness ratio, $6/\dfrac{1}{2} = 12 < 16$ $\underline{O.K.}$

Check minimum width $= 2'' + \dfrac{h}{30} = 2'' + \dfrac{120}{30} = 6''$ $\underline{O.K.}$

1. Bending Strength

An examination of the stress distribution in the cross section of a girder subjected to pure bending provides insight into the way in which the girder sustains the applied moment. Figure 8.13 shows the distribution when the applied moment approaches a value causing first yielding. The broken line represents the stress distribution obtained from beam theory while the solid line represents that obtained from a typical test girder.[8.4] The compressed

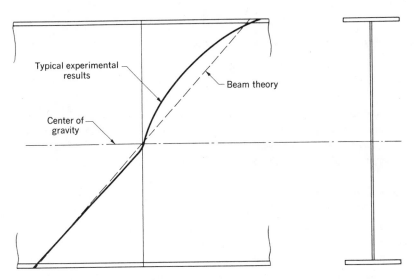

Fig. 8.13 Stress distribution due to bending.

portion of the web does not carry the stress predicted by beam theory, due to a gradual lateral deflection from the initial configuration. The stress in the compression flange therefore exceeds the value obtained using beam theory.

Since some of the compressive force assigned by beam theory to the web is redistributed to the compression flange, it is assumed that the contribution of the compressed portion of the web can be disregarded except for an effective strip along the compression flange.[7.18] The strength of a girder subjected to bending is then dependent on the strength of the compression flange, which may be treated as an isolated column.

The compression flange column may fail either by buckling or by yielding. The three typical buckling modes are shown in Fig. 8.14. Lateral buckling

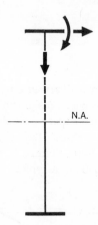

N.A.

Fig. 8.14 Typical buckling modes of compression flange column.

has been discussed previously in Art. 7.6, and Eqs. 7.38, 7.39, and 7.40 are the AISC Specification formulas which are intended to avoid the possibility of failure by this mode. Torsional buckling is essentially a local buckling problem which can be avoided by limiting the ratio of width to thickness of the compression flange, as discussed previously in Art. 8.2. The AISC rule is that the ratio of one-half the flange width to its thickness should not exceed $95.0/\sqrt{F_y}$. (See Art. 17.4.)

Vertical Buckling of the Compression Flange. The curvature of the flanges of a girder subjected to bending causes the web to be subjected to vertical compressive forces. By requiring the maximum value of the vertical compressive forces to be less than the resisting strength of the web as determined by Euler buckling, a limiting web slenderness ratio (web depth to thickness ratio) to prevent vertical buckling of the compression flange into the web can be determined. If the resisting strength of the web is not sufficient to resist vertical buckling of the flange, the result will be similar to that shown in Fig. 8.15, a photograph of a test girder after compression flange failure (Ref. 8.4).

A section with exaggerated curvature is shown in Fig. 8.16. With the aid of the vector diagram at the right of the figure, the vertical component of the

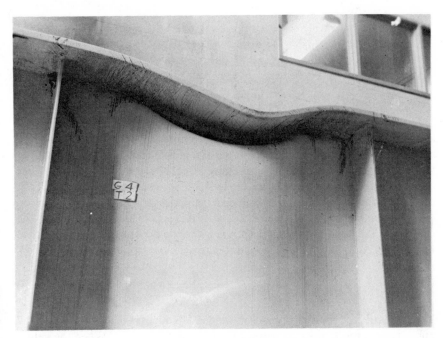

Fig. 8.15 Vertical buckling of compression flange of a test girder.

flange force in a small portion of the girder is evaluated as $A_f f_f \sin \phi$ or $A_f f_f \phi$ for small values of ϕ. Substitution of $\phi = 2\epsilon_f (dx/h)$ results in $2A_f f_f \epsilon_f (dx/h)$ for the flange force. The resisting force of the web is $F_e \cdot t \cdot dx$ where F_e is the Euler buckling stress of the web strip of width dx,

$$F_e = \frac{\pi^2 E}{12(1 - \mu)^2} \left(\frac{t}{h}\right)^2$$

The requirement that the applied force be less than the resisting force results in

$$\frac{h}{t} < \sqrt{\frac{\pi^2 E}{24(1 - \mu^2)} \cdot \frac{A_w}{A_f} \cdot \frac{1}{f_f \epsilon_f}} \tag{8.16}$$

when $A_w = h \cdot t$, the area of the web.

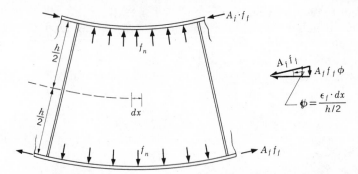

Fig. 8.16 Compressive forces on a girder web due to curvature of flanges.

Two assumptions can be made to simplify Eq. 8.16.[7.18] The first is that the distribution of residual stresses in the compression flange is a linear one as shown in Fig. 11.5 and that the maximum tensile residual stress is equal in magnitude to the maximum compressive residual stress. This means that the flange strain must be equal to $(F_y + F_r)/E$ for every flange fiber to reach the yield point. The second assumption is that for practical plate girders the minimum value of A_w/A_f is 0.5. Using these two assumptions and the value $\mu = 0.3$, and requiring f_f to reach the yield point F_y before vertical buckling occurs, Inequality 8.16 reduces to

$$\frac{h}{t} < \frac{0.48E}{\sqrt{F_y(F_y + F_r)}} \qquad (8.17)$$

The formation of residual stresses in welded H-shapes is discussed in Art. 9.7 and a typical residual stress pattern is shown in Fig. 9.21. When the values of $E = 29,000$ ksi and $F_r = 16.5$ ksi are introduced in the Inequality 8.17, the limiting web slenderness ratio specified by the AISC Specification is obtained,

$$\frac{h}{t} \leqslant \frac{14,000}{\sqrt{F_y(F_y + 16.5)}} \qquad (8.18)$$

A higher value of F_y will result in a lower value of h/t since flange force and curvature increase in proportion to F_y. It is emphasized that F_y in Inequality 8.18 is the yield point of the flange material. This would be of importance in the design of a hybrid girder where, for example, A441 steel is used for flanges and A36 steel is used for the web.

Flange Stress Reduction. The stress redistribution from the compression portion of the web to the compression flange (Fig. 8.13) increases the stress in the compression flange and thus the allowable stress may have to be reduced.

If it is assumed that the compression flange is prevented from buckling, the ultimate bending moment can be expressed as a function of the yield point, the web slenderness ratio h/t, and the ratio of web area to flange area A_w/A_f. This relationship is approximated for $F_y = 33$ ksi in Fig. 8.17 with the ordinate non-dimensionalized by dividing M_u by M_y, the moment required to

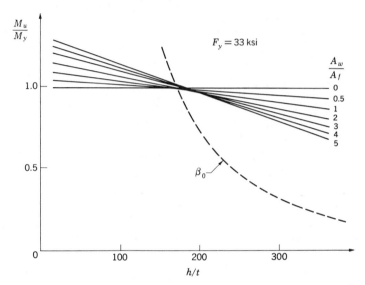

Fig. 8.17 Influence of parameters h/t and A_w/A_f on ultimate bending moment.

initiate yielding at the centroid of the compression flange. The equation for the set of lines in the figure can be approximated[7.18] by

$$M_u = M_y\left[1 - 0.0005\,\frac{A_w}{A_f}\left(\frac{h}{t} - \beta_0\right)\right] \tag{8.19}$$

where β_0 is the abscissa of the buckling curve for a web plate subjected to bending (shown by a broken line).

Since $M_u/M_y = F_u/F_y$, Eq. 8.19 can be expressed in terms of stress. The reduced bending stress F_b' is obtained by dividing by a factor of safety of 1.65,

$$F_b' = \frac{F_u}{\text{F.S.}} = \frac{F_y}{\text{F.S.}}\left[1 - 0.0005\,\frac{A_w}{A_f}\left(\frac{h}{t} - \beta_0\right)\right]$$

or

$$F_b' = F_b\left[1 - 0.0005\,\frac{A_w}{A_f}\left(\frac{h}{t} - \beta_0\right)\right] \tag{8.20}$$

Equation 8.20 was derived by assuming that instability of the compression flange does not influence the carrying capacity. The influence of lateral buckling can be included by using F_b as determined by Eq. 7.38, 7.39, or 7.40 when it is less than $F_y/\text{F.S.} = 0.6F_y$.

The value of β_0 depends on the amount of restraint provided to the web by the flanges. Assuming that the flange offers partial restraint, a value of $\beta_0 = 5.7\sqrt{E/F_{cr}}$ has been suggested,[7.18] where F_{cr} is the web-buckling stress. If F_{cr} is replaced by the product of the allowable bending stress F_b and a safety factor, β_0 is found to be $760/\sqrt{F_b}$. Thus, Eq. 8.20 can be written as

$$F_b' = F_b\left[1.0 - 0.0005\frac{A_w}{A_f}\left(\frac{h}{t} - \frac{760}{\sqrt{F_b}}\right)\right] \qquad (8.21)$$

which is the AISC Formula [1.10-5] for flange stress reduction. The reduction in allowable bending stress is expressed by the second term in the equation. If the girder web is thick enough, no reduction is necessary. The limit is expressed in the inner bracket of Eq. 8.21; thus, if

$$\frac{h}{t} < \frac{760}{\sqrt{F_b}} \qquad (F_b \text{ in ksi})$$

no reduction is required.

2. Shear Strength

Although it is physically impossible for a portion of a beam to be subjected to shear unaccompanied by bending, the theoretical case of a girder subjected to pure shear is considered in this section. Interaction between bending and shear is discussed below.

The stress redistribution in a plate girder web after web buckling due to shear has taken place results in the development of tension field action. A girder panel including its framing elements (the flanges and transverse stiffeners) acts in a manner which is analogous to a panel of a Pratt truss; that is, a diagonal strip of the web acts as a tension member while the transverse stiffeners act as compression struts. Figure 8.18 illustrates the Pratt truss analogy, whereas the yield lines on the test girder of Fig. 8.19 (Ref. 8.4) illustrate the development of tension field action in an actual girder.

Webs which are stocky enough will not buckle before yielding occurs. In this case all of the applied shear force will be taken by beam action shear and no tension field action will occur. For girders with slender webs, it is assumed that an applied shear force is carried completely by beam action shear until the theoretical web buckling stress is reached, and that subsequently, the additional applied shear is carried by tension field action. The ultimate shear which a panel can sustain is determined then by adding the beam

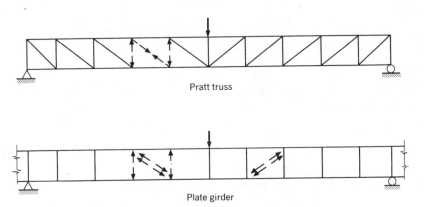

Fig. 8.18 Pratt-truss analogy.

Fig. 8.19 Development of tension field in a test girder.

action contribution V_b to the tension field contribution V_t,

$$V_u = V_b + V_t \tag{8.22}$$

The contribution of beam action shear is already known: it is the shear carried by the girder web at the theoretical web buckling stress and is given by

$$V_b = F_{cr} h t = V_p \frac{F_{cr}}{F_{ys}} \tag{8.23}$$

where F_{ys} is defined as the yield point in shear and taken to be $F_y/\sqrt{3}$,[8.2] and V_p is the shear force causing complete plastification of the web,

$$V_p = htF_{ys} \tag{8.24}$$

F_{cr} is the critical web-buckling stress given by Eq. 8.2 for values of F_{cr} less than the proportional limit (assumed as $0.8F_{ys}$) and expressed by

$$F_{cr} = \sqrt{\frac{0.8F_{ys}\pi^2 Ek}{12(1 - \mu^2)}\left(\frac{t}{h}\right)^2} \quad \text{(for } F_{cr} > 0.8F_{ys}) \tag{8.25}$$

The applicable buckling coefficient k is that given by Eq. 8.9.

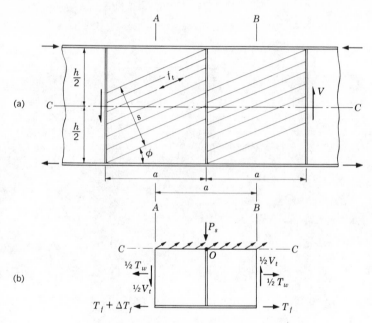

Fig. 8.20 Tension field action shear.

Tension Field Action. The contribution of tension field action to the ultimate shear force can be evaluated from the geometry and equilibrium of the tension field. First, to determine the orientation of the field, the tension strip shown in Fig. 8.20a is considered. The strip is capable of carrying a shear force of $f_t \cdot s \cdot t \cdot \sin\phi$ where s, the width of the strip, is equal to $h \cdot \cos\phi - a \cdot \sin\phi$ and f_t is the tension field stress. By differentiating this shear force with respect to the angle ϕ and setting the resulting expression equal to zero, the optimum value of ϕ is found:

$$\phi = \tan^{-1}\left[\sqrt{1 + (a/h)^2} - \frac{a}{h}\right] \tag{8.26}$$

It is reasonable to expect that the tension field will be oriented in the position given by Eq. 8.26, since it is the position which will provide the maximum shear resistance for direct tension action and hence is the most efficient position.

Next, the free-body diagram shown in Fig. 8.20b is used to evaluate V_t. The total force due to tension field stresses acting at the angle ϕ with the horizontal is $f_t \cdot t \cdot a \cdot \sin \phi$. The change in flange force, ΔT_f, is determined by summation of forces in the horizontal direction and is equal to $f_t \cdot t \cdot a \cdot \sin \phi \cdot \cos \phi$. Finally, by summation of moments about point O, the total shear force V_t is $h \cdot (\Delta T_f/a)$. When the value of ΔT_f given above along with ϕ from Eq. 8.26 is substituted in this expression, the desired expression for the contribution of tension field action to the ultimate shear force is obtained,

$$V_t = f_t \cdot t \cdot h \cdot \frac{1}{2\sqrt{1 + (a/h)^2}} \tag{8.27}$$

Allowable Shear Stress and Stiffener Spacing. The expression for the ultimate shear force, Eq. 8.22, can be rewritten by substitution of Eqs. 8.23, 8.24, and 8.27. The resulting expression is

$$V_u = V_p\left[\frac{F_{cr}}{F_{ys}} + \frac{\sqrt{3}}{2} \cdot \frac{f_t}{F_y} \cdot \frac{1}{\sqrt{1 + (a/h)^2}}\right] \tag{8.28}$$

An approximation of Mises' yield condition,[8.2]

$$\frac{f_t}{F_y} = 1 - \frac{F_{cr}}{F_{ys}} \tag{8.29}$$

can be inserted in Eq. 8.28, resulting in

$$V_u = V_p\left[\frac{F_{cr}}{F_{ys}} + \frac{1 - (F_{cr}/F_{ys})}{1.15\sqrt{1 + (a/h)^2}}\right] \tag{8.30}$$

Dividing by the web area, $h \cdot t$, and a safety factor of 1.67, and using the definition $F_{ys} = F_y/\sqrt{3}$, the AISC allowable shear stress formula is obtained:

$$F_v = \frac{F_y}{2.89}\left[C_v + \frac{1 - C_v}{1.15\sqrt{1 + (a/h)^2}}\right] \qquad \text{(for } C_v < 1\text{)} \tag{8.31}$$

where C_v is used to designate the ratio F_{cr}/F_{ys}. The second term in the brackets is the tension field contribution. A value of $C_v > 1$ would mean that the web is stocky enough to carry all the applied force by beam action shear; hence for this situation only the first term of Eq. 8.31 is used,

$$F_v = \frac{F_y}{2.89} C_v \quad \text{(for } C_v > 1) \tag{8.32}$$

For completeness, the values of C_v to be used in Eqs. 8.31 and 8.32 are listed below. They are obtained by rewriting Eqs. 8.2, 8.9, and 8.25 in terms of $C_v = F_{cr}/F_{ys}$ and using $E = 29,000$ ksi and $\mu = 0.3$.

$$\left. \begin{array}{ll} C_v = \dfrac{45,000k}{F_y(h/t)^2} & \text{(when } C_v < 0.8) \\[2ex] C_v = \dfrac{190}{h/t} \sqrt{\dfrac{k}{F_y}} & \text{(when } C_v > 0.8) \\[2ex] k = 4.00 + \dfrac{5.34}{(a/h)^2} & \left(\text{when } \dfrac{a}{h} < 1.0\right) \\[2ex] k = 5.34 + \dfrac{4.00}{(a/h)^2} & \left(\text{when } \dfrac{a}{h} > 1.0\right) \end{array} \right\} \tag{8.33}$$

The equations for allowable shear stress are now complete. However, two additional requirements for the panel size are specified. The first of these requirements has been selected somewhat arbitrarily to limit a panel size for the cases where the shear stresses are small. This was done to facilitate fabrication, handling, and erection of girders. The requirement is expressed in the form

$$\left. \begin{array}{l} \dfrac{a}{h} \leqslant \left(\dfrac{260}{h/t}\right)^2 \\[2ex] \dfrac{a}{h} \leqslant 3 \end{array} \right\} \tag{8.34}$$

The second requirement pertains to the size of end panels. The tension field in one panel depends on the neighboring panels to absorb the horizontal component of the tension field force. At the ends of a girder, where there is no neighboring panel, two methods can be used to provide anchorage for the horizontal component. The first is to provide an end plate which forms a strong end post over the support to resist the horizontal pull (see Fig. 8.21). The second method, which is used in the AISC Specification, is to limit the width of the end panel so that tension field action is eliminated and only beam action takes place. This is accomplished by using a limitation similar to that of Eq. 8.14, that is, the end panel size should be such that the smaller of the two dimensions, a and h, is less than $348/\sqrt{f_v}$ times the web thickness. The

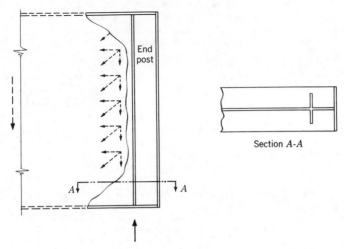

Fig. 8.21 End post as anchorage for tension field.

consequences of failing to provide adequate anchorage for the tension field in the end panel can be seen in Fig. 8.22, which shows a yielded end post on a test girder (Ref. 8.4).

Equations 8.31 through 8.34, which give the allowable values of shear stress or panel size and web thickness, depending on what design procedure is used, are certainly an imposing array of formulas and would be very burdensome to use in design. In the Appendix of the AISC Specification, tables are available which incorporate all of these formulas and thus simplify the designer's work. The table for steel having a yield point of $F_y = 36$ ksi is reproduced in Table 8.5. The use of this table is illustrated in Example 8.2.

Stiffener Requirements. Referring to Fig. 8.20b, summation of vertical forces leads to the expression $P_s = f_t \cdot t \cdot a \cdot \sin^2 \phi$ for the compressive force on a transverse stiffener due to the tension field force. Substitution of the value of ϕ given by Eq. 8.26 results in:

$$P_s = \frac{f_t}{2}\left[\frac{a}{h} - \frac{(a/h)^2}{\sqrt{1 + (a/h)^2}}\right]h \cdot t \tag{8.35}$$

A further substitution of Eq. 8.29 for f_t and division by F_y leads to the AISC Formula [1.10-3] for required stiffener area A_{st}:

$$A_{st} = \frac{1 - C_v}{2}\left[\frac{a}{h} - \frac{(a/h)^2}{\sqrt{1 + (a/h)^2}}\right]Y \cdot D \cdot h \cdot t \tag{8.36}$$

where D has been introduced to reflect the efficiency of stiffeners furnished in pairs as opposed to one-sided stiffeners, and Y provides for the possibility

Table 8.5 Allowable Shear Stresses (F_v) in Plate Girders (ksi) for 36-ksi Specified Yield Stress Steel

(*Italic* values indicate gross area, as percent of web area, required for pairs of intermediate stiffeners of 36-ksi yield stress steel)*

Aspect Ratio a/h: Stiffener Spacing to Web Depth — Slenderness Ratio h/t: Web Depth to Web Thickness

h/t	0.5	0.6	0.7	0.8	0.9	1.0	1.2	1.4	1.6	1.8	2.0	2.5	3.0	over 3
60										14.5	14.5	14.5	14.5	14.5
70							14.5	14.5	14.5	14.4	14.2	13.8	13.6	13.0
80					14.5	14.5	14.0	13.4	13.0	12.6	12.4	12.2 *0.3*	12.0 *0.4*	11.4
90				14.5	14.3	13.4	12.5	12.2 *0.6*	11.9 *0.9*	11.8 *1.1*	11.6 *1.2*	11.3 *1.2*	11.1 *1.2*	10.1
100			14.5	13.9	12.8	12.3 *0.5*	11.9 *1.4*	11.6 *1.8*	11.3 *2.0*	11.1 *2.1*	10.9 *2.2*	10.3 *2.3*	10.0 *2.1*	8.3
110		14.5	13.8	12.6	12.2 *0.9*	11.9 *1.8*	11.5 *2.5*	11.0 *3.1*	10.5 *3.5*	10.1 *3.6*	9.8 *3.6*	9.2 *3.4*	8.8 *3.1*	6.9
120		14.3	12.7	12.2 *1.1*	11.8 *2.1*	11.5 *2.8*	10.8 *4.1*	10.2 *4.7*	9.8 *4.9*	9.4 *4.9*	9.0 *4.7*	8.4 *4.3*	8.0 *3.8*	5.8
130	14.5	13.2	12.2 *0.9*	11.9 *2.2*	11.5 *3.2*	11.0 *4.5*	10.3 *5.6*	9.7 *5.9*	9.2 *6.0*	8.8 *5.8*	8.4 *5.6*	7.8 *5.0*	7.3 *4.4*	4.9
140	14.2	12.4 *0.3*	12.0 *1.9*	11.6 *3.2*	11.0 *4.8*	10.5 *5.9*	9.8 *6.7*	9.2 *6.9*	8.7 *6.8*	8.3 *6.6*	7.9 *6.3*	7.2 *5.5*	6.8 *4.9*	4.2
150	13.2	12.2 *1.2*	11.8 *2.8*	11.2 *4.7*	10.6 *6.1*	10.1 *7.0*	9.4 *7.6*	8.8 *7.7*	8.3 *7.5*	7.9 *7.2*	7.5 *6.8*	6.8 *6.0*	6.3 *5.2*	3.7
160	12.4	12.0 *2.1*	11.5 *4.1*	10.9 *6.0*	10.3 *7.2*	9.8 *8.0*	9.1 *8.4*	8.5 *8.3*	8.0 *8.1*	7.6 *7.7*	7.2 *7.3*	6.5 *6.3*		3.2
170	12.3 *0.9*	11.8 *2.8*	11.2 *5.3*	10.6 *7.0*	10.1 *8.1*	9.6 *8.7*	8.9 *9.0*	8.3 *8.9*	7.7 *8.5*	7.3 *8.1*	6.9 *7.7*			2.9
180	12.1 *1.6*	11.6 *4.0*	10.9 *6.3*	10.4 *7.9*	9.9 *8.8*	9.4 *9.4*	8.7 *9.6*	8.1 *9.3*	7.5 *8.9*	7.1 *8.5*	6.7 *8.0*			2.6
200	11.9 *2.9*	11.2 *6.0*	10.5 *8.0*	10.0 *9.2*	9.5 *10.0*	9.1 *10.4*	8.3 *10.4*	7.7 *10.0*	7.2 *9.5*					2.1
220	11.5 *4.8*	10.8 *7.5*	10.3 *9.2*	9.7 *10.2*	9.3 *10.8*	8.8 *11.1*	8.1 *11.0*	7.5 *10.6*						1.7
240	11.2 *6.2*	10.6 *8.6*	10.0 *10.1*	9.5 *11.0*	9.1 *11.5*	8.6 *11.7*								1.4
260	11.0 *7.3*	10.4 *9.5*	9.9 *10.8*	9.4 *11.6*	8.9 *12.0*	8.5 *12.1*								1.2
280	10.8 *8.2*	10.2 *10.2*	9.7 *11.4*	9.2 *12.1*										
300	10.7 *9.0*	10.1 *10.8*	9.6 *11.8*											
320	10.5 *9.5*	10.0 *11.2*												

Girders so proportioned that the computed shear is less than that given in right-hand column do not require intermediate stiffeners.

* For single angle stiffeners, multiply by 1.8; for single plate stiffeners, multiply by 2.4.

EXAMPLE 8.2

PROBLEM:
  A building girder is to support the loads shown. Lateral support to the compression flange is available at the ends and at the 76 kip concentrated loads. Headroom requirements limit the depth to 6'-0". A36 steel and the AISC specification are to be used.

SOLUTION:

Section:

    Try – web $\text{R} \frac{5}{16} \times 70$

        flange $\text{R} \frac{3}{4} \times 20$

    for flanges, $\frac{b}{2t} = \frac{10}{0.75} = 13.3 < 15.8$ <u>O.K.</u>

    for web, $\frac{h}{t} = \frac{70}{0.312} = 224 < 322$ <u>O.K.</u>

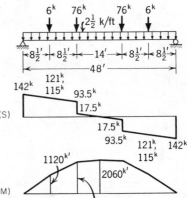

Section properties:

    $I = 2(15)(35.4)^2 + \frac{1}{12}(0.312)(70)^3$

      $= 46,500 \text{ in.}^4$

    $S = \frac{46,500}{35.75} = 1300 \text{ in.}^3$

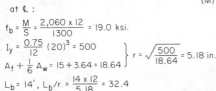

Bending stress:

    at C_L :

    $f_b = \frac{M}{S} = \frac{2,060 \times 12}{1300} = 19.0 \text{ ksi.}$

    $\left. \begin{aligned} I_y &= \frac{0.75}{12}(20)^3 = 500 \\ A_f + \frac{1}{6}A_w &= 15 + 3.64 = 18.64 \end{aligned} \right\} \, r = \sqrt{\frac{500}{18.64}} = 5.18 \text{ in.}$

    $L_b = 14'$, $L_b/r = \frac{14 \times 12}{5.18} = 32.4$

    Since moment is larger within the unbraced length than at both ends, M_1/M_2 is to be taken as unity, thus $C_b = 1$

    Since $L_b/r < 53\sqrt{C_b}$, $F_b = 22$ ksi.

At load points:

    $f_b = \frac{2000 \times 12}{1300} = 18.5 \text{ ksi}$

    $L_b = 17'$, $L_b/r = \frac{17 \times 12}{5.18} = 39.4$

    $M_1/M_2 = 0$, $\therefore C_b = 1.75$

    Since $L_b/r < 53\sqrt{C_b}$, $F_b = 22$ ksi

Flange stress reduction:

    $F_b' = F_b \left[1.0 - 0.0005 \frac{A_w}{A_f} \left(\frac{h}{t} - \frac{760}{\sqrt{F_b}} \right) \right]$

    checking center section only,

    $F_b' = 22 \left[1.0 - 0.0005 \frac{21.9}{15} \left(224 - \frac{760}{\sqrt{22}} \right) \right]$

      $= 21.0 \text{ ksi} > 19.0 \text{ ksi}$ <u>O.K.</u>

Maximum permissible stiffener spacing:

  End panels:

    $f_v = V/A_w = \frac{142}{21.9} = 6.48 \text{ ksi}$

    $\frac{348t}{\sqrt{f_v}} = \frac{348(0.312)}{\sqrt{6.48}} = 42.7 \text{ in.}$ use $a = 42$ in.

EXAMPLE 8.2 (Continued)

Interior panels:

$$a/h \leq \left(\frac{260}{h/t}\right)^2 = \left(\frac{260}{224}\right)^2 = 1.35$$

$$\therefore a_{max.} = 1.35h = 1.35(70) = 94.5 \text{ in.}$$

Since bearing stiffeners will be used under the 76 kip loads and at the supports, try 2 panels with a = 84 in. in the center section and 2 panels with a = 81 in. between the end panels and the 76 kip loads.

center section: a/h = 84/70 = 1.20

end sections: a/h = 81/70 = 1.16

Shear stress:

At 42 in. from ends, $V = 142 - 3\frac{1}{2} \times 2\frac{1}{2} = 133$ kips

$$f_v = V/A_w = \frac{133}{21.9} = 6.07 \text{ ksi}$$

From Table 8.5,   with a/h = 1.16 and h/t = 224

$$F_v = 8.2 \text{ ksi} > 6.07 \text{ ksi} \qquad \underline{O.K.}$$

At 76 kip loads, $f_v = \frac{93.5}{21.9} = 4.27$ ksi

From Table 8.5,   with a/h = 1.20 and h/t = 224

$$F_v = 8.1 \text{ ksi} > 4.27 \text{ ksi} \qquad \underline{O.K.}$$

Resulting stiffener arrangement:

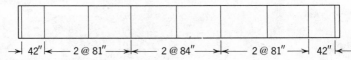

→| 42" |← — 2 @ 81" —→|← — 2 @ 84" —→|← — 2 @ 81" —→| 42" |←

Check interaction:

Under 76 kip loads, $\frac{f_v}{F_v} = \frac{4.27}{8.1} = 0.53 < 0.60$ $\underline{O.K.}$

Check web crippling:

$\underline{O.K.}$ (as checked in Example 17.4)

Intermediate stiffeners:

From Table 8.5 with a/h = 1.20 and h/t = 224

req'd $A_{st} = 0.11 A_w = 0.11 (21.9) = 2.41$ in.^{2}

Try 2 $\mathcal{R}$'s $\frac{5}{16} \times 4$, $A_{st} = 2.50$ in.^{2}

Check width-thickness ratio: $\frac{4.0}{0.312} = 12.8 < 15.8$ $\underline{O.K.}$

Check moment of inertia: req'd $I_s = (h/50)^4 = (1.4)^4 = 3.84$ in.^{4}

$$I_s = \frac{1}{12}\left(\frac{5}{16}\right)(8)^3 = 13.3 \text{ in}^4 > 3.84 \qquad \underline{O.K.}$$

Use: 2 $\mathcal{R}$'s, $\frac{5}{16} \times 4$, bearing on compression flg. and cut 1 in. short of tension flg. ($\frac{5}{16} \times 4 = 1.25 > 1$ in.)

Bearing stiffeners:

Since bearing stiffeners must extend approximately to edges of flg. plates, try 2 $\mathcal{R}$'s, 1/2 x 8

Check width-thickness ratio: $\frac{8.0}{0.5} = 16$ $\sim \underline{O.K.}$

Check compressive stress:

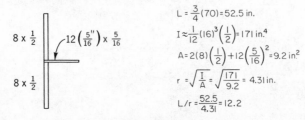

$8 \times \frac{1}{2}$

$12\left(\frac{5''}{16}\right) \times \frac{5}{16}$

$8 \times \frac{1}{2}$

$$L = \frac{3}{4}(70) = 52.5 \text{ in.}$$

$$I \approx \frac{1}{12}(16)^3 \left(\frac{1}{2}\right) = 171 \text{ in.}^4$$

$$A = 2(8)\left(\frac{1}{2}\right) + 12\left(\frac{5}{16}\right)^2 = 9.2 \text{ in}^2$$

$$r = \sqrt{\frac{I}{A}} = \sqrt{\frac{171}{9.2}} = 4.31 \text{ in.}$$

$$L/r = \frac{52.5}{4.31} = 12.2$$

EXAMPLE 8.2 (Continued)

at supports, $f_a = \dfrac{142}{9.2} = 15.4$ ksi

From table [1-36] for $L/r = 12.2$, $F_a = 21$ ksi > 15.4 ksi   <u>O. K.</u>

The bearing stress is obviously less than the allowable value of 33 ksi

Use: 2 ℞'s, 1/2 x 8 at supports and 76 kip loads, bearing on both flanges.

Fig. 8.22   End post failure in a test girder

of using different materials for web stiffeners. Y is defined as the ratio of the yield point of the web steel to that of the stiffener steel.

One-sided stiffeners will be subjected to moment as well as axial force since they will be loaded eccentrically. They will thus be less efficient than a stiffener pair. By allowing the one-sided stiffener to become fully yielded under the combined moment and axial force, and using the case of a pair of stiffener plates as a reference ($D = 1.0$), the following values of D can be determined:[8.2]

$D = 1.8$ for single angle stiffeners, one leg flat against the web
$D = 2.4$ for single plate stiffeners.

As previously mentioned, Eq. 8.36 is presented in tabular form in the Appendix of the AISC Specification. The stiffener area given in the tables may be reduced by the ratio of the actual shear stress in a panel to the allowable shear stress. The resulting required area often will be very small. In such cases, a stiffness requirement will govern the design of the transverse stiffeners, that is, the stiffener will be designed to maintain the shape of the cross section. This requirement, $I_{min} \geqslant (h/50)^4$, is similar to Eq. 8.15, the AASHO formula.

It can be shown, by partial differentiation of Eq. 8.36 with respect to the variables a/h and b/t, that the maximum stiffener force is

$$P_s = 0.015 F_y h^2 \sqrt{\varepsilon_y}$$

The method of specifying the fastening required between a transverse stiffener and the web can be simplified considerably if the maximum stiffener force is used. Assuming that the stiffener force is built up over a third of the depth and incorporating the AISC safety factor of 1.67, the minimum value of the total shear transfer between intermediate stiffeners and the web may be shown to be

$$f_{vs} = h \sqrt{\left(\frac{F_y}{340}\right)^3} \tag{8.37}$$

This requirement is based on the tension field action force and may be less than the amount of fastening required to transfer the shear force from a concentrated reaction or force applied externally at the location of a stiffener. In any case, the fastening requirement is usually so small that it can be taken into consideration by the minimum amount of welding or riveting that might be desired.

Additional provisions on details of transverse stiffeners are given in the AISC Specification. It has been verified experimentally[8.4] that stiffeners can be stopped short of the tension flange without harmful effect. (Actually, if the stiffeners are welded to the tension flange, a fatigue problem will be created.) They should be carried to within a distance of four times the web thickness from the tension flange to avoid local buckling of the web.[8.11] Single stiffeners must be attached to the compression flange to prevent the flange from twisting. Stiffeners shall also be attached to the compression flange when lateral supports are connected to the stiffeners. It has been decided arbitrarily that in the latter case the connection should be sufficient to transfer 1 per cent of the total force in the flange. When the flange is composed only of angles, this provision is not required.

3. Combined Bending and Shear

For the case of a girder section subjected to a high shear combined with a high bending moment, the allowable stresses may have to be reduced. This situation is now examined with the aid of an interaction diagram (Fig. 8.23).

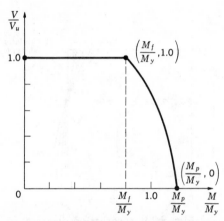

Fig. 8.23 Development of the interaction diagram.

The applied shear force V is non-dimensionalized by dividing by the ultimate shear force V_u and the resulting ratio plotted as the ordinate of the interaction diagram. Similarly, the applied bending moment M is non-dimensionalized by dividing by the yield moment M_y and the resulting ratio plotted as the abscissa of the diagram. M_y is defined as the moment required to initiate yielding at the centroid of the compression flange.

When no moment is present, V can be as large as V_u; when no shear is present, M can be as large as M_p, where M_p is the moment required for complete yielding of the cross section. If the flange moment M_f is now defined as the moment carried by the flanges alone when the stresses over the entire flange are equal to the yield point, it is seen that, for $M = M_f$, V can still be as large as V_u since V_u is assumed to be carried only by the web. For values of M greater than M_f, the applied shear will have to be reduced.

The limits of the diagram are now determined with the exception of the portion between the points $(1, M_f/M_y)$ and $(0, M_p/M_y)$ where interaction takes place. It is convenient to express the diagram in terms of stresses rather than moments and shears, using the relationships $f_b = M/S$, $F_y = M_y/S$, $f_v = V/A_w$, and $F_{vu} = V_u/A_w$, where F_{vu} is defined as the ultimate shear stress. The resulting diagram is shown in Fig. 8.24. It can be shown[8.3] that the undetermined portion of the diagram depends on the ratio A_w/A_f. If the maximum value of this ratio is taken to be equal to or less than 2, which is an upper limit to the practical range of cross section proportions, and it is further specified that the maximum value of f_b is F_y, the following restrictions are placed on the diagram: $f_v \leqslant 0.6F_{vu}$ for $f_b = F_y$ and $f_b \leqslant 0.75F_b$ for $f_v = F_{vu}$. The curve between points (0.75, 1.0) and (1.0, 0.6) is taken to be a straight line.

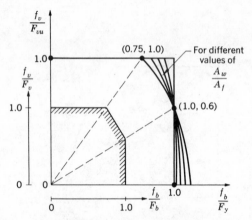

Fig. 8.24 Interaction diagram.

The desired interaction diagram is obtained by connecting the labeled points in Fig. 8.24 with straight lines. If this diagram is converted to allowable stress values, using the 1.67 safety factor, the interaction diagram used in the AISC Specification is determined (cross-hatched lines in Fig. 8.24). The limitation on the bending stresses in the web due to interaction is that they must not exceed $0.6F_y$ and they must satisfy the expression

$$f_b \leqslant \left(0.825 - 0.375 \frac{f_v}{F_v}\right) F_y \qquad (8.38)$$

which is the equation of the inclined line in the allowable stress design.

In summary, no interaction check is required if the web is proportioned on the basis of either of these two criteria:

1. Maximum permissible bending stress when the shear stress is not greater than 0.6 times the full permissible value
2. Full permissible shear stress when the bending stress is not more than 0.75 times the maximum allowable value

When neither of these conditions are satisfied, Eq. 8.38 must be checked.

The application of the AISC Specification provisions on plate girder design which have been discussed in this article is illustrated in Example 8.2.

4. Experimental Results

An indication of the correlation of test results with the ultimate strength theory presented in the previous sections is shown in Fig. 8.25 in the interaction diagram used in the AISC Specification. All of the test points shown on the

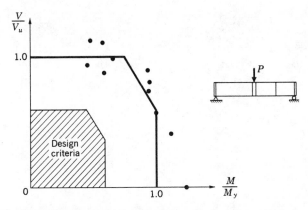

Fig. 8.25 Correlation of test results with ultimate-strength theory.

diagram were obtained from tests on girders which has a span length of 26 ft 6 in. between supports and a 50-in. web depth. The set-up used for these tests is sketched at the right of the figure. A detailed discussion of the test girders and test results is presented in Ref. 8.4.

It was assumed that $M \leqslant M_y$ and $A_w/A_f = 2$ in deriving the interaction curve of Fig. 8.25 in order to simplify the design procedure. If these assumptions had not been made, the correlation of the test points near the vertical and inclined portions of the diagram would be even better (see Fig. 4 of Ref. 8.3).

A comparison of other test results with the theory is presented for the loading conditions of pure bending and high shear in Refs. 7.18 and 8.2 respectively.

PROBLEMS

8.1. Determine the transverse stiffener spacing and number of stiffeners required for Example 8.1 if no longitudinal stiffener is used. How much can the size of the transverse stiffener be reduced in this case?

8.2. For the girder of Example 8.1, at what intervals would the compression flange have to be laterally supported if the size of both the tension and compression flanges were to be reduced to 1″ × 16″ plates? How could this lateral support spacing be achieved practically?

8.3. Redesign the flanges and web of the girder of Example 8.2 for a situation where the girder depth is limited to 5 ft 2 in.

8.4. Redesign the girder of Example 8.2 using a 5 ft 0 in. over-all depth, A441 steel, and the AASHO Specifications. Compare the total weight of the girder obtained with that of Example 8.2.

COMPRESSION MEMBERS

9.1 INTRODUCTION

A compression member in its simplest form is a column. A column may be defined as a compression member whose length is considerably greater than its cross-sectional dimensions.

However, columns are not the only compression members; any structure or part of a structure is a compression member if it is under a compressive load, either alone, or in conjunction with other loadings. Under this category could be included beam-columns, plates, component parts of frames, and, for instance, the compression flange of beams or plate girders.

This chapter will be restricted to columns under centrally applied loads with no end moments, such as the one shown in Fig. 9.1. A column is a basic member of most structures, and a knowledge of column behavior is necessary in the interpretation and understanding of specification requirements.

9.2 HISTORICAL REVIEW

Columns have been in use in one form or another since time immemorial. Even the use of iron columns may be traced back many centuries. However, it was not until 1729 that van Musschenbroek published the first paper concerning the strength of columns.[9.1] An empirical column curve was presented in the form

$$P = k \frac{bd^2}{L^2} \tag{9.1}$$

Fig. 9.1 Simple column under centrally applied load.

where P is column strength, k is an empirical factor, b and d are the breadth and width of the rectangular section, and L is the length of column. Such a column formula is not really too far removed from those in use today.

The development of the differential and integral calculus in the second half of the seventeenth century gave an impetus to the formulation of many natural phenomena, including that of column buckling. Although the theory of beams and bending had been studied some years previously, it was not until 1759 that Euler published his now famous treatise on the buckling of columns.[9.1,9.2] Euler was the first to realize that column strength could also be a problem of stability and not merely a matter of crushing. The original buckling load determined by Euler was for a column with one end built in and the other free:

$$P = \frac{C\pi^2}{4L^2} \tag{9.2}$$

where P is the buckling strength and C is the "absolute elasticity," which was defined merely as depending upon the elastic properties of the material. Euler investigated the purely elastic phenomenon of buckling, which implied that the elastic limit was not exceeded in any fiber in the column cross section.

Elastic instability of columns occurs only with very slender columns, and it was not until one hundred years after Euler that theories were introduced to define inelastic column strength.

Engesser presented his original tangent modulus theory in 1889, and some years later, under the influence of Considere's predictions and Jasinski's criticisms, presented the reduced modulus theory for the inelastic buckling strength of columns.[7.6,9.1,9.3] These two theories provide a basis of the modern concepts of column buckling and are considered in Art. 9.6.

The reduced modulus load had been accepted as the correct buckling theory for columns in the inelastic range until 1947, when Shanley published a paper giving the buckling load of a centrally loaded column as the tangent modulus load.[9.4]

It became apparent in the late 1940's that the key to the application of the tangent modulus concept to the steel column lay in the inclusion of the effect of residual stresses which existed in the cross section of the column even before the application of external load. Since then, a considerable number of studies have been made on that problem, many of these at Lehigh University, the results being summarized in Refs. 2.4, 9.3, 9.5, 9.6, and 9.7.

9.3 COLUMN STRENGTH

The buckling strength of a column may be defined by either the buckling load or the ultimate (or maximum) load. These definitions are applied to an overall column failure as distinct from local failure, such as local buckling. The "buckling load" may be defined as the lowest load at which the theoretically straight column can assume a deflected position. The "ultimate load" is the maximum load a column can carry; it marks the boundary between stable and unstable deflected positions of the column and is reached gradually, unlike the buckling load, which is an instantaneous phenomenon. (However, the attainment of the maximum load of a column, which is part of a frame or subassemblage, does not mean necessarily that the frame cannot carry more load. See Arts. 11.2 and 22.3.)

Buckling will occur only for a centrally loaded, perfectly straight column. The strength of practical columns, however, depends upon whether there is initial out-of-straightness, eccentricity of load, transverse load, end fixity, local or lateral buckling, or residual stress. Most tests on columns did not isolate these various effects, and so a scatter band for column curves resulted from observation of the maximum load and not the buckling load. The usual procedure for defining a column curve (see Art. 9.5) was much the same eighty years ago as now; the column curve was taken as the line of best fit through the test points.[9.1,9.8] Figure 9.2 shows a scatter band of test results obtained in 1915; the maximum stress attained in the test column is plotted against the slenderness ratio of the column.

To take into account the transition in the column curve from the Euler curve to the yield line (see Fig. 9.7), complicated correction factors were evolved, using estimated eccentricities or initial deflections; this approach is still followed in some codes. In the late 1940's, it was shown that, for the

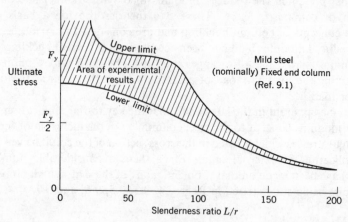

Fig. 9.2 Column strength in 1915.[9.1] (Courtesy of Oxford University Press.)

hypothetical case of straight, centrally loaded, pinned-end columns, the transition curve is due entirely to the presence of residual stresses in the cross section.[2.4] This effect of residual stresses is considered in Art. 9.7.

9.4 COLUMN FORMULAS FOR ELASTIC BUCKLING

1. Euler Load

For a pinned-end column, perfectly elastic, perfectly straight and loaded centrally, the buckling load is known as the Euler load,

$$P_e = \frac{\pi^2 EI}{L^2} \tag{9.3}$$

where P_e is the Euler load, E the modulus of elasticity, I the moment of inertia of the cross section, and L the length of the column between pin ends. The derivation of this formula may be obtained from equilibrium considerations.

Consider the column in Fig. 9.3; it is initially straight, and centrally loaded through pin ends. When the load on the column is of such a magnitude that the column is indifferent as to whether it is straight or deflected, the column is said to be unstable. Consider the column at the first such load to cause instability.

The column is deflected with an assumed mid-height deflection of v_0. Consider the equilibrium of the column in its bent position. At any cross section, the external and internal moments may be expressed as

$$M_{\text{ext}} = Pv \tag{9.4}$$

$$M_{\text{int}} = \frac{EI}{R} \tag{9.5}$$

where R is the radius of curvature at the cross section under consideration. For the column to retain its deflected shape, the internal moment must equal the external moment,

$$Pv = \frac{EI}{R} \tag{9.6}$$

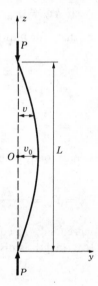

Fig. 9.3 Centrally loaded pinned-end column.

The curvature, for small deflections, may be defined in terms of the deflection by

$$\frac{1}{R} = -\frac{d^2v}{dz^2} \tag{9.7}$$

Applying Eq. 9.7 to Eq. 9.6, and rearranging, the following differential equation results:

$$\frac{d^2v}{dz^2} + k^2v = 0 \tag{9.8}$$

where

$$k^2 = \frac{P}{EI} \tag{9.9}$$

The solution of Eq. 9.8 is

$$v = A \sin kz + B \cos kz \tag{9.10}$$

where A and B are constants, the values of which can be determined from the boundary conditions of the column. When

$$z = 0, \qquad v = 0$$

and hence

$$B = 0 \tag{9.11}$$

when

$$z = L, \qquad v = 0$$

and hence

$$\sin kL = 0 \tag{9.12}$$

The solution of Eq. 9.12 implies that, for non-trivial solutions,

$$k = \frac{n\pi}{L} \qquad \text{for} \quad n = 1, 2, 3, \ldots \tag{9.13}$$

Hence, from Eq. 9.9,

$$P = \frac{n^2\pi^2}{L^2} EI \tag{9.14}$$

The critical buckling load, the Euler load, is the smallest load possible, and corresponds to $n = 1$:

$$P_{cr} = P_e = \frac{\pi^2 EI}{L^2} \tag{9.3}$$

When the column is not pinned-ended, the equivalent length of the column is used. (Equivalent, or effective, length is discussed in Chapter 10.)

2. Initial Curvature

For a hypothetical pinned-end column centrally loaded, with perfectly elastic material, and with an initial curvature which is assumed as a cosine curve, the following equations hold true:

$$\left. \begin{array}{l} v = \dfrac{e}{1 - \dfrac{P}{P_e}} \cos \dfrac{\pi z}{L} \\[3em] v_0 = \dfrac{P_e}{P_e - P} e \\[2em] f_{\max} = \dfrac{P}{A}\left[1 + \dfrac{ec}{r^2} \cdot \dfrac{P_e}{P_e - P}\right] \end{array} \right\} \tag{9.15}$$

where z is the distance along the column axis from the origin at the mid-height, e is the initial mid-height deflection of the column, assuming that the deflection is sinusoidal, and remains so under load, and v is the deflection of any point along the column from the straight position.   The nomenclature is illustrated in Fig. 9.4.

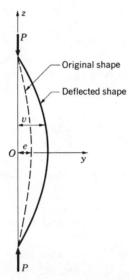

Fig. 9.4 Column with initial curvature.

3. Eccentrically Applied Load

For a pinned-end column with an eccentricity of load e, the maximum compressive stress is defined by the following equation, better known as the secant formula, which was and continues to be the basis of many rational attempts at column design.

$$f_{\max} = \frac{P}{A}\left[1 + \frac{ec}{r^2}\sec\left(\frac{\pi}{2}\sqrt{\frac{P}{P_e}}\right)\right]$$

or

$$f_{\max} = \frac{P}{A}\left[1 + \frac{ec}{r^2}\sec\left(\frac{L}{2r}\sqrt{\frac{P}{AE}}\right)\right]$$

(9.16)

Equations 9.16 are derived in Chapter 11 as Eq. 11.11.   Figure 9.5 illustrates an eccentrically applied load.

Fig. 9.5 Column with eccentrically applied load.

4. Combined Initial Curvature and Eccentrically Applied Load

For a pinned-end column with an initial deflection of e' at the mid-height of the column, and an eccentricity of load e'', the following equation holds true:

$$v_0 = e'\left(1 + \frac{1}{2}\cdot\frac{\pi^2}{4}\cdot\frac{P}{P_e} + \frac{5}{24}\cdot\frac{\pi^4}{16}\cdot\frac{P^2}{P_e{}^2} + \cdots\right)$$

$$+ e''\left(1 + \frac{5}{12}\cdot\frac{\pi^2}{4}\cdot\frac{P}{P_e} + \frac{61}{360}\cdot\frac{\pi^4}{16}\cdot\frac{P^2}{P_e{}^2} + \cdots\right) \quad (9.17)$$

When $P/P_e < 1/5$ the following equations result from Eq. 9.17 and have a negligible inaccuracy.

$$\left.\begin{aligned} v_0 &= (e' + e'')\left(1 + \frac{3}{2}\frac{P}{P_e}\right) \\[2mm] f_{\max} &= \frac{P}{A}\left[1 + \left(1 + \frac{3}{2}\frac{P}{P_e}\right)\frac{c}{r^2}(e' + e'')\right] \end{aligned}\right\} \quad (9.18)$$

Figure 9.6 shows a column with an initial curvature and an eccentrically applied load.

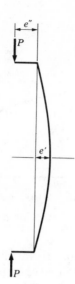

Fig. 9.6 Column with initial curvature and eccentrically applied load.

9.5 THE COLUMN CURVE

Equation 9.1 was an early empirical attempt to express the concept that the strength of a column depends on its length and on its shape, and on the size of the cross section. Equations 9.3, 9.15, 9.16, and 9.18 were correct theoretical solutions to specific cases of column bending or buckling.

These equations and column strength equations in general are based on a column with pinned ends. Such a column is hypothetical, yet its strength is the basis for the design of practical columns and is the reference or anchor point for beam-column behavior.

The strength of a column usually is indicated by a "column curve" which shows graphically the relationship between the buckling stress and the corresponding slenderness ratio, as in Fig. 9.7.

The maximum stress in the column curve may be the absolute maximum stress corresponding to the total failure of the column at the buckling load; it also may be defined as the stress corresponding to the first yield of an extreme fiber in the column cross section.

In the nineteenth century, three important empirical and semi-empirical column formulas were developed:[9.1]

1. Rankine-Gordon (which becomes the secant formula when eccentricity is introduced)
2. Straight-line
3. Johnson parabola

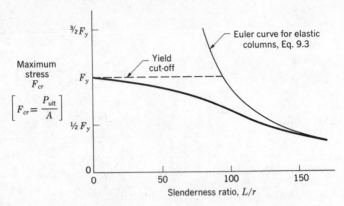

Fig. 9.7 The column curve.

These basic formulas are still in use in many parts of the world. The Johnson parabola, quite similar to the secant formula, is used by the AASHO specifications, and was used in the AISC Specification prior to 1961. The straight-line formula was specified by AREA, although the Johnson parabola was used earlier.

These formulas contain constants which are a function of end-conditions, accidental out-of-straightness, and initial crookedness with different values being used in the different specifications. Figure 9.8 shows these three main column curves; the Johnson parabola and the Rankine-Gordon curve are almost coincident. They are discussed further in Art. 9.9.

As shown in Art. 9.7, recent developments have resulted in a column curve not unlike the Johnson parabola, but based on a more complete understanding of column behavior.

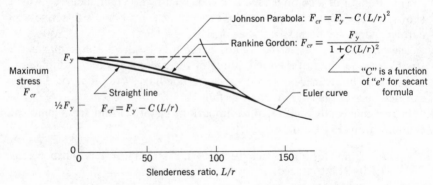

Fig. 9.8 Basic column curves of the nineteenth century.

Column curves generally are such empirical or semi-empirical relationships defining a set of experimental test data. Recent European efforts[9.9] have applied statistical analysis to hundreds of test data to obtain a column curve.

Examples 9.1 and 9.2 illustrate the design of columns using the column curves specified by design specifications. Design criteria specify allowable stresses for a given slenderness ratio.

9.6 COLUMN THEORIES FOR INELASTIC BUCKLING

Column equations for elastic buckling have a limited use, since such buckling occurs only for very slender columns. Most practical columns fail in the inelastic range. The theories which have been developed for inelastic buckling analysis are:

1. Tangent modulus theory
2. Reduced modulus theory

1. Early Development of Inelastic Column Theories

In 1889 Engesser stated his first theory, the original tangent modulus theory. He assumed that the column "remained straight up to the moment of failure, and the (tangent) modulus of elasticity remained constant right across the cross section."[9.1]   He modified Euler's buckling formula by replacing E by E_t. This original concept of the tangent modulus buckling load is illustrated in Fig. 9.9. The generalized stress-strain relationship assumed implies that stress is not directly proportional to strain, as is the case with elastic buckling.

2. Reduced Modulus Theory

In 1895, under the influence of Considere and Jasinski, Engesser realized that when the slightest column deflection occurs the material on the convex side would unload, while the material on the concave side would continue loading.[9.1] He declared his original tangent modulus theory to be invalid, and replaced it by the reduced modulus theory. The reduced modulus theory assumes that strain reversal of fibers takes place on the convex side of the bent column when it passes from the straight to the deflected configuration.[7.6]   This is illustrated in Fig. 9.10.

As was done in the derivation of Eq. 9.3, the equation of internal and external forces on the critical cross section of the deflected column leads to the critical or buckling stress.

$$F_{cr} = F_r = \frac{\pi^2 E_r}{(L/r)^2} \tag{9.19}$$

EXAMPLE 9.1

PROBLEM:
Design the column shown. The unbraced length is 20 ft, fixed at the bottom, pinned at the top. Compare the designs obtained with A36, A 572 (65) and A 514 steels. Compare AISC, AREA and AASHO.

SOLUTION:

AISC: (A36)

TRY W 14 x150 : r_y = 3.99 in., A = 44.1 in^2

$$\frac{KL}{r} = \frac{0.8 \times 20 \times 12}{3.99} = 48 \quad \therefore \quad F_a = 18.53 \, \text{ksi} \quad \{ \text{Table} \, [\text{I}-36] \}$$

A x F_a = 44.1 x 18.53 = 819 kip > 800 kip   OK

$\therefore$ USE W 14 x150 (A36)

AISC: (A 572 (65))

Try W 14 x103 : r_y = 3.72 in., A = 30.3 in^2

$$\frac{KL}{r} = \frac{0.8 \times 20 \times 12}{3.72} = 52 \quad \therefore F_a = 28.8 \, \text{ksi}$$

A x F_a = 30.3 x 28.8 = 870 kip > 800 kip   O.K.

$\therefore$ USE W 14 x 103 (A 572 (65))

AISC: (A 514)

Try W 14 x 84 : r_y = 3.02 in., A = 24.7 in^2

$$\frac{KL}{r_y} = \frac{0.8 \times 20 \times 12}{3.02} = 63.5 \quad \therefore F_a = 33.7 \, \text{ksi}$$

A x F_a = 24.7 x 33.7 = 832 kip > 800 kip   O.K.

$\therefore$ USE W 14 x 84 (A 514)

AASHO: (A 36)

For riveted ends:

$$F_a = 16.0 - \frac{0.3}{1000} \left(\frac{L}{r} \right)^2$$

Trial stress = 15 ksi

$A_{req'd}$ = 53.2 in^2

Try W 14 x 184 : r_y = 4.04 in.

A = 54.1 in^2

$\frac{L}{r_y}$ = 59.5

$\therefore F_a$ = 14.94 ksi

A x F_a = 54.1 x 14.94 = 808 kip > 800 kip   O.K.

$\therefore$ USE W 14 x184

EXAMPLE 9.1 (Continued)

AREA: (A 36)

For riveted ends:

$$F_a = 21,500 - 75 \frac{L}{r}$$

Try W 14 x 167: r_y = 4.01 in.

               A = 49.1 in?

               $\frac{L}{r_y}$ = 59.8

            ∴ F_a = 17.0 ksi

          A x F_a = 835 kip > 800 kip   <u>O.K.</u>

       ∴ USE W 14 x 167

SUMMARY:

AISC SPECIFICATION			AASHO	AREA
A 36	A 572(65)	(A 514)	A 36	A36
W 14 x150	W 14 x103	W 14 x 84	W 14 x184	W 14 x167

The derivation may be shown in its simplest form by considering the cross section of the column to be rectangular, as shown in Fig. 9.11. In this case, equilibrium leads to expressions for external and internal moment given by

$$M_{\text{ext}} = Pv \tag{9.4}$$

$$M_{\text{int}} = \underbrace{[(\phi E_t)d_1]}_{\text{stress}} \underbrace{\left[\frac{bd_1}{2}\right]}_{\text{area}} \underbrace{[\tfrac{2}{3}d_1]}_{\substack{\text{lever} \\ \text{arm}}} + [(\phi E)d_2]\left[\frac{bd_2}{2}\right][\tfrac{2}{3}d_2]$$

$$= \phi \frac{b}{3}(E_t d_1{}^3 + E d_2{}^3)$$

Now, $\phi = 1/R$. Hence

$$M_{\text{int}} = \left(\frac{1}{R}\right)\frac{b}{3}(E_t d_1{}^3 + E d_2{}^3) = \frac{E_r I}{R} \tag{9.20}$$

where the reduced modulus E_r is defined by

$$E_r = \left(\frac{1}{I}\right)\frac{b}{3}(E_t d_1{}^3 + E d_2{}^3) \tag{9.21}$$

Equations 9.4 and 9.20 lead to Eq. 9.19 in the same manner as Eq. 9.3 was obtained.

The reduced modulus E_r depends upon the shape of the cross section, and is a complicated expression for most shapes. For the case of the rectangular section, however, a simple expression (Eq. 9.23) results. Equating forces for

EXAMPLE 9.2

PROBLEM:
It is desired to design alternatives for a column using A36 and A572(42) steels. The column has an unbraced length of 12ft and carries a load of 2100 kip. For reasons of convenience in detailing, the section has to conform with the 16 in. ($\pm1/4$ in.) W sections used in the upper floors. Make use of cover plates if necessary. Use AISC specification.

SOLUTION:
A column of A572(42) material will be designed first and revised by adding cover plates to the section obtained to make use of the lower strength materials.

A572(42)

For a 16 in. W section for columns (14 x 16 variety)

$$r \cong 4.1$$

$$\frac{L}{r} = \frac{144}{4.1} = 35$$

Now, $F_y = 42.0\text{ksi}$ $(1\frac{1}{2}" \le t_{max} \le 4")$

From Table [1-42] $F_a = 22.59\text{ksi}$

$$A_{req'd} = \frac{2100}{22.59} = 93 \text{ in.}^2$$

TRY W 14x314 $r = 4.20$ in., $A = 92.3$ in.^{2}

$$\frac{L}{r} = \frac{144}{4.20} = 34 \quad \therefore F_a = 22.69\text{ksi} \quad (\text{Table [1.42]})$$

$A \times F_a = 92.3 \times 22.69 = 2090\text{kip} \approx 2100 \text{ kip}$ OK

$$\underline{\text{USE W 14x314 (A 572(42))}}$$

A36

Trial stress (for $\frac{L}{r} \approx 34$) = 19.65 ksi   (Table [1-36])

$$A_{req'd} = \frac{2100}{19.65} = 107 \text{ in.}^2$$

Instead of placing cover plates on the W 14 x 314 which will require a fairly narrow plate (for which fillers will be required later in splicing) a lighter section will be used and reinforced with cover plates to obtain a plate thickness comparable to the thickness of the flanges.

TRY W14x264 w/2 Cover plates 15"x1"
Check width to thickness requirements, Section [1.9] (assume cover pl. is welded to W-sect.)
   Section[1.9.1] – Since the cover plate width $\cong$ flange width, projecting element is not critical.
   Section[1.9.2] – Unsupported width of cover plate between lines of welds $\le 42t$.

$$\frac{15}{1} = 15 < 42$$ OK

$I_y = 1331.2 + 2(281.3) = 1893.8$ in.^{4}
$A = 77.63 + 2(15.0) = 107.63$ in.^{2}

$$r_y = \sqrt{\frac{1894}{108}} = 4.2 \text{ in.}$$

$$\frac{L}{r} = \frac{144}{4.2} = 34.5 \quad \therefore F_a = 19.61 \quad (\text{Table [1-36]})$$

$A \times F_a = 108 \times 19.61 = 2110\text{kip} > 2100 \text{ kip}$ OK

$$\underline{\text{USE W 14x264 w. 2 Cover Pls 15"x1" (A36)}}$$

SUMMARY:

AISC	SECTION	WEIGHT lb/ft
A36	W 14x264 w/15"x1"Cov. Pls	366.0
A 572 (42)	W 14 x 314	314.0

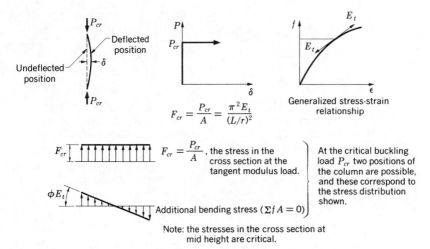

Fig. 9.9   Original Engesser theory.

equilibrium at the critical section,

$$[(\phi E_t)d_1]\underbrace{\left[\frac{bd_1}{2}\right]}_{} = [(\phi E)d_2]\left[\frac{bd_2}{2}\right]$$
$$\quad\underbrace{}_{\text{stress}}\underbrace{\phantom{\left[\frac{bd_1}{2}\right]}}_{\text{area}}$$

or

$$d_1{}^2 = \frac{E}{E_t}d_2{}^2 \tag{9.22}$$

Substituting Eq. 9.22 in Eq. 9.21 and utilizing $I = \frac{1}{12}b(d_1 + d_2)^3$ gives

$$E_r = \frac{4E_t E}{(E^{1/2} + E_t^{1/2})^2} \tag{9.23}$$

which holds true for rectangular shapes.

The reduced modulus theory appeared reasonable due to the recognition of the simultaneous loading and unloading of fibers.   However, experimental results on columns always tended to be much closer to the tangent modulus than to the reduced-modulus-predicted curves.

3. The Shanley Contribution

Shanley drew attention to the assumption in the reduced modulus theory of a column remaining perfectly straight up to the reduced modulus load. (The criterion of perfect straightness is an assumption for any stability problem.)

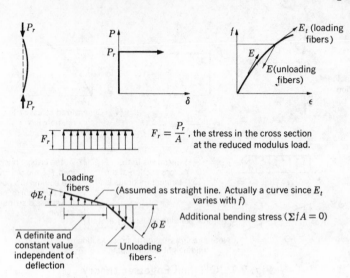

$F_r = \dfrac{P_r}{A}$, the stress in the cross section at the reduced modulus load.

Fig. 9.10 Reduced-modulus concept.

Shanley showed that an initially straight column will buckle at the tangent modulus load, and then will continue to bend with increasing axial load. He showed this as the result of experiments and also confirmed the experiments with a simple analytical study on a column model, the "Shanley" model.[7.6,9.4] The Shanley model is a column with a concentrated flexibility which makes analytical considerations comparatively simple. It is shown in Fig. 9.12. The two rigid ends of the column are connected by springs at the

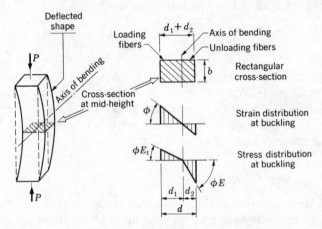

Fig. 9.11 Column buckling, reduced-modulus concept, rectangular cross section.

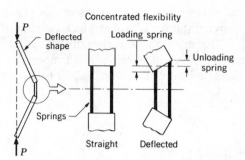

Fig. 9.12 Shanley model column.

center; hence only the behavior of the springs need be considered to study the behavior of the column.

He wrote:[9.4] "Upon reaching the critical tangent modulus load, there is nothing to prevent the column from bending simultaneously with increasing axial load." This quotation may be regarded as the introduction of a completely new concept of column behavior.

It is not necessary to use a Shanley model to illustrate this behavior analytically—the model merely creates simpler handling of the mathematics. Analytically, other investigators have illustrated the Shanley column behavior using actual column properties.[9.10,9.11,9.12,9.13]

The behavior of a centrally loaded column is typified by the stress-deflection curve shown in Fig. 9.13. In order to have such a curve which reaches a maximum and then decreases, it is necessary to have a distribution of strain, as shown in the figure.

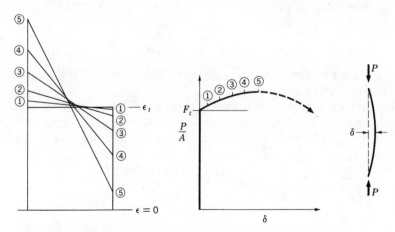

Fig. 9.13 Strain distribution and column deflection.

With the Shanley column concept the tangent modulus load is the lower bound for column strength; it is the load at which an initially straight column will start to bend.   The upper bound is the reduced modulus load since it is the maximum load a column will sustain if it is temporarily supported up to that load.   The ultimate load of a column will lie between these two limits, as shown in Fig. 9.14.

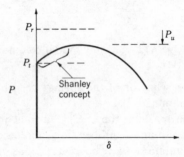

Fig. 9.14 Deflection of initially straight centrally loaded column.

4. Tangent Modulus Theory

Under the influence of Shanley's contributions the tangent modulus concept was reinstated and may be stated as "no strain reversal takes place on the convex side of the bent column when it passes from the straight form to the adjacent deflected configuration."[7.6]   The tangent modulus buckling phenomenon may be explained by reference to Fig. 9.15.

By equating internal and external forces acting on the critical cross section of the deflected column, it may be shown that the critical or buckling stress is

$$F_{cr} = F_t = \frac{\pi^2 E_t}{(L/r)^2} \tag{9.24}$$

where E_t is the tangent modulus of the stress-strain relationship of the material at the critical stress.

The derivation follows the lines of that for the Euler buckling load and the reduced modulus buckling load.   For the rectangular cross section considered in Fig. 9.11, with the additional infinitesimal bending stress shown in Fig. 9.15,

$$M_{\text{int}} = \underbrace{\left[\frac{(\phi E_t)d}{2}\right]}_{\text{stress}} \underbrace{\left[\frac{bd}{4}\right]}_{\text{area}} \underbrace{\left[\frac{2}{3}d\right]}_{\substack{\text{lever} \\ \text{arm}}}$$

$$= \phi E_t \cdot \frac{bd^3}{12}$$

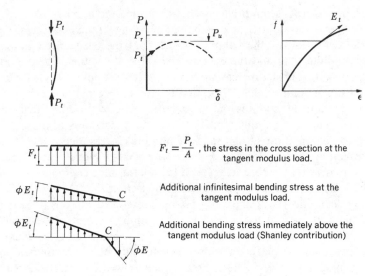

$F_t = \dfrac{P_t}{A}$, the stress in the cross section at the tangent modulus load.

Additional infinitesimal bending stress at the tangent modulus load.

Additional bending stress immediately above the tangent modulus load (Shanley contribution)

Fig. 9.15 Tangent-modulus concept.

that is,

$$M_{\text{int}} = \frac{E_t I}{R} \tag{9.25}$$

The equilibrium of forces leads to Eq. 9.24.

Equation 9.24 has been used as the basis for column strength since it is simple to use and since it is a lower bound to column strength. The Column Research Council in 1952 issued a memorandum[9.14] stating that "the tangent modulus formula for the buckling strength affords a proper basis for the establishment of working load formulas."

In Fig. 9.15, consider the position of the point C. (C is the position of the fiber which has a zero increase in strain between any two adjacent deflected positions of the columns.) When the column is straight, point C is at infinity. At the tangent modulus buckling load, point C may be regarded as moving from infinity to the edge of the cross section, instantaneously. In Fig. 9.15 this limiting condition is shown by the infinitesimal stress distribution. As soon as the column is no longer straight, the tangent modulus load has been exceeded and point C is inside the cross section. As soon as this happens, unloading of some fibers occurs, with loading of other fibers of the cross section.

It is necessary to realize that the tangent modulus load is an instantaneous phase only: as soon as the column buckles, unloading of some fibers takes place, and the column load increases with increase in column deflection. In Fig. 9.13 at the onset of deflection position 1, only a small amount of the cross section has a relief of strain, and hence of stress. This part of the cross section (the part having unloading strains) increases in area until the deflection of the column is such that the tensile force above the tangent modulus load and the compressive force above the tangent modulus load produce no further increase in load. After this point, 5, the deflection is in unstable equilibrium. (Note that the forces referred to are those existing after the tangent modulus load; they correspond to the additional stresses above the tangent modulus stress. The actual stress distribution of any load is the summation of the tangent modulus stress plus these bending stresses, unloading and loading.) With any further movement, the added tensile forces will increase at a greater rate than the added compressive forces, and the load will decrease with increased deflection.

Figure 9.16 illustrates the progressive stress distributions in a simple rectangular cross section as the load on the column increases, causes buckling, and continues to increase under increasing deflection until the maximum load is reached. (For the purposes of illustration, the simplified stress-strain relationship of Fig. 9.18 is used, rather than the generalized relationship; this implies that $E_t = E$ over the complete elastic cross section, whether loading or unloading.)

The stress distribution shown is not necessarily the same for all columns, since cross-sectional area, slenderness ratio, and the presence of residual stresses all play a role in practical columns.[9.11,9.12] In particular, the stresses at the extreme loading fiber will reach the yield at some stage of loading. The stress distribution in practical columns under load is discussed in Ref. 9.13.

9.7 RESIDUAL STRESS AND COLUMN STRENGTH

1. Residual Stresses

Residual stresses are formed in a structural member as a result of plastic deformations; they are stresses which exist in the cross section even before the application of an external load. These plastic deformations may be due to cooling after hot-rolling or welding, or due to fabrication operations such as flame-cutting, cold-bending, or cambering. In rolled shapes, these deformations always occur during the process of cooling from the rolling temperature to air temperature; the plastic deformations result from the fact that some parts of the shape cool much more rapidly than others, causing inelastic deformations in the slower-cooling portions. (The flange tips of a wide-flange shape, for example, would cool more rapidly than the juncture of flange and web.) As explained in Art. 14.5, residual stresses are also introduced during the welding operation as a result of the localized heat input and

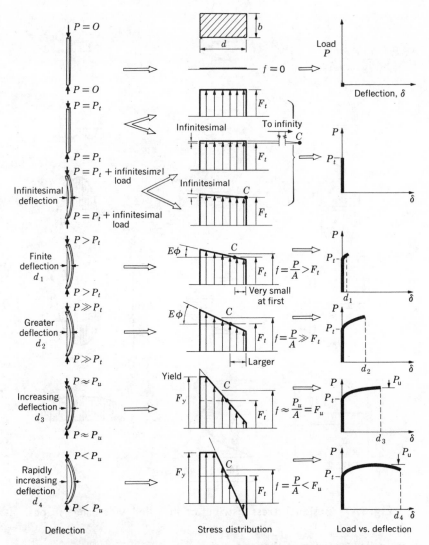

Fig. 9.16 Progressive stress distribution as column is loaded.

resultant plastic deformation.   Reference 9.15 describes one of the common means of measuring residual stresses, and Ref. 9.16 compares a number of measuring techniques.

The magnitude and distribution of residual stresses depend on the shape of the cross section, rolling or welding procedures, cooling conditions, and material properties.[2.4,9.11,9.17,9.18,9.19] Typical cooling residual stresses for

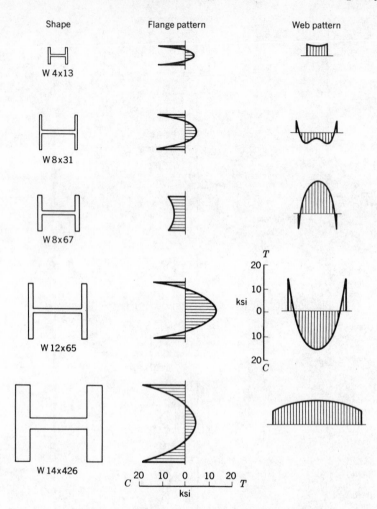

Fig. 9.17 Residual stress distribution in rolled wide-flange shapes.

some rolled wide-flange shapes are shown in Fig. 9.17. The average compressive residual stress at the flange tips of small to medium size shapes is about 13 ksi for ASTM A36 structural carbon steel and for ASTM A242 (50 ksi yield) steels.[2.4,9.20] Residual stresses in welded shapes, in shapes of higher strength steels, and in heavy shapes are considered below.

2. Rolled H-Shapes

The theories discussed in Art. 9.6 were based on the material being homogeneous. Actually, mechanical property tests on a coupon from a portion of the cross section of a column are not necessarily indicative of the mechanical properties of the cross section as a whole.

Research in the 1950's and early 1960's has shown that the main factor influencing the strength of straight centrally loaded steel columns is the magnitude and distribution of residual stress within the column cross section.[2.4,7.14,9.5,9.6,9.7,9.11,9.15,9.17,9.20–9.29] Except for certain modifications, to be discussed below, the general discussion in the preceding section applies also to columns which contain residual stresses.

The stress-strain curve for a complete cross section of steel reflects the presence of residual stress, and so the application of the tangent modulus concept to the stress-strain curve of a coupon from the cross section would be

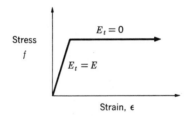

Fig. 9.18　Idealized stress-strain curve.

erroneous. A coupon will be essentially free of residual stresses. It will be seen below that the tangent modulus concept applied to cross sections containing residual stresses results in equations for column strength which are functions of E_t rather than utilizing E_t directly, as in Eq. 9.24.

For column cross sections containing residual stresses, the tangent modulus and reduced modulus theories for column buckling define loads differing from those for the same columns free of residual stress.[9.27]

A column cross section containing residual stresses will have certain fibers yield before others when the column is loaded. Compressive residual stresses exist in these fibers even before the load is applied. The material of the cross section is no longer homogeneous, and the general equations of the preceding section no longer apply. A major difficulty in the analytical solution of column strength utilizing the tangent modulus or reduced modulus concepts is that, whenever the stress-strain relationship is non-linear, the superposition of stresses no longer holds true. However, a comparatively simple solution for column instability with the tangent modulus concept may be obtained when it is assumed that every fiber in the cross section has an idealized elastic, perfectly plastic stress-strain relationship.

With an idealized stress-strain relationship as given in Fig. 9.18 it may be shown that, for a column of symmetrical cross section containing residual stresses in a symmetrical distribution, the critical stress at the buckling load as defined by the tangent modulus concept is given by the following equation.[9.15]

$$\frac{P_{cr}}{A} = F_{cr} = \pi^2 \frac{EI_e/I}{(L/r)^2} \tag{9.26}$$

where EI_e is the effective bending rigidity of the column, derived through the realization that the yielded portion of a structural shape offers no additional resistance to bending. The buckling strength is thus a function of the moment of inertia of the elastic part.[9.22]

The derivation of Eq. 9.26 utilizes again the equilibrium of internal and external moments; in this case, due to the effect of residual stresses, the internal moment must consider the summation of the moments acting on each individual fiber in the cross section. For one fiber

$$M_{\text{int}} = \underbrace{[(\phi E_t)y]}_{\substack{\text{stress}}} \underbrace{[dA]}_{\substack{\text{area} \\ \text{of} \\ \text{fiber}}} \underbrace{[y]}_{\substack{\text{lever} \\ \text{arm}}} \tag{9.27}$$

and for the complete cross section

$$M_{\text{int}} = \int_A \phi E_t y^2 \, dA \tag{9.28}$$

Hence

$$E' = \frac{1}{I} \int_A E_t y^2 \, dA \tag{9.29}$$

where E' is the effective modulus for the cross section. For the idealized stress-strain relationship of Fig. 9.18,

$$\left. \begin{array}{ll} E_t = E & \text{for } f < F_y \\ E_t = 0 & \text{for } f = F_y \end{array} \right\} \tag{9.30}$$

and, substituting in Eq. 9.29

$$E = \frac{E}{I} \int_{A_e} y^2 \, dA = E \frac{I_e}{I} \tag{9.31}$$

where I_e is the moment of inertia about the extreme fiber of the elastic part of the cross section A_e.

The use of equilibrium and a differential equation the same as Eq. 9.8 results in Eq. 9.26.

The solution of Eq. 9.26 requires the function relating I_e to F_{cr}. This can be accomplished as indicated in Refs. 9.15 and 9.20. For the small rolled H-shape, and using the E_t determined from the stress-strain relationship of a stub column, Eqs. 9.32 hold true.[9.15]

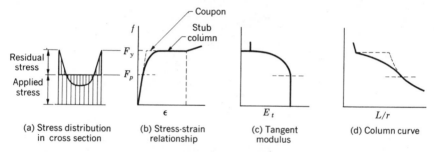

$$E\frac{I_{ex}}{I_x} = \frac{AE_t - \frac{2}{3}A_wE}{2A_f + (A_w/3)} \qquad \left[\approx E\frac{E_t}{E}\right]$$

$$E\frac{I_{ey}}{I_y} = E\left(\frac{AE_t}{A_tE} - \frac{A_w}{2A_f}\right)^3 \qquad \left[\approx E\left(\frac{E_t}{E}\right)^3\right] \tag{9.32}$$

Equations 9.32 enable the direct solution of Eq. 9.26 to be obtained. The relationship between the stub column stress-strain curve and the column curve is typified in Fig. 9.19. The stub column stress-strain relationship

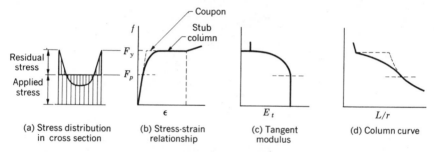

Fig. 9.19 Stress-strain and column curve.

(a) Stress distribution in cross section   (b) Stress-strain relationship   (c) Tangent modulus   (d) Column curve

reflects the presence of residual stresses. This is evident from Fig. 9.19a where, for any fiber, when the sum of the applied stress and the compressive residual stress acting on that fiber becomes equal to the yield stress, yielding will commence in that fiber. The beginning of yielding implies that the stress-strain relationship for the complete cross section is no longer linear, or elastic, as would be the case for a tensile specimen which contains no residual stresses. (The stub column tests is an important control test in experimental investigations of columns, and is described further in Ref. 9.30.)

The column curve in Fig. 9.19d results from the use of the stress-strain relationship (Fig. 9.19b) and the tangent modulus curve (Fig. 9.19c) in Eqs. 9.26 and 9.32.

Equations 9.32 apply only to small and medium-size rolled shapes, and to values of E_t obtained from a stub column stress-strain curve. For all other shapes and fabrication processes, the relationship is more complex, and requires a mathematical formulation of the stress-strain relationship and the residual stress distribution.

From the column curves prepared for small rolled H-shapes, it will be seen that straight-line and parabolic curves give satisfactory predictions for column strength in the weak and strong axes respectively. The maximum loads carried by such columns do not exceed the tangent modulus load enough to warrant the use of a post-buckling analysis.

Test results[2.4] for rolled H-shapes are given in Fig. 9.20 to illustrate the efficacy of the straight-line and parabolic assumptions. The column curves are cut off at $L/r = 20$, to take account of the effect of strain-hardening. The CRC Basic Column Curve is an average (parabolic) curve which is used for bending about both axes; the curve is a compromise, being the average of test results for bending about both axes. The CRC column curve was developed for small rolled H-shapes.[7.14] It is the first column curve based

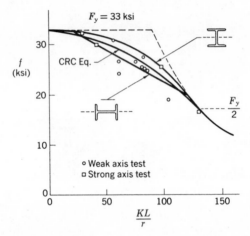

Fig. 9.20 Column curves and test results of rolled H-shapes.

on a theoretical study reflecting actual conditions. The curve was adopted by AISC as the maximum strength column curve used as the basis for the design curves. Even though based on the strength of small rolled shapes, the AISC column design curve currently is used for all shapes, materials and fabrication processes.

3. Welded Shapes

The residual stress distribution set up in a cross section due to welding may be vastly different from that set up in a rolled shape due to cooling, as seen from a comparison of the residual stress distributions in Fig. 9.21. The fact that welding induces a different distribution of residual stress implies that welding may induce different column strength properties.

Welded columns have high residual stresses with the tensile residual stresses approaching the yield point at the weld.[9.6,9.11,9.19,9.24,9.31,9.32,9.33]

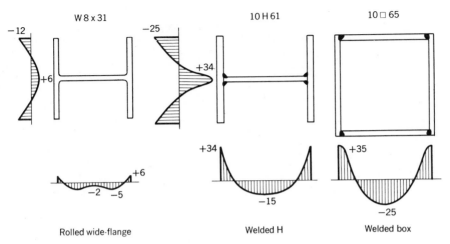

Fig. 9.21   Residual stresses in small welded shapes.

The magnitude and distribution of residual stresses in welded shapes are markedly influenced by the geometry of the cross-sectional shape.[9.11,9.19] The column strength of centrally loaded built-up members can be predicted by the same techniques as for rolled shapes with symmetrical cooling residual stresses. However, it has been shown that the use of the tangent modulus concept is not realistic for the prediction of the strength of welded columns.[9.11,9.24,9.34] Because of the large magnitude of the residual stresses, a maximum strength analysis is necessary. Welded columns will tend to have greater out-of-straightness than rolled shapes and the effect of this on column strength is so great that it cannot be neglected.[9.31] Hence, the study of the strength of centrally loaded welded columns differs essentially from that of centrally loaded rolled columns. Welded columns need a maximum strength study whereas the tangent modulus buckling load presents a realistic figure for rolled columns. These findings hold true for small and medium-size welded shapes built up from UM (universal-mill) plates; studies at Lehigh University are presently under way (1973) on very heavy welded shapes.

For centrally loaded columns, the maximum strength is a load in excess of the tangent modulus load for the column in the deflected position, and hence is a post-buckling problem for the column in the inelastic range. It is difficult to obtain a perfectly general solution except for very simple column shapes which do not contain residual stresses and where the stress-strain relationship of the material may be expressed in simple form.[9.11]

References 9.11, 9.24, and 9.31 present analytical and experimental studies on the strength of small welded box and welded H-columns. The results are summarized in Fig. 9.22.

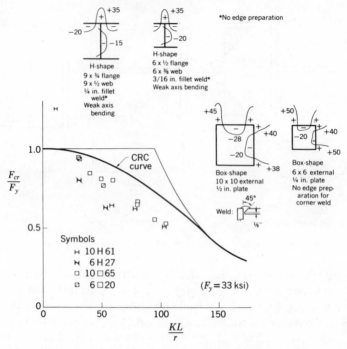

Fig. 9.22 Welded column strength (small shapes).

The reason for the somewhat lower strengths of welded columns of small and medium cross section is twofold: the effect of residual stresses due to welding, and the effect of initial out-of-straightness. The maximum effect of out-of-straightness occurs for the longer columns, L/r from 60 to 120, and this effect will be discussed further below. For shorter columns, the box shapes tend to be stronger than the H-shapes, since the box shapes retain the corners in the elastic condition throughout the bending history of the columns, due to the favorable residual stress distribution there as a result of the weld. (For the same reason, box shapes are able to sustain the maximum load for much larger deflections than the H-shapes.[9.11]) On the other hand, H-shapes, with the compressive residual stress at the flange tips, lose a major part of their rigidity very early under load, since the flange tips yield first.

4. Welded Shapes from Flame-Cut Plates

The above discussion has considered welded columns built up from universal mill (UM) plates. Some residual stress distributions are more "favorable" for column strength than are others. A "favorable" residual stress distribution is one which leads to improved column strength, such as, for example, when the tensile residual stresses are farthest from the axis of bending. Thus, a welded box shape is compared with a welded H-shape above.

For H-shapes, the use of flame-cut (FC) plates generally leads to improved column strength because of the tensile residual stresses set up on the flange tips by the operation of flame-cutting. In practice, most welded shapes are

Fig. 9.23 Automatic flame cutting.

fabricated from FC plates, since rolled UM plates often do not satisfy straightness requirements.

Figure 9.23 shows a wide plate being flame-cut to furnish a number of narrow plates for use as components of welded structural shapes. The residual stress distribution in two typical FC plates is given in Fig. 9.24, while Fig. 9.25 shows the residual stress distribution in a welded H-shape fabricated from these same plates.[9.32] The flame-cutting operation is a heat input, similar to welding, and results in a tensile residual stress equal in magnitude to the yield stress of the material. When additional sources of heat, such as welding, are applied, the additional residual stress will affect

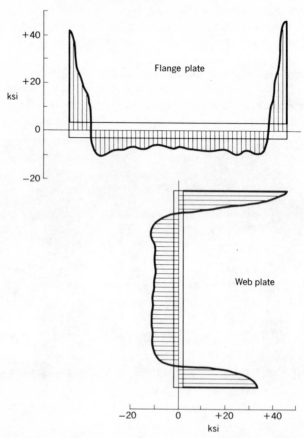

Fig. 9.24 Residual stresses in component FC plates.

that due to flame-cutting, as is shown in Fig. 9.25, where the high value of tension at the FC flange tips is reduced considerably by the welding.

Figure 9.26 compares both test results and theoretical tangent modulus predictions for similar column shapes welded from rolled UM plates and from FC plates. The strength of small-to-medium-size FC welded columns has been shown to be much the same as that of similar rolled shapes, for structural carbon steel.[9.32]

Some early studies[9.35,9.36] showed similar results—improved column strength resulted from both the laying of a weld bead on the flange tips of a rolled shape, and the reinforcement of a rolled shape by the welding of cover plates. Both of these methods are methods of reinforcement since they lead to increased column capacities, useful when loading conditions have changed. (Note that both methods may require support of the column and its loads

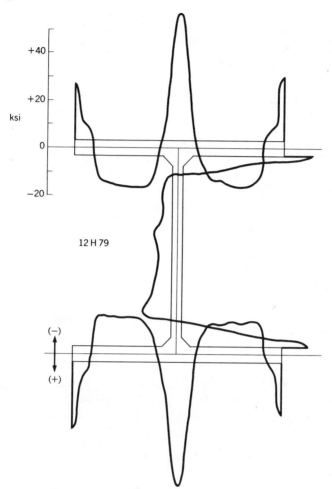

Fig. 9.25 Residual stresses in welded shape.

during welding.) Figure 9.27, (a) and (b), compares the strength of columns before and after reinforcement by welding alone, and by welding of cover plates, respectively.

5. Heavy Shapes

While most shapes in practice are small to medium in size, such as those discussed above, there are many situations requiring the use of "heavy" shapes. A "heavy" shape may be defined as one where the thinnest component plate exceeds $1\frac{1}{2}''$ in thickness. Heavy shapes are used in the lower

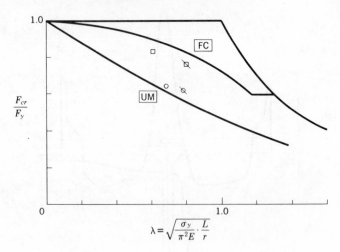

$$\lambda = \sqrt{\frac{\sigma_y}{\pi^2 E} \cdot \frac{L}{r}}$$

Fig. 9.26 Strength of small welded shapes.

stories of multi-story buildings, in major bridges, and in such special buildings as the Apollo Assembly Building at the Cape Kennedy Space Station. Figure 9.28 shows the cross sections of some columns used in large buildings, and compares these with cross sections considered in a research program currently (1973) under way at Lehigh University.[9·19]

Figure 9.29 illustrates the smallest and largest welded column shapes considered in the overall research program. It would appear that the behavior of the two is different; however, specification requirements do not take the size effect into account.

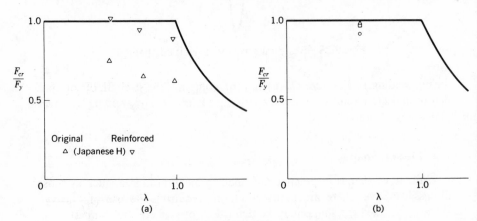

Fig. 9.27 Reinforcement of columns.

Heavy column shapes in existing structures

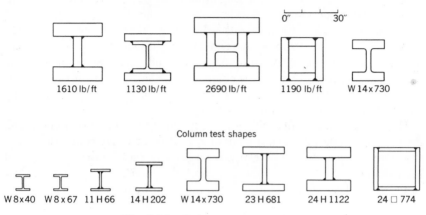

1610 lb/ft   1130 lb/ft   2690 lb/ft   1190 lb/ft   W 14 x 730

Column test shapes

W 8x40   W 8 x 67   11 H 66   14 H 202   W 14 x 730   23 H 681   24 H 1122   24 □ 774

Fig. 9.28   Column cross sections.

Fig. 9.29   Largest and smallest column shapes in test program.

The residual stress distribution in a heavy shape differs from that in a small shape by two major factors—the magnitudes may be considerably larger, and there is also a considerable variation through the thickness. Figure 9.30 shows residual stress measurements in a rolled shape,[9.18] and in a welded shape.[9.19] Figure 9.31 shows the residual stress distribution in thick plates; the distribution is shown for both non-welded and welded plates of the same size. The influence of welding is comparatively small; most of the residual stress distribution is created during the operations of cooling after rolling and cooling after flame-cutting.[9.18,9.19]

The large magnitudes of residual stress, combined with the variation through the thickness indicate reduced column strengths for heavy rolled shapes, as shown in Fig. 9.32.[9.18,9.34] Relatively large slenderness ratios show most of the reduction—for the usual short columns in practice, the strength is close to the yield value, although lower than indicated by the CRC curve.

6. Effect of Yield Strength

Except for the very slender columns, higher column strengths are obtained most simply by using steel of higher yield strength. The results of tests have indicated that the residual stresses arising in shapes of high-strength steels are of the same order of magnitude as in structural carbon steels—hence, the effect of residual stresses becomes comparatively smaller for steels of higher yield strengths.[9.20,9.21,9.33,9.37] Residual stresses are mainly a function of geometry.[9.38]

Figures 9.33,[9.5] 9.34, and 9.35[9.33,9.39] compare the strength of rolled and welded columns for the same shape, for low yield point and high yield point steels. The comparison is not quite complete in the case of rolled A514 steel shapes, since A514 steel is quenched and tempered, and thus contains very small magnitudes of residual stress due to the tempering operation.[9.39,9.40] Reference 9.41 gives some information on the behavior of a rolled column of a 130-ksi-yield quenched and tempered steel.

Results of column tests for structural carbon steel (yield strengths of 33 and 36 ksi respectively), and for A514 steel (quenched and tempered steel, yield strength 100 ksi) are compared in Fig. 9.36[9.39] for both rolled and welded shapes.

7. Other Factors

A number of other factors must be considered in any general study of columns. These factors may play an important part in design. Those to be considered here are the shape of the cross section, riveting, annealing, out-of-straightness, cold-straightening, and effective length.

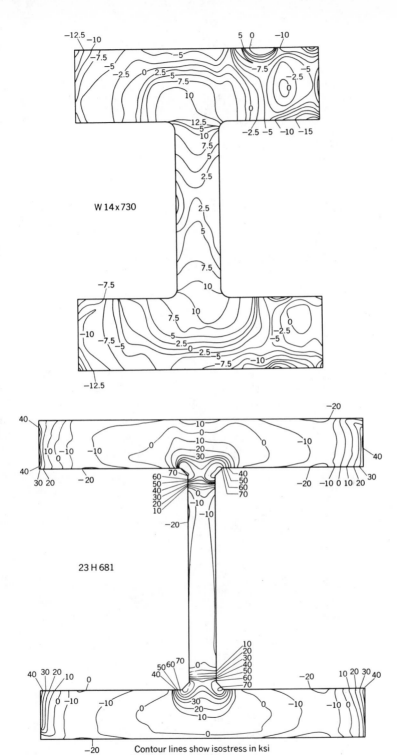

Fig. 9.30 Residual stresses in rolled and welded heavy shapes.

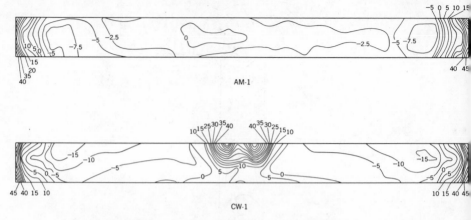

AM-1

CW-1

Fig. 9.31   Residual stresses in 24 × 2 in. plates.

In general, no particular form of cross-sectional shape can be regarded as being the best for column use.   Every situation requires its own evaluation. Box shapes, however, are somewhat stronger than corresponding H-shapes; a cost-strength study would be required for any final decisions.   For the low slenderness ratios (L/r up to 60) when out-of-straightness is not an important factor, columns with favorable residual stress distribution will be stronger than columns with unfavorable distribution.   Whether the residual stress is due to welding or to cooling from hot-rolling, if the material furthest from the axis of bending is in a state of residual compression, then this

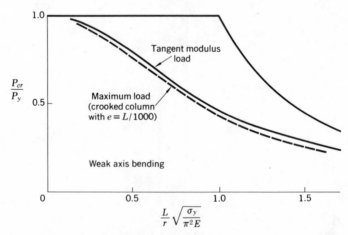

Fig. 9.32   Tangent modulus column strength for W14 × 730 shape based on assumed residual stress distribution.

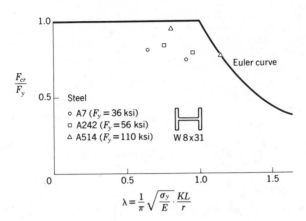

Fig. 9.33 Rolled wide-flange columns and yield point.

material will yield first under load, leading to column failure at a lower load than would otherwise be expected.

Riveted columns exhibit strengths similar to those of rolled columns.[9.24] The reason is that the cooling residual stresses of the component rolled parts have not been changed by the process of riveting, as they would have been by welding.

Annealing reduces the residual stress magnitude to a comparatively low value, so that there is an accompanying increase in column strength; see Fig. 9.37.

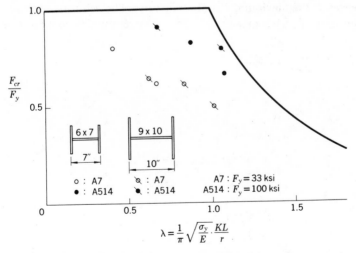

Fig. 9.34 Welded H-columns and yield point.

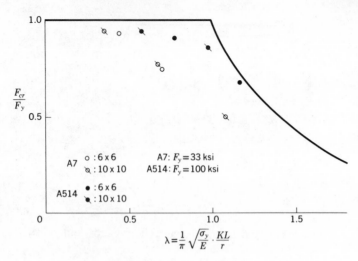

Fig. 9.35 Welded box columns and yield point.

Out-of-straightness is unavoidable and, together with the effect of residual stresses, is a significant factor involved in the strength of columns. Out-of-straightness is used here to refer to all deviations which result in an eccentrically loaded column: initial curvature, eccentric application of load, and unsymmetrical residual stress distributions.

Reference 7.6 gives a general study of eccentrically loaded columns. References 9.11, 9.12, and 9.32 take into account the effect of residual stress. In general, the maximum out-of-straightness allowed for rolled and welded columns is of such a magnitude that the corresponding column strength will

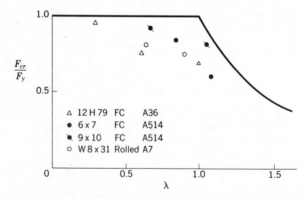

Fig. 9.36 Rolled columns, welded H-columns (FC), and yield point.

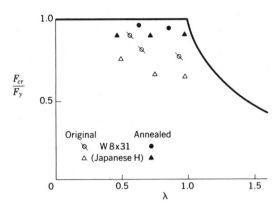

Fig. 9.37 Annealed column strength.

lie on the lower boundary of the test results shown in Fig. 9.38b. The expected maximum out-of-straightness (for example, of the order of ¼ in. in a 20-ft column, AISC Section [1.23.8.1]) will reduce the column strength about 25 per cent below that indicated by the CRC curve in the medium slenderness ratios (*L/r* from 50 to 120). The effect of out-of-straightness on shorter columns is not quite as marked.

Most columns will be either straightened, or else will be framed columns, and have a deflected shape other than the simple single-wave deflection curve so that the problem of out-of-straightness usually is not a factor to be checked in design.

The usual structural columns, rolled or welded, will be cold-straightened to the specified tolerances. The process of cold-straightening induces residual stresses which are of a magnitude similar to, although different in distribution from, cooling residual stresses.[9.17] Findings based on members with cooling residual stress patterns will be conservative when applied to straight members whose cooling patterns have been modified by cold bending. This is borne out by the test result shown in Fig. 9.38. At present, there are no test data on the effect of cold-straightening on welded columns. The effect of cold-straightening depends on the method by which it is carried out, whether by "gagging," or by "rotorizing." "Gagging" concentrates the straightening at some sections, leaving the rest of the column with its initial state of residual stress. "Rotorizing" is a continuous straightening process and changes the residual stress pattern completely. In any case, column strength based on cooling or welding residual stresses will normally be considered, as there is no assurance that these residual stresses will be changed to a more favorable distribution.

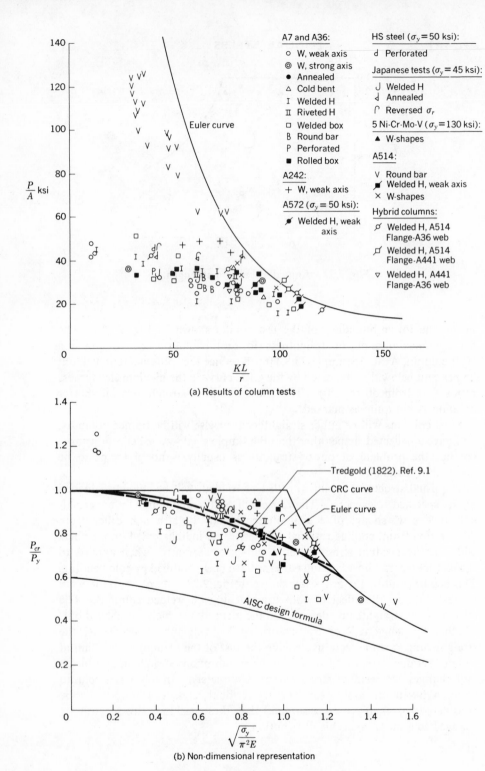

Fig. 9.38 Column test results.

The strength of a column is dependent to a great extent on its effective length, rather than its actual length.　It is the effective length which is used in the equations of design.　(See Chapter 10.)

A pinned-end column does not usually exist in structures.　In earlier days pinned-joint trusses and other structures were used often, their use following analytical and design assumptions.　This simplified the column design, although in most cases a perfect pin-joint did not exist.　Columns which are members of frames (framed columns) make up most of the columns today. The problem of end fixity (the effect of adjacent members) must also be taken into account.　Why, then, even consider a pinned-end column?　The reason is that a pinned-end column may be regarded as a basic or limiting condition, and a knowledge of its behavior under load is necessary in the study of column strength in general; the strength of the pinned-end column is the reference or anchor point for the strength of beam-columns.

The factors which influence the effective length of a column are: (1) flexural column end restraint, (2) torsional column end restraint, and (3) translational column end restraint.　These restraints are usually considered as spring constants in the analytical literature—although values for these spring constants are never given.

Chapter 10 considers effective length and summarizes a number of methods available to determine the K-value.　In particular, the use of Fig. 10.7 is a convenient method.

8. Test Data and the Column Curve

Figure 9.38 presents test data for columns of different shapes, yield strengths, and manufacturing and fabrication methods.　The test data come from references given in this chapter.

The scatter of test results is pronounced, and reflects the residual stress distribution, out-of-straightness, yield strength, and shape of cross section.

It is interesting to compare the modern CRC curve with Tredgold's column curve of 1822.　(In Fig. 9.38, the constants of Tredgold's formula have been determined from correlation with the present column tests.[9.1])

Article 9.5 refers to various column design curves; Fig. 9.38 draws attention to the scatter of results.　Recently, in both the United States and Europe, attention has been given to the possibility of having more than one design curve—one design curve tends to penalize the strength of some columns, while the factor of safety of other columns is reduced. [9.42]

A possible set of three column design curves is sketched in Fig. 9.39 as curves A, B, and C.　Curve A may be appropriate for annealed shapes, rolled tubes, and rolled shapes of very high strength steels; curve B for small rolled shapes, and welded shapes (FC plates); and curve C for welded shapes (UM plates), and for very heavy rolled and welded shapes.

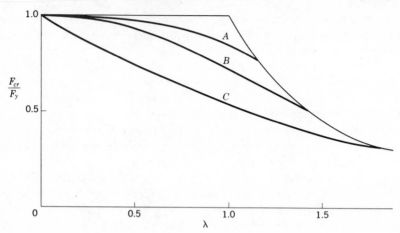

Fig. 9.39 Possible basic column curves.

Example 9.3 illustrates the design of a welded column. The following comments refer to the various methods considered.

For the same material the different design methods give practically the same section except for the two based on the tangent modulus load; in particular, the simple residual stress distribution No. 1 results in a very heavy section. The method utilizing the tables of the AISC Specification is by far the most convenient. Although the secant formula and the equation for initial curvature result in almost the same design section as the AISC formulas, it should be pointed out that while the former were based on incorrect assumptions of failure at first yield, the AISC Formula reflects the actual behavior of a column at its maximum load.

In this problem the tangent modulus solution for residual stress distribution No. 1 considerably underestimates the capacity of the section. This is due primarily to the wide range of slenderness ratio (L/r) where the discontinuity of the column curve occurs, which in turn is due to the idealized residual stress distribution assumed.

A note on the design of the welds is in order here. The welds can actually be designed theoretically by performing a maximum load analysis which will give the column deflection at maximum capacity. Consequently, the stress conditions at the joint of the flange and the web can be obtained by using either the familiar expression for horizontal shear stress VQ/It, or some nominal expression for shear stress such as that found in Ref. 7.14. This analysis, however, yields weld requirements well below the minimum requirements based on the thickness of the plates to be welded. Thus, for instance, the recommended weld sizes in AISC Section [1.17.5] are the appropriate values.

EXAMPLE 9.3

PROBLEM:

Design a welded H-shaped column. L = 15 ft, P = 250 kip

(a) Compare designs for A36 and A572 (50) steels, AISC Spec.

(b) Prepare the design on the basis of the classical secant formula.

(c) Prepare the design using the simple theoretical expression for column bending from initial curvature.

(d) Prepare the design on the basis of the tangent modulus load. Assume both an idealized shape and an idealized residual stress distribution, № 1 and № 2:

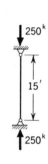

 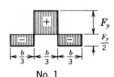 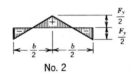

No. 1   No. 2

Compare the designs.

AISC

(a) A36

TRY $\frac{1}{2}$ x 11 flg. $A = 2(\frac{1}{2})11 + \frac{3}{8}(11) = 15.12$ in^2

     $\frac{3}{8}$ x 11 web $I_y = 2(55.5) = 111$ in^4

                  $r_y = 2.82$ in.

                  $\frac{L}{r} = \frac{15 \times 12}{2.82} = 64$ $\therefore$ $F_a = 17.04$ ksi (Table $[1-36]$)

                  $A \times F_a = 258$ kip > 250 kip   <u>O.K.</u>

                           flg. $\frac{b}{t_f} = \frac{5.5}{0.5} = 11 < 16$

       Check $\frac{b}{t}$ ratios (Section $[1.9]$)   <u>O.K.</u>

                          web $\frac{b}{t} = \frac{11}{0.375} = 30 < 42$

USE $\left.\begin{matrix} \frac{1}{2} \text{ x 11 flg.} \\ \frac{3}{8} \text{ x 11 web} \end{matrix}\right.$

A36

AISC

(a) A572 (50)

TRY $\frac{1}{2}$ x 10 flg $A = 13.37$ in^2

     $\frac{3}{8}$ x 10 web $I_y = 83.3$ in^4

                $r_y = 2.5$ in.

                $\frac{L}{r} = 72$ $\therefore$ $F_a = 20.56$ ksi

             $A \times F_a = 274$ kip > 250 kip

                   flg $\frac{b}{t} = \frac{5}{0.5} < 13.5$ <u>O.K.</u>

       Check $\frac{b}{t}$ ratios

                   web $\frac{b}{t} = \frac{9}{0.375} < 36$ <u>O.K.</u>

USE $\left.\begin{matrix} \frac{1}{2} \text{ x 10 flg} \\ \frac{3}{8} \text{ x 9 web} \end{matrix}\right.$

A572 (50)

EXAMPLE 9.3 (Continued)

(b) Classical Secant Formula

The secant formula is based on an assumed initial eccentricity. The carrying capacity P is defined as the load producing the yield stress in any fiber. For this problem assume the maximum allowed out-of-straightness is:

$$\frac{1}{1000}(15 \times 12) = 0.18 \text{ in.} \qquad \text{(Section [1.23.8.i])}$$

The secant formula is:

$$f = \frac{P}{A}\left(1 + \frac{ec}{r^2}\sec\frac{L}{2r}\sqrt{\frac{P}{AE}}\right) \qquad \cdots (9.16)$$

where P = load on the column

Check shape obtained in (a)(A36)

$$r = 2.82 \text{ in.}$$

$$\frac{L}{r} = 64$$

Assume F.S. of AISC Specs. as being applicable,

$$\text{F.S.} = \frac{5}{3} + \frac{3\frac{L/r}{8C_c} - \frac{(L/r)^3}{8C_c^3}}{} \qquad \text{(Section [1.5.1.3] ; for A36 , } C_c = 128)$$

$$\text{F.S.} = \frac{5}{3} + \frac{3}{8}\frac{.64}{128} - \frac{(64)^3}{8(128)^3} = 1.84$$

$$\frac{ec}{r^2} = \frac{0.18(5.5)}{(2.82)^2} = 0.123 \; ; \; \frac{L}{2r} = \frac{15 \times 12}{2(2.82)} = 31.9$$

Design load = 250 kip

Capacity of column, P = 250 x F.S. = 250 (1.84) = 458 kip

$$\sqrt{\frac{P}{AE}} = \sqrt{\frac{458}{15.1 \times 30,000}} = 0.0304$$

Check stress: $f = \dfrac{458}{15.1}\left[1 + 0.123\sec(31.9 \times 0.0304)\right]$

$$f = 30.2\left[1 + 0.216\right] = 36.5 \text{ ksi} \approx F_y = 36 \text{ ksi} \qquad \underline{OK}$$

$$P_{all} = \frac{458}{1.84} = 250 \text{ kip} \qquad \underline{OK}$$

USE A36	$\frac{1}{2}$ x 11 flg.
	$\frac{3}{8}$ x 11 web

(c) Theoretical Solution to Columns with an Initial Curvature

$$f_{max} = \frac{P}{A}\left(1 + \frac{ec}{r^2} \cdot \frac{P_e}{P_e - P}\right)$$

$$\frac{ec}{r^2} = 0.123 \qquad [\text{from (b)}]$$

$$P_e = \frac{\pi^2 EI}{L^2} = \frac{\pi^2(30,000)\,111}{(180)^2} = 1015 \text{ kip}$$

Assuming again $P_{ult} = 458$ kip $\qquad [\text{from (b)}]$

$$f_{max} = \frac{458}{15.1}\left[1 + 0.123\left(\frac{1015}{1015 - 458}\right)\right]$$

$$= 30.3\left[1.226\right] = 36.1 \text{ ksi} \approx F_y = 36 \text{ ksi} \qquad \underline{OK}$$

$$P_{all} = \frac{458}{1.836} = 250 \text{ kip} \qquad \underline{OK}$$

USE A36	$\frac{1}{2}$ x 11 flg.
	$\frac{3}{8}$ x 11 web

EXAMPLE 9.3 (Continued)

(d) Tangent Modulus Solution

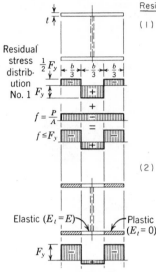

Residual stress distribution No. 1

Residual Stress Distribution No. 1

(1) Elastic case $(f = \frac{P}{A} \leq \frac{F_y}{2})$

Euler's Equation:

$$P_e = \frac{\pi^2 EI}{L^2}$$

$I = Ar^2$ so that $F_{cr} = \frac{\pi^2 E}{(\frac{L}{r})^2}$

Lower Limit: $f \to 0$, $\frac{L}{r} \to \infty$

Upper Limit: $f = \frac{1}{2} F_y$, $\frac{L}{r} = \sqrt{\frac{2\pi^2 E}{F_y}} = 128$ (for A36)

(2) Elastic–plastic case $(f = \frac{P}{A} \geq \frac{1}{2} F_y)$

From eq. 9.26

$$F_{cr} = \frac{\pi^2 \bar{E}}{(\frac{L}{r})^2} \quad \text{where } \bar{E} = \frac{I_e}{I} E$$

and I_e = mom. of inertia of elastic portion

$I = \frac{1}{12} tb^3 \times 2$; $I_e = \frac{1}{12} t (\frac{b}{3})^2 \times 2$

$\frac{I_e}{I} = \frac{1}{27}$; $F_{cr} = \frac{\frac{1}{27} \pi^2 E}{(\frac{L}{r})^2}$

Lower Limit ; $f = \frac{1}{2} F_y$, $\frac{L}{r} = \sqrt{\frac{2\pi^2 E}{27 F_y}} = 24.7$

Upper Limit; $f = F_y$, $\frac{L}{r} = \sqrt{\frac{\pi^2 E}{27 F_y}} = 17.4$

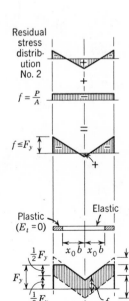

Residual stress distribution No. 2

Residual Stress Distribution No. 2

(1) Elastic case $(f = \frac{P}{A} \leq \frac{F_y}{2})$

Euler's Eq.: $F_{cr} = \frac{\pi^2 E}{(\frac{L}{r})^2}$

Lower Limit: $f \to 0$, $\frac{L}{r} \to \infty$

Upper Limit: $f = \frac{1}{2} F_y$, $\frac{L}{r} = 128$ (for A36)

(2) Elastic – Plastic case $(f = \frac{P}{A} \geq \frac{1}{2} F_y)$

again, $F_{cr} = \frac{\pi^2 \bar{E}}{(\frac{L}{r})^2}$

$I = \frac{1}{12} tb^3 \times 2$

$I_e = \frac{1}{12} t (2 x_0 b)^3 \times 2$

$\frac{I_e}{I} = 8(x_0)^3$; $\boxed{f_{cr} = \frac{\pi^2 E (8 x_0^3)}{(\frac{L}{r})^2}}$ Eq. I

$P_{cr} = 2 \left[f(bt) - 2(\frac{1}{2})(f - \frac{1}{2} F_y)(\frac{1}{2} - x_0) bt \right]$

EXAMPLE 9.3 (Continued)

but,
$$\frac{f - \frac{1}{2}F_y}{(\frac{1}{2} - x_o)b} = \frac{F_y}{\frac{b}{2}}$$

that is,
$$f = (\frac{3}{2} - 2x_o)F_y$$

$$P_{cr} = F_{cr}(2bt) = 2bt\left[(\frac{3}{2} - 2x_o)F_y - (1 - 2x_o)^2(\frac{1}{2}F_y)\right]$$

that is $\boxed{f_{cr} = F_y\left[1 - 2x_o^2\right]}$ Eq.2

METHOD OF SOLUTION: Specify x_o, compute F_{cr} from Eq.(2) then, substitute value of F_{cr} and x_o into Eq.(1) to obtain $\frac{L}{r}$.

Lower Limit: $x_o = \frac{1}{2}$; $f_{cr} = \frac{1}{2}F_y$, $\frac{L}{r} = \sqrt{\frac{2\pi^2 E}{F_y}} = 128$ (A36)

Upper Limit: $x_o = 0$; $f_{cr} = F_y$; $\frac{L}{r} = 0$

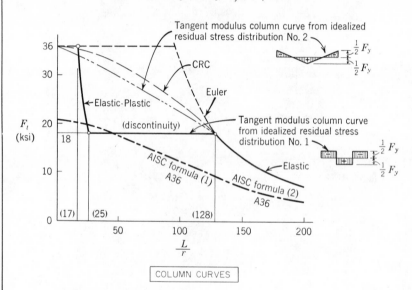

COLUMN CURVES

DESIGN OF SECTION

   Residual Stress Distribution No. I

     Check section obtained in (a)(A36)

$$\frac{L}{r} = 64; \text{ from fig., } F_u = 18.0 \text{ ksi}$$

$$F_a = \frac{18.0}{F.S.} ; \text{ assume F.S.} = 1.836 \left[\text{from (b)}\right]$$

$$F_a = \frac{18.0}{1.836} = 9.8 \text{ ksi}$$

$$A \times F_a = 149 \text{ kip} \ll 250 \text{ kip} \hspace{3cm} \text{N.G}$$

TRY $\frac{7}{8} \times 12$ flg. $A = 2(12)\frac{7}{8} + \frac{1}{2}(14) = 28.0 \text{ in.}^2$

   $\frac{1}{2} \times 14$ web $I_y = 2(126.0) = 252.0 \text{ in.}^4$

           $r_y = 3.00 \text{ in.}$

           $\frac{L}{r} = \frac{15 \times 12}{3.00} = 60$; $F_u = 18 \text{ ksi (from fig.)}$

EXAMPLE 9.3 (Continued)

Assume F.S. of AISC Spec. as applicable:

$$F.S. = \frac{5}{3} + \frac{3(60)}{8(128)} - \frac{(60)^3}{8(128)^3} \quad \text{(Section [I.5.1.3])}$$

F.S. = 1.833

$$F_a = \frac{18}{1.833} = 9.81 \text{ ksi}$$

USE $\begin{cases} \frac{7}{8} \times 12 \text{ flg.} \\ \text{A36} \\ \frac{1}{2} \times 14 \text{ web} \end{cases}$

A x F_a = 28.0 x 9.81 = 274 kip > 250 kip   OK

b/t_f = 13.7 < 16   OK

d/t = 28 < 42

Residual Stress Distribution No. 2   The design is not shown here.

SUMMARY:

Mat'l	Design Method	Section	Computed Design Load	wt/ft
A36	Secant Formula	1/2 x 11 flg	250 kip	51.4 lb
	Eq. of Initially Curved Column	3/8 x 11 web	≈ 250 kip	
	AISC Specs	1/2 x 11 flg 3/8 x 11 web	258 kip	51.4 lb
	Tangent Modulus (F_r No.1)	7/8 x 12 flg 1/2 x 14 web	274 kip	95.2 lb
	Tangent Modulus (F_r No.2)	1/2 x 12 flg 3/8 x 12 web	252 kip	56.1 lb
A572 (50)	AISC Specs	1/2 x 10 flg 3/8 x 10 web	274 kip	45.4 lb

9.8 SHEAR FORCE AND BUILT-UP COLUMNS

When a column buckles, or is initially curved, shearing forces come into effect on the cross section of the column as shown in Fig. 9.40.   The effect of this shearing force on column strength may be seen by a comparison of the differential equations, below.

Taking into account the modulus of steel at the buckling stress (the tangent modulus E_t), the differential equation for buckling of an initially straight, centrally loaded column is

$$\frac{d^2v}{dz^2} + \frac{P}{E_t I} v = 0 \qquad (9.33)$$

with the solution

$$P_{cr} = \frac{\pi^2 E_t I}{L^2} \qquad (9.34)$$

(Equations 9.33 and 9.34 correspond directly to Eqs. 9.8 and 9.24.)

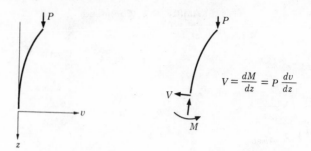

Fig. 9.40 Shear force on columns.

When the effect of shear is taken into account, the equation becomes

$$\frac{d^2v}{dz^2} + \frac{1}{[1 - (\beta P/G_tA)]} \cdot \frac{P}{E_tI} v = 0 \qquad (9.35)$$

The derivation of Eq. 9.35 may be obtained by considering the additional deflection on the column due to the shear force V.

The slope θ, due to the additional deflection, is

$$\theta = \frac{V}{GA} \qquad (9.36)$$

which may be obtained from the geometry of the sheared section shown in Fig. 9.41, since $\theta = \tau/G = (V/A)(1/G)$.

Now, the shear stress τ is not actually uniform throughout the cross section, and a factor β is introduced into Eq. 9.36 to make it valid, since the geometry had assumed uniform shear.[7.17] The value of β depends upon the shape of the cross section.

$$\theta = \beta \frac{V}{GA} \qquad (9.36a)$$

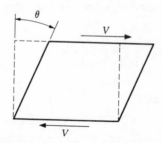

Fig. 9.41 Shear force on fiber.

The change in slope, $d\theta/dz$, due to the additional deflection is

$$\frac{d\theta}{dz} = \beta \frac{1}{GA} \cdot \frac{dV}{dz}$$

and, from Fig. 9.40

$$\frac{d\theta}{dz} = \frac{P\beta}{GA} \cdot \frac{d^2v}{dz^2} \tag{9.37}$$

The total curvature of any point in the column when it buckles is the sum of the curvature due to the buckling (see Eq. 9.8), plus the additional curvature due to the effect of shear:

$$\left.\frac{d\theta}{dz}\right|_{\text{total}} = \left.\frac{d^2v}{dz^2}\right|_{\text{total}} = \underbrace{-\frac{P}{E_t I} v}_{\substack{\text{from} \\ \text{Eq. 9.8}}} + \underbrace{\frac{P\beta}{G_t A} \cdot \frac{d^2v}{dz^2}}_{\substack{\text{from Eq.} \\ 9.37}} \tag{9.38}$$

or, rearranging,

$$\frac{d^2v}{dz^2} + \frac{1}{[1 - (\beta P/G_t A)]} \cdot \frac{P}{E_t I} v = 0 \tag{9.35}$$

This differential equation, Eq. 9.35, is of exactly the same form as Eq. 9.8, and its solution is Eq. 9.13, which here is

$$\frac{1}{[1 - (\beta P/G_t A)]} \cdot \frac{P}{E_t I} = \frac{n^2 \pi^2}{L^2} \tag{9.39}$$

Rearranging Eq. 9.39 gives the solution to Eq. 9.35:

$$P_{cr} = \frac{\pi^2 E_t I}{L^2} \cdot \frac{1}{1 + (\beta/G_t A)(\pi^2 E_t I/L^2)} \tag{9.40}$$

In Eq. 9.35, the term $[1 - (\beta P/G_t A)]$ indicates the influence of shear, where G_t is the tangent modulus in shear.

Equation 9.40 may be put into standard form by modifying the slenderness ratio, or by modifying E_t, that is:

or

$$\left. \begin{aligned} P_{cr} &= \frac{\pi^2 \bar{E} I}{L^2} \\[2em] P_{cr} &= \frac{\pi^2 E_t I}{(k_i L)^2} \end{aligned} \right\} \begin{aligned} &\text{when } \bar{E} \text{ or } k_1 \text{ account} \\ &\text{for the modifications} \\ &\text{due to shear} \end{aligned}$$

When writing in terms of the slenderness ratio,[7.6] and assuming $E_t/G_t = E/G = \gamma$ where $\gamma = 2(1 + \mu)$, then

$$P_{cr} = \pi^2 E_t I \frac{1}{[1 + \pi^2 \beta \gamma (r/L)^2] L^2}$$

or

$$P_{cr} = \frac{\pi^2 E_t}{[k_1(L/r)]^2} \tag{9.41}$$

where $k_1 = \sqrt{1 + \pi^2 \beta \gamma (r/L)^2}$. ($k_1$ should not be confused with K, the end-fixity constant.)

The effect of shear is so small for the usual solid columns that it may be neglected. However, the effect of shear is quite important for latticed columns.

1. Built-Up Columns

Since the available shapes in rolled structural sections are limited, quite often special shapes are built up to suit particular requirements. This is particularly true when very heavy cross sections are needed. These shapes may be built up by welding, riveting, or bolting. The cross-sectional shapes may be similar to those of rolled shapes, or, on the other hand, may be a non-uniform type such as latticed columns. A built-up column may utilize rolled shapes in its cross section. The main vertical members of latticed columns are usually channels or corner angles. Figure 9.42 illustrates latticed and welded columns.

The welded built-up H and box shapes have already been considered. Other welded shapes, such as those indicated above, may be designed in the same manner.

The analysis and design of latticed columns deserve a special study since they differ from other columns in one important aspect. This is the effect of shearing forces. Whereas the effect of the shearing stresses on the usual cross-sectional shape is negligible, this is definitely not the case with latticed columns.

If the number of panels in a laced or battened column is greater than four, the column can be considered as an ordinary column, rather than as a framework, and the general equation, Eq. 9.40, can be utilized for design. Rewritten,

$$P_{cr} = \frac{P_t}{1 + (P_t/P_d)} \tag{9.42}$$

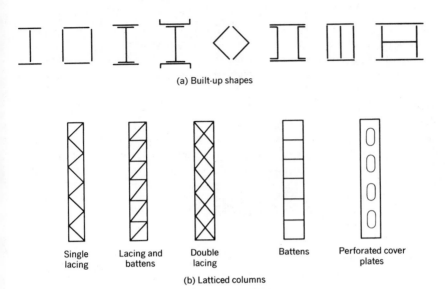

(a) Built-up shapes

| Single lacing | Lacing and battens | Double lacing | Battens | Perforated cover plates |

(b) Latticed columns

Fig. 9.42   Built-up columns.

where

$$P_t = \frac{\pi^2 E_t I}{L^2}$$

$$\frac{1}{P_d} = \frac{\beta}{G_t A}$$

where the value $1/P_d$ is obtained rather simply by considering the forces acting on the deflected panel.[7.6,7.17] The expressions for $1/P_d$ are given in Fig. 9.43 for a laced column and a batten-plated column.

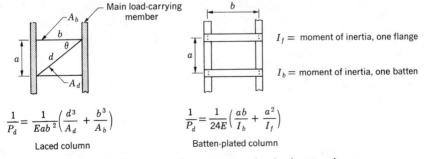

Main load-carrying member

$I_f =$ moment of inertia, one flange

$I_b =$ moment of inertia, one batten

$$\frac{1}{P_d} = \frac{1}{Eab^2}\left(\frac{d^3}{A_d} + \frac{b^3}{A_b}\right)$$

Laced column

$$\frac{1}{P_d} = \frac{1}{24E}\left(\frac{ab}{I_b} + \frac{a^2}{I_f}\right)$$

Batten-plated column

Fig. 9.43   Theoretical expressions for built-up columns.

When the perforations of a perforated cover-plated column are made to the dimensions advised in Refs. 7.14 and 9.43, then, for L/r less than 40 or 50, the perforations do not diminish the critical column strength.

Figure 3.9 of Ref. 7.14 (CRC Guide) presents a convenient manner of determining the shear action on the column cross section; this is needed in the design of the component members. Furthermore, the usual design of laced columns follows a rule-of-thumb method, such as the following from Reference 7.14. A generally conservative estimate of the influence of standard 60 or 45° lacing specified in usual bridge design practice may be made by modifying the effective length factor K, determined by restraint conditions, to a new factor K', as follows:

$$K' = K\left(1 + \frac{L}{4000r}\right)$$

2. Residual Stresses in Latticed Columns

There has been no research conducted on the effect of residual stress on columns which are built up by latticing.

However, in the case of laced and battened columns, it is realized that residual stresses will have only a small effect when corner angles are used. Corner angles do not have high residual stresses so it would be expected that any loss in strength due to their existence may be safely ignored, when compared to their general column strength.[2.4] When the flange shape is much larger, for example, a channel, the effect of residual stress does become more pronounced, but in no case would it be greater than that in a normal solid welded column. Hence, for laced and battened columns, the allowable column stresses may be used with the reduced column lengths with no fear of the adverse effects of residual stresses.

For columns with perforated cover plates, the same general conclusions may be made when the members are joined by structural fasteners, such as rivets. However, when the plates are joined by welding, the situation is completely different. Welded columns with perforated cover plates may be treated directly as welded built-up columns, provided the slenderness ratio of the column is less than about 50. That is, the effect of the perforations may be neglected in such a case. Hence, as a welded column, the magnitudes of residual stress are great enough to warrant the same precautions as with welded columns in general.

Example 9.4 compares the design of a rolled column with that of laced, battened, and perforated coverplated columns.

9.9 DESIGN SPECIFICATIONS

The design specifications of AISC, AASHO, and AREA are compared in the following sections.

EXAMPLE 9.4

PROBLEM:
   Design the column using A36 steel. Consider the design as a:
       (a) rolled shape
       (b) laced column
       (c) battened column (not covered by AISC, AREA or AASHO Specs)
       (d) column with perforated cover plates
   Compare the designs, with AISC Specification.

600^k

$25'$

600^k

SOLUTION:

(a) ROLLED SHAPE

Trial stress = 15.5 ksi

$$A_{req'd} = \frac{600}{15.5} = 38.7 \text{ in.}^2$$

TRY W 14×136 $r_y = 3.77'$,   A = 40.0 in.^{2}

$$\frac{L}{r_y} = \frac{25 \times 12}{3.77} = 79.5 \quad (\text{Section}[1.8.2])$$

$F_a = 15.41$ ksi

$A \times F_a = 616$ kip > 600 kip   <u>OK</u>

     ∴ USE W 14×136

(b) LACED COLUMN

   TRY 4 Ls 6×6×$\frac{3}{4}$ 20 × 20 o. to o.

$$I_x = I_y = 4\left[28.2 + 8.44(8.22)^2\right] = 2400 \text{ in.}^4$$

$$A = 4(8.44) = 33.76 \text{ in.}^2$$

$$r_y = \sqrt{\frac{2400}{33.76}} = 8.45 \text{ in.}$$

$$\frac{L}{r_y} = \frac{25 \times 12}{8.45} = 35.5 \;\therefore F_a = 19.54 \text{ ksi}$$

$A \times F_a = 33.76 \times 19.54 = 660$ kip > 600 kip   <u>OK</u>

<u>Design of Lacings (Section $[1.18.2.6]$)</u>

Distance between lines of rivets = 13" < 15"   <u>OK</u>

Use single lacing inclined at 60°

$\frac{L}{r}$ of L between lacing = $\frac{15}{1.17} = 13 \overset{?}{<} (\frac{L}{r})_{member}$

                              13 < 35.5   <u>OK</u>

Force on lacing bar:

V = 0.02P = 0.02(600) = 12 kip
      (Section $[1.18.26]$)
or 6 kip/lacing on each side

$$P_L = \frac{6}{\cos 30°} = 6.93 \text{ kip}$$

$L_{bar} = 15"$; $\frac{L}{r} \leq 140$ (for single lacing)

r = 0.288t for rectangular section

$$\therefore t_{min} = \frac{15}{140(.288)} = 0.372 \text{ in.}$$

TRY $\frac{3}{8}$ in. flat bar $\frac{L}{r} = \frac{15}{.375 \times 0.288} = 139$ (Secondary member) < 140   <u>OK</u>

$\therefore F_a = 8.54$ ksi

$$A_{req'd} = \frac{6.93}{8.54} = 0.81 \text{ in.}^2$$

$$b_{req'd} = \frac{0.81}{0.375} = 2.16 \text{ in.} \quad \text{use } 2\frac{1}{4} \text{ in.}$$

Use $\frac{3}{8}$ × $2\frac{1}{4}$ flat bar for lacing

EXAMPLE 9.4 (Continued)

<u>Design of tie plates at the ends</u>   (Section $\boxed{\text{J.18.2.5}}$)

Minimum length = 13 in.

$t \geqq \dfrac{1}{50}$ (distance between lines of rivets) (Section $\boxed{\text{J.18.2.5}}$)

$t_{min} = \dfrac{13}{50} = 0.26$ in. $\therefore$ Use a $\dfrac{3}{8}$ in. $\text{P}\!\!\!\text{L} - 15$ in. long

FINAL SECTION:

4 Ls 6×6×$\dfrac{3}{4}$ (20" o. to o.)
$\dfrac{3}{8}$ × $2\dfrac{1}{4}$ Single lacing inclined at 60°
$\dfrac{3}{8}$ ×15 ×1'–8" Tie plates at the ends

wt. per ft. 4 Ls 6×6×$\dfrac{3}{4}$ 114.8 lb/ft

bar $\dfrac{3}{8}$ × $2\dfrac{1}{4}$ $\dfrac{11.5 \text{ lb/ft}}{126.3 \text{ lb/ft}}$

(c) BATTENED COLUMN: (Not covered by AISC Specs.)

TRY section in (b)

4 Ls 6×6×$\dfrac{3}{4}$ (20 in. o.to o.)

Spacing between battens will be governed by the slenderness ratio (L/r) of the single angle between battens.

$$\max \left(\dfrac{L_o}{r_o}\right)_{1L} \leq 0.6\left(\dfrac{L}{r}\right)_{col.} \quad \left[\begin{array}{l}\text{British Standard 449} \\ \text{Ref 7.14}\end{array}\right]$$

$$\left(\dfrac{L}{r}\right)_{col} = 35.5 \quad \left[\text{from (b)}\right]$$

$$\max \left(\dfrac{L_o}{r_o}\right)_{1L} = 0.6 \times 35.5 = 21.3$$

$$\max L_o = 21.3(1.17) = 25 \text{ in.}$$

Try batten spacing = 27 in. c. to c.

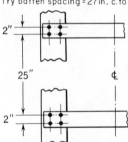

$K_s L$ – "Effective" column length: takes account of the possible reduction in the buckling load of the column caused by the presence of appreciable shear deformation in latticed column (Ref. 7.6)

$$\dfrac{K_s L}{r} = \sqrt{\left(\dfrac{L}{r}\right)^2 + \dfrac{\pi^2}{12}\left(\dfrac{L_o}{r_o}\right)^2}$$

$$= \sqrt{(35.5)^2 + \dfrac{\pi^2}{12}(21.3)^2} = 40.5$$

$F_a = 19.15$ ksi

$A \times F_a = 33.76 \times 19.15 = 645 > 600$ kip   say <u>OK</u>

<u>Design of battens:</u>

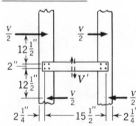

From Ref. 7.6 p.181

$$V = 2\pi(F_y - F_{cr})A\dfrac{r}{L}$$

Where F_y = yield stress

F_{cr} = the critical column stress

A = area of one chord

From Figure 92 of Ref. 7.6

$$\dfrac{V_{max}}{A} = 180$$

$$V_{max} = 180(16.88) = 3040 \text{ lb}$$

$$V'_{max} = \dfrac{1}{2}(3040)\left(\dfrac{27}{15.5}\right) = 2650 \text{ lb}$$

$$M = 2650\left(\dfrac{15.5}{2}\right) = 20,500 \text{ in. lb}$$

EXAMPLE 9.4 (Continued)

The design will be based on the yield strength (Ref. 7.6)

$$F_b = \frac{F_y}{F.S.} = \frac{36,000}{1.65} = 22,000 \text{ psi}$$

$$t_{req'd} = \frac{6M}{F_b(d)^2} = \frac{6(20,500)}{22,000(4)^2} = 0.38 \text{ in.}$$

USE $4 \times \frac{1}{2}$ batten plates @ 27 in. c. to c.

The design of the rivets has not been shown.

Check chord for moment from battens (Ref. 7.6)

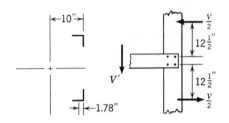

Properties:

$$A = 2(8.44)$$
$$I = 2(28.2)$$

$$M = \frac{1}{2}(12.5)3040$$

$$f_b = \frac{\frac{1}{2}(12.5)3040(6.0-1.78)}{2(28.2)1000} = 1.65 \text{ ksi}$$

$$f = \frac{600}{4(8.44)} + 1.65 = 19.5 \text{ ksi}$$

$$F_{a\left(\frac{L}{r}=0\right)} = 22 \text{ ksi} > 19.5 \text{ ksi} \qquad \underline{OK}$$

(CRC Recommendation)

FINAL SECTION:

4 Ls 6 × 6 × $\frac{3}{4}$ (20" o. to o.)	114.8 lb/ft
$\frac{1}{2}$ × 4 PL (27" c. to c.)	12.1 lb/ft
	126.9 lb/ft wt per ft

(d) PERFORATED COLUMN (A 36)

TRY 4 Ls 4 × 4 × $\frac{5}{16}$

w/ $\frac{3}{8}$ × 20 solid cover plate

and $\frac{5}{16}$ × 21 PL w/perforations (manholes)

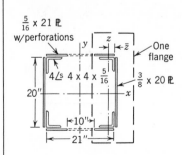

$\frac{5}{16}$ × 21 PL w/perforations

4 Ls 4 × 4 × $\frac{5}{16}$

$\frac{3}{8}$ × 20 PL

One flange

Assuming 10 in. wide perforations,

$$I_{min} = I_x = 4\left[3.7 + 2.4(8.9)^2\right] + 2(250) + 4(5.5)\frac{5}{16}(10.2)$$

$$= 1990.8 \text{ in.}^4$$

$$A = 4(2.4) + \frac{3}{8}(20)2 + \frac{5}{16}(11)2 = 31.5 \text{ in.}^2$$

$$r_{min} = \sqrt{\frac{1991}{31.5}} = 7.95 \text{ in.}$$

$$\frac{L}{r_{min}} = \frac{300}{7.95} = 37.8 \therefore F_a = 19.37 \text{ ksi (Table 1-36)}$$

$$A \times F_a = 31.5 \times 19.37 = 610 \text{ kip} > 600 \text{ kip} \qquad \underline{OK}$$

Note:
    The CRC (Ref. 7.14) recommends that before the design can be based on net area, a check must be made on the unsupported length of the flange at the perforations:

$$\left.\begin{array}{l}\left(\frac{L}{r}\right)_{flg} \quad \begin{array}{l}\leq \frac{1}{3}\left(\frac{L}{r}\right)_{col} \\ \leq 20\end{array}\end{array}\right\} \quad \text{Ref. 7.14 p 50}$$

This is not required by the AISC Specs; however the check will be made here for illustration.

EXAMPLE 9.4 (Continued)

Design of perforations:

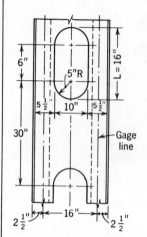

FINAL SECTION:

Properties of one flange (refer to cross section):

From computation of areas,

$\bar{z} = 0.76$ in., $I_z = 30.8$ in.^{4}, $r_z = 1.40$ in., $A = 15.75$ in.^{2}

Unsupported length of flange = 16 in.

$$\frac{L}{r_z} = \frac{16}{1.40} = 11.2 \begin{cases} < \frac{1}{3}(37.8) \\ < 20 \end{cases}$$ <u>OK</u>

b/t ratio:

Perforated cover ℙ: $b/t = \dfrac{16}{5/16} = 51.2 < 53$

solid cover ℙ: $b/t = \dfrac{15}{3/8} = 40 < 42$ } Sect(1.9)   <u>OK</u>

From Section $\boxed{1.18.2.7}$

Length to width of hole: $\dfrac{16}{10} = 1.6 < 2$

Clear distance between holes: 20" > 16"

Min. radius of periphery of holes: 5" > $1\frac{1}{2}$ <u>OK</u>

4 Ls $4 \times 4 \times \frac{5}{16}$ (21"×20" o. to o.)	32.8 lb/ft
$\frac{5}{16} \times 21$ Perforated ℙ	44.6
$\frac{3}{8} \times 20$ ℙ	51.0
	128.4 lb/ft. wt per ft

Type of Column	Member	Design Capacity	Weight (lb/ft)
Rolled Shape (A 36)	W 14×136	616 kip	136.0
Laced column (A 36)	4 Ls 6×6×$\frac{3}{4}$ Bar $\frac{3}{8} \times 2\frac{1}{4}$ ℙs $\frac{3}{8} \times 15 \times 1'\text{-}8"$	660 kip	126.3
Battened Column (A36)	4 Ls 6×6×$\frac{3}{4}$ Bar $\frac{1}{2} \times 4$	596 kip	126.9
Perforated Column (A36)	4 Ls 4×4×$\frac{5}{16}$ ℙs $\frac{5}{16} \times 21$ ℙs $\frac{3}{8} \times 20$	610 kip	128.4

1. Allowable Stresses

Figure 9.44 compares the column curves of AISC, AREA, and AASHO, for ASTM A36 steel.   There is an obvious difference between the allowable stresses.   The equations for AISC, AASHO, and AREA are given below.

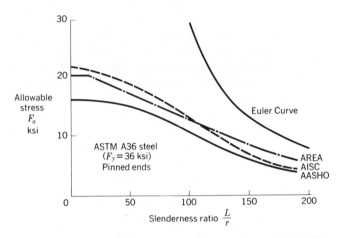

Fig. 9.44   Column formulas for AISC, AREA, and AASHO.

AISC.   The CRC basic column curve[7.14] is used as the basic equation for the maximum strength of a column, and is divided by a factor of safety, F.S., to give the equation for allowable axial stress

$$F_a = \frac{F_y\left(1 - \dfrac{(KL/r)^2}{2C_c^{\,2}}\right)}{\text{F.S.}} \tag{9.43}$$

where C_c is the L/r corresponding to $F_y/2$, and KL/r is the effective slenderness ratio.   Equation 9.43 is valid for slenderness ratios less than C_c,

$$C_c = \sqrt{\frac{2\pi^2 E}{F_y}} \tag{9.44}$$

The factor of safety is given by

$$\text{F.S.} = \frac{5}{3} + \frac{3(KL/r)}{8C_c} - \frac{(KL/r)^3}{8C_c^{\,3}} \tag{9.45}$$

When the effective slenderness ratio exceeds C_c, the Euler curve modified by a factor of safety is used:

$$F_a = \frac{12\pi^2 E}{23(KL/r)^2} \quad \text{(in ksi)} \tag{9.46}$$

For axially loaded bracing and secondary members, when L/r exceeds 120,* the allowable stress

$$F_{as} = \frac{F_a\dagger}{1.6 - (L/200r)} \tag{9.47}$$

When a compression member is composed of plate elements whose width-thickness ratio exceeds the applicable limits of AISC Sect. [1.9], the strength of the column will be reduced.   This is taken into account by the introduction of a reduction factor into Eq. 9.43.   (See Art. 12.4 and AISC Appendix [C].)

AASHO. The Johnson parabola (Fig. 9.8) is used to define the actual allowable design stress for certain slenderness ratios:

For riveted ends: $$F_a = \frac{0.55F_y}{1.25} \left\{ 1 - \frac{\left(0.75\dfrac{L}{r}\right)^2 F_y}{4\pi^2 E} \right\} \quad \text{(in psi)} \tag{9.48}$$

For pinned ends: $$F_a = \frac{0.55F_y}{1.25} \left\{ 1 - \frac{\left(0.875\dfrac{L}{r}\right)^2 F_y}{4\pi^2 E} \right\} \quad \text{(in psi)} \tag{9.49}$$

Equations 9.48 and 9.49 are given in the following form,

$$F_a = C_1 - \frac{C_2}{1000}\left(\frac{L}{r}\right)^2 \quad \text{(riveted)}$$

$$F_a = C_1 - \frac{C_3}{1000}\left(\frac{L}{r}\right)^2 \quad \text{(pinned)}$$

with the constants given below for the different yield strengths of the specification:

F_y (ksi)	C_1	C_2	C_3	Limiting L/r
36	16.0	0.30	0.38	130
42	18.0	0.39	0.52	125
46	20.0	0.46	0.62	125
50	22.0	0.56	0.74	125
55	24.0	0.65	0.88	120
60	26.0	0.77	1.04	115
65	29.0	0.93	1.26	110
90	40.0	1.8	2.4	90
100	44.0	2.2	2.9	85

* For this case, K is taken as unity.
† From Eq. 9.43 or Eq. 9.46.

When a "more exact formula" is desired, or when the slenderness ratio is greater than the limiting, or for compression members of known eccentricity, the following secant formula (Art. 9.4) is used:

$$F_a = \frac{\dfrac{F_y}{\text{F.S.}}}{1 + \left(0.25 + \dfrac{e_g c}{r^2}\right) B \csc \Phi} \tag{9.50}$$

where F.S. = factor of safety varying from 1.80 to 1.87, depending on yield
point
c = distance from neutral axis to the extreme fiber in compression

$$\Phi = \frac{L}{r} \sqrt{\frac{(\text{F.S.})F_a}{E}}$$

L = 75 per cent of the total length of a column having riveted end
connections
= 87.5 per cent of the total length of a column having pinned-end
connections

$$B = \sqrt{\alpha^2 - 2\alpha \cos \Phi + 1}$$

$$\alpha = \frac{(e_s c/r^2) + 0.25}{(e_g c/r^2) + 0.25}$$

e_g = eccentricity of applied load at the end of column having the
greater computed moment, in inches
e_s = eccentricity at opposite end
When the slenderness ratio is equal to or less than

$$(\cos^{-1} \alpha) \left[E\left(1 + 0.25 + \frac{e_g c}{r^2}\right) \Big/ F_y \right]^{\frac{1}{2}}$$

then the allowable stress F_a is defined by

$$F_a = \frac{F_y/\text{F.S.}}{1 + 0.25 + (e_g c/r^2)} \tag{9.51}$$

The value of 0.25 in Eqs. 9.50 and 9.51 "provides for inherent crookedness and unknown eccentricity."

AREA. The straight line column curve is used: For A36 steel,

$$\frac{KL}{r} \leqslant 15 \quad F_a = 20{,}000 \text{ psi}$$

$$15 \leqslant \frac{KL}{r} \leqslant 143 \quad F_a = 21{,}500 - 100\frac{KL}{r} \tag{9.52}$$

$$\frac{KL}{r} \geqslant 143 \quad F_a = \frac{147{,}000{,}000}{\left(\dfrac{KL}{r}\right)^2}$$

For high-strength steels up to 60-ksi yield strength, the corresponding equations are

$$\frac{KL}{r} \leqslant \frac{3388}{\sqrt{F_y}} \quad F_a = 0.55F_y$$

$$\frac{3388}{\sqrt{F_y}} \leqslant \frac{KL}{r} \leqslant \frac{27{,}111}{\sqrt{F_y}} \quad F_a = 0.60F_y - \left(\frac{F_y}{1662}\right)^{3/2}\frac{KL}{r} \tag{9.53}$$

$$\frac{KL}{r} \geqslant \frac{27{,}111}{\sqrt{F_y}} \quad F_a = \frac{147{,}000{,}000}{\left(\dfrac{KL}{r}\right)^2}$$

For all steels, under usual conditions:

$K = 0.875$ for members with pin-end connections
$K = 0.75$ for members with riveted, bolted, or welded end connections.

2. Factor of Safety

Factors of safety have existed as long as the concept of the column curve has been used. Indeed, the same controversy existed one hundred years ago as now: whether the factor of safety should be constant, or variable with L/r.

In 1898 Barth[9.1] suggested that the factor of safety be $\frac{9}{4}$, whereas the value most commonly used with columns then was approximately 4. Today the value is usually around 2. The factor of safety was either based on a maximum strength, or on an allowable stress basis. Quite often a factor of safety was used to estimate the imperfections of the column (end conditions, etc.) and a second factor of safety was used on the allowable stress. A study of the factor of safety is incomplete without an allusion to statistical reasoning. Reference 7.14 also gives much information on the topic. The factor of safety has been defined by Eq. 1.2.

From the test results of Fig. 9.38, it is obvious that a scatter exists for column strength for any L/r. Indeed, if sufficient test data are available, a frequency curve could be drawn for strength at any L/r, and the standard deviation $\bar{s}$ of

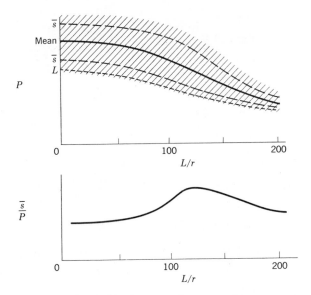

Fig. 9.45 Scatter of test results.

test data could be determined.   In fact, this is the approach used for the column curve of the European Convention of Constructional Steelwork.[9.9] See Fig. 9.48.

Figure 9.45 shows a sketch of a scatter of test results for a column curve, as well as a mean curve and curves for the standard deviation.   The deviation of the scatter of results is not constant.   For small L/r, this deviation is due entirely to variation in material strength; and for large L/r, almost entirely due to the condition of the end fixity.   Medium column lengths (L/r from 50 to 150) reflect the effect of residual stress and initial out-of-straightness. The lower figure shows a plot of the deviation, $\bar{s}/P$ ($P = $ mean value), against the slenderness ratio.

The factor of safety can be based directly on the standard deviation, or else upon a curve which represents the lowest values expected (that is, curve L in Fig. 9.45).   Above such a curve would lie 90 per cent, or 99 per cent, etc., of the column strength.   A constant factor of safety appears desirable for simplicity, yet such a factor proves too conservative for low L/r and high L/r, and is insufficient for medium L/r.   For this reason a variable factor of safety may be introduced.   As a design aid, the factor of safety normally would be in the form of tables or graphs, as in the case of the AISC Specification.

It appears not unlikely that in the future a rational design of a structure will involve load analysis, structural analysis, and safety analysis.   The analysis of the structure could be on the component elements as well as on the whole, to ensure uniform economy and safety.

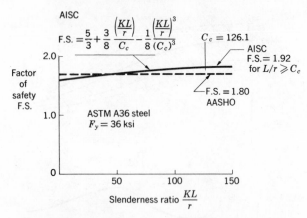

Fig. 9.46 Factor of safety for AISC and AASHO.

Figure 9.46 compares the factor of safety used with the column formulas of AISC and AASHO. The controversy mentioned above is obvious: both constant and variable factors of safety are used.

The AISC Formula takes account of the greater variation in column strength for the medium slenderness ratios. The constant values of AASHO vary from 1.80 to 1.87, depending on the yield point.

3. Slenderness Ratio

The AISC Specification provides for effective slenderness ratios which take into account the end conditions of the column. (See Chapter 10.) The AASHO and AREA Specifications provide indirectly, in a limited manner, for effective slenderness ratios by allowing different allowable stresses for either riveted ends or pinned ends, that is, the use of Eqs. 9.48, 9.49, 9.52, and 9.53. The effect of end restraint becomes much less pronounced for short columns, as is shown in Fig. 9.47.

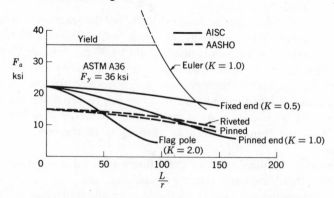

Fig. 9.47 Effect of end fixity on allowable stress.

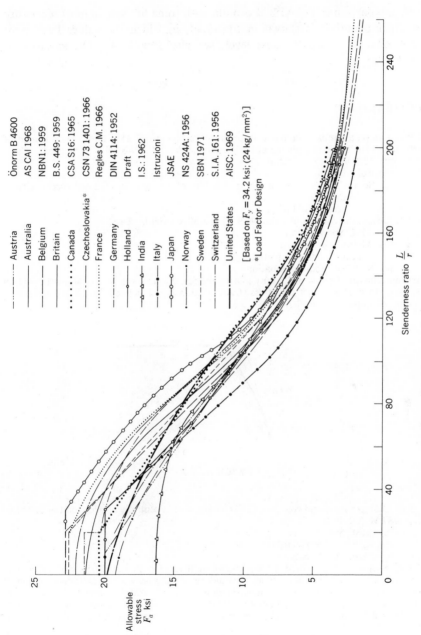

——————— Austria	Önorm B 4600
——————— Australia	AS CAI 1968
——————— Belgium	NBN1: 1959
——————— Britain	B.S. 449: 1959
·············· Canada	CSA S16: 1965
——·——·—— Czechoslovakia*	CSN 73 1401: 1966
·············· France	Regles C.M. 1966
——————— Germany	DIN 4114: 1952
——————— Holland	Draft
——◇——◇—— India	I.S.: 1962
——△——△—— Italy	Istruzioni
——●——●—— Japan	JSAE
——○——○—— Norway	NS 424A: 1956
·············· Sweden	SBN 1971
——————— Switzerland	S.I.A. 161: 1956
——————— United States	AISC: 1969
[Based on F_y = 34.2 ksi; (24 kg/mm²)]	
*Load Factor Design	

Slenderness ratio $\frac{L}{r}$

Allowable stress F_a ksi

Fig. 9.48 Column formulas around the world.

4. Column Formulas around the World

A comparison of the AISC Formula with some of those in most recent use in other countries is shown in Fig. 9.48, in which the values have been adjusted to a common stress level; see also Ref. 9.44. The variation is significant.

PROBLEMS

9.1. Design a simple column, with an unbraced length of 15 ft and a central load of 500 kips. The column is pinned at both ends and ASTM A36 steel is to be used. Use
  (a) The AISC Specification
  (b) Euler equation
  (c) Secant formula with a factor of safety of 1.8, and an assumed eccentricity of 0.25 in.
Comment on the results obtained.
9.2. Design the column of Example 9.1 for a 50-ksi yield. Use:
  (a) AISC Specification
  (b) AREA Specification
  (c) Tangent modulus equation
  (d) Reduced modulus equation
Assume a factor of safety of 1.8 for items (c) and (d), that the cross section does not contain residual stresses, and that the stress-strain relationship is defined by the accompanying graph. Compare the results.

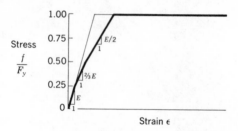

Problem 9.2

9.3. Design a fixed-end column of 25-ft length, with a centrally applied load of 750 kips. Use ASTM A36 steel, and
  (a) AISC Specification
  (b) AASHO Specifications
  (c) AREA Specifications
9.4. A design has required a medium-size column of a structural shape in A36 steel. For the particular situation in mind, it would be good to have some additional margin of safety above that covered by the standard factor of safety. Rather than to increase the size of the member, comment on which of the following shapes would be best to use, and which would be advisable for the special circumstances. (A knowledge of residual stress distributions is needed.)
  (a) Structural rolled wide-flange shape

    (b) Welded H-shape
    (c) Welded box shape
    (d) Riveted shape
    (e) Rolled hollow-tube shape
    (f) Latticed member
    (g) Perforated cover-plated member
    (h) Reinforced, cover-plated rolled wide-flange shape

9.5. Using the AISC Specification, design a column with one end fixed, and the other pinned, with a load of 600 kips and a length of 20 ft. Use a 46 ksi yield steel.

    (a) Design the column as made up from four equal-size angles, made into a cross shape (see accompanying figure a).

    (b) Using the same four angles, weld them into a box shape (b) and compute the design load which can be carried by the member.

    (c) Comment on the strength of the two shapes. Which would be expected to have the greater maximum load? Explain the reasoning.

    (d) Compare the probable price for each column for both riveted and welded construction.

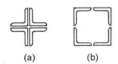

    (a)   (b)

Problem 9.5

9.6. Construct a column curve for the wide-flange shape shown. Assume weak axis bending, and neglect the effect of the web. Assume an idealized stress-strain relationship for each fiber; each flange has a residual stress distribution as shown in the sketch.

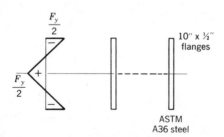

Problem 9.6

9.7. Assume that the column curve of the previous problem, Problem 9.6, holds true for an H-shape: 10-by-$\frac{1}{2}$-in. flanges, and 9-by-$\frac{3}{8}$-in. web (ASTM A36 steel).

    (a) What design load would a 15-ft pinned-end column of this shape support?

    (b) Compare this design load with those obtained from AISC, AASHO, and AREA Specifications.

9.8. Design a laced column to carry a load of 700 kips in a length of 20 ft between pinned ends. The maximum external size of the column is 18 by 18 in. Use any relevant specification. Use A36 steel.

9.9. Repeat the previous problem, Problem 9.8, for a perforated cover plated column with
   (a) plates riveted with angles at corners
   (b) plates welded at the corners

9.10. Repeat the design of Example 9.3 for AASHO and AISC Specifications.

10

COMPRESSION MEMBERS IN FRAMES AND TRUSSES

10.1 INTRODUCTION

The preceding chapter has described in detail the procedures of design for various types of individual compression members subjected to a concentric axial thrust. In the development of these design procedures it was assumed that the ends of the members are perfectly pinned; this means that the members are free to rotate but are not permitted to translate at the supports. This is a highly idealized situation and is not likely to occur in engineering practice. The compression members in a structural framework (such as the top chords of a roof truss) are usually connected to the adjacent members by welding or bolting and thus are provided with rotational restraints at their ends. Furthermore, in certain types of structures, such as a rigid frame, the column top sometimes may translate with respect to the base when buckling occurs. To achieve a completely rational design of compression members, it is necessary to treat these members as integral parts of the structure, rather than as isolated members.

In this chapter, a general discussion of the phenomena of "frame stability" and the importance of considering this type of failure in column design is first presented. This is followed by a summary of the various methods of estimating the "effective length" of framed columns. The procedure of proportioning columns in plastically designed frames which are not braced against possible sidesway is then explained. Finally, a brief description of the buckling of compression members in trusses is given and design methods are outlined.

A structure may buckle either in or out of its plane. Throughout this chapter, it will be assumed that sufficient bracing is provided perpendicular to the plane of the frame or truss to prevent out-of-plane buckling.

10.2 PHENOMENA OF FRAME INSTABILITY

Consider the frame shown in Fig. 10.1. Associated with such a frame there exist two possible modes of instability:

1. The symmetrical mode
2. The antisymmetrical or sidesway mode

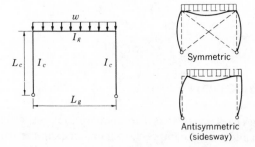

Fig. 10.1 Instability modes of a portal frame.

In the first mode, the frame remains in the symmetrically deformed shape before and after the attainment of the critical load. This type of failure induces no sudden change of the deformed configuration and can occur only when positive bracing is provided to prevent the frame from swaying sideways. In the second mode, the frame shifts suddenly from a symmetrical configuration to an antisymmetrical configuration at the instant when the corresponding critical load is reached. The frame as a whole also moves laterally after buckling. The load-deformation relationships of these two types of instability failure are illustrated in Fig. 10.2. The sidesway mode is always associated with a lower critical load than the symmetrical mode.[10.1]

The phenomena of frame instability described above may take place at a load level below that which produces initial yielding (elastic instability), but more frequently would occur after the applied load has caused yielding in parts of the frame (inelastic instability). Both of these cases will be discussed.

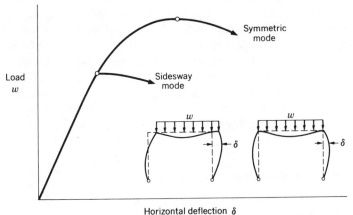

Fig. 10.2 Load-deflection relationships corresponding to two types of instability.

10.3 BUCKLING OF ELASTIC FRAMES

In analyzing the critical strength of the portal frame loaded as shown in Fig. 10.1, it is often convenient to replace the distributed beam load by two equivalent loads P, each applied directly at the tops of the columns.   Such a situation is shown in Fig. 10.3.   This replacement of load usually results in a considerable simplification in the calculations and at the same time leads to solutions with an acceptable degree of accuracy.   However, it should be recognized that there is a basic difference in the nature of the two types of

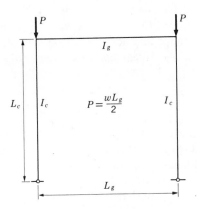

$$P = \frac{wL_g}{2}$$

Fig. 10.3 Simplified loading condition.

problems.   For the case in which the frame is loaded by the distributed load, all the members in the frame are subjected to combined bending moment and axial force; whereas in the case with the simplified loading condition axial forces only are present in the members at the instant of failure.   The presence of bending moment in the members tends to cause some reduction in the critical load for portal frames considered here.   Detailed discussion of the effect of bending moments on elastic frame instability can be found in Refs. 10.1 and 10.2.   For most practical applications, the simplified loading condition is quite satisfactory in determining the elastic critical load of frames.

1. Methods of Buckling Analysis

Numerous methods have been developed and effectively used in solving stability problems associated with various types of structural frameworks. They resemble very much those methods ordinarily used in analyzing statically indeterminate structures, except that some modification is made to include the bending moment produced by the axial forces after buckling.   Essentially there are three different avenues of approach that are now in common use. These are: (1) the analytical methods, including the methods of slope deflection and four-moment equation, (2) the moment-distribution method, and (3) the energy method.   Each method is based on a particular type of stability criterion.   For complete descriptions of the various methods, the reader is referred to Refs. 10.1 and 10.3.   In the following discussion, the results obtained for some simple frames by the application of these methods are examined.

2. Solutions for Two Simple Frames

Effective Length of Columns.   In treating various structural stability problems, it is often useful to express the buckling load in the form of the Euler formula for a pinned-end column with a suitable modification of the column length.   Thus, the critical load of a framed column may be expressed as

$$P_{cr} = \frac{\pi^2 EI_c}{(KL_c)^2} \tag{10.1}$$

in which EI_c is the flexural rigidity of the column and KL_c denotes the effective or equivalent length.   Figure 10.4 gives the K values for six columns with various idealized end conditions.*   It is important to note that the effective length factor of columns can be greater than 1.0 if translational movement of the ends is permitted.

---

* A more detailed discussion of Fig. 10.4 is given in Art. 10.4.

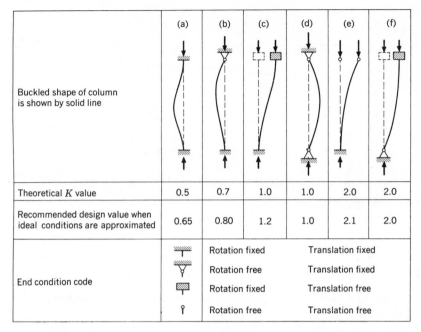

	(a)	(b)	(c)	(d)	(e)	(f)
Buckled shape of column is shown by solid line						
Theoretical K value	0.5	0.7	1.0	1.0	2.0	2.0
Recommended design value when ideal conditions are approximated	0.65	0.80	1.2	1.0	2.1	2.0

End condition code		
	Rotation fixed	Translation fixed
	Rotation free	Translation fixed
	Rotation fixed	Translation free
	Rotation free	Translation free

Fig. 10.4 Effective lengths of centrally loaded columns with various idealized end conditions.[10.5]

Equation 10.1 can also be used to express the critical load of the frame shown in Fig. 10.3. In general, the value of K is a function of the flexural and translational stiffnesses of all the members and can be determined exactly only by solving the complete frame. K depends entirely on the particular buckling mode assumed and should not be confused with the coefficient that gives the distance between the points of inflection (or points of zero moment) in a column as determined by ordinary structural analysis.

Symmetrical or Non-Sway Buckling. Symmetrical buckling will occur for frames which are braced in such a manner that translation of the tops of the columns is fully prevented. Figure 10.5 shows the theoretical effective length factor K corresponding to this type of buckling for a simple portal frame. Two types of column base conditions are considered: the pinned condition (curve a) and the fixed condition (curve b). In both cases the value of K is found to depend only on the non-dimensional parameter G, which is the ratio of the stiffness of the column to that of the beam. It may be seen from Fig. 10.5 that for all values of G the effective length factor K is always less than 1.0. Therefore, if it is assumed that the effective length KL_c is

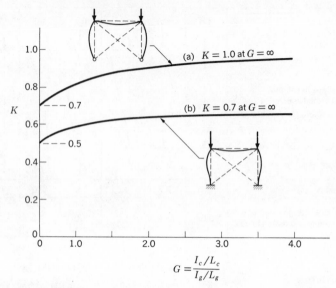

Fig. 10.5 Effective-length factor of columns (non-sway buckling).

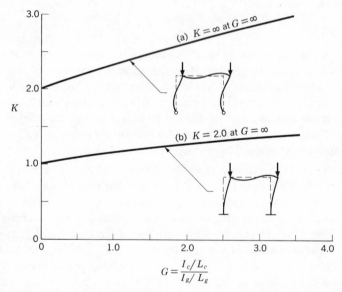

Fig. 10.6 Effective-length factor of columns (sidesway buckling).

equal to the actual length in calculating the buckling strength by using Eq. 10.1, the results will be on the safe side. This explains AISC Section [1.8.2], which follows:

In frames where lateral stability is provided by adequate attachment to diagonal bracing, shear walls, an adjacent structure having adequate lateral stability, or to floor slabs or roof decks secured horizontally by walls or bracing systems parallel to the plane of the frame, and in trusses,* the effective length factor K for the compression members shall be taken as unity unless analysis shows that a smaller value may be used.

Antisymmetrical or Sidesway Buckling. For frames which are not braced to prevent lateral displacement at the tops of the columns, the prevailing buckling mode will be that of sidesway. The curves relating the effective length factor K and the non-dimensional ratio G defined previously are given in Fig. 10.6. Again, two types of column base conditions are considered. For sidesway buckling the value of K is always greater than 1.0. This is in contrast to K values of less than 1.0 for the case of symmetrical buckling (Fig. 10.5). On the basis of this observation, AISC Section [1.8.3] states:

In frames where lateral stability is dependent upon the bending stiffness of rigidly connected beams and columns, the effective length KL of compression members shall be determined by a rational method and shall not be less than the actual unbraced length.

Rational methods for determining the effective length of columns in continuous frames will be discussed in Art. 10.4.

Effect of Partial Base Fixity. Referring to curve a of Fig. 10.6, it can be seen that the coefficient K for pinned-base frames varies from $K = 2.0$ at $G = 0$ to $K = \infty$ at $G = \infty$. These values of K are determined for perfectly pinned-base conditions which can never be realized in actual practice. Investigations have shown that the rotational restraint offered by the so-called "pinned" column bases used in building construction can be counted upon to increase considerably the buckling strength of frames.[10.4] For this reason, a G value of 10 has been recommended for pinned column bases in the Commentary on the AISC Specification,[7.9] and, therefore, a K value smaller than that indicated by curve a may be used in design computations.

10.4 DETERMINATION OF THE EFFECTIVE LENGTH OF COLUMNS IN FRAMES

The effective length factor K of columns (defined in connection with Eq. 10.1) may be determined (1) by interpolating between the K values found

---

* The effective length of compression members in trusses is discussed in Art. 10.6.

for columns with idealized end conditions, (2) by using charts or tables which consider varying end conditions, and (3) by performing a stability analysis of the entire structure.   The first procedure gives only approximate K factors, which must be used with engineering judgment.   A more accurate estimate of the effective length factor may be obtained from chart solutions.   The third method generally furnishes exact solutions, but the numerical work involved is often excessive for practical purposes.

1. Interpolation Method

In this method the effective length factor K for the columns in question is determined by simple interpolation between the K values tabulated for a number of columns with idealized end conditions.   Figure 10.4 gives the theoretical K values for six conditions in which the rotation and/or restraints at the ends of the column are either fully realized or non-existent.[10.5]   For conditions a, b, c, and e the lower end of the column is assumed to be fixed. The condition of full fixity can be approached only when the column is anchored securely by adequate embedment or by a moment-resisting bolted connection to a footing that is designed to resist overturning moment and for which the rotation is negligible.*   Column condition a is approached when the top of the column is integrally framed to a heavy girder many times more rigid than the column under consideration.   Condition c is the same as a except that in this case the top of the column is allowed to translate.   This condition can be realized for columns in an unbraced frame with heavy beam members (corresponding to the fixed base frame shown in Fig. 10.6 with $G = 0$).   Similarly, condition f is approached for columns in a pinned-base frame with heavy beam.   Also shown in Fig. 10.4 are the design values recommended by the Column Research Council.[10.5]   These values are a modification of the ideal values, taking account of the fact that full fixity is impossible to attain, and, likewise, that a perfectly rotation-free condition is not attained in practice.

2. Chart Method

At present numerous charts, tables, and approximate formulas are available for use in estimating the effective length of framed columns.[10.1,10.5,10.6] They were developed from the exact solutions obtained for certain typical structures.   Reference 10.6 contains a summary of table and formulas which are applicable to columns in building frames.

Figure 10.7 shows two convenient alignment charts for determining the effective length factor K of columns in multi-story frames for cases in which

---

* A method for estimating the moment-resisting capacity of column bases can be found in Ref. 10.4.

sideway is prevented or in which sidesway occurs.[10.5]   These charts can be used when the stiffness I/L of the adjacent girders is known or can be closely estimated.   The alignment chart for the case with no sidesway (Fig. 10.7a) is based on the following equation:

$$\frac{G_A G_B}{4}\left(\frac{\pi}{K}\right)^2 + \left(\frac{G_A + G_B}{2}\right)\left(1 - \frac{\pi/K}{\tan(\pi/K)}\right) + \frac{2\tan(\pi/2K)}{\pi/K} = 1 \quad (10.2)$$

The subscripts A and B in this equation refer to the joints at the two ends of the column section being considered. G is defined as follows:

$$G = \frac{\sum(I_c/L_c)}{\sum(I_g/L_g)} \quad (10.3)$$

in which $\sum$ indicates a summation for all members connected to that joint and lying in the plane in which buckling of the column is to occur, I_c is the moment of inertia and L_c the unsupported length of a column section, and I_g is the moment of inertia and L_g the unsupported length of a girder or other restraining member. I_c and I_g are taken about the axes perpendicular to the plane of buckling.

Equation 10.2 was derived on the basis of the following assumptions:

1. All the columns in the structure reach their individual critical loads simultaneously.

2. At the initiation of buckling the rotation at the far end of the girders is equal to that at the near end (adjacent to the column under consideration), but of opposite sense.   (See the buckled shape of the girders of the frames shown in Fig. 10.5.)

3. The restraining moment at a joint provided by the girders is distributed between the columns in proportion to their stiffness.

The above assumptions were also made in developing the alignment chart for the case with sidesway permitted (Fig. 10.7b), except that the rotation at the far end of the girders was assumed to have the same sense as that at the near end.   (See the girders of the frames shown in Fig. 10.6.)   The resulting equation for sidesway buckling is

$$\frac{G_A G_B(\pi/K)^2 - 36}{6(G_A + G_B)} = \frac{\pi/K}{\tan(\pi/K)} \quad (10.4)$$

For the simple portal frames considered previously, Eqs. 10.2 and 10.4 can be shown to yield K values identical to those given in Figs. 10.5 and 10.6, respectively.

In applying the alignment charts of Fig. 10.7, the following considerations may prove useful:

1. For a column base connected to a footing by a frictionless hinge, G is theoretically infinity but should be taken as 10 for practical design.[10.5]   If

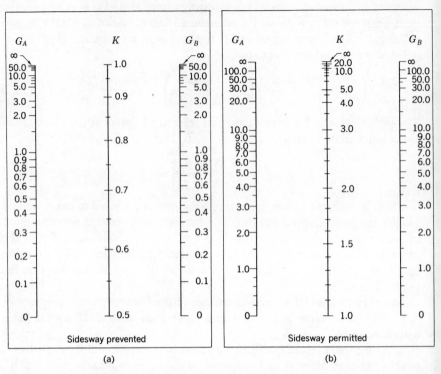

Fig. 10.7 Alignment charts for effective length of columns in continuous frames.[10.5]

the column base is rigidly attached to a properly designed footing, G approaches a theoretical value of zero but should be taken as 10. Other values may be used if justified by analysis.

2. Refinements in girder stiffness I_g/L_g may be made when conditions at the far end of any particular girder are known. For the case with no sidesway (Fig. 10.7a) multiply girder stiffness by the following factors:

1.5 for far end of the girder hinged
2.0 for far end of girder fixed against rotation.

For the case with sidesway permitted (Fig. 10.7b) multiply girder stiffness by:

0.5 for far end of girder hinged
0.67 for far end of girder fixed.

The use of the alignment charts is illustrated by Examples 10.1 and 10.2.

EXAMPLE 10.1

PROBLEM:
Find K values of columns C1, C2, C3 and C4 of the frame shown in the figure. All the columns are oriented with webs parallel to the plane of the drawing and are assumed to be braced in the weak direction. Sidesway is prevented by the diagonal members.

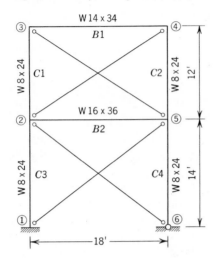

SOLUTION:
Moment of inertia of the members:

$$W\ 8 \times 24 \quad I_x = 82.5\ \text{in.}^4, \quad r_x = 3.42\ \text{in.}$$
$$W\ 14 \times 34 \quad I_x = 340\ \text{in.}^4$$
$$W\ 16 \times 36 \quad I_x = 447\ \text{in.}^4$$

Beams I/L

$$B1 \quad \frac{340}{216} = 1.570$$
$$B2 \quad \frac{447}{216} = 2.066$$

Column I/L

$$C1 \quad \frac{82.5}{144} = 0.573$$
$$C2 \quad \frac{82.5}{144} = 0.573$$
$$C3 \quad \frac{82.5}{168} = 0.491$$
$$C4 \quad \frac{82.5}{168} = 0.491$$

G values from Eq. 10.3:

$$\text{Column C1}: G_3 = \frac{0.573}{1.570} = 0.365 \qquad G_2 = \frac{0.573 + 0.491}{2.066} = 0.515$$

$$C2 : G_4 = G_3 = 0.365 \qquad G_5 = G_2 = 0.515$$
$$C3 : G_2 = 0.515 \qquad\qquad G_1 = 1.0\ (\text{fixed end})$$
$$C4 : G_5 = 0.515 \qquad\qquad G_6 = 10.0\ (\text{hinged end})$$

For a given pair of values of G_A and G_B, the effective length factor K can be obtained by constructing a straight line between the appropriate points on the scales for G_A and G_B. For example, for column C1

$$G_A = G_3 = 0.365$$
and
$$G_B = G_2 = 0.515$$

EXAMPLE 10.1 (Continued)

K is found to be 0.67 from Fig. 10.7a. Similarly, K values for other columns are obtained as follows:

	K	$\dfrac{L}{r_x}$	$\dfrac{KL}{r_x}$
C 1	0.67	42.1	28.2
C 2	0.67	42.1	28.2
C 3	0.73	49.1	35.8
C 4	0.81	49.1	39.8

3. Theoretical Methods

In general, any of the theoretical methods mentioned in Art. 10.3 may be used to determine the critical load of a given structure. The critical load thus determined can then be expressed in the form of Eq. 10.1, from which the effective length factor K may be obtained. In some cases of practical application, the critical load, after being modified by a suitable factor of safety, can be used directly in selecting the proper column size.

Of the theoretical methods that are in use for buckling analysis, the method of moment distribution has shown promise in recent years.[10.7,10.8] Various special techniques have been developed and applied successfully to multistory building frames with or without sidesway. Some of these techniques are discussed in Refs. 10.9, 10.10, 10.11.

4. Use of K Factors in Column Design

In the design of centrally loaded compression members, the effective length factor K determines the equivalent slenderness ratio KL/r to be used in the column formulas of the AISC Specification. For bracing and secondary members, the full unbraced length ($K = 1.0$) should be used.

In designing members subject to combined axial compression and bending stresses (beam-columns),* it is necessary to select the member sizes to meet the requirements of Eqs. 11.24 and 11.25. In Eq. 11.24 the quantities F_a and F_e' are functions of the slenderness ratio of the member being considered. F_a is determined from the column formulas of the Specification by using effective column slenderness ratio, and F_e' is given by

$$F_e' = \frac{12\pi^2 E}{23(KL_b/r_b)^2} \tag{10.5}$$

in which KL_b/r_b represents the effective slenderness ratio of the member in the plane of bending. The use of K factors in the design of beam-columns is illustrated in Examples 11.6, 11.7, and 11.8.

EXAMPLE 10.2

PROBLEM:
  Solve Example 10.1, for the case when the columns are braced in the weak direction but sidesway buckling in the plane of the frame is not prevented.

SOLUTION:
  The G values that are needed for determining the K factors of the columns in this case are the same as those used in Example 10.1. The alignment chart of Fig. 10.7b should be used and the following results are obtained:

	K	$\dfrac{L}{r_x}$	$\dfrac{KL}{r_x}$
C 1	1.15	42.1	48.4
C 2	1.15	42.1	48.4
C 3	1.25	49.1	61.3
C 4	1.78	49.1	87.4

  The effective slenderness ratios given above are considerably higher than those obtained in the previous example. This indicates that larger member sizes must be used for the columns in building frames that are not braced to prevent sidesway buckling.

10.5 BUCKLING OF PARTIALLY YIELDED FRAMES

The discussion in the previous sections was restricted to the buckling of perfectly elastic frames. Such a restriction implies that the methods of analysis and the results obtained are applicable only in designing structures by the allowable-stress method (Part 1 of the AISC Specification). As was mentioned earlier in Art. 10.2, frame instability is more likely to take place after the stresses at some parts of the frame have exceeded the yield limit. In general, the stiffness of a frame decreases as yielding progresses in its members and, eventually, failure will occur when the limit of stability is reached.

Since in plastic methods of designing structures the basis of design is always the attainment of maximum plastic strength, it is required that all types of instability which may cause reduction in this strength should be taken into account. Frame instability constitutes one of the possible types. A considerable amount of research has been done on this subject in recent years,[10.12,10.13,10.14,10.15,10.16] and extensive studies are still under way at several institutions. The following discussion is intended to clarify the problem and briefly to discuss the column design procedure in Part 2 of the AISC Specification.

---

* The design of beam-columns is discussed in Chapter 11.

1. Ultimate Strength of Steel Frames

Depending upon the mode of failure assumed in the analysis, different theoretical ultimate loads may be obtained for the same frame. Figure 10.8 shows the ultimate strength curves for frames corresponding to four possible modes of failure. The symmetrical frames considered here are assumed to have a constant span length L_g and a varying column height L_c. Two types of loads are applied to the frames: a distributed load of intensity w and

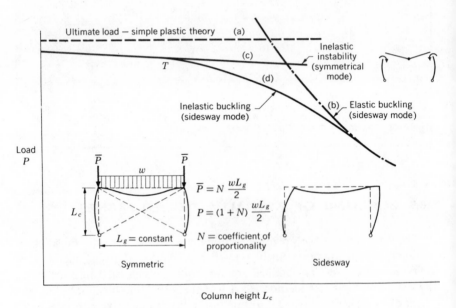

Fig. 10.8 Load-carrying capacity of frames.

concentrated loads $\bar{P}$ on the columns. These loads are related to each other by the coefficient of proportionality N as indicated in the figure. Thus, the total axial force P in the columns is $(1 + N)(wL_g/2)$. This condition of loading simulates that which exists in the lower stories of a tall building.

At a specified column height on the chart of Fig. 10.8, line a gives the carrying capacity according to simple plastic theory. This theory assumes failure by symmetrical bending (beam mechanism) and ignores any reduction in strength due to instability.

Curve b represents the limiting strength corresponding to sideway buckling in the elastic range. Points on this curve can be determined by the methods mentioned in Art. 10.3.

Curves *a* and *b* assume two idealized conditions, the simple plastic failure and the purely elastic buckling failure. Neither of these idealized conditions can actually exist in practical frames. Usually the true ultimate load is determined by considering the combined effects of instability and yielding.

Curve *c* in Fig. 10.8 represents the true ultimate strength of the frames when they are properly braced to prevent sidesway movement. Points on this curve are obtained by considering the reduction of plastic moment capacity at the tops of the columns due to beam-column action. (This is treated in Art. 11.5.)

For unbraced frames, inelastic sidesway buckling will usually take place before the applied load has reached the value given by curve *c*. The limiting strength for this type of failure is shown as curve *d* in Fig. 10.8. To determine any point on this curve, it is necessary to consider the decrease in overall stability in the lateral direction caused by yielding in all the members. The problem is further complicated in that the secondary moments in the columns resulting from their deformations must also be considered. Although a general method has been developed for determining the inelastic sidesway buckling load,[10.14] it is not feasible to base any frame design on this type of limit load. In fact, it is often desirable to proportion the columns in such a way that sidesway buckling does not occur until the applied load has reached the computed ultimate load (corresponding to curve *c* of Fig. 10.8). In certain cases, this may require a slight increase in column size.

It is interesting to note in Fig. 10.8 that curve *d* becomes coincident with curve *c* after passing through point *T*. Therefore, for frames with column height less than that indicated by point *T*, the reduction in strength due to sidesway buckling will be negligibly small. In other words, within this limit of column height (or slenderness ratio) the possibility of sidesway buckling may not affect the final design of the structure.

2. Column Design in Rigid Frames

The 1969 AISC Specification permits the application of plastic theory to the design of the following types of rigid frames:
1. Braced and unbraced frames of one and two stories in height.
2. Braced multi-story frames.

The columns in these frames are to be proportioned to satisfy Eqs. 11.32 and 11.33. For braced frames, a *K* value of unity may be used in computing the terms P_0 and P_e in Eq. 11.32. For low unbraced frames, studies have shown that the possible reduction in strength due to overall frame instability is usually very small and that a nominal amount of base restraints or stiffening effects provided by the non-structural elements can improve the stability of the frames significantly.[10.16] For this reason, it has been found possible to include, in an approximate manner, the effect of frame instability by using

K values determined for the sway mode from an elastic buckling analysis in computing P_0 and P_e. The proportioning of columns in plastically designed frames is illustrated in Example 11.10.

10.6 COMPRESSION MEMBERS IN TRUSSES

From elementary structural theory, it is generally known that the members in a truss framework are subjected only to axial forces if the joints are pinned. The compression members in such a structure therefore may be designed as pinned-end columns. On the other hand, if the joints are rigidly connected by welding or by bolting, some secondary moments are induced. Furthermore, the rigidity of the joints often provides rotational restraints to the members. (This is especially true when heavy gusset plates are used to form the joints.) The effect of the secondary moments on the stability of truss framework has been found to be small and may be neglected in buckling analyses.[10.17,10.18] Thus, the compression members in rigid-jointed trusses may be treated as columns elastically restrained against rotation at both ends.

1. Approximate Determination of Effective Length Factor

Consider the top chord AB of the simple truss as shown in Fig. 10.9a. To determine the effective length of this member, it is necessary to solve the complete truss buckling problem, using one of the methods mentioned in Art. 10.3. This is usually a time-consuming task. For practical applications, a satisfactory estimate of the effective length may be made by considering the member as a simple column subject to rotational restraint offered by the adjacent members. Such a situation is shown in Fig. 10.9b. Note that for simplicity the members not directly connected to AB are disregarded and the far ends of the restraining members are assumed to be hinged. The restraint at end A or B is equal to the sum of the rotational stiffness, $3EI/L$, of all the adjacent members framed to that end. (In cases where some of the adjacent members are also subjected to axial compression, the contribution of these members may be neglected.) These restraints may be regarded as elastic springs attached to the ends of the member as shown in Fig. 10.9c. The amount of end restraint is represented by the spring constant R (in units such as ft-lb per radian).

To determine the K value of member AB in Fig. 10.9c, a convenient chart developed in Ref. 10.19 may be used. The chart is reproduced in Fig. 10.10. The effective length may also be calculated from the following approximate formula:[10.20]

$$K = \left[\frac{(\pi^2 C_A + 2)(\pi^2 C_B + 2)}{(\pi^2 C_A + 4)(\pi^2 C_B + 4)} \right]^{1/2} \tag{10.7}$$

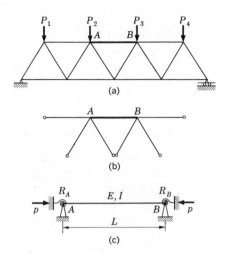

Fig. 10.9 Estimation of end restraints.

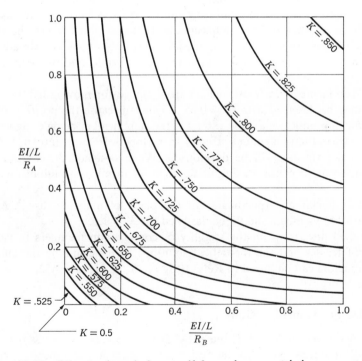

Fig. 10.10 Effective-length factors K for columns with known rotational restraints R_A and R_B.[10.19]

in which

$$C_A = \left(\frac{EI}{L}\right)\left(\frac{1}{R_A}\right)$$

$$C_B = \left(\frac{EI}{L}\right)\left(\frac{1}{R_B}\right)$$

2. Effective Length Factors Recommended for Use in Design

The Column Research Council has recommended a number of K values to be used in the design of various compression members in trusses.[10.5] Some of the recommendations are cited below:

1. In a truss of balanced design, buckling stresses in compression chords and yield forces in tension members will be approached almost simultaneously and the restraints of the ends of the compression members would be greatly reduced at the ultimate load. Therefore, a K value of 1.0 may be assumed in designing the compression chords.

2. In a roof truss of nearly constant depth, where a single chord of constant cross section is used for the full length of the truss, K may be taken as 0.9.

3. In a continuous truss, adjacent to the panel point where the chord stress changes in sign, for the compression chord continuous with the adjacent tension chord, K may be taken as 0.85, if the lengths and make-up of the two members are similar.

4. Web members in trusses which are designed for moving live load systems may be designed with $K = 0.85$. This is because the position of live load which produces maximum stress in the web member being designed will result in less-than-maximum stresses in members framing into it, so that rotational restraints will be developed. When designed for a fixed load system, so as to cause maximum stress in all members simultaneously, K is to be taken as unity.

5. For buckling perpendicular to the plane of a main truss, the web compression members should be designed for $K = 1.0$.

Detailed discussion on the stability of compression members in various types of trusses and the theoretical solutions for determining the effective length factors can be found in Refs. 10.1, 10.18, 10.21, and 4.32.

3. Design Considerations

As mentioned in Art. 10.3, the AISC Specification recommends the use of the full unbraced length in the design of compression members in trusses. In certain cases where an approximate buckling analysis is feasible the effective length factor determined by the analysis may be used. For routine design, the K values given in the preceding section are considered to be adequate.

The use of K factors less than unity will result generally in smaller cross sections for the compression members. This effect, however, may become much less pronounced for short members (this has been illustrated in Fig. 9.47). For structural carbon steel columns with slenderness ratios less than about 50 the allowable stress for a fixed-end column ($K = 0.50$) is only 10 per cent higher than that for a column with pinned ends ($K = 1.0$) (based on the AISC Specification). Thus, the choice of K values has but little effect on the design of these members. In the case of high-strength steel ($F_y = 50$ ksi) columns the figure of 50 may be changed to about 42 for the same difference in allowable stresses. Figure 9.47 also shows that the difference in allowable stresses for the fixed- and pinned-end columns increases steadily as the slenderness ratio increases. Therefore, the use of K values less than unity would result in some reduction in member sizes for long columns.

PROBLEMS

10.1. Select a wide-flange section for the columns of the portal frame shown in the accompanying figure, using the AISC Specification. The members are to be arranged in such a manner that the web is in the plane of the frame. Use A36 steel; assume that sidesway of the frame is permitted.

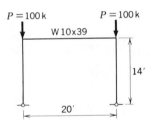

Problem 10.1

Suggested procedure:
    (a) Select a trial section (say, W 8 × 31).
    (b) Compute the ratio G and determine the corresponding effective length factor K from Fig. 10.6.
    (c) If the effective slenderness ratio KL_c/r_x is less than C_c, the column will buckle in the inelastic range, and the critical stress can be computed by

$$F_{cr} = \left[1 - \frac{(KL_c/r_x)^2}{2C_c^2}\right]F_y$$

    (d) For inelastic buckling the flexural rigidity of the column is $E_t I_c$, in which E_t is the tangent modulus of the cross section. The ratio G should therefore be defined as

$$G = \frac{E_t I_c/L_c}{EI_g/L_g} = \frac{\tau I_c/L_c}{I_g/L_g}$$

An approximation for τ may be obtained as follows:

$$\tau = \frac{E_t}{E} = \frac{\pi^2 E_t/(KL_c/r_x)^2}{\pi^2 E/(KL_c/r_x)^2} = \frac{F_{cr}}{F_e}$$

in which F_e may be identified as the elastic buckling stress of the column.

(e) A new value of G is computed and Fig. 10.6 is again used to determine the K factor. Generally, the K value obtained at the end of the second cycle is sufficiently accurate for design use.

(f) The allowable stress F_a corresponding to the new effective slenderness ratio is then determined from the column formula. If the allowable load $P_a = F_a \times A$ is greater than the applied load P, then the selected trial section is satisfactory; otherwise, a new trial section should be selected and the above steps repeated.

10.2. Solve Problem 10.1, assuming that the frame is properly braced to prevent sidesway movement.

10.3. Using the charts in Fig. 10.7, determine the effective slenderness ratios of columns C1, C2, C3, C4, and C5 of the frame shown in the accompanying figure.

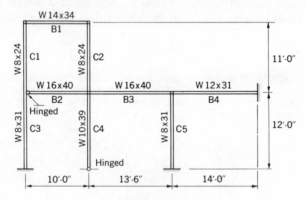

Problem 10.3

10.4. Determine the rotational restraints R_A and R_B of the compression member shown in the figure. All the members are oriented with webs parallel to the plane of the structure and are assumed to be braced in the weak direction.

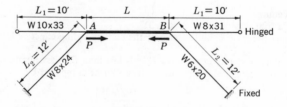

Problem 10.4

10.5. Design the compression member of Problem 10.4, if $P = 200$ kips and $L = 12$ ft. Use A36 steel. (Note: The K factor can be determined either from the chart in Fig. 10.10 or by applying Eq. 10.7. The design procedure outlined in Problem 10.1 is to be followed.)

11

COMBINED BENDING AND COMPRESSION

11.1 INTRODUCTION

Beam-columns are members which are subjected to forces producing significant amounts of both bending and compression.

Several typical beam-columns are shown in Fig. 11.1. Each beam-column is subjected to a concentric force P and to end moments M. In Fig. 11.1a these end moments are applied about one of the principal axes of the cross

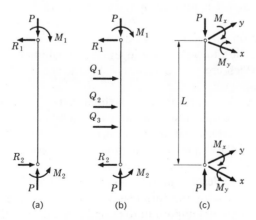

Fig. 11.1 Typical beam-columns.

section.   This case represents the situation occurring in the design of the vertical members of planar frames.   The member in Fig. 11.1b is subjected to transverse forces in addition to the axial force and the end moments.   A more general case of loading is shown in Fig. 11.1c, where biaxial bending is produced about both principal axes.

The analysis of the behavior of beam-columns is not a straightforward matter.   It is complicated by the fact that the so-called "second-order" bending moments (due to the axial force times the deflection) may, in general, not be ignored.   Another complicating feature is that maximum capacity is attained only after some portions of the member have become yielded. Because of these and other factors, the design of beam-columns is based either on simplifying assumptions or on formulas developed from theoretical and experimental research.

There have been essentially three major approaches to the design of beam-columns in steel design specifications.

1. The "secant formula" method is based on rationally developed analytical expressions obtained by assuming that initiation of yielding, caused by the applied loads and including the effect of an assumed initial crookedness or an initial accidental end eccentricity, will terminate the usefulness of the member.   Historically this is the earliest approach; it has served as the basis of the beam-column design provisions of the bridge specifications of both AASHO[11.1] and AREA.[11.2]   This method is also permitted as an alternate in the German buckling specifications.[11.3]

2. Empirically determined and experimentally tested "interaction equations" frequently are used for their simplicity and versatility.   These formulas relate the interaction between the axial force, the bending moments, and the member geometry at the limiting working stresses by means of simple algebraic expressions.   This approach is by far the most popular in steel building specifications, and it is utilized by the AISC specifications,[11.4] the German buckling specification DIN 4114,[11.3] the British specification BS 449,[11.5] the AISI specification for light gage cold-formed construction,[11.6] and the 1969 AREA specification for railroad bridges.[1.25]   In addition to these, there are many others which use this concept, notably the specifications for structures of aluminum alloys.[11.7]

3. The theoretically developed maximum strength interaction curves have been used as the basis for design.   The 1963 AISC beam-column design rules for plastic design used this method.[11.4]   The 1969 version of this specification uses the empirical interaction equation concept for the design of columns in plastically designed frames.

All three of these design philosophies are based on the maximum strength regardless of the use of the formulas in either allowable stress or plastic design.   It is, therefore, important to examine first the mechanisms of failure before starting a detailed explanation of the design methods.

11.2 THE MAXIMUM STRENGTH OF BEAM-COLUMNS

1. Load-Deflection Behavior

For the sake of illustration, the following description is restricted to members which are bent by end moments about the major axis of the cross section. It is stipulated further that these members are structural steel wide-flange shapes. Other types of beam-columns fail in a similar manner.

Failure is influenced by the following factors: (1) yielding, (2) lateral-torsional buckling, and (3) local buckling. The effect of two of these factors is shown in Fig. 11.2 where the curve shows the theoretically determined relationship existing between the end-moment M and the end slope θ for the particular beam-column and axial load shown.

The solid line curve in Fig. 11.2 represents the optimal behavior of the W 8 × 31 member; no lateral-torsional or local buckling effects influence the situation in this case. This means that all deformations take place in the plane of bending, and that the only weakening effect is due to yielding. A constant axial force $P = 90.4$ kips acts on this particular beam-column. The $M - \theta$ curve consists of the following two parts: (1) the ascending part, where an increase of deflection results in an increase in the moment, and (2) the descending part, where an increase of deformation is accompanied by a

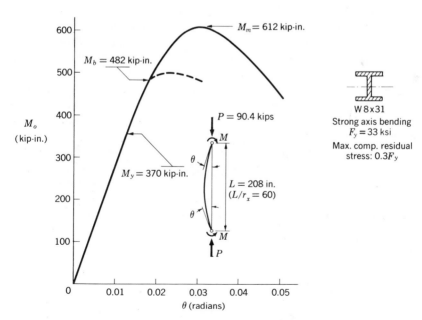

Fig. 11.2 End moment vs. end slope curve for wide-flange beam-column.

decrease in M. The peak point ($M_m = 612$ kip-in.) corresponds to the maximum moment, and the attainment of this moment constitutes failure for an isolated beam-column. However, if this member is part of a frame system (see Example 11.3), then the unloading portion of the curve is also of importance.

Under relatively high axial loads and for relatively long members, unloading occurs rather rapidly after the maximum moment is reached, as illustrated by the curve in Fig. 11.2. However, for many practical combinations of axial load and geometry, the unloading process is gradual and the top of the M-θ curve exhibits a more or less flat yield plateau.

The M-θ curve is a straight line if the axial force is maintained constant and the material is elastic. Due to the presence of residual stresses, yielding commences as soon as the sum of the applied compressive stress and the preexisting maximum compressive residual stress becomes equal to the yield stress. This situation occurs at $M_y = 370$ kip-in. in the example of Fig. 11.2, where it can be seen that the M-θ curve begins to become non-linear. Because of the increased amount of yielding with each increment in M, the stiffness of the member is reduced progressively until finally the over-all member stiffness becomes zero and no additional moment can be supported. Because of the non-linear post-yielding characteristics of metals, the direct analytical computation of the maximum load is not practical and recourse must usually be made to numerical integration.

Failure in the plane of bending is possible only if the member is bent about its weak axis or if it is adequately braced against lateral-torsional buckling in case the member is bent about its strong axis. In the absence of adequate bracing an initially straight member will start to deflect laterally, as well as to twist, at a certain "critical" moment M_b which is lower than the maximum moment M_m ($M_b = 482$ kip-in. in the example of Fig. 11.2 if the member is unbraced between ends). If beam-columns cannot be designed with sufficient bracing, then they must be designed on the basis of the reduced moment capacity due to lateral-torsional buckling.

In addition to lateral-torsional buckling, it is also possible that the M-θ relationship, and thus the maximum strength, is influenced by local buckling. The problem of local buckling, and its effective prevention until after the peak point on the M-θ curve is reached, is discussed in Chapter 17. In the design routine it is necessary to check whether the width-thickness ratios of the plate elements, of which the cross section is composed, do not exceed certain maximum values specified in the appropriate specifications.

2. Determination of Maximum Strength

There are a variety of methods available for the determination of the maximum strength of beam-columns failing by excessive bending in the plane of the applied moments. The starting point for any one of these is the relationship existing between the bending moment M and the thrust P acting on the cross section and the resulting curvature. The moment-thrust-curvature relationship (M-P-ϕ curves) is dependent on the values of P and M, the cross-sectional dimensions, the stress-strain curve of the material, and the residual stress distribution.

One particular strain and stress distribution resulting from the presence of P and M is shown in Fig. 11.3 for a steel wide-flange shape where the material

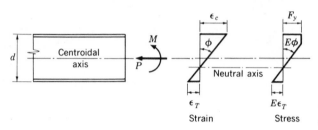

Fig. 11.3 Strain and stress distribution in the cross section at partial yielding.

exhibits the idealized stress-strain law of Fig. 1.9. The corresponding curvature is computed by assuming that plane sections before bending remain plane after bending, and by making use of the equilibrium conditions that the sum of the stresses must equal P, and that the sum of the moments due to the stresses must equal M.

Formulas for both strong and weak axis bending of wide-flange shapes are tabulated in Ref. 11.8. Because of the complexity of these equations, it is easier to present the results in the form of curves. One such set of M-P-ϕ curves is shown in Fig. 11.4 where the non-dimensionalized moment-curvature plots are presented for constant axial loads of 0, $0.2P_y$, $0.4P_y$, $0.6P_y$, and $0.8P_y$. The moment is non-dimensionalized by the yield moment $M_y = SF_y$, and the curvature $\phi_y = 2F_y/dE$ is the elastic curvature corresponding to M_y when $P = 0$. The solid line curves in Fig. 11.4 are for sections without residual stress, and the dashed line curves show the influence of the residual stress pattern of Fig. 11.5. All of the M-ϕ curves form a plastic hinge upon the complete yielding of the section.

The maximum strength of beam-columns may be determined by any one of the available methods if the M-P-ϕ curves for the given cross section and type of material are available. Several of these methods are based on the assumption that the deflection of the member will conform to a certain shape,[11.9,11.10] and others employ some type of numerical integration.[11.11,11.12,11.13,11.14] The latter methods are of importance because they

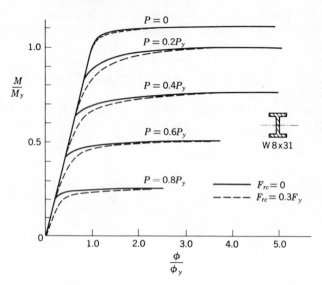

Fig. 11.4 M-P-ϕ curves for W 8 × 31 section.

not only furnish the maximum strength to any desired degree of accuracy, but they also permit the determination of the deformations (such as the M-θ curve of Fig. 11.2).

One of the numerical methods employs Newmark's numerical integration procedure.[11.12] The method consists of the computation of the deflections at evenly spaced points along the length of the beam-column by an iterative integration of the curvature diagram for given values of P, M, and L, as illustrated in Example 11.1. In this example the deflections at the eighth points are determined for a W8 × 31 beam-column which is bent by an end moment $M = 0.2M_y$ about the strong axis of the section. An axial force of

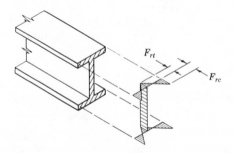

Fig. 11.5 Residual stress distribution.

EXAMPLE 11.1

PROBLEM:

Determine the maximum strength of the following beam–column:

Material: A 36 ; $F_y = 33$ ksi; $E = 30,000$ ksi
Stress–strain curve: Figure 1.4, no strain hardening
Residual stress distribution, Figure 11.5

Section: W 8 x 31, $P_y = AF_y = 9.12 \times 33 = 301$ kip.
$M_y = SF_y = 27.4 \times 33 = 904$ kip-in.

M-P-ϕ curves: Figure 11.4; $\phi_y = \dfrac{2F_y}{dE} = \dfrac{2 \times 33}{8 \times 30,000} = 0.000275$

Length: 138.8 in., $L/r_x = 40$

Axial load: 240.8 kip. $P/P_y = 0.8$

Loading condition: One end moment only

SOLUTION:
1) Subdivide member into 8 equal segments, $\lambda = L/8$
2) Specify M = 181 kip-in, $M/M_y = 0.200$
3) Determine end slope θ by numerical integration (see below)
   Fourth cycle provides sufficient accuracy.
   Assuming that the deflected shape through the last three
   points is a parabola, $\theta = \dfrac{4V_1 - V_2}{2\lambda} = \dfrac{4 \times 0.070 - 0.113}{2 \times 17.35} = 0.0048$ rad.

										Multi. Factor	Remarks
a	181	158	136	113	90	68	45	23	0		Moment due to M, kip-in.
b	0	0.026	0.049	0.068	0.047	0.044	0.032	0.008	0		Assumed deflection, in. (guessed)
c	0	5	10	14	10	9	7	2	0		P x v, kip-in.
d	181	163	146	127	100	77	52	25	0		Total moment, kip-in.,(a+c)
e	0.200	0.180	0.161	0.140	0.111	0.085	0.058	0.028	0		M/My
f	0.350	0.290	0.250	0.210	0.151	0.119	0.083	0.045	0	ϕ_y	ϕ/ϕ_y (from Fig. 11.4)
g		0.350	0.640	0.890	1.100	1.251	1.370	1.453	1.498	$\lambda\phi_y$	Slope
h	0	0.350	0.990	1.880	2.980	4.231	5.601	7.045	8.552	$\lambda^2\phi_y$	Deflection
i	0	1.069	2.138	3.207	4.276	5.345	6.411	7.483	8.552	$\lambda^2\phi_y$	Deflection correction*
j	0	0.719	1.148	1.327	1.296	1.114	0.810	0.438	0	$\lambda^2\phi_y$	Corrected deflection (i-h)
k	0	0.060	0.095	0.110	0.107	0.092	0.045	0.036	0		Corrected deflection, in.

End of first cycle
Second and third cycle

Fourth cycle

a	0	0.069	0.112	0.131	0.129	0.112	0.081	0.043	0		Starting deflection, in.
k	0	0.070	0.113	0.132	0.130	0.112	0.082	0.043	0		Final deflection, in.

*This deflection correction is made by setting the deflection at the right end equal to zero through rotating the member through an angle equal to 8.552/L and subtracting the computed deflection in line h from the rigid–body rotation represented by the deflection in line i.

EXAMPLE 11.1 (Continued)

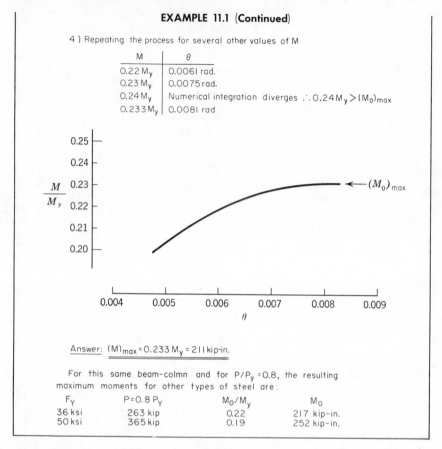

4) Repeating the process for several other values of M

M	θ
0.22 M_y	0.0061 rad.
0.23 M_y	0.0075 rad.
0.24 M_y	Numerical integration diverges $\therefore$ 0.24 M_y > $(M_0)_{max}$
0.233 M_y	0.0081 rad

Answer: $(M)_{max} = 0.233\,M_y = 211$ kip-in.

For this same beam-colmn and for $P/P_y = 0.8$, the resulting maximum moments for other types of steel are:

F_Y	$P = 0.8\,P_Y$	M_0/M_y	M_0
36 ksi	263 kip	0.22	217 kip-in.
50 ksi	365 kip	0.19	252 kip-in.

$P = 0.8P_y$ is also applied to the member. A sufficiently accurate deflected shape is found after four cycles of numerical integration, and the corresponding end slope θ is found to be 0.0048 radian.

By repeating the calculations for the same member and same axial force, but for different values of the end-moment, the inelastic portion of the ascending branch of the M-θ curve can be constructed, as shown in Example 11.1. The maximum point on this curve is the maximum moment of this particular beam-column. For the problem in Example 11.1 the combination of $P = 0.8P_y$ and $M = 0.233M_y$ will result in the failure of the member, due to inelastic instability. This example also shows the effect of varying the yield stress of the steel. As F_y increases, M_0/M_y decreases for a constant ratio P/P_y, but the actual values of P and M_0 increase.

Whereas Newmark's numerical integration procedure is well suited to determining the maximum load of beam-columns, it furnishes only the ascending branch of the M-θ curve. In many applications, as is shown in Chapter

EXAMPLE 11.2

PROBLEM:
Develop the $M_0 - \theta$ curve for the following beam-column.

Material: A 36 steel, $F_y = 36$ ksi; $E = 30,000$ ksi
Stress-strain curve of Figure 1.4, no strain hardening

Section: W 8 x 31
M-P-ϕ curves; Figure 11.4
Axial load P and length L are given. Solution presented below is only schematic:

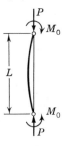

SOLUTION:
1) Because of symmetry it is sufficient to work with only one half of the member. At the center, the slope is zero.
2) Subdivide the halflength into n equal sublengths λ (usually take $\lambda = 3r$ or $4r$, where r is radius of gyration).
3) Assume that the segment in each sublength is a circular arc.
4) Assume a deflection at the center, v_0.
5) The corresponding moment $M_1 = P v_0$.
6) From the M-P-ϕ curves, find ϕ_1 corresponding to P and M.
7) For small deflections

$$v_1 = \frac{\lambda^2 \phi_1}{2}; \quad \theta_1 = \lambda \phi_1$$

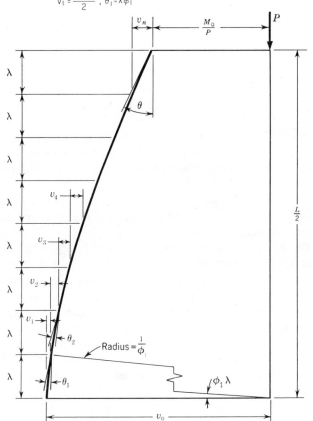

EXAMPLE 11.2 (Continued)

8) At the next point, $M_2 = P(v_0 - v_1)$; find corresponding ϕ_2 from Figure 11.4 and compute $v_2 = \dfrac{\lambda^2 \phi_2}{2}$; $\theta_2 = \lambda \phi_2$.

9) Continue until the end of the member, computing $v_3, v_4 \cdots v_n$ and $\theta_3, \theta_4 \cdots \theta_n$.

10) The end slope θ and the end moment M_0 is computed from geometry (see sketch):

$$M_0 = P\left[v_0 - (v_1 + v_2 + \cdots + v_n)\right]$$

$$\theta = \theta_1 + \theta_2 + \ldots + \theta_n$$

11) Repeat for other assumed values of v_0 until complete $M_0 - \theta$ relationship is determined, then draw $M_0 - \theta$ curve.

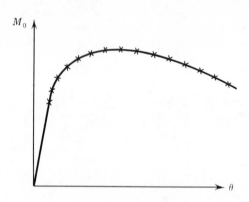

22, it is important to know also the descending branch of this curve. The complete load-deformation curve is obtained by another numerical method which uses a stepwise integration procedure.[11.14]   This method is illustrated in Example 11.2, where the M-θ curve is found for a wide-flange member which is subjected to an axial force P and to two equal end-moments M. This example illustrates the procedure schematically only, and no numerical answers are obtained.

The deflected shape and the values of M and θ for given values of L and P are obtained by first assuming a center deflection v_0, and then developing the deflection of evenly spaced points along the length of the column from the assumption that the deflected shape in the small interval λ is a circular segment with a radius $1/\phi$, where ϕ is the curvature corresponding to the moment at the last computed point.   This curvature is taken from M-P-ϕ curves such as are given in Fig. 11.4 for wide-flange shapes.

By repeating the integration process for different values of the center deflection v_0, the complete M-θ curve can be constructed, as shown in Example 11.2.

This procedure has been illustrated in Example 11.2 on a very simple example. The method has been generalized and further expanded in Ref. 11.11, where it is shown how it is possible to obtain any desired load-deformation relationship for simply supported and restrained beam-columns. Numerous curves for a variety of loading cases are given in Ref. 11.11 for use in plastic design.

The discussion up to now has been concerned with the behavior of individual beam-columns, and it has been shown that one can determine the complete load-deformation history of such members. However, beam-columns never occur as isolated members; they are part of a structure consisting of members joined together in a unified load-carrying system. A proper analysis of such systems can be performed if the load-deformation behavior of the individual components is known. Such analyses are performed in Chapter 22 in connection with studies of multi-story frames. It is shown there how the maximum strength of structural subassemblages can be determined with the aid of the beam-column M-θ curves discussed here.

3. Maximum Strength Interaction Curves

The results of the maximum strength calculations for beam-columns are represented best in the form of P-M-L interaction curves. A set of these is given in Fig. 11.6.[11.12] These particular interaction curves are for the case of two equal end moments causing single-curvature deformation about the strong axis of W8 × 31 members.

Each curve in Fig. 11.6 shows the relationship between P (non-dimensionalized by P_y) and M (non-dimensionalized by M_p) for a given

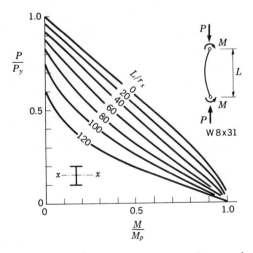

Fig. 11.6 Ultimate-strength interaction curves for equal end moments.

slenderness ratio L/r. The particular curves in Fig. 11.6 include the effect of residual stress.

The following observations can be made about these interaction curves:

1. When $P = 0$, the member is a beam and can support a moment equal to M_p.

2. When $M = 0$, the member is a column which is able to carry a load equal to its own critical load, as discussed in Chapter 9. (The tangent modulus load has been used in the curves in Fig. 11.6 for the residual stress pattern shown in Figs. 11.4 and 11.5.)

3. The situations when $P = 0$ and $M = 0$ are extreme cases. Between these extremes, beam-column action takes place.

4. For a given value of P, the member for which $L/r_x = 0$ can carry considerably more moment than the member with $L/r_x = 120$. Thus, short members are stronger than long members.

5. Up to $L/r_x = 60$ the interaction curves are nearly straight lines. For higher slenderness ratios the curves sag downward, thus showing the larger influence of secondary moments due to deflection.

Fortunately, the variation between the M-P-ϕ curves, and thus the interaction curves, is almost negligible for different wide-flange shapes, and therefore the interaction curves developed for the W8 × 31 shapes can be

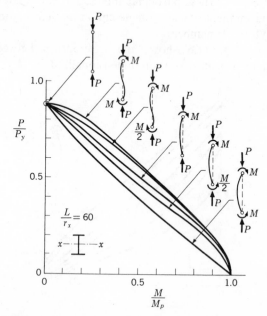

Fig. 11.7 Ultimate-strength interaction curves for various end-moment ratios.

used for other sections also.   It has been shown that the curves are slightly
on the conservative side for all other wide-flange shapes.[11.12]

Curves for other loading conditions have also been computed.[11.12,11.13]
The influence of the variation in the end-moment ratio is illustrated in Fig.
11.7 where interaction curves are presented for $L/r_x = 60$.   A comparison
of the curves shows that the most severe case of loading is that of two equal
end moments causing single-curvature deflection.   The least severe case is
that for which two equal end moments cause double-curvature deflection.

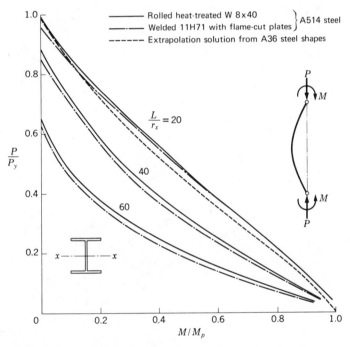

Fig. 11.8 Interaction curves for A514 steel beam-columns.

These two loadings represent the extremes.   The three additional curves in
Fig. 11.7 give intermediate situations.

Similar interaction curves can be determined for weak axis bending,[11.11] for
lateral-torsional buckling,[11.15] for other types of cross sections, and for other
types of steel.[11.17]   The effect of residual stress patterns of different types
from that shown in Fig. 11.5 is relatively small.[11.16]   Figure 11.8 shows
interaction curves for three slenderness ratios of A514 beam-columns.
Three curves are shown; two of these were determined by numerical integration for the measured residual stress of the rolled and heat-treated rolled

shape (relatively small residual stress of the general shape shown in Fig. 9.21) and for the welded shape (very high residual stresses at the flange centers). The third curve was obtained from interaction curves for the residual stress pattern of Fig. 11.5 and A36 steel by using an adjustment to the slenderness ratio:

$$(L/r)_{\text{adj}} = L/r \sqrt{\frac{F_y}{36}}.$$

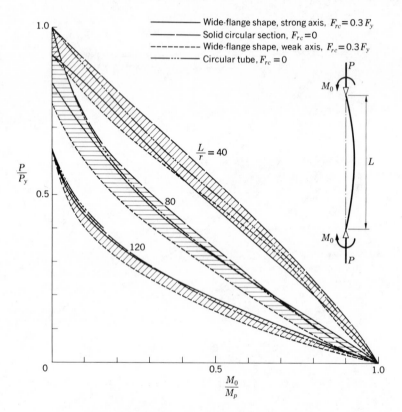

Fig. 11.9 Effect of cross-sectional shape.

This adjustment for different values of F_y is theoretically correct only if the residual stress pattern and the ratio F_r/F_y remain constant. Despite these differences in the residual stress, the differences in strength are relatively small.

The effect of the shape of the cross section is shown in Fig. 11.9 where the differences between widely different shapes are not too large. From the evidence in Figs. 11.8 and 11.9 it is obvious that any design approximation which is tested against the "standard" interaction curves of Figs. 11.6 and 11.7 could not be far wrong. It is for this reason that the approximate interaction equations given in Art. 11.4 are compared with those curves. The validity of these ultimate strength interaction curves has been verified by extensive experiments,[11.12,11.17–11.31] and the correlation between the theoretical and experimental results has, in general, been excellent. This can be seen in Fig. 11.10 where the ratio of experimental to theoretical strength has been plotted as a histogram. The average of this ratio for these 82 tests is 1.005, with most of the values lying between ±10 per cent of 1.00. These tests were performed in several laboratories over a span of 30 years. There

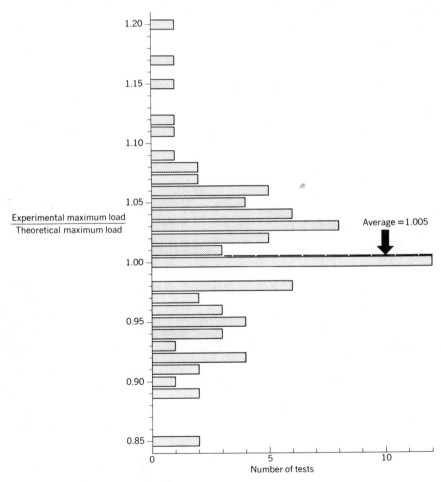

Fig. 11.10 Experimental correlation.

is no doubt that the ultimate strength of beam-columns can be determined with precision, provided enough time is taken to account for all the controlling factors.

It is obvious that the methods discussed in this article can hardly be expected to be used in design office practice. However, the results of the research are used as an index against which the various design methods are measured. The treatment of beam-column behavior is brief and concise in this chapter. More thorough and general discussions are available in other texts.[11.25,11.31]

11.3 SECANT FORMULA METHOD

The secant formula method specifies that the maximum stress in the beam-column, modified by a factor of safety, may not exceed the yield stress of the

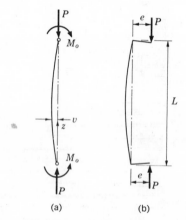

Fig. 11.11   Notation for eccentrically loaded members.

material. This maximum stress is computed by assuming a perfectly elastic material and by including the bending contribution of the axial force times the deflection, but ignoring residual stresses.

The example of the beam-column shown in Fig. 11.11a will be used to illustrate the secant formula approach. For simplicity it is assumed that the member is symmetrical and that bending occurs about an axis of symmetry. From Fig. 11.11a it is seen that the bending moment at any point, a distance z away from the origin, is equal to

$$M = M_0 + Pv = -EI \frac{d^2v}{dz^2} \qquad (11.1)$$

The right side of this equation represents the relationship $M/EI = -d^2v/dz^2$. If Eq. 11.1 is differentiated twice and rearranged, then the resulting differential equation[11.19] is

$$EI \frac{d^4v}{dz^4} + P \frac{d^2v}{dz^2} = 0 \tag{11.2}$$

which has the solution

$$v = A + Bz + C \cos kz + D \sin kz \tag{11.3}$$

where A, B, C, D are constants of integration and

$$k = \sqrt{\frac{P}{EI}} \tag{11.4}$$

Substitution of the boundary conditions $v = 0$ and $d^2v/dz^2 = -M_0/EI$ at $z = 0$ and $z = L$ in Eq. 11.3 leads to the following expression for the deflection v:

$$v = \frac{M_0}{P}\left[\left(\frac{1 - \cos kL}{\sin kL}\right) \sin kz + \cos kz - 1 \right] \tag{11.5}$$

The moment is obtained by differentiating Eq. 11.5 twice, or

$$M = -EI \frac{d^2v}{dz^2} = M_0 \left[\left(\frac{1 - \cos kL}{\sin kL}\right) \sin kz + \cos kz \right] \tag{11.6}$$

Noting that the maximum moment occurs at $z = L/2$:

$$M_{\max} = M_0 \sec \frac{kL}{2} \tag{11.7}$$

The maximum fiber stress in the member is equal to

$$f_{\max} = \frac{P}{A} + \frac{M_{\max}}{S} \tag{11.8}$$

or, with Eq. 11.7 substituted in Eq. 11.8,

$$f_{\max} = \frac{P}{A} + \frac{M_0}{S} \sec \frac{kL}{2} \tag{11.9}$$

Letting $f_{\max} = F_y$ and dividing both sides of Eq. 11.9 by F_y, the following interaction equation is obtained,

$$\frac{P}{P_y} + \left(\frac{M_0}{M_p}\right)\left[\frac{Z}{S} \sec \frac{L}{2r} \sqrt{\left(\frac{P}{P_y}\right)\left(\frac{F_y}{E}\right)}\right] = 1.0 \tag{11.10}$$

A comparison between the maximum strength interaction curve of Fig. 11.6 and the elastic limit curve of Eq. 11.10 is shown in Fig. 11.12 for a W8 × 31 member of $L/r = 60$, which is bent about its strong axis. It can be observed that the elastic method is conservative over most of the range of the curve except where the bending moment is very small. In that case the secant formula is not conservative, leading always to $P/P_y = 1.0$ for this member regardless of the length when $M = 0$. The reason for this is that the secant curve gives $P = P_y$ when $M = 0$, whereas the inelastic interaction curve gives the tangent modulus load, which is always less than P_y. To overcome this defect, most specifications provide for a fictitious initial curvature, as is

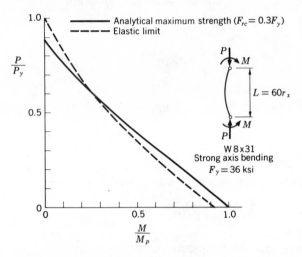

Fig. 11.12 Comparison of an elastic-limit interaction curve with a curve for ultimate strength.

shown below. They replace the end-moment M by an eccentrically applied axial force (see Fig. 11.11b) such that Eq. 11.9 becomes

$$f_{\max} = F_y = \frac{P}{A}\left[1 + \frac{eA}{S}\sec\frac{kL}{2}\right] \tag{11.11}$$

Rearranging Eq. 11.11, as well as introducing an initial eccentricity ratio $(eA/S)_{\text{initial}} = 0.25$ (used in the AASHO Specifications), the following formula is obtained for the axial stress P/A:

$$\frac{P}{A} = \frac{F_y}{1 + [0.25 + (eA/S)]\sec[(L/2r)\sqrt{P/EA}]} \tag{11.12}$$

In the AASHO Specifications the length L is reduced to an effective length of $0.75L$ for columns with riveted ends and to $0.875L$ for intentionally pinned ends. This was done to account for the beneficial effect of the end restraint existing in trusses, as discussed in Art. 10.6.

EXAMPLE 11.3

PROBLEM:
  Check adequacy of cross section by 1969 AASHO specification.

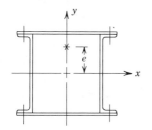

C 10 x 30 Channels
1/2"x 12" plates
Length of member: 30'—0"
Eccentricity in the y-plane: 3 in.
(equal at both ends)
End conditions: riveted
Material: A 440 steel
Axial load: 300 kip

SOLUTION:

Cross-sectional properties:

$A = 2 \times 8.80 + 2 \times 12 \times 1/2 = 29.6 \text{ in.}^2$

$I_x = 2 \times 103 + 2 \times 12 \times 1/2 \times (5.25)^2 = 537 \text{ in.}^4$

$r_x = \sqrt{\dfrac{I_x}{A}} = \sqrt{\dfrac{536}{29.6}} = 4.26 \text{ in.}$

$S = \dfrac{I}{c} = \dfrac{536}{5.5} = 97.6 \text{ in.}^3$

$L/r = \dfrac{30 \times 12}{4.25} = 84.5$

$\dfrac{eA}{S} = \dfrac{3 \times 29.6}{97.5} = 0.910$

$P/A = \dfrac{300}{29.6} = 10.1 \text{ ksi}$

From AASHO specifications (Appendix C):
$F_y = 50 \text{ ksi}$; $E = 29,000 \text{ ksi}$; F.S. = 1.82; Effective length: 0.75L

Design check: (Eq. 11.13)

$$F_a = \frac{F_y/F.S.}{1.00 + \left(0.25 + \dfrac{eA}{S}\right)\sec \dfrac{0.75L}{2r}\sqrt{\dfrac{F_a \times (F.S.)}{E}}}$$

Solving this equation for $\dfrac{eA}{S}$ for $F_a = 10.1 \text{ ksi}$ and $L/r = 84.6$,

$\left(\dfrac{eA}{S}\right)_{required} = 0.951 > 0.910$

∴ Section is O. K.

Equation 11.12 is in terms of the yield-producing stresses. For a working stress formula the factors of safety (F.S.) are introduced as shown in Eq. 11.13,

$$F_a = \frac{F_y/(\text{F.S.})}{1 + [0.25 + (eA/S)] \sec [(L/2r)\sqrt{F_a(\text{F.S.})/E}]} \tag{11.13}$$

The AASHO formula (Eq. 9.48) is somewhat more complicated because it is expressed in terms of any ratio of the eccentricities at the top and at the bottom of the member. In the special and most critical case of equal end eccentricities (Fig. 11.11b), the AASHO formula reduces to Eq. 11.13.

Equation 11.13 has been discussed in Art. 9.9 in connection with its use for columns with an unavoidable eccentricity. Article 9.5 has considered the development and use of column curves, in particular, the Rankine-Gordon curve. The Rankine-Gordon curve was developed from test results and is equivalent to the secant formula when an eccentricity of load is introduced into the column.

In the secant formula approach to the design of beam-columns the usually specified initial eccentricities take care of the effect of the residual stresses in an approximate manner. Comparison with test results and with the ultimate strength methods have shown that the secant formula approach is conservative.[11.20] Direct solutions for F_a in Eq. 11.13 or similar formulas in the design routine are cumbersome. However, charts have been made available (notably in Appendix C of the AASHO Specifications) to alleviate this shortcoming. The secant formula method is illustrated in Example 11.3.

Another utilization of the philosophy of elastic design can be made in the analysis of laterally loaded beam-columns by noting that the deflection of such a member is magnified by the axial force approximately as the ratio[11.19] $1/(1 - P/P_e)$, that is,

$$v = \frac{\bar{v}}{[1 - (P/P_e)]} \tag{11.14}$$

where $\bar{v}$ is the deflection due to the transverse forces only and P_e is the Euler load (also given as Eq. 9.3),

$$P_e = \frac{\pi^2 EI}{L^2} \tag{11.15}$$

The maximum stress can be computed from Eq. 11.8 as

$$f_{max} = \frac{P}{A} + \frac{\bar{M}_{max}}{S} + \frac{P\bar{v}}{S[1 - (P/P_e)]} \tag{11.16}$$

In Eq. 11.16, $\bar{M}_{max}$ is the maximum moment under the transverse forces alone and $\bar{v}$ is the maximum deflection. The third term in this equation corresponds to the contribution of the axial force to the moment. Due to the approximate nature of this solution, it is usually sufficient to substitute $\bar{M}_{max}$ and $\bar{v}_{max}$ in Eq. 11.16, even if they do not occur exactly on the same location along the length of the beam.

The approach discussed above is not used directly in any current specifications; however, it is applied indirectly in the "interaction equation" method, which is the topic of Art. 11.4. The application of the approximate elastic method is illustrated in Example 11.4.

EXAMPLE 11.4

PROBLEM:
  Compute the maximum allowable value of w_0 such that $F_{max} \leqslant 24.0\,ksi$; $F_y = 36\,ksi$.

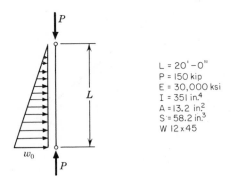

L = 20' – 0"
P = 150 kip
E = 30,000 ksi
I = 351 in.^{4}
A = 13.2 in.^{2}
S = 58.2 in.^{3}
W 12 x 45

SOLUTION:

Eq. 11.16: $F_{max} = \dfrac{P}{A} + \dfrac{M_{max}}{S} + \dfrac{P v_0}{S(1 - P/P_e)}$

$P/A = \dfrac{150}{13.24} = 11.4\ ksi$;

M_{max} for $P = 0$: $M_{max} = 0.0642 w_0 L^2 = 3,700 w_0$ *

v_0 max for $P = 0$: $v_0 = 0.00652 \dfrac{w_0 L^4}{EI} = 2.06 w_0$ *

$P_e = \dfrac{\pi^2 E I}{L^2} = 1,806\ kip$; $1 - P/P_e = 1 - \dfrac{150}{1,806} = 0.917$

$F_{max} = 24.0 = 11.4 + \dfrac{3,700 w_0 + (150 \times 2.06 w_0)\left(\dfrac{1}{0.917}\right)}{58.2}$

Solving for w_0: $(w_0)_{max} = 0.18\ kip/in. = 2.2\ kip/ft$

*From AISC Steel Construction Manual or any other common tabulation of beam solutions.

11.4 INTERACTION EQUATIONS

The relatively simple, empirically determined beam-column interaction equations have proven themselves more popular with specification writers than the elastic method described in Art. 11.3, because the interaction equations are easily adaptable to a multitude of situations. Most significantly, these formulas permit ready inclusion of provisions for lateral-torsional buckling and biaxial bending.

In the following discussion, maximum-strength interaction equations are derived from the analytically determined maximum-strength interaction curves such as those shown in Figs. 11.6 and 11.7.

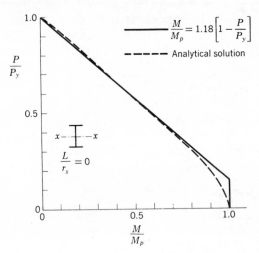

Fig. 11.13 Interaction curves for "zero-length" members.

The maximum-strength interaction relationship for strong axis bending of a "zero length" wide-flange member is shown in Fig. 11.13.[11.18] The dashed line represents the analytically developed curve, and the solid line corresponds to the approximation

$$\frac{M}{M_p} = 1.18\left(1 - \frac{P}{P_y}\right) \tag{11.17}$$

Equation 11.17 corresponds to the case where the whole cross section is yielded under the axial force P and the moment M. This plastic hinge condition is shown in Fig. 11.14. The moment M_0 is the largest moment which the cross section can support under a given value of P. Full plastification can occur usually only for very short members, and thus, the curve in Fig. 11.13 is an upper limit to the capacity of any beam-column.

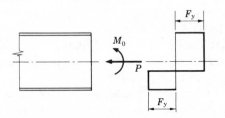

Fig. 11.14 Fully plastic stress distribution with axial force and moment.

The two dashed curves shown in Fig. 11.15 represent the maximum strength for $L/r_x = 40$ and 120 when the loading is that shown in the figure. A very simple interaction formula would result from drawing straight lines between the end points of the curves (solid lines in Fig. 11.15). In equation form this would be

$$\frac{P}{P_0} + \frac{M}{M_p} = 1.0 \tag{11.18}$$

In this equation the new term P_0 is the axial force which would cause failure if no moment were present (Eq. 9.24). A look at Fig. 11.15 indicates that Eq. 11.18 would be an excellent approximation for $L/r_x = 40$. However,

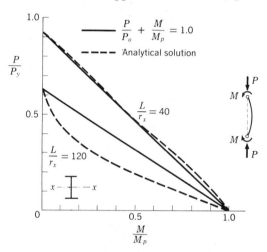

Fig. 11.15 The straight-line interaction formula.

for $L/r_x = 120$ the straight line approximation is not adequate. Therefore, Eq. 11.18 must be further modified.

The dashed curves in Fig. 11.16 show analytically obtained interaction curves for $L/r_x = 40$, 80, and 120. The solid lines represent the formula

$$\frac{P}{P_0} + \frac{M}{M_p}\left[\frac{1}{1 - (P/P_e)}\right] = 1.0 \tag{11.19}$$

This equation is the same as Eq. 11.18, except that the moment term has been multiplied by the amplification factor $1/(1 - P/P_e)$. In this factor the term P_e is the elastic buckling load of the member in the plane of bending, or in non-dimensional form,

$$\frac{P_e}{P_y} = \frac{\pi^2 E}{F_y}\left(\frac{1}{L/r_x}\right)^2 \tag{11.20}$$

The amplification factor takes care of the increased contribution of the axial force to the bending moment for longer members. Indeed, for small

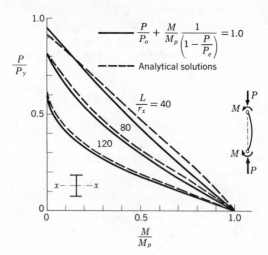

Fig. 11.16 The interaction formula with the amplification factor.

values of L/r_x, P_e is large, and thus $1/(1 - P/P_e)$ deviates very little from 1.0 (as witnessed by the nearly straight line for $L/r_x = 40$ in Fig. 11.16). On the other hand, for longer members P_e becomes small, and, therefore, the influence of the amplification factor is increased (see curve for $L/r_x = 120$ in Fig. 11.16).

The interaction equation (Eq. 11.19) would be safe to use in design. However, as is shown in Fig. 11.17 for $L/r_x = 120$, it would be quite conservative for loading cases other than that of two equal end moments causing

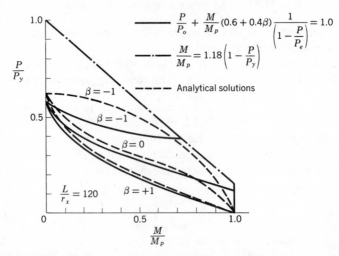

Fig. 11.17 The interaction formula for unequal end moments.

single-curvature deformation. Therefore, an additional modification is introduced:

$$\frac{P}{P_0} + \frac{M}{M_p}\left[\frac{1}{1 - (P/P_e)}\right](0.6 + 0.4\beta) = 1.0 \qquad (11.21)$$

As may be seen in Fig. 11.17, the new term $(0.6 + 0.4\beta)$ merely causes the curve to sweep upwards with a pivotal point $P = P_0$. The term β represents the ratio of the smaller to the larger end moment, and it may vary from -1 to $+1$. The correlation between the exact and the empirical formulas is good for $\beta \geqslant 0$. For $\beta < 0$ the correspondence is not as good; but the error is on the conservative side. It should be noted that the curve of Eq. 11.21 does not go through $M_0 = M_p$ when $P = 0$, unless $\beta = +1.0$. Thus, when Eq. 11.21 reaches beyond the line of Eq. 11.17, this latter equation governs.

The interaction equation (Eq. 11.21) is a reasonable approximation of the maximum strength curves for failure by excessive bending. The influence of lateral-torsional buckling can be included by replacing the term M_p by M', where M' is the maximum bending moment which could be carried by the member before lateral buckling occurred in the $P = 0$ case. (See Art. 7.6.)

The complete maximum-strength interaction equations are given as Eqs. 11.22 and 11.23.

$$\frac{P}{P_0} + \frac{M}{M'}\left[\frac{1}{1 - (P/P_e)}\right](0.6 + 0.4\beta) = 1.0 \qquad (11.22)$$

$$\frac{M}{M_p} = 1.18\left[1 - \frac{P}{P_y}\right] \leqslant 1.0 \qquad (11.23)$$

In any given situation both equations need to be checked. Equation 11.22 can be thought of providing for the condition when the column fails by inelastic instability due to excessive bending within the span of the beam-column, and Eq. 11.23 indicates that a plastic hinge has formed at the end of the member.

These interaction equations have been verified by experiments for various types of cross sections and loading conditions, and they were found to be in satisfactory agreement with the test results.[11.17,11.21,11.22,11.25,11.27]

The interaction equations (Eqs. 11.22, 11.23) are in terms of maximum loads. It is usual to work in terms of allowable stresses. In AISC Section [1.6.1] the interaction equations are in the form[11.4]

$$\frac{f_a}{F_a} + \frac{C_m f_b}{\left(1 - \frac{f_a}{F'_e}\right)F_b} \leqslant 1.0 \qquad (11.24)$$

$$\frac{f_a}{0.6F_y} + \frac{f_b}{F_b} \leqslant 1.0 \qquad (11.25)$$

Equation 11.24 is exactly the same as Eq. 11.22 if $F_b = M'/S$(F.S.); $C_m = 0.6 + 0.4\beta$; $f_a = P_w/A$; $F_a = P_0/A$(F.S.); $f_b = M_w/S$; $F_e' = P_e/A$(F.S.) (where the subscript w refers to working loads and F.S. is the safety factor). Equation 11.25 is the same as Eq. 11.23 if the term 1.18 is rounded off to 1.00 (which is a conservative assumption).

The interaction equations as presented above permit checking a selected member if that member is an individual beam-column subjected to a given combination of end moments and axial force. However, beam-columns are not individual members. As members in a rigid frame, the effect of the members framing into them must also be considered. It has been suggested in Ref. 11.23 that this can be done by computing the value of F_a and F_e' for an effective column length KL.* In case sidesway buckling of the frame is prevented by shear-walls or diagonal bracing, the effective length is less than the story height. If this condition is true, Eqs. 11.24 and 11.25 apply, with $C_m = 0.6 + 0.4\beta$. However, C_m may not become less than 0.4 in the AISC Specification.

For the case in which no bracing against sidesway is provided, the effective length KL is larger than the story height. This must be accounted for in computing F_a and F_e'. The value of $C_m = 0.85$ is to be used in this case.

When the beams framing into the beam-columns in a multi-story frame cause bending about both principal axes of the cross section, the design interaction equation is written as[11.4,11.17]

$$\frac{f_a}{F_a} + \frac{C_{mx}f_{bx}}{[1 - (f_a/F_{ex}')]F_{bx}} + \frac{C_{my}f_{by}}{[1 - (f_a/F_{ey}')]F_{by}} \leqslant 1.0 \qquad (11.26)$$

and

$$\frac{f_a}{0.6F_y} + \frac{f_{bx}}{F_{bx}} + \frac{f_{by}}{F_{by}} \leqslant 1.0 \qquad (11.27)$$

where the subscripts, x and y, refer to the x or the y axis.

In this article it has been shown how an interaction equation can be developed which represents the maximum strength of beam-columns to a reasonable accuracy (Eqs. 11.22 and 11.23), and it has been shown how the AISC interaction equation (Eqs. 11.24 and 11.25) relates to this ultimate strength equation. The discussion here does not provide a detailed discussion of all the provisions of the AISC Specification on beam-columns. These details are presented in the AISC Specification as well as in its Commentary. Further details of the AISC beam-column provisions are also given in Chapters 19 and 20, where the design of rigid frames is discussed.

Five examples illustrating the interaction formula concept are presented. The AISC Specification has been used as the basis of the computations, and

---

* The concept of the effective length is discussed in Chapter 10.

EXAMPLE 11.5

PROBLEM:

A 15 ft. beam −column is subjected to an end moment of 63.5 kip-ft at the top and to an end moment of 58.0 kip-ft. at the bottom such that these moments cause double curvature deformation about the strong axis of the member. The axial force is 152 kip, and the frame is braced against sidesway buckling in the plane of the loading and in the plane perpendicular to it. Design the member by AISC Spec. Material A 36 steel

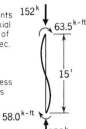

SOLUTION:

Effective length – Because of diagonal bracing, the effective length will be less than the story height in both the weak and strong axes. An independent analysis (not shown here, but based on the theory presented in Chapter 10) shows that

$$K_x = 0.86 \text{ and } K_y = 0.91$$

Try W 10x60

Interaction equations: $\dfrac{f_a}{F_a} + \dfrac{C_m f_b}{\left(1 - \dfrac{f_a}{F'_e}\right) F_b} \leq 1.00$ and $\dfrac{f_a}{0.6F_y} + \dfrac{f_b}{F_b} \leq 1.00$ $\left(\begin{array}{l}\text{Eqs. 11.24}\\ \text{and 11.25}\end{array}\right)$

Cross−sectional properties: $A = 17.7 \text{ in.}^2$, $S = 67.1 \text{ in.}^3$, $r_x = 4.41 \text{ in.}$, $r_y = 2.57 \text{ in.}$

Slenderness ratios: $\dfrac{L}{r_x} = \dfrac{15 \times 12}{4.41} = 40.8$; $\dfrac{L}{r_y} = \dfrac{15 \times 12}{2.57} = 70.0$

Effective slenderness ratios: $\dfrac{KL}{r_x} = 0.86 \times 40.8 = 35$; $\dfrac{KL}{r_y} = 0.91 \times 70.1 = 64$

Allowable axial stress:

(Table [1-36] in AISC Spec.): $F_{ax} = 19.58 \text{ ksi}$; $F_{ay} = 17.04 \text{ ksi}$ ◄—— governs

Axial stress: $f_a = \dfrac{P}{A} = \dfrac{152}{17.7} = 8.59 \text{ ksi}$;

Reduction factor: $C_m = 0.6 - 0.4 \left(\dfrac{58.0}{63.5}\right) = 0.235 < 0.4$ ∴ use $C_m = 0.4$

Amplification factor: for $\dfrac{KL}{r_x} = 35$, $F'_e = 121.9 \text{ ksi}$ (table 2 in. AISC Spec.)

$$1 - \dfrac{f_a}{F'_e} = 1 - \dfrac{8.61}{121.9} = 0.930$$

Allowable bending stress: [AICS, 1969, Sec. 1.5.1.4.6] ∴ $F_b = 0.6 F_y = 22 \text{ ksi}$

Bending stress: $f_b = \dfrac{M_{max}}{S} = \dfrac{63.5 \times 12}{67.1} = 11.36 \text{ ksi}$

Check interaction formulas Eq. 11.24 : $\dfrac{8.59}{17.04} + \dfrac{0.4(11.36)}{0.930(22.0)} = 0.504 + 0.222 = 0.726 < 1.00$

Eq. 11. 25 : $\dfrac{8.59}{22} + \dfrac{11.36}{22} = 0.390 + 0.516 = 0.906 < 1.00$

∴ W 10x60 is O. K.

---

use has been made of tabulated values from it. Examples 11.5 and 11.6 illustrate the use of Eqs. 11.24 and 11.25, and Example 11.7 discusses a biaxial bending problem using Eqs. 11.26 and 11.27. Examples 11.8 and 11.9 show the use of the AISC Specification for beam-columns with lateral loads.

EXAMPLE 11.6

PROBLEM:
 A 15 ft. beam-column is pinned at one end and subjected to a strong axis end moment of 94.0 kip-ft at the other. The frame is not braced in the plane of loading, such that the effective length factor is 1.95*. Bracing is provided in the perpendicular plane, with $K_y = 0.87$. The axial force is 192 kip. Design the member by AISC Spec.
 Material: A 572-50 steel, $F_y = 50$ ksi

192^k

94^{k-ft}

15'

192^k

SOLUTION:
Try W 12 x 65
 Cross-sectional properties: $A = 19.1$ in.^{2}, $r_x = 5.28$ in.,
 $S = 88.0$ in.^{3}, $r_y = 3.02$ in.

 Effective slenderness ratios: $\dfrac{KL}{r_x} = \dfrac{1.95 \times 15 \times 12}{5.28} = 67$; $\dfrac{KL}{r_y} = \dfrac{0.87 \times 15 \times 12}{3.02} = 52$

 Allowable axial stress: from table $\left[1-50\right]$ AISC spec. $F_a = 21.49$ ksi (for $\dfrac{KL}{r_x} = 67$ which is governing)

 Axial stress: $f_a = \dfrac{192}{19.1} = 10.05$ ksi

 Reduction factor: $C_m = 0.85$

 Amplification factor: for $\dfrac{KL}{r_x} = 67$, $F_e' = 33.27$ (table 2 in AISC spec.)

$$1 - \frac{f_a}{F_e'} = 1 - \frac{10.05}{33.27} = 0.698$$

 Allowable bending stress. $\left[\text{AISC, 1969, 1.5.1.4.6}\right]$

$$F_b = 0.6 F_y = 30 \text{ ksi}$$

 Check interaction formulas: Eq.11.24: $\dfrac{10.05}{21.49} + \dfrac{0.85(12.81)}{0.698(30)} = 0.466 + 0.520 = 0.986$

$0.986 < 1.00$ \underline{O. K.}

Eq.11.25: $\dfrac{10.05}{30.0} + \dfrac{12.81}{30.0} = 0.334 + 0.427 = 0.761$

$0.761 < 1.00$ \underline{O. K.}

 $\therefore$ W 12 x 65 is O. K.

*See chapter 10 for determination of effective length.

11.5 BEAM-COLUMNS IN PLASTIC DESIGN

If plastic design is used as the basis for proportioning a frame, the available research information on the maximum strength of beam-columns could be utilized directly. One could use the available interaction curves[11.12,11.13] or one could use tables abstracted from these curves.[11.24,11.26] Although this would be desirable from the standpoint of greater accuracy in design, it is not possible to incorporate many charts and tables into what is usually a relatively minor portion of a specification. In the 1963 AISC Specification,[11.4] therefore, the maximum-strength interaction curves for certain

EXAMPLE 11.7

PROBLEM:
  A beam-column is subjected to biaxial moments as shown in the sketch. Sidesway is not prevented in the x-plane, but bracing is present in the y-plane. Design the member for A36 steel and use AISC spec.

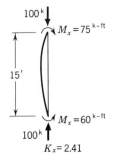
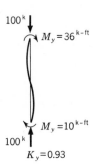

$M_x = 75^{k-ft}$ $M_y = 36^{k-ft}$

$M_x = 60^{k-ft}$ $M_y = 10^{k-ft}$

$K_x = 2.41$ $K_y = 0.93$

SOLUTION:
  Try W 12 x 85
  Cross-sectional properties: $A = 25.0$ in.^{2} $S_x = 116$ in.^{3} $r_x = 5.38$ in.
                                                      $S_y = 38.9$ in.^{3} $r_y = 3.07$ in.

  Effective slenderness ratios: $\dfrac{K_x L}{r_x} = \dfrac{2.41 \times 15 \times 12}{5.38} = 81$; $\dfrac{K_y L}{r_y} = \dfrac{0.93 \times 15 \times 12}{3.07} = 55$

  Allowable axial stress: from table [1-36] AISC spec, $F_a = 15.24$ ksi (for $\frac{KL}{r} = 81$)

  Axial stress: $f_a = \dfrac{100}{25} \approx 4.00$ ksi

  Reduction factors: $C_{mx} = 0.85$ $C_{my} = 0.6 - 0.4\left(\dfrac{10}{36}\right) = 0.489 > 0.40$ use $C_{my} = 0.489$

  Amplification factor: $F'_{ex} = 22.76$ ksi ($L/r_x = 81$, table [2] in AISC spec.)
                        $F'_{ey} = 49.37$ ksi ($L/r_y = 55$)

  $1 - \dfrac{f_a}{F'_{ex}} = 1 - \dfrac{4.00}{22.76} = 0.824$; $1 - \dfrac{f_a}{F'_{ey}} = 1 - \dfrac{4.00}{49.37} = 0.919$

  Allowable bending stress: $=$ [AISC, 1969, Sec. 1.5.1.4.6] $\therefore$ $F_{bx} = 0.6\, F_y = 22.0$ ksi
                            [AISC, 1969, Sec. 1.5.4.6] $\therefore$ $F_{by} = 0.75\, F_y = 27$ ksi

  Bending stress: $f_{bx} \dfrac{75(12)}{116} = 7.76$ ksi ; $f_{by} = \dfrac{36(12)}{38.9} = 11.11$ ksi

  Check interaction formulas: Eqs. 11.26 and 11.27

  $\dfrac{4.00}{15.24} + \dfrac{(0.85)7.76}{(0.824)22.0} + \dfrac{(0.489)11.11}{(0.919)27.0} = 0.262 + 0.364 + 0.219 = 0.845$
  $0.896 < 1.00$ O.K.

  $\dfrac{f_a}{0.6 F_y} + \dfrac{f_{bx}}{F_{bx}} + \dfrac{f_{by}}{F_{by}} = \dfrac{4.00}{22.0} + \dfrac{7.76}{22.0} + \dfrac{11.11}{27} = 0.946$ O.K.

  $\therefore$ W 12 x 85 is O.K.

EXAMPLE 11.8

PROBLEM:
Determine the amount of uniformly distributed load which the beam-column can carry in addition to the axial load. Use AISC spec.

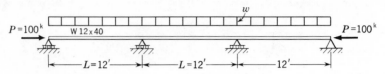

Section: W 12 x 40, A 36 steel

Axial load : P = 100 kip

The transverse load is applied through a concrete slab such that the member can be assumed to be continuously supported in the lateral direction.

SOLUTION:
Cross-sectional properties: $A = 11.8 \text{ in}^2$, $S = 51.9 \text{ in}^3$, $r_x = 5.13 \text{in}$.

$$f_a = \frac{P}{A} = \frac{100}{11.8} = 8.47 \text{ ksi}$$

$$\frac{L}{r_x} = \frac{12 \times 12}{5.13} = 28$$

The effective length would be less than L , however, very little advantage can be gained by considering this factor for such a short column as this, and therefore F_a will be determined for for the actual slenderness ratio.

$$F_a = 20.08 \text{ksi (table } [1-36])$$

Unbraced length of compression flange is zero (continuous lateral support).
From sect. [1.5.1.4.1] of AISC spec.

$$(b/t)_{max} = \frac{52.2}{\sqrt{F_y}} \times 2 = 17.4$$

$$(d/w)_{max} = \frac{257}{\sqrt{F_y}} = 42.8$$

For W 12 x 40 $b/t = \frac{8}{0.516} = 15.5 < 17.4$

$$d/w = \frac{11.94}{0.294} = 40.6 < 42.8$$

$\therefore$ Section is "compact" and $F_b = 0.66 F_y = 24 \text{ksi}$.

Moment diagram (exclusive of increased moment due to deflections)

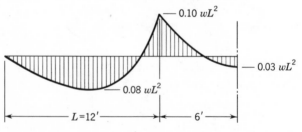

$F_e' = 190.47 \text{ kip (table} [2], \text{ AISC spec. for } L/r = 28)$

$1 - \dfrac{f_a}{F_e'} = 0.955$

Outside span: $C_m = 1 + \psi \dfrac{f_a}{F_e'}$ (sect. [1.6] of AISC commentary)

EXAMPLE 11.8 (Continued)

$\psi = -0.3$ (table [C1.6.1.2] commentary)

$C_m = 1 - 0.3 \left(\dfrac{8.5}{190.5}\right) = 0.987$

Interaction equation: $\dfrac{f_a}{F_a} + \dfrac{C_m f_b}{\left(1 - \dfrac{f_a}{F_e'}\right)F_b} \leq 1.00$

$\dfrac{8.47}{20.08} + \dfrac{0.987}{0.955(24.0)}\left[\dfrac{0.08\ w\ (144)^2}{51.9}\right] = 1.0$

$\therefore W_{allow} = 0.420\ kip/in. = 5.04\ kip/ft$

Support point: $\dfrac{f_a}{0.6F_y} + \dfrac{f_b}{F_b} = 1.00$; $\dfrac{8.47}{22.0} + \dfrac{0.10\ w\ (144)^2}{51.9 \times 24.0} = 1.00$

$\therefore W_{allow} = 0.369\ kip/in. = 4.43\ kip/ft$

Center span: $\psi = -0.4$ (table [C.1.6.1.2] commentary)

$C_m = 1 - 0.4\left(\dfrac{8.5}{190.5}\right) = 0.982$

$\dfrac{f_a}{F_a} + \dfrac{C_m\ f_b}{1 - \left(\dfrac{f_a}{F_e'}\right)F_b} = 1.00$

$\dfrac{8.47}{20.08} + \dfrac{0.982}{0.955(24.0)}\left[\dfrac{0.03\ w\ (144)^2}{51.9}\right] = 1.00$

$\therefore W_{allow} = 1.126\ kip/in. = 13.51\ kip/ft$

The critical point is at the support:

$\underline{W_{allow} = 4.43\ kip-ft}$

extreme loading cases were fitted into equations, and these equations serve as the basis for design.[11.12] For situations falling between the extremes, the more severe extreme should be used to ensure a conservative design. In the 1969 revised version of the AISC Specification, a departure toward a more useful and general method of design has been achieved by using the maximum-strength interaction equations (Eqs. 11.22 and 11.23) directly.

These equations are expressed in the following form:

$$\frac{P}{P_0} + \frac{M}{M'}\left[\frac{C_m}{1 - P/P_e}\right] \leq 1.0 \tag{11.28}$$

$$\frac{P}{P_y} + \frac{M}{1.18M_p} \leq 1.0 \tag{11.29}$$

The value of M in Eq. 11.29 may not exceed M_p. The terms in these equations are identical to their previous definitions, except that the 1969

EXAMPLE 11.9

PROBLEM:
Check the adequacy of the beam-column in Example 11.4 by the 1969 AISC Specifications, Part I.

SOLUTION:
Assume lateral support in the weak direction.
Since the end moments are zero, only AISC Formula 1.6-1a need be checked (Eq. 11.24)

$$\frac{f_a}{F_a} + \frac{C_m\, f_b}{\left(1 - \frac{f_a}{F_e'}\right) F_b} \leq 1.0$$

$$L/r_x = \frac{20 \times 12}{5.15} = 47 \; ; \quad F_a = 18.61 \text{ ksi (Table 1-36)}$$

$$F_e' = 68.60 \text{ ksi (Table 2)}$$

$$f_a = \frac{P}{A} = \frac{150}{13.2} = 11.36 \text{ ksi} \; ; \; \frac{f_a}{F_a} = \frac{11.36}{18.61} = 0.610$$

$$M_{max} = 0.0642\, w_0 L^2 = 3700\, w_0 = 3700 \times 0.18 \text{ kip/in.} = 666 \text{ in.-kips}$$

$$f_b = \frac{M_{max}}{S} = \frac{666}{58.2} = 11.44 \text{ ksi}$$

$$F_b = 24 \text{ ksi} \; ; \; 1 - \frac{f_a}{F_e'} = 1 - \frac{11.36}{68.60} = 0.834$$

The value of C_m can be computed from a formula given in Sec. 1.6 of the AISC Commentary, where

$$C_m = 1 + \psi \frac{f_a}{F_e}, \text{ and } \psi = \frac{\pi^2 \delta_0 E I}{M_0 L^2} - 1$$

δ_0 = maximum deflection due to trasverse loading

$$\delta_0 = 0.00652 \frac{w_0 L^4}{E I} = 2.06\, w_0 = 2.06 \times 0.18 = 0.371 \text{ in.}$$

M_0 = maximum moment between supports due to transverse loading

$M_0 = M_{max} = 666$ in.-kips

$$\psi = \frac{\pi^2 \delta_0 E I}{M_0 L^2} - 1 = \frac{\pi^2 \times 0.371 \times 29,000 \times 351}{666 \times 240^2} - 1 = -0.030$$

$$C_m = 1 - 0.03 \frac{f_a}{F_e'} = 1 - \frac{0.03 \times 11.36}{68.60} = 0.995$$

Check interaction equation

$$0.610 + \frac{0.995 \times 11.44}{0.834 \times 24} = 0.610 + 0.569 = 1.18 > 1.0$$

ANSWER:
Beam-column of Example 11.4 is not quite adequate to support the given load according to the AISC Specification.

AISC Specification defines the following terms thus:

$$P_0 = 1.70 A F_a \tag{11.30}$$

$$P_e = 1.92 A F_{e'} \tag{11.31}$$

$$M' = \left[1.07 - \frac{\sqrt{F_y}(L/r_y)}{06\ 31}\right] M_p < M_p \tag{11.32}$$

where P_0 is the axial load which can be supported in the absence of bending moment, P_e is the elastic buckling load of the column and M' is the bending moment which can be supported in the absence of axial load. The expression

EXAMPLE 11.10

PROBLEM:
 Design the beam-columns in the following structure by plastic design. Use AISC spec.
Assume that bracing against sway buckling is provided in an adjacent frame.

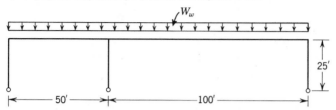

Working loads: W_w = 2.00 kip/ft.
Material: A 572 steel (F_y = 50 ksi)

SOLUTION:
 Plastic design W_u = 1.70 x W_w = 1.70 x 2.00 = 3.40 kip/ft.

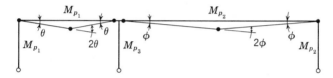

Structure will fail by the formation of two beam mechanisms:
Virtual work equations: $3.4 \times 50 \times 1/2 \times 25\theta = 4M_{p1}\theta$ M_{p1} = 531 $^{kip-ft}$

$3.4 \times 100 \times 1/2 \times 50\phi = 4M_{p2}\phi$ M_{p2} = 2,125 $^{kip-ft}$

From equilibrium $M_{p3} = M_{p2} - M_{p1} = 2,125 - 531 = 1,594$ $^{kip-ft}$

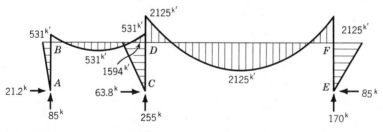

Design beam AB
 Assume continuous lateral support from slab;

$$M = 531 \text{ kip-ft} = 531 \times 12 = 6372 \text{ kip-in.}$$

$$Z_{req} = \frac{6360}{50} = 127 \text{ in.}^3$$

Try W 21x62 $Z = 144 \text{ in.}^3 > 127 \text{ in.}^3$

$\dfrac{d}{w} = 52.5$; $\dfrac{b}{t} = \dfrac{8.24}{0.615} = 13.4 < 14.0$ O.K.

Web local buckling, Sec. 2.7, AISC specifications

$$\left(\frac{d}{w}\right)_{req} = \frac{412}{\sqrt{F_y}} = \frac{412}{\sqrt{50}} = 58.2 > 52.5$$ O.K.

Use W 21x62 - A 572 - 50

EXAMPLE 11.10 (Continued)

Design beam DF

Assume continuous lateral support;

$$M = 2125 \text{ kip-ft} = 25,500 \text{ kip-in.}$$
$$Z_{req} = 510 \text{ in.}^3$$

Try W 33 x 141

$$Z = 514 \text{ in}^3 > 510 \text{ in}^3$$
$$\frac{d}{w} = 55.1 < 58.2 \qquad \text{O.K.}$$
$$\frac{b}{t} = 12.0 < 14.0 \qquad \text{O.K.}$$

Use W 33 x 141 - A 572-50

Design column AB

Assume lateral bracing;

$$M = 6372 \text{ kip-in.}, \ P = 85 \text{ kip}, \ L = 300 \text{ in.}$$

Try W 18 x 70

$$A = 20.6 \text{ in.}^2 \ , \ \frac{d}{w} = 41.1$$
$$P_y = A F_y = 20.6 \times 50 = 1030 \text{ kip}$$
$$\frac{P}{P_y} = \frac{85}{1030} = 0.083$$

Check local buckling Sec. 2.7 in AISC specifications

$$\frac{d}{w} = \frac{412}{\sqrt{F_y}} \left(1 - 1.4 \frac{P}{P_y}\right) = 51.4 > 41.1 \qquad \text{O.K.}$$

Check flange local buckling:

$$\frac{b}{t} = \frac{8.75}{0.751} = 11.7 < 14 \qquad \text{O.K.}$$

Lateral bracing spacing (formula 2.9 - 1b , AISC specifications)

$$\frac{l_{cr}}{r_y} = \frac{1375}{F_y} = 27.5$$
$$r_y = 2.02 \ , \ l_{cr} = 2.02 \times 27.5 = 56 \text{ in.}$$

First lateral brace is 56 in. below joint, or 300 - 56 = 244 in. above base.

Moment at that level: $M = 6375 \times \dfrac{244}{300} = 5185 \text{ kip-in.}$

Reduce this moment to working load level:

$$\frac{5185}{1.7} = 3050 \text{ kips}$$

Corresponding bending stress:

$$f_b = \frac{M}{S} = \frac{3050}{129} = 23.64 \text{ ksi}$$

According to AISC Sec. 1.5.1.4.6, the slenderness ratio of the elastic length is

$$\frac{l}{r_t} = \frac{244}{2.34} = 104$$

From formula 1.5 - 6a of AISC specification:

$$C_b = 1.75$$

$$\sqrt{\frac{510 \times 10^3 C_b}{F_y}} = 134 > 104 ; \quad \sqrt{\frac{102 \times 10^3 C_b}{F_y}} = 60 < 104 ;$$

$$\therefore F_b = F_y \left[\frac{2}{3} - \frac{F_y (l/r)^2}{1530 \times 10^3 C_b}\right] = 50 \left[\frac{2}{3} - \frac{50 \times 104^2}{1530 \times 10^3 \times 1.75}\right] = 23.2 \text{ ksi}$$

EXAMPLE 11.10 (Continued)

From formula 1.5–7 of AISC specification:

$$F_b = \frac{12 \times 10^3 C_b}{ld/A_f} = \frac{12,000 \times 1.75}{244 \times 2.74} = 31.4 \text{ ksi} > 0.6\,F_y = 30 \text{ ksi}$$

$$\therefore F_b = 30 \text{ ksi} > f_b = 23.6 \text{ ksi}$$

$\therefore$ No additional brace is required.

Check formula 2.4–3 in AISC specifications (Eq. II.33)

$$M_p = F_y Z = 50 \times 145 = 7250 \text{ kip-in.}$$

$$\frac{M}{M_p} = \frac{6375}{7250} = 0.879 < 1.000 \qquad \underline{\text{O.K.}}$$

$$\frac{P}{P_y} + \frac{M}{1.18\,M_p} = 0.083 + 0.745 = 0.828 < 1.000 \qquad \underline{\text{O.K.}}$$

Check formula 2.4–2 in AISC specifications (Eq. II.32)

$$\frac{L}{r_x} = \frac{300}{7.50} = 40 \; ; \; \frac{L}{r_y} = \frac{56}{2.02} = 28$$

$$F_a = 25.83 \text{ ksi (Table 1-50, AISC specification for } L/r = 40)$$

$$P_o = 1.70\,A\,F_a = 1.70 \times 20.6 \times 25.83 = 905 \text{ kip}$$

$$F_e' = 93.33 \text{ ksi (Table 2, AISC specification for } L/r = 40)$$

$$P_e = \left(\frac{23}{12}\right) A \, F_e' = 3685 \text{ kip}$$

$$C_m = 0.6$$

$$\frac{P}{P_o} + \frac{C_m M}{M_p\left(1 - \frac{P}{P_e}\right)} = \frac{85}{905} + \frac{0.6 \times 6375}{7250 \times 0.977} = 0.094 + 0.540 = 0.636 < 1 \qquad \underline{\text{O.K.}}$$

Use W 18 x 70 – A 572–50
Lateral brace 20'-6" above base

Design column CD

Assume no lateral bracing

$$M = 1594 \text{ kip-ft} = 19,128 \text{ kip-in.}$$

$$P = 255 \text{ kip}$$

For zero axial force, $\quad Z_{req} = \dfrac{19,128}{50} \cong 383 \text{ in}^3$

Try W 30 x124 $\qquad A = 36.5 \text{ in}^2 \; ; \; \dfrac{d}{w} = 51.6 \; ; \; Z = 408 \text{ in}^3$

$$P_y = 1825 \text{ kip}$$

$$\frac{P}{P_y} = 0.140$$

$$\frac{412}{\sqrt{F_y}}\left(1 - 1.4\frac{P}{P_y}\right) = 46.9 < 51.6 \qquad \text{N.G.}$$

Try W 24 x145 $\qquad A = 42.7 \text{ in}^2 \; ; \; \dfrac{d}{w} = 40.3 \; ; \; Z = 417 \; ; \; \dfrac{P}{P_y} = 0.119$

$$\frac{412}{\sqrt{F_y}}\left(1 - 1.4\frac{P}{P_y}\right) = 48.6 > 40.3 \qquad \underline{\text{O.K.}}$$

$$\frac{b}{t} = 13.8 < 14 \qquad \underline{\text{O.K.}}$$

$$M_p = 20,800 \text{ kip-in.}$$

Formula 2.4–3 $\quad 0.119 + 0.777 = 0.896 \qquad \underline{\text{O.K.}}$

$$\frac{L}{r_y} = \frac{300}{3.32} = 90.4 \; ; \; F_a = 16.85 \text{ ksi} \; ; \; P_o = 1223 \text{ kip}$$

$$\frac{L}{r_x} = \frac{300}{10.34} = 29 \; ; \; F_e' = 177.56 \text{ ksi} \; ; \; P_e = 14,437 \text{ kip} \quad 1 - \frac{P}{P_e} = 0.982$$

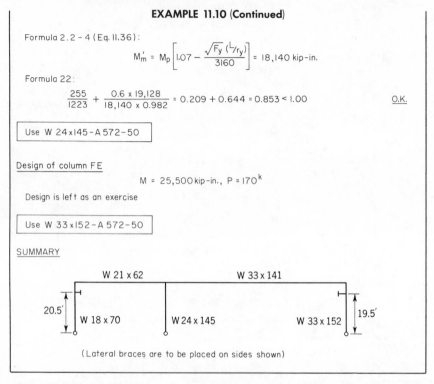

EXAMPLE 11.10 (Continued)

Formula 2.2 – 4 (Eq. II.36):

$$M_m' = M_p\left[1.07 - \frac{\sqrt{F_y}\,(L/r_y)}{3160}\right] = 18,140 \text{ kip-in.}$$

Formula 22:

$$\frac{255}{1223} + \frac{0.6 \times 19,128}{18,140 \times 0.982} = 0.209 + 0.644 = 0.853 < 1.00 \qquad \underline{\text{O.K.}}$$

> Use W 24×145 – A 572–50

Design of column F E

$$M = 25,500 \text{ kip-in.}, \quad P = 170^k$$

Design is left as an exercise

> Use W 33×152 – A 572–50

SUMMARY

(Lateral braces are to be placed on sides shown)

of Eq. 11.36 is an approximation to this moment and it includes the possible effect of lateral-torsional buckling. If lateral bracing is provided, then $M' = M_p$.

The AISC beam-column formulas give an estimate of the maximum capacity of beam-columns; they do not include any provisions concerning rotation capacity. This problem is treated in Chapter 22 where the complete $M_0\text{-}\theta$ curves are used in the determination of the maximum strength of subassemblages. The AISC method of beam-column design is illustrated in Example 11.10, where the 1969 AISC Specification is used as the basis for design. It should be noted in this example that the selection of member sizes for the beam-columns was affected strongly by the web width-thickness requirement (see Chapter 17). In fact, it is advisable to make the local buckling check first so that the more time-consuming beam-column formulas will not need to be solved as frequently.

PROBLEMS

11.1. In Example 11.3 it is found that the allowable eccentricity was larger than the actual one for the given cross section. Assuming that all other conditions remain the same, determine the thickness of the plate such that the allowable eccentricity becomes just equal to 2 in.

11.2. For the cross section and material of Example 11.3, compute the maximum allowable axial force P. Use the AREA Specifications and assume that the eccentricity acts only on one end.

Problem 11.2

11.3. Solve for the maximum value of w_0 in Example 11.4 if the axial load is in tension instead of compression.

11.4. Compare the answer obtained in Example 11.4 with an exact answer obtained from solving the differential equation of this case. This equation is

$$EI\frac{d^4v}{dz^4} + P\frac{d^2v}{dz^2} = \frac{w_0 z}{L}$$

and the boundary conditions are $v = d^2v/dz^2 = 0$ at $z = 0$ and $z = L$.

11.5. Using the approximate relationship established by Eq. 11.14, develop an interaction equation between P/P_y, QL/M_y, and L/r for the beam-column shown. Plot the interaction curves P/P_y versus QL/M_y for $L/r = 20, 60, 120$ to study effect of changing L/r.

Given: $F_{max} = F_y = 36$ ksi
    $E = 29,000$ ksi
    $P_y = AF_y$
    $M_y = SF_y$

Problem 11.5

11.6. Design a beam-column which is subjected to the following conditions:

End moment on top: 3650 kip-in.
End moment on the bottom: 5670 kip-in.

These end moments are applied in such a manner that the moments cause single-curvature deformation. The axial load is 510 kips and the length of the member is 25 ft. Assume that no sidesway buckling occurs, and use the total length as the effective length. Design this member for the following situations, using AISC Section [1.6.2]:
    (a) Rolled wide-flange shape, strong axis bending, A36 steel
    (b) Same as (a), but use A441 steel
    (c) Rolled wide-flange shape, weak axis bending, A36 steel

COLD-FORMED MEMBERS

12.1 INTRODUCTION

Structural members cold-formed from strip or sheet steel are used widely in construction. They are actually complementary to the heavier rolled and built-up sections. They fill the need where hot-rolled sections would be uneconomical to support light loads and also where the members are intended to create surfaces such as roofs, floors, and walls or formwork for concrete in addition to supporting light loads. Often both types of members are incorporated in the same structure, each serving the purpose for which it is best suited. A favorable characteristic of cold-formed members is that they can be formed into a large variety of cross-sectional shapes by using simple equipment, in contrast to the monumental installations needed for the production of hot-rolled shapes. Additional shapes can be obtained by connecting simpler members; spot welding is used widely for this purpose. The protection of cold-formed members against corrosion is greatly facilitated by galvanizing or otherwise precoating the sheet metal before fabrication.

Figure 12.1 shows some of the common sections (more types can be seen in Ref. 12.1). Three groups may be distinguished:

1. Sections a to d—simple, load-carrying members
2. Sections e to h—built-up, load-carrying members
3. Sections i to k—panels or decks

Figures 3.25 and 3.27 illustrate the application of such members in roof and

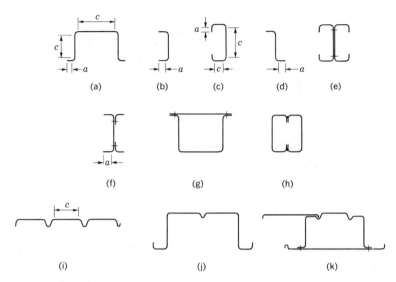

Fig. 12.1 Light gage cold-formed sections.

floor construction.   Apart from developing the necessary strength for carrying the vertical load, floor or roof panels can be used as shear diaphragms to resist and transmit horizontal forces resulting from wind, earthquake, or similar actions.

The principles of design of cold-formed members are basically similar to those described in other chapters.   There are, however, some important additional considerations which make it necessary to give a separate presentation for these members.   The most important feature is that the component plate elements usually have rather high width-thickness ratios and, normally, their post-buckling strength is utilized.   This arises from the use of relatively thin metal (as thin as 0.012 in.) in members whose other dimensions are essentially the same as those of hot-rolled members.   The concepts of effective width (applied both to post-buckling strength and "shear lag") and effective area are employed in designing these members.

The stress and stability analyses of single members of thin-walled open cross section require consideration of torsional-flexural buckling, a problem that was presented only briefly in the analysis of rolled sections (Art. 7.6).

The process of cold-forming introduces changes in the material properties, usually raising the yield point.   Figure 12.2 illustrates the effect of cold-forming in a lipped channel (this figure is based on Fig. 38 of Ref. 12.2). Shown plotted is the material hardness in Diamond Penetration Numbers (DPN), which is indicative of the yield point.   A greater amount of cold-working at the corners is seen to increase the hardness considerably; for

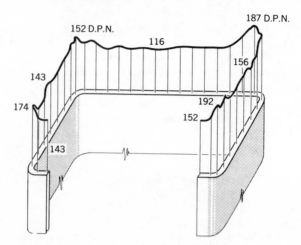

Fig. 12.2 Effects of cold-forming on material hardness. (Courtesy of The Institution of Civil Engineers.)

example, 187 DPN at a corner vs. 116 DPN in the flat portion. The changes in material properties due to cold-forming depend on many factors: chemical composition, previous mechanical treatment, amount of cold-working in the forming process, type of cold-working (tension or compression), original material properties, and so on. A detailed study of this effect led to the following simplified formula for utilizing a higher average yield point for design purposes (Refs. 12.3, 12.4, 12.5, AISI Sec. [3.1.1.1]).

$$F_{ya} = CF_{yc} + (1 - C)F_y \qquad (12.1)$$

where F_{ya}, the average yield point of the cross section, is a function of the ratio C of the sum of the corner areas to the total area and the raised yield point in the corner F_{yc}. F_{yc} is a complex function of the ultimate strength of the material and of the radius of curvature in the corners (see AISI Sec. [3.1.1.1]).

Design procedures also give consideration to the fact that the wide range of steels used for light gage members includes steels with different stress-strain characteristics (strength and ductility). Figure 12.3 shows two extreme types: a steel with a yield level (curve *a*) and a gradually yielding steel (curve *b*).

Since the purpose of this chapter is to explain primarily the aspects of the design of cold-formed members which are different from those of conventional hot-rolled members, some knowledge of the content of Chapters 6 to 10 and 17 to 19 is needed. The emphasis will be on explaining phenomena involved and only the principal equations will be presented for developing basic

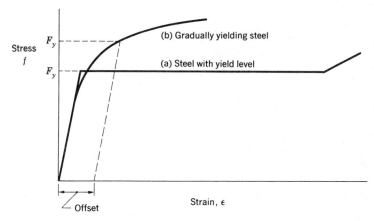

Fig. 12.3 Stress-strain diagrams of steels used for light gage cold-formed structures.

design skills. Many additional detailed provisions and limitations can be found in the Specification.[1.24]

Light gage cold-formed members have been used to a limited extent for as long a time as hot-rolled members. Their wide acceptance in building construction, however, took place mainly after the Second World War, especially after the issuance of the first edition of the *Specification for the Design of Light Gage Steel Structural Members* by AISI in 1946. Most of the research for the development of this specification was conducted at Cornell University. This chapter is based largely on the findings of this and subsequent research as described in the *Commentary on the 1968 Edition of the Specification*[1.19] and also on the *Specification for the Design of Cold-Formed Steel Structural Members*[1.24] which contains the Specification, and on other publications such as Refs. 12.1 and 12.6.

12.2 STRENGTH OF PLATE COMPONENTS, EFFECTIVE WIDTH, AND ALLOWABLE STRESSES

Cold-formed members have as their components flat thin plates, and the behavior of these influences the member strength.

The strength of an element in tension usually is considered to be limited by the yield point stress acting over the full element area. The allowable stress is obtained by dividing the yield point by a factor of safety (AISI Section [3.1]):

$$F_b = 0.60F_y \tag{12.2}$$

(Use F_{ya} instead of F_y in Eq. 12.2 when the effect of cold-forming is utilized.) Very wide plates serving as tension flanges, however, may have a non-uniform stress distribution due to shear lag, and therefore, in their analysis the concept of the effective width is employed (see Art. 12.3).

When a wide plate element is in compression, its buckling or post-buckling strength controls the design. Such plates are usually investigated as long plates, supported along one or two edges.

A plate in compression may be a part of a column cross section or of the compression flange of a member subjected to flexure. Two types of compression elements can be distinguished: *unstiffened plates*, plates supported at one edge only, and *stiffened plates*, plates supported along two edges, and perhaps also at some intermediate points, by webs or specially formed longitudinal stiffeners. Often these stiffeners are folds of the same plate or just a narrow lip at the edge of the plate. The elements designated as a and c in Fig. 12.1 are examples of unstiffened and stiffened elements, respectively.* A distinction between these two types is needed because unstiffened elements have considerably less postbuckling strength than stiffened elements and also are subject to excessive deformations after buckling.

1. Unstiffened Compression Elements

Unstiffened plate elements have considerable deformation in the post-buckling range and their strength usually is considered to be governed by the buckling stress.

The development of the allowable compressive stresses for unstiffened elements, AISI Section [3.2], is illustrated in Fig. 12.4. The light solid curve (curve a) depicts the buckling strength. This curve is analogous to the plate buckling curves shown in Figs. 17.9 and 17.12 and consists of three regions.

$0 < b/t < 63.3/\sqrt{F_y}$: $F_u = F_y$, that is, the plate does not buckle before reaching the yield point.

$63.3/\sqrt{F_y} < b/t < 144/\sqrt{F_y}$: A straight-line transition curve. In contrast to hot-rolled sections, the reduction is influenced not so much by residual stresses as by the fact that many sheet steels are of the gradually yielding type (see Fig. 12.3). In this region plates buckle inelastically and have essentially no post-buckling strength (see Chapter 17).

$b/t > 144/\sqrt{F_y}$: Plates buckle elastically. Since in most cases plates restrain each other, a plate buckling coefficient of 0.5 is used here (see Fig. 17.9). This is somewhat higher than the 0.425 applicable to

---

* Note that only the width of the flat portion of the plate is considered as the controlling dimension.

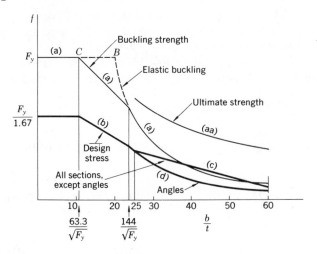

Fig. 12.4 Allowable stresses for unstiffened compression elements. (Courtesy of AISI.)

plates simply-supported at one edge (Fig. 17.6). In this range plates exhibit an increasing post-buckling strength as shown by the difference between curves a and aa. The ultimate strength (curve aa) is computed from Eq. 12.3, which gives the effective width (Eq. 12.3 is the same as Eq. 17.14):

$$\frac{b_e}{b} = 1.19 \sqrt{\frac{F_{cr}}{f_{\max}}} \left(1 - 0.30 \sqrt{\frac{F_{cr}}{f_{\max}}} \right) \tag{12.3}$$

The allowable stress for b/t less than 25 is given by the heavy curve b and is obtained by dividing the buckling stress (curve a) by a factor of safety (F.S. = 1.67).   Curve b defines three portions: $0 < b/t < 63.3/\sqrt{F_y}$, $63.3/\sqrt{F_y} < b/t < 144/\sqrt{F_y}$, and $144/\sqrt{F_y} < b/t < 25.$*

For $b/t > 25$ two design curves are recommended—one for angles (curve d) and one for unstiffened plate elements of other cross sections (curve c).

The design curve for angles (curve d) continues to be the buckling stress divided by the factor of safety.   Plate elements of an equal-leg angle column buckle together and behave as if they were simply supported at the corner. The little post-buckling strength that they possess is compensated for by the use of the higher buckling coefficient as stated above.

To take advantage of the post-buckling strength of cross sections other than angles, the allowable stress is given by curve c.   This straight-line relationship actually allows buckling to occur at working stresses, as indicated by the overlap of curves c and a.

* The very rare case when $144/\sqrt{F_y} > 25$, which occurs when $F_y < 33$ ksi, requires a special formula given in the footnote of AISI Section [3.2].

Unstiffened elements with b/t greater than 60 tend to distort prohibitively at relatively low stresses and, in general, are not used in load-carrying members where distortion would be objectionable.

Analytical expressions for the allowable compressive stresses (in ksi) for unstiffened plate elements are as follows (curves b, c, and d, and AISI Section [3.2]):

$b/t \leqslant 63.3/\sqrt{F_y}$: $\qquad F_c = 0.60F_y = F_b$ $\qquad$ (curve b) $\quad$ (12.4)

$63.3/\sqrt{F_y} < b/t \leqslant 144/\sqrt{F_y}$: $\quad F_c = F_y[0.767 - (2.64/10^3)(b/t)\sqrt{F_y}]$

$\qquad\qquad\qquad\qquad\qquad\qquad\qquad\qquad\qquad$ (curve b) $\quad$ (12.5)

$144/\sqrt{F_y} < b/t < 25$: $\qquad F_c = 8000/(b/t)^2$ $\qquad$ (curve b) $\quad$ (12.6)

$25 < b/t < 60$:

$\quad$ 1. Angles. $\qquad F_c = 8000/(b/t)^2$ $\qquad$ (curve d) $\quad$ (12.7a)

$\quad$ 2. Other sections. $\quad F_c = 19.8 - 0.28(b/t)$ $\qquad$ (curve c) $\quad$ (12.7b)

The application of the formulas is demonstrated by Example 12.1.

2. Stiffened Compression Elements

Effective Width. In contrast to unstiffened elements, stiffened plate elements (plates supported at both edges) develop considerable post-buckling strength without excessive distortions, and their design is based on this post-buckling strength using the effective width concept. (A plate element is assumed to have only a portion of its width effective in carrying the applied load.)* The formula for the computation of effective width is given by Eq. 17.13 which, with the plate buckling coefficient $k = 4.0$, can be written as

$$\frac{b_e}{t} = 1.9\sqrt{\frac{E}{f_{max}}}\left(1 - 0.415\frac{t}{b}\sqrt{\frac{E}{f_{max}}}\right) \qquad (12.8)$$

where f_{max} is the stress at the edges. f_{max} is taken as uniformly distributed over the effective width (see Fig. 17.8).†

As Eq. 12.8 shows, the effective width depends not only on the width-thickness ratio b/t but also on the edge stress f_{max}. This fact poses some difficulty in the evaluation of cross-sectional properties of members, since the stress and the effective cross section are interdependent. When the stress is specified, the effective width and the cross-sectional properties can be

---

* The effective width concept for post-buckling plate behavior is explained in Art. 17.3.
† For closed rectangular sections, Eqs. 12.5 and 12.8 are somewhat liberalized (see AISI Section [2.3.1.1]).

EXAMPLE 12.1

PROBLEM:

Compute the moment capacity for the I-section shown. $F_y = 36$ ksi

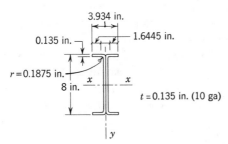

SOLUTION:

$$b = \frac{3.934}{2} - (0.1875 + 0.135) = 1.967 - 0.3225 = 1.6445 \text{ in.}$$

$$\frac{b}{t} = \frac{1.6445}{0.135} = 12.2 \quad \begin{matrix} > 63/\sqrt{F_y} = 10.55 \\ < 144/\sqrt{F_y} = 24 \end{matrix}$$

Allowable compressive stress, Eq. 12.5 (AISI Section [3.2.6])

$$F_c = 36\left(0.767 - \frac{2.64}{1000}(12.2)\sqrt{36}\right)$$

$$= 20.6 \text{ ksi}$$

CROSS-SECTIONAL PROPERTIES (assuming square corners)

Flange: $2 (0.135 \times 3.934)(4 - \frac{0.135}{2})^2 = 16.41$

Web: $2 \frac{(8 - 2 \times 0.135)^3}{12} \times 0.135 \qquad = 10.40$

$$\qquad\qquad\qquad\qquad\qquad I_x = 26.81 \text{ in.}^4$$

$$S_x = \frac{I_x}{c} = \frac{26.81}{4.0} = 6.7 \text{ in}^3$$

Resisting $M = F_c\, S_x = \frac{20.6 \times 6.7}{12} = 11.5$ kip-ft

Answer: Resisting M = 11.5 kip-ft

---

computed directly. AISI Section [2.3.1.1] provides a means of doing just that, using $E = 29,500$ ksi and $f_{max} = 1.67 f_{actual}$ in deriving the effective width formula for load determination (thus, this formula already incorporates a factor of safety). For deflection computations, when cross-sectional properties due to a given loading are needed, AISI Section [2.3.1.1] has another formula, which was obtained from Eq. 12.8 by substituting the values of E and $f_{max} = f_{actual}$ into it.

The limiting value of the width-thickness ratio below which the plate is fully effective for a given stress is found from Eq. 12.8 by setting $b_e = b$,

$$\left(\frac{b}{t}\right)_{\text{lim}} = 1.288\sqrt{\frac{E}{f_{\text{max}}}} \tag{12.9}$$

Multiply-Stiffened Compression Elements. Multiply-stiffened compression elements and stiffened elements with only one edge supported by a web, having b/t greater than approximately 60, have smaller effective widths than those given by Eq. 12.8.[1.19] From AISI Section [2.3.1.2] a design approximation for the modified effective width b_e' to account for this effect is

$$\frac{b_e'}{t} = \frac{b_e}{t} - 0.10\left(\frac{b}{t} - 60\right) \tag{12.10}$$

where b_e is the effective width for the given b/t as computed from Eq. 12.8.

Intermediate stiffeners in this case are unable to develop stresses as high as at the web and thus should be considered to have a reduced effective area as given by the following expressions:

$60 < b/t < 90$:

$$A_{\text{eff}} = \left[\left(3 - 2\frac{b_e'}{b}\right) - \frac{1}{30}\left(1 - \frac{b_e'}{b}\right)\left(\frac{b}{t}\right)\right]A \tag{12.11}$$

$b/t > 90$:

$$A_{\text{eff}} = \left(\frac{b_e'}{b}\right)A \tag{12.12}$$

where A is the full actual area of the intermediate (or edge) stiffener. The centroid and moment of inertia of the stiffener about its own axis are assumed to be those of the full stiffener area.

Stiffeners. The post-buckling strength of a plate can be fully developed only if the edge supports fulfill certain rigidity requirements and do not fail before the plate does. A theoretical analysis verified by tests indicates that a stiffener will not buckle prematurely if its moment of inertia about its centroidal axis parallel to the stiffened plate has the following value[1.19,7.17] (see AISI Section [2.3.2.1]):

$$9.2t^4 \leqslant I_{\text{min}} = 1.83t^4\sqrt{(b/t)^2 - 4000/F_y} \tag{12.13}$$

Since properly designed webs are sufficiently rigid, Eq. 12.13 applies in particular to the edge stiffeners, such as simple lips (Fig. 12.1c, e, j, and Fig. 12.2).

Intermediate stiffeners support two portions of the plate and, therefore, should have twice the moment of inertia given by Eq. 12.13. Figure 12.1j shows a section with an intermediate stiffener in the form of a fold.

3. Comparison of Stiffened and Unstiffened Plate Elements

A comparison of the allowable stresses in stiffened and unstiffened plate elements can be made by studying Fig. 12.5, where design curves are plotted for $F_y = 36$ ksi. The heavy line gives the allowable average stress for stiffened elements which for this discussion is defined as the value of $(b_e/b)(F_y/\text{F.S.})$ and computed by using F.S. $= 1.67$ and Eq. 12.8. The broken heavy lines represent the plot of the design curves of Fig. 12.4 for unstiffened elements.

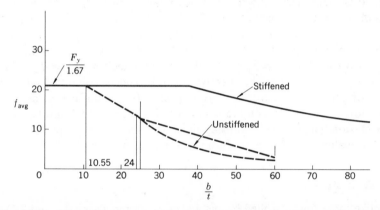

Fig. 12.5 Allowable average stresses for stiffened and unstiffened plate elements.

For any b/t value above $63.3/\sqrt{F_y}$, and especially greater than 25, a stiffened element is seen to be decidedly more efficient than an unstiffened one. For example, for $b/t \approx 50$ the allowable average stress on a stiffened element is 17.9 ksi and on an unstiffened element, 5.8 ksi. Thus, the stiffened element is more than three times as strong as an unstiffened one. This is the major reason why stiffened elements are preferred in cold-formed members. Usually, the formation of an edge lip or of a fold introduces little complication in the fabrication process, yet often more than doubles the strength of the plate.

This figure also illustrates the advantage of introducing an intermediate stiffener. A plate with $b/t = 80$ has the allowable average stress of 12.5 ksi. If an intermediate stiffener is introduced, b/t becomes approximately 40 and the allowable average stress becomes 20.6 ksi, thus giving an increase of about 65 per cent.

12.3 BEAMS

The design of cold-formed beams follows the principles outlined in Chapter 7 for conventional beams. There are, however, a number of additional considerations of which the important ones are described in this article.

1. Properties of a Cross Section

Since part of a beam cross section is in compression, it is necessary, when b/t exceeds certain limits, either to reduce the allowable stress on an unstiffened element (Eqs. 12.5 to 12.7) or to consider only the effective width for stiffened plate elements (Eq. 12.8). In the latter case the ineffective portion of the plate is assumed removed and further analysis is performed on the remaining effective parts. For a strength analysis, the allowable compressive stress is $F_c = F_y/1.67$ and b_e is computed for $f_{max} = F_y$. There are, however, cases when the effective width should be computed by a trial-and-error procedure for a lower stress; such cases would include deflection evaluation and when the compression flange is closer to the neutral axis than is the tension flange. At each trial, all the cross-sectional properties needed must be evaluated. The procedure is illustrated in Example 12.2.

2. Web Buckling

Unlike those for hot-rolled beams and plate girders (Arts. 7.4 and 8.2) the allowable shear and bending stresses for webs in cold-formed beams are variable, depending on the web dimensions given. The approach is analogous to that used for column stresses or stresses in unstiffened plate elements.

Buckling in Shear. Shear is limited by web buckling and thus depends on the depth of the beam. The critical shear stress is obtained from Eq. 17.9 by substituting $E = 29,500$ ksi and the plate buckling coefficient $k = 5.34$ from Eq. 8.8 ($a/h = \infty$), since intermediate or bearing stiffeners are economically impracticable in cold-formed members. h, the clear distance between flanges, is taken as the controlling dimension. Then,

$$F_{cr} = 142,000 \left(\frac{t}{h}\right)^2 \quad \text{(in ksi)} \tag{12.14}$$

For small h/t values the web plate will not buckle but will yield when the shear stress becomes $f_v = F_{ys} = F_y/\sqrt{3}$ (according to the Huber-Hencky-Mises theory of yielding). Usually, this is approximated by

$$f_v = F_{ys} = \tfrac{2}{3}F_y \tag{12.15}$$

EXAMPLE 12.2

PROBLEM:

Given an 18-foot beam with a hat cross section shown and made of a material with F_y = 40 ksi, determine:

   a) Allowable distributed loading w(including dead load)

and b) The resulting mid-span deflection.

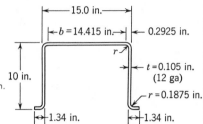

SOLUTION:

a) F_b= 0.6x40.0 =24.0 ksi

Cross-sectional properties will depend on the effective width of the top flange which in turn depends on the stress in it. Neutral axis will be closer to the top flange and therefore the stress in it will be less than F_b.

1st TRIAL

Assume $f = \frac{1}{3}F_b$ = 8 ksi , $b/_t$ = 14.415/0.105 = 137.4

Eq. 12.7 gives for f_{MAX} = 8 x 1.67 ksi and E = 29.5 x 10^3 ksi

$$\frac{b_e}{t} = 1.9\sqrt{\frac{29.5 \times 10^3}{1.67 \times 8}}\left(1 - \frac{0.415}{137.4}\sqrt{\frac{29.5 \times 10^3}{1.67 \times 8}}\right) = 76.6$$

b_e = 76.6 x 0.105 = 8.05 in. , b_e /2 = 4.03 in.

Cross-sectional properties are computed employing so-called "linear method", that is, computing the properties from the plate centerline dimensions and only at the end including the plate thickness.

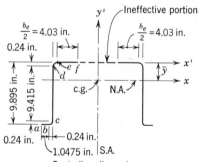

Centerline dimensions

FOR HALF SECTION

PORTION	LENGTH L [in.]	y'	Ly'	L(y')^2	I_{oo}
a b	1.0475	-9.895	-10.35	102.2	—
b c	0.377	-9.808	-3.70	36.3	—
c d	9.415	-4.9475	-46.50	230.3	69.5
d e	0.377	-0.087	-0.03	0.003	—
Subsum	11.2165		-60.58	368.8	69.5
e f	4.03	0	0	0	—
Σ	15.1745		-60.58	368.8	69.5 +368.8

$$\bar{y} = \frac{-60.58}{15.1745} = -4.0$$

$$-(-4.0)(-60.58) = \frac{438.3}{-242.5}$$

Half section I_x = 195.8 (0.105) =20.55 in.^{4}

EXAMPLE 12.2 (Continued)

For maximum tension in the bottom flange of $F_b = 24.0$ ksi compression in the top flange

$$f_c = 24.0 \frac{4.0}{9.895 - 4.0} = 16.3 \text{ ksi} > 8 \text{ ksi (assumed)}$$

2nd TRIAL

Assume   f = 19 ksi

$$b_e = (0.105)(1.9)\sqrt{\frac{29.5 \times 10^3}{1.67 \times 19}} \left(1 - \frac{0.415}{137.4}\sqrt{\frac{29.5 \times 10^3}{1.67 \times 19}}\right) = 5.52 \text{ in.}$$

FOR HALF SECTION

PORTION	LENGTH	Ly'	$Ly'^2 + I_{oo}$
Subsum	11.2165	−60.58	438.3
e f	2.7600	0	0
Σ	13.9765	−60.58	438.3

$$\bar{y} = \frac{-60.58}{13.9765} = -4.34 \qquad \begin{array}{l} -(-4.34)(60.58) = -263.0 \\ I_x = 175.3(0.105) = 18.4 \text{ in.}^4 \end{array}$$

Compression. $f = \dfrac{4.34 \times 24.0}{9.895 - 4.34} = 19$ ksi ≈ 19 ksi   <u>O.K</u>

Both halves: $I_x = 2 \times 18.4 = 36.8 \text{ in.}^4$

     Bottom $S_x = \dfrac{36.8}{5.555 + 0.0525} = 6.56 \text{ in.}^3$

     Top $S_x = \dfrac{36.8}{4.34 + 0.0525} = 8.39 \text{ in.}^3$

Resisting Moment $M = Sf = \dfrac{6.56 \times 24.0}{12} = 13.14 \text{ kip-ft}$

Allowable Loading $w = \dfrac{8M}{L^2} = \dfrac{8 \times 13.14}{18^2} = \underline{0.324 \text{ kip/ft}}$ —   answer

Answer: w = 0.324 kip/ft

b) Mid-span deflection

$$M = 13.14 \text{ kip-ft}, \quad b/_t = 137.4, \quad E = 29.5 \times 10^3 \text{ ksi}$$

1st TRIAL

Assume   f = 19 ksi

     Using AISI section [2.3.1.1]   (same as Eq. 12.8 with f_{max} = actual stress)

$$\frac{b_e}{t} = \frac{326}{\sqrt{f}} \left(1 - \frac{71.3}{(b/_t)\sqrt{f}}\right)$$

$$b_e = \frac{0.105(326)}{\sqrt{19}} \left(1 - \frac{71.3}{137.4\sqrt{19}}\right) = 6.93 \text{ in.}$$

Following the same procedure as in a) and utilizing the available subsums:

$$\bar{y} = -4.14 \text{ in.} \qquad\qquad I_x = 2 \times 19.7 = 39.4 \text{ in.}^4$$

Stress in the top flange

$$f = \frac{Mc}{I} = \frac{3.14(12)(4.14)}{39.4} = 16.51 \text{ ksi} < 19 \text{ ksi (assumed)} \text{ N.G.}$$

EXAMPLE 12.2 (Continued)

2nd TRIAL

Assume f = 16.5 ksi

b_e = 7.35 in. , $\bar{y}$ = 4.08 in. , I_x = 39.9 in.^{4}

Stress in the top flange

$$f = \frac{3.14(12)(4.08)}{39.9} = 16.21 \text{ ksi} \approx 16.5 \text{ ksi (assumed)} \quad \text{close enough}$$

DEFLECTION

$$\delta = \frac{5}{384} \frac{w L^4}{E I} = \frac{5(0.324)(18^4)(12^4)}{384(12)(29.5 \times 10^3)(39.9)} = 0.648 \text{ in.} - \text{Answer}$$

Answer: δ = 0.653 in.

The shear stress is computed as an average stress

$$f_v = \frac{V}{ht} \tag{12.16}$$

where V is the total shear.

With a factor of safety of 1.7 in the elastic range ($h/t > 548/\sqrt{F_y}$), the allowable shear stress is then given (AISI Section [3.4.1]) as

$$F_v = 83,200\left(\frac{t}{h}\right)^2 \quad \text{(in ksi)} \tag{12.17}$$

For h/t less than about 45–50 the web will fail in yielding due to shear, and the allowable shear stress is specified (AISI Section [3.4.1]) as

$$F_v = 0.40 F_y \tag{12.18}$$

which is the shear yield stress divided by a factor of safety of 1.67. In the inelastic range between the values given by Eqs. 12.17 and 12.18, the following transition equation is used:

$$F_v = \frac{152\sqrt{F_y}}{h/t} \tag{12.19}$$

Buckling in Bending. The buckling stress of a web under bending is given by Eq. 17.9 with the buckling coefficient $k = 23.9$. Plates in this case have considerable post-buckling strength, and a lower factor of safety against buckling is justified (F.S. = 1.23). The allowable compressive stress in a web under bending thus becomes (AISI Section [3.4.2])

$$F_{cw} = 520,000\left(\frac{t}{h}\right)^2 \quad \text{(in ksi)} \tag{12.20}$$

This stress, of course, may not be greater than F_b.

Buckling Due to Combined Bending and Shear. The buckling of the web where shearing and normal compressive stresses act simultaneously is described in Ref. 7.6. Introducing factors of safety as stated above for the two types of stresses, an interaction formula is obtained (AISI Section [3.4.3]),

$$\left(\frac{f_b}{F_{cw}}\right)^2 + \left(\frac{f_v}{F_v}\right)^2 \leqslant 1.0 \tag{12.21}$$

The web should be checked not only for buckling but also for yielding. The following interaction formula derived from the Huber-Hencky-Mises yielding criterion is used for this purpose:

$$\left(\frac{f_b}{F_{cw}}\right)^2 + \left(\frac{f_v}{\frac{2}{3}F_b}\right)^2 \leqslant 1.0 \tag{12.22}$$

Usually it is adequate to check the shearing and normal stresses separately against their allowable values, because the shearing stress is maximum at the neutral axis where the bending stress is zero and the normal stress is maximum at the flange where the shearing stress is smaller than the average.

Transverse Load. When a transverse load is applied to the flange in the area of high bending and shearing stresses, the stress condition is more complicated and should be investigated for yielding by the following equation, developed analogously to Eq. 12.22,

$$\left(\frac{f_b}{F_b}\right)^2 + \left(\frac{f_e}{F_b}\right)^2 - \frac{f_e f_b}{F_b^2} + \left(\frac{f_v}{\frac{2}{3}F_b}\right)^2 \leqslant 1.0 \tag{12.23}$$

where f_e is the stress at the edge produced by the load. The web in this area must be checked also for crippling.

3. Web Crippling

Reactions and concentrated loads may produce local failure in the webs of beams. This phenomenon, web crippling, is considerably more complicated in cold-formed sections than in hot-rolled or built-up sections. As explained in Art. 17.6, web crippling in the latter case is essentially elastic-plastic buckling of a *flat plate* under a concentrated edge load. Webs of cold-formed shapes, however, consist of one or two *bent plates* and the edge load is transmitted through a rounded plate portion at the web-to-flange junction, as shown in Fig. 12.6.

Figure 12.6a shows a member whose single web is not restrained from rotation. The crippling of such a web is facilitated by the curvature at the root of the web, as can be seen from the deformed shape. An I-beam made of two channels connected back to back, shown in Fig. 12.6b, has considerably more resistance against crippling because of the mutual restraint furnished by the two webs.

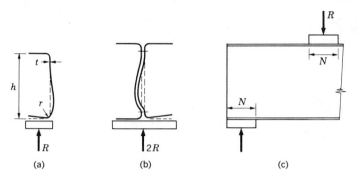

Fig. 12.6 Web crippling in cold-formed members.

Theoretical analysis being impractical, the following design formulas have been developed on the basis of experimental results[12.7] (AISI Section [3.5]):

Single Web (unrestrained web).
End:

$$R = t^2 \left[98 + 4.2\left(\frac{N}{t}\right) - 0.022\left(\frac{N}{t}\right)\left(\frac{h}{t}\right) - 0.011\left(\frac{h}{t}\right) \right]$$

$$\times \left(1.15 - 0.15\frac{r}{t}\right)\left(1.33 - \frac{0.33F_y}{33}\right)\frac{F_y}{33} \quad (12.24)$$

Interior:

$$R = t^2 \left[305 + 2.3\left(\frac{N}{t}\right) - 0.009\left(\frac{N}{t}\right)\left(\frac{h}{t}\right) - 0.5\left(\frac{h}{t}\right) \right]$$

$$\times \left(1.06 - 0.06\frac{r}{t}\right)\left(1.22 - 0.22\frac{F_y}{33}\right)\frac{F_y}{33} \quad (12.25)$$

Tests should be conducted if $r/t > 4.0$ in Eqs. 12.24 or 12.25.
Double Web (restrained web).
End:

$$R = t^2 F_y(4.44 + 0.558\sqrt{N/t}) \quad (12.26)$$

Interior:

$$R = t^2 F_y(6.66 + 1.446\sqrt{N/t}) \quad (12.27)$$

where R = allowable reaction or concentrated load in kips for each component plate of the web (for example, the total reaction in Fig. 12.6b is $2R$),

t = web thickness, N = the length of bearing but not greater than h, h = clear distance between flanges, and r = inside radius of the bend.

If the allowable load R is exceeded, the only adjustment that can be made is to increase the length of bearing N. Bearing stiffeners are used on hot-rolled sections in such cases, but they are very impractical for cold-formed members. Only in special cases are bearing stiffeners cold formed during fabrication.

Example 12.3 illustrates the application of some of the formulas given above.

4. Lateral Buckling

The phenomenon of lateral buckling of flexural members is explained in Art. 7.6.

Of the wide range of cold-formed shapes used in construction, it is primarily the single-web shapes (I-beams, channels, and zees) that may be susceptible to lateral buckling. The box and hat-shaped sections, and especially the panels, have much more lateral stability.

I-Beams and Channels (AISI Section [3.3]). Equation 7.28 gives the lateral buckling stress of I-sections and it may also be used for channels.[12.8] However, since it was derived for I-sections with equal flanges, it is not rigorously applicable to sections which are not symmetrical about the axis of bending. Since such unsymmetrical sections are popular in cold-formed construction, a slightly different equation serves as a basis for deriving the design formulas.[7.19]

As in Eq. 7.28, two contributions appear; one due to warping torsion and the other due to the St. Venant (pure) torsion. Neglecting the contribution of the St. Venant torsion as insignificant for the commonly-used cold-formed sections and making a series of simplifying assumptions analogous to those in Art. 7.6, the allowable stress for the compression flange is found to be, with F.S. = 1.67,

$$F_c' = 0.6\pi^2 E C_b \frac{dI_{yc}}{L_b^2 S_{xc}} \quad \text{(in ksi)} \tag{12.28}$$

where E = 29,500 ksi, C_b is a factor which takes into consideration the variation of the moment along the beam (see Eq. 7.33), d is the beam depth (in.), I_{yc} is the moment of inertia of the compressed portion of the cross section about the y-axis (in.⁴), L_b is the unbraced length of the member (in.), and S_{xc} is the section modulus for the compression fiber of the entire section about the bending axis.

Equation 12.28 is valid only when $(L_b^2 S_{xc}/dI_{yc})$ is such that the critical stress F_{cr} is below the proportional limit. The limiting value of $(L_b^2 S_{xc}/dI_{yc})$ is

$$\frac{L_b^2 S_{xc}}{dI_{yc}} = \frac{1.8\pi^2 E C_b}{F_y} \quad (F_y \text{ in ksi}) \tag{12.29}$$

EXAMPLE 12.3

PROBLEM:

Given the beam of Example 12.2 under the loading $w = 0.314$ kip-ft, with length of bearing 1.5in. and $F_y = 40$ ksi, check the web for buckling and crippling.

SOLUTION:

a) Buckling in shear.

End shear: $V = \dfrac{wL}{2} = \dfrac{0.324 \times 18}{2} = 2.92$ kip

Average shearing stress in each web:

$f_v = \dfrac{V}{ht} = \dfrac{2.92}{2 \times 10 \times 0.105} = 1.39$ ksi

Allowable shearing stress:

Eq. 12.17 $F_v = 83,200 \times (t/h)^2 = \dfrac{83,200 \times 0.105^2}{10^2} = 9.16$ ksi > 1.39 ksi   O.K.

b) Buckling in bending.

Allowable compressive stress:

Eq. 12.20 $F_{cw} = 520 \times 10^3 \, (t/h)^2 = 520 \times 10^3 \times \left(\dfrac{0.105}{10}\right)^2 = 57.4$ ksi > 19 ksi   O.K.

c) Crippling.

Reaction per web $R = \dfrac{2.92}{2} = 1.46$ kip

Unrestrained web: Use Eq 12.24 for the allowable stress.

$N/t = 1.5 / 0.105 = 14.3$ $h/t = 10/0.105 = 95.3$

$r/t = 0.1875 / 0.105 = 1.79$

$R = 0.105^2 \left[98 + 4.2 \times 14.3 - 0.022 \times 14.3 \times 95.3 - 0.011 \times 95.3 \right] \times$

$\times (1.15 - 0.15 \times 1.79) \left(1.33 - \dfrac{0.33 \times 40}{33}\right) \dfrac{40}{33} = 1390$ lb

$R = 1.39$ kip < 1.46 kip   N.G.

Increase N to 2.0 in.

$N/t = 2.0 / 0.105 = 19.05$

$R = 1.5$ kip > 1.46 ksi   O.K.

Answer   a) Buckling in shear   O.K.

b) Buckling in bending   O.K.

c) Crippling O.K. if bearing increased from 1.5in. to 2.0in.

A design approximation for the lower values of $(L_b^2 S_{xc}/dI_{yc})$ is taken as a parabolic-type transition curve analogous to that of Eq. 7.35 and 7.36,

$$F_c' = F_y \left[\frac{2}{3} - \frac{F_y}{5.4\pi^2 E C_b} \left(\frac{L_b^2 S_{xc}}{dI_{yc}} \right) \right] \qquad \text{(in ksi)} \qquad (12.30)$$

F_c' may, of course, not exceed F_b, which should be used for

$$\frac{L_b^2 S_{xc}}{dI_{yc}} \leqslant \frac{0.36\pi^2 E C_b}{F_y} \qquad (12.31)$$

Z-Sections. Z-beams tend to buckle at a smaller stress than I-beams or channels.[12-8] The relationship is quite complicated, especially in view of the twisting of Z-beams under loads acting in the plane of the web. The allowable stress is approximated conservatively by one-half the stress for I-beams in the elastic range and a correspondingly modified transition curve. AISI Section [3.3] gives the pertinent equations.

Box and Hat-Shaped Beams. Box beams have considerable torsional stiffness and seldom are subject to lateral buckling; there is no need to limit the compressive stress as long as the ratio of the length to the distance between webs is below $2500/F_y$ (AISI Section [5.3]).

For hat-shaped sections with $I_y < I_x$, it may be necessary to check for lateral buckling using an approximate formula:

$$F_c' = \frac{152,000}{(L_b/r_y)^2} \qquad \text{(in ksi)} \qquad (12.32)$$

where r_y is the radius of gyration of that portion of the cross section which is in compression.[1-19]

In Example 12.4 an I-beam with the section used in Example 12.1 is investigated for lateral buckling.

5. Bracing of Channels and Z-Beams

Beams with unsymmetrical sections such as channel and Z-beams are susceptible to transverse distortions, in addition to lateral buckling, when subjected to the usual loading in the plane of the web. Research indicates that for most types of practical loading, distributed or concentrated, the lateral braces to counteract such distortion should be located at quarter-points of the beam span.[12-9,12-10] Additional braces may be needed at heavy concentrated loads (AISI Section [5.2.1]).

The forces for the design of the lateral braces can be computed as reactions of a continuous beam consisting of one half of the cross section and subjected to forces T at the level of the braces (AISI Section [5.2.2]).

EXAMPLE 12.4

<u>PROBLEM</u>

Given a 13 ft. simply supported beam of an I cross section as in Example 12.1 and loaded with $w = 0.5$ kip/ft ($F_y = 36$ ksi, $F_b = 21.6$ ksi), investigate the lateral buckling and the spacing of bracing needed to prevent it.

<u>SOLUTION</u>

$$M = \frac{0.5 \times 13^2}{8} = 10.56 \text{ kip-ft}$$

Maximum stress in top flange.

$$f_c = \frac{M}{S} = \frac{10.56 \times 12}{6.7} = 18.91 \text{ ksi} < F_c = 20.6 \text{ ksi (Example 12.1)} \qquad \underline{O.K.}$$

Cross-sectional properties (assuming square corners)

$$A = 2(0.135 \times 3.934) + 2(8 - 2 \times 0.135) \times 0.135 = 3.145 \text{ in.}^2$$

$$I_{yc} = \frac{3.934^3}{12} \times 0.135 = 0.684 \text{ in.}^4 \text{ (one flange)}$$

$$d = 8 \text{ in.}$$

$$S_{xc} = 6.7 \text{ in.}^3 \text{ (Example 12.1)}$$

Actual (full beam length)

$$\left(\frac{L_b^2 S_{xc}}{d I_{yc}}\right) = \frac{(13 \times 12)^2 \times 6.7}{8 \times 0.684} = 29,900$$

Limiting $\left(\frac{L_b^2 S_{xc}}{d I_{yc}}\right)$, Eq. 12.29 (AISI Section [3.3])

$$\left(\frac{L_b^2 S_{xc}}{d I_{yc}}\right) = \frac{1.8 \pi^2 E C_b}{F_y} = \frac{1.8 \pi^2 29.5 \times 10^3 \times 1.0}{36} = 14,560 < 29,900$$

($C_b = 1.0$ for the full beam length or its portion at the mid-span)

Eq. 12.28 should be used

$$F_c' = 0.6 \pi^2 \times 29.5 \times 10^3 \times 1.0 \frac{1}{29,900} = 5.85 \text{ ksi} < f_c \text{ Bracing is required}$$

Required spacing of bracing for $F_c' = f_c = 18.91$ ksi

Use Eq. 12.30 (AISI Section [3.3])

$$f_c = F_c' = F_y \left[\frac{2}{3} - \frac{F_y}{5.4 \pi^2 E C_b}\left(\frac{L_b^2 S_{xc}}{d I_{yc}}\right)\right]$$

$$L_b = \sqrt{\left(\frac{2}{3} - \frac{f_c}{F_y}\right)\frac{5.4 \pi^2 E C_b d I_{yc}}{F_y S_{xc}}}$$

$$= \sqrt{\left(\frac{2}{3} - \frac{18.91}{36}\right)\frac{5.4 \pi^2 \times 29.5 \times 10^3 \times 1.0 \times 8.0 \times 0.684}{36 \times 6.7}} = 71 \text{ in.} = 5.9 \text{ ft}$$

Use bracing at the third points, that is, $L_b = 4'-4''$

<u>Check</u> whether one brace at the mid-span would be sufficient.

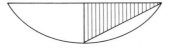

$$L_b = \frac{13}{2} = 6.5 \text{ ft}$$

Assume that the parabola of the M-diagram may be approximated by a triangle (although somewhat unconservatively). Then, from Eq. 7.33,

$$C_b = 1.75 \text{ and from Eq. 12.30}$$

EXAMPLE 12.4 (Continued)

$$F_c' = 36\left[\frac{2}{3} - \frac{36}{5.4\pi^2 \times 29.5 \times 10^3 \times 1.75}\left(\frac{(6.5 \times 12)^2 \times 6.7}{8.0 \times 0.684}\right)\right] = 20.4 \text{ ksi}$$

$$F_c' = 20.4 \text{ ksi} \quad \begin{array}{l} < F_c = 20.6 \text{ ksi} \\ > f_c = 18.91 \text{ ksi} \end{array} \qquad \underline{\text{O.K.}}$$

Use one brace at the mid-span, that is, $L_b = 6'-6''$.

Consider the channel twisting about the shear center due to loads P acting on the web (Fig. 12.7a). The forces T which would produce the same amount of twisting if load P were shifted to the shear center are given by

$$T = \frac{Pm}{c} \tag{12.33}$$

Z-beams, having inclined principal axes 1-1 and 2-2, deflect vertically and laterally due to loads P in the plane of the web (see Fig. 12.7b). The forces T are then obtained from

$$T = P\frac{I_{xy}}{I_x} \tag{12.34}$$

where P = vertical loading on the beam, T = horizontal forces for the computation of forces in braces, m = the distance from the shear center to the

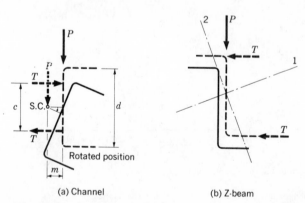

(a) Channel　　　　　　　　　(b) Z-beam

Fig. 12.7 Channel and Z-beams under vertical load.

mid-thickness of the channel web, c = vertical distance between the lateral braces, I_{xy} = product of inertia, and I_x = moment of inertia about the horizontal axis x.

A distributed loading is treated analogously.

Two Channels Forming an I-Beam (AISI Section [4.3]). The longitudinal spacing of connectors (usually spot welds) to form an I-beam from two channels (Figs. 12.1e and 12.6b) is computed from the following equation, which is based on Eq. 12.33:

$$s = \frac{2cS_w}{mw} \tag{12.35}$$

where the additional notation is: s = spacing of connectors, S_w = design capacity of one connector in tension, and w = distributed loading applied to the beam.

6. Buckling of Unbraced Compression Flanges

Instead of the overall lateral buckling discussed previously and shown in Fig. 12.8a, some thin-walled flexural members may be endangered by the lateral buckling of the compression flanges alone as shown in Fig. 12.8b. The analysis of this type of buckling can be simplified to a lateral-torsional buckling analysis of a column on an elastic foundation, the compression flange being elastically restrained by the portion of the cross section in

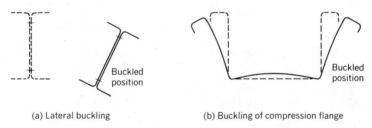

(a) Lateral buckling   (b) Buckling of compression flange

Fig. 12.8 Buckling of compression flanges. (Courtesy of AISI.)

tension.[12.11] The appendix in Ref. 1.24 describes a nine-step procedure for the determination of the critical stress in such a flange.

7. Wide Flanges

Shear Lag. The conventional beam theory (Chapter 7), by assuming that sections plane before deformation remain plane after deformation (Navier-Bernoulli hypothesis), in effect neglects deformations caused by shearing stresses. For beams of ordinary proportions this assumption is acceptable. For unusually deep or wide beams, however, the error becomes significant, and a more refined analysis is necessary. Deep beams (length/depth < 5) are treated elsewhere.[12.12]   Here, a discussion is given of beams having very wide flanges of the type often encountered in cold-formed construction.

An insight into the phenomenon can be gained by considering a simply supported beam subjected to a concentrated load at the mid-span (Fig. 12.9a).[12.13]   If the flanges are allowed to distort freely due to shearing stresses computed by the conventional formulas (Art. 7.4), the compression flange will assume the configuration shown by the dotted line in Fig. 12.9b.   This

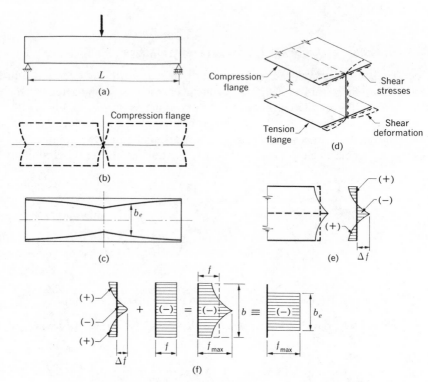

Fig. 12.9 Shear lag in sections with wide flanges.

configuration, however, is obviously inconsistent with the real behavior of the flange since no separation can take place at the mid-span—the material must be continuous.   Restoration of the continuity requires self-equilibrating secondary normal stresses Δf (Fig. 12.9e), which change the original stress condition to a non-uniform stress pattern as shown in Fig. 12.9f.   The effect is the same as if the flanges were less effective in transmitting shear from the web to the flange edges, whence the name of this phenomenon—*shear lag*.

The maximum stress at the web, f_{max}, controls the design of the beam and affects its elastic behavior.

The non-uniform stress pattern usually is assumed to be replaced by a rectangular stress block of intensity f_{max} and width b_e (effective width) such that the areas of the two stress patterns are equal (see Fig. 12.9f).* Figure 12.9c shows the variation of the effective width as computed using the theory of elasticity.[12.12] b_e is seen to be a minimum at the point of load application.

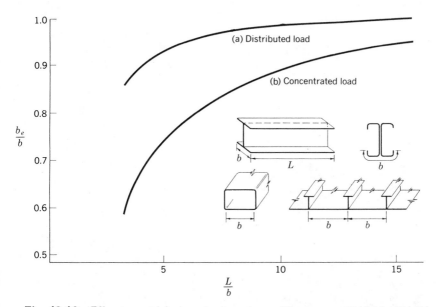

Fig. 12.10   Effective width due to shear lag. (Courtesy of NASA: NACA TN 784. Prepared by George Winter, Cornell University, for National Advisory Committee for Aeronautics.)

For the purposes of design it is usually sufficient to know the effective width only in the region of the maximum moment. Figure 12.10 gives two curves relating such an effective width to the beam span for typical members.[12.14]   Curve a shows that, except for unusually short beams ($L/b < 5$), the effective width for a distributed load may be taken to be equal to the actual width ($b_e/b = 1.0$). For concentrated loads, however, the reduction may be quite substantial, as shown by curve b. The specification gives b_e values based on curve b (AISI Table [2.3.5]).

* This is the same effective width concept as employed in handling the non-uniform stress distribution in the post-buckling range.   Note, however, that b in this discussion is defined as shown in Fig. 12.10, which is somewhat different from elements a and c in Fig. 12.1.

The above reasoning applies equally to tension and compression flanges and assumes that the flanges remain plane. Wide compression flanges, however, may buckle, and then the effective width based on the post-buckling strength should be used rather than that based on the shear lag.

Distortion of Cross Section Due to Bending (Flange Curling). The assumption of an undistorted cross section as used for rolled members is in some cases not valid for cold-formed members. An approximate yet adequate analysis of the phenomenon involved is as follows.[12.14,12.15]

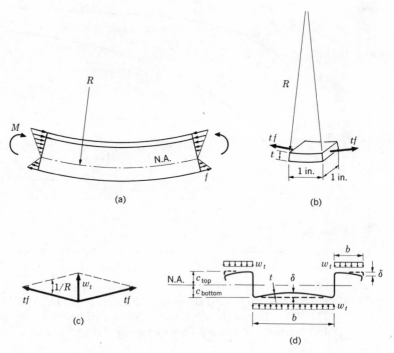

(a)

(b)

(c)

(d)

Fig. 12.11 Distortion of cross section due to bending (flange curling).

When a beam is subjected to bending, the flange stresses give rise to transverse stress components acting toward the neutral axis due to longitudinal curvature. Figure 12.11b shows a 1-in.² portion of the tension flange of the bent beam of Fig. 12.11a. The stress resultants tf on the sides of the element act at an angle equal to $1/R$ to each other (where R is the radius of the longitudinal beam curvature due to bending) and produce a transverse component w_t (see Fig. 12.11c):

$$w_t = \frac{tf}{R} \tag{12.36}$$

(The same occurs in the compression flange.) The effect of this transverse component is the same as if external distributed loading of intensity w_t were acting on the beam flanges as shown in the beam cross section in Fig. 12.11d. w_t gives rise to the deflection of the flanges toward the neutral axis (curling). For thin, wide flanges the maximum deflection δ is computed by assuming that the webs furnish simple support to flanges supported at both edges and fixed support to flanges supported at one edge (intermediate or edge stiffeners are considered as parts of the flanges). With $K = \frac{5}{384}$ for support at both edges and $\frac{1}{8}$ for support at one edge, and $1/R = M/EI = f/cE$ (where c is the distance to the neutral axis), the deflection is given by

$$\delta = K \frac{w_t b^4}{Et^3/12(1 - \mu^2)} = K \frac{12(1 - \mu^2)}{c}\left(\frac{fb^2}{tE}\right)^2 \qquad (12.37)$$

The reduction of the bending stress in the flange is directly proportional to the reduction in the distance to the neutral axis due to flange deflection. If deflections are kept to within about 5 per cent of the depth, the corresponding change in the stress is of little significance. The deflection, however, should be checked.

AISI Section [2.3.3(d)] gives the value of maximum b from Eq. 12.37 in terms of f and permissible δ, by assuming that $c = h/2$ and $K = \frac{1}{8}$ for both cases (b for flanges with webs at both edges is taken as half the actual width).

12.4 COMPRESSION MEMBERS

1. Compression Members Not Subject to Torsional-Flexural Buckling

General column behavior is discussed in Chapter 9. Basically, the strength of cold-formed columns is controlled by the same considerations. The importance of the post-buckling behavior of plate elements is, however, greater than in hot-rolled columns (Q-factor). Also, for many typical columns, there is the possibility of torsional-flexural buckling. The effect of the gradually yielding materials often used in cold-formed structures is counterbalanced by the absence of residual stresses of high intensity encountered in hot-rolled sections.

The design philosophy for cold-formed columns which are not subject to torsional-flexural buckling is essentially the same as explained in Art. 9.5, and Eq. 9.43 is the basis for developing the formulas for these columns. The extremely complex problem of the interaction between the post-buckling plate behavior and the over-all (primary) buckling of a column is simplified by introducing a form factor Q into this equation. It is assumed that individual plate elements are independent of each other, and that their strength (or effective width) can be satisfactorily given by Eqs. 12.4 to 12.8. The influence of plate buckling on column strength is different, depending on whether the cross section is composed of unstiffened or stiffened elements or a combination of both.

The strength of unstiffened elements is taken as limited by their buckling stress F_{cr}. Thus, the total load that may be put on a column cannot exceed $AF_{cr} = AQ_sF_y$, where the factor Q_s is defined by

$$Q_s = \frac{F_{cr}}{F_y} \quad \text{or} \quad \frac{F_c}{F_b} \qquad (12.38)$$

In stiffened elements the stress may reach the yield point, but at this stage only the effective width of the plates can be included in the load-carrying area of the column. Thus, for this case the maximum column load will be $A_{eff}F_y = AQ_aF_y$. Q_a may also be treated as if it were influencing the average stress rather than the cross-sectional area.

$$Q_a = \frac{f_{eff}}{F_y} = \frac{A_{eff}}{A} \qquad (12.39)$$

Column strength is affected in two ways when the cross section consists of both types of elements: the area is reduced to a smaller effective width of the stiffened elements, and the stress is reduced to the lower buckling stress of the unstiffened elements. Thus,

$$P = AQ_aQ_sF_y \qquad (12.40)$$

The column formula (Eqs. 9.43) is then modified to include the effect of plate buckling by using QF_y instead of F_y, where

$$Q = Q_aQ_s \qquad (12.41)$$

The allowable compressive stress F_c of the weakest unstiffened plate element in the cross section should be used for the computation of Q_s. The same allowable stress should be used for the evaluation of Q_a in sections having both types of elements ($f_{max} = 1.67F_c$). When the section has only stiffened elements, Q_a is computed using the ultimate condition, that is, $f_{max} = F_y$. For columns with KL/r above the limiting value (Eq. 9.44) no reduction of the column stress due to local buckling is made.

The use of the factor of safety recommended by the AISI Specification is somewhat different from the use by the AISC. The variable factor of safety given by Eq. 9.45 is applied only to the sections with $Q = 1.0$, that is, only in cases when the plate elements are not expected to buckle (and $KL/r < C_c$). When $Q < 1.0$, the factor of safety should be F.S. = 1.92. The average axial stress in columns not subject to flexural-torsional buckling is thus given by the following formulas, where appropriate values of Q and F.S. should be used.

For

$$\frac{KL}{r} < \frac{C_c}{\sqrt{Q}} = \sqrt{\frac{2\pi^2 E}{QF_y}} \tag{12.42}$$

$$F_a = \frac{QF_y}{(\text{F.S.})}\left[1 - \frac{QF_y}{4\pi^2 E}\left(\frac{KL}{r}\right)^2\right] \tag{12.43}$$

and for $KL/r \geqslant C_c/\sqrt{Q}$

$$F_a = \frac{\pi^2 E}{(\text{F.S.})\left(\dfrac{KL}{r}\right)^2} \tag{12.44}$$

The radius of gyration r is computed on the basis of a full cross section.

The substitution of F.S. = 1.92 and $E = 29.5 \times 10^3$ ksi in Eqs. 12.43 and 12.44 results in the formulas of AISI Section [3.6.1.1].

The design of a column is illustrated by Example 12.5.

2. Compression Members Subject to Torsional-Flexural Buckling

Often cold-formed compression members are susceptible not only to lateral flexural buckling (discussed in the preceding section), but also to torsional buckling (see the definition for P_T after Eq. 12.63). The lowest load then controls the design. For members with unsymmetrical or singly-symmetrical sections, torsional buckling combines with lateral buckling and the phenomenon is quite similar to the lateral-torsional buckling of beams discussed in Chapter 7. When referred to columns in this chapter, this mode is called torsional-flexural buckling. A general analysis is very complex and in Art. 12.7 only the elastic solutions for simply supported members are presented, Eq. 12.62 for an unsymmetrical section and Eq. 12.63 for a singly symmetrical section subjected to an eccentric load. The buckling load is obtained as the smallest root of a cubic (Eq. 12.62) or quadratic (Eq. 12.63) equation. For engineering application these equations must be modified to account for other than simple support end conditions, for stresses in the inelastic range, and for the possibility of local buckling.

An adequately accurate approximation is obtained for other end conditions by using KL for L in the equations for P_{ex}, P_{ey}, and P_T following Eq. 12.63. K is here the effective length factor equal to 1.0 for $u = v = \phi = M_1 = M_2 = M_T = 0$ at both ends (both ends simply supported), or to 0.7 for $u = v = \phi = M_1 = M_2 = M_T = 0$ at one end and $u = v = \phi = du/dz = dv/dz = d\phi/dz = 0$ at the other end (one end simple, the other fixed) or to 0.5 for $u = v = \phi = du/dz = dv/dz = d\phi/dz = 0$ at both ends (both ends fixed).

EXAMPLE 12.5

PROBLEM:

Given a channel as shown to be used as a 15-ft wall stud (F_y = 45 ksi, F_b = 45/1.67 = 27.0 ksi) compute the column capacity for buckling about x-axis (wall material prevents the column from buckling about the y-axis). Assume square corners.

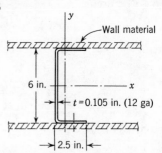

6 in.

t = 0.105 in. (12 ga)

2.5 in.

SOLUTION:

Check whether flanges have adequate rigidity as edge stiffeners of the web. Required (Eq. 12.13):

$$I_{min} = 1.83 \times 0.105^4 \sqrt{\left(\frac{5.8}{0.105}\right)^2 - \frac{4000}{45}} = 0.012 \text{ in.}^4$$

and I_{min} = $9.2 t^4$ = $9.2 (0.105)^4$ = 0.00112 in.⁴

$$\text{Available } I = \frac{2.5^3 \times 0.105}{12} = 0.137 \text{ in.}^4 \quad \begin{array}{l} > 0.012 \text{ in.}^4 \\ > 0.0011 \text{ in.}^4 \end{array} \quad \underline{\text{O.K.}}$$

Cross-sectional properties: (assuming square corners)

$$A = 0.105 \left[2 \times 2.5 + (6 - 2 \times 0.105) \right] = 1.132 \text{ in.}^2$$

$$I_x = 2 \times 2.5 \left(3 - \frac{0.105}{2} \right)^2 \times 0.105 + \frac{5.79^3 \times 0.105}{12} = 6.26 \text{ in.}^4$$

$$r_x = \sqrt{6.26/1.132} = 2.35 \text{ in.}$$

K = 1.0 for pin-ended column

$$KL/r = 1.0 \frac{15 \times 12}{2.35} = 76.6$$

COLUMN FACTOR Q

Unstiffened elements: Flange, $b/t = \dfrac{2.5 - 0.105}{0.105} = 22.8 \begin{cases} > \dfrac{144}{\sqrt{45}} = 21.5 \\ < 25 \end{cases}$

Use Eq. 12.6

$$F_c = \frac{8000}{22.8^2} = 15.4 \text{ ksi}$$

Eq. 12.38

$$Q_s = \frac{F_c}{F_b} = \frac{1.67 \times 15.4}{45} = 0.57$$

Stiffened elements: Web, $\dfrac{b}{t} = \dfrac{6 - 2 \times 0.105}{0.105} = 55.2$

Use Eq. 12.8

$$b_e = 0.105 \times 1.9 \sqrt{\frac{29.5 \times 10^3}{15.4 \times 1.67}} \left(1 - 0.415 \times \frac{1}{55.2} \sqrt{\frac{29.5 \times 10^3}{15.4 \times 1.67}} \right)$$

$$= 5.04 \text{ in.}$$

$$A_{eff} = 0.105 (2 \times 2.5 + 5.04) = 1.056 \text{ in.}^2$$

Use Eq. 12.39

$$Q_a = A_{eff}/A = \frac{1.056}{1.132} = 0.933$$

Use Eq. 12.41

$$Q = Q_s Q_a = 0.57 \times 0.933 = 0.532$$

EXAMPLE 12.5 (Continued)

ALLOWABLE COLUMN STRESS:

Limiting KL/r

Eq. 12.42 $KL/r = \sqrt{\dfrac{2\pi^2 E}{QF_y}} = \pi\sqrt{\dfrac{2 \times 29.5 \times 10^3}{0.532 \times 45}} = 156 > 76.6$

$\therefore$ Use Eq. 12.43

$$F_a = \frac{QF_y}{F.S.}\left(1 - \frac{QF_y}{4\pi^2 E}\left(\frac{KL}{r}\right)^2\right)$$

$$= \frac{0.532 \times 45}{1.95}\left(1 - \frac{0.532 \times 45}{4\pi^2 \times 29.5 \times 10^3} \times 76.6^2\right) = 10.8 \text{ ksi}$$

Column capacity

$P = F_a A = 10.8 \times 1.132 = 12.2$ kip

Column capacity $P = 12.2$ kip

It should be noted that a member may have different end support conditions for buckling about the x', y', or z axis and thus should have different values for K for P_{ex}, P_{ey} or P_T.*

The parabolic transition curve used for other buckling problems has been also recommended for this case to account for the inelastic range.

Considering only singly symmetrical sections subjected to concentric compression ($e_y = 0$), the critical stress for torsional-flexural buckling is obtained from Eq. 12.63.

$$f_{tf} = \frac{1}{2[1 - (y_s'/r_0)^2]}$$

$$\times \; [(P_{ey} - P_T) - \sqrt{(P_{ey} + P_T)^2 + 4[1 - (y_s'/r_0)^2] \times P_{ey}P_T}] \quad (12.45)$$

The design equations then become

For $f_{tf} > 0.5F_y$

$$F_{atf} = \frac{F_y}{(F.S.)}\left[1 - \frac{F_y}{4f_{tf}}\right] \quad (12.46)$$

and for $f_{tf} \leqslant 0.5F_y$

$$F_{atf} = \frac{f_{tf}}{(F.S.)} \quad (12.47)$$

Although the problem of considering the effect of local buckling upon torsional-flexural buckling has not yet been solved (1973), the effect can be taken into account conservatively by replacing F_y by QF_y in Eqs. 12.46 and 12.47. The formulas of AISI Section [3.6.1.2] can be obtained from Eqs.

* More accurate values of K can be found in Ref. 7.3 or Ref. 7.15.

12.45, 12.46, and 12.47 by incorporating F.S. = 1.92 and using a somewhat different notation.

The analysis of thin-walled members for torsional-flexural buckling is complicated by the time-consuming computations of the cross-sectional properties. To simplify the process, many tables and charts have been developed for the most typical sections, for example: Refs. 12.16 to 12.20.

3. Wall Studs

When columns are used as studs to support wall sheeting (wall board panels and the like) besides carrying axial loads, it is appropriate to take advantage of the strengthening effect of the sheeting. There are two ways in which the sheeting is beneficial. It contributes to the effective area of the cross section and thus may increase the capacity of the column for buckling in the direction perpendicular to the wall surface. The degree of strengthening is, however, rather small and uncertain and this effect is disregarded conventionally. Of much greater importance is the bracing restraint of the sheeting for buckling in the direction of the wall surface. With an adequate restraint in this direction, a column may be designed for stresses based only on the slenderness ratio in the direction perpendicular to the wall surface even if the column (by itself) is quite weak parallel to the wall.

Being attached to the sheeting, the stud behaves as a column elastically restrained in the plane of the wall. The restraint may be continuous or concentrated at a series of discrete points of fasteners (bolts, clips, screws, nails).

A design procedure should be concerned with the adequacy of the following items:

1. Strength and rigidity of the wall material.
2. Spacing of fasteners. It should be such that the column would not buckle in the wall direction even if one of the fasteners is fully ineffective.
3. The strength of individual fasteners to transmit restraining forces from the wall sheeting to the stud.

Wall studs faced with wall material on both sides have been investigated both experimentally and theoretically, as described in Ref. 12.16. The research resulted in the design rules of AISI Section [5.1]. The appendix in Ref. 1.24 outlines a test procedure for the determination of the restraint characteristics of wall sheeting and also lists values for some typical materials.

When only one flange of the stud is laterally supported by the wall sheeting, the allowable stress may be controlled by the lateral buckling of the unsupported flange. This stress can be determined by the procedure applicable to the buckling of unbraced compression flanges in beams (Art. 12.3).

12.5 COMBINED BENDING AND COMPRESSION

The behavior of rolled members subjected to a combined action of bending and axial compression is discussed in detail in Art. 11.4. The design of doubly symmetrical cold-formed members is based on the same principles, and the AISC interaction equations (Eqs. 11.26 and 11.27) may be used with some modifications to account for the effect of local buckling (AISI Section [3.7.1]). Both equations, Eq. 12.48 and 12.49, must be satisfied.

$$\frac{f_a}{F_a} + \frac{C_{mx}f_{bx}}{\left(1 - \dfrac{f_a}{F_{ex}'}\right)F_{bx}} + \frac{C_{my}f_{by}}{\left(1 - \dfrac{f_a}{F_{ey}'}\right)F_{by}} \leqslant 1.0 \qquad (12.48)$$

where the subscripts x and y designate quantities pertaining to bending about x' and y' principal axes.

F_a, f_b, and F_b are defined as in Eqs. 11.24 and 11.25, but computed with due consideration of local and lateral buckling, as discussed in the preceding articles of this chapter.

f_a and F_e' ($F_e' = F_a$ in Eq. 12.44) are based on the full cross section.

$C_m = 0.85$, except: (1) when $f_a/F_a \leqslant 0.15$, Eq. 12.48 is replaced with $f_a/F_a + f_{bx}/F_{bx} + f_{by}/F_{by} \leqslant 1.0$; (2) for members without transverse loading C_m may be computed from $0.4 \leqslant C_m = 0.6 + 0.4\beta$ as explained in Art. 11.4; (3) for members with transverse loading, a rational analysis may be used for the computation of C_m—for example, see the Commentary of Ref. 1.21.

$$\frac{f_a}{0.522QF_y} + \frac{f_{bx}}{F_{bx}} + \frac{f_{by}}{F_{by}} \leqslant 1.0 \qquad (12.49)$$

In Eq. 12.49, F_b is computed considering only local buckling (no lateral buckling) and the factor 0.522 was obtained from $1/(\text{F.S.} = 1.92)$.

Members with singly symmetrical cross section may buckle in the torsional-flexural mode when subjected to both axial compression and bending in the plane of symmetry. AISI Section [3.7.2] gives relationships analogous to Eqs. 12.48 and 12.49 which take into consideration the possibility of torsional-flexural buckling under different positions of the eccentricity of the compression force.

12.6 CONNECTIONS

Bolted and welded connections for rolled sections described in Chapters 18 and 19 are used for light gage members with some additional considerations brought about by the thinness of the material. Some special types of connection such as spot (resistance), plug, and puddle welds, self-tapping screws, stitching, and special rivets have also found wide application.

Bolting. The shear strength of finished and high-tension bolts is determined in accordance with the AISC Specification[1.21] (see also Chapter 18). It has been shown by research that the following additional rules should be used in the design of light gage bolted connections.[1.19,12.22]

For the edge distance in the line of stress less than about 3.5 times the bolt diameter d, the strength should be checked against the shear-out of the sheet at the sides of the bolt. The limiting edge distance is then, with F.S. ≈ 2.3,

$$e = \frac{2.3P}{0.7F_y 2t} \approx \frac{P}{0.6F_y t} \tag{12.50}$$

where P is the force transmitted by the bolt, $0.7F_y$ is the shearing strength of the material, and e is the edge distance.

For e greater than $3.5d$, bearing stresses greater than $4.8F_y$ have been found to cause cutting of the sheet. Taking F.S. ≈ 2.3 the allowable bearing force is given by

$$P \leqslant \frac{4.8}{2.3} F_y td \approx 2.1 F_y td \tag{12.51}$$

The tensile force on the net section at which tearing occurs depends on the ultimate stress of the sheet material F_u, relative spacing of the bolts (d/s) perpendicular to the line of force, and the ratio (r) of the force transmitted by the bolt (or bolts) to the tension force in the member at that section (AISI Section [4.5.2]). The allowable tension stress F_t can be determined as follows:

$$0.6F_y \geqslant F_t \geqslant \left(0.1 + 3\frac{d}{s}\right)\frac{F_u}{1.35} \tag{12.52}$$

For medium strength steels $F_u \approx 1.35F_y$.

Self-Tapping Screws. The design values for self-tapping screws usually are furnished by the manufacturer. As an approximate guide, the design procedures applicable to finished bolts may be followed.

Welding. Fusion welding of cold-formed members is designed by the same methods as the welding of hot-rolled members (see Chapter 19).

Resistance welding, especially spot welding, is of much greater importance for thin material. This method is particularly efficient in the fabrication of built-up sections, such as an I-section formed from two channels. The process consists of clamping the sheets to be welded between two electrodes. Strong current is passed through the electrodes, and the heat resulting from the resistance of the contact surfaces melts the metal and fuses the sheets together in the spot under the electrodes. The strength of such spot welds depends mainly on the thickness of the thinnest plate connected. The allowable values with a factor of safety of about 2.5 are tabulated in AISI Section [4.2.2].

The connection of cold-formed members to heavy sections is often accomplished by means of "puddle" welds. A hole is burned through the thin sheet to the face of the adjoining member and filled with the weld metal, all in one operation. The design of these welds is based on the same principles as that of the plug welds, that is, on the shearing strength around the periphery of the fused zone.

Spacing of Stitch Fasteners.   The spacing of non-staggered stitch fasteners (spot welds, bolts, screws) in cover plates is determined to prevent local buckling of the plate between the fasteners as outlined in Art. 17.5. The formula is

$$s = 200 \frac{t}{\sqrt{F_c}} \tag{12.53}$$

where F_c is the allowable compressive stress for the plate in ksi. AISI Section [4.4] gives two additional limitations for s based on fastener strength and the buckling between the longitudinal rows of fasteners.[1.19]

12.7   MEMBERS WITH THIN-WALLED OPEN CROSS SECTIONS

Design procedures presented in the preceding articles are based on conventional methods of stress and buckling analysis and are adequate for the majority of practical structures. In cases of slender members of open cross section subjected to general loading, a more rigorous stress and buckling analysis may be desirable. References 7.3, 7.6, 12.23, 7.15, or 7.17 can be consulted for this purpose.

Below is given a brief summary of the relevant formulas and definitions. They are valid only if no local buckling takes place. Therefore, stresses in flat plate elements should be checked against plate buckling stresses.

1. Stresses in Thin-Walled Open Sections

A thin-walled open cross section is shown in Fig. 12.12; x' and y' are the principal coordinate axes, C is the centroid, and S.C. is the shear center. Normal stress at point i with coordinates (x', y') can be computed from the following general formula, which includes the effect of warping torsion:

$$f = \frac{P}{A} - \frac{M_2}{I_2} x' + \frac{M_1}{I_1} y' + \frac{M_w}{I_w} w \tag{12.54}$$

The shearing stress at point i in the wall is given by another general formula,

$$f_v = \frac{V_2 Q_{1i}}{I_1 t_i} + \frac{V_1 Q_{2i}}{I_2 t_i} + \frac{M_z{}^w Q_{wi}}{I_w t_i} + \frac{M_z{}^T t_i}{K_T} \tag{12.55}$$

With $dA = t \, ds$, the symbols of the two foregoing equations have the interpretation which follows.

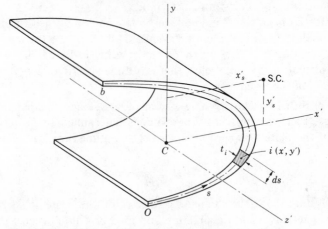

Fig. 12.12 Thin-walled member with open cross section.

Section Properties:

$$A = \int_0^b dA$$

[in.²] Area of the cross section

$$I_1 = \int_0^b (y')^2 \, dA$$

[in.⁴] Moment of inertia about x'-axis

$$I_2 = \int_0^b (x')^2 \, dA$$

[in.⁴] Moment of inertia about y'-axis

$$I_w = \int_0^b w^2 \, dA$$

[in.⁶] Warping torsion constant

$$w$$

[in.²] Warping coordinate defined below

$$Q_{1i} = \int_0^i y' \, dA = \int_i^b y' \, dA$$

[in.³] First moment of area about x'-axis

$$Q_{2i} = \int_0^i x' \, dA = \int_i^b x' \, dA$$

[in.³] First moment of area about y'-axis

$$Q_{wi} = \int_0^i w \, dA = \int_i^b w \, dA$$

[in.⁴] "Warping first moment of area"

$$t_i$$

[in.] Wall thickness at point i

$$K_T$$

[in.⁴] St. Venant torsion constant, for thin-walled open sections $K_T = \frac{1}{3} \int_0^b t^3 \, ds$

Forces:

$$P = \int_0^b f \, dA \qquad\qquad \text{[lb]} \qquad \text{Axial load}$$

$$M_1 = \int_0^b y'f \, dA = -EI_1 \frac{d^2v}{dz^2} \qquad \text{[lb-in.]} \qquad \begin{array}{l}\text{Bending moment} \\ \quad\text{about } x'\text{-axis}\end{array}$$

$$M_2 = \int_0^b x'f \, dA = -EI_2 \frac{d^2u}{dz^2} \qquad \text{[lb-in.]} \qquad \begin{array}{l}\text{Bending moment} \\ \quad\text{about } y'\text{-axis}\end{array}$$

$$M_w = \int_0^b wf \, dA = -EI_w \frac{d^2\phi}{dz^2} \qquad \text{[lb-in.}^2\text{]} \qquad \begin{array}{l}\text{"Warping bending} \\ \quad\text{moment"}\end{array}$$

$$V_1 = \int_0^b f_v t \, dx' = \frac{dM_2}{dz} = -EI_2 \frac{d^3u}{dz^3} \qquad \text{[lb]} \qquad \begin{array}{l}\text{Transverse shear in} \\ \quad\text{direction of } x'\text{-axis}\end{array}$$

$$V_2 = \int_0^b f_v t \, dy' = \frac{dM_1}{dz} = -EI_1 \frac{d^3v}{dz^3} \qquad \text{[lb]} \qquad \begin{array}{l}\text{Transverse shear in} \\ \quad\text{direction of } y'\text{-axis}\end{array}$$

$$M_z{}^w = \int_0^b f_v t \, dw = \frac{dM_w}{dz} = -EI_w \frac{d^3\phi}{dz^3} \qquad \text{[lb-in.]} \qquad \begin{array}{l}\text{"Warping twisting} \\ \quad\text{moment" (twisting} \\ \quad\text{moment carried by} \\ \quad\text{warping torsion)}\end{array}$$

$$M_z{}^T = \int_0^b f_v r \, dA = GK_T \frac{d\phi}{dz} \qquad \text{[lb-in.]} \qquad \begin{array}{l}\text{St. Venant twisting} \\ \quad\text{moment (twisting} \\ \quad\text{moment carried by} \\ \quad\text{pure torsion)}\end{array}$$

Note: $M_z{}^w + M_z{}^T = M_z = $ total twisting moment about the shear center axis (torque).

The warping coordinate w is defined as a function of the coordinate s by

$$w = \frac{1}{A} \int_0^b w_s \, dA - w_s \qquad (12.56)$$

where

$$w_s = \int_0^i r \, ds \qquad (12.57)$$

and r is positive when the tangent at the point (directed in positive s) produces counterclockwise moment about the shear center (in Fig. 12.13, r is negative).

The coordinates of the shear center, x_s' and y_s', can be computed from

$$x_s' = \frac{1}{I_1} \int_0^b w_c y' \, dA \qquad (12.58)$$

and

$$y_s' = \frac{1}{I_2} \int_0^b w_c x' \, dA \qquad (12.59)$$

where w_c is computed by using Eq. 12.57, except that r should be measured from the centroid instead of the shear center. Determination of the values of M_w, $M_z{}^w$, and $M_z{}^T$ is accomplished by solving the following differential equation:

$$EI_w \frac{d^4\phi}{dz^4} - GK_T \frac{d^2\phi}{dz^2} = m_z \tag{12.60}$$

where $m_z = -dM_z/dz$. With $\lambda = \sqrt{GK_T/EI_w}$ the solution is

$$\phi = C_1 + C_2 z + C_3 \sinh \lambda z + C_4 \cosh \lambda z + \phi_{\text{particular}} \tag{12.61}$$

C_1 to C_4, the integration constants, are found from the boundary conditions.

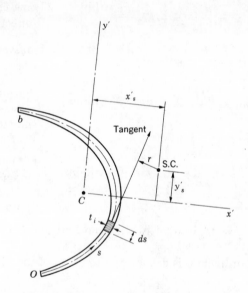

Fig. 12.13 Open cross section.

Derivatives of ϕ are then used to compute M_w, $M_z{}^w$, and $M_z{}^T$ according to the above-listed formulas.

2. Buckling of Columns Having Thin-Walled Open Cross Section

A pinned-end column* with a general thin-walled open cross section will buckle under a concentric axial load P_{cr} when this load takes the value of the

---

* "Pinned-end" in this case means that, at the ends, $u = v = \phi = d^2u/dz^2 = d^2v/dz^2 = d^2\phi/dz^2 = 0$, and thus, the ends are free to warp.

smallest root P of the following cubic equation:

$$\frac{I_1 + I_2}{r_0^2 A} P^3 + \left[\frac{1}{r_0^2}\{(x_s')^2 P_{ey} + (y_s')^2 P_{ex}\} - \{P_{ex} + P_{ey} + P_T\}\right]P^2$$

$$+ (P_{ex}P_{ey} + P_{ex}P_T + P_{ey}P_T)P - P_{ex}P_{ey}P_T = 0 \quad (12.62)$$

When such a column has one axis of symmetry (for example, y'-axis) and the load is applied eccentrically in that plane ($e_y \neq 0$), the buckling load is found as the smallest root of Eq. 12.63 (see Fig. 12.14).

$$(P_{ey} - P)[r_0^2(P_T - P) - e_y\beta_y P] - P^2(y_s' - e_y)^2 = 0 \quad (12.63)$$

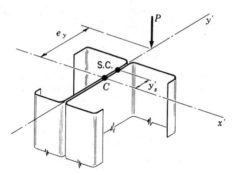

Fig. 12.14 Eccentrically loaded singly symmetrical column.

The additional notation in Eqs. 12.62 and 12.63 is

$$P_{ex} = \frac{\pi^2 EI_1}{L^2} \qquad \text{[lb]} \quad \text{Euler buckling load about } x'\text{-axis}$$

$$P_{ey} = \frac{\pi^2 EI_2}{L^2} \qquad \text{[lb]} \quad \text{Euler buckling load about } y'\text{-axis}$$

$$P_T = \frac{1}{r_0^2}\left(GK_T + EI_w \frac{\pi^2}{L^2}\right) \qquad \text{[lb]} \quad \text{Torsional buckling load about } z\text{-axis}$$

$$r_0^2 = (x_s')^2 + (y_s')^2 + \frac{I_1 + I_2}{A} \qquad \text{[in.}^2\text{]}$$

$$\beta_y = \frac{1}{I_1}\int_0^b y'\{(x')^2 + (y')^2\}\,dA - 2y_s' \qquad \text{[in.]}$$

e_y = eccentricity of the load with respect to x'-axis.   [in.]

The load should be also checked against the load producing flexural yielding. The one which is smaller will be controlling the design.

PROBLEMS

12.1. Determine the resisting moment of the cross section shown; $F_y = 40$ ksi.

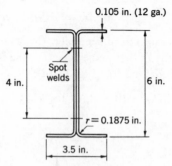

Problem 12.1

12.2. Assuming adequate bracing to prevent lateral buckling, compute the intensity of distributed loading w which can be supported safely by a 10-ft beam with the cross section shown in the sketch. Also compute the resulting mid-span deflection; $F_y = 50$ ksi.

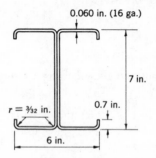

Problem 12.2

12.3. Investigate the capacity and deflection of the beam of Example 12.2 if the cross section is changed to include a fold in the middle of the flat portion as an intermediate stiffener.

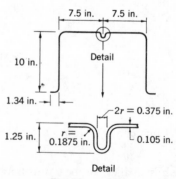

Problem 12.3

12.4. A beam with the cross section and material of Problem 12.1 has a span of 12 ft.

(a) Determine the spacing of bracing against lateral buckling required to develop the maximum capacity of the beam.

(b) What will be the allowable distributed loading if the bracing is located at the third points only?

12.5. Determine the maximum longitudinal spacing of spot welds in the beam of Problem 12.4(b) if they are located as shown in the sketch for Problem 12.1.

12.6. A 10-ft-long beam having a hat cross section and material as in Example 12.2, but with the flat portion for the bottom (tension) flange, is subjected to equal concentrated loads at the quarter-points. Assuming that the top flange has adequate lateral bracing to prevent buckling, compute the allowable value of the concentrated loads. Shear lag, flange curling, and web strength should be considered.

12.7. Check web buckling and crippling for the beam of Problem 12.2.

12.8. A 14-ft column with the cross section as in Problem 12.2 is to be investigated for its load-carrying capacity; $F_y = 45$ ksi.

12.9. If the beam of Problem 12.4 is subjected to the axial load $P = 6000$ lb, what will be the permissible lateral loading w?

12.10. A 16-ft beam of cross section as shown carries five equally spaced concentrated loads; $F_y = 40$ ksi. Determine permissible loads and perform a complete check of the design, including spacing of the spot welds in the cover plate. Note that spot welds should be designed for the channel bracing forces acting together with the forces due to shear.

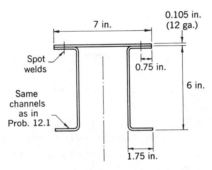

Problem 12.10

12.11. A 16-ft beam is to carry 0.45 kip/ft. Use $F_y = 50$ ksi and select an appropriate lipped I-section from Table 5 of the AISI Manual. The bracing is located at the quarter-points.

12.12. Select a hat cross section from Table 9 of the AISI Manual to serve as a 12-ft column supporting an axial load $P = 16$ kips; $F_y = 36$ ksi.

COMPOSITE STEEL-CONCRETE MEMBERS

13.1 INTRODUCTION AND HISTORICAL REVIEW

Bridges and buildings were designed for many years with each component having only one primary function. Gradually a structure has come to be viewed as a composite system. As high-strength steels, having the same elastic properties as structural carbon steel, became available, it became even more necessary to combine the parts of a structure more effectively to limit deflections. One of the vital steps in this trend was the realization that concrete slabs could be connected to steel beams to produce a composite floor system.

The first appreciation of this type of composite construction probably originated from observations of bridges in service. It was noticed that the deflection of decks was considerably less than the values predicted by considering the steel beams alone. Considerable investigation went into the effort of evaluating the composite action of members in which the only shear connection was the bond between concrete and steel, and allowable bond stresses for these members were developed. Investigations by the Truscon Steel Company in the United States in 1921, by the National Physical Laboratory in England beginning about 1922, and tests by the Dominion Bridge Company of Canada in 1922 were among the first of such investigations.[13.1] These early investigations considered steel beams which were either partially or completely encased in concrete. In the early 1930's, investigations of members in which various types of mechanical shear connectors were used to supplement bond in transferring shear were begun in

Switzerland, and were followed by other investigations in Europe and the United States.[13.1,13.2]

The use of composite construction lagged far behind experimental knowledge and was used only in a few isolated projects prior to 1940. Progress in construction was hampered by the lack of design specifications on the one hand and the lack of suitable economical shear connectors on the other. With the development of confidence in welding techniques, greater prospects of over-all economy in composite construction developed. Thus, in the early 1940's composite construction began to be more widely considered and accepted. The real impetus to composite construction in the United States came with the adoption of the 1944 AASHO Specifications.[13.3] In spite of the somewhat limited design provisions of these specifications, composite design rapidly became accepted for highway bridges.

The development of composite building designs followed somewhat later, and the first building design provisions were not adopted in the United States until the 1952 edition of the AISC Specification was issued.[13.4] In buildings, composite construction was considered at first only where heavy floor loads were encountered. General use in buildings came about with the search for greater economy in all types of steel construction which preceded the formulation of the 1961 edition of the AISC Specification.

In general the advantages of composite construction include a saving in the weight of steel required, a reduction in the depth of members, increased stiffness of the floor system, and an increase in the overload capacity. The relative economy of composite construction is often enhanced by the use of structural lightweight concrete and metal decking. Composite construction becomes particularly economical where the framing plan requires long spans that cannot be achieved satisfactorily with non-composite members.

13.2 BEAM THEORY FOR COMPOSITE BEAMS

Composite floor systems may include both the concrete fireproofing on the steel beams as well as the concrete floor slab as part of the structural member. If steel beams are encased, shear between concrete and steel normally is transmitted entirely by bond and friction. When only a concrete slab exists, shear is transmitted only by properly designed mechanical shear connectors. The type of shear connection is not pertinent to the design of the cross section. It is assumed here that the shear connection will be adequate for transmitting the required shear forces.

A composite beam may be considered as a steel member to which a cover plate has been added on the top flange. Since this cover plate is concrete rather than steel, it is effective only when the top flange is in compression. In the case of continuous beams, only the longitudinal reinforcement in the concrete slab may be used in computing section properties throughout the negative moment region.

Members are proportioned by elastic analysis, using simple beam theory and assuming that the cover plate is fully effective and that plane sections before bending remain plane after bending. As in the case of reinforced concrete design by the transformed area method, it is necessary to transform the cross-sectional area of one material into an equivalent area of the other. Unlike reinforced concrete design, it is usually customary to design in terms of steel area instead of concrete area. Therefore, the concrete area is reduced by dividing by the factor n, where n is defined as the modulus of elasticity of steel divided by the modulus of elasticity of concrete, or

$$n = \frac{E_s}{E_c} \tag{13.1}$$

The value of n to be used in design usually is given in the applicable building code or specification. The application of this method in determining the section properties of a composite beam is illustrated in Example 13.1. The notation required for the design of composite members becomes more involved than for non-composite members. The notation given in Table 13.1 will be used throughout this chapter. The cross section of Example 13.1 will be used in the bridge member of Example 13.2.

Since the neutral axis of the member in Example 13.1 falls below the bottom of the concrete slab, the entire concrete slab is in compression and therefore is included in computing section properties. If the neutral axis of a composite section falls within the concrete slab, only the portion of the concrete slab above the neutral axis should be included in the calculation of section properties. The area of slab reinforcing steel is neglected in the calculations. This is normal design practice and is conservative.

The effective width or the width of the slab that may be considered to function as part of the composite section, must be determined. Theoretically, the effective width of a T-beam is a function of (1) the span of the member, (2) Poisson's ratio for the material, and (3) the shape of the moment diagram.[13.5] The effective width to be used in determining section properties of a beam usually is controlled by the building code or specification. Thus, simple beam theory is extended to composite steel and concrete members in the same manner as it is extended to the reinforced concrete tee beams.

13.3 INFLUENCE OF CONSTRUCTION METHODS

The same allowable stresses which apply to the design of non-composite steel members also apply to the design of composite members. However, the method of construction alters the actual stresses in members, and these

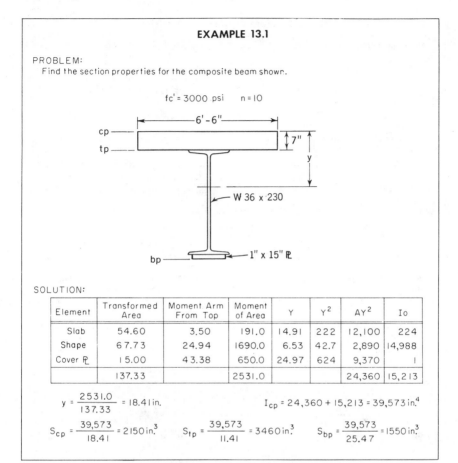

EXAMPLE 13.1

PROBLEM:
Find the section properties for the composite beam shown.

$f_c' = 3000$ psi $n = 10$

SOLUTION:

Element	Transformed Area	Moment Arm From Top	Moment of Area	Y	Y²	AY²	Io
Slab	54.60	3.50	191.0	14.91	222	12,100	224
Shape	67.73	24.94	1690.0	6.53	42.7	2,890	14,988
Cover ℟	15.00	43.38	650.0	24.97	624	9,370	1
	137.33		2531.0			24,360	15,213

$$y = \frac{2531.0}{137.33} = 18.41 \text{ in.} \qquad I_{cp} = 24,360 + 15,213 = 39,573 \text{ in.}^4$$

$$S_{cp} = \frac{39,573}{18.41} = 2150 \text{ in.}^3 \qquad S_{tp} = \frac{39,573}{11.41} = 3460 \text{ in.}^3 \qquad S_{bp} = \frac{39,573}{25.47} = 1550 \text{ in.}^3$$

differences must be allowed for in the design. In the case of members designed for loading where fatigue is a consideration, the range of stress is the important consideration in the design. In specifications for members designed for static loading, the factor of safety against ultimate load rather than range of stress is the more important criterion of design, and in this case the design stresses may be higher than in non-composite design, due to an allowance for the greater ultimate strength of composite members.

During casting of the concrete deck, the steel beams may be shored to support the dead weight of the slab. In this type of construction, the shores usually are left in place until the concrete reaches 75 per cent of the 28-day compressive strength. Upon removal of the shores, the dead load of the entire composite member is supported by the composite cross section.

The calculation of actual stresses is complicated by the fact that dead load

Table 13.1 Notation for Composite Design

Steel Section—No Cover Plate	Steel Section—With Cover Plate
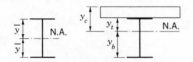	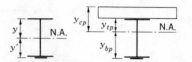
Steel section moment of inertia $= I$	Steel section moment of inertia $= I_p$
Composite section moment of inertia $= I_c$	Composite section moment of inertia $= I_{cp}$

$$n = E_s/E_c = \text{elastic modular ratio}$$
$$n' = E_s/E_c' = \text{plastic modular ratio}$$

When plastic modular ratio is used all properties above are referred to as y_c', I_c', y_{cp}', I_{cp}', etc.

Section Modulus for Calculation of Stresses
Loading To Be Considered

Member	Unshored Dead Load	Live Load	Dead Load on Composite Section	Location of Stress
No cover plate	—	$S_c = \dfrac{I_c}{y_c}$	$S_c' = \dfrac{I_c'}{y_c'}$	Top of concrete
	$S = \dfrac{I}{\bar{y}}$	$S_t = \dfrac{I_c}{y_t}$	$S_t' = \dfrac{I_c'}{y_t'}$	Top of steel
	$S = \dfrac{I}{\bar{y}}$	$S_b = \dfrac{I_c}{y_b}$	$S_b' = \dfrac{I_c'}{y_b'}$	Bottom of steel
With cover plate	—	$S_{cp} = \dfrac{I_{cp}}{y_{cp}}$	$S_{cp}' = \dfrac{I_{cp}'}{y_{cp}'}$	Top of concrete
	$S_{st} = \dfrac{I_p}{y}$	$S_{tp} = \dfrac{I_{cp}}{y_{tp}}$	$S_{tp}' = \dfrac{I_{cp}'}{y_{tp}'}$	Top of steel
	$S_{sb} = \dfrac{I_p}{y'}$	$S_{bp} = \dfrac{I_{cp}}{y_{bp}}$	$S_{bp}' = \dfrac{I_{cp}'}{y_{bp}'}$	Bottom of steel

stresses do not remain constant with time because of the net effect of creep and shrinkage of the concrete. With time, the effectiveness of the concrete slab decreases and the steel stresses due to dead load increase as a result. For this reason, stresses caused by dead load on the composite section usually are determined with section properties calculated using a value of n equal to about 3 times the elastic value of n.[13.6]

Usually beams are not shored and the dead load is then carried by the steel members alone. Stresses due to dead load are calculated using section properties of the steel member alone.

The calculation of actual stresses for unshored construction requires the definition of several sets of section properties for the calculation of bending stress.

Using Table 13.1, it is possible to give equations for the calculation of stresses for both types of construction. M_D, M_L, and M_d are, respectively, moment due to dead load on the steel member, moment due to live load, and moment due to dead load on the composite member. The stresses f_c, f_t, and f_b are the stresses at the top of the concrete slab, top of the steel member, and bottom of the steel member, respectively.

The stress in the top fiber of the concrete slab for shored construction is given by

$$f_c = \frac{M_d}{n'S_c'} = \frac{M_L}{nS_c} \quad \text{or} \quad f_c = \frac{M_d}{n'S_{cp}'} + \frac{M_L}{nS_{cp}} \tag{13.2}$$

The stress in the top fiber of the steel member is given by

$$f_t = \frac{M_d}{S't} + \frac{M_L}{S_t} \quad \text{or} \quad f_t = \frac{M_d}{S_{tp}'} + \frac{M_L}{S_{tp}} \tag{13.3}$$

The bottom fiber stress in the steel section is given by

$$f_b = \frac{M_d}{S_b'} + \frac{M_L}{S_b} \quad \text{or} \quad f_b = \frac{M_d}{S_{bp}'} + \frac{M_L}{S_{bp}} \tag{13.4}$$

Members constructed without shoring may have part of the dead load carried by the steel members and part by the composite section. In bridges, for example, the weight of the slab is carried by the steel beams alone, but the weight of curbs, railing, and wearing surface will be carried by the composite section. Therefore, three terms are required in the equations for calculation of f_t and f_b as given by

$$f_c = \frac{M_d}{n'S_c'} + \frac{M_L}{nS_c} \quad \text{or} \quad f_c = \frac{M_d}{n'S_{cp}'} + \frac{M_L}{nS_{cp}} \tag{13.5}$$

$$f_t = \frac{M_D}{S} + \frac{M_L}{S_t} + \frac{M_d}{S_t'} \quad \text{or} \quad f_t = \frac{M_D}{S_{st}} + \frac{M_L}{S_{tp}} + \frac{M_d}{S_{tp}'} \tag{13.6}$$

$$f_b = \frac{M_D}{S} + \frac{M_L}{S_b} + \frac{M_d}{S_b'} \quad \text{or} \quad f_b = \frac{M_D}{S_{sb}} + \frac{M_L}{S_{bp}} + \frac{M_d}{S_{bp}'} \tag{13.7}$$

Two equations are given for each stress calculation. The first equation of a pair pertains to the portion of a member without a cover plate and the second pertains to the portion of a member with a cover plate.

13.4 BRIDGE DESIGN

A typical composite bridge under construction is shown in Fig. 13.1. Bridges of this type are common throughout the United States and Canada.

Fig. 13.1 Curved plate girders.

Most of these bridges consist of simple spans, but continuous composite bridges are being built in increasing numbers.

Example 13.2 is presented primarily to illustrate the calculation of bending stresses by using Eqs. 13.2 through 13.7. This example presents a preliminary design of a composite bridge for a span of 85 feet center to center of bearings. The deck consists of five steel supporting members spaced 6 ft 6 in. center to center with a 7-in.-thick concrete slab. The deck is designed in accordance with the AASHO Specifications for HS20 loading.

The design of the concrete slab is included in the example to emphasize that it is designed in exactly the same manner as a concrete slab on a non-composite bridge. For this reason, the concrete slab costs exactly the same for both bridges and the only economy in the composite deck is in the steel members. However, the strain distribution in the slabs of the composite and non-composite decks in the longitudinal direction is very much different. The slab of the non-composite deck, although neglected in the design of the longitudinal members, actually is stressed as a member in bending with compression in the top fiber and tension in the bottom. Although these bending stresses are relatively small, the resulting tensile stress in combination with tensile stresses in the concrete due to shrinkage have been advanced as a reason for deterioration of concrete slabs on non-composite bridges. The composite deck slab, on the other hand, is entirely in compression and the cracks are therefore closed rather than opened by the passage of live loads.

EXAMPLE 13.2

PROBLEM:
Design a 85 ft. simple span Highway Bridge for HS 20 loading.

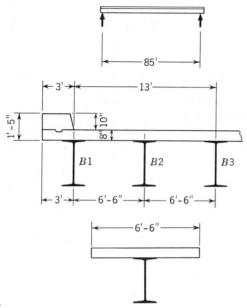

SOLUTION:

Dead load on steel beams

Slab 6.5(100) = 650 lb/ft
Beam   280 lb/ft
                930 lb/ft

Dead load on comp. sections

Wearing surface 25(6.5) = 163 lb/ft
Curb & Rail 450 ÷ 2.5 = 180 lb/ft
                       343 lb/ft

LOADING: MEMBER B2

7" Slab	=	88 psf
2" Asphalt	=	25 psf
Railing	=	75 lb/ft
L.L.	=	H20-S16
Curb	=	375 lb/ft

1" Conc. wearing surface = 12.5 psf

ROADWAY SLAB

Beam spacing = 6'-6" o.c.

S = 78" − 8.25" = 69.75 inches or 5.81 ft

$$M_L = \frac{S+2}{32}\,P_{20} = \frac{7.81}{32}(16) = 3.905 \text{ kip–ft per ft of slab}$$

For slabs continuous on three or more supports

$$M_L = 0.8(3.905)12 = 37.49 \text{ kip–in. per ft of slab}$$

$$M_D = \frac{1}{10}(0.088)(5.81)^2\,12 = 3.57 \text{ kip–in. per ft of slab}$$

$$M_I = 0.30(37.49) = 11.25 \text{ kip–in. per ft of slab}$$

$$M_{WS} = \frac{1}{10}(0.038)(5.81)^2\,12 = 1.54 \text{ kip–in. per ft of slab}$$

$$\text{total } M = 53.85 \text{ kip–in. per ft of slab}$$

For 7" Concrete slab d = 7.0 − 1.50 = 5.50 inches

EXAMPLE 13.2 (Continued)

Main Reinforcement

Try #5 bars @ 6" o.c.　　　　　　　$A_s = 0.62 \text{ in.}^2$ per ft of width

$$P = \frac{0.62}{12(5.5)} = 0.0094$$

$$K = \sqrt{2pn + (pn)^2} - pn$$

$$= \sqrt{2(0.0094)10 + (0.094)^2} - 0.094$$

$$j = 1 - \frac{K}{3}$$

$$= 0.350$$

$$= 1 - \frac{0.350}{3} = 0.883$$

$$f_s = \frac{M}{A_s j d} = \frac{53,850}{0.62(0.883)5.5} = 17,880 \text{ psi}$$

$$f_c = \frac{2pf_s}{K} = \frac{2(0.0094)17,880}{0.350} = 961 \text{ psi}$$

Longitudinal Reinforcement

$$\text{Percentage} = \frac{220}{\sqrt{S}} = \frac{220}{\sqrt{5.81}} = 91\% \qquad \text{Maximum} = 67\%$$

$$P = 0.67(0.0094) = 0.0063$$

Try #5 bars @ 8" o.c.

$$P = \frac{0.460}{12(5.5)} = 0.00697$$

Live Load Distribution:

$$\frac{S}{5.5} = \frac{6.5}{5.5} = 1.182$$

$$\text{Impact Factor } I = \frac{50}{125 + L} = \frac{50}{125 + 85} = 0.238$$

	Shear			Moment		
D.L.	0.930(42.5)	= 39.5 kip	$0.930(85)^2 1.5$	= 10110 kip-in.		
W.S	0.343(42.5)	= 14.6 kip	$0.343(85)^2 1.5$	= 3720 kip-in.		
L.L	$1.182(64.1)\frac{1}{2}$	= 37.9 kip	$1.182(1254.7)\frac{12}{2}$	= 8920 kip-in.		
I	37.9(0.238)	= 9.1 kip	8920(0.238)	= 2120 kip-in.		

Summary of Section Properties (See Table 13.1)

Section	Location	Stress Calculation					
		Initial D.L.		L.L. + I.		Comp D.L.	
no	top concrete			$S_c = 1899 \text{ in.}^3$		$S_c' = 1060 \text{ in.}^3$	
cover	top steel	$S =$	835.5 in.^3	$S_t = 3480 \text{ in.}^3$		$S_t' = 1612 \text{ in.}^3$	
plate	bottom steel	$S =$	835.5 in.^3	$S_b = 1057 \text{ in.}^3$		$S_b' = 963 \text{ in.}^3$	
with	top concrete			$S_{cp} = 2150 \text{ in.}^3$		$S_{cp}' = 1194 \text{ in.}^3$	
cover	top steel	$S_{st} =$	900.0 in.^3	$S_{tp} = 3460 \text{ in.}^3$		$S_{tp}' = 1690 \text{ in.}^3$	
plate	bottom steel	$S_{sb} =$	1230 in.^3	$S_{bp} = 1550 \text{ in.}^3$		$S_{bp}' = 1415 \text{ in.}^3$	

$$I = 14,988.4 \text{ in.}^4, \quad I_s = 19,165.4 \text{ in.}^4, \quad I_c = 29,111.9 \text{ in.}^4, \quad I_{cp} = 39,573 \text{ in.}^4$$

EXAMPLE 13.2 (Continued)

Maximum Bending Stress

	M_D	M_L	M_d
top concrete	—	$-\dfrac{11040}{2150(10)}$	$-\dfrac{3720}{1194(10)}$ = $-$ 0.825 ksi
top steel	$-\dfrac{10110}{900}$	$-\dfrac{11040}{3460}$	$-\dfrac{3720}{1690}$ = $-$ 16.74 ksi
bottom steel	$+\dfrac{10110}{1230}$	$+\dfrac{11040}{1550}$	$+\dfrac{3720}{1415}$ = $+$ 17.99 ksi

Length of Cover Plate

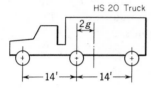

HS 20 Truck

$$g = \frac{1}{2}\left[\frac{14(32)+28(32)}{72}-14\right]$$

$$= 2.33 \, \text{ft}$$

$$L' = g + (L-g)\sqrt{1-\frac{S_b}{S_{bp}}}$$

$$= 2.33 + (82.67)\sqrt{1-\frac{1057}{1550}}$$

$$= 48.90 \, \text{ft}$$

Check stresses if cover plate is 48 ft.

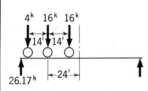

$$\frac{4.5}{85}(4) = 0.21$$

$$\frac{18.5}{85}(16) = 3.49$$

$$\frac{32.5}{85}(16) = \frac{6.13}{9.83 \, \text{kip}}$$

$M_{LL} = [26.17(18.5)-4(14)]12(1.182)$

     $= 6050 \, \text{kip-in.}$

$I \quad = 6090(0.238) = 1450 \, \text{kip-in.}$

$M_L = 7540 \, \text{kip-in.}$

$$M_D = \left(1-\frac{l^2}{L^2}\right) M_D \text{ maximum}$$

$$= \left(1-\frac{(48)^2}{(85)^2}\right)10,110 = 6890 \, \text{kip-in.}$$

$M_d = (0.681)3720 = 2530 \, \text{kip-in.}$

	M_D	M_L	M_d
Bottom flange stress without cover plate	$\dfrac{6900}{835.5}$ +	$\dfrac{7540}{1057}$ +	$\dfrac{2530}{963}$ = 18.02 ksi

Approximate Force in Cover plate at cut-off point

	M_D	M_L	M_d
Bottom flange stress in cover plate	$+\dfrac{6900}{1230}$ +	$\dfrac{7540}{1550}$ +	$\dfrac{2530}{1415}$ = + 12.26

     Force in Cover plate = 12.26(15) = 183.9 kip

     Length of $\dfrac{3}{8}''$ fillet weld = $\dfrac{183.9}{10(0.707)0.375}$ = 69.4 in.

Make Cover plates 1"x 15" x 54 ft - 0 in.

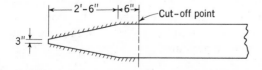

Length of weld = 74 in.

Although concrete shrinkage will create cracks in composite decks which are not shored, these cracks do not open as wide as corresponding cracks in non-composite decks.

The calculations in Example 13.2 require little further explanation. Section properties other than those calculated in Example 13.1 are determined in a similar manner. It should be pointed out that the proportioning of a composite cross section is a trial-and-error process. The work is greatly simplified by the use of tables of section properties.[13.6,13.7,13.8]

The calculation for the length of cover plates requires some explanation since it is somewhat different for a composite member than for a non-composite member. The approximate formula used in Example 13.2 to determine the theoretical length of cover plate is [13.9]

$$L' = g + (L - g)\sqrt{1 - \frac{S_{sb}}{S_{bp}}} \tag{13.8}$$

where L' is the length of the cover plate, L is the span length, and g is the distance from the center of the span to the point of maximum moment. (See the calculation of the cover plate length in Example 13.2.) Equation 13.8 is derived with the following assumptions:

1. The moment diagrams for both live load and dead load moments are parabolic between the point of maximum moment and the support.
2. The ratios of S/S_{sb}, S_b/S_{bp}, and S_b'/S_{bp}' are all approximately equal.
3. At the theoretical cut-off point, the stress in the bottom flange of the composite section without the cover plate is the maximum allowable stress.

The approximate cover plate length given by Eq. 13.8 is checked in Example 13.2 by the calculation of stresses with the cover plate omitted to show that these stresses are less than the allowable stresses.

The actual cover plate length is determined by the amount of weld required to fully develop the stress in the cover plate at the theoretical cut-off point. The cover plate stress is determined approximately by calculations based upon the section properties of the composite section with the cover plate included.

13.5 SHEAR STRESSES

The horizontal shearing stress between concrete slab and steel member must be developed by a suitable means so that the composite section acts as a monolithic section. Ideally, these shear connectors should be stiff enough to prevent any slip between the two parts of the member. This would require an infinitely rigid connection—which cannot be achieved in practice. However, a small amount of slip can be permitted.

The shear stress per unit length may be determined by

$$v = \frac{VQ_c}{I_c} \qquad \text{or} \qquad v = \frac{VQ_{cp}}{I_{cp}} \qquad (13.9)$$

where v is the shear per unit length and Q_c or Q_{cp} are the statical moment of the portion of the concrete slab above the interface between steel and concrete.

The shear thus computed must be resisted entirely by properly designed shear connectors. The bond between the concrete slab and the steel beam cannot be depended upon because this will be destroyed either by shrinkage of the concrete slab or by live load vibrations.

A method of designing connectors based solely upon working stresses and the mechanics of materials is not possible because of the complicated nature of the problem. The connector essentially acts as a dowel embedded in an elastic medium. If this dowel yields under load or if the concrete around it deforms, the dowel may not fail but the slip between concrete slab and beam may become excessive. An important property of connectors in beams is their load-slip behavior. A typical load-slip curve obtained from testing a pushout specimen is shown in Fig. 13.2.

Extensive research has been carried out on the behavior of various types of shear connectors for either static or fatigue loading. The design of

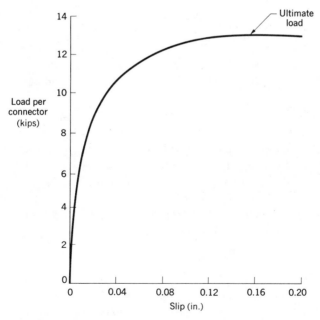

Fig. 13.2 Typical load vs. slip curve from a pushout test.

connectors for bridges is based on providing a sufficient number of shear connectors to prevent fatigue failure of the shear connection. As a secondary requirement a sufficient number of shear connectors must also be provided to develop the static ultimate strength of the composite section.

As a load moves across a bridge a horizontal force is exerted on the shear connectors. The direction of this force is reversed as the wheel load passes the connector. Therefore, in a typical simple span member the range of the force on a connector must be computed from a complete shear diagram showing not only the maximum shear but also the minimum shear for points along the span. A typical diagram is shown in Fig. 13.3. The force on

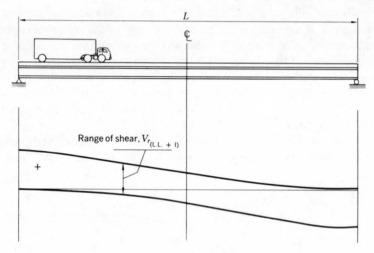

Fig. 13.3 Range of shear diagram for a simple-span beam.

connectors near the supports is applied in only one direcion, but near midspan a complete reversal of stress occurs. It is the range of shear indicated in Fig. 13.3 which is important in fatigue loading. The range of shear does not vary greatly across the span, and consequently the spacing of connectors can be nearly uniform in simple span bridge members.

Equation 13.9 can be adapted for the calculation of the range of shear per linear inch at the junction of the slab and beam by revising it as follows:

$$S_r = \frac{V_r Q_c}{I_c} \quad \text{or} \quad \frac{V_r Q_{cp}}{I_{cp}} \tag{13.10}$$

where S_r is the range of stress per linear inch and V_r is the range of shear. The spacing of shear connectors is determined by dividing the allowable range of shear of a connector or group of connectors occurring at a cross section of the beam by S_r computed for that region of the member.

Table 13.2 Allowable Range of Shear for Connectors

Type of Connector	Formula for Allowable Range in Pounds	Coefficient (α or β)		
		100,000 cycles	500,000 cycles	2,000,000 cycles
Welded studs	$Z_r = \alpha d^2$	13,000	10,600	7,850
Channels	$Z_r = \beta w$	4,000	3,000	2,400

d is diameter of stud in inches. w is width of channel in inches.

The allowable values for the range of shear on connectors are given in Table 13.2. This table contains allowable values for the two types of connectors most commonly used. The values presented in Table 13.2 were derived from a large number of test results on these two types of connectors.[13.10] Figure 13.4 shows the results of this investigation in terms of the shear stress on the weld plotted as ordinate with the number of cycles to failure as abscissa on a log scale. It is obvious from this figure that shear stress on the weld is the important variable, and the allowable loads for other types of connectors could be derived from Fig. 13.4 since fatigue failure of the weld governs.

The design of shear connectors to satisfy the requirements of fatigue loading is illustrated in Example 13.3 in which the shear connectors for the

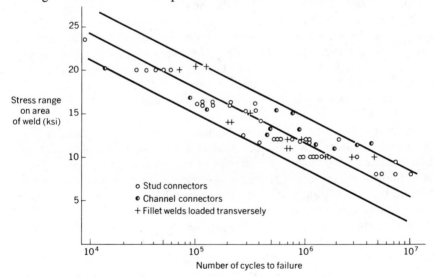

Fig. 13.4 Relationship between number of cycles to failure and the unit stress on the weld for shear connectors and transversely loaded fillet welds.

member of Example 13.2 are designed. In many instances the arrangements of connectors obtained by the procedure of Example 13.3 becomes the final design. However, a check of the requirements for static ultimate strength is also required. The ultimate strength requirement may govern the design in continuous members or in members where the design for fatigue is based on 100,000 cycles.

13.6 EFFECT OF SLIP

The shear connector design presented in the previous article ensures a safe design for fatigue loading. However, with the number of connectors provided in the member of Example 13.3, the slip between slab and beam at working load would not be zero. As a result of slip, there exists a discontinuity of stress at the shear plane. The stress distribution in an actual composite beam, therefore, will be intermediate between that in a non-composite beam and that in a composite beam having zero slip. The latter case is referred to as "complete interaction" and the actual case has been termed "incomplete interaction."

Strain distributions for a non-composite member, a member with incomplete interaction, and a member with complete interaction are shown in Fig. 13.5. The discontinuity in strain at the plane between slab and beam is less for incomplete interaction than for a non-composite beam. Therefore, the member with incomplete interaction obviously will be stiffer than a corresponding non-composite member. However, it is equally obvious that a corresponding member with complete interaction would be the stiffest of the three. The bottom flange steel stresses will be highest for the non-composite beam and lowest for a composite beam with complete interaction. A beam with incomplete interaction is an intermediate case.

The strain distribution for the incomplete interaction case given in Fig. 13.5 can be analyzed to show the relationship between complete and incomplete interaction. The discontinuity of strains at the shear plane when summed up along the length of the member equals the slip observed between slab and beam. The amount of slip in a composite beam is always less than that in a non-composite beam.

The magnitude of the stresses in the incomplete interaction case compared to that of the complete interaction case is of interest. To give some insight into this problem, several theories of incomplete interaction have been developed. The differential equation for the situation of incomplete interaction is derived in Ref. 13.11. A solution of the differential equation for one simple loading condition is given and discussed.

EXAMPLE 13.3

PROBLEM:
Design the shear connectors for the bridge member of Example 13.2

SOLUTION:
Use C 4 x 5.4 channels 10 in. in length

Design for 2 million cycles B = 2400

$$Q_r = B_w = 2400(10) = 24,000 \text{ lbs per connector}$$

Theoretical connector spacing

(a) at support

$$V_{L+I} = \frac{64.1}{2}(1.182)(1.25) = 47.4^k$$

Composite section with no cover plate

$$Q = A_c Y = 54.6(11.81) = 646 \text{ in.}^3$$

$$v = \frac{V_{L+I}\,Q}{I_c} = \frac{47.4(646)}{29,119} = 1.052 \text{ k/in.}$$

$$\text{Spacing} = \frac{Q_r}{v} = \frac{24}{1.052} = 22.8 \text{ in.}$$

(b) at midspan

$$V_{L+I} = 2\left[16\left(\frac{42.5}{85}\right) + 16\left(\frac{28.5}{85}\right) + 4\left(\frac{14.5}{85}\right)\right](1.182)(1.25) = 41.5^k$$

Composite section with cover plate

$$Q_p = A_c Y = 54.6(14.91) = 816 \text{ in.}^3$$

$$v = \frac{V_{L+I}\,Q_p}{I_c} = \frac{41.5(816)}{39,573} = 0.856 \text{ k/in.}$$

$$\text{Spacing} = \frac{Q_r}{v} = \frac{24}{0.856} = 28.0 \text{ in. which exceeds maximum allowable spacing.}$$

Use connectors at 22 in. throughout.

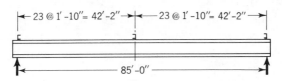

The important assumptions made in the analysis of Ref. 13.11 are as follows:

1. The shear connection is continuous along the length of the member.
2. The amount of slip is directly proportional to the load on a shear connector.
3. There is a linear distribution of strain over the depth of the member.
4. There is no vertical separation of slab and beam at any point along the length of member.

These assumptions create, for the purpose of analysis, an ideal composite beam, but this ideal beam does not differ greatly from actual members. A beam with shear connectors equally spaced with equal loads per connector

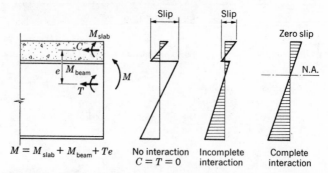

$$M = M_{slab} + M_{beam} + Te$$

No interaction $C = T = 0$ Incomplete interaction   Complete interaction

Fig. 13.5   Strain distribution in steel and concrete beams.

would fulfill essentially the first assumption.   The load-slip curve of Fig. 13.2 reveals that the second assumption is quite good.   The stress distribution in the slab of a composite beam is essentially linear and the linearity assumption is used as in reinforced concrete design.   The slab and beam are held together whenever properly designed shear connectors are used, and hence the third and fourth assumptions are satisfied.

Unfortunately, the solution of the differential equation for incomplete interaction does not yield a solution to the problem of the number of shear connectors required in a composite member.   The question of the number of shear connectors required has been solved empirically for fatigue loading by tests.   If only static loading is considered, the question of how many shear connectors are required may be solved in another way.   Composite construction for buildings requires an economical design with regard to the number of shear connectors.

13.7   ULTIMATE LOAD-CARRYING CAPACITY OF MEMBERS

Another solution to the problem of the design of shear connectors for composite members can be obtained by considering the maximum strength of the member.   All members in which the horizontal shear transfer is accomplished by mechanical shear connectors may be considered from this point of view, but other members where shear forces are resisted only by bond and friction must be excluded.

The concept of ultimate strength design in reinforced concrete when applied to a composite steel-concrete member would involve plastification of the slab and of the steel beam in the positive moment region; it constitutes a practical approach to the design of such composite members.

The ultimate strength concept is especially useful with composite members because it leads to a solution of the difficulties concerning the minimum shear connector requirement.

Normally, a composite beam fails due to crushing of the concrete slab. At this instant a fully plastic state of stress can be assumed for both concrete and steel. The assumed stress distribution is shown in Fig. 13.6. The plastic state of stress is represented by stresses of $0.85f_c'$ in concrete and F_y in

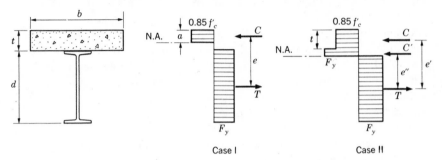

Fig. 13.6 Stress distribution of ultimate moment.

steel. Case I applies when the concrete slab is adequate to resist the total compressive force at ultimate load. Case II applies when the concrete slab is not large enough to resist the total compressive force.

From Fig. 13.6 the equations of equilibrium are:

Case I.

$$C = 0.85f_c'ba \qquad (13.11)$$

$$T = A_sF_y \qquad (13.12)$$

$$C = T \qquad (13.13)$$

$$a = \frac{A_sF_y}{0.85f_c'b} \qquad (13.14)$$

$$M_u = Te = T\left(\frac{d}{2} + t - \frac{a}{2}\right) \qquad (13.15)$$

Case II.

$$C = 0.85f_c'bt \qquad (13.16)$$

$$T = A_sF_y - C' \qquad (13.17)$$

$$T = C + C' \qquad (13.18)$$

$$M_u = Ce' + C'e'' \qquad (13.19)$$

where the symbols are explained in Fig. 13.6.

The values of e' and e'' must be determined from the stress distribution and geometry of the cross section. The ultimate moment for the negative moment region of continuous members is determined by a similar consideration of only the steel portion of the cross section. The longitudinal slab reinforcing steel may be considered if shear connectors are provided in the negative moment region. If slab steel is not considered, the ultimate moment reduces to the plastic moment of the steel member.

It has been tacitly assumed that the shear connection provided is sufficiently strong to develop the fully plastic state of stress. This assumption is made with regard to the bridge member of Example 13.2 in the calculation of the ultimate strength of that section in Example 13.4. The calculations of Example 13.4 are independent of whether or not construction is shored or unshored. The ultimate strength is not affected by the method of construction. This fact provides the basis for the assumption made in some specifications that the dead load can be assumed to be carried by the composite section regardless of the manner or sequence of construction.

If the number of shear connectors in a member is not sufficient to permit the member to attain the ultimate moment as calculated by either Eq. 13.15 or Eq. 13.19, the ultimate moment of the member can still be determined. The ultimate strength of members with inadequate shear connectors is discussed in Ref. 13.12.

The linear portion of the load slip curve given in Fig. 13.2 was used in the previous treatment of shear connectors. It is obvious from the curve that only a small portion of the complete curve has been considered in this analysis. The ultimate capacity of the connectors and the maximum slip of the connector can also be obtained from Fig. 13.2. A deformation as large as 0.2 to 0.3 in. usually occurs prior to failure. The ductility of mechanical connectors makes it possible to use the ultimate strength of connectors as a possible design value. As in riveted and bolted joints, this ductility allows redistribution of load among individual connectors, and at failure it can be assumed that all connectors carry equal loads regardless of the shape of the shear diagram for the member. Tests of composite members have illustrated this fact conclusively.[13.12]

Because of the redistribution of load between connectors, it is possible to define the shear connector requirements at ultimate moment for composite members based upon the ultimate strength of shear connectors.

Figure 13.7 shows a simple span composite beam with two concentrated loads located symmetrically about mid-span. If the magnitude of these loads is increased such that the ultimate bending moment is reached in the region between load points, a free-body diagram of the concrete slab between the load point and the end of the slab defines the shear connector requirements at ultimate load. The maximum compressive force in the slab is given by Eqs. 13.11 or 13.16. This force must be resisted by the sum of the ultimate strength values of the shear connectors in the shear span, defined as q_u. The number of shear connectors required to develop ultimate moment

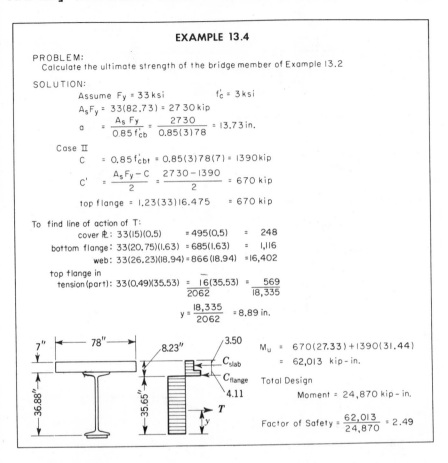

EXAMPLE 13.4

PROBLEM:
Calculate the ultimate strength of the bridge member of Example 13.2

SOLUTION:

Assume $F_y = 33$ ksi $f'_c = 3$ ksi

$A_s F_y = 33(82.73) = 2730$ kip

$a = \dfrac{A_s F_y}{0.85 f'_c b} = \dfrac{2730}{0.85(3)78} = 13.73$ in.

Case II

$C = 0.85 f'_c bt = 0.85(3)78(7) = 1390$ kip

$C' = \dfrac{A_s F_y - C}{2} = \dfrac{2730 - 1390}{2} = 670$ kip

top flange $= 1.23(33)16.475 = 670$ kip

To find line of action of T:

cover ℞: 33(15)(0.5)	= 495(0.5)	= 248
bottom flange: 33(20.75)(1.63)	= 685(1.63)	= 1,116
web: 33(26.23)(18.94)	= 866(18.94)	= 16,402

top flange in
tension (part): $33(0.49)(35.53) = \dfrac{16(35.53)}{2062} = \dfrac{569}{18,335}$

$y = \dfrac{18,335}{2062} = 8.89$ in.

$M_u = 670(27.33) + 1390(31.44)$

$= 62,013$ kip-in.

Total Design

Moment $= 24,870$ kip-in.

Factor of Safety $= \dfrac{62,013}{24,870} = 2.49$

is then obtained as

$$N = \frac{C}{q_u} = \frac{0.85 f'_c\, ba}{q_u} \quad \text{or} \quad N = \frac{C}{q_u} = \frac{0.85 f'_c\, bt}{q_u} \tag{13.20}$$

All flexible connectors including welded studs, channels, spirals, and zees may be treated in this manner. Although connectors generally exhibit a tensile failure in beam tests, connector stresses are given in terms of apparent shear stress. Values of the ultimate strength of shear connectors in terms of apparent shear stress have been developed for flexible connectors.[13.12]

The results of tests to destruction of composite beams have been studied and compared with the foregoing theory. These test results are plotted in Fig. 13.8 where the test moment M divided by the theoretical ultimate moment M_u is plotted as ordinate, and the sum of the ultimate strength values of the shear connectors $\sum q_u$ divided by the theoretical maximum

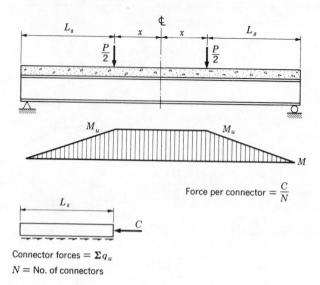

Force per connector $= \dfrac{C}{N}$

Connector forces $= \Sigma q_u$

$N =$ No. of connectors

Fig. 13.7 Shear connector forces at ultimate moment.

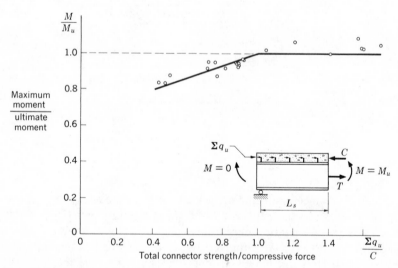

Fig. 13.8 Relationship between shear connector strength and moment capacity.

compressive force C in the slab is plotted as abscissa.   The curve of Fig. 13.8 shows that all members for which $\sum q_u/C$ exceeded 1.00 failed at, or slightly above, the predicted ultimate moment.   All members for which $\sum q_u/C <$ 1.00 failed at values less than the theoretical ultimate moment.   The effect of a weak shear connection on the ultimate strength of the member can be predicted from the slope of the curve in the left portion of Fig. 13.8.

The allowable horizontal shear loads given in AISC Table [1.11.4] were determined from ultimate strength values developed in a research program at Lehigh University.[13.12]   These values are reproduced as Table 13.3.   The

Table 13.3   AISC: Allowable Load Per Connector

Type of Connector	Allowable Horizontal Shear Load* (kips)		
	$f_c' = 3000$ psi	$f_c' = 3500$ psi	$f_c' = 4000$ psi
$\frac{1}{2}''\phi \times 2''$ hooked or headed stud	5.1	5.5	5.9
$\frac{5}{8}''\phi \times 2\frac{1}{2}''$ hooked or headed stud	8.0	8.6	9.2
$\frac{3}{4}''\phi \times 3''$ hooked or headed stud	11.5	12.5	13.3
$\frac{7}{8}''\phi \times 3\frac{1}{2}''$ hooked or headed stud	15.6	16.8	18.0
3″ channel 4.1 lb per inch	4.3w	4.7w	5.0w
4″ channel 5.4 lb per inch	4.6w	5.0w	5.3w
5″ channel 6.7 lb per inch	4.9w	5.3w	5.6w

w = length of channel in inches.
* Applicable only to concrete made with ASTM C33 aggregate.

values specified by AISC include a factor of safety ranging between 2.2 and 2.5, with the higher value of the factor applying to the lower concrete strengths.

Equations for the design of shear connectors by the AISC Specification are given here as Eqs. 13.21 and 13.22:

$$V_h = \frac{0.85f_c'A_c}{2} \tag{13.21}$$

$$V_h = \frac{A_s F_y}{2} \tag{13.22}$$

where A_c is the concrete area, A_s the steel area, and F_y the yield strength of the steel.   Referring to Fig. 13.6, Eq. 13.21 will be recognized as half of the compressive force in the slab for Case II, and Eq. 13.22 will be recognized as half of the compressive force in the slab for Case I.   This presumes that the compressive force in the concrete slab at design load will be equal to half of the compressive force at ultimate load.   In reality, this is a conservative estimate.

In continuous beams, the longitudinal reinforcing steel in the concrete slab will develop a tensile stress as long as the steel is anchored. Shear connectors should be provided in the negative moment region between an interior support and each adjacent point of contraflexure, in accordance with Eq. 13.23,

$$V_h = \frac{A_{sr}F_{yr}}{2} \tag{13.23}$$

where A_{sr} is the total area of the longitudinal reinforcing steel located within the effective slab width, and F_{yr} is the specified minimum yield stress of the longitudinal steel.

The AISC Specification requires that the number of connectors resulting from the application of Eqs. 13.21 and 13.22 be provided between the points of maximum moment and the end of simple span beams or the point of contraflexure in continuous beams. This does not necessarily provide sufficient shear connectors where concentrated loads and uniform loads occur on the same span. In this situation, the designer must investigate the region between the support or the point of contraflexure and the first concentrated load to be sure that the number of connectors in this region is adequate. The number of shear connectors required between a concentrated load and a point of zero moment is given by

$$N_2 = \frac{N_1\left(\dfrac{MB}{M_{\max}} - 1\right)}{B - 1} \tag{13.24}$$

where M is the moment at the concentrated load; $M_{\max}$ is the maximum moment. N_1 is the number of shear connectors given by Eq. 13.20 and

$$B = \frac{S_b}{S} \quad \text{or} \quad \frac{S_{bp}}{S}.$$

The ultimate strength of stud and channel connectors was developed from test data,[13.12] and the following formulas may be used for computing the values:

Channels $\qquad\qquad Q_u = 550(h + t/2)w\sqrt{f_c'} \tag{13.25}$

Welded studs $\qquad\quad Q_u = 930d^2\sqrt{f_c'} \qquad (H/d \geqslant 4) \tag{13.26}$

where Q_u is the ultimate strength of one connector in pounds, h is the average flange thickness of the channel, t is the web thickness of the channel, w is the width of the channel, H is the height of a stud, d is the diameter of a stud, and f_c' is the compressive strength of the concrete.

The ultimate strength of the channel connectors in Example 13.4 is 122.9 kips from Eq. 13.21 for 10-in.-wide connectors. The compressive force in the slab at ultimate load given by Eq. 13.16 is 1392 kips. Therefore the number of connectors required in each half of the span is 1392/122.9 or 11.3 connectors. This is considerably smaller than the number required for fatigue, and the arrangement of connectors found in Example 13.3 is the final design.

13.8 ENCASED BEAMS

Many steel building members are encased in concrete for fire protection. For many years this concrete contributed only fire protection and dead weight to the structure, but more recently this concrete has been designed as a part of the composite member. Because of its low tensile strength, only concrete on the compression side of the elastic neutral axis is considered as part of the composite section. However, the rest of the concrete is useful in developing shear and in stiffening the structure. For calculation of live load deflection, the entire cross section may be used.

The ultimate strength theory does not apply to these members because mechanical shear connectors are not used, and ductility of the shear connection is lacking. A typical encased member is shown in Fig. 13.9. Shear stresses in the encasing material must be checked along lines ab and cd of Fig. 13.9. The shear stress should not exceed $0.12f_c'$ along lines ab and cd.

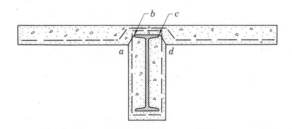

Fig. 13.9 Cross section of an encased composite beam.

Excess shear stress should be developed by steel reinforcement across ab and cd. Wire mesh is required in the encasement to develop bond and friction between steel and concrete even if the allowable concrete stress on lines ab and cd is not exceeded.

Mechanical shear connectors should not be used in encased members to carry excess shear because of the relatively large deformations necessary before mechanical connectors transmit any sizeable shear force.

The design of encased steel composite members is illustrated by Example 13.5. Floor beam B5 is designed as a continuous composite member in

EXAMPLE 13.5

PROBLEM:

Design member B5 as encased composite beam with unshored construction for the following loading:

Live load   100 psf
Ceiling finished floors   25 psf
Dead load   Slab $0.050(8)$ = 0.400 kip/ft
           Beam   = 0.040
           Encasement   = 0.160
                            0.600 kip/ft

steel   A36
concrete f_c' = 3 ksi
4" concrete slab

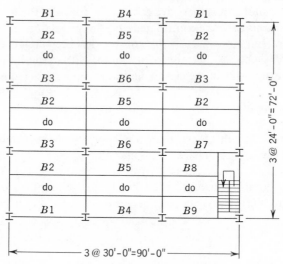

SOLUTION:

D.L.M. = $0.600(30)$ 1.5 = 810 kip-in.

Try W 14x34 steel beam

$$f_b = f_t = \frac{810}{48.5} = 16.70 \text{ ksi}$$

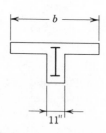

Effective width Tee beam
   b = 1/4 span = 90"
   b = spacing = 96"
   b = 11"+16t = 75"

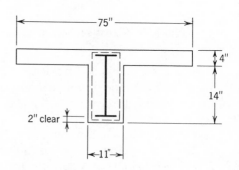

Section at midspan

Assume fixed ends
for live loads
because of encasement
      n = 10

EXAMPLE 13.5 (Continued)

Section	A	y	M	Y	Y²	AY²	I_0
Beam	10.00	9.00	90.0	5.25	27.56	275.6	339.2
Slab	30.00	2.00	60.0	—	—	—	131.8
	40.00		150.0			275.6	471.0

$$I = 746.6$$

$$\bar{y} = \frac{150}{40} = 3.75 \text{ in.} \qquad S_t = \frac{746.6}{3.75} = 199.0 \text{ in.}^3 \qquad S_b = \frac{746.6}{12.25} = 60.95 \text{ in.}^3$$

$$\text{L.L.M} = \frac{0.125(8)(30)^2 12}{24} = 450 \text{ kip-in.}$$

$$f_t = \frac{450}{199(10)} = 0.226 \text{ ksi}$$

$$f_b = \frac{450}{60.95} = 7.38 \text{ ksi}$$

Total stress in bottom flange = 16.70 + 7.38 = 24.08 ksi
Since estimate of dead weight was high this is okay

Check section at support

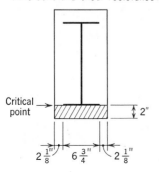

$$\text{L.L.M} = \frac{12}{12}(1.000)(30)^2 = 900 \text{ kip-in.}$$

$$\frac{1}{2}\left(\frac{11}{10}\right)y_b^2 = 10.00(9.00 - y_b)$$

$$0.55 y_b^2 + 10 y_b - 90 = 0$$

$$y_b = 6.60 \text{ in.}$$

6.60 in.

	A	Y	AY²	I_0
steel	10.00	2.40	57.6	339.2
concrete	7.26	—	—	105.4
	17.26		57.6	57.6

$$I = 502.2 \text{ in.}^4$$

$$S_t = \frac{502.2}{9.40} = 53.42 \text{ in.}^3 \qquad f_t = \frac{900}{53.42} = 16.85 \text{ ksi} < 24 \text{ ksi}$$

$$S_b = \frac{502.2}{6.60} = 76.09 \text{ in.}^3 \qquad f_b = \frac{900}{76.09(10)} = 1.18 \text{ ksi} < 1.35 \text{ ksi}$$

Check bond and shear stress for negative moment region. The critical section in a continuous member will be a tendency to slip at the lower steel flange.
Take bond and shear resistance as $0.03 f_c'$ and $0.12 f_c'$

Total resistance to horizontal shear
$$= 0.03 f_c'(6.75) + 0.12 f_c'(4.25) = 2138 \text{ lb/in.}$$

$$V_{support} = 15 \text{ kips}$$

$$Q = 1.1(2.0)(5.60) = 12.32 \text{ in.}^3$$

$$v = \frac{VQ}{I} = \frac{15000(12.32)}{502.2} = 368 \text{ lb/in.}$$

∴ No shear reinforcement required

Critical point   2″

$2\frac{1}{8}″ \qquad 6\frac{3}{4}″ \qquad 2\frac{1}{8}″$

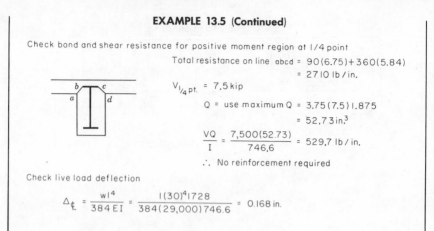

EXAMPLE 13.5 (Continued)

Check bond and shear resistance for positive moment region at 1/4 point

Total resistance on line $abcd$ = $90(6.75)+360(5.84)$

= 2710 lb / in.

$V_{1/4\,pt.}$ = 7.5 kip

Q = use maximum Q = 3.75(7.5)1.875

= 52.73 in.^{3}

$\dfrac{VQ}{I} = \dfrac{7,500(52.73)}{746.6}$ = 529.7 lb / in.

∴ No reinforcement required

Check live load deflection

$$\Delta_{\unicode{x2104}} = \frac{wl^4}{384\,EI} = \frac{1(30)^4 1728}{384(29,000)746.6} = 0.168 \text{ in.}$$

accordance with the AISC Specifications. The member is designed as un-shored construction.

13.9 BUILDING MEMBERS WITH MECHANICAL SHEAR CONNECTORS

The majority of buildings are constructed using mechanical shear connectors rather than concrete encasement. A typical example of this type of building construction is shown in Fig. 13.10, which shows a building frame and floor beams erected with welded stud shear connectors installed. Forms for floor slabs are hung from the steel beams. Floor beams may be shored from the floor below or they may carry the dead load, depending upon the design. The magnitude of the dead load deflection is often the determining factor in deciding whether or not to use shores.

The AISC Specification provides that all loads are considered as applied to the composite section regardless of the method of construction. The method of construction does not alter the ultimate strength of the member but does affect appreciably the actual steel stresses which occur in the member. However, a maximum limit for bottom flange steel stress is fixed by AISC Formula [1.11-2] at approximately $0.82F_y$. This formula, written in terms of the notation of this chapter, is

$$S_b = \left(1.35 - 0.35\frac{M_L}{M_D}\right)S \quad \text{or} \quad S_{bp} = \left(1.35 + 0.35\frac{M_L}{M_D}\right)S_{sb} \quad (13.27)$$

The design of building members is also a trial-and-error process. Fortunately, there are many design tables and charts available.[13.7,13.8] If such

Fig. 13.10   Steel beams for a composite floor (Courtesy of KSM Division, Omark Industries, Inc.).

information is not available, a trial section may be found by assuming that composite action will increase the bending strength of a steel beam by about 30 per cent. The non-composite member determined using a 30-per cent overstress usually will be within one size of the required member.

The design of a composite floor system is illustrated by Example 13.6. This example follows the AISC Specification and needs no further comment. In this example, the design has been achieved using only a rolled beam without cover plates, and the member is designed as a simple beam. A more economical design may be possible by using a lighter rolled section and using a bottom flange cover plate as in the bridge of Example 13.2. A comparison of the weight of steel for various composite sections compared to a non-composite member is given in Example 13.6. The lightest section uses a rolled beam and bottom flange cover plate and this member is 34 per cent lighter in weight than the non-composite member.

Actually, the ends of a floor member such as the one designed in Example 13.6 are not simply supported. The connection to the girder develops some negative moment which can cause cracking of the concrete slab. This is

EXAMPLE 13.6

PROBLEM:
  Design floor member B5 of Example 13.5 using 1/2 in. diameter welded stud shear connectors. Assume simple beam end connections. Use A36 steel AISC specification

SOLUTION:

Loading   L.L. = 125 psf

Slab = 50 psf

= 175 psf

Load per beam = 175(8) = 1400 lb/ft
Estimated weight steel beam = 50 lb/ft

1450 lb/ft

Effective width of flange

b = 1/4 span = 90 in.

b = beam spacing = 96 in.

b = 16 x slab thickness + flange width = 64 in. + 7 = 71 in.

Try W 16 x 36

Steel section properties   A = 10.59 in.^{2} I = 446.3 in.^{2} S = 56.3 in.^{2}

Composite section properties

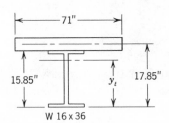

W 16 x 36

$$y_t = \frac{10.59(7.93) + 28.4(17.85)}{10.59 + 28.4} = 15.15 \text{ in.}$$

$$I_{tr} = 446.3 + 10.59(7.22)^2 + \frac{1}{12}(7.1)(4)^3$$
$$+ 28.4(2.7)^2$$
$$= 1242 \text{ in.}^4$$

$$S_{tr} = \frac{1242}{15.15} = 81.9 \text{ in.}^3$$

$$M_D = 0.436(30)^2 1.5 = 589 \text{ kip-in.}$$
$$M_L = 1.000(30)^2 1.5 = 1350 \text{ kip-in.}$$
$$1939 \text{ kip-in.}$$

Maximum section modulus = $S_{tr} = \left(1.35 + 0.35 \dfrac{M_L}{M_D}\right) S$

$$= \left(1.35 + 0.35 \frac{1350}{589}\right) 56.3 = 121.2 \text{ in.}^3$$

$$\therefore S_{tr} = 81.9 \text{ in.}^3 \text{ may be used}$$

Stresses

top of concrete $f_c = \dfrac{1939(4.70)}{10(1242)} = 0.734 \text{ ksi}$

top of steel beam $f_t = \dfrac{1939(0.7)}{1242} = 1.093 \text{ ksi}$

bottom of steel beam $f_b = \dfrac{1939}{81.9} = 23.6 \text{ ksi} < 24 \text{ ksi}$

Check live load deflection

$$\Delta = \frac{5wl^4}{384 E I_c} = \frac{5(125 \times 8)(30)^4 (12)^3}{384(29 \times 10^6)1242} = 0.51 \text{ in.} < \frac{L}{360}$$

EXAMPLE 13.6 (Continued)

Design of shear connectors

$$V_h = \frac{1}{2} \, 0.85 \, f_c' \, A_c = \frac{1}{2}(0.85)3000(284) = 362 \text{ kips}$$

$$V_h = \frac{A_s F_y}{2} = \frac{10.59(36)}{2} = 191 \text{ kip}$$

stud connectors $= \frac{1}{2}'' \, \phi \quad$ by $2\frac{1}{2}''$ in length

$$q = 5.1 \text{ kip /connector}$$

$$\text{number required} = \frac{191}{5.1} = 37.5$$

Use 38 connectors on each side of midspan equally spaced.

It is left as an exercise for the student to prove that the following members may also be used for member B5. The following table compares the relative weights of the various possible solutions in this design problem.

Description of member	total wt of steel	% of non-composite wt
Non-composite steel beams W 16 x 50	1500 lb	100
Encased composite beam – design example 13.5	1020 lb	68.0
Composite rolled beam – design example 13.6	1080 lb	72.0
Composite W 12 x 27 with a 5 1/2" x 1/2" cover plate on bottom flange for 63% of span length of member	988 lb	65.9

often compensated for by adding top slab reinforcing steel, transverse to the girder into which floor beams frame. The design of this steel is somewhat arbitrary, but a typical design suggested in Ref. 13.13 is No. 4 bars at 12 in. on center extending a minimum of 2 ft 0 in. beyond the end of the floor beams. A grid of similar-size bars is also suggested for framing around columns. These members can also be designed as continuous beams as in Example 13.5 for encased beams. This type of design has not been common because it offers little economic advantage. Where welded construction is used and a framing plan permits girders and floor beams of the same depth, a continuous design may offer decided advantages. For a continuous beam, cover plates would be used in the negative moment region to balance a composite section for positive bending moment. Floor beams would, in this case, be spaced to avoid columns, and expensive four-way column connections could be eliminated. The stiffness in the direction of the floor beams is provided by the composite floor system acting as a diaphragm.

It is sometimes inconvenient or uneconomical to provide a complete shear connection as required by Eqs. 13.21, 13.22, and 13.23. The effect of a weak shear connection in reducing slightly the ultimate strength of the member has been illustrated in Fig. 13.8. Members with a weak shear connection may be proportioned by the following equation,

$$S_{\text{eff}} = S + \frac{V_h^1}{V_h}(S_b - S) \quad \text{or} \quad S_{\text{eff}} = S + \frac{V_h^1}{V_h}(S_{bp} - S) \quad (13.28)$$

where V_h^1 is the product of the allowable load per connector times the number of connectors in the shear span, and S_{eff} is the effective section modulus of the composite section with a weak shear connection.

13.10 DEFLECTION OF COMPOSITE MEMBERS

The deflection of a composite member may be calculated by the same means used in calculating deflections of non-composite members. Certain modifications of section properties are necessary in deflection calculations as were also necessary in stress calculations. Dead load deflections for un-shored members are calculated with the properties of the steel beam alone. The deflection for shored dead loads and long-term live loads should be calculated using the composite section and a modular ratio of about 3 times n to allow for creep of the concrete. Finally, for short-term live load the composite section properties determined using the elastic modular ratio are computed. In the case of continuous spans, the members may not be prismatic. However, this can be ignored in initial deflection calculations. If the deflections appear to be critical from these approximate calculations, then the calculations may be refined by considering the non-prismatic member.

Live load deflection calculations are shown for the member of Example 13.6. In this example the live load deflection is approximately one half of the allowable deflection. The total deflection must be provided for in design, either by camber of the steel beams, by using semi-rigid connections, or by varying the thickness of the concrete slab. This is a problem which the structural designer cannot ignore. The problem is somewhat reduced in the case of bridges by a provision in the AASHO Specifications suggesting that the depth-to-span ratio for the steel beam be less than 1/30 and the depth-to-span ratio for the composite section be less than 1/25. These values were developed as limiting values for structural carbon steel and are not necessarily valid when high-strength steels are used or when hybrid girders are specified.

It is important that some of the deflection characteristics of composite members be pointed out so that these can be considered when occasion dictates. In Art. 13.6, the effect of incomplete interaction on stresses was discussed. The incomplete interaction also increases the deflection of members, and the precise amount of the additional deflection is as difficult to determine as are the exact stresses. Figure 13.11 shows load-deflection curves

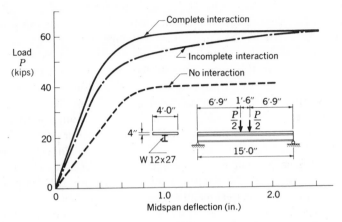

Fig. 13.11 Load-deflection curves for steel and concrete beams.

for a member with complete interaction compared with the same member with incomplete interaction. For comparison, the curve for the same steel member with no interaction is shown. At some load well above design load, the deflection of complete and incomplete interaction members may coincide, but in the region of working load the member with incomplete interaction exhibits greater deflections than the complete interaction case. For members designed in accordance with the AISC Specification with structural carbon steel, the actual deflection of composite members may be expected to be 110 to 120 per cent of theoretical values, depending upon the geometry and loading configuration. At the same time, the deflection of the same member with no interaction would be 200 to 300 per cent of the composite complete interaction deflection.

Deflection due to shear is usually neglected in steel design. In composite members, shear deflection may be appreciable, owing to the fact that the load-carrying capacity of a rolled section has been greatly increased due to composite action. Shear deflection may reach a magnitude of 30 per cent of theoretical bending deflection, and therefore cannot always be ignored. Deflection due to shear may be calculated by simple formulas derived in most strength-of-materials textbooks.

Most of these refinements in the calculation of deflections are not justifiable in general design procedures, and the fact remains that composite floor systems are much stiffer than equivalent non-composite systems.

The problem of the analysis of a building frame which consists of composite members along with non-composite members may occur in design. In this situation, the question arises as to what stiffness should be assumed for the composite member. If a composite member framing between columns bends so as to have positive bending moment throughout its entire length, the stiffness may be taken as I_c/L, using the elastic composite section properties to determine this ratio. The situation for a member having negative bending near the ends and positive bending near mid-span is not as simple. The stiffness to be used in this case is the stiffness of a non-prismatic member consisting of only the steel beam in the negative moment region and the composite section in the positive moment region. The design assumption made in Chapter 21 for the analysis of multi-story buildings is that, due to rotation or deformation of the connections, an inflection point develops in the beams at $0.1L$ from the ends. On this basis, the stiffness of the composite beam for negative bending moment can be determined.

Figure 13.12 expresses the relative stiffness of a composite beam in terms of the value of I/L for the steel members only. In this figure, I_c divided by I of the steel member is plotted as ordinate. Along the abscissa is plotted the relative stiffness of the composite member divided by I/L of the steel beam.

Most floor beams and girders encountered in buildings have a composite moment of inertia between 2.0 and 2.5 times the moment of inertia of the steel member alone. This implies that composite members are on the

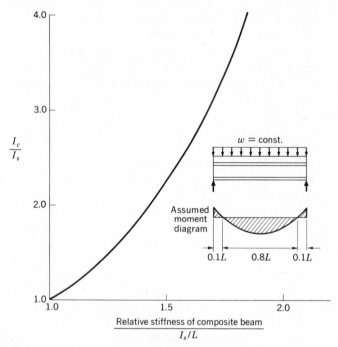

Fig. 13.12 Relative stiffness of non-prismatic composite beams.

average about 1.5 times as stiff as steel beams in frame analysis with semi-rigid connections.

The use of structural lightweight concrete is quite common in composite building construction. The ultimate strength of shear connectors in lightweight concrete has been found to be lower than in regular concrete of the same strength. The allowable shear loads for lightweight concrete are obtained by multiplying the values in Table 13.3 by the reduction coefficients given in Table 13.4.

Table 13.4 Lightweight Concrete Shear Connector Reduction Coefficients

Air-Dry Unit Weight of Concrete, pcf	Reduction Coefficient
90	0.73
95	.76
100	.78
105	.81
110	.83
115	.86
120	0.88

A large amount of composite construction also involves various types of metal decking. When the valleys which contain the shear connectors are less than $1\frac{1}{2}$ inches in depth and at least 2 inches wide, the members may be designed as though the slab were solid. For rib geometries involving deeper ribs and narrower valleys, the volume of concrete in the rib is not sufficient to develop the ultimate strength of the shear connector. In these members it is seldom economical to provide a sufficient number of shear connectors to develop the compressive force in the slab given by Eq. 13.11 or Eq. 13.16. For these members the maximum force in the concrete slab may be taken as the sum of the ultimate strengths of the shear connectors. The ultimate moment may be calculated from Eq. 13.19, and the working load moment may be obtained using a load factor of 2.0.

The value of the ultimate strength of stud shear connectors for various rib geometries has been studied by the manufacturers of metal decking. Suitable values for a wide range of decking geometries and stud connectors in regular concrete may be obtained from Fig. 13.13.[13-14] Suitable values for light weight concrete may be obtained by reducing the values obtained from Fig. 13.13 by the square root of the ratio of the modulus of elasticity for light weight and regular concrete.

Additional information concerning the design of shear connectors with metal decking and allowable shear connector loads may be obtained from Ref. 13.15.

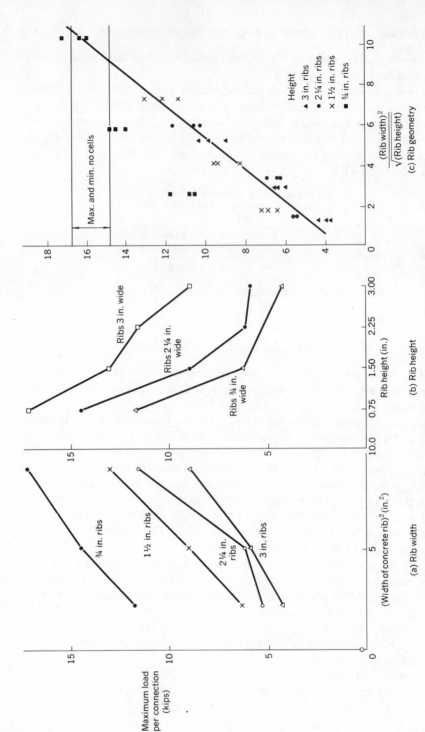

Fig. 13.13 Influence of rib geometry on strength of pushout specimens.

(a) Rib width

(b) Rib height

(c) Rib geometry

PROBLEMS

13.1. Design the bridge of Example 13.2, using shored construction.   Compare member size with the unshored design.

13.2. Design the shear connectors for Problem 13.1 and compare with shear connector design of Example 13.3.

13.3. Design the girder of the floor system used in Example 13.5.   Compare the relative economy of this member with the economy of the floor beam.

13.4. On the basis of ultimate strength, determine the maximum moment which the floor beam of Example 13.6 can carry.   Assume that connections of supports will develop full strength of the steel beam.

13.5. Compare the AASHO and AISC shear connector requirements by determining the number of channel connectors required if the bridge member of Example 13.2 were to be used as a building girder.

13.6. Determine wind moments which could be carried by the composite beams of Examples 13.5 and 13.6.

14

THE WELDING PROCESS

14.1 INTRODUCTION AND HISTORICAL REVIEW

This chapter describes the process of welding and its application to the fabrication of structures. Typical joint preparations are illustrated, and the formation of residual stresses due to welding is discussed briefly.

The term "welding" denotes the process of connecting metals under the application of heat and/or pressure. With this definition, the art of welding encompasses a period from the Iron Age to the present time. The ability to shape and handle iron has been a criterion of the sophistication of primitive peoples. For this book, the term "welding" refers essentially to "arc welding," the joining of metals utilizing the heat generated by an electric arc, with no pressure being necessary.

Arc welding was not possible at all until the discovery of the electric arc by Sir Humphry Davy early in the nineteenth century, and his development of methods of starting and maintaining the arc. The electric arc was not used for welding until 1881, when parts of a storage battery plate were joined by the use of carbon rods.[4.18,14.1] The next development, which followed almost immediately, was the replacement of carbon rods by metallic rods, which, when melted and added to the weld, did not affect the weld as deleteriously as did the carbon. The end of the nineteenth century and the first decade of the twentieth saw a gradual development of the new art, limited to a great extent by the lack of proper electrodes and suitable generators of electricity of the high amperage and low voltage necessary to start and maintain a proper arc.

The first electrodes were bare wire, which resulted in brittle welds and lower strength, as explained below. In the period just before the First World War, the coated electrode was introduced, and, for the next decade, a strong controversy existed among the users of bare wire and the users of covered electrodes. Perhaps the most attractive feature of the bare wire electrode was its lower price, and this played a much greater role with the contemporary welders than the fact that only the coated electrode resulted in a sound and ductile weld.

The purpose of the covering on the electrode was twofold: first, to create an arc shielded from the atmosphere by a gaseous layer resulting from a decomposition of the coating and, second, to act as a flux to cause the impurities to float to the surface of the molten weld. The bare wire electrodes were neither able to discharge impurities from the weld nor to prevent the deleterious formation of oxides and nitrides which occurred readily at elevated temperatures from lack of a shield. It would not be incorrect to define the purity of a weld in terms of its ductility; bare wire welds had very little ductility. Yet, in the first decades of this century, the bare wire was used almost exclusively in the United States.

The period between the two world wars was one of rapid expansion; the covered electrode became accepted, the welding process began to challenge riveting, and many major structures used welding exclusively. The Second World War and its fast production requirements greatly influenced the development of arc welding; so much so, that the use of welding was often ahead of the development of safe welding procedures. In particular, the problem of brittle fracture became critical with reference to the failure of some ships and bridges.

14.2 WELDING PROCESSES

Approximately forty welding processes are classified by the American Welding Society.[14.2] For structural steel, arc welding is used almost exclusively, although other processes, such as electroslag welding, are being used to some extent.

Figure 14.1 shows, in simplified form, the welding circuit basic to almost all processes. A power source is connected on one line by the ground cable

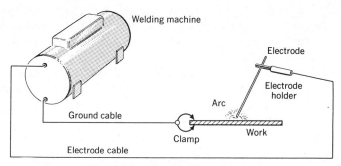

Fig. 14.1 The welding circuit.

Fig. 14.2 Typical manual arc-welding setup. (Courtesy of Lincoln Electric Co.)

to the work to be welded, and on the other line by the electrode cable to the electrode holder and thence to the electrode. The arc is formed when the electrode is touched to and then raised slightly above the work. The arc is maintained by the operator by keeping the arc length constant during the welding. (Inexperienced welders may lose the arc either by pushing the electrode onto the work or else by drawing it away so that the arc collapses.) A typical manual arc-welding setup where three welders are working simultaneously is shown in Fig. 14.2.

An intense heat is generated by the arc, a heat sufficient to reduce steel to the molten state. The temperature in the arc itself has been estimated to be as high as 10,000°F; the temperature in the steel near the arc has been measured to be approximately 3500°F.

1. Shielded Arc-Welding Process

The shielded metal-arc-welding process is shown in Fig. 14.3. The

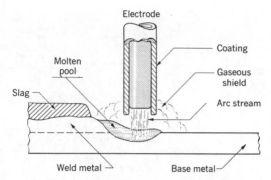

Fig. 14.3 Shielded arc-welding process.

"shield" refers to the gaseous shield which surrounds the arc. The shield is necessary for two reasons: (1) to prevent the molten metal from oxidizing in the atmosphere (and thus becoming porous), as well as from absorbing nitrogen from the atmosphere (and thus becoming embrittled and losing ductility); and (2) to create stability of the arc.

One type of coating which produces the shielding contains a large proportion of pure cellulose; the hydrogen and carbon monoxide released from cellulose protect the molten weld from the oxygen and nitrogen of the atmosphere. Fluxes present in the coating provide a molten covering which floats on top of the weld metal to help shield it from air. Various chemicals present in the coating assist the impurities to float to the surface as a slag. It is this slag which must be knocked off the weld when it has cooled.

Since so many different kinds of steels and metals exist, it is readily appreciated that an even greater variety of electrodes is available. These electrodes have different chemical compositions both in the steel and in the coating, and are designed for use with different steels and with different welding conditions and positions.[14.3,14.4]

The shielded arc-welding process is used generally with manual welding. Manual welding today is used mostly for fitting-up and for short welds on clip angles, and the like. Automatic welding, on the other hand, is used for general structural welding and for the fabrication of welded shapes whenever the joint is linear or regular enough to permit mechanization. Fully automatic welding is used extensively in the *shop* for long welds which require a considerable amount of weld metal. Shorter welds on stiffeners and other attachments may be made with manual shielded metal-arc welding or with a semi-automatic process. In the *field*, increasingly more welding is being carried out with the semi-automatic process such as for column splices (Fig. 14.4) and the flanges of beam-to-column connections. Recently, semi-automatic welding has been used for the vertical welding of the welds of beam-to-column connections (Fig. 14.5) as well as for truss connections.

Fig. 14.4 Semi-automatic welding of column splice. (Courtesy of Lincoln Electric Co.)

2. Submerged Arc Process

Automatic welding frequently uses the submerged arc process.   The word "submerged" refers to the fact that the arc is submerged under a mound of granular flux and, hence, is hidden from view.   Figure 14.6 illustrates the submerged arc process.   The welding circuit is exactly the same as that shown in Fig. 14.1, and the basic process is similar to that shown for the shielded arc-welding process in Fig. 14.3.   In addition, there is a motor which automatically feeds the bare wire electrode into the weld and a voltage control which maintains the desired arc voltage and arc length.   As with the manual shielded arc welding, the arc is formed between the electrode and the work.   The arc is completely and continuously covered by a mound of flux which is deposited from the hopper as the welding progresses along the work.   The heat of the arc melts the electrode, a portion of the parent metal, and part of the flux.   Again, the flux causes a slag covering to form on the weld; this is generally self-cleaning, that is, as the weld cools, the differential in cooling rates between the weld and the slag is sufficient to cause the slag to

Fig. 14.5 Vertical welding of beam-to-column connection. (Courtesy of Lincoln Electric Co.)

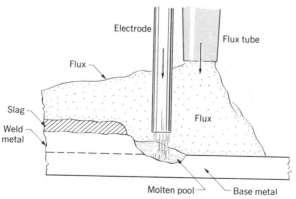

Fig. 14.6 Submerged arc-welding process.

Fig. 14.7 Automatic submerged-arc welding of a building column. (Courtesy of Lincoln Electric Co.)

free itself from the weld on its own accord. Figure 14.7 shows an automatic submerged arc-welding machine in action: the mound of granular flux directly over the weld is clearly visible, as is the electrode tube. (The column dimensions are 24 × 24 in., with 4-in. thick plates.)

Both manual and automatic welding may be operated on either alternating current or direct current, for both shop and field operations. The choice of the machine depends on its availability, the availability of current, the type and finish desired for the weld, the deposition and penetration rates, and even on the availability of skilled labor. Very high welding currents may be used—these will provide high deposition rates. Multiple electrodes also are used—these greatly increase the deposition rate and welding speed. The answer to questions concerning type of machine may be found in handbooks such as Ref. 4.18.

3. Gas Metal-Arc Welding

During the past decade, gas shielded-arc welding has come into increasing use with structural welding. Although this welding process originally had referred to the use of a non-consumable electrode, and although such processes are still in use, it is more common today to use the gas metal-arc welding process where the electrode is of a consumable material.

This process is also known as MIG welding, for metal inert gas; in particular, the so-called CO_2 process often is used in the welding of structural steel because of its many inherent advantages. The welding circuit is basically the same as shown in Fig. 14.1, with the addition of a line for supply of the shielding gas, which may be carbon dioxide, helium, or argon. The gas provides a shield surrounding the arc and weld region. Carbon dioxide gives an oxidizing effect since it decomposes into carbon monoxide and oxygen; thus, the process requires the use of an electrode which contains chemicals to neutralize this effect. It is very important that the gas is not blown away from the arc by a blower or wind; structural codes prohibit its use when the wind exceeds 5 mph unless some protection is built around the operator.

Because of high currents (100 to 400 amps) and the highly concentrated arc, the electrode wire must be thin—it is consumed at rates of up to 800 inches per minute. This requires the mechanical feeding of the wire, even for manual operations; see Fig. 14.8.

4. Flux-Cored Arc Welding

In the flux-cored arc welding process, there is a flux inside the cored wire electrode, which supplies both the shielding and the simple addition of any necessary alloying elements. External gas is not required, a real advantage for structural field welding because the wind will not blow the shielding away. The process provides a very high deposition rate.

5. Electroslag Welding

Essentially, the electroslag welding process is a single-pass vertical welding method for plates.[14.5,14.6] See Fig. 14.9. The two plates to be welded are set up in the electroslag welding machine, spaced about one inch apart. Water-cooled copper shoes are pressed against both sides of the vertical space to produce a "container" in which the welding proceeds and moves vertically upwards. Electrodes are fed down into this container and melted by the electric resistance of the welding current—the bare metal at the edges of the plates are also melted, and the mixture of molten weld metal is covered by a molten slag which is produced, either from small introduced amounts, but

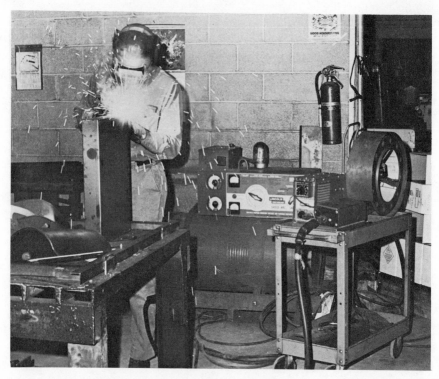

Fig. 14.8 MIG-welding (CO_2 process). (Courtesy of Lincoln Electric Co.)

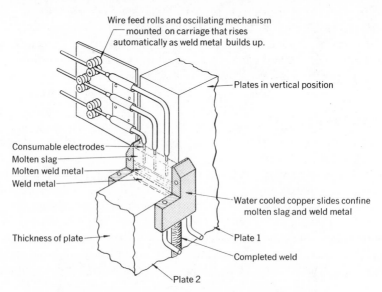

Fig. 14.9 Basic elements of electroslag welding.

usually from the use of flux-cored electrodes. There is no arc in this process. The reinforcement of the weld is dependent on the shape of the copper shoes; the electroslag process can handle plate thicknesses from about 1 in. to over 18 in.

6. Electrogas Welding

The electrogas welding process[14.5,14.6] was developed as an extension of the electroslag process, applicable to relatively thin plates, down to $\frac{1}{2}$ inch in thickness. The process is illustrated in Fig. 14.10. In contrast to the

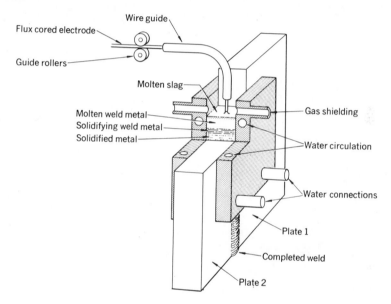

Fig. 14.10 Basic elements of electrogas welding.

electroslag process, an electric arc is maintained between the flux-cored electrode and the surface of the weld. The arc and weld are shielded by a gas introduced through the copper sliding shoes. When the plate thickness is over about $\frac{3}{4}$ inch, the arc is oscillated over the surface of the weld for uniformity.

Because of many similarities in the two processes, combination electroslag/ electrogas welding machines are in use.

14.3 WELDING POSITION

Most welding processes can be adapted to use in any desired position. However, all processes have certain limitations with respect to the position in which they can be most efficiently used.   Certain processes, such as the electroslag and electrogas, can be used in only one position.   There are four common positions: flat, horizontal, vertical, and overhead, and these are illustrated in Fig. 14.11 for both fillet and butt welds.   The welding position is of great importance in welding since the quality of the weld is directly affected by the manner in which it was carried out.   Not all electrodes are suitable for use in all positions; indeed, many electrodes are specified for use in a particular position, and should be used only in that position to obtain the best results.[4.18,14.4]   Obviously, the best position for most arc welding is in the downhand position, and welding in this position can be up to 4 times as fast as welding in the overhead position.   The welding of joints in

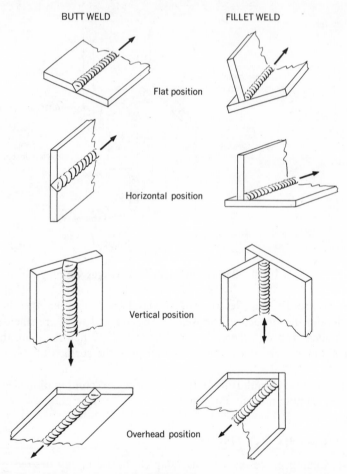

BUTT WELD   FILLET WELD

Flat position

Horizontal position

Vertical position

Overhead position

Fig. 14.11   Positions of welding for fillet and butt welds.   (Courtesy of American Welding Society.)

the overhead or vertical position requires limited arc-energy input and electrodes which can be operated to deposit in this position. These comments apply to those processes which can be used in all positions.

The welded joint should be designed in such a manner that the joint itself is accessible and the welding process can be carried out in the downhand flat position. To do this may require the use of mechanical positioners, such as the simple circular plate shown in Fig. 14.12. Mechanical positioning

Fig. 14.12 Positioning of column for ease in welding. (Courtesy of Bethlehem Steel Corp.)

results in welds which have a minimum cost, since the largest single item of cost is that of labor. Some mechanical positioners, in particular those for the welding of machines, may be very large and complicated:[4.18] their one purpose is to make the welding easy by welding in the downhand position. Through the use of appropriate electrodes, the production of welds of uniform quality in all positions is possible, and very often desirable when welding, for example, a beam-to-column connection in one continuous movement.

14.4 WELD TYPES AND JOINT PREPARATION

There are two basic types of welded joints: those with fillet welds and those with groove welds. These are shown in Fig. 14.13, and may be subdivided

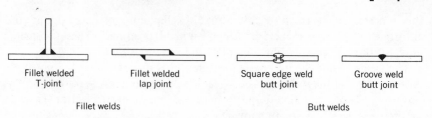

| Fillet welded
T-joint | Fillet welded
lap joint | Square edge weld
butt joint | Groove weld
butt joint |

Fillet welds   Butt welds

Fig. 14.13 Basic welded joints.

further into *joints,* butt, lap, tee, corner, and edge; and *welds or preparation,* fillet, square groove, bevel groove, V-groove, J-groove, and U-groove. Some of these joint classifications are shown in Fig. 14.14.

Joints to be welded may need preparation, that is, the machining or flame-cutting* of the mating parts into the proper shapes to facilitate welding and weld penetration. The preparation of a joint is important: the designer must consider the cost of preparation and need of increased penetration on the other hand. When joints are not beveled, the gap between the parts is somewhat larger than for prepared parts, and the welding procedure is somewhat more complicated.[4.18] Normally, for arc-welding processes, only mechanized welding will produce good welds in joints with no preparation. Welds in joints with no preparation may be the source of fractures due to brittle fracture or fatigue failure. Recommended proportions of grooves in joints to be welded may be found in Ref. 14.2. Figure 14.15 gives joint details for some welded shapes. The joint preparation for the box shape was obtained by flame-cutting. A typical double-torch gas flame cutter is shown

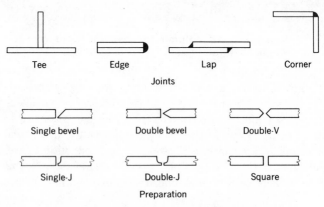

Tee   Edge   Lap   Corner

Joints

Single bevel   Double bevel   Double-V

Single-J   Double-J   Square

Preparation

Fig. 14.14 Joint and weld classification.

* Flame-cutting is also referred to as oxygen-cutting.

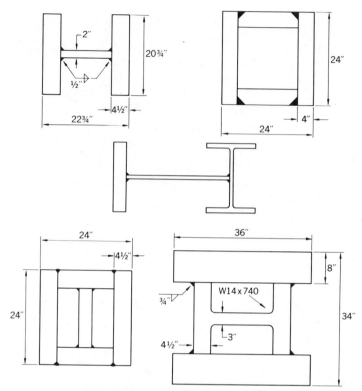

Fig. 14.15 Joint details for some welded columns.

in Fig. 14.16.   The detail of a prepared joint as the assembly is being tacked together prior to welding is shown in Fig. 14.17.   Flame-cutting is fast and accurate, so that joint preparation by machine cutting becomes uneconomical in comparison.

A complete classification of welding symbols is necessary to ensure the correct translation of the designer's ideas into the final fabricated product. Such a classification has been standardized and adopted by the AWS. Figure 14.18 shows some of these standard welding symbols; the complete classification is given in Ref. 14.2.   A complete glossary of standard welding terms is given also in Ref. 14.2.   Figure 14.19 shows the use of welding symbols in a structural drawing; in this case, the structure in the figure is the welded plate girder, part of which is shown under test in Fig. 8.19.

Not all joints are allowed in structural connections; some welded joints may have such poor penetration that they can be used only for fitting-up.   Some joints which are allowed without qualification are given in Fig. 14.20 for bridges.[1.26]   Similar prequalified joints are allowed for buildings.[1.26]   Joints

Fig. 14.16 Joint preparation by flame cutting. (Courtesy of Bethlehem Steel Corp.)

which are not prequalified may not be used without performing a relevant joint welding procedure qualification test.

The completed weld must be checked for soundness and size. Non-destructive tests normally would be used. The usual tests are visual: inspection of the weld to see if there are any obvious faults and use of the welder's gage to measure the size of the weld. Other non-destructive tests are the use of penetrant dyes, magnetic particle inspection, radio-graphic inspection, and ultrasonic inspection. Where the presence of a flaw would be critical, radiographic inspection can be used for its detection; however, the X-ray cannot reveal very small defects in very thick material.[14.2]

The design of welded joints is treated in Art. 19.4.

Fig. 14.17   Box column in tacking jig.   (Courtesy of Lincoln Electric Co.)

14.5 MATERIALS AND ELECTRODES

A judicious choice of material and electrode is necessary to produce a sound weld.   Although any steel can be welded under the right conditions, the mere process of welding does not imply that a satisfactory weld will necessarily result.   A satisfactory weld requires certain material properties and a certain chemical composition in both the steel and the electrode.

1. Steel

The chemical composition of steel to be used for welding has a limitation on the contents of certain elements which may be regarded as impurities because of their deleterious effect on weld strength.   Phosphorus and sulfur are the impurities which give the most trouble.   The weld metal is not made up wholly of electrode metal; on the contrary, it is a mixture of electrode

Basic Weld Symbols and Their Location Significance									
Location Significance	Arc and gas weld symbols								
	Fillet	Plug or slot	Arc-Seam or arc-spot	Groove					
				Square	V	Bevel	U	J	Flare-V
Arrow-side									
Other-side									
Both-sides		Not used	Not used						

Supplementary Symbols			
Weld all around	Field weld	Contour	
		Flush	Convex
◯	●	——	⌒

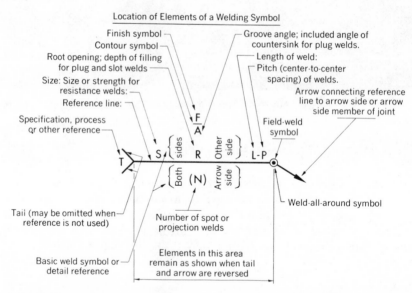

Location of Elements of a Welding Symbol

Fig. 14.18   Some standard welding symbols (AWS).   (Courtesy of American Welding Society.)

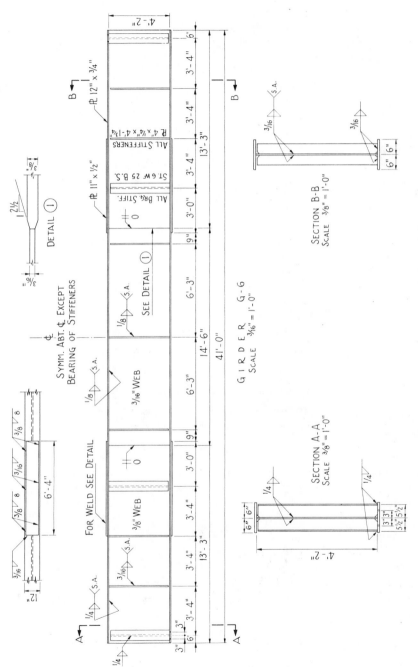

Fig. 14.19 Welding symbols on a structural drawing.

Fig. 2.9.1a of Ref. 1.26
Prequalified Manual Shielded Metal-Arc Welded Joints—
Material of **Limited** Thickness (L)

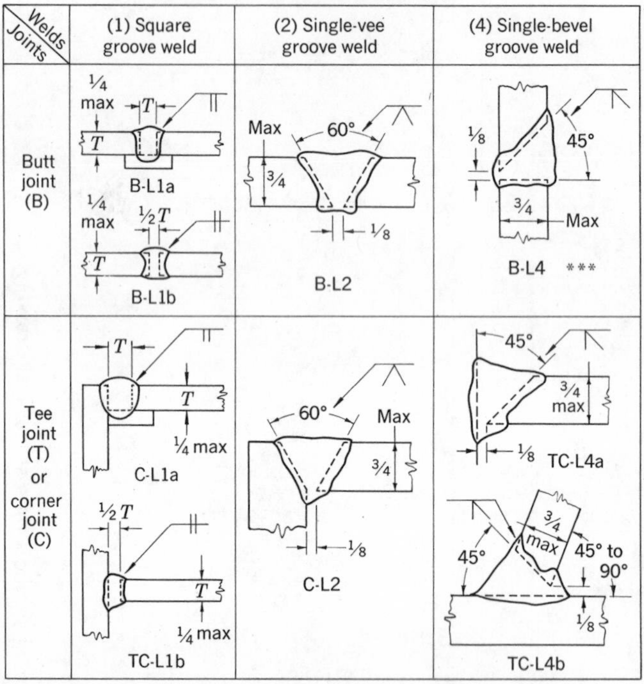

Method of joint and weld classi-
cation for prequalified joints in
Figs. 2.9.1a through 2.14.1 of
Ref. 1.26

1st — Joint type
  B — butt joint
  C — corner joint
  T — tee joint
  TC — tee and corner joint

2nd — Material thickness
  L — limited
  U — unlimited

3rd — Weld type
  1 — square groove
  2 — single-Vee groove
  3 — double-Vee groove
  4 — single-bevel groove
  5 — double-bevel groove
  6 — single-U groove
  7 — double-U groove
  8 — single-J groove
  9 — double-J groove

4th — Welding process, if
  not manual shield-
  ed metal-arc

  S — submerged arc
  G — gas metal-arc
  F — flux cored arc

Notes: 1. For all joints, except B-L1a, gouge root before welding second side (see 4.10.8).
  2. See 2.9.2 for allowable variation of dimensions and 3.3.4 for workmanship tolerances.
  3. If fillet welds are used in buildings to reinforce groove welds in tee and corner joints, they shall
     be equal to T/4 but not more than 3/8 in. Groove welds in tee and corner joints of bridges
     shall be reinforced with fillet welds equal to T/4 but not more than 3/8 in. T is the thickness
     of the groove weld.
  ***Bridge application limits the use of this joint to the horizontal position. (See 9.12.1.5).

Fig. 14.20 Some AWS bridge joints allowed without qualification.[1.26]
(Courtesy of American Welding Society.)

metal and parent metal, the mixture varying from the center of the weld to
the outside edge of the heat-affected zone. For this reason, it is necessary
that both parent metal and electrode be as pure as possible. In practice,
it is the electrode which is manufactured to very strict specifications to
exclude impurities.

The effect of some elements on the properties of steel is shown in Table
14.1. In particular, the effect of carbon is important, both for material
properties and for ease in welding. The higher the carbon content of steel,
the higher are the yield strength and tensile strength and the lower the

Table 14.1   Effect of Elements Present in Steel

Element	Effect	Preferred Content for Good Weldability (%)
Carbon	As C content increases, steel becomes both harder and stronger; a tendency to crack for over 0.25% C.	0.13–0.20
Manganese	With increase in Mn, steel becomes harder and stronger; tendency for cracking when too high.	0.40–0.60
Silicon	Produces uniformity in steel.	<0.10
Sulfur	When in right amounts, improves machinability.	<0.035
Phosphorus	Almost always an impurity.	<0.03

ductility; hence, the greater is the possibility of the occurrence of cracking on cooling after welding.

The term "weldability" basically is the ease of making a sound and service-able joint.[14.7] Some steels are said to be more weldable than others. A steel which is weldable is one which enables easy and economical welding and in which the welded joint meets all the necessary strength requirements. These factors, ease, cost, and strength, depend upon many variables, such as the design of the joint, the position of welding, and the chemical and material properties of both parent material and electrode.[14.7]

Most high-strength steels (high carbon content), or thick plates of structural carbon steel requiring heavy welds, may not be weldable without additional precautions; cracking is liable to occur in the weld upon cooling. Such a situation requires the use of preheating, which is the raising of the temperature of the area to be welded. Preheating is a very simple device to reduce the cooling rate and to reduce the temperature gradient between the molten metal of the weld and the cooler material of the parent metal; a more uniform temperature gradient ensures cooling without the possibility of cracking. The temperature used for preheating varies, depending on the thickness and chemical composition of the steel; however, temperatures to use and procedures to follow are given in any welding handbook.

2. Electrode

The chemical and mechanical properties of the deposited weld metal should be as close as possible to those of the parent material. Also, these properties should be such that it is possible to deposit sound welds under service conditions. Hence, a variety of electrodes is necessary for all the variations existing in job and material.

AWS has specifications for electrodes, for instance AWS A5.1,[14.8] for shielded arc-welding. These specifications provide, among other data and requirements, a numbering system which is in itself a classification of electrodes.

The numbering system is of the following form:

$$E \quad A \quad B \quad C \quad D$$

where E indicates metal arc-welding electrode; AB, the minimum tensile strength in ksi; C the welding position: 1 = all positions, 2 = flat and horizontal, 3 = only flat; and D indicates such items as the current supply and welding technique variables. An example of the numbering system is E6012, the common electrode used with structural steel; it is defined as: all position, either DC straight polarity or AC machine, and a minimum tensile strength of 60 ksi.

There are literally hundreds of electrodes. Each one is designed for a specific purpose.

The selection of an electrode is not usually difficult since most electrodes are much stronger than the corresponding parent material. The first step in the selection is to decide on the minimum tensile strength and yield point; next, on the current supply and application (this may depend on the availability of a welding machine); and finally, to decide on the position for the welding.[4.18,14.3,14.4] When problems (such as corrosion) may exist which require an exact control of chemical content, the manufacturer's composition of the electrode or deposited weld metal should be checked. It should be noted that the welding characteristics of an electrode are important and influence the speed of welding, and the appearance and strength of the finished weld.

For submerged arc welding, gas metal-arc welding, flux-cored arc welding, electroslag, and electrogas welding, a wide variety of electrodes is available in all sizes, and similarly, a wide variety of fluxes. Any desired requirement for the welding process or for the finished weld can be obtained by careful combination of electrode and flux types.[1.21,4.18,14.3] The flux may be open as in submerged-arc welding, or it may be inside a cored electrode, as in flux-cored arc welding.

The AWS has specified[1.26,14.5] that combinations of electrode, flux, and/or shielding gas will produce weld metal having specified mechanical properties; for example:

Exx   manual shielded metal arc welding
Fxx   submerged arc welding
ExxS   gas metal arc welding
ExxT   flux-cored arc welding

indicate the grades of the welding processes which correspond to structural carbon steel. For example, E90S, has a minimum tensile strength of 90 ksi, a minimum yield strength of 78 ksi, and a minimum elongation in 2 in. of 17 per cent.

14.6   HEAT INPUT AND RESIDUAL STRESSES

1. Heat Input and Temperature Distribution

The instant the circuit is closed and the welding process commenced, a source of heat is set up. The heat input creates a number of changes, the most important consequence of which is the plastic deformations which result and which lead to the formation of residual stresses. The marked effect of residual stresses on column strength is treated in Art. 9.7; the effect of residual stresses is considered also in Arts. 8.3, 11.2, and 17.3.

The first obvious effect of a heat input is the increase in the temperature at and around the weld. The properties of steel at elevated temperatures are not the same as those at room temperatures. Indeed, with rising temperatures, there is a decrease in both the yield point and Young's modulus, until, at about 1500°F, both values are zero. At this temperature, the material is plastic. (The variation of the properties of steel with temperature are discussed further in Ref. 14.9.)

2. Residual Stresses

Residual stresses are formed as the result of plastic deformations. Heat is not necessary in the formation of residual stresses; the plastic deformation from cold-bending also produces residual stresses. Residual stresses due to welding cannot be formed until the temperature is approximately 1400°F, when the material has little or no capacity to resist loading induced by differential movements due to the temperature distribution.

The formation of the residual stresses is described in Ref. 14.9. In brief, residual stresses due to welding are formed as a result of the differential in the heating and cooling rates of the various fibers of the material. At any time during and after the instant of welding, some plastic fibers in the cross section will be cooling and other elastic fibers will be heating. The action of the loads on the elastic fibers due to the changes in temperature is not counteracted by the plastic fibers until the plastic fibers have cooled sufficiently to become elastic. The interaction between the different fibers in the cross section results in the presence of internal stresses after the material has cooled. These internal stresses are called "residual stresses." The stresses may be quite large, and in any event, the residual stress in the weld will always be tensile and approximately equal to the yield point of the weld metal. Typical residual stress distributions in welded shapes are shown in Fig. 9.21.

Away from the weld, the magnitude of the residual stresses in the shape is a function of the magnitude of the heat input.[14.9] In general, the smaller the compressive residual stress is, the better for column strength (see Art. 9.7). To minimize residual stress, the heat input should be lowered. This can be accomplished by decreasing the effective current and voltage or increasing the welding speed.[14.10] In the future, it may be possible to develop welding procedures wherein the welding is conducted at so fast a rate that the heat input at any section is reduced, but the penetration is sufficient to create a proper weld. The submerged arc-welding process does result in welding at high speed, with a concentrated arc giving high penetration; but this combination does not lead to low residual stresses, as witnessed by the distribution shown in Fig. 9.21.

Residual stresses should be kept as low as possible in magnitude by a strict control of fabrication procedures. These include preheating, avoidance of intersecting welds which may lead to high stresses at the intersections, and the use of a welding sequence allowing some movement during welding. Other precautions are enumerated in Ref. 14.11.

3. Distortion

Unless care is taken, fabrication by welding may result in some deformations in the final shape. Since a distorted shape is undesirable, methods must be used to limit these deformations. For instance, the sequence of welding should be designed so that the distortions resulting after each welding pass tend to cancel each other, resulting in a straight and square end product. Most structural members seldom present distortion problems since a proper welding sequence is not difficult to obtain.

Figure 14.21 shows the welding sequence of the 10-by-10-in. box column referred to in Fig. 14.12; diagonally opposite corners are welded in sequence, approximately 8 ft at a time. Figure 14.22 shows the straightness of a fabricated length of approximately 60 ft; no mechanical straightening was

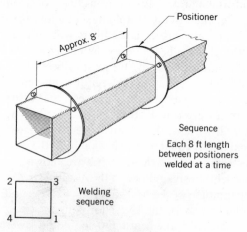

Fig. 14.21 Welding sequence.

Fig. 14.22   Column after welding. (Courtesy of Bethlehem Steel Corp.)

required. The usual practice, however, is to weld the whole length of a column continuously.

Another example of the deliberate use of distortions is in the fabrication of structural shapes such as H-shapes. Their fabrication requires the pre-bending of the component parts so that after welding and cooling the shape is perfectly square. As with the principle of welding sequence, this method also uses a process of cancellation of effects. This is illustrated in Fig. 14.23. Depending on the practice of the shop, the plate either may be pre-bent by clamping or else may be pre-bent by plastically deforming the plate along its center line. Rule-of-thumb methods exist in each welding shop for the amount of pre-bending required. For very thick plates, it is not possible to have any prebending at all. Automatic welding machines, where the shape is fed through the machine, accomplish the same effect as in Fig. 14.23 through the use of positioning wheels.

The problem of distortions has been studied theoretically in great detail in Ref. 14.12. Reference 14.13 has shown theoretically that high residual

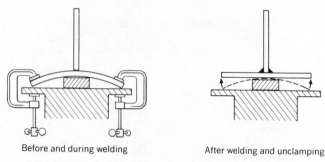

Before and during welding   After welding and unclamping

Fig. 14.23 Pre-bending of shapes for fabrication.

stresses set up in plates due to welding are the cause of the distortion of some plates after cooling and before the application of any load.

14.7 WELDING IN STRUCTURAL DESIGN

The use of welding is an accepted practice for all structural fabrication; as with any other fabrication process, there are both advantages and disadvantages to the use of welding.

The important advantages to be gained by welding are:

1. A decrease in the weight of the structural member, both by the use of simple joints and the utilization of the full section, unimpeded by rivet or bolt holes.
2. A saving in time, and consequently of money, since much less preparation is needed than for bolting; there are no punching and drilling of holes, assembling for fit, etc.
3. The ease of repairing or strengthening of old structures with a minimum of inconvenience.
4. The design time is shortened since there is less detailing of the fabrication.
5. Good appearance is obtained, leading to a greater architectural versatility.
6. The noise of fabrication is kept at a low level.
7. The joints can develop their full moment resistance for both elastic and plastic design, leading to a more economical use of continuous structures.

The main disadvantages of welding are the problems of brittle fracture and fatigue. These problems arise from the change in the properties of, or the introduction of notches into, the parent metal. Careful design and fabrication can eliminate most of these effects; structural failures due to fatigue or brittle fracture are rare. The topics of brittle fracture and fatigue are treated in Chapters 15 and 16 respectively.

BRITTLE FRACTURE

15.1 INTRODUCTION

The service life of a structure may be terminated by an unexpected, rapidly propagating brittle fracture. Brittle fracture in service is, fortunately, rare but can be catastrophic because it may occur at low nominal stresses, demonstrates no visible plastic deformation or signs of distress prior to fracture and, once initiated, may propagate at a high velocity to complete fracture. Brittle fractures have been observed in bridges, ships, natural gas transmission pipe lines, tanks, stacks, penstocks, turbo-generators and galvanized transmission towers.[15.1–15.4]

Non-ship brittle fractures generally have occurred in wide-plate structures such as tanks made of carbon steel. The loading at the time of fracture was usually static and not impact. The steel was often of an unsatisfactory metallurgical structure and chemical composition. Fractures originated at fabrication defects, weld cracks, incomplete weld penetration, gouges in the plates, or at punched rivet holes. Cracks did not follow welds or heat-affected zones except in the presence of gross weld defects. Defective welds were responsible for crack initiation but crack propagation was associated with brittleness of the base material. Ship failures were caused by defective welds combined with poor design practice. Although brittle fractures have occurred in riveted structures, complete fractures are more likely to occur in welded structures due to an absence of crack-arresting discontinuities such as lap joints.

Fractures are generally classified by the amount of deformation prior to fracture, crystallographic mechanism, or the appearance of the fracture surfaces, as in Table 15.1. Steel has a crystalline atomic structure and may

Table 15.1   Fracture Classification[15·5]

Characteristic	Type of Fracture	
Amount of visible deformation prior to fracture	Ductile	Brittle
Crystallographic mechanism	Shear	Cleavage
Fracture surface appearance	Fibrous	Granular (or crystalline)

fracture along shear (or slip) planes after considerable plastic flow or by cleavage between planes of atoms with little or no visible plastic flow. In most service brittle fractures, some crystals undergo sufficient plastic deformation to produce alternate regions of shear and cleavage fracture.

A brittle fracture in a plate is modeled in Fig. 15.1. After sufficient local plastic deformation at the notch root, a fibrous (shear) crack is initiated and is visible, after fracture, as a "thumbnail" region. The shear crack sharpens the notch producing constraint to further plastic flow and, accelerating, increases the strain rate until the value of the constrained yield strength is driven above the fracture strength. The crack then propagates by a continuous process of cleavage microcrack initiations ahead of the main crack front. These microcracks spread radially until they link with other spreading microcracks or the advancing fracture front by shear fracture. The regions

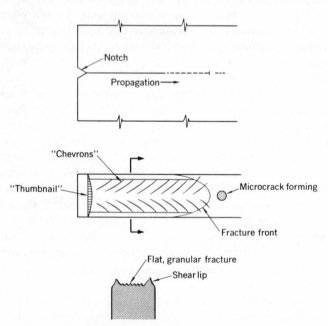

Fig. 15.1   Crack propagation.[15·6]

of alternate shear and cleavage give rise to the widely observed "chevrons" which are orthogonal to the fracture front. It should be noted that "chevrons" may not be visible at low temperatures and/or high crack propagation velocities. The flat fracture surface may intersect the surface of the plate in 45° shear-lips, which decrease in width with an increase in brittleness and/or thickness. The brittle crack will continue to propagate as long as the elastic strain energy released by crack growth exceeds the work required to form and drive the crack. If the required work surpasses the released strain energy, the crack will be arrested in a ductile manner.

15.2 FACTORS INFLUENCING NOTCH TOUGHNESS

Notch toughness or the resistance to brittle fracture is affected by the interactions of many factors, for example:

Notch acuity	Grain size
Service temperature	Chemical composition
Strain rate	Prior strain history such as cold-working and
Thickness	strain-aging.

Brittle fracture will occur when the yield strength is elevated above the fracture strength. Not only is the stress at the notch root increased by an increase in notch acuity but the yield strength at the notch root is also elevated. The Hencky-Mises yield criterion states that plastic flow will be initiated when the maximum principal stress equals

$$f_1 = F_y \left[1 + \left(\frac{f_2}{f_1}\right)^2 + \left(\frac{f_3}{f_1}\right)^2 - \frac{f_2}{f_1} - \frac{f_3}{f_1} - \left(\frac{f_2}{f_1}\right)\left(\frac{f_3}{f_1}\right) \right]^{-1/2} \quad (15.1)$$

where f_1, f_2, and f_3 are the principal stresses and F_y is the uniaxial yield strength. At the notch root, f_2, which is parallel to the notch axis, is zero but is a maximum just ahead of the notch root. The combination of f_1, perpendicular to the plane of the notch, and f_2 cause a decrease in thickness just ahead of the notch which is resisted by the less stressed material adjacent to the notch root. A through-the-thickness stress f_3 is developed by this constraint to lateral contraction. Inspection of Eq. 15.1 reveals that the flow stress f_1 exceeds F_y by an increasing amount as f_2 and f_3 increase. An increase in section thickness, decrease in service temperature and/or an increase in strain rate will also act in combination with notch acuity to elevate the yield strength. The elevation of yield strength due to notch acuity and temperature decrease is illustrated in Fig. 15.2.

A decrease in grain size increases the fracture strength and retards the growth of cleavage cracks. Grain size is controlled by the deoxidation

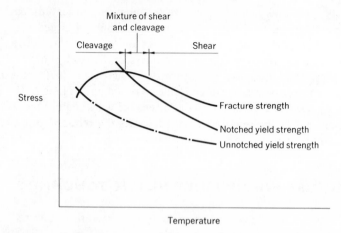

Fig. 15.2 Effect of temperature on yield and fracture strength.

practice used in steel-making, the finishing temperature of the rolling operation and the rate of cooling after rolling.   Aluminum-killing, low finishing temperature, fast cooling rate, and increasing the manganese-to-carbon content ratio promote grain refinement.   Thicker plates cool more slowly, allowing time for grain growth and a concomitant reduction in notch toughness.

Residual stresses alone will not cause brittle fracture.   However, residual stresses combined with sharp notches, defective welds, and the like can facilitate crack initiation.   When plastic flow is unimpeded, residual stresses are of little consequence.

Factors influencing notch toughness as well as fracture mechanisms have been reviewed elsewhere in detail.[15.5,15.7–15.13]

15.3 FRACTURE CRITERIA

1. Transition Temperature

The most widely used fracture criterion has been the ductility transition temperature (DTT) derived from the Charpy V-notch impact test (see Chapter 2).   The energy required to fracture small notched beams with a pendulum-like hammer is plotted versus test temperature (Fig. 15.3). Cleavage fractures occur below the DTT and shear or ductile fractures occur at the upper shelf energy (shear energy or tear energy) level.   The fracture appearance transition temperature (FATT) is the temperature at which the Charpy fracture surface appearance is 50 per cent fibrous (shear) and 50 per cent granular (cleavage).

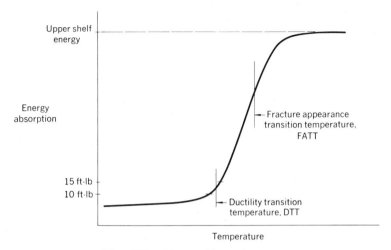

Fig. 15.3   Charpy V-notch test.

The Charpy V-notch test has been correlated with the World War II ship failures.   Extensive impact tests[15.2] on casualty plates in which fractures were initiated showed that the service temperature at fracture was, approximately, the DTT.   Tests on casualty plates in which fractures were arrested showed that the fracture arrest temperature in service was, approximately, the FATT. The correlation between the Charpy V-notch 10 to 15 ft-lb DTT and the ship failures was developed for wartime rimmed and semi-killed carbon steels and may not be generally valid for service evaluation of steels developed since that time, or for structures substantially different from ships.

Improvement in the 15 ft-lb DTT may result in a reduction of the upper shelf energy and, instead of ductile arrest, a crack may be propagated by a "low energy" shear or tearing.   Therefore, the upper shelf energy should be considered as well as the DTT when using the Charpy impact test for comparison purposes.[15.14]   The effects of various parameters on both the DTT and the upper shelf energy of the Charpy test are shown in Table 15.2.

2. Fracture Analysis Diagram

A generalized stress-temperature relation,[15.3] the fracture analysis diagram, for both fracture initiation and crack arrest is shown in Fig. 15.4a.   The characteristics are defined by the nil-ductility temperature (NDT), the fracture transition elastic (FTE) temperature and the fracture transition plastic (FTP) temperature.   At the NDT (measured in the drop weight test), the tensile ductility is decreased, essentially, to zero.   In the presence of a small sharp flaw, the nominal fracture stress equals the yield strength when

Table 15.2   Factors Affecting Charpy
Impact Curve

Factor	Ductility Transition Temperature	Upper Shelf Energy
Increase content of		
Carbon	Increase	Decrease
Manganese	Decrease	Increase
Nickel	Decrease	Increase
Phosphorus	Increase	*
Sulfur	Increase	*
Increase		
Grain size	Increase	*
Thickness	Increase	Decrease
Finishing Temp.	Increase	*
Cooling rate	Decrease	*
Strain aging	Increase	Decrease
Neutron irradiation	Increase	Decrease

* Little change.

the temperature is below the NDT.   As the flaw size increases, the nominal fracture stress decreases approximately as the square root of the flaw size. The FTE is the highest temperature at which fractures will propagate for purely elastic loads.   The FTP is the temperature above which fractures are 100 per cent shear and the fracture strength approaches the tensile strength. The crack arrest temperature (CAT) curve shows the temperatures for the arrest of brittle cracks at various levels of nominal stress.   The FTE and the FTP are measured in the explosion bulge test while the CAT curve was developed from Robertson, Esso, and Illinois wide-plate studies.   Correlation between the fracture analysis diagram and service failures has been claimed.[15.3]

The importance of the upper shelf energy was mentioned in the discussion of the Charpy test.   The fracture propagation by low energy shear or tearing is illustrated in Fig. 15.4b.

3. Fracture Mechanics

Fracture mechanics offers a quantitative relationship between stress, flaw or crack size, and material and has been developed from linear elastic continuum mechanics and extensive experimentation.[15.15–15.19]

The basis of fracture mechanics is the hypothesis that a stable growing crack will become unstable and propagate rapidly with no further increase in load when:

The rate of strain energy released by crack growth equals the rate of plastic work required to advance the crack.

The plastic work goes into deformation of the crystals just beneath the flat fracture surface and into formation of shear lips.   For a given crack length,

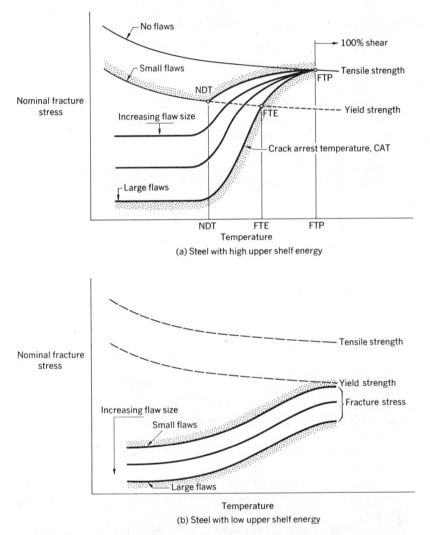

Fig. 15.4 Fracture-analysis diagram.

the plastic work required for crack extension decreases with an increase in section thickness, a decrease in temperature, or an increase in strain rate.

At the instant of crack instability, the critical values of the crack length and stress (remote from the crack) are related by a parameter called fracture toughness K_c. K_c is the critical value of the stress intensity factor K. K, a function of flaw or crack size, stress distribution away from the flaw, and geometry of the member, is derived from linear elastic analysis of a continuum containing a sharp crack (zero root radius). For example, K for a wide plate containing a very short crack (compared to plate width) is

$$K = f\sqrt{\pi a} \qquad (15.2)$$

where f is the stress applied to the plate at a considerable distance from the crack and a is one-half the crack length. Within certain limitations, K_c may be assumed to be a constant for a given material, metallurgical structure and section thickness, regardless of geometry. Therefore, K_c for complex geometries may be determined from small, relatively simple laboratory specimens.

As section thickness increases, K_c becomes smaller approaching a minimum value K_{Ic}, the plane strain fracture toughness (Fig. 15.5). In general, K_{Ic}

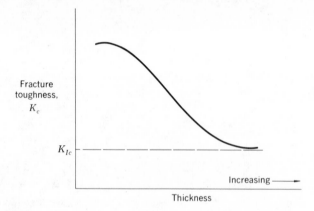

Fig. 15.5 Effect of thickness on fracture toughness.

is the value of stress intensity factor K at which the crack will begin to grow in a stable manner with increasing stress f. The crack becomes unstable and propagates rapidly with no further increase in stress when K equals K_c. For very thick sections, K_c equals K_{Ic} and the crack is unstable upon initiation.

Fracture mechanics analysis is valid for steels with yield strengths greater than 150 ksi, providing certain criteria concerning section thickness, crack length, and net section stress (uncracked material in the plane of the crack) are met. For lower strength materials, fracture mechanics is valid when the yield strength of the material is elevated by high strain rates and very low temperatures. Extensive research[15.18,15.19] is under way to extend the application of fracture mechanics to the lower-strength materials at strain rates and temperatures of practical interest to structural engineers.

The fracture analysis diagram has been interpreted and expanded in terms of fracture mechanics parameters.[15.20]

Efforts to correlate Charpy V-notch impact data with K_{Ic}[15.21,15.22] have received considerable notice among structural engineers. For example, one correlation is[15.22]

$$K_{Ic} = 15.5\sqrt{\text{CVN}} \qquad (15.3)$$

where K_{Ic} is the plain strain fracture toughness, in ksi$\sqrt{\text{in}}$., and CVN is the Charpy V-notch impact energy at the temperature of interest, in ft-lb. The correlations relate impact energy at low temperatures with values of K_{Ic} measured at those temperatures. However, when the correlation equations are used to predict K_{Ic} for structural steels at service temperatures usually experienced by structures, the stress–crack length combinations are quite conservative. Generally, the stresses exceed the yield strength of the material, a situation in which linear elastic fracture mechanics is not applicable. The use of Eq. 15.3 is illustrated for a $\frac{3}{4}$-in.-thick A36 carbon structural steel at 0°F. The range of Charpy V-notch impact energies at 0°F for this material is 15 to 34 ft-lb and the comparable range of K_{Ic}, from Eq. 15.3, is 60 to 90 ksi$\sqrt{\text{in}}$. The crack length, $2a$, computed from Eq. 15.2 with a working stress $f = 20$ ksi, ranges from 6 to 13 inches. A common fracture design concept ("leak before burst") requires that a metal tolerate a through-thickness crack of a length twice the metal's thickness before fracture occurs. Combining Eqs. 15.2 and 15.3, with plate thickness t, gives

$$f_{cr} = 15.5\sqrt{\frac{\text{CVN}}{t}} \qquad (15.4)$$

For $\frac{3}{4}$-in.-thick A36 at 0°F, with CVN from 15 to 34 ft-lb, the critical fracture stress (Eq. 15.4) varies from 39 to 59 ksi.

General yielding fracture mechanics, which is applicable to structural steels, is still in the development stage. However, three areas of research show promise. One, the COD or crack opening displacement method[15.23], is examining the strain near the crack tip as a fracture criterion. The second approach[15.24] is studying the flaw tolerance of a metal in terms of gross strain to maximum, or fracture, load. The third method[15.25] proposes to establish the upper and lower bounds on fracture toughness by means of an equivalent energy relationship between a model (cracked test specimen) and the prototype (a cracked structural member or component).

15.4 DESIGN OF FRACTURE-RESISTANT STRUCTURES

There are two approaches or concepts to the design of fracture-resistant structures:

1. Minimize the probability of fracture initiation, or "safe-life" concept.*
2. Minimize the probability of major fracture propagation, or "fail-safe" concept.†

---

* See Chapter 16 for a discussion of "safe-life" and "fail-safe" design concepts.
† See previous footnote.

1. Safe-Life Design Concept

Since crack initiation requires a much higher nominal stress than propagation, the structure is designed and fabricated so that notch severity is reduced and initiation does not occur under the expected loading. Weld details producing high triaxial constraint to yielding are eliminated. During fabrication, intensive inspection is used to try to detect and eliminate flaws. A material is selected which is notch tough at or below the expected service temperature and which is not susceptible to extreme strain aging effects.

The safe-life approach can be extremely expensive and, although apparently quite conservative, could result in a catastrophic failure due either to an accidental overload surmounting the crack initiation barrier, or to critical flaws that were undetected during inspection.

2. Fail-Safe Design Concept

Crack initiation may be inevitable and may occur early in the life of the structure. In addition, performance of non-destructive tests (such as ultrasonic, magnetic particle, eddy current techniques) during fabrication is no guarantee that critical flaws will be detected. Therefore, the structure is designed with crack arresters and redundancies to prevent crack growth to critical proportions. Materials with sufficient ductility to retard or arrest crack growth are selected. High notch toughness and relatively long critical crack lengths are synonymous.

The fail-safe concept will also reduce the probability of fatigue cracks triggering brittle fractures.

3. Application of Impact Tests to Design

The safe-life concept is illustrated by the American Bureau of Shipping (ABS) steel selection methods, which have reduced (but not eliminated) the incidence of ship fractures. The ABS specifies material composition and steel-making practice. Fine-grain practice (aluminum-killing), decrease in carbon content, and an increase in manganese content are specified for thick plates. ABS also sets a minimum value for Charpy V-notch energy absorption at a specified temperature. ABS concluded from the World War II studies that fractures were unlikely in plates with an energy absorption of 10 ft-lb or more at the expected service temperature. The energy value was later raised to 15 ft-lb to cover a wider range of steels.

The reliability of the 15 ft-lb DTT criterion when applied to steels other than those of the wartime studies is questionable. It is often economically impractical to select a material with a DTT less than the expected service temperature and, in addition, impact tests give no information about fracture initiation under static loading or about the critical combination of stress and flaw size.

The real value of impact tests is as a quality assurance or acceptance test to screen steels. Until a critical test is available that will reliably predict the fracture behavior of steel structures in service, impact tests and the enormous amount of available impact test data will be used as a standard for acceptance or rejection. However, upper shelf energy as well as the DTT should be included in criteria based on impact tests.

4. The Fracture Analysis Diagram

The fracture analysis diagram permits a more extensive design criterion. Using the values of yield strength F_y and tensile strength F_u from a tensile test, the NDT from a drop weight test, and the following equalities:

$$FTE = NDT + 60°F \qquad (15.3)$$

$$FTP = NDT + 120°F \qquad (15.4)$$

the fracture analysis diagram is constructed for a particular steel (Fig. 15.6).[15.3,15.14] The initiation stress-temperature curves for various flaw sizes are based on the interpretation of cracked plate tests.[15.3] In Fig. 15.6, values of temperature and stress for $\frac{3}{4}$-in. thick A441 plate have been added for illustration.

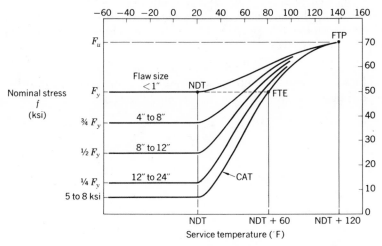

Fig. 15.6 Fracture-analysis diagram for $\frac{3}{4}$-in.-thick A441 plate.

Given an expected temperature, the designer may use the fracture analysis diagram to select a nominal working stress:

1. To the right of or below the CAT curve
2. To the right of the NDT and CAT curve
3. To the left of the NDT or CAT curve, provided that the maximum flaw size is considered

Alternatives 1 and 2 follow the safe-life design concept and may be conservative. Alternative 3, a fail-safe approach, is of unknown reliability because the required stress–flaw size–temperature test data is not readily available for low-strength materials. It is also recognized that the CAT curve may be shifted to the right in the fracture analysis diagram for low-strength steel thicknesses over 2 in. In general, however, the diagram has been used in design and experience has indicated that the method is satisfactory albeit conservative.

5. Design Principles

As in fatigue-resistant design, a brittle-fracture-resistant structure is not assured by only the selection of a working stress satisfying some criterion. It is also necessary to follow certain principles in order that the probability of brittle fracture is minimized. They are:

1. Avoid severe stress concentrations.
2. If no crack-arresting discontinuities exist, introduce crack arresters and/or redundant stress paths to localize fractures.
3. Avoid welded details with high restraint to plastic flow.
4. Avoid rapid weld cooling. Use preheat or postheat where required.
5. Weld with low-hydrogen electrodes and the submerged-arc process.
6. Avoid arc strikes.
7. Avoid excessive chipping and hammering of welds.

It should also be remembered that bad welds may result from inaccessibility due to poor design as well as from inexperienced welders.

FATIGUE

16.1 INTRODUCTION

If a structural member or connection is subjected to a cyclically varying load, it may fail after a certain number of load applications even if the maximum nominal stress in a single cycle is much less than the yield stress of the material, weld metal, or fastener. Figure 16.1 shows two members subjected to cyclic loads; the variation of stress with time between maximum and minimum limits, S_{max} and S_{min}, is indicated.

A crack may be initiated at some mechanical or metallurgical discontinuity (stress concentration) and be propagated through the material with successive load repetitions until the affected part loses its ability to carry load. This fracture phenomenon is known as "fatigue."

If the stress ratio R (algebraic ratio of minimum stress to maximum stress) is held constant and the maximum stress S_{max}* is varied, the number of cycles to failure (the fatigue or service life) will also vary. This relationship is plotted diagrammatically in a semilog or log-log "S-N curve." The semi-log S-N curve for axially loaded structural carbon steel (A36) specimens with ground surfaces is shown in Fig. 16.2. The S-N curve that best describes the central tendency of the data at a given maximum stress defines the fatigue life prior to which 50 per cent of the specimens failed at that stress

---

* S denotes the critical value of the stress f in a stress spectrum sufficient to cause a failure for a given number of stress cycles N. Notation in this chapter conforms to that of ASTM.[16.1]

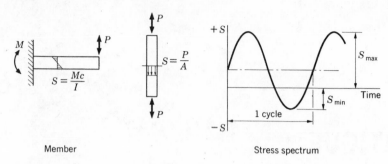

Fig. 16.1 Structural member subjected to cyclically varying load.

level. The family of parallel lines define, for each maximum stress, fatigue lives for per cent failures other than 50 per cent. The lines shown in Fig. 16.2 apply only to the tested sample. If it is desired to estimate fatigue lives for various per cent failures of the population from which the sample was taken, statistical measures of uncertainty must be considered.[16·2]

The fatigue life of ductile metals increases with a decrease in maximum stress until a region is reached in which some proportion of the specimens tested at a given maximum stress survive an arbitrarily large number of cycles. This proportion of survivals increases with a further decrease in maximum

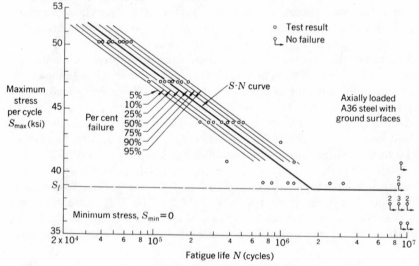

Fig. 16.2 S-N curve.[16·2]

stress. The fatigue limit S_f is defined as the maximum stress at which 50 per cent of the specimens will survive some large number of cycles, for example 10 million cycles.

At the present time, there is no analytical method available by which the fatigue strength of a structural member or connection may be predicted for a desired fatigue or service life. Instead, fatigue tests must be conducted on the prototype or on models, and the prediction is then based on the experimentally determined S-N curve (an empirical model.)

16.2 FACTORS AFFECTING FATIGUE OR SERVICE LIFE

The prediction of fatigue resistance is complicated by the fact that citation of maximum stress alone does not define a unique service life. In general, the stress spectrum to which a member or connection will be subjected, the nature and condition of the part, and the environment in which it will function will all influence the service life.

The factors influencing the fatigue resistance of a structural part or laboratory specimen are:

A. Load spectrum
   1. Stress ratio
   2. Maximum stress
   3. Stress range
   4. State of stress
   5. Repetition of stress (regular or random)
B. Nature and condition of member
   1. Prior stress history
      a. Residual stresses
      b. Work-hardening
   2. Size and shape of member
      a. Size effect (simulation of a member by a small specimen)
      b. Stress gradient
      c. Presence of notches
      d. Surface condition
   3. Metallurgical structure
      a. Microstructure, grain size, and chemical composition
      b. Mechanical properties
   4. Welding
      a. Metallurgical
      b. Mechanical
C. Environmental
   1. Temperature
   2. Atmosphere

1. Effects of Load Spectrum

Stress Ratio.   It was mentioned in Art. 16.1 that the stress ratio R may be held constant during the construction of a given S-N curve.   If R is changed from one S-N curve to the next, the plots will appear displaced as in Fig. 16.3. It can be seen in Fig. 16.3 that different values of R will each provide a different S-N curve.

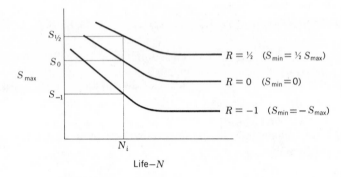

Life$-N$

Fig. 16.3   S-N curves — variation of R.

A more convenient graphical presentation of fatigue behavior is the constant-life diagram recommended by the Welding Research Council and adopted by the AWS.   This diagram is presented in Fig. 16.4 and is constructed from the S-N curves in Fig. 16.3.

In Fig. 16.4, the ordinate of a general point in this diagram is the maximum stress f_{max} in a single stress cycle and the abscissa is the minimum stress f_{min} in that cycle.   The curve $ABCD$ represents the critical combination of maximum and minimum stresses, S_{max} and S_{min}, that will cause fatigue failure at N_i cycles.   If the point (f_{max}, f_{min}) falls below the curve $ABCD$, fatigue failure will probably not occur at N_i cycles.   If, on the other hand, the point falls on or above the curve $ABCD$, failure is likely to occur at or prior to N_i cycles.   The usual set of fatigue test data often reflects considerable scatter, and it must be remembered that the S-N curve fitted to that data is a measure of the mean values only.   In Fig. 16.4, the critical range of stress S_r is the vertical distance between the "N_i" curve and the sloping line $R = 1$.   It is the range of stress that will produce failure in N_i cycles for a given minimum (or maximum) stress.   The critical range of stress S_r decreases with an increase in maximum stress per cycle, S_{max}.   Note that a separate curve is required for each fatigue life (Fig. 16.5).

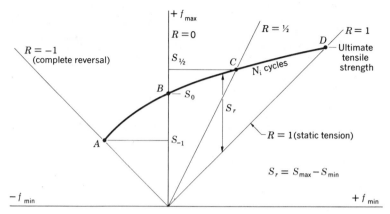

Fig. 16.4 AWS–WRC (Roš) diagram.

State of Stress. It has been assumed in the previous discussion that the stresses are uniaxial. The theories of static failure under combined stresses have been applied to the condition of cyclic combined stresses and have been found to give reasonably good results. The distortion energy theory is used to approximate the behavior of ductile materials under combined cyclic stresses.

In Fig. 16.6, an S-N curve AB has been obtained for a material loaded uniaxially. If a combined cyclical stress state in which R remains unchanged is imposed on the material, an equivalent uniaxial stress $(S^c)_{\max}$ may be calculated from the modified distortion energy equation.[16.3]

$$(S^c)_{\max} = \sqrt{(f_x)_{\max}^2 - (f_x)_{\max} \cdot (f_y)_{\max} + (f_y)_{\max}^2 + 3(f_v)_{\max}^2} \qquad (16.1)$$

When $(S^c)_{\max}$ is greater than $(S_x)_{\max}$, the S-N curve of Fig. 16.6 shows that the fatigue life has been reduced by replacing a uniaxial stress state with a

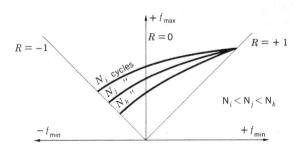

Fig. 16.5 AWS–WRC diagram for different fatigue lives.

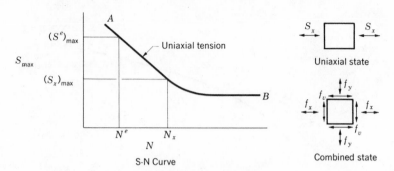

Fig. 16.6 Stress state.

particular combined stress state. It can also be shown that certain combined stress states may increase fatigue resistance. Equation 16.1 is valid strictly only for complete reversal and requires modification for other stress ratios.

Repetition of Stress. The situation in which the maximum stress is constant throughout the life of a member is more critical than the more realistic case in which this maximum stress is achieved only in a certain percentage of cycles. The member loaded with the former spectrum will, in general, have a fatigue life that is shorter than a member loaded with the latter spectrum.[16.4]

Constant stress-amplitude fatigue tests adequately rank structural details as to their relative fatigue resistance and may reveal the sections of a structure that are critical in cyclic loading. Constant-amplitude tests, however, cannot be used to forecast the service lives of structures with any confidence, unless they are related through cumulative-damage hypotheses to the variable-amplitude loading experienced by structures. Many cumulative-damage hypotheses have been proposed but the simplest and most widely used is the Miner-Palmgren hypothesis, which states

$$\frac{n_1}{N_1} + \frac{n_2}{N_2} + \cdots + \frac{n_i}{N_i} + \cdots + \frac{n_m}{N_m} = 1 \qquad i = 1, \ldots, m \quad (16.2)$$

where n_i = number of cycles of application of the ith load, moment, or stress level

N_i = fatigue life (cycles) if only the ith load, moment, or stress level were applied during the life of the member (constant-amplitude test)

m = number of load, moment, or stress levels

The use of this hypothesis is shown in Fig. 16.7 for a two-stress-level fatigue test.   If stress level S_1 were applied for the entire test, the fatigue life would be N_1. N_2 is the fatigue life if only S_2 were applied cyclically.   In the two-level test of Fig. 16.7, the total number of applications of S_1 is $n_1 = \frac{3}{4}N_1$.   It is predicted that failure would occur when the number of cycles of S_2 equals $n_2 = \frac{1}{4}N_2$, or

$$\frac{n_2}{N_2} = 1 - \frac{n_2}{N_1} = 1 - \tfrac{3}{4} = \tfrac{1}{4} \qquad (16.3)$$

In this case, the total life N was greater than N_2 but less than N_1, or

$$N_2 < (N = \tfrac{3}{4}N_1 + \tfrac{1}{4}N_2) < N_1$$

Experimental verification of Miner's hypothesis, using successive blocks of constant-amplitude loading applied in either increasing or decreasing magnitude, has been, in general, less than satisfactory.   However, service stresses do not occur in blocks of either increasing or decreasing stress magnitude.   Instead, the various magnitudes occur in some random or near-random sequence.   It has been shown recently[16.5 – 16.8] that Miner's hypothesis, used in conjunction with the cyclic stress-strain behavior of a material and the unnotched, strain-controlled fatigue data for the material, will adequately predict the variable-amplitude fatigue life of that material in a notched condition.   It may be concluded that the application of Miner's hypothesis is not unacceptable as a design or investigative tool.

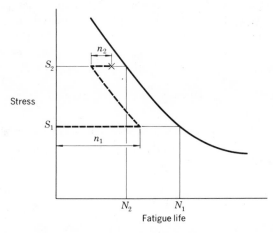

Fig. 16.7   Miner's cumulative-damage hypothesis.

Miner's hypothesis may be used to estimate the service life N of structural members and details subjected to stresses of cyclically varying intensity. The following assumptions are made :

1. Fatigue damage is both stress-independent and interaction-free.
2. For the case of pulsating stress, the critical stress range (maximum stress — minimum stress) for failure at N cycles is a constant regardless of stress ratio.
3. The relative frequencies (per cent occurrence) of each cyclic stress level are known.

Miner's hypothesis may be expressed as

$$\sum_{i=1}^{m} \frac{\alpha_i N}{N} = 1 \qquad (16.4)$$

where α_i = the relative frequency of occurrence of the ith stress range

N = service life (cycles) of member or detail when subjected to m levels of stress range

N_i = the fatigue life (cycles) of a similar member or detail when subjected to a constant-amplitude stress range S_{r_i}.

The foregoing equation may be rearranged to express service life N as

$$N = \frac{1}{\displaystyle\sum_{i=1}^{m} \frac{\alpha_i}{N_i}} \qquad (16.5)$$

Other cumulative-damage hypotheses, more complicated than Miner's hypothesis, have been presented. The primary differences among the various hypotheses are whether or not a particular hypothesis is stress-independent or stress-dependent and whether or not it is interaction-free. A hypothesis is stress-independent if its damage function is the same for all stress levels. For example, the damage function of the stress-independent Miner's hypothesis is

$$D_k = \sum_{i=1}^{k} \frac{n_i}{N_1} \qquad k < m \qquad (16.6)$$

which has the same form regardless of stress S_i. On the other hand, the damage function of a stress-dependent hypothesis (for example, Grover's hypothesis) may be

$$D_k = \sum_{i=1}^{k} \frac{n_i}{a_i N_i} \qquad k < m \qquad (16.7)$$

where a_i is a parameter that is a function of the stress level S_i. A damage hypothesis is interaction-free when its damage function is independent of

---

EXAMPLE 16.1

PROBLEM:
The service stress ranges S_{r_i} (nondimensionalized with respect to the material's tensile strength S_u), and their relative frequencies of occurrence α_i for a given structural member, are listed in the table below. Also listed in the table are the constant-amplitude fatigue lives N_i from laboratory fatigue tests of the given structural member. Find the service life of the member.

SOLUTION:
The sum of the quotients α_i/N_i, from the table, is
$$\sum_{i=1}^{5} \frac{\alpha_i}{N_i} = 3.472 \times 10^{-6}$$
The service life of the member or component is, by substitution into Eq. 16.5,
$$N = \frac{1}{3.472 \times 10^{-6}} = 0.288 \times 10^6 \text{ cycles}$$
The constant-amplitude fatigue lives N_i may be either the mean values expressed by an S-N curve or values expressed by the lower tolerance limit [16.2] to that S-N curve.

level, i	$\dfrac{S_{r_i}}{S_u}$	α_i	N_i	$\dfrac{\alpha_i}{N_i}$
1	0.20	0.30	∞	0
2	0.25	0.20	1.4×10^6	0.143×10^{-6}
3	0.30	0.18	0.34×10^6	0.529×10^{-6}
4	0.35	0.17	0.15×10^6	1.131×10^{-6}
5	0.40	0.15	0.09×10^6	1.669×10^{-6}
		$\Sigma \alpha_i = 1.00$		$\Sigma \dfrac{\alpha_i}{N_i} = 3.472 \times 10^{-6}$

prior stress history or the order in which the various stress levels were applied.

If the second assumption listed above is not acceptable or invalid for a particular detail or member, the equivalent zero-to-tension maximum stress $S^0_{\max_i}$ may be computed from the modified Goodman constant-life relation

$$\frac{S^0_{\max_i}}{s_u} = \frac{(1 - R_i)\dfrac{S_{\max i}}{s_u}}{(1 - R_i)\dfrac{S_{\max i}}{s_u}} \tag{16.8}$$

where the stress ratio R_i is

$$R_i = \frac{S_{\min i}}{S_{\max i}} \tag{16.9}$$

Cumulative-fatigue damage analysis and the various hypotheses are discussed in detail in Refs. 16.9, 16.10, and 16.11.

The use of Eq. 16.4 to predict service life N is demonstrated in Example 16.1.

2. Effects of Nature and Condition

Prior Stress History. Residual compressive surface stresses will increase the fatigue life of a machine part or structural member and residual tensile surface stresses will lower the fatigue life to some extent.[16.3,16.12]

Residual compressive surface stresses can be induced in steel components through shot peening, rolling, nitriding, carburizing, and spot-heating. These treatments have increased the fatigue strength of steel parts by as much as 100 per cent.

Thermal stress relief or proof loading (overloading by a single cycle) may be used in certain cases to reduce detrimental residual tensile stresses and increase the fatigue life. Proof loading also may be used to introduce compressive residual stresses at notches. Thermal stress relief, however, has not proved to be beneficial to the fatigue resistance of structural carbon steel weldments.

Size and Shape of a Member. Notches or stress raisers may be classified as metallurgical, mechanical, or service. Metallurgical notches are the result of segregation, inclusions, blowholes, laminations, and quench-cracks. Sudden changes in contour are considered as mechanical notches. Service notches include nicks, erosion, and corrosion pitting.

Maximum stresses at notches, sufficient to cause fatigue failure, generally exceed the yield point of the structural material and cause inelastic stress redistribution. The presence of residual stresses in weldments cause inelastic behavior at nominal stresses considerably below the yield point. The inelastic stress distributions in the vicinity of notches are described in detail in Chapter IV of Ref. 16.13.

Studies on aluminum aircraft alloys and steels[16.6,16.7,16.14,16.15] indicate that the fatigue life of notched specimens may be predicted from unnotched specimen data. If the cyclic strains at the notch root are the same as cyclic strains in an unnotched specimen, similar lives are observed.

Stress raisers or notches are one of the most serious causes of reduced fatigue strength.

Metallurgical Structure. There has been much interest in the effects of chemical composition and microstructure on the fatigue properties of metals. The fatigue resistance of unnotched heat-treated steels may be enhanced by relatively rapid cooling from above the critical temperature to produce finely dispersed hardening constituents. The fatigue limit of unnotched specimens is raised when the microstructure consists mainly of tempered martensite but is lowered by the presence of areas of coarse pearlite or free ferrite. Inclusions may or may not be detrimental to the fatigue resistance of steels. Angular or elongated inclusions decrease fatigue strength whereas small rounded inclusions do not appear to affect significantly the fatigue strength. Any alloying element, carbon, manganese, nickel, chromium, molybdenum, vanadium, or phosphorus, that increases the tensile strength will also increase the unnotched fatigue limit.

An empirical relationship between fatigue limit, strain-hardening exponent (slope of the true stress–true strain curve) and tensile strength for polished, rotating beam tests[16.16] has been extended to axially loaded fatigue tests of plain material.[16.17] The latter relation describes tests on specimens with both ground and as-rolled surfaces ($R = 0$):

$$\log S_f = \frac{0.01487}{n} + 0.7889 \log (F_u) \left(\frac{e}{n}\right)^n \qquad (16.10)$$

where n is the strain-hardening exponent and e is the Napierian base.

Another mechanical property that has been studied in conjunction with fatigue is the brittle-fracture transition temperature. A decrease in temperature will increase the fatigue resistance of a steel although it may decrease impact resistance. Cyclic stressing below the fatigue limit can, however, raise the brittle-fracture transition temperature significantly. Some service brittle-fracture failures have been initiated at fatigue cracks which served as severe notches. Although both fatigue failures and brittle fractures may be initiated at stress raisers, in regions of high restraint or of tensile residual stresses, they are entirely different fracture phenomena. Brittle fracture is discussed in Chapter 15.

Welding. Deposition of weld metal in a joint is similar to the casting of steel (see Chapter 14). Welds are subject to such internal heterogeneities as shrinkage cracks, lack of fusion, porosity, and inclusions. The heat-affected zone* of the base metal is heated by the welding procedure to a temperature above the lower critical temperature (the temperature at which a steel changes phase), and the rate of cooling, governed by the conduction of heat away from the fusion zone, will produce metallurgical transformations similar to those produced by deliberate heat treatment. The grain size of the heat-affected zone is coarsened adjacent to the weld metal but refined in the outer region. These metallurgical notches and the geometrical stress raisers due to surface contour, undercutting, surface ripples, and lack of penetration reduce the fatigue resistance of metals when welded.

Experimental techniques and a lack of uniform criteria for judging the severity of heterogeneities make it difficult to discuss, quantitatively, the effects on fatigue strength of internal weld defects. It is possible, however, to draw some general conclusions from reviews of existing tests.[16.18] Lack of penetration lowers the fatigue strength of transverse welds significantly but has relatively little effect on longitudinal welds. Porosity and slag inclusions

---

* The heat-affected zone is that portion of the base metal that has not been melted but in which the mechanical properties and microstructure have been changed by welding.

decrease fatigue resistance in proportion to the decrease in effective weld area of transverse welds. Severe quenching due to the sudden extinguishing of the welding arc may initiate a fatigue crack at the point of change of electrode. Severe quenching also results from stray flashes and weld spatter. Geometric stress raisers may be reduced by removal of weld reinforcement and by control of the welding procedure.

Effect of Notches and Tensile Strength on Fatigue-Crack Initiation and Propagation. The insensitivity of the fatigue resistance of riveted joints and weldments (with reinforcement intact) (Art. 16.3) to variation in tensile strength of the base material is a special case of the more general truism— severe notches reduce the fatigue resistance of metals to a common value regardless of tensile strength.

The fatigue life of a test specimen, or structural member, consists of three phases:

1. Initiation of a microscopic crack.
2. Propagation of the crack to a critical size.
3. Exceeding the residual strength of the cracked element, causing complete fracture.

The bulk of the fatigue life consists of the first two phases. The effects of tensile strength and notch severity on their duration are examined below.

Cycles to crack initiation, as a function of cyclic range of stress intensity factor ΔK^* and tensile strength, is presented in Fig. 16.8 for carbon steel,[16.19–16.21] constructional alloy steels,[16.22] and maraging steel ($S_u = 253$ and 414 ksi)[16.23]. For a given ΔK, the cycles to crack initiation decreases with notch-root radius r until $r = 0.010$ in.[16.23] Below $r = 0.010$ in., crack initiation is a function only of ΔK. At root radii $r < 0.010$ in., the behavior of the constructional alloy steels and that of the maraging steels are identical. However, the mild steels do exhibit shorter lives to crack initiation at a given ΔK than do the higher-strength steels.

Extensive measurements of toe radii in transverse butt weld reinforcement showed that the average radius was about 0.010 in.[16.25] The range of stress-intensity factor is, for $S_r = 20$ ksi, of the order of 16. Entering Fig. 16.8 with $\Delta K = 16$ ksi$\sqrt{\text{in}}$., it may be seen that crack initiation at the transverse butt weld toe would be expected to occur at lives from 1000 to 10,000 cycles. Crack initiation consumes less than 1 per cent of the total life (see

---

* See Chapter 15 for a discussion of the stress-intensity factor. It has been shown[16.23,16.24] that the stress field ahead of a notch with a small but finite radius is approximately described by the stress-intensity-factor concept. For example, the error decreases from 10 per cent for a notch-root radius of 0.180 in. to about 4 per cent for notch radii less than 0.040 in.

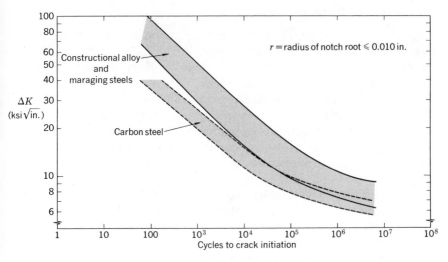

Fig. 16.8 Fatigue-crack initiation.

Fig. 16.11), and differences in crack-initiation time due to differences in tensile strength have no significant effect on total life to fracture. However, it should be noted that, as the notch severity, that is, stress concentration, decreases, the crack-initiation period increases markedly and that differences in tensile strength significantly influence such periods.

The second phase of the fatigue process, and the one of greatest duration in the presence of severe notches, is propagation of the crack to critical size. Crack-growth data for carbon steel,[16.19,16.26—16.28] low-alloy high-strength steel,[16.26] heat-treated constructional alloy steels,[16.26,16.29,16.30] weld metal, and heat-affected zones[16.31,16.32] are summarized in Fig. 16.9. For a given range of stress-intensity factor ΔK, the crack-growth rate is insensitive to grade of steel (that is, tensile strength) until the growth rate exceeds 10^{-5} to 10^{-4} inches per cycle. However, the growth rate, for up to 95 per cent of the life, never exceeds 10^{-6} to 4×10^{-5} inches per cycle, depending on the magnitude of the stress range.[16.19] Therefore, a crack, once initiated, will propagate at a rate that is basically a function of stress range and crack length *only* until achievement of the critical crack length causes fracture.

The foregoing arguments indicate that any differences in fatigue resistance due to differences in tensile strength arise, primarily, in the crack-initiation phase. However, once initiated, the life from initiation to fracture is insensitive to tensile strength. It follows that any improvement in fatigue life due to an increase in tensile strength will be realized only when notches are reduced in severity or are eliminated.

3. Effects of Environment

Fatigue strength varies inversely with temperature and, at high temperatures, the effect of creep must be considered with fatigue. If a structural member or connection is subjected to cyclic temperature changes during service, it may experience thermal fatigue. The stresses caused by differential thermal expansion or differences in thermal conductivity encountered when dissimilar metals are used together may cause fatigue when superimposed on the live- and dead-load stresses carried by the structure. The life of a thermally stressed member may be considerably less than that of an identical member mechanically stressed for the same strain spectrum at a constant temperature.[16.33]

The presence of oxygen, water vapor, sulfides, or other corrosive agents in the atmosphere lowers fatigue resistance. The failure of a metal caused by the simultaneous action of cyclic stresses and corrosive agents is known as "corrosion fatigue."

The effect of neutron irradiation on fatigue of low-carbon and low-alloy high strength steels used in nuclear reactor containment vessels has been studied to a limited extent. Although neutron irradiation lowers the ductility, the changes in fatigue life appear to be slight.

16.3 FATIGUE TESTS OF STRUCTURAL STEEL CONNECTIONS AND MEMBERS

In the beginning of this chapter, it was pointed out that, at present, no analytical method exists by which safe or critical stress combinations can be predicted. Since designs and design specifications must be based upon test data, this article presents some typical results of fatigue tests on structural steels commonly used in the United States. The previously mentioned factors affecting fatigue resistance preclude the extrapolation of test results from small coupon specimens to a structural member or connection in which the only similarity is the type of steel. For this reason, fatigue tests have been conducted on relatively large specimens, the size of which is limited only by the capacity of the testing machine.

Extensive compilations[16.18,16.34,16.35] of fatigue tests on plain material and welded joints are available and references are made only to particular papers that appeared subsequently.

1. Plain Material

Fatigue test data for axially loaded plain carbon, low-alloy high-strength and constructional alloy steels with as-rolled surfaces are shown in Fig. 16.10.

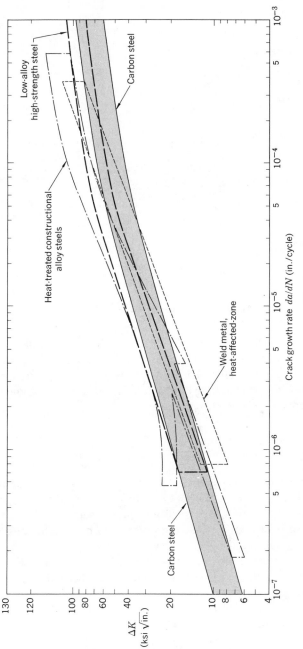

Fig. 16.9 Fatigue-crack propagation in structural steels.

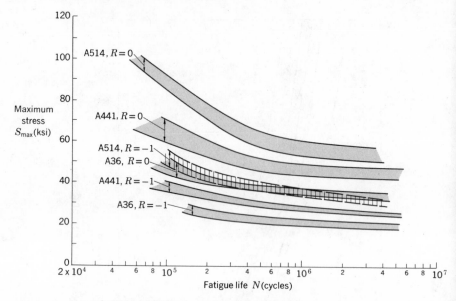

Fig. 16.10 Plain, as-rolled structural steel.

Scatter bands rather than the individual data points, summarized in Ref. 16.35, show the trends for zero-to-tension and complete reversal, $R = 0$ and $R = -1$, respectively. The differences between the various grades become less pronounced at long lives and at stress ratios where the minimum stress is compressive $(R < 0)$.

2. Butt Welds

The fatigue resistance of structural steels as well as the difference in fatigue resistance between various grades is reduced by the presence of transverse butt welds (Fig. 16.11). The effect of stress ratio is also less pronounced in the case of butt-welds. Removal of reinforcement from transverse butt-welded carbon steel may increase the fatigue life by a factor of 4 or 5.[16.36] Similar increases have also been observed for butt-welded constructional alloy steel.[16.35] Stress relief of butt-welded carbon steel has no significant effect on fatigue resistance[16.36] but may be slightly beneficial in the case of the alloy steels at long lives.

The fatigue resistance of carbon steel is little affected by the presence of longitudinal butt welds. On the other hand, limited tests show that longitudinal butt welds reduce the fatigue life of low-alloy high-strength steels. This is probably due to the fact that the weld metal, in the latter case, was of a lower strength than the base material.

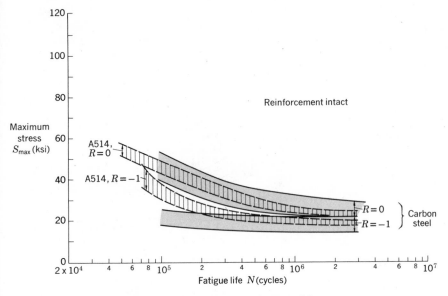

Fig. 16.11 Transverse butt welds.

With the reinforcement intact, the transverse welded specimens failed at the edge of the weld reinforcement where a geometric stress raiser existed. With the reinforcement removed, the specimens failed either in the base metal or in the weld at inclusions or gas pockets. Although internal heterogeneities probably existed in the longitudinal welds also, they are less critical when aligned with the principal stress. The longitudinal butt welds failed due to surface imperfections on the weld bead such as ripples or at a point of change of electrode.

3. Fillet Welds

A continuous, longitudinal fillet welded specimen has been developed to simulate the flange-web weldment of built-up beams, axially loaded box members, or attachments.[16.37] Results of fatigue tests utilizing this specimen are shown in Fig. 16.12 for carbon steel[16.36] and constructional alloy steel.[16.38] The reduction in life of a continuous, longitudinal fillet weld is due, generally, to a metallurgical notch. On the other hand, transverse or intermittent longitudinal fillet welds combine the detrimental effects of both geometrical and metallurgical notches. The fatigue strengths for a given life of transverse and intermittent longitudinal fillet welds are about one-half of those for continuous longitudinal fillet welds shown in Fig. 16.12.

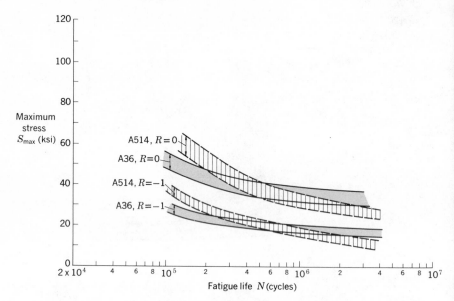

Fig. 16.12   Longitudinal fillet welds.

4. Comparison of Plain Material and Welds

The trends of fatigue data for carbon and constructional alloy steels, plain material, transverse butt welds and longitudinal fillet welds, are shown in Fig. 16.13 for $N = 100,000$ cycles, and Fig. 16.14 for 2 million cycles.

The welded constructional alloy steel demonstrates an advantage over welded carbon steel when either the stress ratio R is greater than 1/4 to 1/2 or the expected life is short.   However, at long lives or in the case of alternating stresses ($R < 0$), the welded constructional alloy and carbon steels are similar in behavior.   The sensitivity of alloy steels to notches (both metallurgical and mechanical) is greater than in carbon steels and increases with increases in life.   Comparison of Figs. 16.13 and 16.14 shows that the ratio of the fatigue strength at two million cycles to the fatigue strength at 100,000 cycles is greater for the welded carbon steels than for the welded constructional alloy steels.

The fatigue strength of fillet welded carbon steel appears to be greater than that of plain carbon steel in Fig. 16.13.   This is due to the higher tensile strength of the carbon steel used in the fillet weldment tests.

5. Rolled and Welded Beams

The fatigue resistance of rolled beams is similar to that of plain, as-rolled material but greater than that of welded, built-up beams (Fig. 16.15).

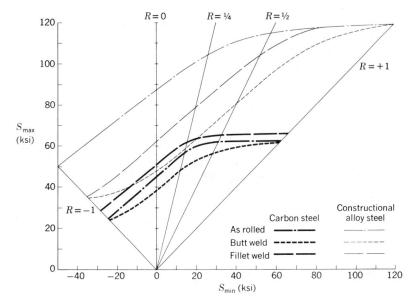

Fig. 16.13 Constant-life diagram, $N = 10^5$ cycles.

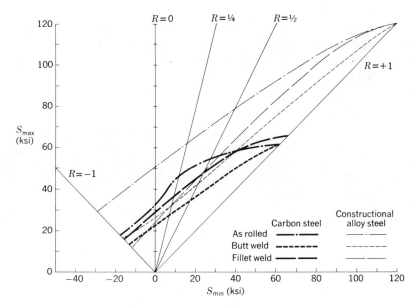

Fig. 16.14 Constant-life diagram, $N = 2 \times 10^6$ cycles.

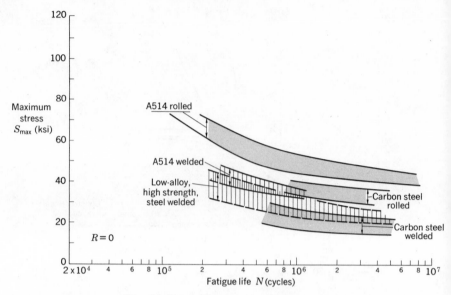

Fig. 16.15   Rolled and welded beams.

Fatigue failures in manually welded built-up beams are generally initiated at craters due to electrode changes in the flange-web fillet weld and propagate into both the flange and web.   Built-up beams of BS968, a steel similar to A441, with fillet welds laid both manually and automatically, have been cyclically loaded; the fatigue strength at 2 million cycles was increased by about 35 per cent when automatic submerged arc welding was used.[16·18]

6. Welded Beam Details

Partial-length cover plates significantly reduce the fatigue strength of welded beams but, on the other hand, full-length cover plates have no effect (Fig. 16.16).   The scatter band for partial-length cover plates includes various details and butt welded flange transitions.   It has been suggested that cover plates should be extended past the theoretical cut-off point a sufficient distance so that the stress at the end of the cover plate is 40 per cent of the stress at its center.   Intermittent fillet welds on cover plates should be avoided.

Beam splices (Fig. 16.17) have little or no effect on the fatigue resistance of welded beams.   Most failures were initiated at the weld toes or discontinuities in the weld.

Various types of stiffeners, some welded to the tension regions of beams and some not, have been studied.   In the case of stiffeners welded to the

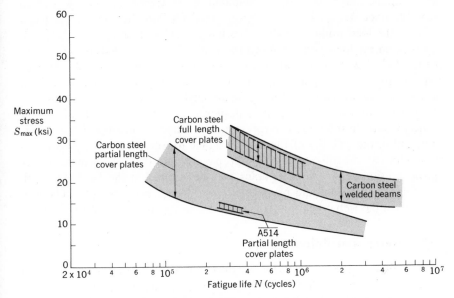

Fig. 16.16 Welded cover plates.

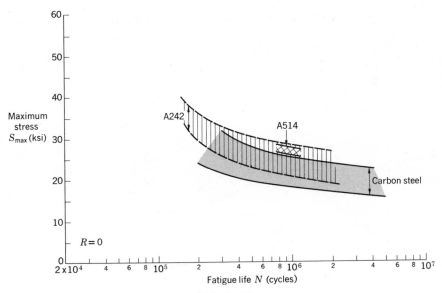

Fig. 16.17 Welded beam splices.

web, fatigue failure usually was initiated at the termination of the web to stiffener fillet weld. The crack propagated up into the web in the panel toward the load point and along the flange-to-web fillet weld in the panel away from the load point. When the stiffeners were welded to the tension flange, failure was initiated at the toe of the fillet weld on the flange. No one type of stiffener has a particular advantage over another although welding to a tension zone in the web or flange could lead to an early fatigue failure.

In deep, thin-web plate girders designed according to the tension field concept, fatigue fractures may occur in the following manner. A crack is initiated in the web toe of the tension flange to web fillet weld, propagates toward a vertical stiffener and then propagates into the web perpendicularly to the tension field. Another crack may be initiated in the toe of the web to stiffener fillet weld and propagates toward the tension flange. The web fails when the web crack joins the crack along the toe of the stiffener weld.

7. Riveted and Bolted Joints

Important factors affecting the fatigue strength of riveted and bolted joints are clamping force, grip length, type of fastener, fastener pattern, tension-shear ratio, tension-bearing ratio, and type of steel in fastener or main material.

It has been shown that high-strength bolted joints are superior to riveted joints in fatigue resistance and that the fatigue strength of bolted joints approaches the yield strength of the connected material[16.39] (Fig. 16.18).

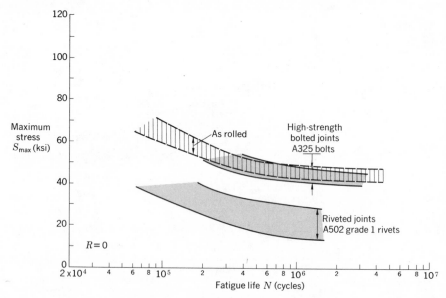

Fig. 16.18 Carbon steel structural joints.

Failure through the net section occurs in both bolted and riveted joints, but an increase in clamping force of the bolts can cause failure in the gross section.

Some research indicates that no steel shows a superiority over carbon steel as plate material in riveted joints subjected to cyclic loading. It has been suggested that the effect of variation in clamping force within a joint or from one joint to another subordinates the relative superiority of one steel over another and produces the great experimental scatter. In one series of tests, the clamping forces of individual carbon steel rivets were measured after fatigue failure of the connection. The clamping force varied from 1 to 20 kips, with 50 per cent less than 5 kips and 20 per cent greater than 11 kips. The average clamping stress of hot-driven rivets increases with the grip length and, at a grip length of $3\frac{1}{16}$ in., can approach the yield strength of the undriven rivet material. In comparison, high-strength bolts have a uniformly high clamping force, although the average clamping force of bolts has been observed to decrease about 20 per cent after a few cycles of loading.

16.4 CONSIDERATION OF FATIGUE IN STRUCTURAL SPECIFICATIONS

The current structural specifications in the United States consider, in general, the effects of maximum and minimum stress or stress range, ultimate tensile strength, and the stress concentration effect of various details on the service life of a structure.

These specifications present equations expressing maximum allowable cyclic stress in the form

$$f_{\max} = \frac{k_1 f_0}{1 - k_2 R} \qquad (16.11)$$

The maximum allowable cyclic stress may not exceed $f_{\max}$ of Eq. 16.11, the maximum allowable static tensile stress F_t, or the maximum allowable compressive stress F_a (in the case of a compressive member). Equation 16.11 was obtained in the following manner. Figure 16.19 is a constant-life diagram showing the test points for the fatigue strength at N cycles for a certain steel that was cyclically loaded in complete reversal ($R = -1$), zero-to-tension ($R = 0$) and half-to-full tension ($R = \frac{1}{2}$). The fatigue strengths, $S_{\frac{1}{2}}$, S_0, and S_{-1}, are mean test values. It was decided that line AB would be the design curve for cases in which $f_{\min}$ is negative and line BC would be the design curve for the static stress case. Reference 16.34

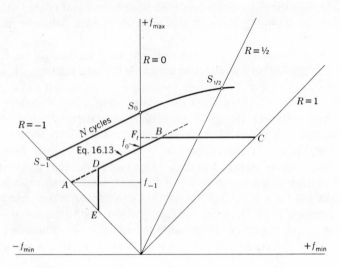

Fig. 16.19 Construction of design specifications from tests.

indicates the thoughts behind the selection of a design curve such as
AB. Point A coincides with the permissible value of maximum stress in
complete reversal, f_{-1}, and point B coincides with the maximum allowable
stress f_0. The slope of line AB is

$$\frac{f_0 - f_{-1}}{f_{-1}} = k_2 \tag{16.12}$$

and the equation for line AB is

$$f_{\max} = f_0 + k_2 \cdot f_{\min}{}^* \tag{16.13}$$

In addition

$$f_0 = \frac{S_0}{(\text{F.S.})_0} \tag{16.14}$$

and

$$f_{-1} = \frac{S_{-1}}{(\text{F.S.})_{-1}} \tag{16.15}$$

where $(\text{F.S.})_0$ and $(\text{F.S.})_{-1}$ are the safety factors at $R = 0$ and $R = -1$
respectively.   Since

$$f_{\min} = R \cdot f_{\max}$$

* $f_{\min}$ can be either tension (+) or compression (−).

Eq. 16.13 becomes

$$f_{max} = \frac{f_0}{1 - k_2 R} \qquad (16.16)$$

If line AB intersects the ray $R = 1$ (Fig. 16.19), $k_2 = \frac{1}{2}$ and Eq. 16.16 becomes

$$f_{max} = \frac{f_0}{1 - \frac{1}{2}R} \qquad (16.17)$$

If the safety factors $(F.S.)_0$ and $(F.S.)_{-1}$ are equal and the constant life curve through the test values is parabolic $(S_0 \approx 5/3 S_{-1})$, Eq. 16.16 becomes

$$f_{max} = \frac{f_0}{1 - \frac{2}{3}R} \qquad (16.18)$$

If the line AB is parallel to the ray, $R = +1$, Eq. 16.16 becomes

$$f_{max} = \frac{f_0}{1 - R} \qquad (16.19)$$

Equation 16.19 is a rather important special case which states that, for a given fatigue life, the allowable stress range is a constant value regardless of stress ratio R.

If f_0 is the fatigue strength of carbon steel, Eq. 16.16 may be amended for application to high strength steels by the factor k_1 or

$$f_{max} = \frac{k_1 f_0}{1 - k_2 R} \qquad (16.11)$$

where

$$k_1 = 1 + \alpha \left(\frac{F_u}{(F_u)_{c.s.}} - 1 \right) \qquad (16.20)$$

F_u, $(F_u)_{c.s.}$, and α are, respectively, the tensile strength of the higher-strength steel, the tensile strength of carbon steel, and a parameter that is a function of expected life and type of joint detail. In specifications presenting fatigue design criteria in the form of Eqs. 16.11 and 16.20 (AASHO[1.23] and AWS[1.26]), k_2, f_0, and α are tabulated for various cycle lives, joint details, location of material (away from or adjacent to a detail) and type of stress (tension, compression, or shear). The AREA[1.25] provisions are, basically, those expressed by Eq. 16.11 but the individual values of k_2, f_0, and α are not tabulated.

On the other hand, the AISC[1.21] fatigue provisions are based on Eq. 16.19 when the stress ratio is equal to or greater than zero. The allowable stress range F_{sr} where

$$F_{sr} \geqslant f_{\max} - f_{\min},\qquad(16.21)$$

is tabulated for various details and loading conditions (frequency of load repetition). When R is less than zero, the AISC allowable stress range F_{sr} is multiplied by the factor

$$\frac{1 - R}{1 - 0.6R}.$$

All materials except constructional alloy steel plain material are proportioned for the allowable stress ranges of carbon steel plain material and details.

16.5 CONSIDERATION OF FATIGUE IN STRUCTURAL DESIGN

1. Design Philosophies

There are two philosophies in the design of cyclically loaded members and structures, the "safe-life" and "fail-safe" approaches.[16.9,16.11,16.13]

In the "safe-life" approach, the fatigue life of a member or detail is evaluated experimentally. This minimum expected life is divided by a safety margin (or scatter factor) to obtain the maximum service or permissible life. The service life is terminated by the initiation of fatigue cracks and the question of residual strength of cracked elements never arises. The "safe-life" approach is wasteful of material and does not guard against catastrophic failure caused by accidental or unexpected damage.

On the other hand, the "fail-safe" approach accepts a specified amount of fatigue cracking but prevents catastrophic failure through inspection and repair. The member may fail locally but the rate of fatigue crack propagation is low enough and the residual strength of the cracked member is high enough that the structure will continue to perform satisfactorily until the cracks are discovered at the next scheduled inspection. Large sections are interrupted by crack-arresting details or are replaced by smaller, parallel members so that failure of one member will not jeopardize the safety of the complete structure. The validity of the "fail-safe" approach is based on the presence of redundancies and *periodic, rigorous* inspections.

Lack of information on the growth of fatigue cracks in structural steel and residual strength of cracked structural steel members has prevented the general application of "fail-safe" techniques to civil engineering structures. Instead, "safe-stress," a variant of the "safe-life" approach, is used.

The safety factor and margin of safety are illustrated in Fig. 16.20 where the frequency distributions of the applied or service stresses and the fatigue

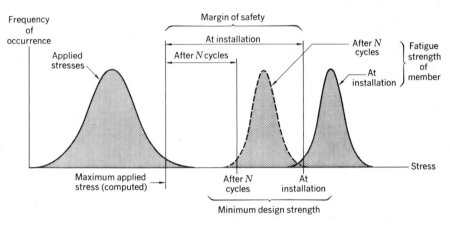

Fig. 16.20 Margin of safety: required service life of *N* cycles.

strength of a member or detail are shown. The minimum design strength is selected on the basis of the lower limit of scatter (shown in Figures 16.10 to 16.18) or some probability of failure.[16·2] The margin of safety is the difference between the minimum design strength and the maximum applied stress. The safety factor is the ratio

$$\frac{\text{Minimum design strength}}{\text{Maximum applied stress}}$$

As service cycles accumulate on the member or detail, the distribution of strengths is displaced to the left and the margin of safety is diminished. It is also possible that the distribution of applied stresses shifts to the right as permissible live loads increase with time. The safety factor at *N* cycles is that used in fatigue design criteria.

2. Fatigue Safety Factors

Safety factors account for:

1. Statistical variation of fatigue strengths of details and materials.
2. Inadequacy of stress analysis.

3. Incomplete simulation of service stress spectra in laboratory tests.
4. Effects of environment.
5. Unknown fabrication and residual stresses.
6. Lack of a satisfactory theory of cumulative damage to predict adequately the fatigue life of a structure subjected to random loading.

Representative structural carbon steel safety factors from current fatigue provisions are shown in Table 16.1. The ranges of safety factors in Table 16.1 were determined from minimum and average fatigue strengths listed in Ref. 16.34.

Table 16.1 Safety Factors—Carbon Steel

Base Metal	N Cycles	Safety Factor $R = 0$	Safety Factor $R = -1$
Away from	10^5	2.0 to 2.3	1.7 to 1.9
weld	2×10^6	1.3 to 1.5	1.0 to 1.3
Adjacent to	10^5	1.2 to 1.8	1.3 to 1.8
butt weld	2×10^6	1.0 to 1.5	1.0 to 1.4

3. Consideration of Fatigue in Design

When the consideration of fatigue in a particular detail is required, the maximum and minimum load carried by a structural member must be established, as well as the number of load repetitions to which that member will be subjected during its service life.

A more rational design of bridges would utilize traffic load surveys made in various regions, rural, urban, highly industrialized, throughout the country. These surveys would analyze data statistically on the number and magnitude of load repetitions. In addition, higher working stresses would be justified if rigorous periodic inspections of structures, similar to those on aircraft, were instituted.

In the case of industrial structures subjected to repeated loads, such as ore bridges and crane runways, a good estimate of the magnitude and number of load repetitions can be determined from surveys of existing installations. These structures generally are designed in accordance with AISC Specification which permits the increase of allowable stress range for a corresponding decrease in load repetitions.

Consideration of fatigue in structural design utilizes the standard methods of proportioning members. The only departure from the customary static design procedures is the increase of the maximum load by some fraction of the minimum load. The point is illustrated in Examples 16.2, 16.3, and 16.4.

EXAMPLE 16.2

PROBLEM:
The stress sheet for a camel back Pratt truss railway bridge, see figure 4.3f, indicates, for a vertical near the midspan:

Dead Load	Live Load	Impact
	+ 320 kip	+ 125 kip
− 20 kip	− 260 kip	− 82 kip

Proportion this member in A36 steel with high-strength bolted ends. Assume $L/r = 80$.
400,000 cycles of load are expected to occur.

SOLUTION:

From AREA (1969):

$$F_t = 20 \text{ ksi}$$

$$F_a = 21.5 \text{ ksi} - 0.100 \ (3/4)(80) \text{ ksi} = 15.5 \text{ ksi}$$

with $P_{max} = -20 + 320 + 125 = 425$ kip and

$$P_{min} = -20 - 260 - 82 = -362 \text{ kip}$$

$$R = P_{min}/P_{max} = -362/425 = -0.851$$

Allowable cyclic unit stress in tension =

$$\frac{20.5 \text{ ksi}}{1 - 0.55R} = \frac{20.5}{1 + 0.468} = 13.96 \text{ ksi}$$

Allowable cyclic unit stress in compression =

$$\frac{20 \text{ ksi}}{1 - 0.50R} = \frac{20}{1 + 0.425} = 14.02 \text{ ksi}$$

$$A_{net} = \frac{425 \text{ kip}}{13.96 \text{ ksi}} = 30.5 \text{ in.}^2$$

$$A_{gross} = \frac{362 \text{ kip}}{14.02 \text{ ksi}} = 25.8 \text{ in.}^2$$

The vertical is proportioned for a net area equal to or greater than 30.5 in.^{2}

Reference 16.40 contains several design examples illustrating the AASHO criteria.

Good detail practice does much to improve the fatigue resistance of a structure subjected to repeated loads. Sections should be changed gradually to reduce the stress concentration effect. Joints or details with a large variation in stiffness should be avoided. High restraint in localized zones may cause high secondary stresses that are not considered in the design calculations.

An example of fatigue failure due to a large variation of stiffness in a detail is shown in Fig. 16.21. The diaphragm connecting a crane runway

EXAMPLE 16.3

PROBLEM:
   For the railroad bridge of Example 16.2, proportion the welded connection of the vertical to joint U4. The bridge is single-tracked, 300 ft long. A36 steel and E70 electrodes are to be used.

Joint $U4$

$U_4 L_4$ $L = 50'-0''$

2-web $\mathbb{L}$s $\frac{7}{8}$ x 18

2-flg. $\mathbb{L}$s 1 x 20

SOLUTION:
   For $U_4 L_4$, try the cross section indicated above

      I_{min} = 4208 in^4 A = 71.5 in^2 r_{min} = 7.68 in.   L/r = 78

      from AREA (1969)

$$F_a = 21.5 \text{ ksi} - 0.100 \ (3/4)(78) \text{ ksi} = 15.7 \text{ ksi}$$

   for base material adjacent to fillet welds, maximum stress tension, AWS, (1969),

   Loading Case No. 2.

$$F_r = \frac{K_1 f_{ro}}{1 - K_2 R} = \frac{1 \ (10.5 \text{ ksi})}{1 - 0.8 \ (-0.851)} = 6.25 \text{ ksi}$$

$$A_{req'd} = \frac{425 \text{ kip}}{6.25 \text{ ksi}} = 68 \text{ in}^2 < 71.5 \text{ in}^2 \qquad\qquad \text{O.K}$$

   for fillet welds, allowable unit stress in shear, AWS (1969)

$$F_r = \frac{K_1 f_{ro}}{1 - K_2 R} = \frac{1 \ (10.8 \text{ ksi})}{1 - 0.55 \ (-0.851)} = 7.36 \text{ ksi}$$

$$A_{req'd} = \frac{425 \text{ kip}}{7.36 \text{ ksi}} = 57.7 \text{ in}^2$$

   The total length of fillet weld is

      4 (34 in.) = 136 in.

   The weld throat is

$$\frac{57.7 \text{ in}^2}{136 \text{ in.}} = 0.425 \text{ in. say } 0.5 \text{ in.}$$

   Use 1/2 in. fillet welds to connect the vertical to the gusset plates.

Note: the allowable cyclic stresses in the base material and the fillet welds permitted by AREA (1969) are identical to those computed above.

EXAMPLE 16.4

PROBLEM:
The beam shown below is to be designed for over 2 million cycles of load from zero to some maximum. The beam is to be fabricated from A36 steel. Using the AISC specification, determine the maximum allowable load on the beam. What is the minimum safety factor against fatigue failure?

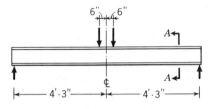
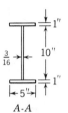

A-A

SOLUTION:
From the AISC Specification (1969) for over 2 million cycles of load applied to a welded built-up beam, the allowable stress range is 15 ksi

$$I = 2(5)(5.5)^2 + \frac{2(5)(1)^3}{12} + \frac{0.1875(10)^3}{12} = 349 \text{ in.}^4$$

$$M_{max} = 22.5\, P_{max}, \quad M_{min} = 0, \quad \therefore P_{min} = 0, \quad C = 6 \text{ in.}$$

$$f_{max} = \frac{22.5\, P_{max}(6)}{349} = 0.387\, P_{max}$$

$$f_{min} = 0$$

$$\text{stress range} = f_{max} - f_{min} = 0.387\, P_{max} \leq 15 \text{ ksi}$$

$$\therefore P_{max} \leq 38.8 \text{ kip}$$

From Fig. 16.15, the fatigue strength at 2 million cycles for welded carbon steel beams varies from 15 to 24 ksi. The factor of safety against fatigue failure is

$$\frac{15}{15} < \text{S.F.} < \frac{24}{15}$$

or $1.0 < \text{S.F.} < 1.6$

---

to a roof column did not restrain the webs of the runway beams over their full depth and thus permitted lateral bending of the upper flange and part of the web about the top of the diaphragm. This bending, caused by the lateral force of the crane on the crane rail, produced critical bending stresses in the web and eventually led to a fatigue crack that propagated toward the center of the beam. A more favorable detail in this case would be the extension of the diaphragm angles over the full depth of the web as end stiffeners.

Partial-length cover plates on girders should extend well beyond the theoretical cut-off point and girder splices should be located in regions of low bending stress.

In welded fabrication, procedures should be used that will eliminate such internal defects as gas pockets and slag inclusions. Undercutting, incomplete penetration, weld spatter, and promiscuous arc striking outside of the weld area must be prohibited. Intermittent fillet welds should be avoided.

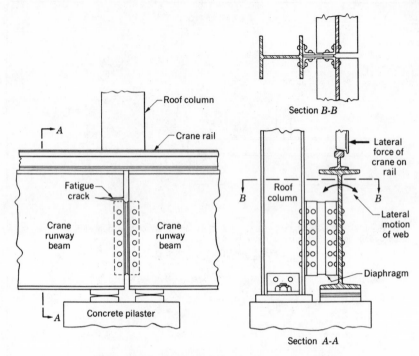

Fig. 16.21   Fatigue failure in crane runway.

It is evident that there are some cases of fatigue loading that are more severe than those normally encountered in the design of structures which are governed by the existing specifications.   For such special cases, available sources of test information should be consulted, keeping in mind the effects of the various parameters discussed in this chapter, the scatter inherent to fatigue testing, and the number of specimens studied in a given test program. For a thorough coverage of the fatigue phenomenon in general, attention is called again to Refs. 16.3 and 16.9 to 16.11.   Fatigue tests in the United States and abroad on welded joints are reviewed in Refs. 16.18, 16.34, 16.35, and 16.41.   Bolted and riveted joint tests are summarized in Ref. 16.39.   An example using published data is presented in Example 16.4.

PROBLEMS

16.1. The following data for axially loaded transverse butt welds in structural carbon steel were established from fatigue tests.   Construct the three *S-N* curves on a semilog plot (stress, arithmetic; life, logarithmic). Notice the rather great difference in life for a small change in maximum stress.

$R = \frac{1}{2}$		$R = 0$		$R = -1$	
Fatigue Strength (ksi)	Life (kilocycles)	Fatigue Strength (ksi)	Life (kilocycles)	Fatigue Strength (ksi)	Life (kilocycles)
49	184	32.5	135	22	150
48	300	29.5	211	17.5	400
44.5	370	28	430	17	1200
40	1100	24	730	15	1600
43	2210	23	1690	14.5	4300*
42	1150	22	3650*	13	3600*
36.5	4000*				
35	4400*				

Fatigue strength = maximum stress per cycle.
* Run-out, that is, no failure.

16.2. Construct an AWS-WRC constant life diagram from S-N curves of Problem 16.1 for lives of 10^5, 6×10^5, 10^6, and 2×10^6 cycles. Assume the tensile strength is 61 ksi. Discuss the effect of maximum stress on the stress range in the alternating stress region.

16.3. Construct an AWS-WRC constant life diagram showing the allowable stresses for the current specifications:

(a) AASHO. Axially loaded A441 steel member with thickness varying from $\frac{3}{4}$ to $1\frac{1}{2}$ in. Use $L/r = 75$, 100, and 125.

(b) AISC. Extreme fibers of a rolled beam of A441 steel. Construct curves for 50,000, 175,000, and 3,000,000 stress repetitions.

(c) AWS. Fillet welds in shear for 100,000 and 2,000,000 cycles of stress, A36 steel.

16.4. A crane runway, two-span continuous, 25 ft per span, supports two traveling wheel loads 5 ft apart and 15 kips each.

(a) Select a wide-flange section according to the AISC Specifications. Assume that the crane traverses the runway 10 times per day and that the useful life is 25 years.

(b) Select a wide-flange section for the runway assuming that the crane rail is secured to the top flange by short fillet-welded attachments.

16.5. For the data in the table of Example 16.1, compute the variable-amplitude service life, using the modified Goodman constant-life relationship, Eq. 16.8. Compare the service life to that computed in Example 16.1 for the constant-stress-range criterion. The minimum stress is, at all levels i, 10 per cent of the ultimate tensile strength.

17

LOCAL BUCKLING

17.1 INTRODUCTION

With a few exceptional cases of solid sections, metal structural members are composed of flat plate elements. They are either rolled into standard shapes or assembled from individual plates by welding, riveting, or bolting, or cold formed from sheet material. Typical sections are shown in Fig. 17.1a to 17.1i. Sections a, b, and c are rolled; sections d to g are built-up; and sections h and i are cold formed.

The strength of a structural member may be influenced by the buckling of a component plate element in two major ways: the buckling may precipitate an over-all failure by making the plate element ineffective, or it may produce a redistribution of stresses and thus influence the carrying capacity and behavior of the member. In rolled sections and sections built up from thick plates, the maximum plate strength is usually coincident with the yield stress of the material. The width-thickness ratio is selected such that the plate does not buckle at a stress below this level. Cold-formed members, however, quite often have such large b/t ratios that their strength is based on the post-buckling strength—which may be below the yield stress. The thin web plate in a plate girder also, for example, has considerable post-buckling strength.

The maximum average stress which can be applied to a plate element depends on the width-thickness ratio of the plate, on the boundary conditions, and on the stress distribution. It may be larger than, or smaller than, the theoretical elastic buckling stress. An important consideration in arriving at the limiting width-thickness ratios is whether the member is to be designed on an allowable stress basis or on a plastic design basis.

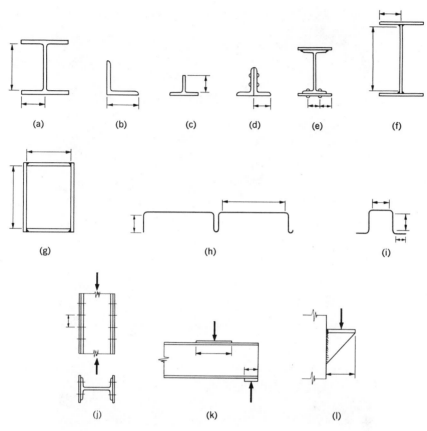

Fig. 17.1 Structural sections and controlling dimensions.

There are four basic types of plate elements, the buckling of which controls the strength of a member:

1. Long plates connected at one or both longitudinal edges and subjected to compression on the short edges. The strength of such plate elements is limited either by buckling or by the post-buckling strength. The controlling dimension is the width of the plate. These plates are indicated by dimension lines in Fig. 17.1a to i. Figure 17.2 shows a plate girder flange, the buckling of which led to the failure of the whole member.
2. Portions of outside plate elements in built-up members between transverse lines of fasteners, such as the dimension in Fig. 17.1j.
3. Web plates which may buckle or cripple due to a transverse load or reaction. This type is shown in Fig. 17.1k.

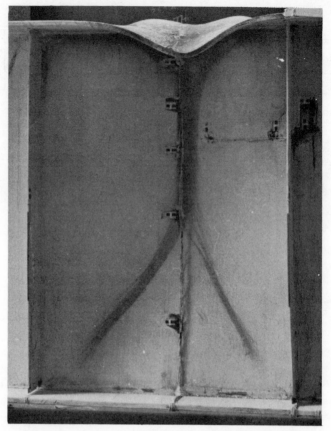

Fig. 17.2 Local buckling of a plate girder flange.

4. Triangular plate brackets whose strength is controlled by plate buckling are represented by Fig. 17.1*l*.

The limiting width-thickness ratios and the post-buckling strength are obtained from a study of the plate behavior under edge loading.

17.2 HISTORICAL REVIEW

Only a few of the important milestones leading to the present-day theories of plate buckling are described below. Detailed accounts of the historical development and many additional references can be found in Refs. 7.6 and 17.1.

Navier, in 1823, was the first to formulate the correct differential equation of a buckled plate. His equation is applicable to rectangular plates subjected to equal edge pressure in two directions. Actually, his equation was just a byproduct of the then current interest in plate vibration—sound produced by a vibrating plate—and it was apparently forgotten. It was in 1888 that Bryan solved the problem of plate buckling by deriving the following differential equation for a simply supported rectangular plate under edge compression in one direction:

$$\frac{Et^3}{12(1-\mu^2)}\left(\frac{\partial^4 w}{\partial x^4} + 2\frac{\partial^4 w}{\partial x^2 \partial y^2} + \frac{\partial^4 w}{\partial y^4}\right) = -f_x t \frac{\partial^2 w}{\partial x^2} \tag{17.1}$$

where E is the modulus of elasticity, μ is Poisson's ratio, t is the plate thickness, w is the lateral deflection of the plate, and f_x is the edge compression. Later, Timoshenko (1907) and H. Reissner (1909) analyzed plates with various other boundary conditions. Timoshenko also investigated the influence of plate buckling on the column strength.

The first treatment of inelastic plate buckling was given by F. Bleich in 1924. Roš and Eichinger made important contributions in 1932. Further development of plate buckling theories in the inelastic range was spurred by the requirements of the aircraft and shipbuilding industries.

The strength of the girder and beam webs subjected to a concentrated transverse load or reaction (Type 3 in Art. 17.1) was assumed to be controlled by elastic buckling. Only in 1935 was it recognized through experimental studies that two types of failure may occur in this case: *plate buckling* in the webs with a large depth-thickness ratio such as used in plate girders; and *plate crippling*, which is essentially plastic buckling due to local yielding. The AISC included web crippling as a design criterion in its specification in 1936.[17.2] The present AISC Specification considers both types of buckling.[1.21]

The post-buckling strength of plates was first noted in experiments in 1930. Empirical approaches then developed for its determination were, however, unsuitable for any practical use. In 1932 von Kármán introduced the concept of the effective width to handle this problem with ease, and derived an approximate formula for simply supported plates. For structural engineering an important contribution was made in 1947 when, on the basis of extensive tests, the effective width formulas were proposed which with minor modifications have been used in the design of cold-formed members.[1.19,1.24]

The buckling of structural steel plates in the strain-hardening range has been studied since 1956.[17.3,17.4,17.5] The consideration of this type of buckling is essential to plastic design in which members should be able to undergo considerable plastic deformation without local failure. The effect of residual stresses on the buckling of plates in the elastic and plastic ranges has been intensively studied since 1962.[14.13]

17.3 PLATES UNDER EDGE COMPRESSION

1. Behavior of Plates Under Edge Compression

As an example of plate behavior under edge compression, a perfectly flat rectangular plate made of an ideal material and subjected to edge compression in one direction will be considered. As shown by the sketch in Fig. 17.3, the load is applied through rigid end blocks and the edges thus remain straight in the process of loading. The width of the plate is b and the thickness is t. The material has a stress-strain relationship as given by another sketch in Fig. 17.3. A diagram of plate behavior is obtained by plotting the average compressive stress P/bt versus the average strain ϵ.*

The heavy solid line $OABC$ in Fig. 17.3 is a typical relationship for a plate with a large width-thickness ratio b/t. Several stages can be observed. At first, the strain increases linearly with the increasing average stress P/bt.

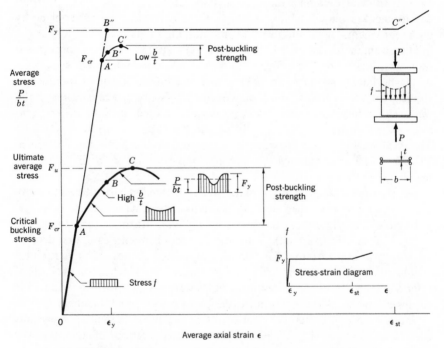

Fig. 17.3 Behavior of plates under edge compression.

* Plate behavior before and after buckling can be also illustrated by plotting the average stress P/bt versus lateral plate deflection w (see Figs. 17.5 and 8.4d).

The stress f, which is the stress intensity at a particular point, is distributed uniformly across the width, and there is no lateral deflection of the plate. Then, when the average stress P/bt reaches a certain magnitude F_{cr} (point A), the plate starts deflecting laterally—it buckles. For columns, buckling is often synonymous with failure (see Art. 9.3). This is not so for a plate; the buckles are restrained by the plate spanning in the transverse direction, and the plate is able to carry loads beyond buckling. After the point of buckling, the uniform stress distribution changes to a non-uniform pattern as shown for portion AB. Further increase in load leads to a greater stress redistribution. The stress along the edges gradually increases and the edges start yielding at point B on the curve. Yielding then spreads quickly until the ultimate load is reached at point C on the curve. The average stress at this load is defined as the ultimate stress F_u. The increase in the average stress beyond buckling may be quite substantial for high b/t ratios.

For plates with lower width-thickness ratios, the critical stress is close to F_y and yielding starts almost immediately after buckling. The ultimate stress is then only insignificantly above the critical stress, as shown by the thin curve $OA'B'C'$ in Fig. 17.3.

In plates having a width-thickness ratio b/t below some specific value, the average stress P/bt reaches the yield point F_y without buckling. The plate may then undergo further straining at the same stress level as shown by the dash-dot line $OB''C''$ in Fig. 17.3. (Point C'', at which the stress-strain curve starts changing, signifies the strain-hardening strain). Eventually, the plate will buckle and fail at a certain strain between points B'' and C'' or even at a strain greater than C'', depending on the b/t ratio.

Plates with various edge support conditions and different distribution of edge stresses behave, qualitatively, in the same manner as just described. The difference lies in the numerical magnitude of the critical buckling stress, and the amount of the post-buckling strength.

The phenomena of the buckling (point A) and post-buckling (ABC) behavior of plates, as well as other related considerations, are described in more detail in the following sections.

2. Plate Buckling

The buckling load is the load at which a structure may assume more than one deflected position without disturbing equilibrium. A perfectly flat rectangular plate with all edges simply supported, and subjected to uniformly distributed edge compression in one direction, will be in this state of indifferent equilibrium when the intensity of the edge compression assumes the value of the buckling stress. (See Fig. 17.4.) This stress will remain constant while the plate may deflect in either direction, as shown by point A in Fig. 17.5. The magnitude of the deflection at the beginning is not defined.

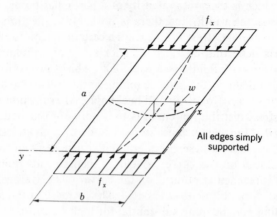

Fig. 17.4 Buckled plate.

Only for a somewhat larger deflection will an increase in the edge stress be necessary to produce additional deflection, but this will already be the range of the post-buckling behavior.

The buckling phenomenon for a plate under compression in one direction is described by Eq. 17.1. The solution is obtained when the deflection function w and the value of f_x are found which satisfy this equation at the boundaries and over the whole plate. A popular approach is to assume the deflection w to be given by a series, each term of which satisfies the conditions at the boundaries, and the coefficients and f_x are determined to satisfy the differential equation. A convenient series for a simply supported plate is

$$w = w(x, y) = \sum_{m=1,2,3\ldots}^{\infty} \sum_{n=1,2,3,\ldots}^{\infty} A_{mn} \sin \frac{m\pi x}{a} \sin \frac{n\pi y}{b} \qquad (17.2)$$

where A_{mn} are the unknown coefficients. A substitution of this series into both sides of Eq. 17.1 leads to an equation in which a term in the left-side

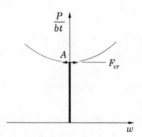

Fig. 17.5 Lateral deflection of a buckled plate.

series for some particular value of m and of n must be equal to a corresponding term in the right-side series. Thus,

$$\frac{Et^3}{12(1 - \mu^2)} A_{mn}\left[\left(\frac{m\pi}{a}\right)^2 + \left(\frac{n\pi}{b}\right)^2\right]^2 \sin\frac{m\pi x}{a} \sin\frac{n\pi y}{b}$$

$$= f_x t A_{mn}\left(\frac{m\pi}{a}\right)^2 \sin\frac{m\pi x}{a} \sin\frac{n\pi y}{b} \quad (17.3)$$

The fact that the constant A_{mn} cancels in the above equality indicates that the magnitude of the plate deflection defined by each set of the m and n values is indefinite and that f_x is equal to the buckling stress F_{cr}.   This stress is found from Eq. 17.3 for each set of m and n to be

$$F_{cr} = f_x = \left[m\frac{b}{a} + \frac{a}{b}\frac{n^2}{m}\right]^2 \frac{\pi^2 E}{12(1 - \mu^2)}\left(\frac{t}{b}\right)^2 \quad (17.4)$$

The bracketed expression is defined as the plate buckling coefficient k.

$$k = \left[m\frac{b}{a} + \frac{a}{b}\frac{n^2}{m}\right]^2 \quad (17.5)$$

Of primary interest is the smallest value of F_{cr}.   It is given when $n = 1$. The value of m is found by minimizing Eq. 17.5 with respect to m:

$$\frac{\partial}{\partial m}\left[m\frac{b}{a} + \frac{a}{bm}\right]^2 = 2\left(\frac{mb}{a} + \frac{a}{bm}\right)\left(\frac{b}{a} - \frac{a}{bm^2}\right) = 0$$

or $b/a - a/bm^2 = 0$ and, thus, $m = a/b$.   The substitution of $n = 1$ and $m = a/b$ in Eq. 17.5 leads to

$$k_{\min} = 4 \quad (17.6)$$

for simply supported rectangular plates.

It may be of interest to compare Eq. 17.4 with a column buckling formula, for example, Eq. 9.3.   For a very wide plate, that is, when b/a is very large, $a/b \to 0$, and Eq. 17.4 can be simplified to

$$F_{cr} = m^2 \frac{\pi^2 E}{12(1 - \mu^2)}\left(\frac{t}{a}\right)^2 \quad (17.7)$$

in which $k = m^2$ and is a minimum when $m = 1$.

$$k_{\min} = 1 \quad (17.8a)$$

Noting that the radius of gyration of a rectangular section of depth t is $r = t/\sqrt{12}$, Eq. 17.7 becomes

$$F_{cr} = \frac{\pi^2 E}{(1 - \mu^2)\left(\frac{a}{t/\sqrt{12}}\right)^2} = \frac{\pi^2 E}{(1 - \mu^2)\left(\frac{a}{r}\right)^2} \quad (17.8b)$$

Except for $(1 - \mu^2)$, which reflects the effect of plate action due to Poisson's ratio, Eq. 17.8b is identical with the Euler column buckling formula, Eq. 9.3, since $KL = a$ for simply supported ends.

In general, the plate buckling stress is conveniently given by

$$F_{cr} = k \frac{\pi^2 E}{12(1 - \mu^2)} \left(\frac{t}{b}\right)^2 \tag{17.9}$$

k, the plate buckling coefficient, should be determined for each particular case of plate geometry, boundary conditions, material, and edge loading. For long and narrow plates ($a \gg b$, Eq. 17.6) $k = 4$, and for wide and short plates ($b \gg a$, Eq. 17.8b) $k = 1$. b in Eq. 17.9 is some significant dimension of the plate; for rectangular plates it is the width, or the length, whichever is smaller.

Figure 17.6 gives the values of the buckling coefficient k for long rectangular plates having typical boundary conditions at the side edges.   The buckling

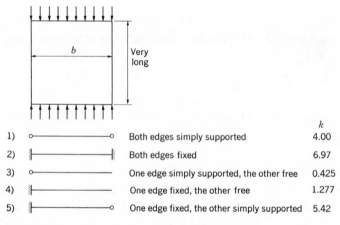

$$k$$

1) Both edges simply supported   4.00

2) Both edges fixed   6.97

3) One edge simply supported, the other free   0.425

4) One edge fixed, the other free   1.277

5) One edge fixed, the other simply supported   5.42

Fig. 17.6 Plate buckling coefficients, k.   (Courtesy of CRC.)

coefficient k, and thus the critical stress, is seen to vary considerably.   The buckling of plates due to non-uniformly distributed edge compression and due to edge shear is discussed briefly in connection with web buckling in plate girders in Art. 8.2.

The plate buckling stress in the inelastic range can be shown to be given simply and conservatively by[7.6]

$$F_{cr} = k \frac{\pi^2 \sqrt{EE_t}}{12(1 - \mu^2)} \left(\frac{t}{b}\right)^2 \tag{17.10}$$

where E_t is the tangent modulus of elasticity of the material. An even simpler and more conservative expression is obtained by approximating $\sqrt{EE_t}$ by E_t[17.6]

$$F_{cr} = k \frac{\pi^2 E_t}{12(1 - \mu^2)} \left(\frac{t}{b}\right)^2 \qquad (17.11)$$

3. Post-Buckling Behavior

An exact analysis of the post-buckling behavior of plates can be performed only by using the large-deflection theory of plates. The differential equations for this theory were derived by von Kármán in 1910, but they found little application due to their complexity.[7.6] An engineering simplification of the theory was made by von Kármán in 1932, when he introduced the concept of the "effective width."[17.7]

The physical nature of the post-buckling plate behavior can be explained best by means of a model. The plate can be thought of as being replaced by a system of straight bars in the horizontal and vertical directions, as shown in Fig. 17.7. After buckling, the vertical bars are enabled to carry additional loads because of the support provided by the bent horizontal bars (horizontal plate elements). With the loaded horizontal edges remaining straight, the load on the vertical bars closer to the side edges will be greater than on

Fig. 17.7 Model of plate in post-buckling range. (Courtesy of George Winter.)

the bars in the middle, for these bars deflect less. Thus, the stress distribution on the plate becomes non-uniform, the maximum stress being at the edges and the lowest stress in the middle. With an increasing axial load, the buckles are amplified and the difference between the stresses becomes more and more pronounced. Eventually, the stress at the edges reaches the yield point. The yielded edge zone widens concurrently with a reduction of the stress in the middle portion until the plate is no longer able to carry any additional load. The load at first yielding usually is considered as the ultimate load because any increase beyond it is relatively small in most cases.

Since the over-all strain of the plate (such as that shown plotted in Fig. 17.3) is controlled by the edge stress, this stress is considered the significant value. It is convenient to describe the behavior of a plate by assuming that the maximum (edge) stress is distributed uniformly over some reduced, "effective" width* b_e of the plate such that the total load is still the same. In Fig. 17.8 the non-uniform stress diagram is replaced by two rectangles of the width $b_e/2$ and having the same area as the original stress diagram. Thus, only a part of the plate is considered effective in carrying the load.

Von Kármán proposed the following approximate formula for the effective width of simply supported plates:[17.7]

$$\frac{b_e}{b} = \sqrt{\frac{F_{cr}}{f_{\max}}} \qquad (17.12)$$

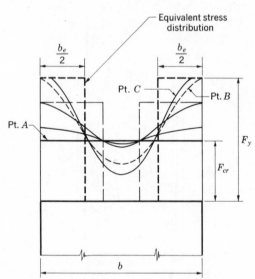

Fig. 17.8 Stress distribution and effective width b_e.

---

* The concept of the effective width is also used in connection with shear lag which is not associated with buckling. Shear lag is discussed in Art. 12.3.

where F_{cr} is the buckling stress for a simply supported plate ($k = 4$) and f_{max} is the edge (maximum) stress.

Several formulas have since been developed for plates having other boundary conditions.[17.8,17.9]   Usually, they are applicable to some particular field of engineering: aircraft, shipbuilding, building construction.   Each formula incorporates requirements and material properties pertaining to that particular field.

For structural design in steel von Kármán's formula was slightly modified to fit the experimental results.*   Thus, the effects of initial out-of-flatness and of other deviations from ideal conditions were taken into account.   Two formulas for the effective width of steel plates supported along two edges and one edge, respectively, are as follows:

$$\frac{b_e}{b} = \sqrt{\frac{F_{cr}}{f_{max}}}\left(1 - 0.218\sqrt{\frac{F_{cr}}{f_{max}}}\right) \tag{17.13}$$

and

$$\frac{b_e}{b} = 1.19\sqrt{\frac{F_{cr}}{f_{max}}}\left(1 - 0.30\sqrt{\frac{F_{cr}}{f_{max}}}\right) \tag{17.14}$$

4. Experimental Results

Numerous experiments have been conducted on plates under edge compression.[7.6,7.17,14.13,17.1,17.7–17.12]   In order to have a meaningful comparison of the results, it is convenient to plot them non-dimensionally since material and edge conditions often differ.   Figure 17.9 gives such a plot of F_{cr}/F_y or F_u/F_y for the ordinate, and the plate slenderness parameter for the abscissa.

$$\lambda = \sqrt{\frac{F_y}{F_{cr}}} = \frac{b}{t}\sqrt{\frac{F_y}{E} \cdot \frac{12(1 - \mu^2)}{k\pi^2}} \tag{17.15}$$

For a specific material and edge condition, this curve would correspond to a curve of F_{cr} or F_u vs. b/t.   Note that this representation is of the same nature as the one used for columns (see, for example, Fig. 9.38).   The available test results, if plotted in this figure, would fall approximately within the shaded zones.   A deviation from the theoretical curves could be due to initial imperfections, residual stresses, and other factors, which either were not considered or were approximated in the computation of the theoretical critical stress F_{cr}.

Curve a is the theoretical elastic buckling curve which is analogous to the Euler column curve.   Curves b and c are the theoretical buckling curves in the strain-hardening range for plates supported at both edges, and at one edge, respectively.[17.3,17.4]

* A series of references is given in Ref. 1.24.

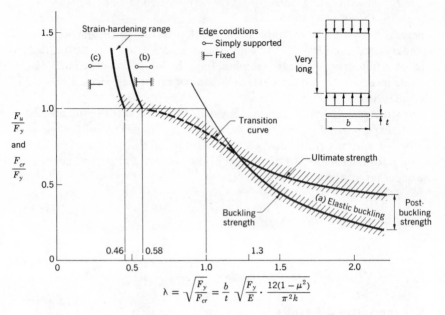

Fig. 17.9 Test results—plate-buckling curve.

The increasing separation of the ultimate and buckling strengths for values of λ higher than about 1.3 indicates an increasing post-buckling strength. A fact of great importance for steel design is that for λ less than 1.3, the ultimate and buckling strengths are for all practical purposes identical. This substantiates the previous observation of the plate behavior in Fig. 17.3 for low b/t values (curve $OA'B'C'$) for which the ultimate average stress was reached very soon after buckling.

The conclusion therefore follows that, in the range of λ below about 1.3, the plate strength is limited by buckling. This buckling, however, does not correspond to the theoretical buckling curve (curve a); rather, it forms a reasonably smooth transition curve between curve a at $\lambda \approx 1.3$ and the yield strength curve, $F_{cr}/F_y = 1.0$, in the vicinity of curves b and c. The reduction is ascribed to the effect of residual stresses and other imperfections.

5. Effect of Residual Stresses on Plate Buckling

Although residual stresses* were not considered in the previous discussion, they have a definite influence on the axial loading required to initiate plate

---

* The formation of residual stresses was discussed in Arts. 9.7 and 14.6.

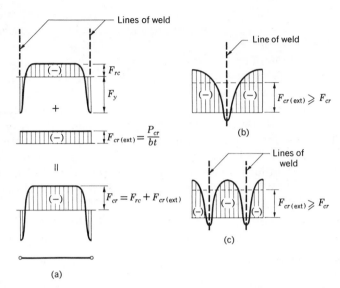

Fig. 17.10   Residual stress patterns in plates due to welding.

buckling. This influence was first recognized in tests on welded crane girders in 1941.[17.13]

Figure 17.10a shows that if the compressive residual stress F_{rc} is nearly uniformly distributed over the width of the plate (such as due to welding at the edges), it may be added to the stress from the external loading. Thus, the external stress $F_{cr(ext)}$ required to produce buckling would be approximately equal to the buckling stress F_{cr} less the residual stress F_{rc}.[14.13,17.14]

Curves a and b of Fig. 17.11 show, respectively, the buckling stress of a plate without residual stresses and of a plate with residual stresses having the

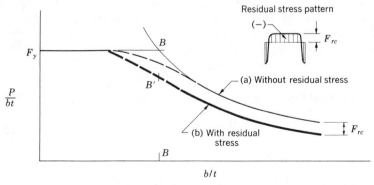

Fig. 17.11   Effect of residual stresses on plate buckling.

pattern in Fig. 17.10a. Note that for b/t ratios larger than indicated by points B and B', curve b lies below curve a, almost exactly by the magnitude of the residual compressive stress F_{rc}.[14.13]

Some other patterns of residual stress, such as those due to a weld or welds along the middle of the plate width (Fig. 17.10b and 17.10c), may increase the buckling strength of a plate, so that $F_{cr(\text{ext})} > F_{cr}$.[17.15]

The plate buckling problem becomes more involved when compressive residual stresses induce local yielding before the buckling stress is reached. Narrow plates with low b/t ratios have a higher critical stress, and the effect of residual stresses on local yielding is more pronounced. Examples of this behavior are the transition curve formed by the test results in Fig. 17.9, and the heavy dashed curve in Fig. 17.11, which was computed theoretically.[14.13]

17.4 WIDTH-THICKNESS RATIO REQUIREMENTS

1. Member Strength As Controlled by Plate Buckling

In this article, the width-thickness requirements of long plate elements subjected to edge compression are derived.

In order that a member be able to carry its design load, the strength of its component plates must be greater than or at least equal to the strength of the member itself. Since the capacity of a structural member is based on a stress which is close to or equal to the yield point, the plate elements should be controlled by their buckling or yield strength, and their post-buckling strength (which is very small at such stress level) may be neglected.*

In a properly designed centrally loaded column, the buckling strength of the plate components should be at least equal to the buckling strength of the column itself. In other words

$$(F_{cr})_{\text{plate}} \geqslant (F_{cr})_{\text{column}} \tag{17.16}$$

Considering a general case of the elastic or inelastic buckling, and substituting Eq. 17.11 for the plate stress and Eq. 9.24 for the column stress, the following inequality is obtained:

$$k\frac{\pi^2 E_t}{12(1 - \mu^2)}\left(\frac{t}{b}\right)^2 \geqslant \frac{\pi^2 E_t}{(KL/r)^2} \tag{17.17}$$

The requirement for b/t for steel ($\mu = 0.3$) is, from Eq. 17.17,

$$\frac{b}{t} \leqslant 0.3\,\frac{KL}{r}\,\sqrt{k} \tag{17.18}$$

where KL/r is the effective column slenderness ratio (see Chapter 10).

---

* Cold-formed members in which the post-buckling strength is not negligible are treated in Chapter 12.

Compressed plate elements in beams and beam-columns can be proportioned following the reasoning applied to columns, that is, by equating the plate buckling stress to the stress expected. The required b/t will then be a function not only of KL/r but also of the axial load and the moment.

The procedure described leads to specific width-thickness ratios for each particular design. Some specifications (for example, AISI,[1.24] Aluminum Alloy,[17.16,17.17] German,[8.9] and Russian[17.18]) follow it in one form or another. In order to ensure definitely that failure will be general rather than local, the permissible width-thickness ratios are usually limited to 2/3 to 3/4 of the critical ratios. This eliminates the uncertainty about imperfections in boundary conditions, in material, and in flatness.

For structural steel design in the United States, it was found more convenient to stipulate the yield point (the onset of strain-hardening, in plastic design) as the critical condition for the determination of the required width-thickness ratios. Requirements are then uniform for all members and depend only on the material, edge conditions, and on whether the member is designed for allowable stress or is designed plastically. When the b/t ratio exceeds these requirements, the effective b/t may be computed from Eqs. 17.13 and 17.14. The AISC specification formulates this approach in terms of reduction factors Q_s and Q_a described in Sect. 1 of Art. 12.4 (Appendix C of the AISC Specification[1.21]).

2. Width-Thickness Ratio Requirements in Allowable Stress Design

In allowable stress design, the capacity of a member is assumed to be limited by the state at which stresses just reach the yield point. Elastic plate buckling will be prevented if the buckling stress is taken equal to or greater than the yield point.

$$(F_{cr})_{\text{plate}} \geqslant F_y \quad \text{or} \quad \frac{F_y}{F_{cr}} \leqslant 1.0 \tag{17.19}$$

In Fig. 17.12 this condition corresponds to point B, that is, the point where the elastic buckling hyperbola intersects the line corresponding to the yield point F_y ($F_{cr}/F_y = 1.0$). Substituting Eq. 17.9 in Eq. 17.19 and taking $\mu = 0.3$ leads to the following equation for the limiting width-thickness ratio:

$$\frac{b}{t} \leqslant 0.951 \sqrt{\frac{kE}{F_y}} \tag{17.20}$$

Point B, however, lying above the transition curve, overestimates the actual plate strength. To be more certain of safe b/t ratios, point C corresponding to

$$\lambda = \sqrt{\frac{F_y}{F_{cr}}} = 0.7 \tag{17.21}$$

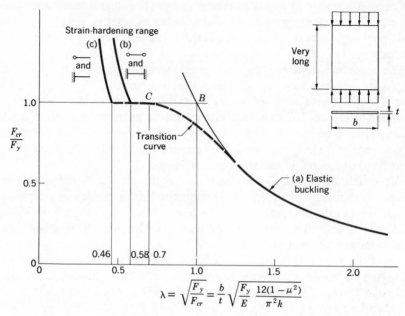

Fig. 17.12 Plate-buckling curve.

is used. This point, 0.7, is approximately where the transition curve reaches F_y and also is between 2/3 and 3/4, which have been used by various specifications. The general expression for computing the width-thickness ratios is obtained by substituting Eq. 17.9 and Eq. 17.21 and taking for steel $\mu = 0.3$ and $E = 29 \times 10^3$ ksi; thus

$$\frac{b}{t} \leqslant \frac{114\sqrt{k}}{\sqrt{F_y}} \tag{17.22}$$

where the yield point F_y is in ksi.

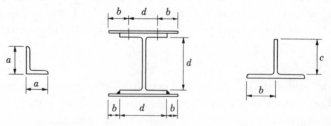

Fig. 17.13 Critical plate dimensions.

The following may be considered as typical of plate elements and of their k values in hot-rolled and built-up structural members:

1. Single angles (dimension a in Fig. 17.13): The worst condition occurs in an equal-legged angle subject to concentric thrust; both legs buckle simultaneously and thus each leg behaves as a plate simply supported along one edge. Figure 17.6 gives $k = 0.425$ for this condition.
2. Flanges or other projecting plate elements partially restrained along one edge, such as double angles in contact, angles or plates projecting from girders and columns, stiffeners on plate girders (dimension b in Fig. 17.13): The degree of restraint of these plate elements may vary considerably. A study led to the conclusion that a value of $k = 0.70$ may be considered as the lowest occurring in practical members. Procedures for computing k from the geometry of a member are discussed in Ref. 7.6 and 17.6. Figure 17.2 (buckled flange of a plate girder) illustrates the buckling of such plate elements.
3. Stems of tees (dimension c in Fig. 17.13): The edge restraint of the stem by the flange is essentially complete for structural tees, and therefore, $k = 1.277$ applicable to a plate fixed along one edge (case 4 in Fig. 17.6) may be used.
4. Plate elements supported along two edges such as webs, cover plates, diaphragms (dimension d in Fig. 17.13): As in the case of flanges, the edge restraint is partial. The two extremes are: both edges simply supported ($k = 4.0$) and both edges fixed ($k = 6.97$). An investigation of commonly used sections employing the procedure of Ref. 7.6 and 17.6 showed that $k = 5.0$ may be considered as a safe value. Taking at random, for example, a S24 × 100 section gives $k = 5.3$, and a W30 × 210 gives $k = 6.5$. A substantially lower degree of restraint exists in square and rectangular pipe sections, and, correspondingly, an almost simple support is assumed with $k = 4.4$ (AISC Sec. [1.9.2.2]).

Substitution of the plate coefficient values in Eq. 17.22 yields the expressions shown in column 3 of Table 17.1. In the fourth column are given formulas adopted by the AISC specification. They are in very close agreement with the derived formulas of column 3. In the subsequent columns are listed b/t requirements of various specifications for structural carbon steel ($F_y \approx 36$ ksi).

Example 17.1 illustrates the application of the b/t limitations in the design of a built-up column.

3. Width-Thickness Ratio Requirements in Plastic Design

The width-thickness ratios of a profile subject to rotation due to plastic hinge action are more restrictive than similar ratios of a profile designed on an

Table 17.1 Width-Thickness Ratio Requirements

Type	Plate Buckling Coefficient k	b/t (Eq. 17.22)	AISC (Ref. 1.21) General	AISC A36 $F_v = 36$ ksi	AASHO (Ref. 1.23)	AREA (Ref. 1.25)	AISI (Ref. 1.24)	B.S.449 (Ref. 17.19)	DIN 4114 (Ref. 8.9)	Russian (Ref. 17.18)
(1)	(2)	(3)	(4)	(5)	(6)	(7)	(8)	(9)	(10)	(11)
					A36 steel or equivalent*					
1. Projecting plate elements										
(a) Single angles	0.425	$74.3/\sqrt{F_v}$	$76/\sqrt{F_v}$	13	12 (16)†	10	10.6	16	15	14‡
(b) Flanges	0.70	$95.5/\sqrt{F_v}$	$95/\sqrt{F_v}$	16		12 (14)				
(c) Stems of tees	1.277	$129/\sqrt{F_v}$	$127/\sqrt{F_v}$	21						
2. Plate elements supported along two edges										
(a) Webs§	5.0	$254/\sqrt{F_v}$	$253/\sqrt{F_v}$	42	32	32‡	28.5	50	45	40‡
(b) Cover plates	> 5.0				40	40‡			45 and higher	
(c) Perforated cover plates			$317/\sqrt{F_v}$	53	50	40				

* Lower values should be used for higher-strength steels according to additional provisions.
† The value in parentheses is used for secondary members.
‡ Minimum value, otherwise b/t depends on the column slenderness (as in Eq. 17.18) or on the computed stress (Art. 17.4).
§ Buckling of webs in flexural members is discussed in Chapters 8 and 12.

EXAMPLE 17.1

PROBLEM:

The designer selected a W14 x 127 section of ASTM A36 steel to be used as an interior column in an office building. The member was chosen to carry a maximum allowable concentric load of 695 kip. Prior to construction it was learned that the column would be required to carry an additional load of 185 kip. It is decided that the previously selected member will be built-up with $\frac{5}{16}$ in. plates. Unsupported column length is 14 ft. The AISC Specification is to be used.

SOLUTION: WELDED CONSTRUCTION

$A(W14 \times 127) = 37.3$ in.^{2} ; Original design stress $= \dfrac{695}{37.3} = 18.65$ ksi

Trial stress $= 18.7$ ksi ; $A_{\text{R}} = \dfrac{200}{18.7} = 10.7$ in^2

Try 2 Rs, $\frac{5}{16}$ x 16 (A = 10 in^2)

Total applied load $= 695 + 185 = 880$ kip

$I_y = 528 + \dfrac{5}{16}(16^3)(\dfrac{2}{12}) = 741.0$ in.^{4}

$A = 37.3 + 10.0 = 47.3$ in.^{2}

$r_y = \sqrt{\dfrac{741.0}{47.3}} = 3.96$ in.

$\dfrac{KL}{r} = \dfrac{14(12)}{3.96} = 42.4$

$F_a = 19.00$ ksi   (Table [1-36])

$P_{\text{all}} = 47.3(19.00) = 899$ kip

Check $\dfrac{b}{t}$ of plate between welds: $\dfrac{b}{t} = \dfrac{14.69}{5/16} = 47$

Table 17.1(2) $(\dfrac{b}{t})_{\text{all}} = \dfrac{253}{\sqrt{F_y}} = 42.2$

$47 > 42.2$ ∴ N.G.

Thus, the effective width of the cover plate should be reduced and the allowable stress modified. According to (Section [C 3]) the effective width is with $f = 0.6 F_y = 21.6$ ksi

$$b_e = \dfrac{253t}{\sqrt{f}}(1 - \dfrac{44.3}{(b/t)\sqrt{f}})$$

$$= \dfrac{253(5/16)}{\sqrt{19.14}}(1 - \dfrac{44.3}{(47)\sqrt{19.14}})$$

$$= 13.56 \text{ in.}$$

and the effective section is show below. The stress reduction factor (Section [C 4])

$Q_a = \dfrac{\text{effective area}}{\text{actual area}}$

$= \dfrac{47.3 - (5/16)(14.69 - 13.56)}{47.3} = 0.993$

From (Section [C5]), $F_a = 18.97$ ksi

$P_{\text{all}} = F_a Q_a A = 18.97(0.993)47.3$

$= 891^k > 880^k$ O.K.

web, $\dfrac{b}{t} = \dfrac{14.62 - 2(0.998)}{0.61} = 20.7 < 42$ (Table 7.1)   O.K.

Flg(W + R) $\dfrac{b}{t} = \dfrac{8}{1.31} = 6 < 16$

Use W 14 x 127 w/ $\frac{5}{16}$ x 16 Cover Plates Welded

EXAMPLE 17.1 (Continued)

RIVETED CONNECTION:

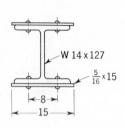

W 14 x 127

$\frac{5}{16}$ x 15

|← 8 →|

|← 15 →|

The unsupported plate width between rivets will not be excessive.

Try 2 ℞, $\frac{5}{16}$ x 15

$I_y = 528 + \frac{5}{16}(15^3)\frac{2}{12} = 704$ in.^{4}

$A = 37.3 + 9.4 = 46.7$ in.^{2}

$r_y = \sqrt{\dfrac{704}{46.7}} = 3.88$

$\dfrac{KL}{r} = \dfrac{14(12)}{3.88} = 43.3 \quad \therefore F_A = 18.92$ ksi

$P = 46.7 \times 18.92 = 884 > 880$ kip   <u>O.K.</u>

Check $\dfrac{b}{t}$; — projecting part of cover plate = $\dfrac{15-8}{2} = 3.50$ in.

$\dfrac{b}{t} = \dfrac{3.50}{5/16} = 11 < 16$ Table 17.1 (Ib)(Section [1.9])   <u>O.K.</u>

All other plate width–thickness ratios have been checked, or more critical ratios have been shown adequate.

> Use W 14x127 $^W/$ $\frac{5}{16}$ x 15 Cover Plates
> Riveted

allowable stress basis.   The plate elements of the latter are required merely to *reach the yield stress*.   To ensure adequate hinge rotation capacity the proportions required for compression elements in regions of maximum moment in plastically designed framing are such that these elements can *compress plastically to strain-hardening without buckling.**

Flanges.   Strain-hardened plates can be analyzed as orthotropic plates, that is, as plates for which the material properties are different in two perpendicular directions.   Equations for such an analysis are developed in Refs. 17.3, 17.4, and 17.6.   The results of this analysis are shown in Figs. 17.9 and 17.12 as curves *b* and *c*.   The theoretical results were substantiated by experiments, as indicated by the shaded area in Fig. 17.9.

For plates with both edges supported, the value of $\lambda = \sqrt{F_y/F_{cr}}† = 0.58$ as given in Fig. 17.12 by curve *b* is the same whether the edges are simply supported or fixed.   The equations for F_{cr} are, however, quite different and thus the maximum width-thickness ratios are also different.   The degree of edge restraint for plates in plastified structural sections was determined both

* This statement is paraphrased from the Commentary on the AISC Specification, Ref. 1.21.

† F_{cr} is defined in this section as the buckling stress in the strain-hardening range.   It is a complicated function of $(b/t)^2$, the degree of edge restraint, and the elastic moduli and Poisson's ratios in both directions (see Ref. 17.3 and 17.6).

analytically and experimentally, and it was found to be essentially equivalent to that of simple supports (Fig. 6.13 of Ref. 1.1).   The resulting requirement is

$$\frac{b}{t} \leqslant \frac{190}{\sqrt{F_y}} \qquad (17.23)$$

Observations similar to the above are applicable also to plates supported at one edge only such as flanges of rolled sections and other projecting plate elements.   Again, $\lambda = \sqrt{F_y/F_{cr}} = 0.46$ (from curve c in Fig. 17.12) is the same for all of them, but the formula for F_{cr} (and thus for b/t) varies with the degree of edge restraint and material properties.   Investigations show that the degree of restraint of flanges in typical sections after plastification is just a little greater than that for simple supports (Fig. 6.12 in Ref. 1.1). This gives, for structural carbon steel ($F_y = 36$ ksi):

$$\frac{b}{t} \leqslant 8.5 \qquad (17.24)$$

The influence of the yield point in this case is more complex; it is not exactly proportional to $1/\sqrt{F_y}$; and b/t for other values of F_y is given in Table 7.3.
   Since there is no uncertainty about residual stresses (they disappear in the range of strain-hardening) and the effect of small imperfections is already included in the material properties, there is no need to introduce any reduction of the b/t values in Eqs. 17.23 and 17.24.[17.3]

   Webs.   A similarly complicated analysis is required to establish the safe width-thickness ratios for beam and column webs subjected to combined axial force and plastic bending moment at ultimate loading.   The plate in this case is stressed and strained non-uniformly across its width, as shown in the sketch in Fig. 17.14.   Since the critical b/t ratio depends to a large extent on the magnitude of the maximum strain, a maximum strain ε_{max} equal to 4 times the yield strain ε_y was assumed as a strain sufficient to furnish adequate rotation capacity of a plastic hinge.   Again, the b/t ratio can be found only for a specific material.   The dashed curve in Fig. 17.14 depicts the theoretical relationship between the maximum b/t ratio and the magnitude of the axial force P/P_y for the following conditions: $F_y = 33$ ksi; total section area equal to twice the web area; and $\varepsilon_{max} = 4\varepsilon_y$.   The design approximation is shown by the solid line.[1.1]   It is generalized for steels with other yield stresses into Eq. 17.25.[1.1,1.21 (Commentary)]

$$\frac{257}{\sqrt{F_y}} \leqslant \frac{b}{t} \leqslant \frac{412}{\sqrt{F_y}}\left(1 - 1.4\,\frac{P}{P_y}\right) \qquad (17.25)$$

The limit for the maximum axial load ($P/P_y = 1.0$) is $257/\sqrt{F_y}$.   It is practically the same as the width-thickness requirement for compressed

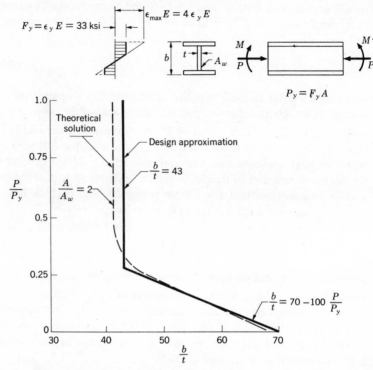

Fig. 17.14 Width-thickness requirements for webs of beam-columns in plastic design. (Courtesy of ASCE.)

plates in allowable stress design (column 4 in Table 17.1). This should be approximately so, since in plastic design the web plates should not be expected to undergo much plastic straining when subjected to uniform compression.

Example 17.2 explains the application of Eqs. 17.23 to 17.25 in the design of a beam-column.

17.5 STITCH FASTENERS IN BUILT-UP COMPRESSION MEMBERS

If a component of a built-up compression member is an outside plate, the spacing of fasteners (bolts, rivets, spot welds) should be sufficiently small to preclude local buckling of the plate between the fasteners. In Fig. 17.15 the controlling dimensions are given as the distance b between the longitudinal

---

EXAMPLE 17.2

PROBLEM:

    The welded built-up beam-column shown below was designed under the provisions of Part 2 of the AISC Specification to carry the loads shown. The material is A36 steel. Check for local buckling.

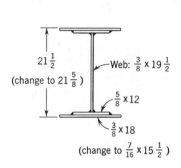

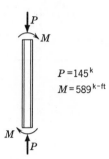

$P = 145^k$

$M = 589^{k\text{-}ft}$

SOLUTION:

Projecting elements: Outer $\mathbb{R}$, $\dfrac{b}{t} = \dfrac{(18-12)}{2(3/8)} = 8 < 8\tfrac{1}{2}$, Eq. 17.24 (Section [2.6])   <u>O.K.</u>

Both $\mathbb{R}$s, $\dfrac{b}{t} = \dfrac{9}{1} = 9 > 8\tfrac{1}{2}$ <u>NG</u>
(assumed 18 in.)

Change outer plate to $\dfrac{7}{16} \times 15\tfrac{1}{2}$. This will provide approximately the same area
                             (6.78 in.^{2} vs. 6.75 in.^{2})

Then, $\dfrac{b}{t} = \dfrac{7.75}{1\frac{1}{16}} = 7.3 < 8\tfrac{1}{2}$ <u>O.K.</u>

Outer plate between welds: $\dfrac{b}{t} = \dfrac{12}{7/16} = 27.4$; $\left(\dfrac{b}{t}\right)_{allow} = \dfrac{190}{\sqrt{36}} = 31.7$, Eq. 17.23 (Section [2.7])

Web plate: The web is governed by Eq. 17.25 (Section [2.7])

    $P_y = F_y A = 36(35.9) = 1292$ kip; $\dfrac{P}{P_y} = \dfrac{145}{1292} = 0.112$

    $\left(\dfrac{d}{w}\right)_{allow} = \dfrac{412}{\sqrt{36}} \left[1 - 1.4(0.112)\right] = 58.0$

    $\dfrac{d}{w} = \dfrac{21^{5}/_{8}}{3/8} = 57.7 < 58.0$ <u>O.K.</u>

There is no danger of buckling assuming the noted change of the outer cover plate.

---

rows of fasteners, the edge distance b_1, and the spacing s of the fasteners in a row. The distance b between rows and the edge distance b_1 are controlled by the requirements established in Art. 17.4. In this article, provisions for the fastener spacing s in a row for staggered and non-staggered fasteners are developed.

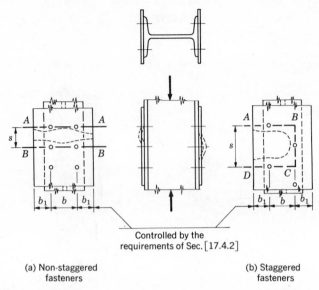

Controlled by the
requirements of Sec. [17.4.2]

(a) Non-staggered
fasteners

(b) Staggered
fasteners

Fig. 17.15 Plate buckling between fasteners in built-up compression members.

Considering first the non-staggered fasteners and referring to Fig. 17.15a, it is seen that the plate portion between the transverse lines of fasteners, that is, lines AA and BB, should be analyzed for buckling. Since the plate is tightly clamped to the member and can deflect only away from it, it may be assumed that lines AA and BB furnish a fixed edge support. With the side edges free, the plate buckling coefficient is $k = 4.0$.* The true value probably is somewhat smaller, but any deviation is accounted for in Eq. 17.22. Replacing b/t by s/t in Eq. 17.22, the following spacing-thickness ratio limitation is derived:

$$\frac{s}{t} \leqslant \frac{114\sqrt{k}}{\sqrt{F_y}} = \frac{228}{\sqrt{F_y}} \qquad (F_y \text{ in ksi}) \qquad (17.26)$$

For staggered fasteners, the plate portion enclosed by line $ABCD$ in Fig. 17.15b should be analyzed. The plate buckling coefficient depends in this case not only on the support conditions, but also on the side ratio of the plate element, $BC/AB = s/(b_1 + b)$. This side ratio for values of s and b

---

* This k value can be obtained from Eq. 17.7 with $m = 2$. The plate in this case is analogous to a fixed-end column with the effective length equal to one-half of the length of a pinned-end column.

encountered in practice is approximately equal to or less than 2 to 3. Assuming lines AB, BC, and CD to furnish fixed-edge support, the k coefficient is found to be approximately equal to 6.5.[17.11] With this k value

$$\frac{s}{t} \leqslant \frac{290}{\sqrt{F_y}} \qquad (F_y \text{ in ksi}) \qquad (17.27)$$

Many specifications are concerned not only with the prevention of local buckling in establishing the spacing-thickness limits, but also with ensuring a close fit-up over the entire faying surface to inhibit corrosion. The latter consideration led to limits which were developed primarily on the basis of judgment and experience with only casual reference to the reasoning involved in the derivation of Eqs. 17.26 and 17.27.* As a result, the rules of some of the specifications may appear quite arbitrary and conservative. A comparison of them is shown in Table 17.2.

Table 17.2 Spacing-Thickness Ratio Requirements (Stitch Fasteners)

	Maximum Allowable Spacing-Thickness Ratios, s/t. ($F_y = 36$ ksi)							
Type	Eqs. 17.26 and 17.27	AISC 1969	AASHO 1965–67	AREA 1969	AISI 1968	B.S.449 (British) 1959	DIN4114 (German) 1959	Russian 1961
(1)	(2)	(3)	(4)	(5)	(6)	(7)	(8)	(9)
Non-staggered fasteners	$\dfrac{228}{\sqrt{F_y}}$	$\dfrac{127}{\sqrt{F_y}}$ but $\leqslant 12$ in.	—	—	$\dfrac{257}{\sqrt{F_y}}$	—	—	—
	38	21	12	12	43	16 but $\leqslant 6$ in.	15	12
Staggered fasteners	$\dfrac{270}{\sqrt{F_y}}$	$\dfrac{190}{\sqrt{F_y}}$ but $\leqslant 18$ in.	—	—	—	—	—	—
	48	32	24	24	—	—	—	18

Columns 4 and 5 (AASHO and AREA) give the most conservative fastener spacing (12 vs. 38). This is well justified in view of the exposure of the bridge structures to weather. The requirements of AISC (column 3) are also tempered considerably by corrosion considerations but not as much as AASHO and AREA (21 vs. 38). The AISI rule (column 6) for light gage cold-formed members is actually more liberal than the rule derived with Eq.

---

* Some specifications give limiting spacing of sealing fasteners for members subjected to tension and bending to prevent corrosion.

EXAMPLE 17.3

PROBLEM:

   Determine the rivet spacing if rivets are used in Design Example 17.1 (AISC).

SOLUTION:

   AISC Specification   NON-STAGGERED RIVETS:
                        Allowable spacing:

   $$s/_t \;=\; \frac{127}{\sqrt{F_y}} \;=\; \frac{127}{\sqrt{36}} \;=\; 21.2 \;,\quad \text{Table 17.2, or (Section } [1.18.2.3])$$

   $s \;=\; 21.2 \times 5/16 \;=\; 6.6 \text{ in.}$ Use: s = 6 ½ in.

                        STAGGERED RIVETS:
                        Allowable spacing:

   $$s/_t \;=\; \frac{190}{\sqrt{F_y}} \;=\; 31.7 \quad \text{Table 17.2, or (Section } [1.18.2.3])$$

   $s \;=\; 31.7 \times 5/16 \;=\; 9.9 \text{ in.}$ Use: s = 10 in.

17.26.   In effect, AISI utilizes $\lambda = \sqrt{F_y/F_{cr}} = 0.833$ for point C in Fig. 17.12 as compared to $\lambda = 0.7$ used in the derivation of Eq. 17.22.

Example 17.3 illustrates the AISC provisions.

17.6 WEB CRIPPLING

1. Allowable Stress Design

Basically, there are two types of web behavior which may be expected at a beam support or at the point of application of a high concentrated load. These are "web crippling" and "web buckling." Web crippling is predominantly a plastic behavior which affects a rather local region by causing the formation of small wrinkles, whereas web buckling is predominantly an elastic behavior which affects a larger part of the web. Specifications usually do not differentiate between these two phenomena and group them under the same heading.

Web Crippling. Early steel specifications made provisions only against web buckling.[17.20] Then it was found that standard sections are not susceptible to web buckling but rather to web crippling.[17.21] The 1936 AISC specification[17.2] introduced the web crippling criterion as the only one controlling the web strength of such sections. The 1969 specification uses the 1936 formulas modified only to cover materials with different yield points. (See AISC Section [1.10.10.1].)

The formulas were obtained empirically on the basis of extensive tests. It was assumed that the transverse load spreads from the edges of a bearing plate under 1:1 slope and that the stress at the toe of the fillet should not exceed $0.75F_y$ at working loads. Figure 17.16 shows the assumed stress distributions for an interior load and an end reaction. The resulting formulas are:

For interior loads:

$$\frac{R}{t(N + 2k)} \leqslant 0.75F_y \tag{17.28}$$

For end reactions:

$$\frac{R}{t(N + k)} \leqslant 0.75F_y \tag{17.29}$$

where R is the concentrated load or reaction, t is the thickness of web, N is the length of bearing, and k is the distance from outer face of flange to web toe of fillet.

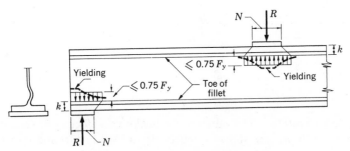

Fig. 17.16 Web crippling.

Although the formulas assume a uniform stress distribution, actually the stresses are distributed approximately as shown by the broken lines. The flat areas of the high stress intensity correspond to the portions of yielding. A little yielding, however, is not detrimental to the strength of a beam or girder even at working loads because considerably more yielding than shown in Fig. 17.16 would be required before crippling ensues.

Other specifications either use essentially the same rules as Eqs. 17.28 and 17.29 or require bearing stiffeners at each concentrated load. For cold-formed members, provisions are more complex and are discussed in Art. 12.3.

Web Buckling Due to Transverse Load. Over-all failure of the web due to transverse loading applied to the flange between stiffeners is presently controlled by the provisions of the AISC specification based on buckling considerations. The buckling strength of a web panel enclosed between the flanges and neighboring intermediate stiffeners is assumed approximately to consist of two contributions: resistance in the direction of the girder depth (parallel to the transverse loading), and resistance in the direction of the spacing of the stiffeners (perpendicular to the transverse loading), that is, dimensions h and a, respectively, in Fig. 17.17a. These two contributions are the two terms in Eq. 17.30.

$$F_u = k_1 \frac{\pi^2 E}{12(1 - \mu^2)}\left(\frac{t}{h}\right)^2 + k_2 \frac{\pi^2 E}{12(1 - \mu^2)}\left(\frac{t}{a}\right)^2 \qquad (17.30)$$

As indicated in Fig. 17.17b and c, two conditions of restraint furnished by the loaded flange are considered: the flange is prevented from rotation and the

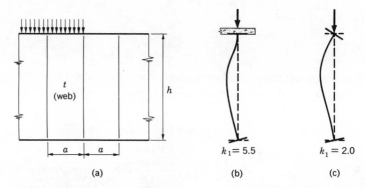

(a)   (b)   (c)

Fig. 17.17 Buckling of girder web due to transverse loading. (Courtesy of AISC.)

flange is free to rotate. The values of the buckling coefficient k_1 assumed for these two conditions are shown respectively in Fig. 17.17b and c to be 5.5 and 2.0. With $k_2 = 4.0$ as a minimum value for the horizontal direction, known values of E and μ, and using a factor of safety F.S. = 2.6, the following two expressions are obtained:

$$F_{\text{edge}} \leqslant \left[5.5 + \frac{4}{(a/h)^2}\right] \frac{10,000}{(h/t)^2} \qquad \text{(in ksi)} \qquad (17.31)$$

when the flange is restrained against rotation, and

$$F_{\text{edge}} \leqslant \left[2 + \frac{4}{(a/h)^2}\right] \frac{10,000}{(h/t)^2} \qquad \text{(in ksi)} \qquad (17.32)$$

when the flange is not so restrained.

In the case of concentrated loads or loads distributed over a partial length of a panel, the allowable edge stress F_{edge} as given by Eqs. 17.31 and 17.32 should be greater than or equal to the stress obtained from dividing the total panel load by the product of the web thickness and the panel length a or the girder depth h, whichever is smaller. Equations 17.31 and 17.32 are used in AISC Section [1.10.10.2] as Formulas [1.10-10] and [1.10-11], respectively. Example 17.4 demonstrates the application of this provision.

EXAMPLE 17.4

PROBLEM:

Check web crippling and determine the length of the bearing plates (if needed) in Example 8.2 (AISC)

SOLUTION:

WEB BUCKLING
The top flange is not restrained from rotation, therefore use Eq. 17.32, or Section [1.10.10.2] for the allowable edge stress.

$$f_{edge} \leq \left[2 + \frac{4}{(a/h)^2}\right]\frac{10^4}{(h/t)^2} = \left[2 + \frac{4}{(81/70)^2}\right]\frac{10^4}{(70/0.3125)^2} = 0.994 \text{ ksi}$$

Actual edge stress:

Distributed loading: $f = \dfrac{2.5}{12 \times 0.3125} = 0.666$ ksi < 0.99 <u>O.K.</u>

6 kip load $f = \dfrac{6}{70 \times 0.3125} = 0.274$ ksi

           $f_{total} = 0.274 + 0.666 = 0.94$ ksi < 0.99 <u>O.K.</u>

76 kip load $f = \dfrac{76}{70 \times 0.3125} = 3.48$ ksi $\gg 0.99$ <u>N.G.</u>

         Bearing stiffeners are required under 76 kip loads acc. to Eq. 17.32

WEB CRIPPLING
Assume ¼ in. weld $k =$ Flange ℞ + Weld $= \frac{3}{4} + \frac{1}{4} = 1$ in.

From Eq. 17.28, or Section [1.10.10]

Length of bearing ℞ $N = \dfrac{R}{0.75t\,F_y} - 2k = \dfrac{6}{0.75 \times 36 \times 0.3125} - 2 \times 1 = -1.3$ in.

<u>No bearing plate is required</u>

Check for 76 kip load according to Eq. 17.33 and Fig. 17.18

f_b (from Example 8.2) $= 18.46$ ksi

         $f_b/f_e = \dfrac{18.46}{3.48} = 5.3$

From Fig. 17.18 for $a/h = 81/70 = 1.16$

         $k_e = 5.3$

Eq. 17.33

$$F_e = \frac{40,000\,k_e}{(h/t)^2\sqrt{a/h}}\sqrt{\frac{F_y}{36}}$$

$$= \frac{40,000}{(70/0.3125)^2\sqrt{1.16}}\sqrt{\frac{36}{36}}\,k_e$$

$$= 0.74\,k_e = 0.74\,(5.3) = 3.94 \text{ ksi} > 3.48 \text{ ksi} \qquad \underline{\text{O.K.}}$$

No bearing stiffeners are needed under 76 -k loads acc. to Eq. 17.33

The actual capacity of this panel is found after several trials to be:

Assume $f_e = 4.75$ ksi, $f_b/f_e = 3.89$, $k_e = 6.5$, $F_e = 4.8$ ksi

Length of bearing ℞ for 76 kip load if no bearing stiffener is used:

$$N = \frac{R}{0.75t\,F_y} - 2k = \frac{76}{0.75 \times 36 \times 0.3125} - 2 \times 1 = 7 \text{ in.}$$

<u>Bearing ℞ should be at least 7 in. long</u>

Other specifications have no provisions for loads applied directly to the web of plate girders without bearing stiffeners. Apparently it is assumed that loading is transmitted to the plate girder by means of transverse members which would require bearing stiffeners in any case.

A more rigorous analysis has shown that Eqs. 17.31 and 17.32 are reasonably accurate in predicting the buckling load when only the edge loading is applied and the moment and shear acting on the panel are zero. Otherwise, the buckling stress may be considerably reduced. On the other hand, these equations do not consider the post-buckling strength of the web, and tests have indicated that the ultimate strength of a panel under practical edge loading and an ordinary design moment will be adequate if Eqs. 17.31 and 17.32 are used.

However, when the bending moment is less than the bending capacity of the girder panel, these equations may lead to overconservative requirements. In this case, the following equation, solvable by trial and error, may be used:[17.22,17.23]

$$F_e = \frac{40,000 k_e}{(h/t)^2 \sqrt{a/h}} \sqrt{F_y/36} \qquad (17.33)$$

where the buckling coefficient k_e is to be taken from Fig. 17.18 as a function of a/h and of the ratio of the bending to edge stresses f_b/F_e. The top flange is

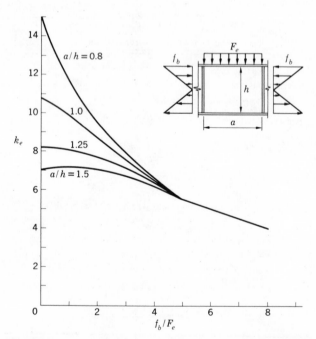

Fig. 17.18 Plate-buckling coefficient k_e for edge loading.

assumed here to be prevented from rotation. This equation is based on theoretical buckling computations and ultimate strength tests and should be at present limited to 34 ksi $< F_y <$ 45 ksi, $\dfrac{f_b}{F_e} <$ 8.0, 0.8 $< a/h <$ 2.0, and 240 $< h/t <$ 300.

2. Web Crippling in Plastic Design

If a load is applied at a point where, according to design, a plastic hinge would form, the plastified web cannot be relied upon to carry this load without buckling, and web stiffeners are required.

A more complicated situation arises when a member is expected to develop a plastic hinge at the point of its connection to another member. A typical combination is of a plastically designed beam framing into a column.

In Fig. 17.19, the effect of a beam moment on a column is shown as a couple composed of the two flange forces, the beam web forces being of secondary importance. Significant effects of these flange forces can occur in

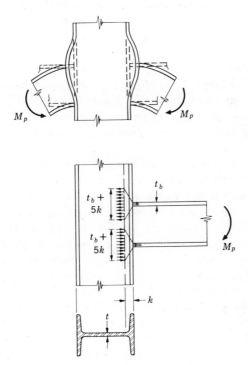

Fig. 17.19 Beam-to-column connection in plastic design, web crippling. (Courtesy of ASCE.)

two regions in the column. The first region is the column web where yielding may be accompanied by crippling due to the beam compression flange force or by fracture due to the beam tension flange force. The second region is the column flange where bending may contribute to the fracture of welds connecting the beam flange to the column flange; this type of failure is treated in Art. 19.10.

The stresses in the column caused by a concentrated beam flange force will spread out as they penetrate into the column, as shown in Fig. 17.19. If the spread of stresses is insufficient to reduce their intensity to the yield level at the toe of the column flange fillet (distance k in Fig. 17.19), the web will not be able to provide sufficient resistance to the beam flange force. Due to the danger of crippling, this effect is more serious in the region stressed by the compression flange. A rational analysis of the spread of stress is difficult and is replaced by a linear assumption based on test results.[17.24] The spread is assumed to be on a 2.5:1 slope from the point of contact to the column "k-line." Then the web area available is equal to $(t_b + 5k)t$, and its capacity must be equal to or exceed the beam flange force. This can be stated as an inequality of the web and beam forces, $F_y(t_b + 5k)t \geqslant F_{yb}A_f$. Then, for a safe design, the column web thickness must satisfy the requirement

$$t \geqslant \frac{C_1 A_f}{t_b + 5k} \qquad (17.34)$$

where t is the column web thickness, A_f is the area of the beam flange, t_b is the beam flange thickness, $C_1 = (F_{yb}/F_y)$ is the ratio of the beam flange yield stress to the column yield stress, and k is the "k-distance" from the outer face of the flange to the web toe of fillet of the column. If the requirement of Eq. 17.34 is not fulfilled, stiffeners are needed opposite the connected beam flanges to carry the excess force.

Equation 17.34 is applicable to a connection of any two members having conditions similar to a beam-to-column connection.

There is an apparent inconsistency between the assumptions made for the spreading of stresses in allowable stress design and in plastic design. In allowable stress design a slope of 1:1 is used (see Fig. 17.16 and Eq. 17.28), and in plastic design the slope is 2.5:1 (see Fig. 17.19 and Eq. 17.34). Theoretical analysis being impractical, both equations are based almost exclusively on experimental results. The explanation of the discrepancy is in the fact that in the case of a beam-to-column connection the high concentrated beam flange force is introduced into the column web and, just a short distance away, is transferred to the other flange. The close interaction between these two opposite actions has a stabilizing effect on the web, and the web is able to sustain a higher load than if only one load were applied. Another reason is that Eq. 17.28 already incorporates a factor of safety, whereas Eq. 17.34 states the ultimate condition.

EXAMPLE 17.5

PROBLEM:

Given the interior beam-to-column connection shown. Design stiffeners to reinforce the column if they are required. Column $F_y = 36$ ksi, beam $F_{yb} = 45$ ksi, stiffeners $F_{ys} = 50$ ksi. Refer to the AISC Specification Part 2.

SOLUTION:

Compression:

$$t_{COL} = 0.47 \text{ in.}, C_1 = 45/36 = 1.25$$

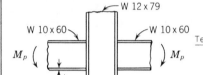

Eq. 17.34 $(t_{col.})_{reqd.} = \dfrac{C_1 A_f}{t_b + 5k} = \dfrac{(1.25)(10.1)(0.683)}{(0.683 + 5 \times 1\frac{7}{16})}$

$$= 1.09 \text{ in.} > 0.47 \text{ in.} \qquad \underline{\text{N.G.}}$$

$$\underline{\text{Stiffeners required}}$$

Tension:

$$t_{f_{COL}} = 0.736 \text{ in.}$$

$$t_{f_{reqd}} = 0.4\sqrt{C_1 A_f} = 0.4\sqrt{1.25(6.90)}$$

$$= 1.18 \text{ in.} > 0.736 \qquad \underline{\text{N.G.}}$$

$$\underline{\text{Stiffeners required}}$$

DESIGN OF STIFFENERS

Considering equilibrium at the beam's compression flange.

$$A_f F_y = F_y \, t\,(t_b + 5k) + F_{ys} A_{st}$$

where A_{st} = area of a pair of stiffeners

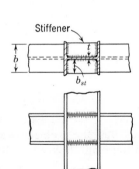

Stiffener

$$A_{st} = \left[C_1 A_f - t\,(t_b + 5k) \right] \frac{F_y}{F_{ys}}$$

$$= \left[(1.25)(6.90) - 3.70 \right](36/50) = 3.55 \text{ in.}^2 \text{ or}$$

$$1.90 \text{ in.}^2 \text{ per stiffener}$$

For best effect the stiffeners should extend as far as the beam flanges.

$$b_{st} = \frac{b - t}{2} = \frac{10.07 - 0.47}{2} = 4.8 \approx 4.75 \text{ in.}$$

$$t_{st} = \frac{3.55}{2 \times 4.75} = 0.37 \text{ in.}$$

For safety against local buckling (Sect. $[2.7]$ for 50 ksi)

$$t_{st} \text{ (min)} = \frac{b}{7.0} = \frac{4.75}{7.0} = 0.68 \quad (\text{Section } [2.6.])$$

A convenient stiffener size, than, would be

$$\underline{\frac{11}{16} \times 4\frac{3}{4} \text{ P.}}$$

$$b/t = 6.9 < 7.0, \; A_{st} = 6.53 \text{ in.}^2 > 3.55 \text{ in.}^2 \qquad \underline{\text{O.K.}}$$

The equilibrium equation at the tension flange is the same as that at the compression flange; therefore, the stiffener requirements will be the same.

$$\underline{\text{Use } \frac{11}{16} \times 4\frac{3}{4} \text{ P. for stiffeners}}$$

The final detail may be as shown above; however other combinations would be acceptable. Design of the welds for this connection follows the procedures given in Example 19.12.

Equation 17.34 together with some additional requirements is given in AISC Section [1.15.5]. Application of the limitation derived is shown in Example 17.5.

17.7 TRIANGULAR BRACKET PLATES

Triangular bracket plates are employed in structures as support brackets, as stiffeners, and as buttresses in heavy bolted or riveted frame joints and column bases.

A bracket can be asssumed to be uniformly loaded on the top edge (dimension b in Fig. 17.20) and supported on the side edge (dimension a). The sloping edge is free. The notation and the arrangement are shown in Fig. 17.20.

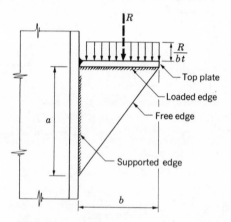

Fig. 17.20　Triangular bracket plate.

In design, dimensions a and b are usually established by the details of the load seat and of the connection to the column, and, thus, the problem is to determine the thickness of the bracket plate. The top plate should be fully developed by the weld at the column.

1. Allowable Stress Design

Since the post-buckling strength of such triangular plates has not been adequately investigated, conditions of buckling or the initiation of yielding are recommended as design criteria in the allowable stress approach.

The average stress on the top loaded edge, which would cause buckling of the bracket plate, is given by Eq. 17.35. This equation is obtained from Eq.

17.9 by substituting $E = 29,000$ ksi and $\mu = 0.3$.

$$F_{cr} = \left[\frac{R(\text{F.S.})}{bt}\right]_{cr} = 26,500k'\left(\frac{t}{b}\right)^2 \qquad (17.35)$$

The buckling coefficient k' is approximately computed from[17.25]

$$k' = 3.2 - 3.0\left(\frac{b}{a}\right) + 1.1\left(\frac{b}{a}\right)^2 \qquad (17.36)$$

Using a factor of safety of 1.67, Eq. 17.35 is transformed into the following plate thickness requirement:

$$t_{cr} = 0.04 \sqrt[3]{\frac{Rb}{k'}} \qquad (17.37)$$

The maximum stress occurs at the free edge and it is related to the average stress on the loaded edge through factor z:[17.25]

$$z = \frac{\text{Average stress-loaded edge}}{\text{Maximum stress-free edge}}$$

$$= 1.39 - 2.20\left(\frac{b}{a}\right) + 1.27\left(\frac{b}{a}\right)^2 - 0.25\left(\frac{b}{a}\right)^3 \qquad (17.38)$$

The smallest plate thickness required to preclude the initiation of yielding is then derived from Eq. 17.38

$$t_y = \frac{R(\text{F.S.})}{F_y bz} \quad \text{or} \quad = \frac{R}{F_{\text{all}} bz} \qquad (17.39)$$

In design, both t_{cr} and t_y should be computed and the larger one is to be used. Equations 17.36 and 17.39 are based on theoretical and experimental research and should be limited to $0.75 \leqslant b/a \leqslant 2.0$.[17.25,17.26]

If dimensions a and b are not fixed, the conditions of buckling and yielding can be equated and the following b/t requirement is derived from Eqs. 17.35 and 17.38:

$$\frac{b}{t} \leqslant 163\sqrt{\frac{k'}{F_y z}} \qquad (17.40)$$

The procedure given above is valid if the loading can be assumed to be distributed uniformly over the top edge, that is, if the resultant falls approximately in the middle or closer to the column. When the resultant moves toward the edge, the approach described may become unsafe. Recourse then can be made to some approximate procedures such as, for example, that given in Ref. 17.27. Here, it is assumed that the bracket acts as an eccentrically loaded column and that the elementary bending formula is applicable. See Fig. 17.21.

Section B-B' passing through the corner B and perpendicular to the free edge CA is assumed to be subjected to the load R', with an eccentricity equal to e'. The relationships between R, e, b, and R', e', and $\overline{BB'}$ in terms of

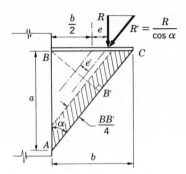

Fig. 17.21 Simplified procedure for design.

the angle α are obvious from Fig. 17.21. The maximum stress, at point B', is found to be

$$f_{\max} = \frac{R}{bt \cos^2 \alpha}\left(1 + \frac{6e}{b}\right) \tag{17.41}$$

This stress should not exceed the allowable stress.

The possibility of buckling can be checked, conservatively, by assuming that the load R' acts concentrically on the strip which is cross-hatched in Fig. 17.21. This strip forms a column of length AC and rectangular cross section $t \cdot (b \cos \alpha)/4$.

Example 17.6 illustrates both procedures.

2. Plastic Design

Triangular plates of sufficiently small b/t value can be loaded to undergo considerable yielding. The material becomes not only anisotropic but also non-homogeneous. A conservative solution can be obtained by assuming that the material is isotropic and its modulus of elasticity in all directions is equal to $\sqrt{EE_{st}}$. The limiting b/t ratios then are found from Eq. 17.35 by setting $F_{cr} = F_y$ and using $\sqrt{EE_{st}}$ according to Eq. 17.10 instead of $E = 29{,}000$ ksi (and using $E_{st} = 900$ ksi).[1.1]

$$\frac{b}{t} \leqslant 68\sqrt{\frac{k'}{F_y}} \tag{17.42}$$

where k' is computed from Eq. 17.36.

EXAMPLE 17.6

PROBLEM:

Determine the thickness of the triangular plate in the stiffened beam seat of Example 19.6. Use A36 steel (F_y = 36 ksi)

SOLUTION:

Assume that the plate behaves as if it had a triangular outline as shown by A-B-C.

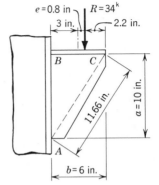

DESIGN FOR ALLOWABLE STRESS

Exact Analysis assuming no eccentricity.

Stability: $\dfrac{b}{a} = \dfrac{6}{10} = 0.6$

From Eq. 17.36 $k' = 3.2 - 3.0\,(0.6) + 1.1\,(0.6)^2$
                        $= 1.796$

From Eq. 17.37 $t_{cr} = 0.04\sqrt[3]{\dfrac{34(6)}{1.796}} = 0.194$ in.

Stress: Use allowable stress $F_a = 36/1.65 \approx 21$ ksi

From Eq. 17.38 $z = 1.39 - 2.20\,(0.6) + 1.27\,(0.6)^2 - 0.25\,(0.6)^3 = 0.473$.

From Eq. 17.39 $t_y = \dfrac{34}{21(6)(0.473)} = 0.57$ in. > 0.194 in.

For allowable stress design <u>USE : 5/8 in. ℞</u>

Approximate Analysis considering eccentricity.

Stress: From Eq. 17.40

$t = \dfrac{R}{f_{max}\,b\,\cos^2 a}\left(1 + \dfrac{6e}{b}\right) = \dfrac{34}{21 \times 6(\frac{10}{11.66})^2}\left(1 + \dfrac{6 \times 0.8}{6}\right) = 0.66$ in.

Use: $^{11}\!/_{16}$ or $^3\!/_4$ in. ℞
(too conservative)

PLASTIC DESIGN

Stability: From Eq. 17.42 $t = \dfrac{6}{68}\sqrt{\dfrac{F_y}{k'}} = \dfrac{6}{68}\sqrt{\dfrac{36}{1.796}} = 0.395$ in.

Strength: From Eq. 17.43

$t = \dfrac{R_{ult} = (F.S.)R}{F_y \cos^2 a\,(-2e + \sqrt{4e^2 + b^2}\,)} = \dfrac{1.65 \times 34}{36(\frac{10}{11.66})^2(-1.6 + \sqrt{1.6^2 + 6^2}\,)} = 0.46$ in.

0.46 in. > 0.42 in.

For plastic design <u>USE : 1/2 in. ℞</u>

Besides establishing the permissible width-thickness ratio, it is necessary to determine the maximum load that may be applied on the bracket. If the resultant R of the loading on the top plate has an eccentricity e, portion BCB' (Fig. 17.22) at the point of plastification may be assumed to have the stresses distributed approximately as shown in the figure. The maximum value of $R = R_{ult}$ consistent with this stress distribution can be computed by equating the resultant of the stresses on line BB' to the component R' which is

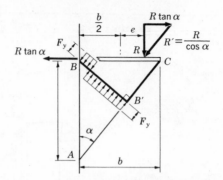

Fig. 17.22 Plastic design of triangular bracket plates.

parallel to the free edge and by taking the equilibrium of moments about point B produced by the stress and R'. The ultimate value of R is then

$$R_{\text{ult}} = F_y t \cos^2 \alpha(-2e + \sqrt{4e^2 + b^2}) \tag{17.43}$$

The area of the top plate must be adequate to carry the horizontal component ($R \tan \alpha$) and thus, is given by

$$A_{\text{top plate}} = \frac{R_{\text{ult}} \tan \alpha}{F_y} \tag{17.44}$$

PROBLEMS

17.1. A specially formed channel section with equal section moduli about x- and y-axes is to be used as a flexural member. In what position (see sketch) is the channel less endangered by local buckling? Assume that the section is prevented from twisting.

Problem 17.1

17.2. A test was conducted on a plate with uncertain boundary conditions, and the stress F_{cr} was measured at the moment of the appearance of the buckles. If the width-thickness ratio of the plate is b/t and the material properties are E, F_y, and μ, determine the plate buckling coefficient k by assuming that the plate buckling curve in Fig. 17.12 may be used for this purpose. Assume that $F_{cr} = 20$ ksi, $E = 29,000$ ksi, $F_y = 36$ ksi, $\mu = 0.3$, and $b/t = 24$.

17.3. The column of Example 17.1 is to be used in a highway structure for a correspondingly different load. Check the local-buckling requirements according to the AASHO Specification.

17.4. What would be the required rivet spacing in Example 17.1 according to the AREA Specification?

17.5. Consider a plate girder having the dimensions and material given in Example 8.1 and determine the maximum concentrated load that may be applied to the top flange within the third panel from the left support, without requiring bearing stiffeners. Also compute the length of the bearing plate. The top flange is restrained from rotation.

17.6. Use Eq. 12.7 and the concepts associated with Fig. 17.12 to show that the b/t requirement in the AISI Specification to prevent buckling before reaching the yield point is more conservative than in the AISC Specification.

RIVETED AND BOLTED
CONNECTIONS

18.1 INTRODUCTION

The design of main structural members involves the use of formulas based upon theories which have been developed and refined so that generally they can be depended upon to give safe and economical results. On the other hand, the behavior of a connection is often so complex that it is impossible to describe by formulas. Concentration on structural analysis and the design of the main members often results in neglect of the design of the connections. Poorly designed connections not only cause weak "links" in the structure but also affect the over-all behavior of the structure and the stress in the main members.

The cost of fabricated steel construction is controlled to a large degree by the connections. Simplicity of connection details may have a greater influence on the total economy of the structure than a reduction in weight of the main members. Before 1950, riveting was the primary method of connecting structural steel, but today, connections utilizing welding and/or high-strength bolting are the most practical; shop riveting is decreasing and field riveting is almost never used.

This chapter deals with connections using mechanical fasteners. A discussion of the types, properties, and installations of rivets and bolts is followed by an analysis of types of connections. Emphasis is placed on understanding connection behavior, and its relationship to design procedures and the provisions of specifications.

The Research Council on Riveted and Bolted Structural Joints has been the focal point of most of the research conducted in the United States during the last twenty years. More specific information on many of the topics discussed in this chapter will be found in the publications of that organization.[18.1,18.2]

18.2 TYPES OF FASTENERS

The mechanical fasteners generally used in connections for buildings and bridges are rivets, square head bolts, high-strength bolts, and interference body bolts, as shown in Fig. 18.1. Each year close to five million tons of structural steel are connected by one hundred and twenty-five million of these fasteners. The dimensions of standard fasteners are specified by the American National Standards Institute (ANSI). The mechanical properties of the fastener materials are specified by ASTM in terms of hardness and tensile requirements.

1. Rivets

Riveting is one of the oldest methods of joining metal, dating as far back as man's first use of ductile materials, but its use in structural steel fabricating is now dying out. Rivets are made from bar stock by either hot- or cold-forming of the manufactured head. The manufactured head is usually of the high button (acorn) head variety although flattened and counter-sunk head rivets are made for applications with limited clearance. Rivets for structural fabricating purposes are made in diameters from $\frac{1}{2}$ to $1\frac{1}{2}$ inches.

Rivet material is usually of the same or a somewhat lesser strength than the steel of the parent structure. Two grades of rivets are specified: ASTM A502—Grade 1, a carbon steel rivet for general purposes, and Grade 2, a rivet for use with high-strength carbon and low alloy structural steels. The A502 specification for rivets defines hardness requirements only; however, these rivets correspond to those formerly made from steel conforming to the strength properties shown in Table 18.1 for full-diameter bar stock.

Driving increases the strength of rivets. The average of many tests of low carbon rivets shows that shop machine driving increases the rivet strength by about 20 per cent compared to a 10 per cent increase for hot rivets driven in the field by a pneumatic hammer.[18.3]

Since rivets usually are loaded in shear, the relationship between shear strength and tensile strength is important. Tests show that the ultimate shear strength of a driven rivet is 70 to 75 per cent of its ultimate tensile strength when driven.[18.3]

Rivets

A 307 square head bolt

A 325 high-strength bolt

Interference body bolt

Fig. 18.1 Mechanical fasteners. (Courtesy of Bethlehem Steel Corp.)

Table 18.1 Properties of Rivet Material*

ASTM Designation	Former ASTM Designation	Type Name	Tensile Strength (ksi)	Yield Point (min) (ksi)	Elong. in 8 in. (min) (%)
A502-65 Grade 1	A141-58	Structural rivet steel	52–62	28	24
A502-65 Grade 2	A195-59	High-strength structural rivet steel	68–82	38	20

* Tensile test of as rolled or annealed material.

594

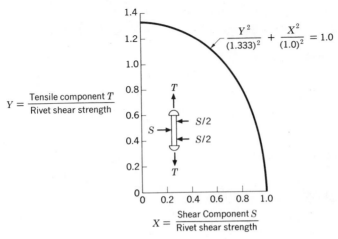

Fig. 18.2   Rivets subjected to combined tension and shear.

In some connections rivets are loaded by combined tension and shear forces. Tests of rivets under various ratios of applied tension and shear forces show that the ultimate strength of a rivet is defined by an elliptical interaction curve as shown in Fig. 18.2.[18.4]

2. Square Bolts

The square bolt, recognized in the field by its square head and nut, is ideally suited for joining low-tonnage, light-framework structures where slip and vibration do not have to be considered.   The square bolt is made of low-carbon steel with physical properties as designated by ASTM A307.   These are shown in Table 18.2.   Unlike rivets, the strength of bolts is specified in

Table 18.2   Properties of Structural Bolts

ASTM Designation	Type Name	Bolt Diameter (in.)	tensile strength stress area* (ksi)	proof load stress area* (ksi)
A307-68	Low carbon steel externally and internally threaded standard fasteners	All	55	None
A325-68	High-strength bolts for structural steel joints	$\frac{1}{2}$–1 $1\frac{1}{8}$–$1\frac{1}{2}$	120 105	85 74
A490-67	Quenched and tempered alloy bolts for structural steel joints	$\frac{1}{2}$–$2\frac{1}{2}$ $2\frac{1}{2}$–4	150–180 140–170	120 105

* Stress area $= 0.785 \left(D - \dfrac{0.9743}{n} \right)^2$ where $D =$ nominal bolt size, and $n =$ threads per inch.

terms of a tensile test of the complete threaded fastener. The weakest section of any bolt is the threaded portion, and the strength of the bolt is usually computed by using the "stress area"—a sort of average area based on the nominal and root diameters.

The nut is an important part of the bolt assembly. Nut dimensions and strengths are specified so that the tensile strength of the bolt is developed before thread stripping or other types of nut failures occur.

3. High-Strength Bolts

Within the last twenty years the A325 high-strength bolt has become the prime field fastener of structural steel. As originally conceived, the high-strength bolting assembly consisted of four pieces—a bolt, a nut, and two washers. However, as a result of continued research the latest specifications now require a heavy hexagon structural bolt, a heavy semi-finished hexagon nut, and either one washer or no washers, depending upon the tightening method used. The bolt is identified on the top of the head by the legend "A325" and often by three radial lines. The standard nut marking consists of three circumferential marks spaced at 120°.

The A325 bolt for structural joints is made by heat-treating medium-carbon steel. Different strength levels are specified for different size bolts (see Table 18.2). In addition to the minimum ultimate tensile strength, a proof load also is specified. The specified proof load represents a lower bound to the proportional limit load as shown in Fig. 18.3. A325—Type 3 bolts are made with corrosion resistance and weathering characteristics comparable to those of A588 and A242 weathering steels.

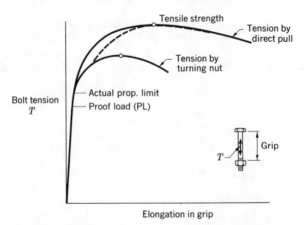

Fig. 18.3 Tension-elongation curves for A325 bolts.

Although the ASTM Specification requires a direct tension test of the bolts, in application tension is induced in the bolt by turning the nut. This operation sets up a combined tension and torsional shear condition that results in development of a maximum tension in the bolt lower than that indicated by a direct tension test. However, if a bolt is prestressed by tightening the nut and is then loaded by an applied direct tensile force, the bolt is able to develop the full tensile strength, as shown by the dotted line in Fig. 18.3.

Tests have shown that the ultimate shearing strength of A325 bolts is about 70 per cent of the ultimate tensile strength. Elliptical interaction curves similar to that in Fig. 18.2 also apply for bolts under combined tension and shear loading. However, these are complicated by the fact that tensile failures occur through the thread, and shear failures may occur through the shank or thread area.

The A490 bolt is an alloy steel bolt that has been developed for structural applications, particularly for use with high-strength steel members. This bolt has "A490" marked on the head, and the nut to be used with it is identified by the mark "2H" or "DH." A490 bolts have mechanical properties as shown in Table 18.2. An upper limit to the tensile strength is specified in order to ensure certain desirable properties. It should be noted that the ratio of proof load to ultimate tensile strength for the A490 bolt is 0.80 whereas for A325 bolts it is about 0.70.

Both the A325 and A490 bolts have shorter thread lengths than on comparable bolts used for other applications in order that the threads will be out of the shearing plane in most instances.

4. Interference Body Bolts

The fasteners discussed so far are standard and non-proprietary in design, manufacture, or application. There are available commercially a few special types of interference body bolts that have proven useful in certain applications such as high television towers and radarscopes.

Interference body bolts have a rivet type head, interrupted ribs on the shank and often use a nut with a locking device. They are designed to be driven into a standard hole, the interrupted ribs producing a tight fit with the gripped material. Interference body bolts conform to the strength of A325 bolts.

18.3 INSTALLATION

In the United States high labor costs make the installation of fasteners an important factor in the over-all costs of connections. As an example, the labor costs of a four- or five-man field riveting crew installing the relatively cheap rivet have surpassed the costs of the more expensive A325 bolt and the two-man bolting crew.   Furthermore, qualified field riveters are in very short supply.   As a result, the high-strength bolt has become the leading field fastener for structural steel.   Studies have shown that economies also may be achieved by shop bolting.[18.5,18.6]   The square head bolt should not be overlooked either, for it can offer economy in certain applications.

1. Riveting

Holes for hot-driven rivets usually are punched or drilled $\frac{1}{16}$ in. larger than the undriven (nominal) rivet diameter.   When rivets are driven, the hole is filled and an almost hemispherical head is formed from the portion of the shank that extends beyond the gripped material.   At the same time the shape of the manufactured head is also changed to the hemispherical form.   All parts of riveted members must be well pinned and bolted together rigidly while being riveted.

Hot-driven rivets are driven when in a cherry-red condition (1000–1950°F). On cooling, the rivet contracts longitudinally, developing a residual tensile stress in itself and a clamping force on the gripped material.   The residual tension developed is unpredictable, ranging from a stress equal to the yield point of the rivet material to practically nothing.[18.7]   This extreme variability has been used to justify the design assumption that no clamping exists in riveted joints.   Also, on cooling, the rivet contracts laterally so that it may not completely fill the hole as is usually assumed.   A rivet of short grip will usually fill its hole more closely than a rivet of long grip.

Rivets are driven by power riveters of either compression or manually operated types employing pneumatic, hydraulic, or electric power.   This chapter will not discuss the relative merits of the various types.

2. Bolts

Bolts are used in holes $\frac{1}{16}$ in. larger than the nominal bolt diameter. Square bolts may be tightened easily using spud wrenches.   The tension induced is low and it is generally considered that these bolts exert no clamping force.

Whereas rivets and ordinary bolts develop residual tensions that vary widely and are unpredictable, tightening of high-strength bolts can produce high and consistent tensions.   Because of this high clamping force, high-strength bolted joints can be depended upon to carry load by friction provided the contact surfaces of the joint are free of oil, paint, lacquer, or galvanizing.

All high-strength bolts (A325 or A490) should be tightened to a tension equal to or greater than a value of 70 per cent of the specified minimum tensile strength of the bolt. For example, a $\frac{7}{8}$ in. A325 bolt should be tightened to 39 kips; a $\frac{7}{8}$ in. A490 bolt to 49 kips. Such tightening requires the use of long-handled hand torque wrenches or powered impact wrenches. Two methods of controlling bolt tension are used, the calibrated wrench and the turn-of-nut methods.

In the calibrated wrench method at least three bolts of the lot to be used in the structure are tightened in a calibrating device that reads the tension in the bolt directly. The wrench is adjusted to stall at a load 5–10 per cent in excess of the prescribed tension and is used thereafter for installation of that lot of bolts. Checking of calibration should be done every few hours or when a new lot of bolts is used. Such a method is essentially a torque control, and success depends on using a hardened washer under the nut in order to limit the variation of the friction between the underside of the nut and the gripped material. If the bolt is tightened by turning the bolt, the washer should be used under the bolt head.

Although early specifications stated a general formula relating torque to bolt tension, it is now known that such relationships are unreliable because of the great variability of thread condition, surface conditions under the nut, lubrication, and other factors that use up the torque energy without inducing tension in the bolt.[18·8] Torque relationships become especially erratic when bolts are tightened into the plastic range and threads begin to deform. Such formulas should not be used for field installation of structural bolts.

The turn-of-nut method utilizes a strain control and, therefore, is ideally suited to controlling tightening in the plastic range. Current practice for A325 bolts is to turn the nut one half a turn from a "snug-tight" position.[18·9] The snug-tight position corresponds to a few blows of the impact wrench or the full effort of a man using an ordinary spud wrench. The half-turn produces tensions well beyond the proof load and into the plastic range as shown for $\frac{7}{8}$ in. A325 bolts in Fig. 18.4. One half-turn from snug-tight is approximately equivalent to one turn from a finger-tight position. Tightening into the plastic range is advantageous because even large variations in bolt elongations will cause only small variations in tension since the tension-elongation curve is flat in that range.

Originally, hardened washers were used under both the head and nut of A325 bolts in order to provide a consistent torque relationship, to prevent galling of the structural material, and to prevent bolt relaxation. It is now known that the washer is not needed for the last two reasons and, since the turn-of-nut method is a strain control rather than a torque control, no washers at all are required when this method of tightening is used. Because greater clamping force is attained with A490 bolts, hardened washers are installed under both the nut and bolt head when A490 bolts are used in steels having a yield point less than 40 ksi, and under the turned element when they are used in higher strength steels.

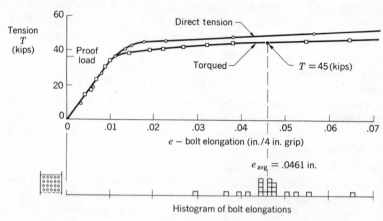

Fig. 18.4 Tension in bolts installed by turn-of-nut method.

18.4 TYPES OF CONNECTIONS AND THEIR BEHAVIOR

There are many types of connections used in joining structural steel members. Some typical ones are shown in Fig. 18.5. In this chapter classification of connections will be made according to the manner of stressing the fasteners, that is, shear, tension, and combined shear and tension. However, consideration of the stress in the connected material also influences the design of connections. For example, the tensile stress on the net section may influence the arrangement of holes. This matter is discussed in Art. 6.6. The compressive stress where the fasteners bear against the sides of the holes may regulate the force transmitted by a fastener. Bearing stress is discussed in Art. 18.6. Shear and bending stresses also determine dimensions of the connected material.

The behavior of connections is very complex. Most connections are statically indeterminate, and the distribution of forces and stresses depends upon the relative deformations of the component parts and the fasteners. Stress concentrations also complicate the situation.

It is practically impossible to analyze most connections by a rigorous and exact mathematical procedure. The analyses used in design are approximate and based upon simplifying assumptions. Tests of prototype connections have made it possible to set allowable stresses that can be applied safely to the approximate design procedures.

In the following discussion of the behavior of connections, it will be assumed that all fasteners in a group have the same cross-sectional area, a situation that usually occurs in practice for reasons of economy.

1. Axial Shear

Double Shear.   The simplest type of structural connection is the splice of a tension member, particularly the flat plate type.   The butt splice (Fig. 18.5a) is preferred because symmetry of the shear planes prevents bending of the plate material.   The load is applied through the centroid of the fastener group in the plan view and has small opposing eccentricities, e, with respect to each of the shear planes.   The fasteners act in double shear, that is, two potential shearing planes cross the fasteners.

Analytical solutions of this type of connection have been made by several investigators for riveted joints stressed elastically.[18.10,18.11]   An analysis of the bearing-type high-strength bolted joint has been made for the plastic range to predict the ultimate strength of the connection.[18.12]   These theoretical analyses are too cumbersome for ordinary design practice and simplifying assumptions must be made.

The usual design assumptions made for riveted joints are:

1. No misalignment of holes.
2. No hole clearance—all driven rivets fill the holes completely.
3. No friction between the faying surfaces.
4. Each rivet carries an equal share of the applied load.

However, perfect hole alignment occurs only when holes are subpunched and reamed or drilled in the assembled position; lateral contraction of the rivet occurs as it cools, and in long grips the driving energy may not be sufficient to force the rivet to fill the hole at the manufactured head end. The residual tension in rivets may be quite high, and as a consequence friction may be developed at least until the rivet undergoes some shear deformation.

The last assumption is valid only if the plates do not deform at all.   This accounts for the design procedure being called "rigid plate theory."   Even if the first three assumptions are true, the loads carried by the rivets usually are not equal because of differential elastic strains in the main plate and in the splice plates in corresponding pitches along the length of the joint.   For short joints the assumption may be justified because the plastic deformation of the more highly strained end rivets allows equalization of load on all rivets. The action in this case is similar in principle to the successive formation of plastic hinges in a continuous beam or frame.

For high-strength bolted joints it is necessary to think in terms of two methods of load transfer: friction, and shear and bearing.

Initially the load is transferred by friction forces concentrated near the ends of the joint.   As the load is increased, the zone of friction extends towards the center of the joint.   Eventually the maximum value of static

1. Axial shear

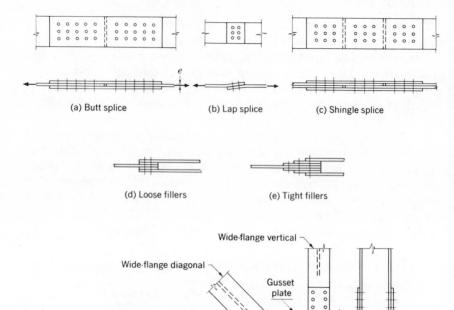

(a) Butt splice (b) Lap splice (c) Shingle splice

(d) Loose fillers (e) Tight fillers

Wide-flange vertical

Wide-flange diagonal

Gusset plate

Built-up bottom chord

Tie plate

Splice plate Butt splice Diaphragm

Bridge panel point and chord splice

(f) Double plane connections

2. Eccentric shear

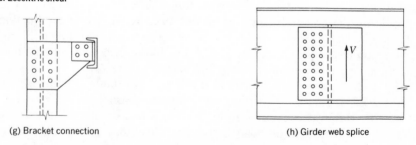

(g) Bracket connection

(h) Girder web splice

Fig. 18.5 Typical riveted and bolted connections.

3. Tension

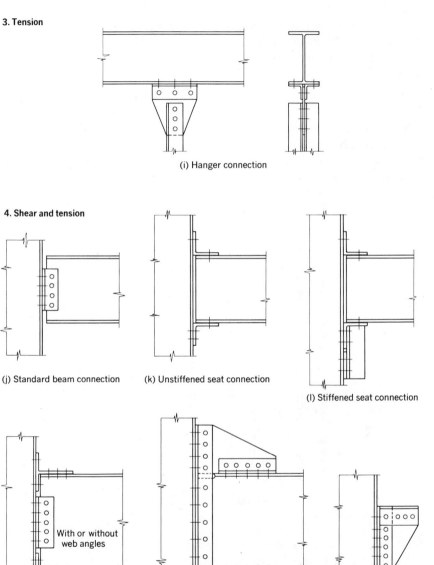

(i) Hanger connection

4. Shear and tension

(j) Standard beam connection   (k) Unstiffened seat connection

(l) Stiffened seat connection

(m) Structural tee connection

With or without web angles

(n) End plate moment connection

(p) Crane bracket

Fig. 18.5   *Continued*

friction is exceeded at the ends and partial slip occurs.   Finally, the limiting value of static friction is exceeded over the whole contact surface and a relative slip of the inner and outer plates occurs, taking up the hole clearance. This range of loading has been investigated theoretically.[18·13]   The slip load may be computed as

$$P_{\mathrm{slip}} = K_s m \sum T_i \qquad (18.1)$$

where K_s = coefficient of slip, m = number of slip planes, and $\sum T_i$ = sum of the initial bolt tensions.

When major slip occurs, only the end bolts come into bearing against the main plate and the splice plates as shown in Fig. 18.6.   As the load increases,

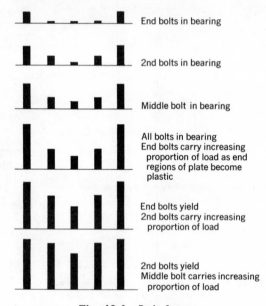

End bolts in bearing

2nd bolts in bearing

Middle bolt in bearing

All bolts in bearing
End bolts carry increasing
  proportion of load as end
  regions of plate become
  plastic

End bolts yield
2nd bolts carry increasing
  proportion of load

2nd bolts yield
Middle bolt carries increasing
  proportion of load

Fig. 18.6   Bolt forces

these fasteners deform until the next interior bolts are bearing.   The process continues until all the bolts bear against the plates on both sides.   This does not mean that each fastener is carrying an equal share of the total load.   As load is increased, the fastener forces change as shown diagrammatically by the height of the bars in Fig. 18.6, and finally an end fastener fails because it is overstrained.

In short connections, with a few fasteners in line, almost complete equalization of load may occur before the end fasteners fail.   But in longer connections, the end fasteners reach a critical shear deformation and fail before the full strength of each fastener can be utilized.   Such a connection is shown in Fig. 18.7 where large shearing deformation of the end bolts and the greater elongation of the end holes can be seen.   The "premature" sequential failure of fasteners, progressing from the ends of the joint inward, has been called "unbuttoning."   The phenomenon has been witnessed in tests of riveted[18.14] and high-strength bolted joints[18.15,18.16] and has been predicted by theoretical analysis.[18.12,18.16]

The shingle splice (Fig. 18.5c) is often used for bridge members consisting of several plies of material.   It has usually been assumed that a more gradual transfer of load occurs with this staggered splice than if both main

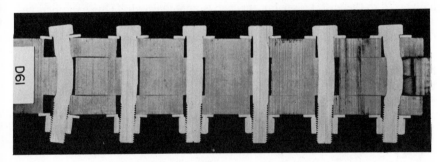

Fig. 18.7 Sawed section of bolted joint.

plates are terminated at the same location; however, recent research indicates that the end fasteners carry most of the load and that the shingle splice is not much different in behavior from a butt joint.[18.17]

When the members being spliced lie in planes that are separated, it is necessary to use filler plates.   According to most specifications loose fillers (Fig. 18.5d) may be used as packing pieces if they do not exceed a thickness of $\frac{1}{4}$ in.   For greater thicknesses, except in friction type connections assembled with high strength bolts, tight fillers are required.   Tight fillers (Fig. 18.5e) are fastened by additional fasteners to the spliced member outside of the connection, thereby becoming a load-carrying part of the member.   This results in a more uniform loading on the fastener and reduces the bending of the fastener.   As an alternative, the additional fasteners may be put in the connection without extending the filler plates.

Single Shear.   In the lap splice (Fig. 18.5b) the fasteners act in single shear. For practical purposes, the load-carrying capacity of a given pattern of fasteners in single shear is equal to one-half that for the same pattern in double shear.

A lap splice is not a desirable structural joint and should be used for minor connections only.   The eccentricity of the loads pulling on the connected members causes bending of the members as the loads attempt to align themselves axially. The combined axial and bending stresses are critical at the net section.

There are other types of connections in which the fasteners are stressed in single shear but in which bending is prevented because of symmetry of two shearing planes. The wide-flange bridge connections shown in Fig. 18.5f are of this type. Because of symmetry of the shearing planes and the diaphragm action of the web, bending of the main material does not take place.

Often in connections of this type only a portion of the actively stressed material is actually connected. In the illustration just cited, the flanges are connected to the gusset plate but the web is unconnected. Due to a phenomenon known as "shear lag," it takes a considerable length along the member for stress to flow from the unconnected regions to the fasteners. Whereas tests of plate splices,[18.15] where all the material is connected, have indicated the desirability of making the connection as short as possible to minimize unbuttoning of the fasteners, the shear lag phenomenon requires that a reasonable length of connection be provided in double-plane connections in order to utilize effectively the cross-sectional area of the main material.[18.18] (Shear lag is also discussed in Art. 12.3.)

Although the fasteners in this section have been classified under axial shear, it is quite obvious that the fasteners also bend and in some cases have secondary tensile components due to bending of the joint.

2. Eccentric Shear

The line of action of the force acting on a connection should pass through the centroid of the fastener group, but this is not always practicable and eccentric forces must be tolerated. Examples in practice are certain types of bracket connections (Fig. 18.5g) and web splices of plate girders (Fig. 18.5h).

Once the static friction has been overcome, the fasteners slip into bearing and are stressed in single or double shear by an axial component and by a torsional component. It is convenient to visualize the behavior of the eccentrically loaded fastener group by replacing the original eccentric force by a force acting through the centroid and a couple as shown in Fig. 18.8.

If it is assumed that the connected parts are rigid and that the resisting force of a fastener is proportional to the applied strain, the resisting forces of the fasteners may be determined. Due to the axial component, each of the fasteners provides an equal resisting force $R_a = P/n$. Due to the torsional component, each fastener provides a resisting force R_t proportional to its distance from the centroid.

The latter force may be computed by a variation of the elementary torsion formula. Thus, the average shearing stress on a fastener ρ units from the

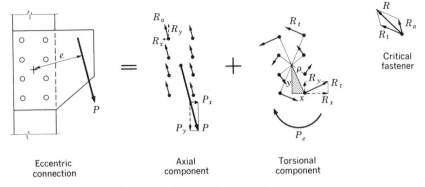

Fig. 18.8   Resolution of eccentric connection.

centroid of the group is given by

$$f_v = \frac{T\rho}{J} = \frac{Pe\rho}{\sum A_b \rho^2}$$

and when the area of each fastener is constant, the force on that same fastener equals

$$R_t = f_v A_b = \frac{Pe\rho A_b}{A_b \sum \rho^2} = \frac{Pe\rho}{\sum (x^2 + y^2)} \tag{18.2}$$

The resistance force of the fastener acts normal to the direction of ρ and opposes the direction of the applied torque Pe.

The total resisting force R of any fastener is the vector resultant of the components R_a and R_t.

A useful variation of Eq. 18.2 gives the x and y components of R_t.

$$R_x = \frac{Pey}{\sum (x^2 + y^2)} \quad \text{and} \quad R_y = \frac{Pex}{\sum (x^2 + y^2)} \tag{18.3}$$

Although the above method of analysis has been used for many years and has provided apparently satisfactory designs, the actual factor of safety against failure has been unknown. The application of the method to a bracket connection is illustrated in Example 18.1.

A limited experimental study[18.19] of riveted framing angles has shown factors of safety of about $4\frac{1}{2}$, and empirical values of "effective eccentricities" have been recommended to reduce the factor of safety to about 3, a value more consistent with other connection design. These values are used in the tables for eccentric loads on fastener groups in the AISC Manual. A more recent study of high-strength bolted connections has developed an analytical solution[18.20] that gives results comparable to those determined by use of the AISC effective eccentricities. It must be noted that these studies have been limited to loads applied parallel to lines of equally spaced fasteners and no ultimate strength information is available for eccentric inclined loads and unequal fastener spacing.

3. Tension

There are not many cases where rivets or bolts are loaded in direct tension only. One example is the connection of a platform hanger shown in Fig. 18.5i. Usually it is assumed that the flange fasteners in the top tee of the split-T moment connection (Fig. 18.5m) are stressed in tension only.

The cooling of hot-driven rivets and the tightening of a nut on a bolt result in a residual tension in the fastener. Inasmuch as this stress exists prior to the application of any primary loading, the fastener may be considered as prestressed. The question has often been asked, "What happens to the fastener when external tensile load is applied?" Because of this concern over the additive effects of prestress and applied tensile stress the use of rivets in tension often has been avoided. The following analysis should serve to answer the question posed above.

If a prestressed bolt grips two plies of material (Fig. 18.9a) the prestress force T_i and the total contact pressure C_i are equal. If an external tensile load P is applied at the contact surface (Fig. 18.9b) it tends to reduce the thickness of the plates, but this is simultaneously offset by a corresponding increase in thickness because the contact pressure has been decreased. The

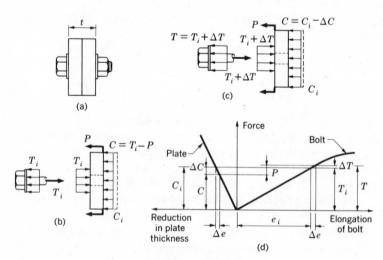

Fig. 18.9 Prestressed fastener.

EXAMPLE 18.1

PROBLEM:

    A bracket supporting a crane
runway is connected to the flange
of the supporting column by a
group of ten fasteners of like size.
What is the force on the critical fastener?

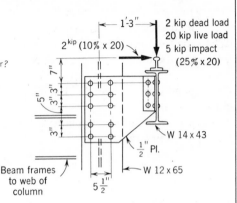

SOLUTION:
  Properties of fastener group.

   b
$$\bar{y} = \frac{\Sigma A_b y}{\Sigma A_b} = \frac{3+8+11+14}{5} = 7.2'' \text{ above bottom row}$$

    $\bar{x}$ = centroid on column centerline by symmetry.

    $\Sigma = (x^2+y^2) = 10 \times 2.75^2 + 2(7.2^2 + 4.2^2 + 0.8^2 + 3.8^2 + 6.8^2) = 337 \text{ in.}^2$

  Axial Shear Force

$$R_x = \frac{P_x}{n} = \frac{2}{10} = 0.20^{kip} \longleftarrow \qquad R_y = \frac{P_y}{n} = \frac{27}{10} = 2.70^{kip} \uparrow$$

  Torsional Shear Force

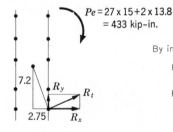

$Pe = 27 \times 15 + 2 \times 13.8$
   $= 433$ kip-in.

        By inspection lower right fastener is critical.

$$R_x = \frac{Pey}{\Sigma x^2 + y^2} = \frac{433 \times 7.2}{337} = 9.25 \text{ kip}$$

$$R_y = \frac{Pex}{\Sigma x^2 + y^2} = \frac{433 \times 2.75}{337} = 3.53 \text{ kip}$$

  Combined Force

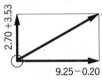

$R = \sqrt{9.05^2 + 6.23^2} = 11.0$ kip

    Note: The AISC effective eccentricity formulas are
         not strictly applicable to this case but if
         applied to the vertical load only the torsional
         moment becomes

$$P_e = 27 \left[15 - \left(\frac{1+5}{2} \right) \right] + 2 \times 13.8$$

              = 352 kip-in.
    thus reducing R

bolt does not elongate and the tension remains T_i. When $P = T_i$ the contact pressure reduces to zero and the plates are on the verge of separation. If $P > T_i$ the plates are separated and the bolt tension thereafter is always equal to P. Thus, in this case there is no increase in bolt tension unless the working load exceeds the prestress.

When the load P is applied at the outer surfaces (Fig. 18.9c) the bolt will stretch immediately and the precompressed plates will expand. If the plates do not separate, the elongation of the bolt Δe and the amount by which the plates expand must be equal. Figure 18.9d illustrates this case on a force-deformation diagram. The increase in the bolt tension ΔT depends on the stiffnesses of the bolts and plates. For elastic conditions and bolts and plates of the same material

$$\Delta T = \frac{P}{1 + (A_p/A_b)} \tag{18.4}$$

In an actual connection the area of plate that is compressed, A_p, is not clearly defined. If it is assumed that A_p is a circular zone whose diameter is three bolt diameters, then $T = 0.11P$. Thus, in this case, until the applied load exceeds the bolt prestress the increase in bolt tension will be rather small, only about $\frac{1}{10}$ of the applied tensile load. If the bolt is installed by the turn-of-nut method which produces initial tensions on the flat part of the tension-elongation curve, the increase in tension will be negligible.

In practice, the application of load is somewhere between the cases discussed. Although the distribution of pressure between the plates is not uniform and the contact area is doubtful, this approximate analysis does help to allay fears concerning the use of rivets or bolts in tension.

In ordinary design calculations fastener stresses are computed as though caused by externally applied loads alone and the allowable tensile stresses have been established on this basis.

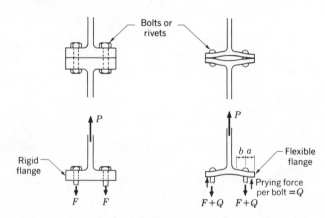

Fig. 18.10 Prying action of fasteners.

Prying Action. Another factor, prying action, also enters into consideration of practical cases involving rivets or bolts under tension. Tests of riveted and bolted T-type tension connections as shown in Fig. 18.10 have indicated that the main factors influencing prying action are the flange thickness and the ratio b/a.[18.21] When flanges are thick and narrow they act with almost infinite rigidity so that the fasteners are loaded in axial tension only. However, when the flanges are more flexible the fasteners must carry additional tension due to prying caused by the bending of the flange. In connections with four gage lines of fasteners, the inner lines of fasteners carry all the load initially. Even at ultimate load the outer lines are not very effective.

Further analytical and experimental studies have been made.[18.22,18.23] The latter reference proposes the following equation for computing the prying force per bolt, Q, for A325 bolts.

$$Q = F\left[\frac{100b(d_b)^2 - 18w(t_f)^2}{70a(d_b)^2 + 21w(t_f)^2}\right] \tag{18.5}$$

where Q = the prying force per bolt, F = the direct tension per bolt due to external load, d_b = nominal bolt diameter, a = distance from center of bolt to flange tip but not to exceed $2t_f$, b = distance from center of bolt to edge of fillet, t_f = thickness of flange, and w = length of flange tributary to each bolt. For A490 bolts the coefficients 18 and 70 become 14 and 62 respectively.

A simplified design equation

$$Q = F\left(\frac{9b}{16a} - \frac{t_f^3}{14}\right) \tag{18.6}$$

has also been suggested.[18.6] The notation is the same as for Eq. 18.5 but it is specified that $a \leqslant 1.25b$.

4. Shear and Tension

In building frames and bridge deck systems connections may be subjected to shears and bending moments resulting from continuous structural action. The amount of continuity achieved depends upon the ability of the connections to resist moment.

Brackets (Fig. 18.5p) and the various types of moment connections (Figs. 18.5m, n) present essentially the same basic conditions with the upper fasteners in the outstanding legs being loaded in shear by the vertical reaction and loaded in tension by the end moment. This is true even in the case of the standard double angle beam web connection (Fig. 18.5j) which is commonly assumed to be a shear connection only.

The AISC Specification defines three basic types of construction:

Type 1. Rigid-frame (continuous frame)
Type 2. Simple framing (unrestrained)
Type 3. Semi-rigid framing (partially restrained)

Articles 21.2 and 21.4 discuss these types and their influence on the design of multi-story frames. In this chapter the behavior of various framing connections is examined in order to determine the categories in which they belong.

Web Angles (Type 2). The conventional framing angles that fasten to the web of the beam (Fig. 18.5j) are usually considered to be completely flexible and to carry shear only. Actually, they have a limited ability to resist moment[18,25] and thus they oppose to some extent the rotation of the end of the beam. On the other hand, good rotation capacity is important whenever a joint is assumed to be completely flexible. Such connections should be able to rotate through the end rotation of a simply supported beam

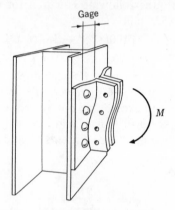

Fig. 18.11 Deformation of standard beam connection.

without fracturing. The relationship between the applied moment and the rotation of a connection is determined by experiment. Tests have shown that standard web angle connections have good rotation capacity. (A typical moment-rotation curve for a pair of web angles is shown as curve *A* in Fig. 18.14.)

When rotation takes place, the upper part of the connection is in tension while the lower part is compressed against the column. The rotation is accommodated by deformation of the angles and, once the prestress has been overcome, by elongation of the fasteners, as pictured in Fig. 18.11. To minimize the resistance to rotation the angles should be as thin as practicable and the gage in the outstanding legs large. The AREA Specifications help to assure the necessary flexibility by giving an empirical formula for this gage.

Example 18.2 illustrates the design of a web connection for a bridge stringer.

The AISC has developed standard framing connections and these are tabulated in the AISC Manual. The shear-carrying capacity of these

EXAMPLE 18.2

PROBLEM:
Referring to Example 4.1 design the end connections for the W 21 x 55 stringers. Use 7/8" A501 Grade I rivets, A36 steel members and connections, and AASHO specs.

SOLUTION:

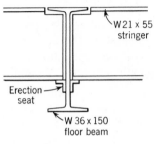

Stringers designed as
simple beams — no continuity
Therefore standard web angles
will be satisfactory.

Rivet values:
S.S. .601 x 13.5 = 8.1 kip
D.S. = 16.2 kip
Brg. Str. - .875 x .375 x 40 = 13.1 kip
Brg. Fl.Bm .875 x .625 x 40 = 21.91 kip

Loads
Design shear (Ex. 4.1) = 37.3^{kip}

Shear strength of W 21x55 = 20.80 x .375 x 12 = 93.6 kip

Average = $\dfrac{37.3 + 93.6}{2}$ = 65.5 kip

75% of strength = .75 x 93.6 = 70.2^{kip} ◀── Use; AASHO section $\left[1.7.21\right]$

No. of Rivets
Stringer web $\dfrac{70.2}{13.1}$ = 5.4 use 6

To floor beam – SS $\dfrac{70.2}{8.1}$ = 8.7

To floor beam – Brg $\dfrac{2 \times 70.2}{21.9}$ = 6.4 use 8

Use 2L^S 4 x 3$\frac{1}{2}$ x $\frac{3}{8}$ x 1'-6"

Min thickness = $\frac{3}{8}$

Check shear on angle

$$f_v = \dfrac{70.2}{2 \times 18 \times \frac{3}{8}} = 5.2 \text{ ksi} < F_v = 12$$

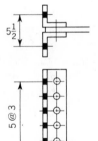

To insure flexibility of simple connection use AREA rule for gage in outstanding legs (flexing legs)

Min. gage $\geq \sqrt{\dfrac{lt}{8}}$ where l = length of stringer, inches
t = thickness of angles, inches

$$= \sqrt{\dfrac{26 \times 12 \times \frac{3}{8}}{8}} = \sqrt{14.6} = 3.92 \text{ inches}$$

$5\frac{1}{2} > 3.92$ O.K.

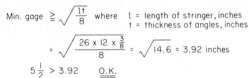

connections may be governed by any one of the following considerations: double shear on fasteners through the web legs, single shear on fasteners through the outstanding legs, or shear on a longitudinal gross section through the angle. In all but friction-type joints bearing stresses also must be considered. In calculating the shearing stress on the fasteners, the eccentric effect of gages less than $2\frac{1}{2}$ in. is neglected. The tensile stress on the upper fasteners in the outstanding legs is neglected also.

If greater eccentricities exist, a more complete analysis is required as illustrated in Examples 18.3 and 18.4. The analysis of the fasteners in the outstanding legs depends upon the conditions of prestress that are assumed. Example 18.3 assumes prestressed high-strength bolts and Example 18.4 assumes rivets with no initial tension.

Seat Angle Connections (Type 2). Unstiffened seat angle connections (Fig. 18.5k) also are used for supporting the ends of unrestrained beams in Type 2 construction. The seat angle usually is connected to the supporting column in the fabricating shop and this simplifies the erection of the beam. Lateral and torsional rigidity must be provided by a top angle or a one-sided clip angle on the beam web. Connections of this type are limited to relatively small reactions.

The behavior of the seat angle may be inferred from Fig. 18.12, which shows the connection after being stressed well above the working range. The outstanding leg acts as a cantilever beam except that it is restrained by the bottom flange of the beam that is connected to it. If the angle is thick and rigid the beam reaction tends to concentrate at the toe of the angle, but if the angle is more flexible the reaction is more uniformly distributed over the bearing length. The length of the outstanding leg is dictated by the end clearance for the beam, usually $\frac{1}{2}$ in., and the bearing length N required to prevent web crippling as discussed in Art. 17.6 and as given by Eq. 17.29. For design purposes the bearing pressure is assumed to be uniformly distributed; hence the total reaction acts at the midpoint of the bearing length.

It might appear that the critical bending section in the seat angle would be just above the top fasteners in the vertical leg. It is found, however, that the bottom flange of the beam elongates under load and pushes the angle back against the column, relieving the bending in the vertical leg. Thus, the usual practice is to consider the toe of the fillet in the horizontal leg as the critical section and to compute the required angle thickness using the bending moment at that section. An allowable flexural stress of $F_b = 0.75F_y$ is used. Justification for use of this higher bending stress is based on the combined behavior of the angle leg and the connected bottom flange of the beam, and the higher plastic shape factor for rectangular sections.

The moment-rotation characteristics of seat angle connections primarily depend on depth of beam (moment arm of couple), stiffness of the top angle, stiffness of the fasteners connecting the top angles to the columns, and stiffness

EXAMPLE 18.3

PROBLEM:
Investigate the stresses on the fasteners in this bracket. AISC specifications. All material is A440 steel.

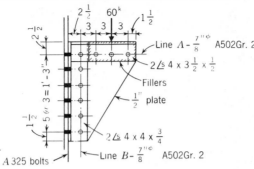

SOLUTION:

Line A - Rivet values

Double shear $= 2 \times 0.601 \times 20 = 24.05$ kip governs.
Bearing $= 0.5 \times 0.875 \times 1.35 \times 50 = 29.53$ kip

$$R_y = \frac{P}{n} = \frac{60}{4} = 15.00 \text{ kip} \qquad R_y = \frac{Pex}{\Sigma x^2} = \frac{60 \times 1.5 \times 4.5}{2(1.5^2 + 4.5^2)} = 9.00 \text{ kip}$$

Max $R_y = 15.00 + 9.00 = 24.00 < 24.05$ O.K.

Fillers should be extended below the angles and riveted to the web plate. Divide load passing thru fillers according to bearing areas.

No. of rivets on line $D \cong 3 \left[\frac{2 \times 3/4}{(2 \times 3/4 + 1/2)} \right]$ say 2

Note: An effective eccentricity of 2.75 could be used here reducing R_x to 12.8

Line B

$$R_y = \frac{P}{n} = \frac{60}{6} = 10.00 \text{ kip} \qquad R_x = \frac{Pey}{\Sigma y^2} = \frac{60 \times 6 \times 7.5}{2(1.5^2 + 4.5^2 + 7.5^2)} = 17.15 \text{ kip}$$

Resultant on bottom rivet

$19.85 < 24.05$ kip O.K.

Line C - Tensile stress on bolts

Negl. to simplify calc.

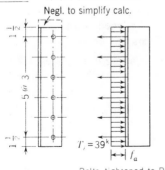

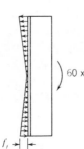

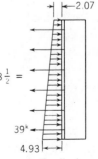

Bolts tightened to P.L.
0.7 tensile strength

$$f_a = \frac{12 \times 39}{8 \times 18} = 3.25 \text{ ksi}$$

Applied M

$$f_t = \frac{60 \times 8\frac{1}{2}}{\frac{1}{6} \times 8 \times 18^2} = 1.18 \text{ ksi}$$

Resultant
Practically the tension in the bolt does not change although the pressure distribution is different.

615

EXAMPLE 18.3 (Continued)

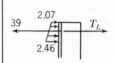

The difference between the resultant pressure on the back of the angle and the unchanged bolt tension represents the portion of the bolt tension caused by the applied moment.

$$T_L = 39 - \left(\frac{2.07 + 2.46}{2}\right)(3 \times 4) = 39 - 27.18 = 11.82$$

$$f_t = \frac{11.82}{0.601} = 19.7 \text{ ksi} < F_t = 40 \quad \text{(No prying action included)} \quad \underline{\text{O. K.}}$$

Actually, the bolt tensile stress is only slightly greater than that corresponding to the initial tightening force.

(Alternate approach)

Applied moment expressed in terms of concentrated forces acting at bolt locations

$$2 \times 2 \left(7.5 \, T_L + \frac{4.5^2}{7.5} T_L + \frac{1.5^2}{7.5} T_L\right) = 60 \times 8\tfrac{1}{2}$$

$$T_L = 12.14 \text{ kip} \ (\cong T_L = 11.82 \text{ kip above})$$

$$f_t = \frac{12.14}{0.601} = 20.3 \text{ ksi} < F_t = 40 \quad \underline{\text{O. K.}}$$

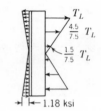

"Shear" stress on bolts – (friction type connection)

$$f_v = \frac{60}{12 \times .601} = 8.32 \text{ ksi} < F_v = 15 \text{ ksi}$$

Notes: 1. The calculation for T_L in the alternate approach is nothing more than $R_x = Pey / \Sigma y^2$ which can be shown by multiplying the left side of the equation by 7.5 / 7.5 and rearranging.
2. There is no need to reduce the allowable "shear" stress in this friction type connection because the reduction in pressure at the top of the connection is balanced by the increase at the bottom. The total frictional resistance remains the same.

of the face of the column to which the top angle is connected. The $M\text{-}\phi$ relationship for a typical seat and top angle connection is shown as curve B in Fig. 18.14. Although this connection is stiffer than the web angle connection it is still considered a simple, flexible connection.

The design of a seat connection is shown in Example 18.5.

For heavier loads, stiffened beam seats, as shown in Fig. 18.5*l*, should be used. The stiffened seat permits extension of the connection so that more fasteners can be used to carry heavier shears, and the stiffener prevents bending of the seat angle. The analysis of stiffened beam seats is similar to that for the rigid bracket illustrated in Examples 18.3 and 18.4.

Semi-Rigid Connections (Type 3). By combining web angles with a seat and top angle it is possible to develop a connection that has a greater moment resistance than either of the previously described flexible connections. An $M\text{-}\phi$ curve is shown in Fig. 18.14 as curve C. Although such a connection is sufficiently stiff to make a substantial reduction in the mid-span moment of the beam it is not rigid enough to prevent all rotation of the end of the beam. The connection is called "semi-rigid."

The actual behavior of the connection is quite complex. The end moment tends to put the upper portions of the connection in tension and the lower portion in compression. The amount of initial tension influences the behavior

---

EXAMPLE 18.4

PROBLEM:
Reinvestigate the bracket of example 18.3 if the fasteners in line C are 7/8" A 502 Gr. 2 rivets assumed to have no initial tension.

SOLUTION:
No initial tension in rivets. Bottom of angles compressed against column by applied moment. Top of angles pulls away from column. Rivets in tension.

Assume location of neutral axis

$$\bar{y} \cong \frac{1}{6} \times 18 = 3''$$

Check location

$$\frac{2 \times 4 \times \bar{y}^2}{2} = 10 \times .601 \, (10.5 - \bar{y})$$

$$\bar{y}^2 + 1.50\,\bar{y} - 15.78 = 0$$

$$\bar{y} = 3.29'' \cong \bar{y} \text{ assumed}$$

Moment of inertia of effective flexural section,

$$I = \frac{1}{3} \times 8 \times 3.29^3 + 2 \times .601 \, (13.21^2 + 10.21^2 + 7.21^2 + 4.21^2 + 1.21^2)$$

$$I = 515.85 \text{ in}^4$$

Tensile stress on top rivet

$$f_t = \frac{Mc}{I} = \frac{60 \times 8\,1/2 \times 13.21}{515.85} = 13.08 \text{ ksi}$$

Shear stress on top rivet

$$f_v = \frac{60}{12 \times .601} = 8.32 \text{ ksi}$$

Check allowable combined stress

$$F_t = 38 - 1.6 \, f_v$$

$$= 38 - 1.6 \times 8.32$$

$$= 24.69 \text{ ksi} > f_t = 13.08 \quad \underline{O.K.}$$

Note: Comparison with example 18.3 shows that the tensile stress on the fastener is lower when no initial tension is assumed than when a prestressed fastener is assumed.

(Figure at right: $18''$ section with N.A., dimensions $4''$ and $4''$, with $M = 60^k \times 8\frac{1}{2}''$, stress distribution f_t and f_c.)

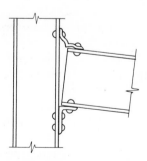

Fig. 18.12 Deformation of seat angle connection.

of the fasteners and their ability to elongate to accommodate rotation. The thickness of the top angle, web angles, and column flange also influences the rotation. The shear is divided among the various fasteners in an indeterminate fashion.

Because of the uncertainties of behavior as just described, a simplified approach is used in the design of the connection. The web angles are considered to carry the shear, and the top and bottom angles the moment. This arbitrary division of function has produced adequately proportioned connections. The fasteners joining the flange of the beam to the top and bottom angles are stressed in single shear by the forces of the moment couple, as shown in Fig. 18.13a. The fasteners joining the top angle to the supporting column must balance this force by pulling in tension. Thus, the angle is subjected to bending (Fig. 18.13b) and its thickness must be determined so that allowable bending stresses are not exceeded. An elastic analysis of the free body shown in Fig. 18.13c may be made.[3,7] However, considering

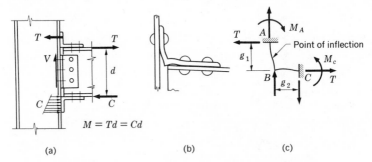

$$M = Td = Cd$$

(a)   (b)   (c)

Fig. 18.13 Assumed behavior of semi-rigid connection.

all the unknown aspects of the problem, especially the extent of the clamping zone around the fasteners, it is sufficiently accurate for design purposes to assume a point of inflection midway between the top face of the horizontal leg and the center line of the top fastener, thus leading to quick computation of the moment at A. For the reasons discussed previously under "Prying Action," probably only one gage line of fasteners should be used in the top angle connection to the column flanges.

The analysis described above is illustrated in Example 18.6.

Structural tees used in place of the top and bottom angles of the flexible seat angle connection result in the most rigid of the semi-rigid-type connections. The increase in the resistance to rotation occurs because the top tee is loaded in tension axially, whereas top angles are loaded eccentrically resulting in large deformations. The slope of the M-ϕ curve for the tee connection (curve D in Fig. 18.14) is much greater than that for the seat angle connection (curve B). In fact, the curve approaches the axis of ordi-

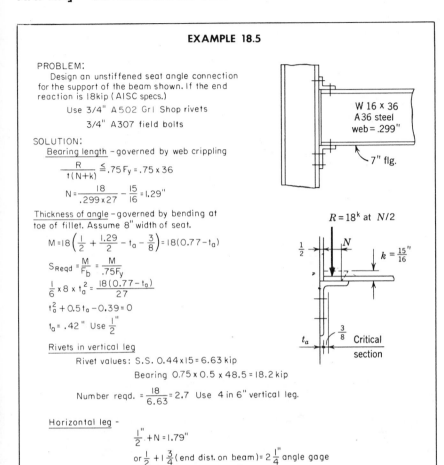

EXAMPLE 18.5

PROBLEM:
    Design an unstiffened seat angle connection for the support of the beam shown. If the end reaction is 18kip (AISC specs.)

    Use 3/4" A 502 Gr I Shop rivets

    3/4" A 307 field bolts

SOLUTION:
    Bearing length – governed by web crippling

$$\frac{R}{t(N+k)} \leqq .75\,F_y = .75 \times 36$$

$$N = \frac{18}{.299 \times 27} - \frac{15}{16} = 1.29''$$

    Thickness of angle – governed by bending at toe of fillet. Assume 8" width of seat.

$$M = 18\left(\frac{1}{2} + \frac{1.29}{2} - t_a - \frac{3}{8}\right) = 18(0.77 - t_a)$$

$$S_{Reqd} = \frac{M}{F_b} = \frac{M}{.75 F_y}$$

$$\frac{1}{6} \times 8 \times t_a^2 = \frac{18(0.77 - t_a)}{27}$$

$$t_a^2 + 0.5\,t_a - 0.39 = 0$$

$$t_a = .42'' \text{ Use } \frac{1}{2}''$$

    Rivets in vertical leg
        Rivet values: S.S. 0.44 x 15 = 6.63 kip
                    Bearing 0.75 x 0.5 x 48.5 = 18.2 kip

        Number reqd. $= \dfrac{18}{6.63} = 2.7$ Use 4 in 6" vertical leg.

    Horizontal leg –

$$\frac{1}{2}'' + N = 1.79''$$

$$\text{or } \frac{1}{2} + 1\frac{3}{4}(\text{end dist. on beam}) = 2\frac{1}{4}'' \text{ angle gage}$$

        Use 2 - A 307 bolts in $3\frac{1}{2}''$ horizontal leg

    Use seat angle $6 \times 3\frac{1}{2} \times \frac{1}{2} \times 8''$ shop rivet to col.

        top angle $3\frac{1}{2} \times 3 \times \frac{5}{16} \times 8''$ bolt to col. for shipment.

nates, which corresponds to a fully rigid connection, and this connection is often considered to be completely rigid.

Because no web angles are used to carry the shearing reaction the fasteners through the tee flanges must act in shear. The fasteners in the upper tee may have combined shear and tensile stresses if they are installed with no initial tension. If designed as a friction joint the tensile force on the top tee reduces the contact force, and the frictional resistance is reduced. On the other hand, the compressive force at the bottom will increase the friction. It is

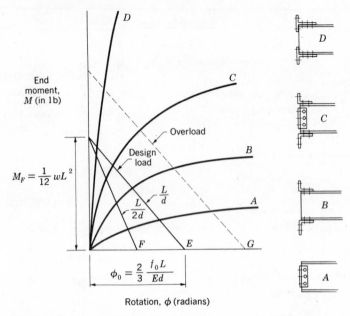

Fig. 18.14 Typical moment-rotation curves and beam-lines.

probable that the major portion of the end shear is resisted by friction at the lower tee.

Beam-Line Concept.   The moment that will be developed by a particular connection when it is used on a beam of a given span and loading may be determined from the M-ϕ curve of that connection by using the beam-line concept.[18.24]

If a beam with a uniformly distributed load w has equal restraining moments M at the ends, the following relationship of end slope ϕ and end moment can be derived.

$$\phi = \frac{1}{24}\frac{wL^3}{EI} - \frac{1}{2}\frac{ML}{EI} \tag{18.7}$$

This is a linear equation in ϕ and M which is easily plotted as line E in Fig. 18.14, using the following values:

$$(\phi = 0) \qquad M_F = \tfrac{1}{12}wL^2 \qquad \text{(fixed-end moment)}$$

$$(M = 0) \qquad \phi_0 = \frac{wL^3}{24EI} \qquad \text{(simple beam end slope)}$$

EXAMPLE 18.6

PROBLEM:

Design a semi-rigid connection of the type shown to carry an end shear of 30 kip and a moment of 35 kip-ft.

   AISC Specifications

   $\frac{3"}{4}$ 502 Gr. I shop rivets

   $\frac{3"}{4}$ A325 field bolts

   A36 main and connection material

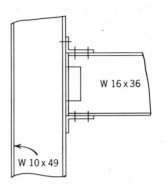

W 16 x 36

W 10 x 49

SOLUTION:

  Web angles –

$$4" \times 3\frac{1}{2} \times t \text{ std.}$$

     Carry 100% of shear – no moment.

   Rivet values: D.S. 0.44 x 21 x 15 = 13.25 kip
                Bearing .75 x .299 x 48.5 = 10.87 kip-governs

$$\frac{30}{10.87} = 2.8 \text{ Use 3-rivets thru web.}$$

   Bolt values – bearing type, threads excluded :
          S.S. 0.44 x 22 = 9.72 kip governs
          Bearing .75 x .558 x 48.5 = 20.6 kip

$$\frac{30}{9.72} = 3.09 \text{ Use 4 bolts O.S. legs}$$

   Angle thickness
      Bearing .75 t x 48.5 $\geq$ 9.72

$$t = .27 \text{ Use } \frac{5}{16}$$

   Shear on gross long. section of $8\frac{1}{2}"$

$$\frac{15}{\frac{5}{16} \times 8\frac{1}{2}} = 5.6 \text{ksi} < F_v = 14.5 \quad \underline{\text{O.K.}}$$

  Flange Angles –

   Top angle critical
   Carry 100% of moment – no shear
   Bolt values – friction type, S.S. 0.44 x 15 = 6.63 kip
               tension   0.44 x 40 = 17.67 kip

$$T = \frac{M}{d} = \frac{35 \times 12}{16} = 26.25^k$$

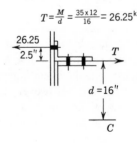

$d = 16"$

$$\frac{26.25}{6.63} = 3.96 \text{ Use 4 bolts thru beam flange}$$

$$\frac{26.25}{17.67} = 1.49 \text{ Use 2 bolts thru col. flange}$$

   Try 7 x 4 x 7/8 angle x 8" wide

   Moment = 26.25 x $\frac{1.62}{2}$ = 21.26 kip-in.

$$S = \frac{M}{F_b} = \frac{21.26}{.75 \times 36} = 0.79 \text{in}^3$$

$$S = \frac{1}{6} bt^2 = 0.79$$

$$t = \sqrt{\frac{0.79 \times 6}{8}} = 0.77$$

  ∴ Angle assumed is O.K.

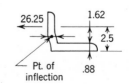

Pt. of inflection

For beams of symmetrical cross section, the value for ϕ_0 may be expressed in other terms by noting that the maximum moment $M = wL^2/8 = f_b I/(d/2)$. Proper substitution gives

$$\phi_0 = \frac{2}{3} \frac{f_b L}{E d} \tag{18.8}$$

Thus it is seen that for a given stress the end rotation of a simple beam depends on L/d.

The intersection of the beam-line E with the curves for the various connections indicates the amount of moment that the connections may be expected to resist. If line E is plotted for the allowable design stress in the beam, it is seen that the standard web angle connection A in this case develops about 25 per cent of the fixed-end moment. The same connection supplemented with top and seat angles C develops about 75 per cent of the fixed-end moment.

If the length of the span is reduced to one-half of that for beam-line E, beam-line F results, running from the same allowable value of M_F to one-half of the value of ϕ_0. Thus, it is seen that the same connections A and C when used on the shorter beam develop only 17 and 60 per cent of the fixed-end moment.

If the allowable beam stress is increased, as for permission of a short-duration overload on a bridge, the beam-line E will be translated to the right to a new position G parallel to its original one. Under these conditions it is important that the rotation capacity of the connection is not exceeded. The limiting position of beam-line G is that which corresponds to $f_b = F_y$.

Rigid Connections (Type 1). In analyzing portal frames and building frames where the resistance to wind depends upon the continuity of beams and columns, it is usually assumed that the connections are completely rigid and capable of maintaining the original angles between beams and columns. Moment-resistant connections that provide practically 100 per cent restraint are typified by the structural tee connection with web angles (Fig. 18.5m) and the end plate moment connection (Fig. 18.5n). The behavior of rigid connections is essentially the same as that described under semi-rigid construction. However, the greater depth of connection and the greater stiffness of the component parts produce very rigid assemblies.

It is difficult to develop moment-rotation curves by analytical methods because of the highly indeterminate interaction of the component parts.[18.25] Only tests or experience can show the amount of rigidity of that connection. In this regard, reliance must be placed on the past experiences and practices of the profession that have established details of construction, approximate methods of analysis and design, and safe allowable stresses.

Example 18.7 illustrates the design of a rigid wind bracing connection.

EXAMPLE 18.7

PROBLEM:
 Design the end connection for the W 14 x 34 beam. Use $\frac{7}{8}''^{\phi}$ A325 bolts.

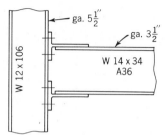

Bolt Values – $\frac{7}{8}''^{\phi}$ A325–F
 S.S – 9.02 kip
 T – 24.05 kip

REACTIONS	D + L	Wind	D + L + W	$\frac{3}{4}$(D + L + W)
Shear, kip	20	6	26	19.5
Moment, kip–ft	60	54	114	85.5

SOLUTION:

 1. Design structural tees to carry moment

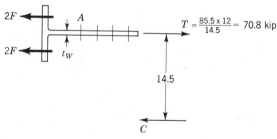

$$T = \frac{85.5 \times 12}{14.5} = 70.8 \text{ kip}$$

Bolts – beam flange to tee

   Use friction type connection for wind connection

   No. reqd. $= \frac{70.8}{9.02} = 7.85$ Use 8 on $3\frac{1}{2}''$ ga

Bolts – Tee to column flange

   Neglecting prying action

   No. reqd. $= \frac{70.8}{24.05} = 2.94$ Use 4

   $F = \frac{70.8}{4} = 17.70$ kip

Thickness of tee web

   Critical net section– Row A

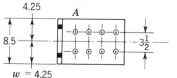

   $t_w = \dfrac{70.8}{(8.5 - 2 \times 1) \times 22} = 0.49$

   Try ST 12 x 50 $t_w = 0.747 > 0.49$

EXAMPLE 18.7 (Continued)

Determine prying force Q – Use Eq 18.5

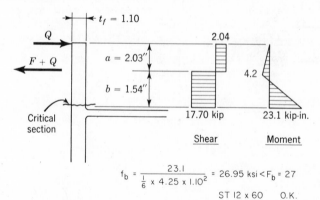

$$Q = 17.70 \left[\frac{100 \times 1.56 \left(\frac{7}{8}\right)^2 - 18 \times 4.25 \,(0.87)^2}{70 \times 1.62 \left(\frac{7}{8}\right)^2 + 21 \times 4.25 \,(0.87)^2} \right]$$

$$= 17.70 \left[\,0.40 \,\right]$$

$$= 7.12 \text{ kip}$$

$$F + Q = 17.70 + 7.12 = 24.82 \text{ kip} \quad 24.05 \text{ N.G}$$

Bolt overstressed due to prying action.   Use stiffer flange. Try ST 12 x 60

$$t_w = 0.798 > 0.49$$

$$Q = 17.70 \left[\frac{100 \times 1.54 \left(\frac{7}{8}\right)^2 - 18 \times 4.25 \,(1.10)^2}{70 \times 2.03 \left(\frac{7}{8}\right)^2 + 21 \times 4.25 \,(1.10)^2} \right]$$

$$= 17.70 \left[\,0.12 \,\right]$$

$$= 2.04$$

$$F + Q = 17.70 + 2.04 = 19.74 < 24.05$$

$$\therefore \ 4 - \tfrac{7}{8}{}''{}^\phi \ \text{A325 Bolts} \quad \text{O.K. with ST 12 x 60}$$

Check bending stress in flange of tee

2.04

4.2

$a = 2.03''$

$b = 1.54''$

$t_f = 1.10$

Critical section

17.70 kip   23.1 kip·in.

Shear   Moment

$$f_b = \frac{23.1}{\frac{1}{6} \times 4.25 \times 1.10^2} = 26.95 \text{ ksi} < F_b = 27$$

$$\text{ST 12 x 60} \quad \text{O.K.}$$

2. Design connection to carry shear

Bolts – Beam to column flg

$$\text{No. reqd.} = \frac{20}{9.02} = 2.2 \quad \text{Use 4}$$

4 bolts through flange of bottom structural tee will be adequate to carry shear. No direct force on bolt – compression side of beam. Web angles could be used to carry shear but not necessary.

Check web of bottom tee for bending as a seat, similar to Example 18.5.   O.K. by comparison to Ex. 18.5

18.5 ALLOWABLE STRESSES FOR FASTENERS

Although the allowable stresses for the design of main members are based on the initiation of unrestricted plastic flow of the steel, the allowable stresses for fasteners usually are based on the ultimate strength of the connection. The following paragraphs examine the allowable stresses for fasteners as given in the AISC Specification and try to relate the resulting factors of safety to those for the main material. In general, AASHO and AREA bridge specifications give allowable stresses about 10 per cent lower than AISC but in some cases they have not recognized items covered by the latter.

Rivet and bolt stresses usually have been adjusted so as to be used with the nominal area of the fastener, even though driven rivets are larger and bolts are smaller in the threaded portion.   This is done in the interest of simplicity. However, tensile stresses are sometimes given for the "stress area" and specification provisions must be read carefully.

1. Rivets in Tension

Beginning with the minimum yield point for A502, Grade 1 rivet bar stock, 28 ksi, and increasing that value 10 per cent due to driving and about 10 per cent more to account for the difference between the driven and nominal rivet size, the adjusted yield point of a nominal rivet becomes about 34 ksi as a minimum.   Thus, an allowable tensile stress of $F_t = 20$ ksi is reasonable and consistent with a factor of safety of 1.67.   In similar fashion it may be shown that the factor of safety against ultimate strength is approximately 3.

Because the A502 Grade 2 high-strength rivet steel has the same ratio of yield point to ultimate strength as A36 steel and A502 Grade 1 rivet material, the factors of safety with a working stress of $F_t = 27$ ksi remain at approximately 1.7 and 3.0.

The AREA Specifications do not permit rivets in direct tension.   The AASHO Specifications permit direct tension rivets only by inference from the interaction formula for rivets under combined tension and shear.

2. Bolts in Tension

A307 bolts are used occasionally in direct tension.   The allowable stress $F_t = 20$ ksi on the "stress area" corresponds to the value of $F_t = 14$ ksi on the nominal area found in the 1963 AISC Specifications.   The factors of safety are comparable to those shown in Table 1.3.

A325 bolts have two strength levels, depending on the size of the bolt. Nevertheless, the allowable tensile stress F_t has been set at 40 ksi (on the nominal area of the bolt) for all sizes. Under tensile loading a section through the thread is critical and the factor of safety against yielding may be computed for the common-size bolts as follows:

$$(F.S.)_y = \frac{\text{Specified proof load}}{\text{Nominal area} \times \text{Working stress}}$$

$$(F.S.)_y \approx \begin{cases} 1.6 \text{ for 1 in. and smaller} \\ 1.4 \text{ for } 1\frac{1}{8} \text{ in. and larger} \end{cases}$$

The factor of safety against ultimate strength is calculated in a similar fashion.

$$(F.S.)_u \approx \begin{cases} 2.3 \text{ for 1 in. and smaller} \\ 2.0 \text{ for } 1\frac{1}{8} \text{ in. and larger} \end{cases}$$

A490 bolts with $F_t = 54$ ksi have corresponding factors of safety of $(F.S.)_y \approx 1.7$ and $(F.S.)_u \approx 2.1$.

As noted previously the allowable tensile stresses are intended for comparison with calculated stresses resulting from applied loads and prying, and the values shown assure that the bolt stress will not greatly exceed the pretension nor will the clamped parts separate.

3. Rivets in Shear: Bearing-Type Connections

It was explained previously that the ultimate shear strength of a single rivet is approximately 75 per cent of its ultimate tensile strength. It follows that for buildings the allowable shear stress for an A502 Grade 1 rivet should be $F_v = \frac{3}{4} \times 20 = 15$ ksi. However, this ignores unbuttoning and the fact that in connections failure takes place by shearing of the end rivets at a somewhat lower average shear stress than the ultimate shear strength of a single rivet. Tests of short connections have shown that hot-driven low-carbon rivets possess enough deformation capacity to allow almost complete equalization of rivet forces. Thus, for short connections designed for $F_v = 15$ ksi, the factor of safety against a shear failure is about 3, resulting in a balanced ultimate design for a connection of A36 steel. However, some reduction in the factor of safety occurs in a long joint, as discussed below for high-strength bolts.

The allowable shear stress for high-strength rivets also has been set at three-quarters of the tensile stress, thus $F_v = 20$ ksi. The unbuttoning phenomenon for A502 Grade 2 steel rivets has not been investigated extensively.

Although the AASHO and AREA Specifications follow their traditional position by setting $F_v = 0.90 \times 15 = 13.5$ ksi for A502 Grade 1 steel rivets, AREA does not specify the use of A502 Grade 2 rivets and AASHO permits $F_v = 20$ ksi, the same as AISC.

4. High-Strength Bolts in Shear: Bearing-Type Connections

There are many structural joints assembled with high-strength bolts in which the parts will be erected in bearing, that is, the dead weight of the parts forces the bolts into bearing against the sides of the holes in the same way that they will bear under service loads. Even though the bolts are tightened so that friction probably carries the working load, frictional resistance is not required, and the design may properly be based on the shearing strength of the bolts. It is the bearing-type connection that can utilize best the high strength of A325 or A490 bolts.

Extensive tests and analytical studies of large plate splices have established the behavior of A325 bolts in long joints.[18·15,18·26,18·27]   The average ultimate shear strength decreases as the joint becomes longer. To maintain a constant factor of safety, the allowable stress would have to be related to the length of the connection. Because this presents a complication never introduced for rivet design where the same phenomenon occurs, the AISC Specification has adopted a constant value of allowable shear stress $F_v = 22$ ksi. This results in a factor of safety of about 3 for net section plate failures in short connections of A36 steel, but when many bolts are in line shearing failure of the end bolts occurs and the factor of safety is only about 2. The factor of safety against failure for many types of steel splices joined with A325 bolts is shown in Fig. 18.15.[18·28]

$F_v = 22$ ksi is permitted only when threading is excluded from the shear plane. Each threaded area occurring at a shear plane results in a 15 per

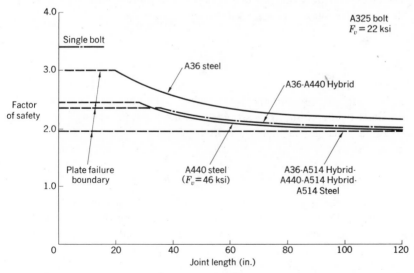

Fig. 18.15   Variation of factor of safety for A325 bolted joints.

cent reduction from the nominal area.   Thus, when a double shear connection has bolt threads at both shearing planes the allowable shear stress based on the nominal area should be $0.70 \times 22 = 15$ ksi.   In the interest of simplicity the AISC Specification does not recognize the case of only one shear plane with threading and requires $F_v = 15$ ksi any time a bolt thread intercepts a shear plane.   Because of the short thread length on structural high-strength bolts this occurs only when the ply of gripped material at the nut end is less than $\frac{3}{8}$ or $\frac{1}{2}$ inch thick depending on the bolt diameter.[18.2]

Analytical and experimental studies of bearing type joints using A490 bolts[18.16,18.28,18.29] and A440 high strength plates have indicated an unbuttoning behavior similar to joints using A325 bolts.   With $F_v = 32$ ksi the factor of safety varies between 2.5 and 2.   However, with A514 plates no unbuttoning occurs for lengths up to 120 inches and plate failures occur at a factor of safety of 1.9.

5. High-Strength Bolts: Friction-Type Connections

In the friction-type connection, bolts are not actually stressed in shear nor is bearing a consideration.   Nevertheless, it is convenient to specify an allowable "shear" stress in order that the proportioning of friction-type connections may be carried out using the same well-established methods of riveted connection design.   However, the important parameters affecting the slip resistance of bolted joints, namely, condition of faying surface and bolt clamping force, must not be forgotten.

The factor of safety that exists when a friction joint is designed on the basis of an allowable shear stress F_v is given by

$$\text{F.S.} = \frac{K_s m \sum T_i}{F_v m \sum A_b}$$

where the numerator is Eq. 18.1, with terms previously defined, $\sum A_b$ is the total bolt shearing area in one plane, and m is the number of slip planes.   If the n bolts in the connection are of the same size and are tightened to the same initial tension, this may be written

$$\text{F.S.} = \frac{K_s m n T_i}{F_v m n A_b} = \frac{K_s T_i}{F_v A_b} \tag{18.9}$$

This expression is shown graphically in Fig. 18.16 for $\frac{7}{8}$ in. bolts and $F_v = 15$ ksi as specified for static loading conditions, and $F_v = 1.33 \times 15$ for static plus wind conditions.   Two values of slip coefficient, $K_s = 0.20$ and 0.35, have been used in plotting the sloping lines.   The value of 0.20 represents the slip coefficient for semipolished steel surfaces where the mill scale has been removed by a power tool.   Although some investigators have

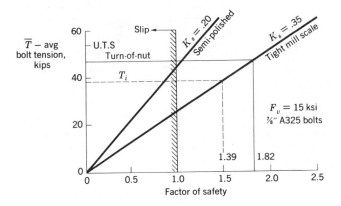

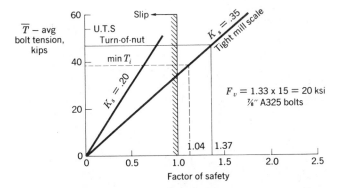

Fig. 18.16 Factor of safety against slip.

reported lower values, the slip coefficient of 0.35 is believed to represent a reasonable lower limit for dry mill scale surfaces of A36 steel. The slip coefficient for painted surfaces is about 0.10[18.30] and for galvanized surfaces it is about 0.16.[18.31]

The other important factor in friction-type joints is the bolt tension or the clamping force. Two significant values have been shown as horizontal lines in Fig. 18.16. The line marked "Min T_i" corresponds to the specified minimum tension. The line marked "Turn-of-Nut" corresponds to tensions achieved by the turn-of-nut method.

Inspection of Fig. 18.16 shows clearly the desirability of using a tightening method that achieves a high clamping force, and the reason for specifying that the contact surfaces of friction-type joints be dry mill scale. The factor of safety against slip need not be as great as that for ultimate strength since slip into bearing is not catastrophic in nature.

Note that the allowable stresses for "shear" in a friction-type joint amounts to a one-for-one substitution of an A325 bolt for an A502 Grade 1 rivet.

It was pointed out in Art. 18.4 that when prestressed fasteners are subjected to applied tensile loads, the initial bolt tension T_i does not change appreciably but the compressive force between the gripped members does decrease. Thus, when a friction-type joint is loaded by a tensile component P the clamping force is reduced to $T_i - P$ (see Fig. 18.9b) and the frictional resistance to the shearing component is also reduced. To acknowledge this fact a reduced allowable "shear" stress $F_v{}'$ must be used. Because the frictional resistance is proportional to the clamping force, the allowable shear stresses should be proportional also. Therefore,

$$\frac{F_v{}'}{F_v} = \frac{T_i - P}{T_i} \quad \text{or} \quad F_v{}' = F_v\left(1 - \frac{P}{T_i}\right) = F_v\left(1 - \frac{f_t A_b}{T_i}\right) \quad (18.10)$$

where T_i is taken conservatively as the specified minimum tension. If $P = T_i$ the gripped material is on the verge of separating and the joint no longer acts as a friction joint.

This provision of the AISC Specification is illustrated by Example 18.8.

The value of $F_v = 20$ ksi for friction type joints using A490 bolts has been adopted because the tightening tension for A490 bolts is about one third greater than that for A325 bolts. The coefficient of slip for high-strength steel surfaces with which A490 bolts normally will be used is comparable to that for A36 material.

6. Combined Shear and Tension

Tests on rivets and bolts subjected to combined shear and tension have shown that elliptical interaction curves (Fig. 18.2) describe the ultimate strength of the fastener. For design purposes, these ellipses can be dimensionalized and reduced to working stresses by dividing by appropriate factors of safety. For rivets of A502 Grade 1 steel used in building work the dashed ellipse shown in Fig. 18.17 results. Thus, a rivet having a computed shear stress of $f_v = 10$ ksi and a computed tensile stress of $f_t = 14$ ksi would plot on the graph as point A. The rivet would be considered satisfactory because the point falls inside the elliptical curve, or because the left-hand side of the equation is less than the right-hand side. This type of interaction equation is specified by AASHO but with the right-hand side equal to 13.5 squared.

The AISC Specification has replaced the ellipse by three straight lines, shown solid in Fig. 18.17. Thus, for rivets of A502 Grade 1 steel, if $f_v < 5$ ksi, $F_v = 20$ ksi, or if $f_t < 4$ ksi, $F_v = 15$ ksi. Between these limits the sloping line interaction equation, $F_t = 28 - 1.6f_v$, must be used. This line is more conservative than the elliptical curve over much of its length. For example, the rivet whose stresses are plotted at point A is seen to be unsatisfactory

EXAMPLE 18.8

PROBLEM:
  Investigate the design of the connection shown. Slip can not be tolerated. All steel A 36 AISC Specifications.

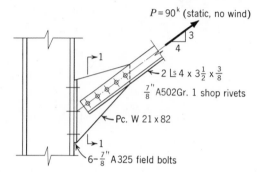

$P = 90^k$ (static, no wind)

2 Ls $4 \times 3\frac{1}{2} \times \frac{3}{8}$

$\frac{7}{8}''$ A502Gr. 1 shop rivets

Pc. W 21 x 82

$6 - \frac{7}{8}''$ A325 field bolts

SOLUTION:

Angles

$$f_t = \frac{P}{A_{net}} = \frac{90}{2 \times \frac{3}{8}\left(4 + 3\frac{1}{2} - \frac{3}{8} - 1\right)} = 19.6 \text{ ksi} < F_t = 22 \text{ ksi} \quad \underline{\text{O.K.}}$$

Rivets

$$f_v = \frac{P}{A_r} = \frac{90}{5 \times 2 \times 0.601} = 15.0 \text{ ksi} = F_v = 15 \text{ ksi} \quad \underline{\text{O. K.}}$$

$$f_p = \frac{P/n}{dt} = \frac{90}{5 \times \frac{7}{8} \times \frac{1}{2}} = 41.2 \text{ ksi} < F_p = 48.5 \text{ ksi} \quad \underline{\text{O. K.}}$$

21WF82

On section I–I $A = (13-1)\frac{1}{2} = 6 \text{ in}^2$

$$f_t = \frac{\frac{4}{5} \times 90}{6} = 12 \text{ ksi} < 22 \text{ ksi} \qquad \underline{\text{O. K.}}$$

$$f_v = \frac{\frac{3}{5} \times 90}{6} = 9 \text{ ksi} < 14.5 \text{ ksi} \qquad \underline{\text{O. K.}}$$

Bolts

Horizontal component of load reduces pressure on friction surface.

$$F_v' \leq 15\left(1 - \frac{f_t A_b}{T_b}\right) ; \quad f_t A_b = \frac{\frac{4}{5} \times 90}{6} = 12 \text{ kip}$$

$$F_v' = 15\left(1 - \frac{12}{39}\right) = 10.3 \text{ ksi}$$

$$f_v = \frac{\frac{3}{5} \times 90}{6 \times 0.601} = 15 \text{ ksi} > F_v = 10.3 \text{ ksi} \qquad \underline{\text{NG}}$$

The original design failed to recognize the loss of frictional resistance caused by the tensile component of the load. Add 2 bolts balanced about the working line to preserve symmetry. Do not introduce eccentricity.

$$f_v = \frac{\frac{3}{5} \times 90}{8 \times 0.601} = 11.2 \text{ ksi} < F_v = 15\left(1 - \frac{9.0}{39}\right) = 11.5 \text{ ksi} \quad \underline{\text{O.K.}}$$

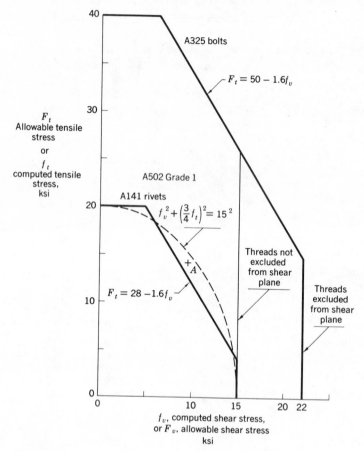

Fig. 18.17 Interaction curves for combined tension and shear.

according to this specification. If the computed shear stress is $f_v = 10$ ksi then the allowable tensile stress is $F_t = 12$ ksi. The computed tensile stress $f_t = 14$ ksi exceeds this value.

A similar approximation to the interaction ellipse has been made for A325 bolts in bearing-type connections (Fig. 18.17) and for each of the other fasteners covered by the AISC Specification.

7. Bearing

When a connection slips, at least one fastener comes into bearing against the side of the hole. A rather complicated distribution of compressive stress is developed in the material adjacent to the hole and in the fastener itself as shown in Fig. 18.18.   Initially this stress may be concentrated at the points of contact, but an increase in load brings local yielding and a larger area of contact together with a more uniform distribution of the compressive stress.   Further loading brings other fasteners into bearing.

Because of the complex nature of the bearing stress at any fastener and the different load on each fastener, as discussed in Art. 18.4, it is the practice to

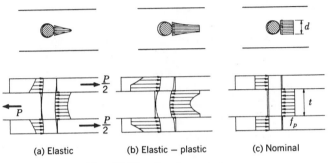

(a) Elastic   (b) Elastic — plastic   (c) Nominal

Fig. 18.18   Bearing stresses.

use a nominal bearing stress.   The nominal bearing stress is calculated by assuming that each fastener carries an equal part of the applied load, and that the bearing stress f_p is uniformly distributed over an area equal to the nominal fastener diameter d times the thickness of the material t (Fig. 18.17c).   Thus

$$f_p = \frac{P/n}{dt} \qquad (18.11)$$

In Fig. 18.18a it is shown that the bearing stress distribution on the outside plates is less uniform than on the center plate.   In a single shear lap joint the bearing stress is even more complex.   Because of this non-uniform distribution some specifications have differentiated between "double shear" or enclosed bearing, and "single shear" or exterior bearing, with a lower allowable stress being used for the latter.   This type of reasoning has not been borne out by tests and the AISC Specification makes no distinction between the two bearing conditions.

Although the fastener itself is subjected to the same magnitude of compressive forces as those acting against the side of the hole, tests have shown that the fastener is not the critical component.   Allowable bearing values are not provided as a protection to the fastener.   Actually, a high bearing stress reduces the effective net section of a tension member as calculated by Eq. 6.5.   Therefore, allowable bearing stresses are provided as a measure of the efficiency of the geometry of the connection, and the same measure is valid for joints connected with rivets or bolts, regardless of the strength of the fasteners or the presence or absence of threads in the bearing zone. Bearing stresses need not be computed for friction-type connections.

Tests on riveted joints of low carbon steel and rivets have shown that the effective net section of tension members is not adversely affected if the bearing stress is no greater than 2.25 times the net section stress.[18.32] On this basis the AISC Specification has set the allowable nominal bearing stress F_p as

$$F_p = 2.25 \times F_t = 2.25 \times 0.60F_y = 1.35F_y \tag{18.12}$$

This value is considerably more liberal than the value previously specified by AISC for single shear bearing or by the bridge specifications for either single or double shear bearing. As a consequence, in building work, allowable rivet and bolt loads will be governed by bearing in only a few cases.

In joints with no more than two bearing-type fasteners in the line of stress, the allowable bearing stress is valid only if the end distance is adequate to prevent splitting out of the fastener. The end distance is the distance from the center line of the end fastener to the unstressed end of the member. As a rule of good practice the AISC specifies that the end distance shall be equal to the cross-sectional area of the rivet divided by the thickness of the material for rivets in single shear or twice the distance for rivets in double shear. For high-strength bolts these distances are multiplied by the ratio of the specified minimum tensile strength of the bolt to the specified minimum tensile strength of the connected part.

PROBLEMS

18.1. What is the maximum static tensile load allowed on this connection (see sketch) according to the AISC and AASHO Specifications? Assume A36 material and A502 Grade 1 rivets.

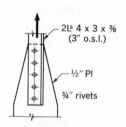

2Ls 4 x 3 x ⅜
(3″ o.s.l.)

½″ Pl

¾″ rivets

Problem 18.1

18.2. Design the splice for the bottom chord of the truss illustrated in Example 6.5, using ¾ in. A325 bolts. Assume a friction-type joint. If the entire connection area is shop-painted with red lead, what is the probable factor of safety against slip?

18.3. The connection shown is field bolted to a column flange, using $\frac{5}{8}$ in. A307 bolts. Which is the most highly stressed bolt? Calculate the shear stress and check against the allowable stress of the AISC Specification. If the bottom bolt is not installed by the steel erector what is the critical bolt stress?

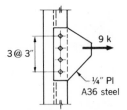

Problem 18.3

18.4. What is the maximum static load P that can be applied to the channel bracket? Use AREA Specifications.

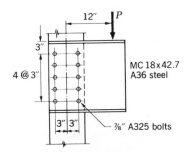

Problem 18.4

18.5. For the configuration shown derive an expression for the number of equal size fasteners, n, required to carry the eccentric load, P, without exceeding the safe carrying capacity, R, of a fastener.

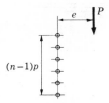

Problem 18.5

18.6. A $1\frac{1}{4}$ in. diameter rod with upset end is connected to a gusset plate by a clevis and pin, as shown. What size pin is required to develop the strength of the rod? What thickness of gusset plate should be used? What size A502 Grade 1 steel rivets are needed for the connection of the gusset plate to the supporting member? All materials are comparable in strength to A36 steel. Use AISC Specifications.

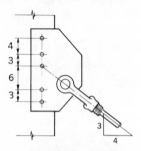

Problem 18.6

18.7. A structural tee is fastened to a column flange by four $\frac{7}{8}$ in. A502 Grade 1 rivets as shown. Assuming that the rivets have an initial tensile stress of 24 ksi due to cooling after driving, plot a graph of rivet tensile stress versus applied tensile load P. Indicate all important values of stress and load. Indicate the working load according to AISC Specifications. Repeat the plot for $\frac{7}{8}$ in. A325 bolts installed at the recommended tension.

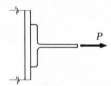

Problem 18.7

18.8. Design the connection of the hanger to the supporting beam (see figure) to develop the strength of the hanger; all material of A36 steel. Use AISC Specifications.

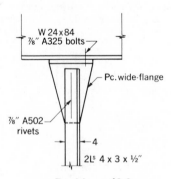

Problem 18.8

18.9. Considering axial shear only and assuming a bearing-type connection, what is the carrying capacity of the standard framed connection (see figure) as governed by the $\frac{7}{8}$ in.

A325 bolts through the beam web? If the eccentricity of the shear is taken into account, is the critical bolt overstressed?

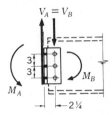

Problem 18.9

18.10. Design a seat angle connection similar to that shown in Example 18.5 to carry the end shear of a uniformly loaded W18 × 50 spanning 20 ft; A36 steel and A502 Grade 1 rivets.

18.11. What end moment can be sustained by the connection of Example 18.5 if the top and bottom angles are replaced by tees, WT7 × 17? See sketch.

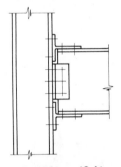

Problem 18.11

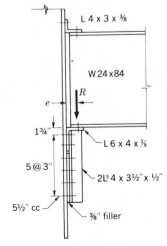

Problem 18.12

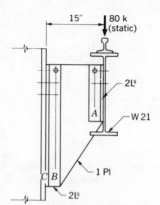

Problem 18.13

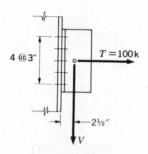

Problem 18.16

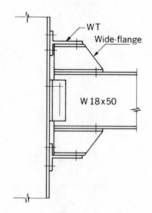

Problem 18.17

18.12. A W24 × 84 beam of A36 steel carries a uniformly distributed live load on a 20-ft span, as shown.   Can the stiffened seat connection safely develop the strength of the beam under the following conditions?

(a) $\frac{3}{4}$-in. A502 Grade 1 rivets; neglect effect of eccentricity.

(b) $\frac{3}{4}$-in. A502 Grade 1 rivets, no initial tension; consider effect of eccentricity.

(c) $\frac{3}{4}$-in. A325 bolts, friction-type connection, initial tension equal to 0.70 tensile strength: consider effect of eccentricity.

18.13. Design a bracket connection to support the crane load shown.   Allow for impact.   Use A440 material, $\frac{7}{8}$-in. A502 Grade 2 shop rivets, and $\frac{7}{8}$-in. A325 field bolts. Use the expression derived in Problem 18.5 to determine the number of fasteners required on lines A, B, and C.   State all assumptions clearly.

18.14. Derive Eq. 18.7.

18.15. Develop the requirements for flexible connections stated in AISC Section [1.15.4].

18.16. A structural tee (see figure) is connected to a column by 10 $\frac{3}{4}$-in. A325 bolts installed to the prescribed tension.   If slip cannot be tolerated what is the maximum allowable shear, V?

18.17. Design a wind-bracing connection of the type shown for a maximum shear of 40 kips and a maximum moment of 150 kip-ft.

19

WELDED CONNECTIONS

19.1 INTRODUCTION

The analysis and behavior of welded connections in many ways parallel that of structural steel connections with mechanical fasteners. Actually, only a different method of transferring the loads between the connected parts is involved. Generally, the welded connection is simpler and more compact because the drilling of holes for rivets or bolts is avoided and gusset plates and framing elements normally can be eliminated. Because of this and the opportunities of using smaller connecting material than could be conveniently fastened with mechanical fasteners, welded structural connections usually are lighter in weight.

Welding can be particularly advantageous when complete continuity between the connected parts is desirable or necessary. Also, the cleaner contours and shapes of welded structures are easier to maintain.

This chapter explains the strength and behavior of welded connections. The discussion in this chapter is primarily intended for manual or automatic submerged arc, shielded metal arc, and gas metal arc welding. Electroslag and Electrogas welding are being increasingly used and the design concepts described below are applicable. Descriptions of the welding process, nomenclature, standard symbols, and the effect of weldments on the behavior of structures can be found in Chapters 14, 15, and 16.

19.2 TYPES OF WELDS

The two basic types of welds are fillet welds and groove welds as shown in Figs. 19.1, 14.13, and 14.14. The basic difference between these two types of

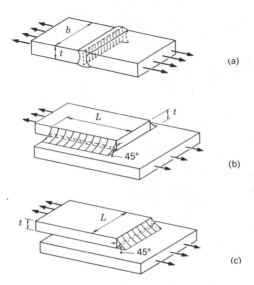

Fig. 19.1 Basic types of welds.

welds is the manner in which the stress transfer takes place. Groove welds usually act in direct tension or compression (although they may sometimes transfer shear), whereas fillet welds normally are subjected to shear as well as tension or compression and flexure since they are placed on the edge or side of the base metal. Fillet welds are nearly all of one type since they are placed in the right angle formed by lapping plates, intersecting plates, or other structural members. Groove welds are classified according to their preparation (see Art. 14.4); for example: square, bevel, vee, J, and U. With the exception of the square, these are further divided into single or double; and further into no backing or backing.

Even though the groove weld possesses greater strength than the fillet weld, many structural connections are joined by fillet welding. This is due mainly to the larger fit-up tolerances which are permitted when fillet welds are used. The butting together of sections necessitates the cutting of members to more or less exact lengths.

Welded joints may be classified as tee, edge, lap, corner, and butt. Another grouping refers to the position during placement of a weld, such as flat, horizontal, overhead, or vertical (see Fig. 14.11). Additional information on the welding process and weld types can be obtained in Chapter 14.

Occasionally, short lengths of fillet welds are distributed along the edges of a joint or along the edges of a cover plate on a girder. These are known as intermittent fillet welds. They are used when the strength required is less than that developed by a continuous fillet weld of the smallest permitted size. Intermittent butt welds are not permitted by the AISC, AASHO, or AWS Specifications.

Plug or slot welds may be used to provide additional strength when there is insufficient room to place the needed length of fillet weld. Their use is permitted by the AISC and AWS Specifications to transmit shear in a lap joint, to prevent buckling of lapped parts, and to join components of built-up members. The effective shearing area of a plug or slot weld is considered to be the nominal cross-sectional area of the hole or slot in the plane of the faying surface. Fillet welds in holes or slots are not to be considered as plug or slot welds. The AWS Specification does not allow plug or slot welds in tension members in bridges except for steady load conditions.

19.3 RESULTS OF TESTS OF WELDS

Tests have indicated that groove welds (Fig. 19.1a) of the same thickness as the connected parts are adequate to develop the full static capacity of these parts.[19.1] There is no need to calculate the stresses in the weld as long as the weld metal matches the plate in physical properties.

The ultimate strength of a fillet weld is dependent upon the direction of the applied load, which may be parallel or transverse to the weld. In both cases the weld fails in shear, but the plane of rupture is not the same. Tests indicate that a fillet weld loaded at right angles to the fillet (transverse weld, Fig. 19.1c) is approximately 30 per cent stronger than when it is loaded in a parallel direction (Fig. 19.1b).[19.2,19.3,19.4] However, this added strength is not considered in design stress provisions since welds may be located at various angles with respect to the applied loads.

In design it is often necessary to use long welds in line with the load. Elastic theory indicates that such welds have higher stresses near the ends, as shown in Fig. 19.2, analogous to long riveted or bolted joints.[19.5,19.6] However, tests have shown that for static loads this concentration is not too serious.[19.3,19.4,19.5] The plastic deformation near the ends is adequate to distribute the shear more or less uniformly before failure.

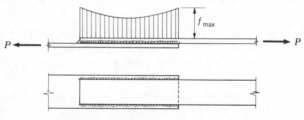

Fig. 19.2 Stresses in long fillet welds.

The increased use of high-strength steels and the need to refer to them in specification provisions resulted in further studies on fillet welded connections.[19.4] Since fillet welds may be made with electrodes whose mechanical properties are not equal to those of the base metal, the study evaluated the influence of type of electrode, size of fillet weld, type of steel, and type of weld. All test specimens were designed to fail in the welds, even though the mechanical properties of the weld metal exceeded those of the base metal.

The study indicated that when longitudinal fillet welds were made with electrodes that "matched" the connected steel, the weld strength varied from 60 to 85 per cent of the electrode tensile strength as illustrated in Fig. 19.3. The study indicated that the failure plane generally was at an angle

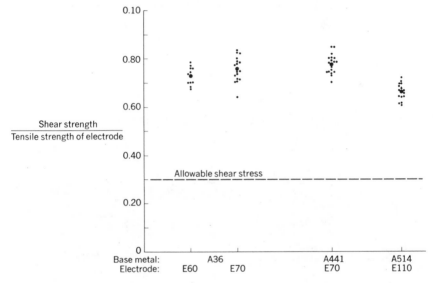

Fig. 19.3 Shear strength of longitudinal fillet welds with matched base metal.

less than 45° to the plane of a leg.   Thus, use of the minimum throat thickness is conservative.

Since weld metal may be deposited on base metal with different mechanical properties, combinations of strong base metal with weaker weld metals and vice-versa were also evaluated.[19.4] The results are summarized in Fig. 19.4. This revealed that the effect of dilution upon weld strength was not great.

Where plate bending is not a problem, tests of welds subjected to combined bending and shear have indicated a varying factor of safety against weld failure.   The results of tests on vertical weld groups are plotted in Fig. 19.5. As the ratio of eccentricity to weld length (e/L) varies from 0.06 to 2.4, the

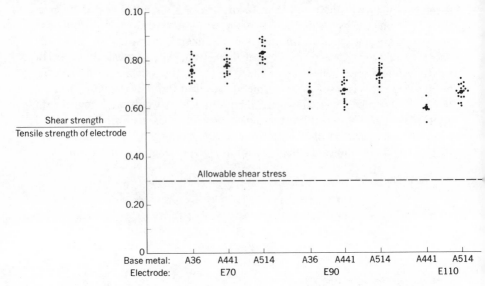

Fig. 19.4 Effect of base metal on the shear strength of fillet welds.

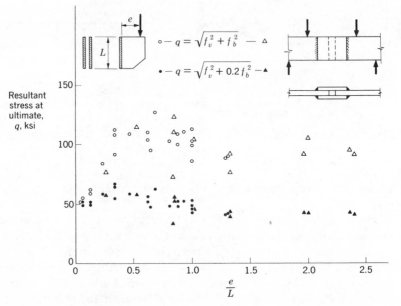

Fig. 19.5 Welds subjected to bending and shear. (Courtesy of *Civil Engineering and Public Works Review.*)

factor of safety against failure varies from 3.6 to 7.6, when the direct vector addition of the shear forces on the weld is used.   The analysis made in Ref. 19.7 has shown that a more uniform factor of safety would result if only 20 per cent of the shear stress component due to bending, f_b, were considered. Similar results were obtained for the weld groups reported in Ref. 19.8, when analyzed in a similar manner.

19.4 STRESSES IN WELDED CONNECTIONS

The weld must transmit tension, compression, or shear, as the case may be. For purposes of allowable stress design the usual direct stress formula $f = P/A$, the flexure formula $f = Mc/I$, the torsion formula $f_v = T_p/J$, and the beam-shear formula $f_v = VQ/It$ are used.   Welds designed to carry ultimate forces and moments are proportioned for stresses at or lower than the yield point of the weld metal but not less than the yield point of the base metal. In groove welds the forces are carried in compression, tension, or shear and are limited only by the resistance of either the base or the weld metal, depending on which is the smaller.   The same combined stresses exist in groove welds that exist in the material they connect, as the weld metal takes the place of an equivalent amount of base metal.

In manually deposited fillet welds, the critical stress is assumed to be a shear stress on the *minimum throat area* regardless of the direction of the applied load.   The throat area is taken as the product of the weld length and its theoretical throat t, as shown in Fig. 19.6.   Macro-etched cross sections of manually made fillet welds (see Fig. 19.7) exhibited very little penetration below the root of the weld.[19.4]

Fillet welds made by the automatic submerged arc process have greater penetration into the base metal, as illustrated schematically in Fig. 19.8 and

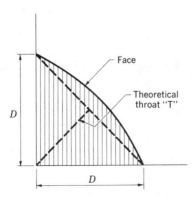

Fig. 19.6 Throat thickness of manual fillet welds.

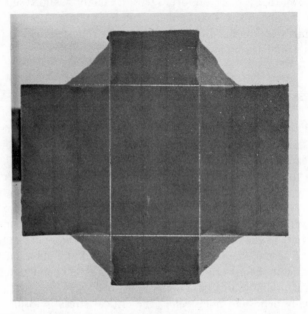

Fig. 19.7　Macro-etched cross section of fillet weld made by manual arc process.　(Courtesy AISC.)

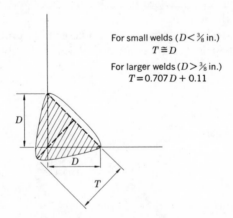

For small welds $(D < \frac{3}{8}$ in.$)$
$$T \cong D$$

For larger welds $(D > \frac{3}{8}$ in.$)$
$$T = 0.707D + 0.11$$

Fig. 19.8　Throat thickness of fillet welds made by automatic submerged-arc process.

in the photograph of the etched surface of Fig. 19.9.   To account for the improved penetration, the throat t may be taken as equal to the leg size D, for small fillet welds up to and including $\frac{3}{8}$ in. in leg size.   Since the added penetration is primarily due to the first pass, automatic submerged arc welds made by more than one pass can count only on the initial pass penetration for added size.   Hence, for fillet welds larger than $\frac{3}{8}$ in. leg size, 0.11 in. may be added to the theoretical throat size as illustrated in Fig. 19.8. Regardless of the welding process, it is good practice to require adequate demonstration of the root pass penetration before it is counted on.

Fig. 19.9   Macro-etched cross section of fillet weld made by automatic submerged-arc process.   (Courtesy of Lincoln Electric Co.)

In computing the stress due to direct loads on longitudinal and transverse fillet welds, it is customary to divide the force acting on the weld by the appropriate area.   Shear, flexure, and axial forces all cause shearing stresses in fillet welds.   Hence, the resultant of the shearing stresses is used to proportion the weld.

Most welded connections will consist of several welds rather than a single weld, and the welds are subjected to combined types of loading.   Nominal stresses are used to evaluate the adequacy of such welds.   The following sections illustrate the analysis of simple welded joints.

1. Axial Stress

The stress in the butt weld, shown in Fig. 19.1a, is computed as

$$f_a = \frac{P}{A_w} \leqslant F_t \tag{19.1}$$

where $A_w = bt$ and F_t is the allowable axial stress in the base metal. For the fillet welds illustrated in Figs. 19.1b and c, the stress is computed as

$$f_v = \frac{P}{A_w} \leqslant F_v \qquad (19.2)$$

or

$$q_v = \frac{P}{L} = \text{shear per inch of weld} \qquad (19.2a)$$

where A_w is $(\sqrt{2}/2)LD$ for manual fillet welds, LD for small automatic submerged arc fillet welds ($\frac{3}{8}$ in. leg or less), and $(\sqrt{2}\, D/2 + 0.11)L$ for automatic submerged arc fillets larger than $\frac{3}{8}$ in. leg. D is the leg size of the weld (Fig. 19.6), L is the length of the weld, and 0.11 is the amount of extra penetration permitted for large automatic submerged arc fillet welds.

2. Bending

When welds are subjected to flexure, the fundamental assumption of a linear strain distribution is made. In the elastic case the stress is proportional to strain, and the relationship $f_b = Mc/I$ is obtained. This is illustrated in Fig. 19.10 for both a butt and a fillet weld.

In the plastic design of the weld the full plastic moment must be resisted,

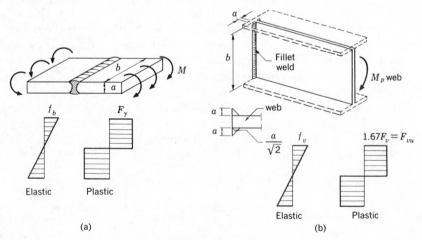

Fig. 19.10 Welds subjected to bending.

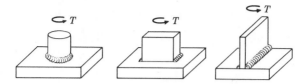

Fig. 19.11 Welds subjected to torsion.

and a uniform stress state is assumed throughout the section except for a change in sign.   This also is illustrated in Fig. 19.10 for both groove and fillet welds.   Reference 19.9 discusses the application of plastic design to welds.

The design of fillet welds subjected to torsion can be considered analogous to welds subjected to bending.   The resulting stresses due to torque are considered as shearing stresses acting on the throat area.   Figure 19.11 shows the applied torque which induces shearing stresses in fillet welds.

3. Combined Stresses in Fillet Welds

Although the same combined stresses exist in groove welds as exist in the base metal, only combined stresses in fillet welds will be considered in this section for illustrative purposes.   Article 7.3 considers the determination of combined stresses in the base metal.

When a bracket of the type illustrated in Fig. 19.12 is subjected to load, the stresses due to twisting are computed from the torsion formula, $f_v = T\rho/J$, where the applied torque is equal to the moment Pe.   The shearing stress is assumed to vary linearly from the centroid of the weld group just as the shear stress varies in a circular member.   Hence, from equilibrium,

$$Pe = T = \int \underbrace{\frac{\rho}{c}\, q_{max}}_{\text{force/in.}}\; \underbrace{d\rho}_{\text{length}} \; \underbrace{\rho}_{\text{arm}} = \frac{q_{max}}{c} J \tag{19.3}$$

where q_{max} = maximum shear per inch for a weld of unit width, c = distance from centroid to extreme fiber, and $J = \int_A \rho^2\, d\rho$, the polar moment of inertia of a unit width of weld.   Since the shear per inch at any distance ρ from the centroid is assumed to vary linearly, then

$$q = q_{max}\frac{\rho}{c} = \frac{T\rho}{J} \tag{19.4}$$

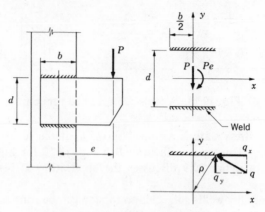

Fig. 19.12 Welded bracket.

It is more convenient to deal with the components q_x and q_y, as other forces may be acting on the weld group. Hence, Eq. 19.4 becomes

$$q_x = \frac{Ty}{J} \tag{19.4a}$$

$$q_y = \frac{Tx}{J} \tag{19.4b}$$

Equations 19.2 and 19.4 are used to determine the shear per inch of weld for the weld group shown in Fig. 19.12. The resulting shear components are

$$q_x = \frac{Ty}{J} = \frac{Pe(d/2)}{J} \qquad \text{(kip/in.)}$$

$$q_y = \frac{Tx}{J} = \frac{Pe(b/2)}{J} \qquad \text{(kip/in.)}$$

$$q_v = \frac{P}{L} = \frac{P}{2b} \qquad \text{(kip/in.)}$$

$$q = \sqrt{(q_v + q_y)^2 + q_x{}^2} \qquad \text{(resultant)}$$

where

$$J = \int (x^2 + y^2)\, dA = I_x + I_y$$

$$= \tfrac{1}{6}(b^3 + 3bd^2)$$

The direct vector addition of shear forces on welds is a conservative approach to the design of several fillet welds, as was illustrated in Fig. 19.5. The factor of safety was found to vary with the ratio of eccentricity to weld length (e/L).

19.5 TYPES OF STRUCTURAL CONNECTIONS

Structural welded connections are frequently combinations of groove and fillet welds. Welded connections are used in rigid, simple, and semi-rigid construction. The most advantageous use of welded construction is realized with rigid or continuous structures.

Beam splices, column splices, beam-to-column connections, and beam-girder connections are typical welded connections that assure continuity in multi-story steel structures. Several of these connections are illustrated in Fig. 19.13.

Rigid frame construction is used in many industrial frames and single-span bridges. Typical of the connections used in such structures are the rigid frame knees, apex joints, valley joints, and knee braces shown in Fig. 19.14.

When simple framing is used, there are generally two methods of obtaining a simple welded connection. In one, web angles straddle the beam web in much the same manner as do riveted and bolted connections. In the other, seat and clip angle connections are used, as is illustrated in Fig. 19.15. These connections are assumed to allow free rotation under load. Tests indicate that this is a reasonable assumption.[19.10]

Frameworks of intermediate rigidity, called "semirigid," can be used if the moment-rotation characteristics are known. Tests have indicated that it is beneficial to the design to take advantage of the partial fixity.[19.11] It is, however, difficult to evaluate the degree of fixity.

Welded structural connections are in common use in both building and bridge construction. Except for differences in allowable stresses, there are no differences in the design procedures used to proportion the connections.

19.6 ALLOWABLE STRESSES

Allowable stresses for structural welds in buildings are given in Ref. 1.21. On complete-penetration groove welds, the allowable tension, compression, shear, and bearing stresses are the same as those allowed in the connected material—when specified electrodes are used that "match" the base metal. The allowable stresses for several electrodes and "matching" base metal are given in Table 19.1.

a. Beam-girder splice

b. Column splice

Fig. 19.13 Rigid welded connections. (Courtesy of Lincoln Electric Co.)

c. Beam-to-column connection

d. Welded beam trusses

Fig. 19.13 *Continued.*

653

a. Rigid-frame knees with bracing and purlins

b. Valley and apex joints

Fig. 19.14 Rigid frame connections. (Courtesy of Lincoln Electric Co.)

Fig. 19.15 Flexible welded connections. (Courtesy of Lincoln Electric Co.)

The allowable stress for fillet, plug, slot, and partial-penetration groove welds is taken as 0.3 of the specified minimum tensile strength of the electrode classification, when the mechanical properties of the electrode and base metal "match". Table 19.2 summarizes the allowable shear stresses for various electrode-matching base metal combinations.

Table 19.1 Allowable Groove-Weld Stresses by AISC Specification

	Electrode	E60 or E70	E70	E110
Type of Stress	Base Metal	A36	A242, A441 ($\frac{3}{4}$ to $1\frac{1}{2}$ in.)	A514
Tension		22 ksi	30 ksi	60 ksi
Compression		22	30	60
Shear		14.5	20	40
Bending		22 or 24	30 or 33	60

Table 19.2 Allowable Fillet-Weld Stresses

Stress	Electrode	Matching Base Metal
18.0	E60xx; F60-Exxxx E60s	A500 Grade A and A570 Grade I
21.0	E70xx; F70-Exxxx E70S, E70T	A36, A242, A441, A572 Grades 42–60
24.0	E80xx; SAW 80; GMAW 80	A572 Grade 65
27.0	E90xx; SAW 90; GMAW 90	A514 Types E and F over $2\frac{1}{2}$ in. thick
30.0	E100xx; SAW 100; GMAW 100	A514 over $2\frac{1}{2}$ in. thick
33.0	E110xx; SAW 110; GMAW 110	A514 $2\frac{1}{2}$ in. or less

When the fillet or partial penetration weld and base metal strengths are different, the AISC specification conservatively requires that the permissible unit stress for the weld be taken from the weaker matching base metal.

The allowable weld stresses for welded highway and railway bridges are given in Refs. 1.23 and 1.25.   For complete penetration butt welds, the allowable stresses are the same as the base metal joined as is the case for building construction.   The most important difference between the provisions for designing welds for bridges and buildings is in their treatment of fatigue affects, an aspect treated in Chapter 16.   The allowable shear stress for fillet welds is 12.4 ksi on A36 steel and 14.7 ksi on base metal having a yield point between 40 and 50 ksi.   Only 12.4 ksi shear stress is permitted on plug welds regardless of the type of base metal.

19.7 WELDED CONNECTIONS: ALLOWABLE STRESS DESIGN

The allowable shear stress for a fillet weld is applicable to the appropriate shear area.   For manually deposited welds, this area is based on the throat dimension T (Fig. 19.6).   For submerged arc welds, the area is based on the leg dimension D (see Fig. 19.8) when D is $\frac{3}{8}$ in. or less.   For larger submerged arc welds, the area is based on the throat dimension T plus an increment of 0.11 in. to account for the improved penetration.   The AASHO and AREA Bridge Specifications do not differentiate between manual and automatic submerged arc fillet welds.   The allowable shear per inch of manual fillet weld is, for E70 electrodes,

$$\text{AISC:} \quad S = 21.0D/\sqrt{2} = 14.85D \quad \text{(kip/in.)} \quad (19.5)$$

or

$$S \approx 0.93 \quad \text{(kip/in.)}$$

if D is expressed in $\frac{1}{16}$ in. increments; and

$$\text{AASHO:} \quad S = 14.7D/\sqrt{2} = 10.4D \quad \text{(kip/in.)} \quad (19.6)$$

1. Design of Welded Angle Connections

Unless single- or double-angle members are subjected to repeated stresses, there is no need to proportion the weld to eliminate the unavoidable eccentricities that occur with this type of connection (AISC Section [1.15.3]). Tests reported in Ref. 19.12 have shown that welded angle connections subjected to static loads do not need to have the weld group balanced.

If the angle member is subjected to repeated variation in stress, the fatigue strength of the connection is critical.   It is then necessary to place the fillet welds so that the forces are balanced about the centroidal axis or axes for end connections of single-angle, double-angle, and similar types of members. In Example 19.1, the total length L would be unchanged if designed in

---

EXAMPLE 19.1

PROBLEM:

Design the welded connection shown using the AISC Specification.

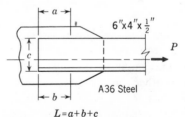

SOLUTION:

If the full strength of the angle is to be developed, the allowable load P is

P = (22 ksi) (4.75 in.²) = 104.5 kip

The maximum weld size, based on the toe of the angle, is

$\frac{1}{2}$ in. $-\frac{1}{16}$ in. $= \frac{7}{16}$ in. (AISC Section [1.17.5])

Using E 70 electrodes, the allowable shear stress is 21.0 ksi (Table 19.1 x). The shear per inch of weld is

$S = \frac{21.0}{2} \times \frac{7}{16} = 6.5$ kip/in.

The total length of weld required is L = P/S = 104.5/6.5 = 16 in. | Use 16 in. |

This length of weld can be placed around the joint at the designer's discretion. However, some portion of the total required weld should be deposited along both longitudinal edges of the member to prevent rotation of the angle about its longitudinal axis.

accordance with the AISC Specification; however, the placement of weld would then have to be considered.

A welded angle connection in which welds are distributed in such a way as to provide better resistance to repeated loading is shown in Fig. 19.16. If designed by the AASHO Specifications, the placement of weld would be similar. Equilibrium of forces and moments acting on the connection leads

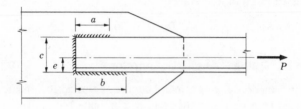

Fig. 19.16 Balanced fillet welds.

to the following if welds a, b, and c are the same size. From equilibrium of the forces,

$$L = \frac{P}{S} = a + b + c \tag{19.7}$$

From equilibrium of the moments about the centroidal axis,

$$a(c - e) + \tfrac{1}{2}(c - e)^2 = be + \frac{e^2}{2}$$

from which is obtained

$$a = \frac{eL}{c} - \frac{c}{2} \quad \text{and} \quad b = \frac{L}{c}(c - e) - \frac{c}{2} \tag{19.8}$$

If no weld is placed along the end of the angle, the equilibrium conditions again lead to expressions for the lengths of welds a and b.

The maximum size of fillet weld to be used along the rolled edge of an angle or other rolled shape usually is taken as $\frac{1}{16}$ inch less than the thickness of the angle or flange. Using a weld of the same thickness as the angle leg or beam flange requires a specially built-out weld in order to ensure full throat thickness and adds to the expense of welding.

2. Design of Lap Joints

Two types of lap joints are in general use. The one type uses longitudinal fillet welds along the edges of the lapped plates parallel to the load. The other type uses transverse fillet welds across the ends of the plate perpendicular to the load. These two types of fillet welds are illustrated in Fig. 19.1. The design of the two types of lap joints is illustrated in Examples 19.2 and 19.3.

EXAMPLE 19.2

PROBLEM:

Determine the size and length of the longitudinal fillet welds which will develop the strength of the smaller plate. Use the AASHO/AWS Specification.

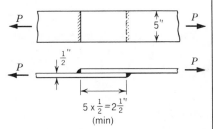

SOLUTION:

The allowable plate load is

$P = (20 \text{ ksi})(5)(\frac{1}{2}) = 50 \text{ kip}$

The size fillet weld is limited by the plate thickness

$D = \frac{1}{2} - \frac{1}{16} = \frac{7}{16} \text{ in.}$

The allowable shear of a $\frac{7}{16}$ in. fillet weld using E60 electrodes is

$S = 8.8D = 3.85 \text{ kip/in.}$

The total length of fillet weld required is $\quad L = P/S = 50/3.85 = 13.0 \text{ in.}$

Since the plate is symmetrical, the weld length should be made the same on each side. Hence, 6.5 inches of weld would be placed along each side of the plate.

Note: To insure full development of the plate despite possible "shear lag", AWS Section-[222] specifies a length of fillet along each side not less than the perpendicular distance between them. This requirement has been met. If the plate were subjected to fatigue loading a reduction in allowable stress may be necessary.

EXAMPLE 19.3

PROBLEM:

Two $\frac{1}{2}$ in. plates are to be lapped and connected by transverse fillet welds. Design the fillet welds for the lap splice using the AISC Specification. Can the strength of the plates be fully developed? A441 steel, E70 electrodes

$5 \times \frac{1}{2} = 2\frac{1}{2}''$
(min)

SOLUTION:

The allowable plate load P is

$P = 30(5)(\frac{1}{2}) = 75 \text{ kip}$ (Plate)

The maximum size fillet weld is limited by the plate thickness

$D = \frac{1}{2} - \frac{1}{16} = \frac{7}{16} \text{ in.}$

The total length of the two transverse fillet welds is 10-inches. Hence, the total load the welds can resist is

$P = 2 \times 5(21)\frac{\sqrt{2}}{2}(\frac{7}{16}) = 65 \text{ kip (weld)}$

Hence, to develop the strength of the plate would require additional welding such as a slot weld.

The efficiency of the joint which is defined as the ratio of the joint strength to the strength of the plate is

$\text{Efficiency} = \dfrac{\text{strength of joint}}{\text{strength of solid member}} \times 100 = \dfrac{65}{75} \times 100 = 87 \%$

Note: A minimum lap of $5t = 2\frac{1}{2}$ in. is required, AISC Section [1.17.8]

3. Web Angles

A welded beam-to-column or beam-girder connection that is analogous to riveted or bolted web angle connections is used occasionally. The angles are shop welded to the web of the beam and field welded to the face of the support. Deflection of the outstanding legs must provide the same cushioning effect against injury to the welds as with other types of fasteners.[19.10] Erection bolts are usually placed near the bottom of the framing angles to ensure flexibility. The connection must be flexible enough to satisfy specification requirements, such, for example, as AISC Section [1.15.4]. Similar connections exist in bridge structures when the floor beams are connected to the trusses.

The welds attaching the angles to the beam web and to the column must be capable of transmitting the load acting on the beam to the column.

A typical web angle connection is shown in Example 19.4. End returns help relieve the effects of high stress concentrations at the end of a length of weld due to prying as the beam end rotates. Hence, they are especially useful when joints are subjected to eccentric loads and impact. Reference 19.13 contains many examples of the elastic design of welds and miscellaneous connections.

4. Beam Seats

Another common beam-to-column connection is the beam seat. This seat may be flexible or stiffened. The flexible beam seat is generally an angle. Typical beam seats are shown in Fig. 19.15. The connection on the left is a stiffened beam seat and the connection on the right is a flexible beam seat. If web angles are not provided, usually a top angle is employed to assure stability of the beam. However, this top angle must allow reasonably free rotation of the connection as shown in Fig. 19.17.

The three principal factors which affect the design are:

1. The location of the beam reaction. (The most common practice is to assume the reaction to be uniformly distributed over such a length of the beam as to just satisfy the web crippling requirement.)
2. The thickness of the beam seat.
3. The size of weld to resist the shear and bending forces.

Flexible Beam Seats. Tests of flexible beam seats have indicated that the critical section is at the root of the inside fillet of the angle as indicated in Fig. 19.18.[19.14] The design of welded seat angles is the same as for riveted or bolted seat angles except for the weld design. Example 19.5 illustrates the weld design.

EXAMPLE 19.4

PROBLEM:
Design the welds for the flexible angle connection shown using AISC Specification.

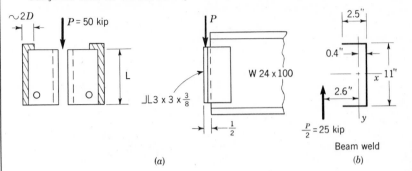

(a)

Beam weld

(b)

The design for combined stresses was covered in Art. 19.4. Each of these welds, the beam web and the column flange, can be considered in the manner illustrated in Fig.19.12.

SOLUTION:
The beam web weld will be assumed to have the dimensions shown in Fig. b

Treating a unit width,

$$\bar{x} = 0.4 \, \text{in.}$$

$$J = I_x + I_y = \frac{(11)^3}{12} + 2(2.5)(5.5)^2 + 2(2.5)(.85)^2 + 11(0.4)^2$$

$$= 270 \, \text{in.}^4/ \text{in. width.}$$

$$q_v = \frac{P/2}{A} = \frac{25}{1[11+2(2.5)]} = 1.56 \, \text{kip/in.}$$

$$q_x = \frac{Ty}{J} = \frac{25(2.6)5.5}{270} = 1.33 \, \text{kip/in.}$$

$$q_y = \frac{Tx}{J} = \frac{25(2.6)2.1}{270} = 0.51 \, \text{kip/in.}$$

$$q = \sqrt{(1.56 + 0.51)^2 + (1.33)^2} = 2.46 \, \text{kip/in.}$$

Using E 70 electrodes, $S = \frac{\sqrt{2}}{2} D \, 21 \, \text{kip/in.}$

Therefore $D = \frac{2.46}{\frac{\sqrt{2}}{2} 21} = 0.165 \, \text{in. or } 3/16 \, \text{in. fillet}$

With 1/2 in. thick web the minimum fillet weld size is 3/16 in., hence **O.K.**

For A36 steel the allowable web shear is 14.5 ksi. Since the weld throat thickness is less than 1/2 the web thickness the full weld size is effective.

The column connection weld is proportioned to resist the combined stresses due to the eccentrically applied load. In recognition of the stiffness of the outstanding angle leg, the bottom 5/6 of the weld is assumed to assist in resisting the eccentric force. This is an arbitrary assumption and is usually made in recognition of the conservativeness of the direct vector addition of stresses in certain instances. It is analogous to decreasing the effect of the bending component as suggested in Art 19.3

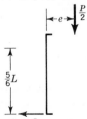

Method I: Assume neutral axis for bending L/6 from top

$$q_v = \frac{P/2}{A} = \frac{25}{11} = 2.27 \, \text{kip/in.}$$

EXAMPLE 19.4 (Continued)

$$q_x = \frac{9P_e}{5L^2} \qquad \text{Note:} \quad \frac{P_e}{2} = \frac{1}{2} q_x \times \frac{5L}{6} \times \frac{2L}{3}$$

$$= \frac{9 \times 50 \times 3}{5 \times 11 \times 11} = 2.23 \text{ kip/in.}$$

$$q = \sqrt{(2.27)^2 + (2.23)^2} = 3.18 \text{ kip/in.}$$

$$\text{Therefore} \quad D = \frac{3.18}{\frac{\sqrt{2}}{2} \, 21} = 0.215 \text{ in. or } 1/4 \text{ in. fillet}$$

Method 2: Reduction of bending contribution

$$q_v = 2.27 \text{ kip/in.}$$

$$q_x = \frac{P/2^e \; L/2}{L^3/12} = \frac{50 \times 3 \times 3}{11 \times 11} = 3.72 \text{ kip/in.}$$

$$q = \sqrt{(2.27)^2 + 0.2(3.72)^2} = 2.82 \text{ kip/in.}$$

$$\text{Therefore} \quad D = \frac{2.82}{\frac{\sqrt{2}}{2} \, 21} = 0.19 \text{ in. or } 1/4 \text{ in. fillet}$$

In either case the end return must be at least 2D (AISC Section [1.17.9])

Stiffened Beam Seats. When the beam reactions are large, the required thickness of the seat angle may exceed that of available sections. It is then convenient to use a stiffened beam seat. The stiffened beam seat can be formed easily by welding two plates together or by using a piece of standard structural tee. The stiffened seat can be used with the web of the beam either parallel or perpendicular to the stiffener of the seat as indicated in Fig. 19.19. When the beam web is parallel to the stiffener, generally the reaction is assumed to act at the midpoint of the length required for bearing, measured from the end of the seat rather than from the end of the beam (Fig. 19.19a). The bracket is usually short enough so that local buckling of the stiffener is not critical, but the thickness of the stem t should not be thinner than the web of the beam which it supports. Article 17.7 indicates a method for checking the stem's stability. The design of a stiffened beam seat is illustrated in Example 19.6.

19.8 COLUMN BASES

There are several points of concern in the design of column bases. One is the design of the base plate area so that the actual bearing stresses on the concrete foundation will be less than the permissible stresses allowed by the specifications. The bending stresses in the plate must be considered. Another consideration is the design of the anchorage. The column must be

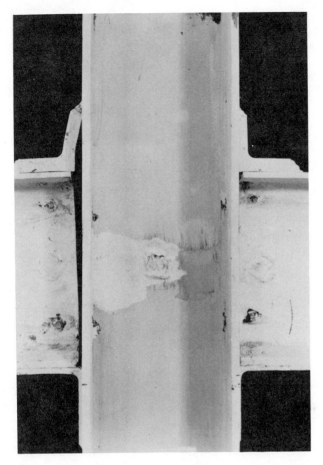

Fig. 19.17 Behavior of framing angles.

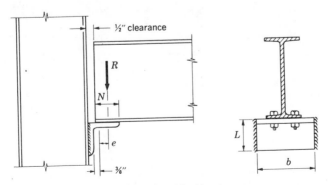

Fig. 19.18 Flexible beam seat.

EXAMPLE 19.5

PROBLEM:

Design the weld required to support the seat angle of Example 18.5. A $\frac{7}{16}$ in. thick angle with a $3\frac{1}{2}$ in. horizontal leg can be used. Determine the length of the vertical leg and the size of weld. Assume the seat is 8 inches wide.

SOLUTION:

The length of the vertical leg and the size of the weld can be obtained from applying the principles described in Article 19.4. The forces on the weld group result from the shear R and the bending moment Re. Hence

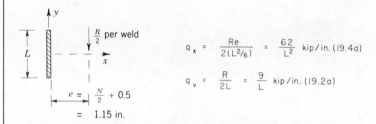

$$q_x = \frac{Re}{2(L^2/6)} = \frac{62}{L^2} \text{ kip/in. (19.4a)}$$

$$q_v = \frac{R}{2L} = \frac{9}{L} \text{ kip/in. (19.2a)}$$

$$e = \frac{N}{2} + 0.5$$

$$= 1.15 \text{ in.}$$

The resultant force on the weld is the vector addition without any decrease in the bending components, because of the flexibility of the angle.

$$q = \sqrt{q_x^2 + q_v^2} = \frac{1}{L^2}\sqrt{62^2 + (9L)^2}$$

The size of fillet weld is given by equating the resultant q to the allowable shear. For E 70 electrodes this becomes

$$\frac{1}{L^2}\sqrt{62^2 + (9L)^2} = 21.0\sqrt{2}\,\frac{D}{2}$$

The equation contained two unknowns L and D. One must be assumed and the other solved for. If L is taken as 5 in.,

$$D = 0.20 \text{ in.}$$

Use 1/4 in. fillet weld.   <u>O.K.</u>

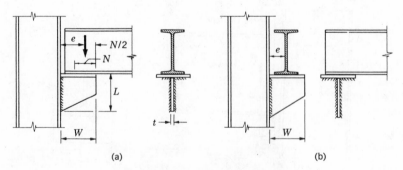

(a)   (b)

Fig. 19.19 Stiffened beam seats.

EXAMPLE 19.6

PROBLEM:

The beam reaction in Example 19.5 is increased to 34 kips. Design a stiffened beam seat for the larger reaction.

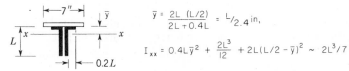

$34\ k = R$

W 10 × 21
$t_w = 0.21$ in.
$k = 0.69$ in.
$A36$ steel

SOLUTION:

From Eq. 17.32

$$N = \frac{R}{27t_w} - k = \frac{34}{27(.24)} - 0.69 = 4.55\,\text{in.}$$

$$W_{min} = N + \tfrac{1}{2} = 5.05\,\text{in.} \qquad \boxed{\text{Use: } W = 6''}$$

The eccentricity of the load is

$$W - N/2 = 6 - 4.55/2 = 3.73\,\text{in.}$$

The length of bearing, N, was chosen so that the beam web was satisfactory as far as web crippling is concerned. The initial web thickness of the web of the beam seat can be assumed as thick as the beam web. Its stability should be checked as indicated in Art. 17.7. Try a 5/8 in. plate.

$$\bar{y} = \frac{2L\ (L/2)}{2L + 0.4L} = \frac{L}{2.4}\,\text{in,}$$

$$I_{xx} = 0.4L\bar{y}^2 + \frac{2L^3}{12} + 2L(L/2 - \bar{y})^2 \sim 2L^3/7$$

The stresses due to bending and shear in the weld can be computed by treating a unit thickness of weld. The width of weld at the top of the tee can be assumed to be about 0.4L. Tables are available which give the allowable loads on such a weld group. As the stem of the seat is bearing against the column, the tensile stresses at the top are considered most critical.

Due to bending $\qquad q_x = \dfrac{Mc}{I_{xx}} = \dfrac{34(3.73)(L/2.4)}{2L^3/7} = \dfrac{185}{L^2}\,\text{kip/in.}$ (19.4)

Due to shear $\qquad q_v = \dfrac{P}{A} = \dfrac{34}{2.4L} = \dfrac{14.2}{L}\,\text{kip/in.}$ (19.2)

The resultant force on the weld is

$$q = \sqrt{q_x^2 + q_v^2} = \frac{1}{L^2}\sqrt{(185)^2 + (14.2L)^2}$$

For E 70 electrodes, the allowable shear stress is 21 ksi. Hence, Eq. 19.5a can be equated to q

$$q = S = \frac{21\sqrt{2}}{2} = \frac{1}{L^2}\sqrt{185^2 + (14.2\,L)^2}$$

This equation contains two unknowns L and D. It is necessary to assume one.

If $L = 8$ in.; $D = 0.23$ in.

$$\boxed{\text{Use } D = \tfrac{1}{4}\text{ in.}}$$

anchored to the plate and the plate in turn anchored to the concrete foundation.

1. Axial Loads

The base plate itself must be designed to distribute the concentrated column forces into the concrete foundation. The thickness and area of the base plate are found by considering the extended portions outside of the column dimensions to act as cantilever beams with their fixed end just inside the column edges. When only axial loads are present the bearing pressure is assumed uniform and equal.

A more sophisticated analysis could be made; but in view of the uncertainties in the actual bearing pressure, added refinement does not seem necessary. The bearing pressure would cause plate bending which would decrease the bearing pressure near the edges of the plate and reduce the bending moments. Hence, this approach can be considered as an upper bound.

The following description illustrates the AISC method of proportioning the plate.[7.9] The allowable bending stress on the extreme fibers of the rectangular bearing plate is taken as $F_b = 0.75F_y$. The increase in plate bending stress is permitted by AISC specification in recognition of the large shape factor of the rectangular cross section and the conservative assumption of a uniform bearing pressure.

The dimensions and reactions are given in Fig. 19.20. The steps in proportioning the plate are as follows:

1. Determine the required area $A = R/p$.
2. Determine B and C so that the dimensions m and n are approximately equal.
3. Determine the thickness of the plate assuming a cantilever beam uniformly loaded.

The resulting plate moment is

$$M = \frac{pm^2}{2} \text{ or } \frac{pn^2}{2} \quad \text{(whichever is larger)} \quad (19.9)$$

The required plate thickness can be obtained from simple beam theory:

$$F_b = \frac{M}{S} = \frac{pm^2}{2}\left(\frac{6}{t^2}\right) \quad (19.10)$$

or

$$t^2 = \frac{3pm^2}{F_b} = \frac{3pm^2}{0.75F_y} \quad (19.11)$$

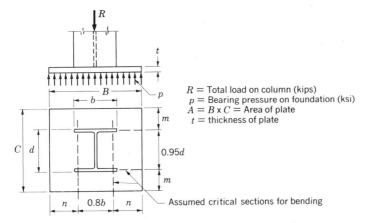

Fig. 19.20 Column base dimensions and loads.

For A36 steel

$$t^2 = \frac{pm^2}{9} \tag{19.12}$$

Typical details for simple column bases are illustrated in Fig. 19.21. These are applicable to columns which are assumed to be pinned at the base. The anchor bolts are used to resist the shearing forces which are acting on the column base. The base may be welded to the column in the shop or in the field. The design of the base plate is illustrated in Example 19.7. Simple column bases will actually carry some bending moment and are useful in resisting frame instability, as was discussed in Chapter 10.

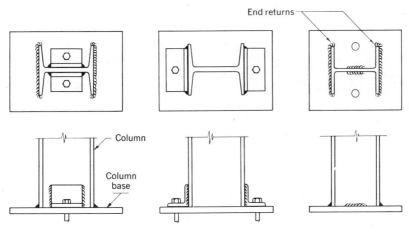

Fig. 19.21 Simple column bases.

EXAMPLE 19.7

PROBLEM:

A W 10 x 49 column has a reaction of 210 kip and rests on a concrete foundation with an allowable bearing stress of 0.650 ksi. Design a steel base plate to support the column.

SOLUTION:

The required bearing area is $A = \dfrac{210 \text{ kip}}{0.65 \text{ ksi}} = 324 \text{ in.}^2$

A plate measuring 18 x 18 in. will provide the necessary bearing area. From Fig. 19.15 the values of m and n can be obtained for this column. Hence,

$$2m = 18 - 0.95d = 18 - 9.5$$
$$m = 4.25 \text{ in.}$$
$$2n = 18 - 0.8b = 18 - 8$$
$$n = 5.0 \text{ in.}$$

Since n is greater than m, the maximum bending moment is along the flange tips

$$t^2 = \frac{1}{9}(0.65)(5)^2 = 1.81 \text{ in.}^2 \qquad\qquad (19.12)$$
$$t = 1.35 \text{ in.} \qquad \boxed{\text{Use } 1\tfrac{1}{2}\text{-inch Plate}}$$

2. Column Bases to Develop Moment

Columns frequently are required to carry end moments because of eccentric loads or base fixity. The column base connection must be designed to transfer these loads to the footing, and must be designed to carry the end moments as well as the axial force. The end moment normally is developed and carried into the concrete footing by means of anchor bolts, as illustrated in Fig. 19.22.

As long as the bending moment is small so that compressive stresses are always present between the column base and the concrete foundation, the base plate design can follow the procedure outlined in Example 19.7.

When the bending moment is large, a continuous pressure distribution is not possible because tensile stresses cannot be developed between the column base plate and the concrete foundation. The tensile force required to maintain equilibrium is provided by the anchor bolt. The problem can be considered analogous to a reinforced concrete beam-column, the anchor bolt being in this case the reinforcing steel.

The base is subjected to the simultaneous application of axial load, bending moment, and preload in the bolts. However, the tension in the bolts is not increased by the bending moment until the stress due to bending exceeds the preload stress. The phenomenon of preloaded bolts is discussed in Art. 18.4.

The design of a moment-resistant base plate is illustrated in Example 19.8.

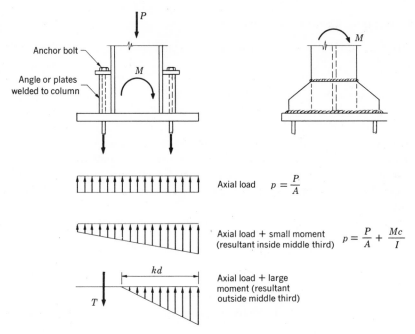

Anchor bolt

Angle or plates
welded to column

$$P$$

$$M$$

$$M$$

Axial load $\quad p = \dfrac{P}{A}$

Axial load + small moment
(resultant inside middle third) $\quad p = \dfrac{P}{A} + \dfrac{Mc}{I}$

kd

T

Axial load + large
moment (resultant
outside middle third)

Fig. 19.22 Moment-resistant column bases.

19.9 RIGID FRAME CONNECTIONS

Rigid frame connections are AISC Type 1 construction, and the AISC Specification recognizes both plastic and allowable stress design methods for the proportioning of the members. Both rectangular and haunched connections are considered in this article.

Rigid frame joints have been used extensively during the past thirty years; however, their design was generally based on rules derived from experience. Several methods of analysis were developed but were tedious to use.

Much research was conducted during the past fifteen years on the behavior of connections. However, the advent of the philosophy of plastic analysis provided a significant impetus toward better design procedures. The plastic design methods are in general simpler and more rational than the elastic methods based on the analysis of plates and plate bending. Reference 1.1 contains a summary of this work. The principal requirements for connections are:

1. Sufficient strength
2. Adequate rotation capacity
3. Over-all stiffness for maintaining the location of all structural units relative to each other
4. Economy of fabrication

1. Straight Corner Connections

Straight corner connections usually are formed by joining two rolled sections. The straight corner connection is easy to fabricate and is used most widely.

Several methods of analysis for the allowable stress design of straight connections are available.[19·15,19·16,19·17] The allowable stress design of straight connections is not covered here; it is given in the references cited. It is recommended that the connections for frames designed on an allowable stress basis be proportioned, using the plastic methods which follow.

The plastic analysis of a rectangular unreinforced corner connection will be considered here because it is more critical than the connection whose members do not meet at right angles.

The following assumptions are made in the analysis of the knee:[19·18]

1. Normal stresses caused by bending moment and thrust are carried by the flanges.
2. Shear stresses are carried by the web.

It is readily apparent from Fig. 19.23 that the tensile force in the outer flange of the beam is carried into the web by shear and similarly for the tensile

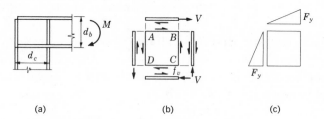

(a)   (b)   (c)

Fig. 19.23 Forces acting on a square knee.

force in the outer flange of the column. In each case, the tensile stress in the flange is assumed to be linearly reduced from F_y at the edge of the corner B or D to zero at the external corner.

If the members AB and CD are the flange sections of the beam, then no further consideration of them is necessary, since their strength is sufficient for the beam, and hence sufficient for the knee. Local flange buckling will be less critical within the knee and no check is necessary if the beam has been checked. For convenience, AD should be either a continuation of the column flange or butt-welded to it. If a portal frame connection has a continuous beam, the member BC transfers the compression force of the column flange into the web of the beam. Normally a half-depth stiffener is used at BC; tests have shown this to be adequate.[19·18,19·19,19·20]

EXAMPLE 19.8

PROBLEM:

A W 14 × 87 column has an axial load of 20 kip and the end moment is 130 kip-ft. Design the anchorage and base plate. Use the AISC Specification and assume $f_c = 1.0$ ksi.

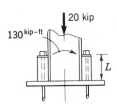

SOLUTION:

Equilibrium leads to

$$\sum F_v = T + 20 - f_c B \frac{kd}{2} = 0$$

$$\sum M = 9.5T + f_c B \frac{kd}{2}\left[13.5 - \frac{kd}{3}\right] - 1560 = 0$$

Substituting $f_c = 1.0$ ksi and $B = 24$ in. gives

$$\sum F_v = T - 12kd + 20 = 0$$
$$\sum M = 9.5T + 162kd - 4(kd)^2 - 1560 = 0$$

The solution of these two simultaneous equations gives.

$$kd = 7.1 \text{ inches and } T = 65 \text{ kip}$$

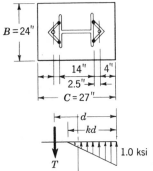

Required area of A307 bolt is

$$A_b = \frac{65}{14} = 4.64 \text{ in.}^2$$

Use $2\frac{1}{2}$ in. diameter bolts, $A_s = 4.91$ in.^{2}

If the plate is assumed critical near the edge of the column flange the maximum moment is

$$M = 0.155 \frac{(6)^2}{2} + 0.845 \frac{(6)^2}{3} = 12.9 \text{ kip-in.}$$

$$t^2 = \frac{6M}{F_b} = \frac{6 \times 12.9}{27} = 2.87, \quad t = 1.70 \text{ in.}$$

Use $1\frac{3}{4}$ in. Plate

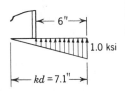

Angles or plates can be used as anchors when welded to the column flange. A 6"×6"×$\frac{1}{2}$" angle will provide sufficient clearance for the bolt. The amount of welding necessary to resist the bolt forces can be determined by considering these forces to act on the weld groups. The force in the bolt is $T = 65$ kip. The weld on each side of the angle can be treated as a line. Hence

$$q_x = \frac{Te}{2\frac{L^2}{6}} = 65(2.5)\frac{3}{L^2} = \frac{487.5}{L^2} \text{ kip/in.} \quad (\quad .4)$$

$$q_v = \frac{T}{2L} = \frac{65}{2L} = \frac{32.5}{L} \text{ kip/in.} \quad (19.2)$$

The resultant shear force on the weld is

$$q = \sqrt{0.2q_x{}^2 + q_v{}^2}$$

$$q = \frac{1}{L^2}\sqrt{0.2(487.5)^2 + (32.5L)^2}$$

Equating the resultant shear force to Eq. 19.5a gives

$$q = S = \frac{21\sqrt{2}}{2} = \frac{1}{L^2}\sqrt{0.2(487.5)^2 + (32.5L)^2}$$

Since there are two unknowns L and D either a weld size D or a length L must be assumed. If $L = 10$ inches, then

$$D = \frac{1}{1485}\sqrt{0.2(487.5)^2 + (325)^2} = 0.264 \text{ inches}$$

Use $\frac{5}{16}$ in. fillet weld

Note: The reduction in the bending component was discussed in Art. 19.3
Since the column flange is 11/16 in. thick, the minimum fillet size is 1/4 in., hence the 5/16 in. fillet is O.K.

From the assumption that moment is carried by the flanges, consideration of equilibrium of the forces acting on the top flange gives an expression for the web thickness necessary to resist shear. The moment M_p is resisted by a couple having a moment arm equal to $0.95d_b$ since the flange thickness is approximately $0.05d_b$.

$$V = M_p/0.95d_b \qquad (19.13)$$

The maximum web shear-resisting force between A and B must not exceed

$$V = \frac{0.95F_y w d_c}{\sqrt{3}} \qquad (19.14)$$

Equating this to the full flange force, there is obtained

$$\frac{M_p}{0.95d_B} = \frac{0.95F_y w d_c}{\sqrt{3}}$$

and

$$w = \frac{\sqrt{3}\,M_p}{0.90d_c d_b F_y} \qquad (19.15)$$

If this required thickness is greater than the thickness provided, it is necessary to provide a diagonal stiffener or doubler plate to prevent undesirable shear deformation in the web. Normally, a stiffener is more practical. It acts like the diagonals of a truss panel in preventing shear deformation, as indicated in Fig. 19.24.

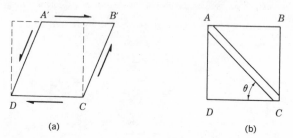

(a)   (b)

Fig. 19.24 Shear distortion in frame connection.

The stiffener area is obtained by considering equilibrium of the top flange. The flange force V must be resisted by the web shear and the force in the stiffener. Hence, equating Eq. 19.13 to Eq. 19.14 gives the resistance of the stiffener:

$$\frac{M_p}{0.95d_b} = \frac{0.95F_y w d_c}{\sqrt{3}} + F_y A_{st} \cos \theta$$

$$A_{st} = \frac{1}{\cos \theta}\left[\frac{M_p}{0.095d_b F_y} - \frac{0.95w d_c}{\sqrt{3}}\right] \qquad (19.16)$$

The results of connections proportioned by this method and then tested are shown in Fig. 19.25. They show the connections to possess adequate strength, stiffness, and rotation capacity. The shaded zone at $M_h/M_y = M_p$ indicates the variation in the section properties of the five wide-flange shapes that were tested. The plastic moment M_p refers to the rolled wide-flange shape.

The design of a straight corner connection is illustrated in Example 19.9. The example shows the design of the member and welds.

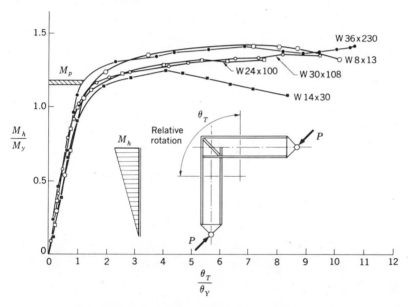

Fig. 19.25 Tests of straight corner connections.

2. Haunched Connections

Haunched corner connections are used primarily for esthetic considerations. However, in many cases the use of haunches may also achieve a more economical design. The increased fabrication costs must be considered together with the apparent savings in weight when considering the use of haunched connections for economical reasons.

Allowable stress design attempts to adapt the haunched member to the shape of the moment diagram. The most effective use is made of the material, with a resulting reduction in weight. In plastic design similar results are obtained as the maximum moments in the rolled sections are reduced and smaller members can be used. The use of typical haunched connections is illustrated in Fig. 19.26.

EXAMPLE 19.9

PROBLEM:
Design the straight corner connection shown and proportion the welds.

Load factor = 1.70

$M = 860$ kip-ft

$M_p = 1.70 \times 860 = 1462$ kip-ft

Use A 36 steel, E 70 electrodes

The AISC specification is applicable:

SOLUTION:
From Eq. 19.15 (AISC Section [2.4])

$$w_{req'd} = \frac{(23,000)\,1462}{(36)(36)\,36,000} = 0.784 \text{ in.}$$

$$w_{furn} = 0.625 \text{ in.} < 0.784 \text{ in.}$$

USE diagonal web stiffener – A 36 steel

Using Eq. 19.16

$$A_{st} = \frac{1}{\cos\theta}\left[\frac{M_p}{0.95\,d_b F_y} - \frac{0.95\,w\,d_c}{\sqrt{3}}\right]$$

$$\theta = 45°; \ \cos\theta = 0.707$$

$$A_{st} = \frac{1}{0.707}\left[\frac{1590(12)}{0.95\,(36)(36)} - \frac{(0.95)\,0.625\,(36)}{\sqrt{3}}\right]$$

$$A_{st} = 4.5 \text{ in.}^2$$

DIAGONAL STIFFENER: USE – 2 ℝs, ¾ x 3 in.
One plate on each side of the web

Design of welds:
Weld design is covered extensively in Ref. 19.8. Only the major details are covered here.

Butt welds – Design to develop full plastic yield strength of column flanges.

Fillet welds – (See welds A, B, C, and E above)

Allowable weld stress = $0.5\sigma_f = 0.5 \times 70 = 35$ ksi

Weld strength (kip) = 35 (D cos 45°)(L) = 24.7 DL

Weld A – Fillet welds for end plate to beam web
The welds must transmit the plastic flange force from the column into the beam web by shear. The plastic flange force is

$F = F_y\,b\,t = 36\,(11.97)\,0.94 = 405$ kip, therefore, size fillet weld required

is $D = \dfrac{405}{24.7\,[2(d-2t)]} = \dfrac{405}{24.7\,[2(33.96)]} = 0.243$ in.   Normal size
5/16 in.*

Weld B – Column web fillet weld to beam flange
These welds are designed to develop the combined tensile forces in bending, the most severe axial force component in the web, and the shear force from the transverse component of load. The direct tensile or compressive force per inch of weld due to axial load and bending is

$$q_D = F_y\left(\frac{W}{2}\right) = 36\,(0.3125) = 11.25 \text{ kip/in.}$$

*Since column flange is 15/16 in. thick, minimum fillet weld size is 5/16 in.

EXAMPLE 19.9 (Continued)

Force per inch due to shear

$$q_v = \frac{V}{2(d-2t)} = \frac{1.85(80)}{2(33.96)} = 2.12 \text{ kip/in.}$$

Resultant force per inch

$$q = \sqrt{q_D{}^2 + q_v{}^2} = \sqrt{(11.25)^2 + (2.12)^2} = 11.45 \text{ kip/in.}$$

Fillet weld size required

$$D = \frac{q}{0.707(35)} = \frac{11.45}{0.707(35)} = 0.466 \text{ in.}$$

| Nominal Size |
| $\frac{1}{2}$ in. |

Note: The use of a complete penetration groove weld may be more economical and would eliminate the need for the above analysis.

Weld C - Fillet welds between beam web and stiffener

These welds must develop stiffener strength   Stiffener = $2\frac{1}{2} \times \frac{1}{2}$ in.

$$24.7 \, DL = 2.5(0.5)36 = 45 \text{ kip}$$

$$D = \frac{45}{24.7\left[2\left(\dfrac{d-2t}{0.707}\right)\right]} = \frac{45}{24.7\left[2(48.1)\right]}$$

$$D = 0.0190 \text{ in.}$$

| Nominal Size |
| $\frac{1}{4}$ in. |
| minimum |

Weld E - Fillet welds between vertical half-depth stiffener and web of beam

Flange force to be carried by the web is

$$P_w = F_y w(t + 2k) = 36(0.625)(0.940 + 3.625) = 103 \text{ kip}$$

The force in the vertical stiffener is

$$F_s = F - P_w = 405 - 103 = 302 \text{ kip}$$

Fillet weld size required is

$$D = \frac{F_s}{4(0.707)\,35\left(\dfrac{d}{2}-k\right)} = \frac{302}{4(0.707)\,35(16.11)} = 0.191 \text{ in.}$$

| Nominal Size |
| $\frac{1}{4}$ in. |

Tests have indicated that haunches designed on the basis of an elastic analysis (tapered or curved) normally have sufficient strength but lack satisfactory rotation capacity to function at ultimate load.[19.18]

A number of theoretical solutions are available for the elastic design of tapered and curved haunches. However, they are usually too unwieldy for everyday use. The reader is referred to Refs. 19.16, 19.17, 19.21, and 19.22 for elastic analysis of haunches.

Only the development and application of the plastic design method is covered here. This method was developed in Ref. 19.23 and has been summarized in Ref. 1.1.

Tapered Haunches. The following items must be considered when proportioning the haunch: (1) resistance to bending in the tapered portion,

a. Courtesy of Lincoln Electric Co.

b. Courtesy of Bethlehem Steel Corp.

Fig. 19.26 Typical application of haunched connections.

(2) resistance to local and lateral buckling, and (3) shear stresses in the web and flange forces in and around the knee panel.

The assumptions made in the analysis are:

1. Plane sections remain plane.
2. Equilibrium of loads and moments.
3. A linear moment relationship over the haunch length.

The influence of axial thrust and shear on the plastic moment usually is neglected.   If $P/P_y > 0.15$ the same reduction can be used as applied to the beam.[19·23]

A typical tapered knee is shown in Fig. 19.27 together with a portion of the moment diagram and the assumed stress distributions.

The moment, thrust, and shear acting on the knee are known from the analysis of the structure.   The web thickness should be made equal to the

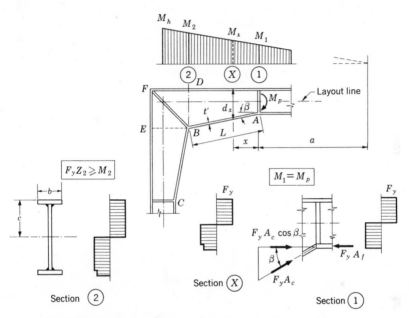

Fig. 19.27 Forces and stresses acting on tapered knees.

adjoining members to ensure that the web is able to carry at least as much shear and axial force as the web of the adjoining member.

The strength of the haunch at any point must equal or exceed the applied moment at that section, determined from the plastic analysis.   Hence,

$$M_{px} = F_y Z_x \qquad (19.17)$$

The stress distribution at any cross section is assumed as shown in Fig. 19.27.   The neutral axis obtained by considering equilibrium is

$$F_y(bt' + cw) = F_y bt' \cos \beta + F_y(d_x - c)w$$

$$C = \frac{d_x}{2} - \frac{bt'}{2w}(1 - \cos \beta) \qquad (19.18)$$

A simplification can be made at this point by observing that an increase of the sloping flange area by the factor $1/\cos\beta$ restores the symmetry of the section and gives $c = d_x/2$. As a result, the plastic modulus defined as $\int_A y\, dA$ can be expressed as

$$Z_x = bt'(d_x - t') + \frac{w}{4}(d_x - 2t')^2 \tag{19.19}$$

The required flange thickness can be obtained from Eq. 19.17 by substitution. The sloping flange should be increased by $1/\cos\beta$. However, this can normally be neglected because, for $\beta < 20°$, a negligible increase is required.

A uniform member will have a constant plastic modulus. However, this is not true for the tapered knee. Hence, the varying plastic modulus furnished must be adequate to resist the varying moment along the haunch. As illustrated in Fig. 19.28, this can be achieved by maintaining the effective flange area the same as that of the rolled section and by varying the angle β.

Under the moment gradient at ultimate load, with $\beta \approx 12°$, the haunch is plastic along most of its length.[19.23] If $\beta < 12°$ the ultimate strength is

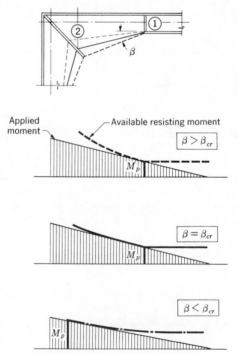

Fig. 19.28 Effect of changing the haunch angle.

exceeded by the varying moment at or near the maximum haunch depth before a hinge can form at the intersection of the haunch and rolled section. If $\beta > 12°$ the haunch is only partially plastic, with the hinge action localized at the intersection of the haunch and rolled section.

In order to attain the required plastic strength, the stability of the compression flange is of utmost importance. Lateral buckling of the tapered haunch depends on the slenderness of the compression flange and on the extent it has yielded and strain-hardened. The web of the haunch provides restraint against buckling about its weak axis; hence the flange must buckle in its strong direction between points of support.

An approximation of the critical buckling length was obtained, using the tangent modulus buckling concept. It was assumed that the flange was uniformly stressed with strains reaching strain-hardening and that it buckled as a pinned-end column.[19·23] This resulted in

$$\left(\frac{L_{cr}}{r_x}\right)^2 = \frac{\pi^2 E_{st}}{F_y} \tag{19.20}$$

Observing that $r_x = b/\sqrt{12}$ for a rectangular plate and that some restraint is offered by adjacent portions of the flange so that an effective length factor, K, of 0.8 can be used, and substituting appropriate values for F_y and E_{st} in Eq. 19.20 gives

$$L_{cr} = \frac{b\pi}{K}\sqrt{\frac{E_{st}}{12F_y}} \tag{19.21}$$

By substituting the mechanical properties for the appropriate grade of steel, the critical length is obtained. For A36 steel, the critical length would be

$$L_{cr} \approx 6b \tag{19.22}$$

Often this critical length may be less than the length desired for the design. Because this is a stability criterion, additional support can be provided or the flange thickness can be increased. Increasing the points of support decreases the buckling length. If the flange thickness is increased, the flange strains are in turn decreased and the buckling condition is not as critical. A means of increasing the flange thickness to control the strain was proposed in Ref. 19.23. It was found that the required plastic strength can be achieved, if L_{cr} is exceeded, by increasing the flange thickness an additional increment Δt. The increment Δt was found to be

$$\Delta t = 0.1\left(\frac{L}{b} - 6\right) \tag{19.23}$$

where L is the length of the compression flange and b is the width.

Equations 19.22 and 19.23 have resulted from the most severe condition to which a haunch can be subjected. The entire compression flange was assumed to yield. If the angle β is greater than approximately 12°, a combination elastic-plastic condition exists along the flange length. The beneficial effect of the decrease in the yielded zone of the compression flange can be taken into account by considering the lateral buckling provision presented in Art. 7.6. Hence,

$$L_{cr} = (60 - 40M/M_p)r_x\sqrt{\frac{36}{F_y}}$$

where r_x is taken as $b/\sqrt{12}$. The moments M and M_p can be expressed as functions of stress and the section properties, which results in

$$L_{cr} \approx \left(17.5 - \frac{11.5}{F_y f}\right)b\sqrt{\frac{36}{F_y}} \qquad (19.24)$$

where f is the computed stress based on the section modulus at the haunch end of the compression flange and is always less than the yield point.

If the bending stress computed on the basis of the section modulus is less than the yield point at all transverse sections, the lateral buckling provisions given in Art. 7.6 should be used.

The stiffeners can be proportioned by assuming that they resist the flange forces in a truss type action. For the most critical condition the resulting stiffener area is approximately 75 per cent of the flange area.

The design of a typical haunched connection is illustrated in Example 19.10.

Curved Haunches. The analysis and design of curved haunches is identical in many respects to that of the tapered haunch. The three main problems to be considered are (1) resistance to bending in the haunch, (2) resistance to local and lateral buckling, and (3) shear deformation. As with tapered knees, the influence of axial thrust and shear on the plastic moment may be neglected.

A typical curved knee is shown in Fig. 19.29 together with the assumed moment and stress diagrams.

If the curved flange thickness is increased at the critical section x by the factor $1/\cos \beta$, the analysis is simplified. This assumption is justifiable because the controlling angle β is small. The plastic moment at the critical location is

$$M_{px} = F_y Z_x \qquad (19.25)$$

where Z_x is equal to $bt_x(d_x - t_x) + w/4(d_x - 2t_x)^2$ and d_x is $d + R(1 - \cos \beta)$. The required flange thickness is obtained by substituting in Eq. 19.25,

$$t = \frac{d_x - \sqrt{d_x^2 \dfrac{b}{b-w} - \dfrac{4M_{px}}{F_y(b-w)}}}{2} \qquad (19.26)$$

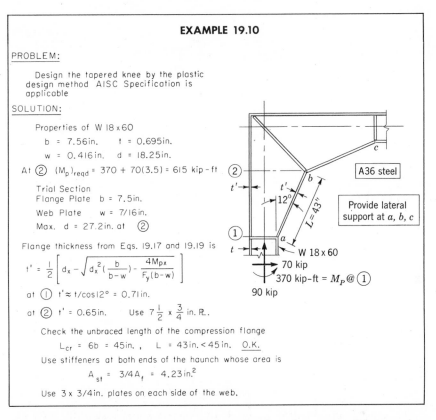

EXAMPLE 19.10

PROBLEM:

    Design the tapered knee by the plastic design method AISC Specification is applicable

SOLUTION:

    Properties of W 18 x 60

       b = 7.56in.   t = 0.695in.

       w = 0.416in.   d = 18.25in.

    At ② $(M_p)_{reqd}$ = 370 + 70(3.5) = 615 kip-ft

    Trial Section
    Flange Plate   b = 7.5in.

    Web Plate   w = 7/16in.

    Max.   d = 27.2in. at   ②

    Flange thickness from Eqs. 19.17 and 19.19 is

$$t' = \frac{1}{2}\left[d_x - \sqrt{d_x^2\left(\frac{b}{b-w}\right) - \frac{4M_{px}}{F_y(b-w)}}\right]$$

    at ① $t' \approx t/\cos 12° = 0.71$in.

    at ② $t' = 0.65$in.   Use $7\frac{1}{2} \times \frac{3}{4}$ in. Pʟ.

    Check the unbraced length of the compression flange

       L_{cr} = 6b = 45in. ,   L = 43in. < 45in.   <u>O.K.</u>

    Use stiffeners at both ends of the haunch whose area is

       A_{st} = 3/4A_f = 4.23 in.^{2}

    Use 3 x 3/4in. plates on each side of the web.

Generally the flange has the same width as the adjacent straight member. If Eq. 19.26 is maximized with respect to the angle β the critical section is found to be at $\beta \approx 12$ degrees, and this holds true for a wide range of moment gradients and haunch radii.

The relationship between moment gradient and flange thickness can be obtained from Eq. 19.26 and the function $\partial t_x/\partial \beta$. If the flange thickness of the rolled section is assumed to be $0.05d$ and the web thickness $0.05b$, the relationship shown in Fig. 19.30 is obtained; t is the thickness of the beam flange and t' the required thickness of the haunch flange, and the dimensions a and d are defined in Fig. 19.29. The required thickness of the haunch is

$$t_x = t' = (1 + m)t \tag{19.27}$$

The lateral stability of the curved flange is treated in much the same manner as the compression flange of the tapered knee.[19.23] The critical arc length is

$$K\frac{R\phi}{r_x} = \pi\sqrt{\frac{E_{st}}{F_y}} \tag{19.28}$$

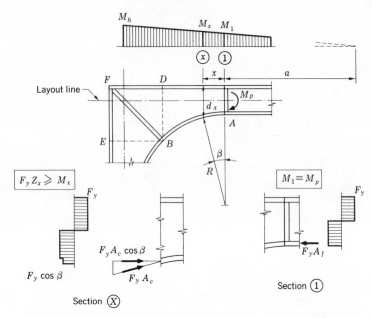

Fig. 19.29 Forces and stresses acting on curved knees.

For A36 steel this results in

$$R\phi = L_{cr} \approx 6b \qquad (19.29)$$

The same approximation as used for tapered knees results in a similar provision for increasing the flange thickness.

$$\Delta t = 0.1\left(\frac{L}{b} - 6\right) \qquad (19.30)$$

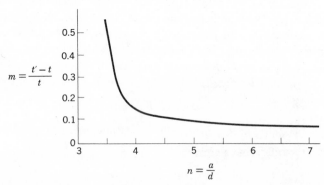

Fig. 19.30 Effect of moment gradient on flange thickness.

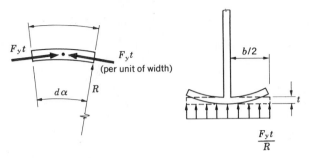

Fig. 19.31   Cross bending in curved flange.

An additional problem not encountered with tapered haunches is the cross-bending of the curved flange which can be considered as indicated in Fig. 19.31; the resulting deformation of the flange is also shown.

When the outstanding flange, acting as a cantilever beam, reaches its plastic moment, the following relationship is obtained:

$$F_y Z = F_y \frac{t}{R} \frac{b^2}{8}$$

where $Z = t^2/4$.   Hence

$$F_y \frac{t^2}{4} = F_y \frac{t}{R} \frac{b^2}{8}$$

or

$$\frac{b^2}{2Rt} \leqslant 1 \tag{19.31}$$

If expressed as a width-thickness ratio, then

$$\frac{b}{t} \leqslant \frac{2R}{b} \tag{19.32}$$

In order to provide restraint to the curved compression flange, stiffeners should be provided at points of tangency with the beam and column.   A stiffener should also be provided midway between the points of tangency to help provide lateral stability as well as restraint for the component of flange force normal to the plane of the web.   Only a nominal thickness is required at the points of tangency.   The midpoint stiffener can be approximated by considering the curved haunch to be analogous to the tapered haunch.   This gives a stiffener area equal to three-quarters of the flange area.

The results of tests of tapered and curved connections are given in Fig. 19.32.   The specimens were designed to test the theoretical concepts and indicated that the connections performed satisfactorily.   They also show the need of adequate lateral support since one specimen (44) was not braced sufficiently.

Example 19.11 shows the design of a typical curved haunch.

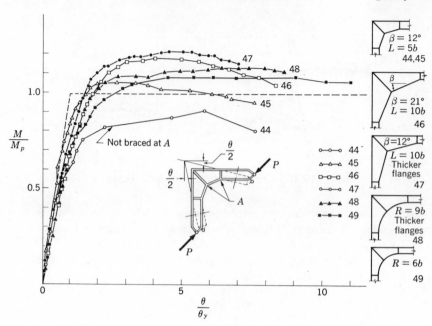

Fig. 19.32   Tests of haunched connections.

19.10   BEAM-TO-COLUMN CONNECTIONS

Several beam-to-column connections are shown in Fig. 19.33. Some of the methods for proportioning beam-to-column connections have evolved from research reported in Ref. 17.24. This study has shown that two factors influence the behavior of beam-to-column connections. These factors are the possible buckling of the column web due to the concentrated action of

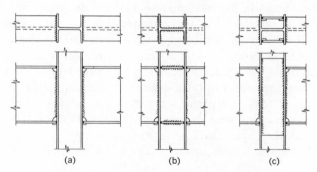

Fig. 19.33   Typical beam-to-column connections.

the compression beam flange and behavior of the tension beam flange butt weld. These two modes of failure are shown in Fig. 19.34.

EXAMPLE 19.11

PROBLEM:

   Proportion the curved knee using the plastic design method. AISC Specification is applicable.

SOLUTION:

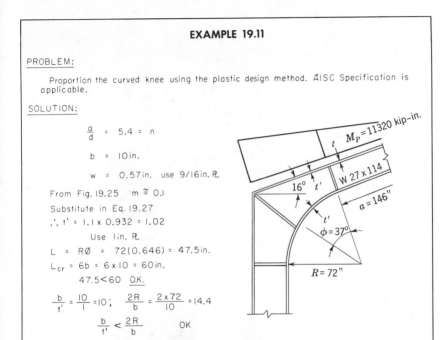

$$\frac{a}{d} = 5.4 = n$$

$$b = 10 \text{ in.}$$

$$w = 0.57 \text{ in.} \text{ use } 9/16 \text{ in. } \mathbb{R}$$

From Fig. 19.25 $m \cong 0.1$

Substitute in Eq. 19.27

$\therefore t' = 1.1 \times 0.932 = 1.02$

   Use 1 in. $\mathbb{R}$

$$L = R\phi = 72(0.646) = 47.5 \text{ in.}$$

$$L_{cr} = 6b = 6 \times 10 = 60 \text{ in.}$$

$$47.5 < 60 \underline{\text{O.K.}}$$

$$\frac{b}{t'} = \frac{10}{1} = 10 ; \quad \frac{2R}{b} = \frac{2 \times 72}{10} = 14.4$$

$$\frac{b}{t'} < \frac{2R}{b} \text{OK}$$

Provide stiffeners 3/4 x 4- 3/4 in. each side of the web at midpoint and at points of tangency.

The buckling of the column web under the compression beam flange is discussed in Art. 17.6. Under the compressive flange force, the column flange was assumed to remain undeformed. This is not necessarily true in the case of the tension flange. The tensile force in the beam flange tends to pull the outstanding column flanges, as shown in Fig. 19.35. The flange bending occurs in two directions, both longitudinal and transverse to the axis of the member. The flange deformation causes the weld to exhaust its ductility and to fracture.

An approximate analysis can be made by assuming the beam flange force to be resisted by the direct resistance of the thick center portion of the column flange and by the column flange bending resistance. Hence

$$F_{yf}A_f = F_{yc}t_b m + F_{yc}C_1 t_c^2 \tag{19.33}$$

The coefficient C_1 was evaluated by analyzing the column flange by means of the yield line theory. The dimensions t_b and m of the beam and column can be approximated by considering these dimensions in common rolled shapes. Upon substituting these values in Eq. 19.33 the following inequality results:

$$t_c \geqslant 0.4 \sqrt{A_f \frac{F_{yf}}{F_{yc}}} \tag{19.34}$$

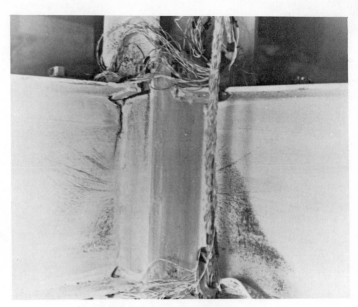

a. Tension flange failure

b. Compression flange failure

Fig. 19.34   Buckling and fracture in beam-to-column connections.

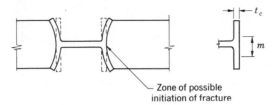

Fig. 19.35 Column flange bending.

Figure 19.36 shows the results of tests on connections designed to satisfy this criterion.

These tests, and tests on connections without stiffening, were used to help evaluate the semi-empirical methods formulated to proportion beam-to-column connections.[17.24] The test results have indicated that this is a reasonable approximation of the connection's strength. Further studies have been made on beam-to-column connections to evaluate the effect of axial load and strain hardening on the connection behavior.[19.24]

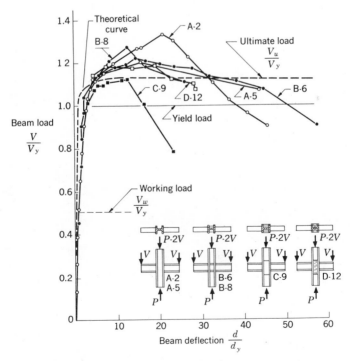

Fig. 19.36 Tests of beam-to-column connections.

EXAMPLE 19.12

PROBLEM:

Design the welds for the beam-to-column connection proportioned in Example 17.5 .

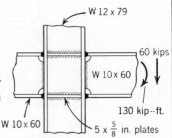

SOLUTION:

(a) If erection can be accomplished with the close tolerances that are required when butt welds are used, the beam flanges can be butt welded to the column flanges to carry the moments, and the fillet welds along the beam web will carry the shear. Only minimum size welds are needed to connect the stiffeners to the column web flanges.

(b) It may be desirable to design the connection welds such that field welding and erection could be done easily and without regard to close tolerances. This can be accomplished by providing stiffened beam seats which are shop welded to the column flanges. The top flange can be provided with a connecting plate which will be field butt welded to the column flanges and field fillet welded to the beam flange. This design is illustrated below.

A 36 Steel
Dimensions W 10 x 60

t_w = 0.415 in.

t_f = 0.683 in.

d = 10.25 in.

b = 10.08 in.

k = 1 $\frac{3}{16}$ in.

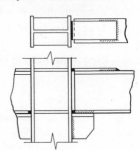

Top flange – Use a connecting plate to resist the flange force

$$F = \frac{M}{d} = \frac{130 \times 12}{10.25} = 152 \text{ kip}$$

area of plate

$$A = \frac{F}{F_b} = \frac{152 \text{ kip}}{22 \text{ ksi}} = 6.9 \text{ in.}^2$$

$\boxed{\text{USE } 1'' \times 7'' \text{ Plate}}$

length of fillet weld required

$$L = F/14.85\,D = 152/14.85\,D = \frac{10.2}{D} \text{ in.}$$

Using 5/8 in. fillet weld would require 16½ inches of weld to connect the plate to the beam flange. The disposition of the weld is shown.

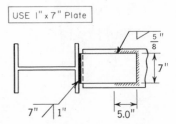

The bottom flange of the beam would be butt welded to the column to resist the bending moment. The stiffened beam seat would be designed to resist the shear force, P as described in Example 19.6.

$$N = \frac{R}{27 t_w} - k = \frac{60}{27(.415)} - 1.19$$

$$= 4.16 \text{ -inches}$$

$$W_{min} = 0.5 + 4.16 = 4.66 \quad \boxed{\text{USE } W = 5 \text{ in.}}$$

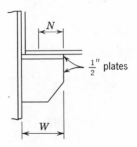

The load eccentricity

$$e = W - N/2 = 5 - 2.08 = 2.92 \text{ in.}$$

The force due to bending, (Eq.19.4)

$$q_x = \frac{60(2.92)\,L/2.4}{2L^3/7} = 256/L^2 \text{ kip /in.}$$

EXAMPLE 19.12 (Continued)

The force due to shear, (Eq.19.2)

$$q_v = \frac{60}{2.4L} = 25/L \quad \text{kips/in.}$$

The resultant force on the weld

$$q = \frac{1}{L^2} \sqrt{(256)^2 + (25L)^2}$$

For E70 electrodes, the resultant q can be equated to Eq. 19.5a

$$q = S = 14.85D = \frac{1}{L^2} \sqrt{(256)^2 + (25L)^2}$$

A length L = 10in. would yield

$$D = 0.241 \text{ in.} \quad \text{Use 1/4 in. fillet welds.}$$

The 3/4 in. thick column flange would require a minimum fillet weld size of 1/4in., hence <u>O.K.</u>

Equations 17.34 and 19.34 indicate whether column stiffeners are required at either the beam compression flange or at the beam tension flange. When stiffeners are required, they can be either horizontal or vertical. Both types may be proportioned by considering the additional amount of resisting force required to achieve equilibrium. They must also maintain stability under their full load.

For horizontal stiffening in both the tension and compression regions, the condition of equilibrium results in[17.24]

$$t_s = \frac{F_{yf}A_f - w(t + 5k)F_{yc}}{b_s F_{ys}} \tag{19.35}$$

For vertical stiffening, tests have indicated that the vertical plates carry about half of the flange forces. Assuming this to hold true, the condition of equilibrium gives

$$t_s = \frac{A_f}{t + 5k} - w \tag{19.36}$$

For many cases local buckling requirements will govern.

For horizontal stiffeners $\quad \dfrac{b_s}{t_s} \leqslant 8.5 \tag{17.24}$

For vertical stiffeners $\quad t = d_c F_y / 190 \tag{17.23}$

The design of a typical beam-to-column connection is illustrated in Example 17.5. The stiffeners were designed to prevent web crippling in the column due to the concentrated flange forces and fracture of the welds connecting the beam tension flange to the column flange. The design of the welds for the beam-to-column connections proportioned in Example 17.5 is illustrated in Example 19.12.

PROBLEMS

19.1. Determine the maximum-size fillet weld and the efficiency of the connection shown; AISC Specification.

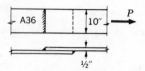

Problem 19.1

19.2. Determine the length of slot a in order to develop the full strength of the channel shown (p. 690); AISC Specification.

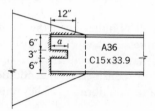

Problem 19.2

19.3. The angle connection shown is to be subjected to alternating loads. Proportion the welds so that the full strength of the angle is realized. Use AASHO Specifications.

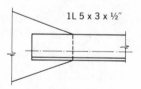

Problem 19.3

19.4. Determine the size of weld required to resist the axial load P (see sketch). Use AISC Specifications.

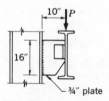

Problem 19.4

19.5. Determine the force P, assuming the weld (see sketch) was made with E70 electrodes; AISC Specification.

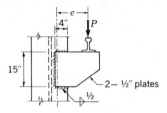

Problem 19.5

19.6. The W 21 × 142 beam shown is subjected to a uniform external load of 17,500 lb/ft and spans 20 ft. A 12-in. cover plate is available to reinforce the beam. Determine the thickness of the cover plate and design the continuous fillet welds connecting the plate to the beam flange. Use AASHO Specifications.

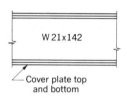

Problem 19.6

19.7. A 4 by 4 by $\frac{3}{8}$ in. steel angle (A36) is used as a tension member in a multi-story building. The angle is connected to a $\frac{1}{2}$-in. gusset plate. Using AISC Specifications, design the welded connection.

19.8. Design a flexible beam seat to support a W 12 × 27 with an end reaction of 27 kips. Use AISC Specifications.

19.9. Design of stiffened beam seat to support an S 20 × 75 beam with an end reaction of 60 kips. The beam is in line with the bracket as shown in Fig. 19.14a. Use AISC Specifications.

19.10. Design a stiffened beam seat to support a S 20 × 85 with an end reaction of 50 kips. The bracket is at right angle to the beam as shown in Fig. 19.14b. Use AISC Specifications.

19.11. Design welded framing angles to support the S 20 × 75 beam having an end reaction of 60 kips. The angles will be shop-welded to the beam and field-welded to the column. Two $\frac{3}{4}$-in. erection bolts will be used. Use AISC Specifications.

19.12. A W 18 × 55 spans 50 ft and supports 550 lb per linear ft. The beam is to frame into W 12 × 79 columns. Design a welded framing connection using AISC Specifications.

19.13. Design a base plate for a W 8 × 31 column having an axial load of 100 kips and supported by a concrete footing with a 28-day strength of 3600 psi. Use AISC Specifications.

19.14. Design a base plate for a W 14 × 87 column having an axial load of 500 kips. The column is supported by a concrete footing with a 28-day strength of 300 psi. Use AISC Specifications.

19.15. Design a column base to support a W 14 × 87 supported by a concrete footing with a 28-day strength of 4000 psi. The column base will be subjected to an axial load of 100 kips and an end moment of 75 kip-ft. (Note: On $\frac{1}{3}$ of area, $F_p = 0.375f_c'$.) Use AISC Specifications.

19.16. Design a square rigid frame knee whose intersecting members are W 24 × 100 (see figure). Design the stiffening and proportion the welds. Use AISC Specifications.

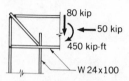

80 kip

50 kip

450 kip-ft

W 24×100

Problem 19.16

19.17. Design a square corner connection for a W 21 × 68 girder to a W 14 × 87 column to develop the ultimate strength of the weaker member. Use AISC Specifications.

19.18. A tapered haunch is to be used in a gable frame (see sketch); the beam and columns are W 14 × 30. Design the tapered knee; find flange thickness and stiffener thickness, using AISC Specifications.

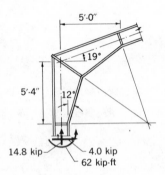

5'-0"

19°

5'-4"

12°

14.8 kip   4.0 kip

62 kip-ft

Problem 19.18

19.19. Redesign the gable frame knee in Problem 19.18, using a curved knee.

20

SINGLE-STORY RIGID FRAMES

20.1 INTRODUCTION

This chapter illustrates the design of a single-story welded rigid frame using both the allowable stress and plastic design provisions of the AISC Specification. Emphasis is given to the behavior of rigid frames and the reasoning involved in applying the AISC Specification.

The advantages of welded fabrication, the pleasing esthetic qualities, the ease and speed of erection, and the research concerned with the plastic behavior of rigid frames, all have contributed to the economy and popularity of this type of structure. Figures 3.15 and 3.16 indicate the flexibility and variety of rigid frame construction. In spite of this variety of form, the ideas and methods used in rigid frame design vary only in minor details. These methods are illustrated in this chapter by the design of a single-span rigid gable frame of the type shown in Fig. 20.1.

Three decades ago the rigid frame was described as a "beam arch" since the rafters and columns were stressed both in bending and axial compression.[20.1] Research has verified that bending is the dominant structural action in single-story frames, up to and including ultimate load. Thus the overload behavior of the rigid frame is similar to that of a continuous beam. When overloads are applied, plastic hinges form at the most highly stressed sections of the frame. If these hinge locations are braced to prevent lateral and torsional movement, they rotate under constant moment. This results in a considerable increase in ultimate load capacity by permitting redistribution of moment to other sections of the frame.

The conscious use of plastic behavior is one of the more significant differences between modern rigid frame design and the "beam arch" design of previous years.

693

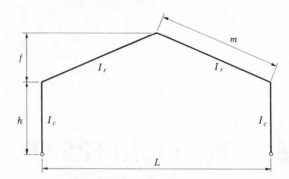

Fig. 20.1   Rigid gable frame.

20.2   DESIGN METHODS

The rigid frame considered in this chapter is classified as Type I construction in AISC Section [1.2].   (The Type I classification is discussed in Art. 21.2.)   Two different design methods are recognized.   These are allowable stress design in Part 1, and plastic design in Part 2 of the Specification.

The allowable stress method is based on working loads and allowable stresses.   Chapter 1 discusses some of the criteria used in establishing allowable stresses.   In many instances, these criteria involve the ultimate load capacity of the member as limited by "unrestrained plastic flow" or by stability.   These criteria represent a better utilization of the strength of steel structures than was possible in designs based upon earlier editions of the AISC Specification, and are indicative of progress in the art of structural engineering.   These criteria also place emphasis on bracing requirements and on careful attention in design, detailing, and fabrication to preserve ductility and stability.

In an allowable stress design an elastic analysis is used to determine the bending moments in the frame.   After modifying this analysis for moment redistribution (see Arts. 7.8 and 20.3), smaller critical design moments are obtained.   As a result, smaller members are required, thus taking partial advantage of the plastic strength available in ductile steel structures.

Methods of elastic analysis are treated in many references and will not be explained here.   The rigid frame equations in Ref. 19.17 are used in the design example.   These equations are valid for single-span frames with rafters and columns of constant section and can be modified to approximate the stiffening effect of haunches.   Superposition of statically indeterminate moments due to different gravity load systems is valid before, but not after, moment redistribution is applied in an allowable stress design.

The plastic design method is based on ultimate loads and ultimate moments. The ultimate loads are obtained by applying an appropriate load factor to the working loads. Article 1.6 considers load factors and indicates that they provide a margin of safety against overload plus several additional items.

A plastic analysis is used to determine the bending moments in the frame at ultimate load. Methods of plastic analysis are reviewed in Art. 7.10 and in Ref. 1.5. Frame charts Nos. 1 and 2 in Ref. 20.2 provide a convenient means for the plastic analysis of single-span frames with rafters and columns of constant section. These same charts can be used to estimate the location of plastic hinges in frames with haunches and thus aid in the analysis of such frames. A set of charts for the plastic analysis of multi-span frames is also available in Ref. 20.2. Superposition of statically indeterminate moments due to different load systems is not valid in a plastic analysis.

The rigid frame considered in this chapter will be designed using both the allowable stress and the plastic design methods. The design sheet numbers for the allowable stress design are prefixed by the letter "A" and those for the plastic design by the letter "P." These design examples are purposely somewhat more detailed than would be necessary in practice, to indicate which factors are important in design and which may be neglected. However, caution should be exercised in reaching general conclusions based only on these examples. For instance, the loading conditions which produce critical moments in these examples may not control the design of taller frames.

Regardless of the design method used to proportion a rigid frame, the intent of the design calculations is the same. These calculations must demonstrate that the frame components provide an adequate margin of safety against the several failure modes which may limit the capacity of the frame.

This article concludes with a brief review of the failure modes which have been observed in tests of rigid frames and their components. The purpose of this review is to provide a check list relating observed structural behavior with factors to be considered in design.

The most frequently observed failure mode in rigid frame tests involves the formation of a sufficient number of plastic hinges to form a mechanism. (See Arts. 7.8 and 20.3.) This mode is preferred at ultimate load because it involves the most economical use of plastic bending strength which a frame can deliver.

The rafters and columns of a single-story rigid frame are stressed in bending and in axial compression. A member loaded in this manner was termed a beam-column in Art. 11.1. Although the axial forces are small, they are regarded as "significant," in that they should not be neglected in proportioning the members. Failure modes for wide-flange beam-columns include (1) inelastic instability (excessive deflection in the plane of bending),

(2) lateral-torsional buckling (deflection plus twisting out of the plane of bending), and (3) local buckling. Usually mode 1 will control the required size of the rafter and column if sufficient lateral bracing can be provided to prevent failure in mode 2. In addition, failure due to frame instability (sidesway) may limit the ultimate strength of slender rigid frames. The shearing stresses within the connection between the rafter and column, the capacity of welds, and the stability of haunch compression flanges should be carefully checked. These are some of the more important factors which must be provided for in the design. Chapters 7, 10, 11, 17, and 19 discuss these failure modes and related Specification provisions.

20.3   MOMENT REDISTRIBUTION

Article 7.8 indicates how the elastic distribution of moment in a beam with fixed ends is altered or redistributed by plastic hinge rotation at the fixed ends. A similar redistribution of moment occurs in an adequately braced rigid frame, as illustrated in Fig. 20.2. Sketch a shows the braced points and the elastic moment diagram under uniform load with maximum negative moments M_D at the eaves. As the uniform load w is increased beyond the working load range,* yielding followed by plastic hinge rotation

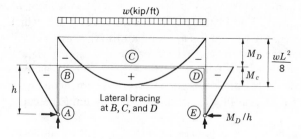

(a) Elastic moment distribution

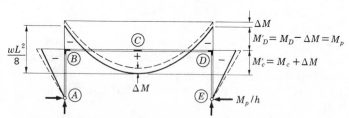

(b) Moment distribution resulting from plastic hinge rotation at B and D

Fig. 20.2 Moment redistribution in a rigid frame.

* If full account is taken of residual stresses resulting from factors other than external loads, some local yielding may occur when working loads are first applied. See Fig. 1.15 and Ref. 1.1, page 3. This local yielding is rarely detrimental to structural behavior in the working load range because of the ductile properties of steel.

occurs at these sections of maximum negative eave moment. This corresponds to the range of contained plastic flow referred to in Art. 1.4. The solid moment diagram in Fig. 20.2b represents the moment distribution which results from plastic hinge rotation at the eaves. The maximum moments in this diagram are limited by the plastic moment capacity M_p of the rafter and column. The dashed moment diagram in Fig. 20.2b shows the hypothetical elastic distribution of moment. The distance ΔM between the solid and dashed moment diagrams indicates how the frame moments are redistributed (from the hypothetical elastic moments) as the result of plastic hinge rotation at the eaves. The change in moment ΔM is termed the "redistributed moment" in the discussion which follows.

Redistribution of moment continues under increasing load until a plastic hinge forms at the center of the rafter in Fig. 20.2, thus producing a mechanism. The mechanism condition corresponds to the ultimate load capacity of the frame and marks the beginning of the range of unrestrained plastic flow mentioned in Art. 1.4. The ductile overload behavior of adequately braced rigid frames described here does not depend on the design method used to proportion the frames and has been demonstrated in numerous full-scale frame tests.[1.1]

A mathematical description of the ductile behavior discussed above yields some interesting results. Consider the uniformly loaded, rectangular, rigid frame with constant rafter and column sections shown in Fig. 20.2. If ΔM is the redistributed moment resulting from plastic hinge rotation, then the positive and negative rafter moments after redistribution are

$$M_C' = M_C + \Delta M$$
$$M_D' = M_D - \Delta M \tag{20.1}$$

where M_C and M_D are the corresponding rafter moments, assuming hypothetical elastic behavior. When a mechanism develops at ultimate load, the positive and negative rafter moments are equal and

$$M_C' = M_D' = M_p = \frac{wL^2}{16} \tag{20.2}$$

This equation represents the plastic analysis of the frame in Fig. 20.2.

Using Eq. 20.1 and 20.2, the elastic moments and the redistributed moment may be related in the form

$$\frac{\Delta M}{M_D} = \frac{1}{2}\left(1 - \frac{M_C}{M_D}\right) \tag{20.3}$$

The ratio $\Delta M/M_D$ is a measure of the effect of moment redistribution in a rectangular rigid frame. It indicates that portion of the negative elastic rafter moment which is redistributed by plastic hinge rotation at the first-formed hinges.

The ratio $\Delta M/M_D$ varies with the frame geometry (h/L ratio) and column base restraint, as indicated in Fig. 20.3. The M_C/M_D values in this figure result from the elastic analysis of rectangular frames with identical column and rafter sections. The $\Delta M/M_D$ values follow from Eq. 20.3.

Figure 20.3 illustrates the point that moment redistribution under overload conditions causes a significant reduction in the hypothetical maximum elastic moments developed in rigid frames. For example, this figure indicates that for a rectangular rigid frame with $h/L = 0.30$ and pinned column bases,

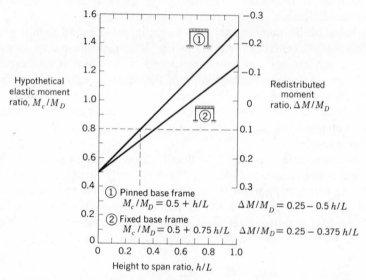

Fig. 20.3 Moment redistribution in rectangular rigid frames (notation from Fig. 20.2).

the hypothetical elastic moment ratio $M_C/M_D = 0.80$ and the redistributed moment ratio $\Delta M/M_D = 0.10$. Thus, 10 per cent of the maximum hypothetical elastic moment M_D at the eaves is transferred to the positive moment region of the rafter at ultimate load by moment redistribution. The negative values of $\Delta M/M_D$ in Fig. 20.3 correspond to the condition that the positive elastic rafter moments are larger than the negative elastic moments at the eaves. This indicates that plastic hinges form first at the center of the rafter. Rotation of these braced hinges redistributes moment from the center of the rafter to the eaves. The maximum hypothetical elastic moment developed in the frame is still reduced by moment redistribution.

The plastic design method makes full use of the benefits of plastic hinge rotation and moment redistribution regardless of the frame geometry, column base restraint, and loading conditions. This is not necessarily true of the allowable stress method. An approximate device for utilizing moment redistribution in an allowable stress design of the frame in Fig. 20.2 is given in AISC Section [1.5.1.4.1]. This section permits a redistribution of 10 per cent of the negative support moment ($\Delta M = 0.1 M_D$) to the positive moment region of the rafter (under certain conditions described below).

In Fig. 20.3 the horizontal dashed line at $\Delta M/M_D = 0.1$ represents the 10 per cent redistribution provision of AISC Section [1.5.1.4.1]. This figure indicates that the 10 per cent redistribution provision makes partial use of moment redistribution if the ratio of hypothetical positive to negative elastic rafter moments, M_C/M_D, is less than 0.8. The condition $M_C/M_D < 0.8$ is satisfied for pinned-base rectangular frames with $h/L < 0.30$ and for fixed-base rectangular frames with $h/L < 0.40$. The height-to-span ratio of many rectangular rigid frames falls within these h/L ranges. When M_C/M_D exceeds 0.8, the approximation, $\Delta M = 0.1 M_D$, overestimates the maximum frame moment after redistribution. For instance, if $M_C/M_D = 0.9$, then Eq. 20.1, with the approximation $\Delta M/M_D = 0.1$, gives

$$M_C'/M_D = 0.9 + 0.1 = 1.0$$
$$M_D'/M_D = 1.0 - 0.1 = 0.9$$

so that the maximum moment $M_C' = 1.0 M_D$ after redistribution. However, Eq. 20.3 or Fig. 20.3 gives $\Delta M/M_D = 0.05$. Then from Eq. 20.1,

$$M_C'/M_D = 0.9 + 0.05 = 0.95$$
$$M_D'/M_D = 1.0 - 0.05 = 0.95$$

and the maximum moment $M_C' = M_D' = 0.95 M_D$ after redistribution. In this example, the approximation $\Delta M/M_D = 0.1$ overestimates the maximum moment after redistribution by $0.05 M_D$. Under this circumstance, the moment-redistribution provisions of AISC Section [1.5.1.4.1] need not be used.

To realize the full potential of moment redistribution in a rigid frame, it is important to preserve stability and rotation capacity at plastic hinge locations by providing adequate bracing. It is also important to preserve ductility, particularly at plastic hinge locations. The several formulas in Part 2 of the AISC Specification provide a means for checking the stability of plastically designed members and frames. Measures to preserve ductility are considered in AISC Section [2.10] and must be provided in the structural details. For example, holes in the region of a member which is subjected to tensile stresses approaching the yield stress should be drilled or subpunched and reamed to remove cold-worked (and thus less ductile) material. The potential loss of ductility at the intersection of triaxial welds (see Art. 15.4) often can be minimized by redesigning the detail or by some form of stress relief such as preheating. Chapters 14 and 15 provide further guidance on the topic of ductility, which is important regardless of the design method used.

AISC Section [1.5.1.4.1] states the conditions which must be satisfied if moment redistribution is to be utilized in an allowable stress design. These conditions are outlined in Table 20.1. The intent of requirements (4) and (5) is to preserve rotation capacity at potential plastic hinge locations for wide-flange shapes. Thus, the lateral bracing requirements may be relaxed in regions of a frame where plastic hinges will not form. In these regions, the

Table 20.1   Moment Redistribution Requirements for Wide Flange Shapes in Allowable Stress Design According to AISC Section [1.5.1.4.1]

Condition	Requirement	Values for A36 Steel
1. Framing	Continuous beams and rigid frames	—
2. Load source	Gravity	—
3. Loading plane	A plane of symmetry	—
4. Lateral bracing*	Unbraced compression flange length must satisfy: $\dfrac{L_y d}{A_f} \leqslant \dfrac{20 \times 10^6}{F_y}$ $\dfrac{L_y}{b_f} \leqslant \dfrac{76.0}{\sqrt{F_y}}$	$\dfrac{L_y d}{A_f} \leqslant 556$ $\dfrac{L_y}{b_f} \leqslant 12.7$
5. Compact section		
Flange	$\dfrac{b_f}{2t_f} \leqslant \dfrac{52.2}{\sqrt{F_y}}$	$\dfrac{b_f}{t_f} \leqslant 8.7$
Web†	$\dfrac{d}{w} \leqslant \dfrac{412(1 - 2.33 f_a/F_y)}{\sqrt{F_y}}$ or $\dfrac{d}{w} \leqslant \dfrac{257}{\sqrt{F_y}}$	$\dfrac{d}{w} \leqslant 68.7 - 4.4 f_a$ or $\dfrac{d}{w} \leqslant 42.8$
6. Axial load	$\dfrac{f_a}{F_a} \leqslant 0.15$	—

* Both requirements must be satisfied.
† The larger d/w value may be used.

allowable bending stresses are governed by AISC Sections [1.5.1.4.6 and 1.6.1].

Moment redistribution is not permitted for loads involving wind or other horizontal components in an allowable stress design. No such limitation is placed on the plastic design method. This makes plastic design advantageous for tall frames and for frames which must resist large horizontal (static) loads.

20.4 PRELIMINARY DESIGN CONSIDERATIONS

The design will follow the steps outlined in Art. 1.1, which begin with the performance requirement. A single-story building is required, providing

25,000 sq ft of floor area for light manufacturing purposes. A minimum clear height of 18 ft and a width of 120 ft, free of interior columns, are desired. No interior cranes are needed but light mechanical equipment may be suspended from the roof. The fire hazard is considered negligible. The building is located in the southern United States on soil suitable for spread footings.

A welded rigid gable frame in A36 steel is well suited to these performance requirements. To keep the design example simple, hot-rolled sections are used although more efficient tapered welded sections find frequent application in this type of frame.

The next step in the design is to make preliminary sketches of the structure and the bracing system. This step is shown in Fig. 20.4. The following paragraphs illustrate some of the factors considered in this step, before calculations are begun. If these factors are carefully reviewed at this stage, the remainder of the design process is better defined. Only those factors which are related directly to the structural design are considered here. Chapter 3 reviews the many additional items which are involved in the preliminary design phase.

The plan dimensions of the building are 120 by 216 ft. The structure consists of ten rigid gable frames with 120-ft span, spaced 24 ft apart. To facilitate shipment, the rafter is spliced for moment and shear at the ridge and near each column. The sides and ends of the building consist of corrugated, insulated siding and window sash supported on channel girts. The girts are connected to the outside flange of the columns.

The column bases utilize a typical base plate and anchor bolt detail with a tie rod between the base plates (below the floor) to resist the horizontal thrust produced by vertical loads. This tie reduces the overturning moment and horizontal thrust on the foundation to that caused by wind loads. While the columns are regarded as hinged at their base, no special provisions to relieve column base moments (such as pins or rockers) are normally required.[19.17,20.1]

The purlins serve two purposes. They support the roof deck and they brace the rafter. Both of these purposes should be considered in establishing the purlin spacing. A purlin spacing of 6 ft is compatible with most of the more economical roof deck sections, and poses no problems with rafter bracing in this building. Figure 3.19 shows how the gravity load on the top flange of a sloping roof purlin is resolved into two force components normal and parallel to the roof and a twisting moment about the shear center of the purlin.

Several types of roof deck are described in Art. 3.5. The building designed in this chapter will use a metal roof deck with interlocking ribs. One characteristic of metal roof decks is their ability to resist shear forces in the plane of the roof if the seams between adjacent deck panels are adequately

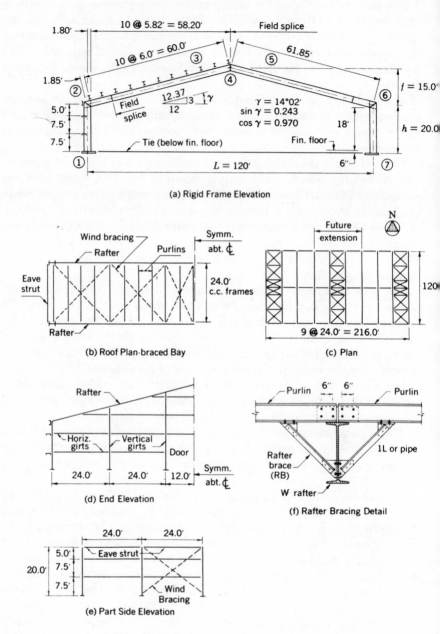

(a) Rigid Frame Elevation

(b) Roof Plan-braced Bay

(c) Plan

(d) End Elevation

(f) Rafter Bracing Detail

(e) Part Side Elevation

Fig. 20.4 Preliminary framing plans.

connected to transmit shear. "Diaphragm action" is the term used to describe this characteristic. Diaphragm action can be used to transmit the parallel component of the roof load to the rafters at each end of a bay. This eliminates weak axis bending and twisting of the purlins. In addition, a metal roof deck also provides continuous lateral support for the top flange of the purlin if the in-plane shear stiffness of the deck and the connectors between the deck and purlin are adequate (which is usually the case).[20.3] This lateral support feature is another result of the diaphragm property of a metal roof deck. Sag rods may be used to provide temporary lateral support during erection of the purlins but are not required once the metal deck is installed.

At this stage it is helpful to visualize the bracing requirements for the frame. The bracing for this building has two functions: (1) to resist wind loads on the ends of the building, and (2) to prevent weak axis buckling and lateral-torsional buckling of the rafters, columns, and corner connection. Article 3.5 discusses the use of wind bracing in the end bays (Figs. 20.4b and e) to resist wind loads on the ends of the building. The combination of the wind bracing and eave struts in Fig. 20.4e prevents the occurrence of the sway buckling modes illustrated in Fig. 10.6. A similar weak axis buckling mode for the rafters is prevented by the combination of purlins and wind bracing in the roof plane. If the wind bracing is the only dependable source of shear resistance in the side and roof planes, the unbraced lengths for lateral buckling of the rafters and columns are equal to the distance between panel points of the wind bracing system.

The diaphragm behavior of a properly installed metal roof deck and of some types of side wall panels also provides shear resistance in the roof and side wall planes. This diaphragm property reduces the unbraced lengths for lateral buckling of the rafters and columns to the distance between purlins and girts respectively. If these unbraced lengths are used, the in-plane shear stiffness of the deck or wall panels and the connectors between these elements and the supporting purlins or girts must be adequate to prevent lateral buckling of the rafters or columns.[20.3]

In the vicinity of the eave, the bottom flange of the rafter and the inside flange of the column are subjected to large compressive stresses. These flanges should be laterally braced to prevent lateral-torsional buckling and to preserve rotation capacity at the plastic hinge which will usually form at the eave under ultimate load conditions. Resistance to lateral-torsional buckling of the eave and the rafter near the eave is provided by the rafter bracing detail* in Fig. 20.4f. A similar detail, utilizing the girts in place of the purlins, can be used to brace the column compression flange below the eave. This detail will be indicated by the symbol "RB" (for rafter brace) in the

---

* Note that the only fabrication required for the purlins in this detail is punching. This is desirable from the standpoint of economy in the fabrication of repetitive members.

design examples.   Note that the first purlin is located approximately above the inside flange of the column so that a rafter brace can be placed close to the inside corner of the eave.   If tapered or curved haunches are used at the eaves of a rigid frame the haunch compression flange must also be braced. This bracing is sometimes provided by a shallow truss in the plane of the haunch diagonal.   Design provisions for haunches are explained in Art. 19.9.

Once the structure and its bracing have been selected, the next step in the design is to establish loads and load combinations.   This step is considered in Art. 3.3.   The minimum snow load listed in Table 3.1 for southern states will be used in the design.   Loads I $(D + S)$ and III $(D + S + W)$ are considered appropriate for this frame because load combinations involving wind usually do not govern the design of rigid frames with a small height-to-width ratio.   For taller frames and for frames with haunches, the load combinations mentioned in Art. 3.3 (including snow drift on one side of the roof) should be investigated.

At this stage, either the allowable stress or the plastic design method may be selected.   The next two articles of this chapter comment on design calculations using each of these methods.

20.5 ALLOWABLE STRESS DESIGN—EXAMPLE 20.1

The allowable stress design example is based on the provisions of Part 1 of the AISC Specification and is discussed in the following sequence: (1) loads, (2) purlin design, (3) rigid frame analysis, (4) statics, (5) frame moments, (6) frame stability, (7) rafter design, and (8) column design. It is presented in five individual sheets, numbered A1 through A5.

1. Loads (Sheet A1)

The dead load on the roof is based on an area in the roof plane while the snow load and uniformly distributed frame load are based on the horizontal projection.   The miscellaneous item in the roof dead load is for "light mechanical equipment."

An estimate of the dead load of the frame can be based on the anticipated rafter size.   The ratio of span* to nominal rafter depth frequently will fall in the range from 35 to 45 for single-story rigid frames with hinged bases using A36 steel.   (This ratio tends to decrease for short spans, tall frames, large frame spacing, and small rafter slopes.)   Once the nominal rafter depth is estimated, the dead load can be based on the average weight of the lighter wide-flange shapes in that depth.   In Example 20.1, a span-to-depth ratio of 40 gives an estimated rafter depth of 3 ft.   A W 36 × 150 rafter is assumed tentatively in the dead load estimate.

---

* The clear span between haunches is appropriate if haunches are used.

EXAMPLE 20.1

SINGLE STORY GABLE FRAME — Allowable Stress Design	Sheet A1

Design Spec. AISC, 1969 Part I, Allowable Stress.

Material A36 Steel

<u>Loads</u>

<u>Roof</u>		<u>Rigid Frame</u>
Roofing	4	Roof 16 x 24 x $\frac{12.37}{12}$ = 0.395
Insulation	3	say 0.155
Metal deck	2	Frame DL = $\overline{0.550}$ kip/ft
Misc.	4	Snow 20 x 24 = 0.480
Purlins	3	TL = 1.030 kip/ft

DL = $\overline{16}$ psf

Snow = 20 psf

Vertical reactions 1.03 x 60' = $\underline{61.8 \text{ kip}}$ ↑

<u>Wind</u> on vertical projection of frame

20 psf x 24' = 0.480 kip/ft

Overturning moment = 0.5 x 0.48 (20 + 15)^2 = 294 kip–ft

Eave thrust T = 294/20 = $\underline{14.7 \text{ kip}}$ →

<u>Purlins</u> : 24 ft. simple span at 6 ft. ctrs.

Top flange braced by metal deck

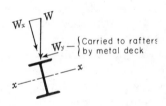

<u>Loads</u>

Dead 16 x 6'   = 96

Snow 20 x 6' x $\frac{12}{12.37}$ = 116

w = 0.212 kip/ft

$w_x = w \cos \gamma$ = 0.206 kip/ft

$w_y = w \sin \gamma$ = 0.052 kip/ft

<u>Moment</u> $M_x = \frac{0.206 (24)^2}{8}$ = 14.8 kip–ft

Try W 10 x 11.5 M_R = 21 kip ft > 14.8   <u>O.K</u>

$$f_b = \frac{14.8 \times 12}{10.5} = 16.9 \text{ ksi}$$

Check deflection : Assume connection $M_c = \frac{5}{24}$ x 14.8 = 3.1 kip–ft

$$\Delta_x = \frac{3}{4} \times \frac{5}{24} \times \frac{f_b L^2}{Ed} = \frac{5}{32} \times \frac{16.9 (24 \times 12)^2}{29 \times 10^6 (9.87)} = 0.77 \text{ in.}$$

$\Delta_x/L = 1/374$ for (D + L) load <u>O.K.</u>

Use W 10 x 11.5 Purlins

The overturning moment due to wind is based on a uniformly distributed horizontal load acting on the vertical projection of the building. This moment is replaced by a statically equivalent force T at the eave. The frame moments produced by this eave thrust are a reasonable design approximation for the variable and uncertain moment distribution produced by wind. The local effects of wind pressure and suction are more important with respect to the connections between the roof deck or siding and the steel structure, as indicated in Ref. 20.4, which summarizes valuable information concerning field observations of wind load failures.

2. Purlin Design (Sheet A1)

The assumptions made in the purlin design concerning the metal roof deck are stated on Sheet A1 and are discussed in Art. 20.4. The purlins are designed to support the normal component of the roof load on a 24-ft simple span. The trial W 10 × 11.5 purlin is selected from the Allowable Stress Design Selection Table in Part 2 of the AISC Manual. This table reduces the allowable resisting moment to account for the large flange $b_f/2t_f$ ratio of the trial shape.

The lighter M 10 × 9 shape also provides an adequate resisting moment but is not selected here because its narrow flange would be quite limber for sloped roof erection. Although this shape is not rolled by most producers, it would be an acceptable substitution if preferred by the erector.

Another possibly more economical alternative would be a cold-formed steel purlin, assuming that suspended mechanical equipment can be accommodated. Instead of designing the cold-formed purlin, the contract documents should provide performance requirements for vendor designs, tailored to vendor fabrication capabilities, and subject to the approval of the engineer. This allows a vendor to propose his most suitable and economical product. These comments suggest some of the many factors, beyond simple allowable stress requirements, that enter into the design process.

The deflection of simple-span purlins is sometimes a source of concern. One way to reduce this deflection is to provide some end restraint at the purlin-to-rafter connection. To estimate the effect of end restraint on deflection in a simple manner, the following reasoning may be used. If M_C is the negative moment developed at the supports of a uniformly loaded beam, the mid-span deflection Δ_x is the difference between the simple-span deflection due to uniform load and the mid-span deflection due to end moments M_C. In symbols

$$\Delta_x = \frac{5}{384} \cdot \frac{wL^4}{EI} - \frac{1}{8} \cdot \frac{M_C L^2}{EI}$$

Let the end moments $M_C = kwL^2/8$ where k is the ratio of the end moment to the maximum simple-span moment. Then

$$\Delta_x = \left(\frac{5}{384} - \frac{k}{64}\right)\frac{wL^4}{EI}$$

This can be written more conveniently in terms of the maximum simple-span bending stress

$$f_b = \frac{M}{I} \cdot \frac{d}{2} = \frac{wL^2}{8} \cdot \frac{d}{2I}$$

with the substitution $wL^4/EI = 16f_bL^2/Ed$. Thus

$$\Delta_x = \left(\frac{5}{24} - \frac{k}{4}\right)\frac{f_bL^2}{Ed}$$

Hence if the purlin connections develop an estimated moment

$$M_C = \frac{5}{24} \cdot \frac{wL^2}{8} \tag{20.4}$$

the estimated purlin deflection is

$$\Delta_x = \frac{3}{4} \cdot \frac{5}{24} \cdot \frac{f_bL^2}{Ed} \tag{20.5}$$

or 75 per cent of the unrestrained simple-span deflection.

Figure 20.4(f) shows one simple method of developing end restraint at the purlin-to-rafter connection, using four bolts in the web of each purlin. The purlin reactions are transmitted to the rafter in bearing so that the web bolts need resist only the connection moment $M_C = 37$ kip-in. estimated on Sheet A1. If the vertical gage lines for these bolts are 6 in. apart, each bolt must resist a shear force of 3 kips without slip. This illustrates that simple web connections at the purlin supports can easily develop the connection moment assumed in Eq. 20.4. The deflection-to-span ratio $\Delta_x/L = 1/374$ for the W 10×11.5 purlin is well within acceptable limits.

Another means of controlling purlin deflection is to use continuous purlins. If lateral bracing is required for the compression (bottom) flange of such purlins in the negative moment region, the advantages of continuity are partially offset by the additional fabrication and erection cost introduced by the bracing. One economical way to provide this bracing is to use sag rods through the web, close to the bottom flange, in the negative moment region of the purlins.

3. Rigid Frame Analysis (Sheet A2)

The redundant horizontal rigid frame reactions are computed using the rigid frame formulas in Ref. 19.17 and the notation in Fig. 20.1. The rafter

and column are assumed to be of equal section.   The results are not affected
very much by this assumption.   Since moment redistribution will be used to
modify the moments, the need for precision in the elastic analysis is open to
question.   The reactions for load III are obtained by superposition, using
the factor 0.75 for loads involving wind.   This is explained in Art. 3.3.

4. Statics

The next phase of the rigid frame design involves the drawing of bending
moment diagrams for the two load combinations.   For this purpose, the
concentrated purlin reactions or uniformly distributed loads may be used.
The latter are used in the examples.   Moments producing tension on the
inside of the frame are considered positive.

The moment diagrams are conveniently drawn in two steps, explained in
Fig. 20.5.   Sketch a shows the structure and loading.   The steps in drawing
the composite moment diagram in part e are:

1. Draw the statically determinate moment diagram d for the frame with
   one reaction replaced by a roller as in b.
2. Draw the redundant moment diagram for the loading shown in c.

The vertical distance between these diagrams is the net moment, positive
when below, and negative when above, the redundant moment line.   The
net moment diagram is cross-hatched in sketch e.   This construction is valid
both for gravity and wind loads.   Only the statically determinate wind
moments are different.

Note the location of P, the pivot point for the redundant moment line in
the rafters.   Every redundant moment diagram, regardless of the value of H
or of the external loads, pivots about the same point.   Sketch c of Fig. 20.5
shows that this is an obvious requirement of statics.

Simple statical aids, based on this pivot idea, are also shown in Fig. 20.5.
For example, the moment in the rafter at point F is the difference between the
statically determinate moment $M_{SF} = wab/2$ and the redundant moment
$M_{RF} = M_B(C + a)/C$.   This last equation is based on similar triangles.

To find the maximum positive moment in the rafter, notice that this
moment is located at the section where the statically determinate and
redundant moment diagrams are parallel in Fig. 20.5e.   If x is the horizontal
distance from the ridge (point C), the slope of the determinate moment diagram in part d is the vertical shear wx.   The slope of the redundant moment
diagram is M_B/C.   At the section of maximum positive moment, the
equation $wx = M_B/C$ can be used to find x.   Then the maximum positive
moment is found as for point F with $a = (L/2) - x$.   A similar procedure
can be used for other loading conditions by including (1) the vertical
statically determinate shear at mid-span (with proper sign) in the equation

EXAMPLE 20.1 (Continued)

SINGLE STORY GABLE FRAME – Allowable Stress Design	Sheet A2

Rigid Frame Analysis – Horizontal Reactions

Assume $\dfrac{I_r}{I_c}=1.0$ $J=\dfrac{I_r h}{I_c m}=1.0 \times \dfrac{20}{61.8}=0.324$ $Q=\dfrac{f}{h}=\dfrac{15}{20}=0.75$

$N=4(J+3+3Q+Q^2)=24.56$

Dead + Snow

$W=1.03\ kip/ft$ $H=\dfrac{WL^2}{8hN}(8+5Q)=44.4\ kip$

Wind $T=14.7$ kip at eave

Lee side $H_7=\dfrac{T}{N}(2J+6+3Q)=5.3$ kip ←

Wind side $H_1=14.7-5.3 \quad =9.4$ kip ←

Vertical $V=Th/L=2.45$ kips ↓↑

Horizontal reactions

	$H_1 \longrightarrow$ (Wind side)	$\longleftarrow H_7$ (Lee side)
Load I (D+S)	= 44.4 kip	44.4 kip

Load III (D+S+W) x 0.75 = (44.4 − 9.4) x 0.75 = 26.2 kip

= (44.4 + 5.3) x 0.75 =   37.2 kip

Frame Moments – Load I (D+S) $W=1.03\ kip/ft$

Case I – Elastic Moments $H=44.4$ kip

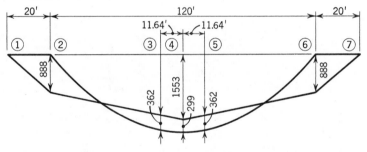

Case 2 – Moment Redistribution AISC Section [1.5.1.4.1]

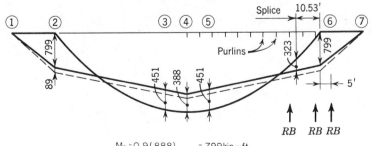

$M_2=0.9(888) \qquad = 799\ kip-ft$

$M_3=362+0.1(888)=451\ kip-ft$

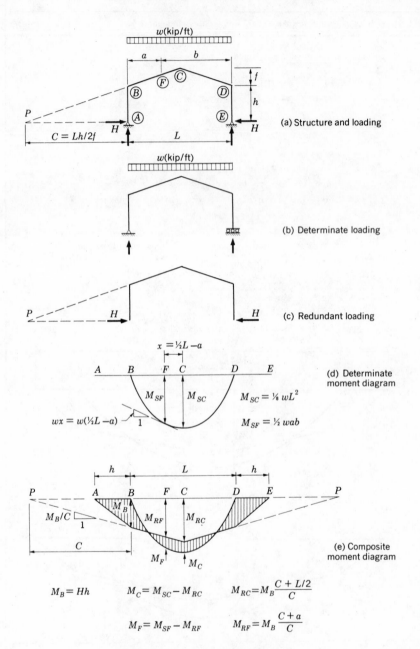

$M_B = Hh$

$M_C = M_{SC} - M_{RC}$

$M_{RC} = M_B \dfrac{C + L/2}{C}$

$M_F = M_{SF} - M_{RF}$

$M_{RF} = M_B \dfrac{C + a}{C}$

Fig. 20.5   Gable frame statics.

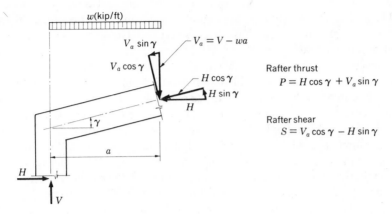

Fig. 20.6 Rafter thrust and shear.

for x and (2) the determinate moment at x in the moment sum.   The axial thrust P and the transverse shear S at any rafter section are found as indicated in Fig. 20.6.

5. Frame Moments (Sheets A2, A3)

Two moment diagrams for load I $(D + S)$ are shown on Sheet A2.   The elastic moment diagram is labeled case 1.   The statically determinate portion of this diagram is the same as the moment diagram for a simply supported beam with a span of 120 ft and carrying a uniform load w.   The redundant moment line is fixed by the moment Hh at the eaves due to the horizontal reactions H.   Maximum moments occur at the eaves.   It is at these locations that plastic hinge action will occur first, followed by redistribution of moment to the center of the rafter.

The moment-redistribution provisions of the AISC Specification are used to obtain the moments in case 2 on Sheet A2.   According to these provisions, the maximum negative moments at points 2 and 6 are reduced by 10 per cent, and the maximum positive moments at points 3 and 5 are increased by 10 per cent, of the negative moments.   For convenient comparison the redundant moment line for case 1 is shown dotted.

It will be observed that the rafter moments for case 2 do not satisfy equilibrium.   This is an unavoidable feature of an engineering approximation intended to cover many different situations.   However, this should not be a source of undue concern if it is realized that an allowable stress design (or any other "elastic" method) uses many such engineering approximations based on plastic behavior.

The load system under investigation satisfies the first three conditions in Table 20.1. The remaining three conditions cannot be checked until trial rafter and column sections have been selected. However, the preliminary consideration of rafter bracing in Art. 20.4 indicates that lateral bracing can be conveniently added, wherever required, by condition 4.

If architectural or functional requirements prohibit any form of column bracing, it may not be feasible to utilize redistributed moments in the column design because of the flange width required for condition 4 in Table 20.1. In this case, redistributed moments can be used in the rafter design and the elastic moments used in the column design if the allowable bending stress for the column is determined from AISC Section [1.5.1.4.6]. The reason for this is that the larger column size required to resist the elastic moment at the eave will tend to force plastic hinges into the rafter. If the unbraced column remains elastic and stable at ultimate load, and if the rafter and eave are braced to preserve rotation capacity, the design is adequate.

The moment diagram for load III $(D + S + W)$ (0.75) is shown on Sheet A3. The statically determinate portion of this diagram is obtained by superposition of the moments for gravity load and eave thrust acting on the frame with the pinned column base at point 7 replaced by a roller. This is treated in Fig. 20.5. Since loads other than gravity load are included in load III, redistribution may not be used.

Comparison of the moment diagrams indicates that the largest moments in the frame occur at the eaves for load I, case 2. This is the condition which controls the rafter and column design. If large unbraced lengths together with large bending moments occur in other regions of the frame (for example the negative moment region in the rafter adjacent to point 6 for load III), additional bracing or an increased member size may be required. The best indicator for bracing requirements is the moment diagram for each loading condition.

6. Frame Stability

The allowable axial stresses for members in axial compression depend on the effective lengths of the members for buckling either in or out of the plane of the frame. Before the rafter and column designs are discussed, the choice of effective length factors for these members will be considered. The rafter and column design on Sheets A3 and A4 will indicate that the effect of axial stress in this example, as in many single-story rigid frames, is small. Frequently it may be assumed, with no significant error, that the effective length of the rafter and column is equal to the unbraced length. The following discussion is intended to interpret the provisions of AISC Section [1.8.3] as they apply to single-story rigid frames so that this assumption can be verified in any particular design.

EXAMPLE 20.1 (Continued)

SINGLE STORY GABLE FRAME – Allowable Stress Design	Sheet A3

<u>Frame Moments – Load III</u> (D+S+W) x 0.75   0.75W = 0.77 kip/ft
0.75T = 11.0 kip →

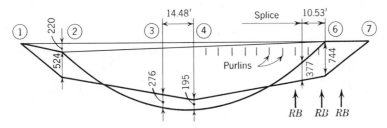

<u>Rafter Design</u> – Load I, Case 2 at Sect. ②

$M_2 = 799$ kip-ft $H \cos \gamma = 43.0$ Unbraced lengths
$H = 44.4$ kip $V \sin \gamma = 15.0$ $L_x = 61.85$ ft
$V = 61.8$ kip $p = 58.0$ kip ↗ $L_y = 6.0$ or 12.0 ft
                                              as required

Estimate $S = \dfrac{799 \times 12}{24.0} = 399 \text{ in}^3$

<u>Try W 36 x 135</u> $S_x = 440 \text{ in}^3$ $A = 39.8 \text{ in}^2$ $r_x = 14.0$ in.
                  $b_f = 11.95$ in. $t = 0.794$ in. $r_y = 2.39$ in.
                  $d = 35.55$ in. $w = 0.598$ in. $d/A_f = 3.75 \text{ in.}^{-1}$

<u>Load I Stresses</u>

$f_a = \dfrac{58.0}{39.8} = \underline{1.46 \text{ ksi}}$ $f_b = \dfrac{799 \times 12}{440} = \underline{21.8 \text{ ksi}}$

<u>Allowable Stresses</u>

$(L/r)_y = 30$ $(L/r)_x = 53$ ← controls F_a
$(L_y = 6.0 \text{ ft})$ $F_a = 18.08$ ksi   AISC Appendix A

For $F_b = 0.66 F_y$ AISC Section $[1.5.1.4.1]$

<u>Check Compact Section</u>

Allow $\dfrac{d}{w} = 68.7 - 4.4 f_a = 62.3$
For W 36 x 135 $d/w = 59.4 < 62.3$ <u>O.K.</u>
           $b_f / 2 t_f = 7.52 < 8.7$ <u>O.K.</u>

<u>Check Lateral Bracing</u>   (Comp. flange on bottom)

Allow $L_y d / A_f = 556$ or $L_y / b_f = 12.7$ for $F_y = 36$ ksi
For 36 WF 135 $L_y d / A_f = 144 \times 3.75 = 540 < 556$ <u>O.K.</u>
With $L_y = 144$ in $L_y / b_f = 144 / 11.95 = 12.0 < 12.7$ <u>O.K.</u>

∴ $F_b = 24.0$ ksi for braced compact section

Resistance to deflection of the columns in the plane of the frame is provided only by the bending stiffness of the rafters and columns. Thus, the sidesway mode of frame instability controls the effective length factor K_x for buckling in this plane. Figure 20.7 considers the basis for this determination. A rectangular frame is included in this figure because simple methods of determining effective length factors for a rectangular frame are provided in the alignment charts of Fig. 10.7. This provides the clue for estimating effective length factors for gable frame columns. Solutions for the critical sway buckling load, based on the axial loading approximations in Figs. 20.7c and d, are given in Appendix B2 of Ref. 10.6. These solutions are identical except that the length L of the horizontal rafter is replaced by the total length $2m$ of the gable rafter. Therefore, chart b in Fig. 10.7 may be used to estimate the effective length factor K_x for both rectangular and gable frame columns. This chart is entered with the restraint factors G_{top} and G_{bot} tabulated in Fig. 20.7g.

One effect of gravity load on the roof of a gable frame is to deflect the eaves outward as shown in Fig. 20.8. This effect is not produced by the axial loading shown in Fig. 20.7d. A qualitative estimate of the influence of eave deflection on the sidesway buckling of a gable frame is available in Ref. 20.5. This reference tentatively suggests that a gable frame may be stiffened against sidesway as the result of "prebuckling deformations."

Since the wind bracing and eave struts shown in Fig. 20.4e resist deflection of the column tops out of the plane of each frame, the effective length factor K_y for weak axis column buckling can be taken as unity. Figure 10.6 indicates the weak axis sidesway buckling mode which may occur if wind bracing and all other sources of stiffness in the side plane (other than the column and eave strut*) are absent. Reference 10.6 contains several solutions for frame stability problems of this type which may arise in buildings with large window areas or many large doors on one side.

Buckling of the girder in Fig. 20.7c under axial force H is restrained by the bending stiffness of the girder and columns. The force H is normally small and the effective length of the girder is close to half its span unless the columns are slender, in which case Fig. 10.7 can be used to estimate an effective length of m for the sloping rafter in Fig. 20.7b.

7. Rafter Design (Sheets A3 and A4)

The maximum moment and axial load for load I, case 2 are recorded first. The unbraced lengths L_x and L_y about the strong and weak axes are indicated also. (The diaphragm stiffness of the metal roof deck is used here

---

* The purlins may also provide some resistance to sway if the torsional stiffness of the rafter and the purlin-to-rafter connections force the purlins to deform in double curvature.

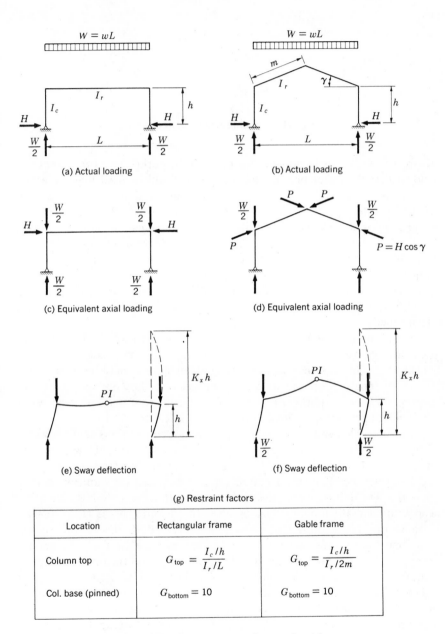

(a) Actual loading

(b) Actual loading

(c) Equivalent axial loading

(d) Equivalent axial loading

(e) Sway deflection

(f) Sway deflection

(g) Restraint factors

Location	Rectangular frame	Gable frame
Column top	$G_{top} = \dfrac{I_c/h}{I_r/L}$	$G_{top} = \dfrac{I_c/h}{I_r/2m}$
Col. base (pinned)	$G_{bottom} = 10$	$G_{bottom} = 10$

Fig. 20.7   Strong axis sidesway buckling.

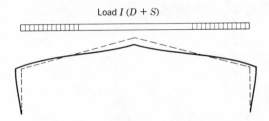

Fig. 20.8 Gable frame deflection.

to prevent weak axis buckling of the rafter at the purlins.)   At this point it is
known that the rafter bracing detail Fig. 20.4f will be used under the first
purlin above the eaves to brace the rafter compression flange.   (See Art.
20.4.)   It is likely that additional bracing for the bottom flange of the rafter
will be required.   Convenient locations for this bracing are under the second
or third purlin from the eave.   This explains the values shown for L_y on
Sheet A3.   The values thus far recorded define the design problem for the
rafter.

Based on an estimated bending stress of $0.66F_y$, a W 36 × 135 is selected
tentatively.   Then the axial load and bending stresses for load I are
determined.

The allowable stress F_a for axial load is controlled by the larger of the
slenderness ratios for weak axis or strong axis (not lateral-torsional)
buckling.*   The distance $L_y = 6.0$ ft between purlins may be used as the
unbraced length for weak axis buckling because the combination of the
purlins and the diaphragm stiffness of the metal roof deck prevents weak
axis buckling under axial load in the rafter.   (See Art. 20.4.)

The allowable bending stress F_b depends on the conditions summarized in
Table 20.1.   The compact section requirements for A36 steel in this table are
met by the trial W 36 × 135 shape as indicated on Sheet A3.   Thus, local
buckling will not limit the rotation capacity required for moment redistribution.   The unbraced compression flange length $L_y = 144$ in. is used in the
check for lateral support.   This assumes that a rafter brace (Fig. 20.4f) is
located under the third purlin from the eave.   Both the lateral support
requirements for A36 steel in Table 20.1 are met by the trial W 36 × 135
shape.   Thus, lateral-torsional buckling will not inhibit the rotation capacity
required for moment redistribution, according to AISC Section [1.5.1.4.1].
Therefore, $F_b = 24.0$ ksi for the braced, compact section.   The compact
section and lateral bracing calculations are recorded at the bottom of Sheet
A3 only to illustrate the application of conditions 4 and 5 in Table 20.1.

---

* Bracing connected to either flange will prevent weak axis buckling.   Compression
flange bracing is usually required to prevent lateral-torsional buckling.

EXAMPLE 20.1 (Continued)

SINGLE STORY GABLE FRAME – Allowable Stress Design	Sheet A4

Check Combined Stress AISC Section $[1.6.1.]$ For $f_a/F_a < 0.15$

$$\frac{f_a}{F_a} + \frac{f_b}{F_b} = \frac{1.46}{18.08} + \frac{21.8}{24.0} = 0.081 + 0.908 = 0.989 < 1.0 \quad \underline{O.K.}$$

Note : Brace bottom flange of rafter $\boxed{\text{Use W 36 x 135 Rafter}}$
       below purlins no.1 and 3 from eaves.

Effective Length Factors for Column AISC Section $[1.8]$

   Weak axis – wind bracing prevents sidesway $\therefore K_y = 1.0$
   Strong axis – sidesway not prevented
   Assume same size rafter and column : $I_r = I_c$

     Restraint $G_{top} = \dfrac{I_c/h}{I_r/(2m)} = \dfrac{2 \times 61.85}{20.0} = 6.19$ $\left.\begin{array}{c} \\ \\ \end{array}\right\}$ $K_x = 2.7$
     Factors
           $G_{bot} = 10$ for non–rigid base   From Fig.10.7 (b)
           AISC $[\text{Fig. C1.8.2}]$

Column Design – Load I, Case 2 at Sec. ②

   $M_2 = 799$ kip-ft   Unbraced Lengths $L_x = 20.0$ ft $K_x = 2.7$
   $P = 61.8$ kip $L_y = 5.0$ and 7.5 ft between girts

Try W 36 x 135 (See Sheet A3)

Load I Stresses $f_a = \dfrac{61.8}{39.8} = \underline{1.55}$ ksi $f_b = \dfrac{799 \times 12}{440} = 21.8$ ksi

Allowable Stresses $(L/r)_y = 38$ $(KL/r)_x = 46 \longleftarrow$ Controls F_a
                          $\underline{F_a = 18.70}$ ksi AISC Appendix A

For $F_b = 0.66 F_y$ AISC $[1.5.1.4.1]$

Check Compact Section (See Sheet A3)

   Allow $\dfrac{d}{w} = 68.7 - 4.4 f_a = 61.9 > 59.4$ $\underline{O.K.}$ b/t $\underline{O.K.}$

Check Lateral Bracing (Comp. flange inside)

   For $L_y = 60 - 36/2 = 42$ in $L_y d/A_f = 158 < 556$ $\underline{O.K.}$
      ↳ Rafter $L_y/b_f = 3.5 < 12.7$ $\underline{O.K.}$
   $\therefore F_b = 24$ ksi for braced compact section

For routine design, the Allowable Stress Design Selection Table in Part 2 of the AISC Manual can be used in place of the compact flange and lateral bracing checks. The compact web check must be made separately because the rafter carries an axial load.

Since the rafter is a beam-column, it must be checked for combined stress, using AISC Section [1.6.1], as shown on Sheet A4. The axial stress ratio f_a/F_a is less than 0.15 so that only the simple AISC interaction equation [1.6-2],

$$\frac{f_a}{F_a} + \frac{f_b}{F_b} \leqslant 1.0 \tag{20.6}$$

need be considered in this example. The relatively minor role of axial stress in the rafter, which is characteristic of single-story rigid frames, is evident from this check.

The W 36 × 135 rafter is adequate for the critical combination of moment and axial load at the eaves caused by load I, case 2, if rafter bracing is provided under the first and third purlins from the eaves. These rafter bracing locations are indicated by the symbol "RB" below the moment diagrams for load I, case 2, and load III. With a rafter brace below the third purlin, the unbraced compression flange lengths in the negative moment region for load III are adequate.

In the positive moment region of the rafter, the rafter compression flange is supported by the purlins and roof deck. The Ld/A_f ratio for a 6-ft unsupported length of the W 36 × 135 rafter is 270 and the maximum bending stress in the positive moment region is 12.3 ksi. Both of these values are considerably smaller than the limits imposed by AISC Section [1.5.1.4.6]. Thus, no additional lateral bracing is required.

8. Column Design (Sheets A4 and A5)

The effective length factor K_x for sidesway buckling in the plane of the frame is obtained from Fig. 10.7b. The assumption $G_{\text{bot}} = 10$ (see Art. 10.4) includes the effect of partial restraint at the column base, and reduces K_x from 3.6 (for a theoretical pinned base frame) to 2.7 for the frame with flat-ended columns. This reflects the conservative nature of frame stability formulas using pinned column bases.

The moment and axial load for load I, case 2, and the unbraced lengths are recorded next. The girt spacing in Fig. 20.4a determines the unbraced lengths L_y on Sheet A4. (The diaphragm stiffness of the siding is used here to prevent weak axis column buckling at the girts.) These moments, forces, and lengths define the design problem for the column.

A W 36 × 135 shape is selected tentatively for the column and the actual and allowable stresses are obtained as for the rafter. The allowable axial stress F_a is controlled by the effective length $K_x L_x$.

EXAMPLE 20.1 (Continued)	
SINGLE STORY GABLE FRAME − Allowable Stress Design	Sheet A5

Check Combined Stress. AISC Section $[1.6.1]$ For $f_a/F_a < 0.15$

$$\frac{f_a}{F_a} + \frac{f_b}{F_b} = \frac{1.55}{18.70} + \frac{21.8}{24.0} = 0.083 + 0.908 = 0.991 < 1.0 \quad \text{O.K.}$$

Check unbraced compression flange below top girt

$L_y = 15\,ft \quad c_b = 1.75 \quad \dfrac{L_y d}{A_f C_b} = \dfrac{180 \times 3.75}{1.75} = 386 < 556 \; \therefore \; F_b = 22\,ksi \quad \text{AISC Section}[1.5.1.4.6]$

At 1st girt $f_b = \dfrac{15}{20} \times 21.8 = 16.4\,ksi$

Combined stress AISC Section $[1.6.1]$ For $f_a/F_a < 0.15$

$$\frac{f_a}{F_a} + \frac{f_b}{F_b} = 0.083 + \frac{16.4}{22.0} = 0.828 < 1.0 \; \therefore \; L_y = 15\,ft \quad \text{O.K.}$$

Note: Brace inside flange of column at girt 5 ft below eave. | Use W 36 x 135 Column |

Illustration 1 showing the influence of sidesway buckling in the plane of the frame.

Check W 36 x 135 Column neglecting sidesway buckling

Say $K_x = 1.0$ $(L/r)_x = 17$ $F_a = 20.78\,ksi$

$$\frac{f_a}{F_a} + \frac{f_b}{F_b} = \frac{1.55}{20.78} + 0.908 = 0.075 + 0.908 = 0.983$$
$$\text{Versus } 0.991 \text{ for } K_x = 2.7$$

Illustration 2 showing the influence of sidesway buckling out of the plane of the frame. Say omit wind bracing and all other sources of stiffness in the side wall planes except columns and eave struts. Estimate $K_y = 2.0$; $L_y = 20\,ft$

Try W 36 x 170 Column:

$$A = 50.0\,in^2 \quad S_x = 580\,in^3$$
$$r_y = 2.53\,in.$$

$$\left(\frac{KL}{r}\right)_y = 190 \quad F_a = 4.14\,ksi \quad \text{Say } F_b = 24\,ksi$$

$$f_a = \frac{61.8}{50.0} = 1.24\,ksi \quad f_b = \frac{799 \times 12}{580} = 16.5\,ksi$$

$$f_a/F_a + f_b/F_b = 0.300 + 0.687 = 0.987 \quad \text{O.K.}$$

The inside (compression) flange of the column must be braced within a distance $12.7b_f = 152$ in. from the inside corner of the eave in order to satisfy the lateral support requirement of Table 20.1. This distance is less than the length of the column. Therefore a detail similar to that shown in Fig. 20.4f is used at the girt 5 ft below the eave to brace the column compression flange. Deducting the half-depth of the rafter from this 5-ft girt spacing gives the unbraced length used in the lateral support check on Sheet A4. The compression flange below this girt is not a region of maximum bending stress and therefore need not be considered in the lateral bracing requirements of AISC Section [1.5.1.4.1].

Equation 20.6 is used to check the column for combined stress on Sheet A5. This check indicates that most of the capacity of the column is used to resist bending, which is typical of single-story rigid frame members.

The inside (compression) flange of the column is unbraced for a length of 15 ft below the top girt. This unbraced span is checked for lateral-torsional buckling next on Sheet A5. The larger value of F_b given by AISC Formulas [1.5-6] and [1.5-7] can be used to determine the allowable bending stress in this span because no plastic hinges form here. Formula [1.5-7] is used on Sheet A5 for simplicity. Note that $C_b = 1.75$ in this formula is acceptable when F_b is used in Formula [1.6-2] as in this example, but $C_b = 1.0$ applies when F_b is used in Formula [1.6-1a] together with $C_m < 1.0$.

The unbraced span under investigation is a beam-column; therefore a combined stress check using Eq. 20.6 is required. The term f_a/F_a was determined previously using $K_x = 2.7$. Since lateral-torsional buckling rather than frame stability is under investigation here, one could determine a new value of F_a which is controlled by the unbraced length $L_y = 7.5$ ft between girts. This gives a small increase in F_a but since the term f_a/F_a is much less than the bending term f_b/F_b, the effort is hardly worthwhile. The sum of the terms in the combined stress check for the unbraced span is 0.828. Thus, no additional bracing is required for the W 36 × 135 column.

In the discussion of frame stability it was indicated that, frequently, the effective length of a single-story rigid frame column may be taken equal to the unbraced length with no significant error. Illustrations 1 and 2 on Sheet A5 qualify what is meant by "frequently."

The first illustration indicates the result of neglecting sidesway buckling in the plane of the frame by using $K_x = 1.0$ instead of 2.7 in the column design. The error of 0.008 in the combined stress check is not significant. The minor role played by axial loads in most single-story rigid frames and the small variation in F_a, as the slenderness ratio changes from $(KL/r)_x = 46$ to $(L/r)_x = 17$, are the factors which contribute to this small error. For the frame in Fig. 20.4 an even smaller error results if $(L/r)_y = 38$ (for the 7.5-ft girt spacing) is used to determine F_a.

The second illustration on Sheet A5 shows the influence of sidesway buckling out of the plane of the frame. It is assumed that the wind bracing and all other sources of shear stiffness in a side wall plane are absent. This might occur, for example, in a warehouse with large doors which fill each bay of one side. In this case the columns and eave struts can be used to form a multi-span rigid frame in the side plane to resist wind loads. Figure 10.6 indicates that the smallest value of K_y for this case is 2.0, corresponding to an infinitely stiff eave strut and pinned column bases. A W 36 × 170 is the smallest wide-flange shape which might be adequate. (Moment-redistribution provisions, lateral-torsional buckling checks, and biaxial bending under wind load would require a further increase in the column size. Only the effective length factor is considered in this illustration, for clarity.) It is evident that bracing about the weak axis of rigid frame columns is an important design factor. If this bracing is prohibited by architectural or functional requirements, the effective length factor K_y is indeed significant.

20.6 PLASTIC DESIGN—EXAMPLE 20.2

The plastic design of the frame follows the provisions of Part 2 of the AISC Specification and is the recommended method for utilizing the plastic strength of single-story rigid frames. The example is discussed in the following sequence: (1) loads, (2) frame moments, (3) rafter design, and (4) column design. It is presented in three sheets, P1, P2, and P3.

1. Loads (Sheet P1)

Load factors of 1.7 for load I $(D + S)$ and 1.3 for load III $(D + S + W)$ are specified in AISC Section [2.1]. The range of rafter span-to-depth ratios mentioned in Art. 20.5 also apply to the plastic design method. However, some reduction in rafter and column weight can be anticipated in many cases using this method. The basis for the eave thrust due to wind is considered in Art. 20.5.

2. Frame Moments (Sheet P1)

Frame Chart No. 1 of Ref. 20.2 is used to obtain the required plastic moment M_p for the rafter and column and to locate the plastic hinges in the rafter for load I. This chart is entered with the values $Q = f/h$ and $C = 0$ since load I excludes wind. The chart gives the ratio $M_p/w_u L^2$ and the horizontal distance αL from the column to the plastic hinges in the rafter near the ridge. This provides the information needed to complete the moment diagram. The failure mechanism for load I is shown on Sheet P1. From statics, the horizontal reactions $H = M_p/h$.

Frame Chart No. 1 is used to investigate load III, involving wind. In this case the eave thrust coefficient $C = 2T_u h/W_u L$, where $T_u h$ is the overturning moment produced by the factored wind load. For any value of Q, the dashed curve in Frame Chart No. 1 indicates the value of C for which loads I (1.70) and III (1.30) require approximately the same plastic moment capacity. (The dashed curve was prepared for load factors of 1.85 and 1.40

but is approximately correct for load factors of 1.70 and 1.30.) For $Q =$ 0.75, this value of $C = 0.125$, which is 3.9 times the C value for the 20-psf wind load in Example 20.2. Hence, load III does not control the rafter or column size.

3. Rafter Design (Sheet P2)

The required plastic moment and axial load for load I are recorded first. The distance L_y is the unbraced length between purlins. (The diaphragm stiffness of the metal roof deck is used here to prevent weak axis buckling of the rafter at the purlins.) Based on the required plastic modulus, neglecting axial load, a W 33 × 118 is tentatively selected.

The flange and web proportions of the trial W-shape are checked against the requirements of AISC Section [2.7] to ensure that local buckling will not inhibit rotation capacity. Next, the section is reviewed for lateral bracing requirements.

The rafter should be laterally and torsionally braced at the first-formed plastic hinges which must rotate to develop the plastic strength of the frame. These hinges form at the eaves under the following conditions:* (1) The eave thrust coefficient C is less than 0.5; (2) the column h/L ratio is less than 0.5; (3) the single-span, pinned-base, rigid frame carries a uniformly distributed gravity load and an eave thrust $T_u = CW_uL/2h$; and (4) the column and (horizontal or sloping) rafter use the same prismatic section. It is good practice to brace the eave even in the relatively infrequent case when first hinges may form near the ridge for tall frames and unusually large horizontal forces. The rafter bracing detail in Fig. 20.4f is used to brace the rafter compression flange under the end purlins of each bay. Additional lateral support requirements in the positive and negative moment regions of the rafter are investigated on Sheet P2.

The moment diagram for load I (Sheet P1) indicates that the bottom flange of the rafter is in compression from the eave to the point of inflection between the third and fourth purlin from the eave. To preserve rotation capacity at the eave hinges below the first purlin, the adjacent unbraced length of the rafter compression flange should not exceed

$$L_{cr} = \left(\frac{1375}{F_y} + 25\right)r_y \quad \text{for} \quad \frac{M}{M_p} > -0.5 \tag{20.7}$$

$$L_{cr} = \left(\frac{1375}{F_y}\right)r_y \quad \text{for} \quad -0.5 > \frac{M}{M_p} > -1.0 \tag{20.8}$$

from AISC Section [2.9] where M/M_p is the end-moment ratio for the unbraced span. In the unbraced rafter span under investigation, $M/M_p = 0$,

---

* Figure 4.19 of Ref. 1.5 extends these conditions.

EXAMPLE 20.2

SINGLE STORY GABLE FRAME – Plastic Design Method	Sheet P1

Design Spec. AISC, 1969 Part 2, Plastic Design

Material A36 Steel

Rigid Frame Loads (Ultimate) AISC Section [2.1]

Load I (D+S) : LF = 1.70

Roof	0.395 k/f
Frame	0.125 k/f
DL	0.520 k/f
Snow	0.480
TL =	1.000 x 1.70 = 1.70 k/f

Vertical Reactions
V = 1.70 x 60 = 102 kip ↑

Load III (D+S+W) : LF = 1.30

DL	0.520 k/f
Snow	0.480
TL	1.010 x 1.30 = 1.31 k/f

Vertical Reaction
V = 1.31 x 60 = 78.6 kip ↑

Wind –on vert. projection
20 psf x 24' x 1.30 = 0.625 k/f

Overturning moment = 0.5 x 0.625 (20 + 15)^2
= 384 kip-ft

Eave thrust T_u = 384/20 = 19.2 kip ⟶

Frame Moments – Use Frame Chart No.1 (Ref.20.2)

$Q = \dfrac{f}{h} = \dfrac{15}{20} = 0.75$ Column base: pinned

Load I (D+S) x 1.70 $w_u L^2 = 2 \times 102 \times 120 = 24{,}480$ kip-ft

For vertical load only ; C = 0

From Frame Chart No.1 $M_p = 0.0464\, w_u L^2 = 1{,}136$ kip-ft

Rafter hinges at : $\alpha L = 0.431\ \underline{L = 51.7\ ft}$ from columns

$w_u = 1.85$ kip-ft   H = 1,136/20 = 56.8 kip

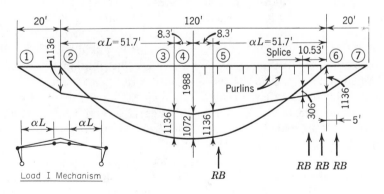

Load I Mechanism

RB

RB RB RB

Load III (D+S+W) x 1.30 $C = \dfrac{2 T_u h}{W_u L} = \dfrac{2 \times 384}{2 \times 78.6 \times 120} = 0.041$

From Frame Chart No.1 : Load III does not control rafter or column size.

so that $L_{cr} = 63.2r_y = 147$ in. for A36 steel. The distance between the first and third purlins exceeds this value so a bottom flange brace is added below the second purlin.

The flat moment gradient adjacent to the plastic hinges in the positive moment regions of the rafter (near the ridge) indicates that Eq. 20.8 may be used to check the 6-ft unbraced compression flange between purlins. However, a different procedure should be used here if the purlin spacing were to exceed the limitations of Eq. 20.8.

It is known from the preceding discussion that the rafter hinges near the ridge are the last to form. This means that the moment at these hinge locations must approach the plastic moment M_p at ultimate load but that no rotation capacity is required because no further redistribution of moment is needed once the ultimate load is reached. Thus lateral bracing at these hinges need only be sufficient to prevent lateral-torsional buckling as the moment approaches M_p. The unbraced length is adequate if the axial force and moments, divided by the applicable load factor, satisfy the allowable stress provisions in Part 1 of the AISC Specification.

Finally, the rafter is checked on Sheet P2 for axial load and bending, using AISC Formulas [2.4-2 and 3]. The second of these formulas is satisfied whenever $P/P_y < 0.15$. Section [2.4] places a limit of C_c on the slenderness ratio L/r in the plane of bending for axially loaded members that would form plastic hinges under factored loads. For the rectangular rigid frame in Fig. 20.7a, the total span would be used to determine rafter slenderness for this comparison with C_c. (This limits L/d to about 50 for an A36 rafter.) For the gable frame in Fig. 20.7b, the gable slenderness is limited to $2m/r_x < C_c$ where m is the length of one sloping rafter.

The axial force in the rafter is small and the column bending stiffness restrains buckling of the rafter (at least until plastic hinges form at the eaves). Hence the effective length of the sloping rafter is taken as m in the calculation of the maximum axial load capacity $P_{cr} = 1.7F_a A$ and the elastic buckling load $P_e = 1.92F_e'A$ for use in AISC Formula [2.4-2]. These loads are discussed in Chapter 11. The value $C_m = 1 - 0.3P/P_e$ is used as a conservative approximation for the moment and curvature distribution produced by distributed gravity load on the end-restrained rafter. The value $C_m = 0.85$ does not apply to the rafter. Both C_m and $1 - P/P_e$ are close to unity in this example and could be dropped from Formula [2.4-2] with little significant error whenever axial loads are small. This formula then reduces to a plastic design counterpart of Formula [1.6-2].

The 6 ft laterally unbraced segments adjacent to plastic hinges govern the slenderness L/r_y in Formula [2.4-4] for the lateral-torsional buckling moment capacity M_m. The rafter compression flange between the point of inflection and the braced point under the second purlin (see moment diagram on Sheet P1) need not be considered in applying this formula because no plastic hinges

EXAMPLE 20.2 (Continued)

SINGLE STORY GABLE FRAME – Plastic Design Method	Sheet P2

Rafter Design – Load I $(D + S)$ x 1.70

$M = 1,136$ kip-ft $H \cos \gamma = 55.0$ kip   Unbraced lengths

$H = 56.8$ kip $V \sin \gamma = 24.8$ kip $L_x = 61.85$ ft

$V = 102$ kip $P = 79.8$ kip $\longrightarrow$ $L_y = 6.0$ or 12.0 ft as required

Estimate $Z = \dfrac{1,136 \times 12}{36} = 379$ in^3

Try W 33 x 118 $Z = 415$ in^3 $A = 34.8$ in^2 $r_x = 13.0$ in.

$\qquad\qquad\quad$ $b = 11.48$ in. $t_f = 0.738$ in. $r_y = 2.32$ in.

$\qquad\qquad\quad$ $d = 32.86$ in. $t = 0.554$ in. $d/A_f = 3.88$ in.$^{-1}$

$\qquad\qquad\quad$ $P/P_y = 79.8 / (1250) = 0.064$

Check width/thickness. AISC Section [2.7]

Allow $\dfrac{d}{t} = 68.7 \left(1 - 1.4 \dfrac{P}{P_y}\right) = 62.5$

For W 33 x 118 $\dfrac{d}{t} = 59.5 < 62.5$ _O.K._ $\dfrac{b}{2t_f} = 7.78 < 8.5$ _O.K._

Check Lateral Bracing. AISC Section [2.9]

For $M/M_p < -0.5$ $L_{cr} = 38.2 \, r_y = 89$ in. $> 6.0 \times 12$ _O.K._

at ridge hinges   Purlin spacing

For $M/M_p > -0.5$ $L_{cr} = 63.2 \, r_y = 147$ in. $> 6.0 \times 12$ _O.K._

at eave hinges

Brace bottom flange of rafter below purlins no. 1 & 2 from eave

Check Axial Load + Bending   AISC Section [2.4]

$M_p = 1250$ kip-ft $M/M_p = 0.909$

Formula [2.4-3] $P/P_y = 0.064 < 0.15$ _O.K._

Gable slenderness: $2(L/r)_x = 114 < C_c = 126.1$ _O.K._

$(L/r)_x = 57$ $P_{cr} = 1.7 \times 17.71 \times 34.8 = 1054$ kip $P/P_{cr} = 0.0757$

$\qquad\qquad\quad$ $P_e = 1.92 \times 45.96 \times 34.8 = 3071$ kip $P/P_e = 0.0260$

$\qquad\qquad\quad$ $C_m = 1 - 0.3 \, P/P_e = 0.992$

$(L/r)_y = 31$ $M_m = \left[1.07 - \dfrac{(L/r)_y}{527}\right] M_p$ $M_m = M_p = 1250$ kip-ft

for $L_y = 6.0$ ft

Formula [2.4-2] $\dfrac{P}{P_{cr}} + \dfrac{C_m}{1 - P/P_e} \times \dfrac{M}{M_m} = 0.076 + \dfrac{0.992}{0.974} \times 0.909$

$\qquad\qquad\qquad\qquad\qquad\qquad = 1.002$ say   _O.K._

Use W 33 x 118 Rafter

Note: Provide web stiffener, AISC Section [2.6], or brace bottom flang below purlin no. 9
from eave

Column Design – Load I $(D + S)$ x 1.70

$M = 1136$ kip-ft   Unbraced lengths

$P = 102$ kip $L_x = 20$ ft $L_y = 5.0$ and 7.5 ft between girts

form in this region. When the lateral bracing requirements of Eq. 20.8 are satisfied, there is no significant reduction in M_m from Formula [2.4-4]. This formula will give $M_m > 0.95M_p$ when the larger bracing spacing permitted by Eq. 20.7 is used.

The sum of the terms in Formula [2.4-2] is 1.002. Thus the W 33 × 118 rafter is adequate. Rafter bracing requirements are indicated by the symbol "RB" below the moment diagram for load I on Sheet P1. To prevent web crippling, AISC Section [2.6] requires the use of web stiffeners at plastic hinge locations such as the rafter hinges near the ridge. The detail in Fig. 20.4f serves the same purpose and provides more positive torsional restraint to the rafter.

If a single-story frame is adequately braced for the critical loading condition which controls the size of the members, it is rarely necessary to investigate lateral bracing requirements for other loading conditions. The negative moment region of the rafter is the only area where load III might require bracing not supplied for load I. The steep moment gradient in this region serves to reduce any tendency for lateral-torsional buckling. There is no question concerning rotation capacity for load III because the bracing for load I ensures rotation capacity.

4. Column Design (Sheets P2 and P3)

The column design follows the same steps used in the rafter design. An effective length factor of 2.7 in the plane of the frame provides for frame stability in the plastic design as described in Section 6 of Art. 20.5. The effective column length is used to determine the loads P_{cr} and P_e for the column but the actual length is used in the column slenderness comparison with C_c as required by AISC Section [2.4]. Since the frame is subject to sidesway in its plane, $C_m = 0.85$ (Section [1.6.1]).

If the inside (compression) flange of the column is not laterally braced between the eave and base, the length L in Formula [2.4-4] is 20 ft giving a minor axis slenderness of 103. This exceeds the limitation of 63.2 from Eq. 20.7 so a bracing detail similar to Fig. 20.4f is added at the girt 5 ft below the eave. This serves to brace the compression flange and preserve rotation capacity for the eave hinge. The minor axis slenderness is reduced to 26, giving $M_m = M_p$ from Formula [2.4-4]. The lower 15 ft of the column compression flange need not be considered in applying this formula because no plastic hinge forms in this segment.

The sum of the terms in AISC Formula 2.4-2 is 0.887 so the W 33 × 118 column is adequate for frame stability and axial load plus bending. A lighter W 30 × 116 lacks adequate plastic moment capacity.

The unbraced compression flange length of 15 ft below the top girt is checked next on Sheet P3. Since this span is an elastic beam-column at

---

EXAMPLE 20.2 (Continued)

SINGLE STORY GABLE FRAME – Plastic Design Method	Sheet P3

Try W 33 x 118   (See Sheet P2)

$$P/P_y = 102/(1250) = 0.082$$

Check Width / Thickness. AISC Section $[2.7]$

Allow $\dfrac{d}{t} = 68.7\,(1 - 1.4\,\dfrac{P}{P_y}) = 60.8 > 59.5$ **O.K.** $\dfrac{b}{2\,t_f} = 7.78 < 8.5$

Effective Length Factors for Column   AISC Section $[1.8]$

see sheet A 4 $K_x = 2.7$ $K_y = 1.0$

Check Axial Load + Bending   AISC Section $[2.4]$

   $M_p = 1250$ kip-ft $M/M_p = 0.908$ $P/P_y < 0.15$ Formula $[2.4-3]$ **O.K.**

   $(L/r)_x = 18.4 < C_c = 126.1$ **O.K.**

   $(KL/r)_x = 2.7 \times 18.4 = 50$ $P_{cr} = 1.7 \times 18.35 \times 34.8 = 1086$ kip $P/P_{cr} = 0.0939$

   $C_m = 0.85\ [1.6.1]$ $P_e = 1.92 \times 59.73 \times 34.8 = 3991$ kip $P/P_e = 0.0256$

   For $L_y = 5.0$ ft below eave

   $(L/r)_y = 26$ $M_m = \left[1.07 - \dfrac{(L/r)_y}{527}\right] M_p$ $M_m = M_p = 1250$ kip-ft

   Formula $\left[2.4-2\right]$ $\dfrac{P}{P_{cr}} + \dfrac{C_m}{1 - P/P_e} \times \dfrac{M}{M_m} = 0.094 + \dfrac{0.85}{0.974} \times 0.909$

                          $= 0.887 < 1.0$ **O.K.**

   Note W 30 x 116 $M_p = 1130$ kip-ft < 1136 **NG**

Check Lateral Bracing   AISC Section $[2.9]$

1. Try brace below purlin no. 1 and girt 5 ft below eave

   $L_y = 60 - 33/2 = 43.5$ in.   At girt $M = (15/20)\,1136 = 852$ kip-ft
      ↖ Rafter

   $M/M_p < -0.5$ $L_{cr} = 38.2\,r_y = 89$ in. > 43.5 in.   **O.K.**

2. Unbraced compression flange (inside) below girt

   $L_y = 15.0$ ft $= 180$ in. This span elastic

   Use Part I of Spec. to check unbraced length

   Load I stresses $P = 102/1.7$ $M = 852/1.7$

   $f_a = \dfrac{102}{34.8 \times 1.70} = 1.73$ ksi $f_b = \dfrac{852 \times 12}{359 \times 1.7} = 16.8$ ksi

   Allowable stresses AISC Sections $[1.5.1.3.1]$ and $[1.5.1.4.6]$

   For F_a : $L_y = 7.5$ ft $(L/r)_y = 39$ $F_a = 19.27$ ksi

              $L_x = 15.0$ ft $(L/r)_x = 14$

   For F_b : $L_y = 180$ in. $\dfrac{L_y d}{A_f C_b} = \dfrac{180 \times 3.88}{1.75} = 399 < 556$

              $C_b = 1.75$ $F_b = 22.0$ ksi

   combined stress AISC Section $[1.6.1]$ $f_a/F_a < 0.15$

   $\dfrac{f_a}{F_a} + \dfrac{f_b}{F_b} = \dfrac{1.73}{19.27} + \dfrac{16.8}{22.0} = 0.090 + 0.764 = 0.854 < 1.0$

                   $\therefore\ L_y = 180$ in.   **O.K.**

> Use W 33 x 118 Column

Note : Brace inside flange of column 5 ft. below eave

ultimate load, it is reviewed using AISC Sections [1.5] and [1.6]. The factor 1.70 is used to convert the ultimate load stresses to an allowable stress basis. Note that the weak axis slenderness ratio for the 7.5-ft girt spacing controls F_a because the girts and siding prevent weak axis buckling under axial load. There is no need to consider effective length factors in this calculation because lateral-torsional buckling rather than frame stability is under investigation. The allowable bending stress F_b determined from AISC Formula [1.5-7] is 22 ksi using $C_b = 1.75$. This C_b value is acceptable since F_b is used in AISC Formula [1.6-2] but should not be used in conjunction with Formula [1.6-1a]. The combined stress check, using Eq. 20.6 for $f_a/F_a < 0.15$, is valid. Therefore, no additional column bracing is required for load I.

The column moments for load III cannot exceed those for load I so load III does not require additional column bracing.

Occasionally, architectural or functional requirements may prohibit any form of bracing for a rigid frame column except at the eave. In this case, the rafter may still be designed plastically if the column is strong enough to force the eave plastic hinge to form in the rafter. The column must then resist the end reaction and plastic moment of the rafter as an elastic and stable beam-column at ultimate load. The unbraced column may be reviewed, using Eq. 20.6 in a manner similar to that illustrated on Sheet P3 for the 15-ft. unbraced elastic column segment. Usually, the unbraced column length about the minor axis will determine F_a. Sidesway buckling in the plane of the frame and out of this plane may be considered in the plastic design by using two effective length factors larger than unity.

20.7 REVIEW OF DESIGN METHODS

The preceding examples illustrate the relative importance of bending versus axial load in proportioning the members of single-story rigid frames. The fact that most of the capacity of the rafter and column sections is used to resist moment implies that plastic behavior in bending is the primary factor governing the strength of adequately braced frames. This plastic behavior is best utilized by the plastic design method, which required a W 33 × 118 rafter and column, as compared with a W 36 × 135 for the allowable stress example.

The allowable stress example indicates that the interaction formula $f_a/F_a + f_b/F_b \leqslant 1.0$ frequently will control the design of single-story rigid frames using Part 1 of the AISC Specification. This results from the fact that f_a/F_a is usually less than 0.15 for these frames. It is also apparent that frame stability is frequently a minor design factor for single-story rigid frames unless the columns are unusually slender or are not braced about the weak axis at the eave.

In the plastic design example, less than a 10 per cent reduction in plastic moment capacity at the eaves is required for axial load in the rafter or column. This indicates that the length and axial load for these members are not sufficient to produce significant inelastic instability (failure by excessive deflection in the plane of bending) at ultimate load in this example. Instead, the size of these members primarily is controlled by their plastic bending capacity at the eaves.

Bracing is required at similar locations in both examples. The best guides to bracing requirements for rigid frames are the moment diagrams. These diagrams locate potential plastic hinges (where rotation capacity is required) and regions where the compression flange of a member is not supported by adjacent framing (where lateral-torsional buckling may occur). The allowable stress parameters $L_y d/A_f C_b \leqslant 556$ and $L_y/b_f \leqslant 12.7$ (for A36 steel W-shapes) are convenient bracing indicators for an unbraced compression flange length L_y. Similar parameters (at hinge locations) in the plastic design method are $L_y/r_y \leqslant 38$ in a region of flat moment gradient and $L_y/r_y \leqslant 63$ in an unbraced span with steep moment.

The emphasis on bracing in this chapter should not be taken to mean that rigid frames cannot be built without all of the bracing indicated in Examples 20.1 and 20.2. Instead, this emphasis is intended to convey the idea that the best advantages of plastic strength can be realized when bracing is supplied at a few critical plastic hinge locations.

The diaphragm properties of the metal roof deck and siding are used in several ways in Examples 20.1 and 20.2. The metal roof deck carries a portion of the roof load to the rafters and provides lateral support for the purlins. The in-plane shear stiffness of the deck and siding is used to provide lateral support for the rafters and columns. In a sense, these uses of the roof deck and siding represent performance requirements which should be verified by the manufacturer.

Factors not considered in these examples include provisions for thermal expansion and the design of the corner connection, rafter field splices, the column base and tie rod, the eave struts and girts, and the wind bracing. Article 19.9 discusses the design of rigid frame corner connections.

Satisfactory deflection behavior under working loads is an essential design criterion. Elastic moments should be used in deflection calculations since moment redistribution does not occur under working load conditions. Figure 20.8 shows the deflected shape of the frame for load I ($D + S$) and Table 20.2 summarizes the working load deflections for Examples 20.1 and 20.2. The dead load deflections are of little concern except to suggest camber and drainage provisions for larger frames with flat roofs. These deflections will occur once the roof is in place and before windows, doors, walls, and suspended equipment are installed.

The vertical deflection at the ridge caused by snow on the plastically designed frame amounts to 1/819 times the span. The leeward eave deflection

Table 20.2   Deflection Summary for Examples 20.1 and 20.2

	Deflection (in.) at Indicated Location				
	Dead Load		Snow Load		0.75 (Snow + Wind)
Frame Size	Vertical at Ridge	Horizontal at Eaves	Vertical at Ridge	Horizontal at Eaves	Horizontal at Leeward Eave
W36 × 135	1.53	0.38	1.33	0.33	0.67
W33 × 118	2.02	0.49	1.76	0.44	0.89

of this frame for the live load portion of load III $(D + S + W) \times 0.75$ is 0.89 in. This corresponds to a deflection index (ratio of horizontal deflection to vertical column height) of 0.0037. These deflections should cause no problems if considered in the architectural details. For example, if the columns are partially encased in masonry, some provision for horizontal movement under live load should be made to avoid cracking the masonry. Thermal contraction of a continuous line of eave struts, particularly if they are erected in the hot sun, should also be considered in masonry details.

The deflections summarized in Table 20.2 are acceptable, but they indicate that stiffness rather than strength may influence the design of taller pinned-base frames or frames using higher-strength steels. In this case, fixed column bases, masonry walls, or other measures may be used to control wind load deflections. (See Art. 20.8.)

It is of interest to compare the weight of steel used by the two design methods and summarized in Table 20.3. The weight reduction for the plastically designed frames represent a saving of 10.9 per cent with no increase in the unit cost of steel fabrication or erection and no additional design effort. Although many factors other than the weight of steel contribute to the cost of a building, it is apparent that the design method, judgement, and assumptions used can be significant cost factors. The economy of rigid frame fabrication, erection, and maintenance should be considered when comparing the weights in Table 20.3 with like values for other framing schemes.

Table 20.3 indicates that the weight differential between a plastically designed frame and a similar frame designed using the allowable stress method

Table 20.3   Weight Summary for Examples 20.1 and 20.2

Item and Size	Weight of Steel (lb/sq ft of floor)
W36 × 135 Rafter and column	7.66
W33 × 118 Rafter and column	6.70
W10 × 11.5 Purlins	2.11
Total: Example 20.1	9.77
Total: Example 20.2	8.81

and hot-rolled shapes may vary but will not reflect a disadvantage for the plastic design. This is because plastic design makes full use of moment redistribution.

Space limitations preclude treatment of rigid frames with haunches or tapered members and of multi-span frames. It can be shown that haunches which extend into the rafter a distance equal to 10 per cent of the span will reduce the plastic modulus of the rafter required for gravity load by more than one-third. Several examples in Ref. 20.2 illustrate the plastic analysis of haunched frames and of multi-span frames using the extensive plastic frame charts of this valuable design reference.

20.8 FUTURE TRENDS

The use of higher-strength steels in large rigid frames, or highly stressed portions of such frames, may serve to further increase the economy and architectural appeal of rigid frame construction by reducing main member sizes and permitting increased spans. Bracing and deflection requirements must be carefully reviewed for rigid frames of such steels.

The ability of a metal roof deck to reduce the deflection of rigid gable frames has received research attention only recently. Reference 20.6 describes how the "cladding" effect (diaphragm behavior) of the roof deck can reduce the live load eaves deflection (and frame moments) of a gable frame to values on the order of 50 per cent of those for the bare frame. This may offer a means for deflection control of high strength steel rigid frames.

The use of web plates welded across the flange tips as a bracing device for W-shapes and the use of box shapes with their superior lateral and torsional stiffness should be mentioned. Article 4.2 of Ref. 7.14 describes one method for estimating the critical inelastic lateral-torsional buckling stress for such members. It is instructive to compare the 50-ft unbraced length of the 37-in.-deep by 10-in.-wide box shape considered in Example 4.1 of Ref. 7.14 with the critical unbraced length of 12 ft for the W36 rafter of similar nominal dimensions in Example 20.1.

An imaginative development in rigid frame construction is their application in clear span framing for round buildings, using radial frames tied with a tension ring at the eaves and a spider or box-type compression ring at the ridge.

PROBLEMS

20.1. (a) What are the failure modes for the components of single-story rigid frames? (b) How are each of these modes provided for in Examples 20.1 and 20.2? (c) What mode controlled the required size of the rafter and column? (d) How does bracing (or its omission) influence the critical mode for design?

20.2. Redesign the purlins in Example 20.1 using a purlin spacing of 8.5 ft. Assume that the roof deck (a) does, (b) does not behave as a diaphragm in the roof plane. In part (b) use two lines of sag rods in each bay to brace the purlin compression flange and design the ridge purlins. Comment on the relative economy of the purlins and the roof deck when the purlin spacing is increased from 6.0 to 8.5 ft.

20.3. Redesign the rafter and rafter bracing (a) in Example 20.1, (b) in Example 20.2, using a purlin spacing of 8.5 ft.

20.4. Assume that the side and end walls of the frame in Fig. 20.4 are to be made of 8-in. concrete blocks faced with 4-in. bricks. (a) Prepare a sketch showing how the columns may be braced against deflection in the weak direction and against twisting using the 8-in. block wall. (*Hint:* Your sketch should provide for deflection of the columns in the plane of the frame to avoid cracking the wall.) Redesign the braced columns using the (b) allowable stress method and (c) the plastic design method.

20.5. The building plan in Fig. 20.4 indicates a future extension 72 ft wide on the north side of the building. When this extension is built, the girts, bracing, and siding below the eave of the existing frames within the 72-ft width will be removed, leaving the north columns of two existing frames unbraced below the eave. Assume that the future extension will not apply any additional load to these columns (by utilizing cantilever framing). Redesign the columns (a) in Example 20.1, and (b) in Example 20.2 to provide for this future extension.

20.6. Design the rafter splices near the eaves in Fig. 20.4 for the rafter in Example 20.1, using A325 bolts. Assume that the flange splices (three plates plus bolts) carry the moment couple and axial load and that the web splices (two plates plus bolts) carry the shear. Should these splices be designed as friction- or as bearing-type connections? Use redistributed moments in the splice design. Compare the effect on the redundant horizontal reaction of (a) plastic hinge action at the eave and (b) slip in the flange splices near the eave. What bearing does this comparison have on the use of redistributed (rather than elastic) moments in the rafter splice design?

20.7. Design the three rafter splices in Fig. 20.4 for the rafter in Example 20.2 using A325 bolts. (*Hint:* Use the statically determinate moment diagram for load III × 1.30 and the redundant moment diagram $M = M_p$ at the leeward eave to estimate the splice moments for load III. Review the ridge connection in Problem 11 of Ref. 20.4.)

20.8. Design the column base and tie rod for (a) Example 20.1 and (b) Example 20.2. Compare the bending action of the base plate for a W36 column with that of the base plate for a column with approximately equal flange width and web depth. How does this comparison influence the choice of critical sections in this design problem? Compare the elongations of an upset tie rod (minimum section outside of the threaded length) and a threaded tie rod (minimum section within the threaded length) if both tie rods have yielded at their minimum sections. If the tie rod in Example 20.2 is designed on the basis of the yield stress on the minimum section, which type of tie rod should be used?

20.9. Design the wind bracing for the building in Fig. 20.4 using the wind load of Example 20.1 on the end of the building. Assume that the diagonal wind bracing acts only in tension and that an initial tension of 5 ksi is produced by fabricating the bracing slightly shorter than its theoretical length to reduce sagging. What shear is transmitted to the anchor bolts of the end frame by the wind bracing? How does this compare with the shear on the anchor bolts due to wind acting on the side of the building? How may the wind shear on the end frame anchor bolts be approximated for a north-east wind?

20.10. In the end bays of the building in Fig. 20.4, the fifth purlin from the eave carries an axial load of 15 kips' compression due to wind on the end of the building. Check the W 10 × 11.5 purlins in Example 20.1 for combined bending and axial load due to (*D* + *S* + *W*) loading. (*Hint:* Review Example 11.6.)

20.11. The owner wants to suspend an air-conditioning unit weighing 1500 lbs from the W 10 × 11.5 purlins in Example 20.1. The unit is mounted on a 2-ft-by-6.5-ft rectangular angle frame which can be welded or bolted to the bottom flange of two purlins. Assume that no other mechanical equipment will be suspended from these purlins, that the center of gravity of the unit is 5 in. below the angle frame, and that the unit must be located at least 8 ft from the rafters. Prepare a sketch and design calculations showing where and how the angle frame for this unit may be suspended from the purlins. (*Hint:* Consider using secondary bracing between purlins to resist torsion.) Estimate the purlin deflection.

20.12. The future extension on the north side of the building in Fig. 20.4 will be 72 ft wide by 144 ft long with a flat metal deck roof and corrugated metal siding to match the existing building. Rectangular rigid frames with a clear height of 18 ft and spaced 20 ft apart will span the 72-ft width. Eight frames will be used, with the south end frame located 4 ft from the existing building. Use cantilever purlins to span this 4-ft space. (This minimizes alterations to the existing building and its foundation.) Prepare preliminary framing plans (similar to Fig. 20.4) for this future extension. Indicate structural provisions for (a) roof drainage, (b) horizontal movement between the existing building and future extension, and (c) field splices. (Hint: The slope required for roof drainage (about $\frac{1}{8}$ in. per ft) may be obtained by (1) using lightweight fill of varying depth, (2) by cambering the rigid frame rafter for dead plus live load plus drainage slope, or (3) by using columns of unequal height.)

20.13. Design continuous rolled-steel purlins for the future extension in Problem 20.12. (*Hint:* The roof loads will depend on provisions for drainage if lightweight fill is used.) Use the AISC Specification, Part 2, and loads appropriate to the local climate.

20.14. Design the rigid frames for the future extension in Problem 20.12 using the AISC Specification and local load provisions. Use (a) plastic design or (b) allowable stress design.

20.15. Redesign the frame shown in Fig. 20.4 using (a) curved haunches or (b) tapered haunches, which join with the rafter under the second purlin from the eave. Use (c) plastic design or (d) allowable stress design and the load provisions of Example 20.1.

21

MULTI-STORY FRAMES I

21.1 INTRODUCTION AND HISTORICAL REVIEW

This chapter discusses the selection of structural members for multi-story building frames proportioned by allowable stress design. It will be shown that there are many possibilities available for alternate choices of members through the use of different types of construction and different types of steel.

Before beginning the study of the design of multi-story frames it is desirable to review the history of steel multi-story buildings. An interesting account of the evolution of such buildings may be found in Refs. 21.1 and 21.2. The following summary is drawn mostly from these references.

Any review of the history of multi-story frames must begin properly with the use of cast iron in buildings. In the late eighteenth century some use of cast iron columns in buildings was made in Europe. Hollow cast iron columns of about 8-in. diameter were used in some buildings in Philadelphia in 1820. An entire wall of cast iron provided a two-story facade for the Miners' Bank at Pottsville, Pennsylvania in 1829.

As early as 1842 another material, wrought iron, was used as a material for building frames. A 34-ft tower of wrought iron framing was used in the Black Rock Lighthouse on Block Island, Rhode Island, in Long Island Sound.

From about 1850 to 1890, multi-story buildings in the United States were constructed using cast iron columns and girders, timber beams and joists, and brick and stone piers and bearing walls. The iron members were bolted together by hand with common bolts through flanges in the ends of the members. The use of cast iron permitted the members to be cast readily into any desired shape.   As a result, columns and facades often followed the ornamental form of classic Greek, Roman, and antique Italian architecture. The cast iron buildings had inadequate capacity for fire resistance and lacked rigidity because of the use of hand-bolted joints.   These factors tended to limit cast iron buildings of the day to about five stories in height.   However, protection from corrosion by adequate painting has permitted some of these buildings to last for at least a century.

The use of wrought iron structural joists rolled as I-beams with a maximum depth of 7 in. began with the Harper & Bros. (now Harper & Row) printing house in New York City in 1854.

The first passenger elevator in the United States was installed in a department store in 1857.   (See Fig. 21.1.)   This development meant that the height of buildings no longer had to be limited to the distance people could climb with reasonable ease.

Fig. 21.1 Haughwout Building, New York City.

Interesting innovations in the Cooper Union Building in New York (1854) were a steam-driven exhaust fan, the forerunner of air conditioning, and a cylindrical elevator and shaft.

In the 1860's, rising real estate prices and intensive land use forced a turning point in New York City building. Improvements in elevators made height increases practical for tenants, and improvements in methods of fireproofing columns and floor beams contributed to the safety of the taller structures.

Another development at this time was the installation of testing machines by iron manufacturers in 1865. Up to this time iron construction was really an adventure, with quality control being almost nonexistent.

With these improvements the height of buildings was increased from the 85 ft of the Wanamaker Building (1859) to 130 ft in the Equitable Life Assurance Building (1868). This was followed by the 230-ft-high Western Union Building and the 260-ft-high Tribune Building in 1875.

The first structure in New York other than a bridge to use main bearing members of steel was the framework of the Statue of Liberty (1883). The steel framework inside the sculptured copper sheeting presented many of the structural problems of a multi-story building. Since the frame has no massive masonry to help resist wind, it had, instead, an extensive system of diagonal bracing.

Advancement of cast iron building in Chicago was stimulated by the rapid growth of industry and commerce beginning about 1860, and by the fire of 1871. Among the pioneers of the new structural developments in Chicago was William LeBaron Jenney. In one of his projects, the Home Insurance Building (1884), the load of all walls and floors was supported on the columns at each level and this minimized the effects of different thermal coefficients of expansion of materials. This building was one of the forerunners of modern framed construction. Another first for this building was the substitution of the first shipment of Bessemer steel beams for the wrought iron beams above the sixth floor.

Riveted connections used in the Tacoma Building in Chicago (1888) provided greater rigidity than was possible with common bolts. The Chamber of Commerce Building built at the same time was the first building in Chicago using an iron and steel frame 13 stories high "without any masonry adjuncts." The extensive use of plate glass in bay-wide windows introduced in this building became a dominant characteristic of the Chicago style.

From 1890 to 1893 a great number of large steel-framed buildings was erected in Chicago. Typical features of the new system of skyscraper or "Chicago" framing were the riveted steel frame, wind bracing, bedrock caissons under column footings, and special framing for projecting windows.

Around the end of the nineteenth century, the rolled H-column began to appear in buildings but the growth of its size was limited by the capacity of rolling mills.

Over a period of twenty years after the Chicago system was introduced in New York (1888–1908) many tall buildings were constructed, with the tallest a 50-story building 700 ft in height.

A significant milestone in the history of multi-story buildings was the 55-story, 760-ft-high Woolworth Tower completed in 1913 (Fig. 21.2). Until 1930 this structure reigned as the world's tallest building. Because of its extreme height, the tower required the use of many types of bracing to resist wind, including knee braces, K-bracing, rigid portal frames, and diagonal bracing in the plane of the floors. At three levels, the straight vertical lines of the building were interrupted by setbacks which required the columns to be supported by floor girders at the level of the offset. In addition to the immense superstructure required for the building, foundation conditions presented an additional challenge. Since the building was built near the lower tip of Manhattan Island on a thick bed of alluvial mud and sand, concrete caissons set 100 ft below grade level were required for the column footings.

The next outstanding development in the multi-story building was the construction of large buildings on air rights over tracks and other railroad facilities of Grand Central Railway Terminal. The terminal was built in 1913, and by 1921 arrangements were made to construct a number of buildings over the underground tracks system of the terminal. The structural and construction problems involved were to install the columns and footings without interfering with the operations of trains and to prevent vibrations due to railroad traffic from being transmitted to the building. The first building completed in the air rights development was the Park-Lexington Building erected north of the terminal property in 1922–23. The original set of air rights buildings was completed with the construction of the Waldorf-Astoria Hotel in 1931 and the New York Central Building in 1932.

The beginning of the thirties saw the construction of the first two buildings to top the 1000-ft mark, the 75-story Chrysler Building and the 102-story, 1250-ft Empire State Building (Fig. 1.2). For nearly 40 years, it was not found necessary to exceed the heights of these great buildings.

Since the successful completion of these tall buildings, the tendency has been to simplify the frame and wind bracing in order to increase the floor area of individual bays. The first structure to achieve this in an efficient manner was the Gulf Oil Building in Pittsburgh (1931). This 30-story building, 108 by 108 ft in plan, had a central service core containing all the needed bracing and providing clear floor areas right out to its exterior walls.

Tall structures built during the post-World War II era have been distinguished by curtain walls of glass and thin steel or aluminum panels, although the continuous window wall was used for industrial buildings as early as 1910. The first steel-framed multi-story structure with glass curtain walls was the research building of the A. O. Smith Corporation built in Milwaukee, Wisconsin, in 1928.

Fig. 21.2 Woolworth Tower, New York City, 1913. (Courtesy of Parker-Brand Corp.)

One significant development in building construction which took considerable time to make its debut was the introduction of electric welding in 1881. (See Art. 14.1.) At first, welding was limited to the joining of small pieces of metal; it was not until 1920 that the factory of the Electric Welding Company of America was built, using the first steel building framework with welded connections. In 1926 the five-story factory of the Westinghouse Electric and Manufacturing Company at Sharon, Pennsylvania, became the first building supported throughout by a multi-story all-welded steel frame. Welding was used to construct, in what is possibly the ultimate use of framing, a column-free interior in the Inland Steel Building in Chicago, Illinois (1955). The entire floor and roof of this 19-story building are carried by 77-ft transverse girders framing into only seven pairs of columns located wholly outside the glass curtain walls of its long elevations (Fig. 21.3).

It is probable that the use of high-speed computers and the availability of the most recent solutions for the behavior of structural members will soon lead to the most refined designs possible for traditional beam and column framing of multi-story frames. For further improvements in design, new structural framing systems may be needed. A promising example of a new system was used in Chicago's John Hancock Center (Fig. 1.3) where a 100-story building was built for the equivalent unit cost of a 40-story framed structure.[21.3,21.4] In this system diagonal members and exterior columns act together as an exterior tube to resist both gravity loads and horizontal wind forces.

The first buildings to exceed the height of the Empire State Building were completed in the early 1970's. The twin 110-story towers of the World Trade Center in New York (Fig. 1.4) rise 1350 feet.[21.5] A system of closely spaced columns and massive spandrel beams in the shape of a vertical tube forms the outer walls and resists both vertical loads and wind. Simply supported floor trusses span between the outer walls and interior columns in a core containing the elevators, providing a large amount of column-free floor space. In Chicago, the 109-story Sears Tower rises 1450 feet. It too uses the vertical tube framing principle to provide large areas of column free space.[21.6]

The development of the plastic analysis and design of welded rigid frames in the 1950's provided a better understanding of the behavior of structures. More recent research has been conducted on the plastic design of multi-story frames, and is described in Chapter 22. The first steps into plastic design of multi-story buildings are seen in the design of an eight-story apartment building (1960) in Toronto, Canada (Fig. 21.4) and in the 17-story Tower Building (1960) in Little Rock, Arkansas (Fig. 21.5). Both of these buildings include plastically designed continuous beams in their floor systems but have columns and diagonal wind bracing proportioned by allowable stress design. Results of the more recent research were applied in the

Fig. 21.3 Inland Steel Building, Chicago. (Courtesy of Bill Engdahl, Hedrich-Blessing.)

Fig. 21.4 Eight-story apartment building—plastic design. (Courtesy of Canadian Institute of Steel Construction.)

design of the Stevenson apartments constructed in Bladensburg, Maryland, where the columns and bracing, as well as the beams, were proportioned plastically.[21.7] (See Fig. 22.1.)

An ever-increasing list of steels available for structural use and the ability to make a variety of shop and field connections by welding, riveting, and high-strength bolting give promise that the future history of multi-story buildings will be limited only by the ingenuity of the architect, structural designer, fabricator and erector.

21.2 TYPES OF CONSTRUCTION

In most large cities the design of multi-story buildings is governed by a local building code. Usually the structural design and allowable stress

Fig. 21.5 Tower Building—plastic design. (Courtesy of F. E. Withrow and H. A. Berry, Architect, E. Saunders, Photographer.)

provisions of the local code parallel those of the AISC Specification, which is the basis for the treatment of multi-story frames in this chapter.

The AISC Specification permits the unconditional use of Type 1 (or "rigid frame") construction for multi-story frames. The effect of rigid frame construction is to reduce the size of girders as determined by gravity loading while requiring that columns and beam-to-column connections be able to carry larger moments than for other types of framing. Splices and beam-to-column connections which have sufficient rigidity to hold "virtually unchanged" the original angles between intersecting members qualify as Type 1.

Tier buildings designed as Type 2 construction are cited by the Specification as the general case. In Type 2 construction, designated as "simple framing," the ends of beams and girders are connected for shear only and are free to rotate under load. This results in using the largest possible girders, but the elimination of end moments makes possible the use of the simplest connections and reduces the bending moments in the columns. Because of the absence of moment-resisting connections, it is necessary to provide lateral stability by means of diagonal bracing, shear walls, or attachment to an adjacent structure having adequate lateral stability.

A paradox is created in a second definition of the term "bracing" as used in tier building. As recommended in AISC Section [1.2], Type 2 construction may be designed with beam-to-column connections assumed to be pinned-end for gravity loading but able to carry moments caused by wind. The wind moment connections which are designed to resist wind effects by means of rigid frame action frequently are termed "wind bracing." This presents a paradox when one tries to imagine connections which will develop a bending moment only when the wind blows. From a purely academic standpoint it would appear that this type of construction might really fit in the category of Type 3 (semi-rigid) construction.

The practical reasons for the use of Type 2 construction arise from long and successful experience, the ductile behavior of steel, and the expense of fabricating riveted or bolted connections capable of fully rigid action. Grotesque connections like the wind-bracing bracket in Fig. 21.6e will probably cost more than the approximately 30-per cent saving in girder material which can be achieved by full continuity. In low-rise buildings and in the top several stories of high-rise buildings, wind moments are likely to be quite small. Girder connections considerably smaller than those necessary to develop the full bending strength of a girder may be adequate to resist the wind moments. Such connections can be of the types shown in Fig. 21.6c and 21.6d. These connections will resist the small wind moments but will also permit end rotation of the girders under gravity loading sufficient to approximate nearly pinned-end conditions. Article 18.4 described how the restraining moment in semi-rigid connections can be determined for beams subjected to gravity loading. Reference 21.8 analyzes the behavior of such connections subjected to wind loads which are reversed in direction, in addition to gravity loads. It is shown that self-limiting plastic rotation of the connections results in static equilibrium under gravity loading with residual moments in the connections.[21.8] After the first full application of wind in each direction, the residual moment at the connections is reduced. All

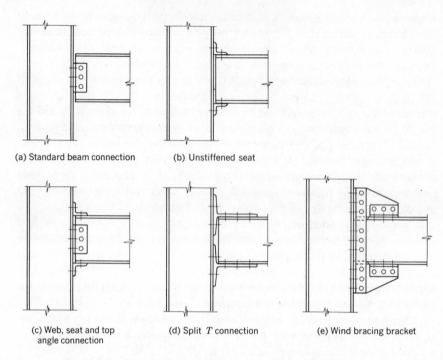

(a) Standard beam connection   (b) Unstiffened seat

(c) Web, seat and top   (d) Split T connection   (e) Wind bracing bracket
angle connection

Fig. 21.6 Types of riveted beam-to-column connections.

further applications of wind loading result in moment-rotation behavior in the elastic range.   Reference 21.8 concludes that such semi-rigid connections are adequate to sustain the necessary combination of gravity and wind loads if certain conditions are met.   These conditions are:

1. The connections must be able to undergo large inelastic deformations without failure.
2. The connections must be adequate to resist the wind moments alone at allowable stresses computed on the assumption that the connections are elastic.

Though welding can facilitate the use of fully continuous beam-to-column connections with less complicated details, the difficulties of making close fits during erection often make it desirable to use flexible types of connections for welded structures.

In the remainder of this chapter, the allowable stress design of multi-story frames for Type 1 and Type 2 construction will be discussed.   Article 21.7 includes material on semi-rigid connections as used in "simple" framing.

Plastic design can be used for braced tier buildings under certain conditions. This subject is covered in Chapter 22.

21.3 DISTRIBUTION OF LOADS TO FRAME

In Chapter 1, the logical steps in the planning and design of structures are outlined. The steps are discussed further in Chapter 3. The procedure for the structural design of a multi-story frame is illustrated in Examples 21.1 through 21.17. Example 21.1 presents the initial steps in the design of the bent sketched in Fig. 21.7 which is the basis for discussion in this chapter. This could be a typical bent from a tier building rectangular in plan designed for office, hospital, or dwelling use. It is assumed that all floors are 5-in. concrete slabs supported on steel beam and girder systems. The weight of exterior walls is assumed to be carried on spandrel beams framing to the outer faces of the exterior columns.

Uniformly distributed wind loads on the vertical surfaces of the building are assumed to be carried as horizontal concentrated loads applied at each floor level.

To design the structural members of the bent, it is necessary to know the amount and distribution of live and dead loads on the bent itself. The manner in which loads are carried from the floor to the bent is indeterminate and depends to a large extent on the type of floor system and framing of the bent. For purposes of designing the steel frame, the designer has a choice of patterns of distribution of floor loads which can be used. The important thing is to provide a path for all floor loads to be carried to the frame.

Figure 21.8 shows a possible system of floor beams and girders for a bay of the sample building. Three possible patterns of distribution of assumed uniform loads are indicated by sectioned areas in the plan views. The girder loadings consistent with the assumed distributions are also shown.

The distribution shown in Fig. 21.8a would be appropriate to use with two-way concrete slabs spanning each panel formed by pairs of floor beams and girders. Lines at an angle of 45° from the corners of panels divide each area into portions to be assigned to the adjacent members. This results in triangular patterns of distributed loads on the girders as well as concentrated loads from the floor beams. With little error, the triangular loads may be regarded as uniformly distributed.[21.9]

Figure 21.8b shows a distribution more consistent with the action of one-way floor slabs. Most of the floor load is carried by the floor beams due to the one-way slab action. Because transverse reinforcement having a length about one-fourth the girder span is provided over each girder, some of the loads might be carried directly to the girder. Reference 4.24 recommends the use of a strip of load covering a width one-fifth of the girder span.

EXAMPLE 21.1

PROBLEM:
State the assumptions and conditions to be used in the design of a multi-story frame in Examples 21.2 through 21.17.

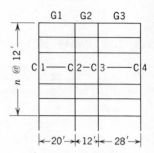

SOLUTION:
Design a multi-story frame for a tier building having column spacing shown and 12 ft. story heights with 24 ft spacing of bents.

Design as many stories as possible starting from the top down. By cutting off the design at any level, a result typical for that height may be achieved. Use AISC Specification.

Live Loads: Roof   30 psf
                  Floor   100 psf including movable partitions.
          Live Load Reduction according to ASA A58.1.

Dead Loads: To facilitate computations
                  Roof   95 psf
                  Floor   120 psf

   Dead loads are assumed to include a 5 in. concrete slab, floor or roof and ceiling finish and the weight of any beam or girder framing plus necessary fireproofing.

   Dead loads of columns will be assumed as a constant per story height including fireproofing.

   An exterior wall with a dead weight of 85 psf including windows and openings will be assumed to be carried by spandrel beams framed to the outside of columns at each floor level. A four foot parapet wall will be assumed on roof spandrels.

Wind Loads:
          20 psf acting horizontally on entire surface.

Materials:
          Design will be attempted with each of the following steels, trying to gain an economical solution.
                  ASTM   A36
                  ASTM   A441

Type of Construction:
          Tier buildings may be designed for all three types of construction. This trial design will investigate:
                  Type 1 – Rigid Frame Construction
                  Type 2 – Simple Framing

Distribution of Wind Moments

   Distribution of wind moments may be made by a recognized empirical method as long as proper provisions are made for connections.

   The cantilever and portal methods are used because they require no knowledge of member sizes beforehand.

   Other methods can be valuable for checking once the sizes have been assumed reasonably.

Distribution of Gravity Moments

   The Specification implies that these should be handled by an exact method. However, this is impractical for Type I framing and reasonable approximations will be made to allow selection of members.

   The preceding assumptions and conditions are used as needed in Examples 21.2 through 21.17.

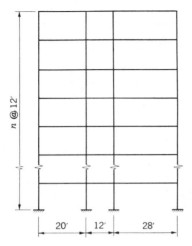

Fig. 21.7   Elevation of typical bent.

Figure 21.8c shows the distribution which will be assumed for examples in this chapter.   All floor loads will be assumed to be carried by the main girders as uniformly distributed loads, even though the floor beams framing into the girders would cause concentrated loads.   It should be noted that the share of load belonging to the spandrel beams is also assigned to the girders.   While this procedure assigns a total load to the girders which is too high, the elimination of consideration of the concentrated loads reduces the severity of the assumption.   The net result is that all three cases would give nearly the same maximum bending moment to the main girders.   The procedure of assigning all the floor loads to the girders reduces some of the arithmetic in calculating column moments.   Example 21.2 shows the calculation of uniform loads on girders based on the assumed distribution of floor loads.

21.4 ANALYSIS OF STRUCTURE DUE TO GRAVITY LOADS

An exact analysis of a rigid frame structure by classical methods depends on prior knowledge of the relative stiffness of members.   To make a preliminary selection of members, it is desirable to use a reasonably approximate method of analysis.

1. Simple Framing

With simple framing, the determination of the complete moment diagram of each girder presents no problem.   Moments are readily determined by

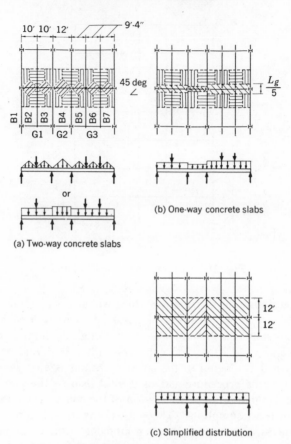

Fig. 21.8   Distribution of floor loads to girders.

statics.   Clear spans of girders may be used if Eq. 21.1 is used for column moments.

Bending moments in columns may be induced by the girder shears applied to the outer flanges of the columns at each floor level.   If the shears at the two faces of a column are unequal as shown in Fig. 21.9a, the unbalanced moment will be equal to the difference in shears times half the column depth, thus:

$$M = (V_2 - V_1)\frac{d_c}{2} \tag{21.1}$$

The deflected shape of the columns in a two-story tier as shown in Fig. 21.9b enables the designer to visualize the proper sense of the moment above and below the joint.

EXAMPLE 21.2

PROBLEM:

Determine girder uniform loads, live load reduction percentages, and concentrated wind loads for the frame of Example 21.1.

SOLUTION:

Roof Loads:

Live Load	w = 30 psf x 24 ft. =	0.72 kip/ft
Dead Load	w = 95 psf x 24 ft. =	2.28 kip/ft
	Total =	3.00 kip/ft

Floor Loads:

Live Load	w = 100 psf x 24 ft. =	2.40 kip/ft
Dead Load	w = 120 psf x 24 ft. =	2.88 kip/ft
	Total =	5.28 kip/ft

Live Load Reduction:

Maximum Reduction

$$R = 100 \times \frac{D+L}{4.33L} = \frac{100 \times 220 \text{ psf}}{4.33 \times 100 \text{ psf}} = 50.9\%$$

Girder No.	G1	G2	G3
Floor Area Served (sq. ft.)	480	288	672
% Reduction of Live Load (0.08 x Area)	38.4	23.04	50.9

In applying maximum reduction formula to columns, consider effect of spandrel dead loads.

Column No.		C1	C2	C3	C4
Floor Area Served (sq. ft.)		240	384	480	336
% Reduction of Live Load	–Roof	0	0	0	0
	–Roof-1	19.2	30.7	38.4	26.9
	–Roof-2	38.4	52.1	51.8	53.8
	–Roof-3	57.6	51.9	51.7	60.0
–All Other Stories		60.0	51.7	51.5	60.0

Wall Loads on Exterior Columns: Roof Story

P = 85 psf x 12 ft x 24 ft = 24.5 kip   4/12 x 24.5 = 8.2 kip

Concentrated Wind Force: Roof Story

H = 20 psf x 12 ft x 24 ft x 3/4 = 4.32 kip   10/12 x 4.32 = 3.6 kip

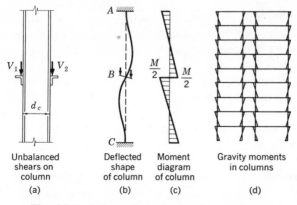

Unbalanced shears on column	Deflected shape of column	Moment diagram of column	Gravity moments in columns
(a)	(b)	(c)	(d)

Fig. 21.9 Effect of gravity loads on columns.

749

The proportion of the column moment to be distributed above and below the joint depends on the stiffness of the two column segments. When selecting a member for segment AB of Fig. 21.9b, it is generally conservative to assume that half the total moment acts on that segment if both columns are of equal length. The column section BC of the lower tier is likely to be heavier than AB, the tier being designed. The actual moment in AB probably would be less than half the total, making the assumption that AB takes half the joint moment more severe than the actual case. For equal column lengths, the moment diagram for gravity moments in the columns for a complete bent would be similar in shape to that in Fig. 21.9d.

The distribution of moments above and below a joint might be changed for a case in which the story below has a greater height. If both columns are assumed to be of the same section, the stiffnesses would be proportional to the reciprocals of the column lengths and the upper column would be called upon to carry more than half of the total moment. By a reasoning similar to that used previously, it can be shown that this assumed moment on the upper column will probably be more severe than the actual moment.

2. Rigid Framing

With rigid framing, the moment distribution in the girders as well as in the columns is indeterminate. By examining the maximum moments and location of inflection points for a number of the known limiting cases of indeterminate beams, an approximation can be made which will result in the selection of reasonable member sizes for girders.[21.10]

Moment diagrams for four cases of a uniformly loaded girder are shown in Fig. 21.10. Figure 21.10a gives the moment diagram for a simply supported girder in which the ends are perfectly free to rotate. The maximum positive moment at mid-span for this case would require the greatest possible member size of any distribution of gravity moments. An inflection point can be assumed to occur at the ends of the member.

Figure 21.10b shows the moment diagram for a member with ends perfectly fixed against rotation. In this case, member size would be controlled by the negative moment at the support, and the inflection points would occur $0.21L$ away from the ends.

A moment diagram for the case of optimum redistribution of moment as obtained from a plastic analysis is given in Fig. 21.10c. With equal negative and positive moments, this case results in the smallest possible member size. The inflection points occur at a distance $0.146L$ from the ends.

The transition from Fig. 21.10b to Fig. 21.10c represents the effect on moment distribution which could occur from rotation of the end joints due to either elastic bending of the adjacent columns or to plastic behavior of the end of the girder. Further rotation of the end joint could result in a distribution

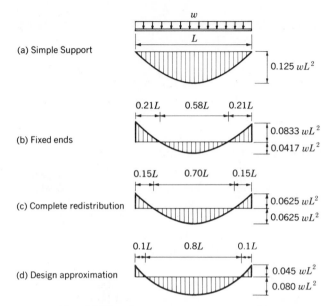

Fig. 21.10 Gravity moments in girders.

of moments similar to that in Fig. 21.10d.   Here the moment controlling member size is the maximum positive moment, and the inflection points would occur at $0.1L$ from the ends.

Comparison of the three moment diagrams 21.10b, c, and d shows that the moment controlling the member size would decrease from $0.0833wL^2$ to $0.0625wL^2$ and increase back up to $0.080wL^2$ as the inflection point moved from $0.21L$ to $0.10L$.   Obviously, a member selected on the basis of the approximation given in Fig. 21.10d would be strong enough to carry the load no matter where the inflection point occurs between these limits.   For this reason, girders will be selected for rigid construction using the moment diagram of Fig. 21.10d.

Due to the girder end moments, additional moments beyond those caused by unbalanced girder shears will be applied to the columns.   These additional moments will be distributed above and below the joint in the same manner as was described for simple framing, as shown above.

Once members have been selected on the basis of the foregoing approximate analysis, relative stiffnesses are available to make an exact analysis by one of the classical methods.   These methods are well documented in structural analysis books and are not discussed here.

In the examples to follow, the members selected by approximate analysis are used as the basis of all further calculations.

21.5 ANALYSIS OF STRUCTURE DUE TO LATERAL LOADS

Lateral loads on a building, such as wind and earthquake forces, are reduced usually to a series of horizontal concentrated loads applied to the bent at each floor level.[4.24] Analysis of the rigid frame bent with unknown member sizes may be made most quickly by the portal method or the cantilever method.

1. Portal and Cantilever Methods[3.7,3.13,4.24,21.10,21.11,21.12]

Both the portal and cantilever methods of analysis are based on a visualization of the deflected shape of a bent under concentrated loads alone. Such a picture shows that inflection points would tend to form near the center of each column segment and girder. For purposes of analysis the inflection points are assumed to occur exactly at the center of each member. One additional assumption as to distribution of force is necessary.

In the portal method one version of the additional assumption is that each column in a story resists a percentage of the total horizontal shear on the story proportional to the width of aisle the column supports. For the bent of Fig. 21.7 the columns would carry 0.17, 0.27, 0.33, and 0.23 of the total-story shear based on aisle widths of 10, 16, 20, and 14 ft respectively. With known shears in each column, the column moments, joint moments, and girder moments can be determined. From these, girder shears, girder thrusts, and column thrusts can be determined by simple statics. The wind analysis of a typical story of the bent by the portal method is illustrated in Example 21.3.

In the cantilever method of wind analysis the building is treated as a cantilever beam standing on end and fixed to the ground. The neutral axis of the cantilever beam is the centroid of the areas of the columns in the bent. (The columns are usually assumed equal in area for this calculation.) The moment of inertia of the cantilever beam depends on the column areas and their distances from the neutral axis. The assumption to be made in the cantilever method is that the direct column stresses are proportional to the distances from the neutral axis. For the bent of Fig. 21.7 the column thrusts would be $0.015M$ tension, $0.004M$ tension, $0.002M$ compression, and $0.017M$ compression, respectively. M is the moment in kip-ft at mid-height of the story considered, caused by all lateral forces above the story. Again, all other forces may be determined by simple statics. Analysis of a typical story of the bent by the cantilever method is illustrated in Example 21.4.

2. Witmer K-Percentage and Spurr Methods

When member sizes are known, a more exact analysis recognizing the elastic properties of the members is possible. Some approximate methods

EXAMPLE 21.3

PROBLEM:
  Make a wind analysis of the 8th story below the roof of the frame of Example 21.1 by the portal method.

SOLUTION:
  (1) Assume inflection points at the middle of all columns and girders.
  (2) Portion of story shear taken by each column.

Column No.	C1	C2	C3	C4
Aisle Width	10	16	20	14
% of Total Shear	16.7	26.7	33.3	23.3

  (3) Shear above floor T-8
$$H = 3.6 \text{ kip} + 7\left(4.32 \text{ kip}\right) = 33.84 \text{ kip}$$
  (4) Shear below floor T-8
$$= 33.84 \text{ kip} + 4.32 \text{ kip} = 38.16 \text{ kip}$$
  (5) Distribution of Shears

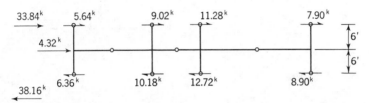

  (6) Moment Diagram (kip-ft.)

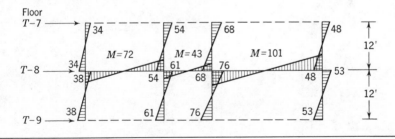

using known elastic properties are able to give results almost as accurate as the slope-deflection and moment distribution methods which are accepted as "exact."

  One of these is the Witmer K-percentage method which considers the relative stiffnesses of the frame and determines vertical wind reactions at all columns. The wind moments may be computed from these reactions.[3.13]

  A second method, the Spurr method, is especially valuable when the height-to-width ratio of a building exceeds about 5. This method assumes

EXAMPLE 21.4

PROBLEM:

Make a wind analysis of the 8th story below the roof of the frame of Example 21.1 by the cantilever method.

SOLUTION:

(1) Assume inflection points at the middle of all columns and girders. Assume column areas equal. (A)

(2) Location of neutral axis.

$$X = \frac{(20 \times A) + (32 \times A) + (60 \times A)}{4A} = 28\,ft.$$

(3) Moment of inertia of column areas.

$$I = \left[(28)^2 + (8)^2 + (4)^2 + (32)^2 \right] A = 1888\,ft.^2 \times A$$

(4) Column thrust = column area × stress caused by cantilever moment at midheight of column.

$$P = A \frac{Mx}{I}$$

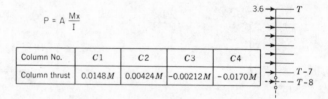

Column No.	C1	C2	C3	C4
Column thrust	$0.0148M$	$0.00424M$	$-0.00212M$	$-0.0170M$

(5) Moment above floor T-8.

$$M = 3.6\,(90) + 7(4.32)(42) = 1594\,kip\text{-}ft.$$

(6) Moment above floor T-9.

$$M = 3.6\,(102) + 8(4.32)(48) = 2026\,kip\text{-}ft.$$

(7) Column thrusts.

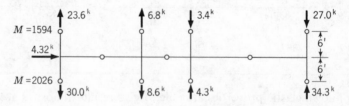

(8) Moments:

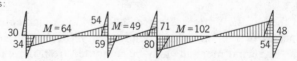

EXAMPLE 21.4 (Continued)

NOTE:
Analysis must be carried from roof step-by-step to obtain column moments, using procedure illustrated in the following table:

Step No.	Operation	Moment M (kip-ft)	Shear H (kip)	C1	C2	C3	C4
		Total Wind		Column Number			
	Roof Story	21.6	3.6				
(1)	Column Thrust = Mc/I (kip)			0.32	0.09	−0.05	−0.36
(2)	Girder Shear (Roof) (kip)			0.32	0.41	0.36	
(3)	Girder Moment (1) (kip-ft)			3.2	2.5	5.0	
(4)	Column Moment (2) (kip-ft)			3.2	5.7	7.5	5.0
(5)	Column Shear (3) (kip)			0.53	0.95	1.25	0.83
	T-1 Story	90.7	7.9				
(6)	Column Thrust = Mc/I (kip)			1.35	0.38	−0.19	−1.54
(7)	Girder Shear (4) (kip)			1.03	1.32	1.18	
(8)	Girder Moment (1) (kip-ft)			10.3	7.9	16.5	
(9)	Column Moment (2) (kip-ft)			7.1	12.5	16.9	11.5
(10)	Column Shear (3) (kip)			1.18	2.08	2.82	1.92
	T-2 Story	211.7	12.2				
(11)	Column Thrust = Mc/I (kip)			3.14	0.90	−0.45	−3.59
(12)	Girder Shear (4) (kip)			1.79	2.31	2.05	
(13)	Girder Moment (1) (kip-ft)			17.9	13.9	28.7	
(14)	Column Moment (2) (kip-ft)			10.8	19.3	25.7	17.2
(15)	Column Shear (3) (kip)			1.80	3.22	4.28	2.87
	T-3 Story	384.5	16.6				
(16)	Column Thrust = Mc/I (kip)			5.71	1.63	−0.82	−6.52
(17)	Girder Shear (4) (kip)			2.57	3.30	2.93	
(18)	Girder Moment (1) (kip-ft)			25.7	19.8	41.0	
(19)	Column Moment (2) (kip-ft)			14.9	26.2	35.1	23.8
(20)	Column Shear (3) (kip)			2.48	4.37	5.85	3.97

(1) Girder Shear x 1/2 Span.

(2) At each joint Σ M = 0.

(3) Column Moment ÷ 1/2 Column Height.
    For check: Σ Column Shears = Total Shear H.

(4) Σ Thrust on this story minus Σ Thrust on story above.

the original area ratios of the columns to be proportional to each column's share of the gravity loads. Using these ratios of areas, column and girder shears are determined in a manner similar to the cantilever method. Girders are selected with relative stiffness proportioned so that end slopes due to bending remain equal throughout the structure. This serves to maintain planarity of the floors under wind loads and to make the structure act as a true cantilever.[3.13]

In 1940 a subcommittee of ASCE presented a comprehensive report evaluating methods of analysis of steel buildings under wind loads.[3.13] The committee reported that the portal method is generally satisfactory for buildings up to 25 stories in height and the cantilever method up to about 35 stories. The committee recommended the K-percentage method and the Spurr method as particularly valuable—the K-percentage method to analyze a structure designed by any method, and the Spurr method to aid in proportioning a structure for uniform behavior.

For the bent discussed in this chapter, it was found that the portal and cantilever methods gave almost identical results for a 14-story building.

3. Other Methods

Two other approximate methods in which the frame can be solved one story at a time are the factor method[21.10] and the Gottshalk method.[21.13] Each of these methods is derived on the basis of deformation considerations and each distributes the forces in a story to the members in proportion to certain combinations of relative stiffness of members.

A so-called "exact" analysis considering only the bending of the members may be made by the method of slope deflection.[1.6,21.5,21.11,21.14,21.15] Though the method makes use of a prohibitive number of simultaneous equations for most applications to multi-story frames, the method may be rendered practical by the use of digital computers and methods of iteration or successive approximation for solution of the equations. The Maney-Goldberg method is a modified slope-deflection method using successive corrections.[21.14]

The method of moment distribution with holding forces by Hardy Cross uses iterative procedures to analyze frames, considering bending only.[1.6,4.24,21.10,21.11,21.14,21.15] For multi-story frames this method also can result in large numbers of simultaneous equations. Simultaneous equations are eliminated by using modified moment-distribution methods, such as the Grinter method,[21.15] the Morris method,[21.11] and the modified Cross method,[4.24] which make successive corrections for shear.

More recent recommendations on the loading and design of structures subjected to wind and earthquake forces are contained in Refs. 21.16 and 3.10.

21.6 DESIGN OF GIRDERS

Once the gravity analysis and wind analysis of the structure are completed, it is possible to design the girders. Girders must be proportioned to carry dead and live loads at the basic allowable stresses or to carry dead plus live plus wind or seismic loads at a $33\frac{1}{3}$-per cent increase in allowable stress. In some cases, the live load on girders supporting a large floor area may be reduced, owing to the improbability of the whole area being loaded at once. Clauses in the AISC Specification provide the added flexibility to the designer of allowing for a range of grades of steel and using composite girders.

1. Live Load Reduction

In accordance with ASA Code A58.1,[3.4] the amount of live load reduction to use in design depends on the floor area supported by the member and the ratio of dead to live load.[21.17]   For the three girders of each floor of the frame of Fig. 21.7, the floor areas of 480, 288, and 672 sq ft result in live load reductions of 38.4, 23.0, and 50.9 per cent respectively as shown in Example 21.2.

2. Design for Gravity Loads

For simple framing, each girder may have a section based on the typical moment diagram of Fig. 21.10a including the effect of dead load and the reduced amount of live load. Since the complete compression flange will be supported by the concrete floor slab, the basic allowable bending stresses for fully supported beams may be used as given in AISC Section [1.5.1.4]. For rigid framing, the positive moment in Fig. 21.10d would control member size.

For both simple and rigid framing, a more economical design might be obtained by using composite action in the positive moment regions of the beams by adding shear connectors. As shown in Art. 403 of Ref. 4.24, the steel member for a trial composite design may be selected by multiplying the allowable stress by a factor of 1.28 for a modular ratio of 10. This means that a composite member may use a steel beam with a section modulus of about 78 per cent of the comparable non-composite member.

3. Design for Wind Plus Gravity Loads

With combinations of wind and other loads, a $33\frac{1}{3}$-per cent increase in allowable stress is permitted by the AISC Specification. This is usually accomplished by multiplying loads by $\frac{3}{4}$ and designing for the basic stress.

In rigid framing, negative moments may be checked by adding the moments at the support due to gravity and wind loadings. The maximum positive moment will vary in magnitude and location depending on the size of the wind moment.

The moments and reactions of a girder are shown in Fig. 21.11. The positive moment M at any point x on the girder is equal to

$$M = [M_w - M_g(\text{neg.})] + x\left(\frac{wL}{2} - \frac{2M_w}{L}\right) - \frac{wx^2}{2} \qquad (21.2)$$

where M_w is the wind moment in the girder, $M_g(\text{neg})$ is the negative gravity moment in the girder, w is the uniform load per unit length, and L is the

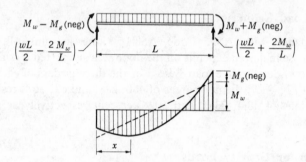

Fig. 21.11 Moments on girder subjected to wind plus gravity loads.

length of the girder. The maximum positive moment occurs at

$$x = \frac{L}{2} - \frac{2M_w}{wL} \qquad (21.3)$$

Its magnitude is

$$M = \frac{wL^2}{8} + \frac{2M_w{}^2}{wL^2} - M_g(\text{neg.}) \qquad (21.4)$$

The design moment using normal working stresses is three-quarters of this value.

For the positive design moment due to gravity plus wind load to exceed that due to gravity loading alone, three-quarters of the wind moment must exceed $0.108wL^2$ for rigid framing designed in accordance with Fig. 21.10d. For the negative design moment due to gravity plus wind load to control design, three-quarters of the wind moment must exceed $0.125wL^2$ for simple framing (with connections adequate for wind moments) and $0.04625wL^2$ for rigid framing.

Figures 21.12a and 21.12b show the design moments for the middle or corridor girder of the sample frame based on simple and rigid framing

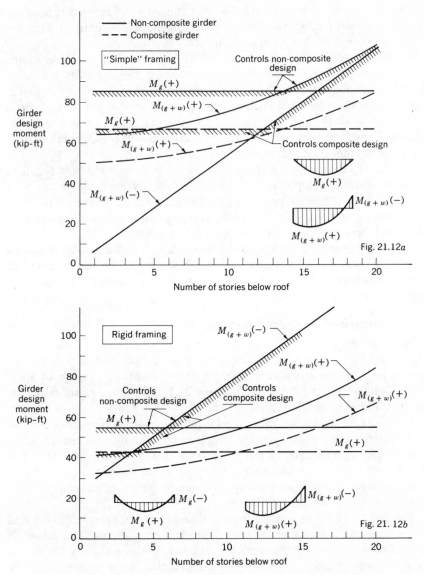

Fig. 21.12 Design moments for corridor girder.

respectively. Solid lines are used to designate non-composite girders and dashed lines to designate composite girders. For each type of framing, the positive moment due to gravity loads is designated as $M_g(+)$. The positive moment due to gravity plus wind loads times $\frac{3}{4}$ is designated as $M_{g+w}(+)$. The negative moment due to gravity plus wind loads times $\frac{3}{4}$ is designated as $M_{g+w}(-)$ and is the same for both non-composite and composite framing. Sectioning marks designate the moments which would control design for composite and non-composite construction.

From Fig. 21.12a it can be observed that gravity loads control the design for the top 13 stories. Then the positive moment due to gravity plus wind load begins to control for the lower stories of non-composite construction. With composite construction the negative wind moment begins to control 13 stories below the roof. From the slopes of the $M_{g+w}(+)$ and $M_{g+w}(-)$ curves it might be expected that negative moment will control both designs somewhere more than twenty stories below the roof.

In Fig. 21.12b it is seen that negative gravity plus wind moment begins to control both non-composite and composite construction only six and four stories below the roof respectively for rigid framing. From the standpoint of member size, there would be no advantage in using composite girders for the lower stories by the allowable stress design method.

For the longer span girders in the two wider bays, the gravity moments control the design for a greater number of stories, but it can be expected that the patterns of design moment are similar to those for the corridor girders.

4. Comparison of Girder Designs

Eight designs of girders for the sample bent were made using both A36 and A441 steel for combinations of simple and rigid framing with non-composite and composite construction. Example 21.5 presents the design of some typical girders for the frame of Example 21.1. Girder sizes are tabulated in the solution to Example 21.5.

Figure 21.13 gives a weight and relative cost of steel comparison for the eight designs. Bar charts give the weight in kips and the cost related to the A36 simple framing case for the girders in a 14-story bent. It is seen that A441 steel results in a lower weight of steel in each case. Rigid framing uses less steel than simple framing, and composite construction saves steel for either rigid or simple framing.

Relative costs are computed from a base price plus average quality and size extras. It is seen that for each type of framing the net cost based on weight of steel is almost the same for A36 and A441 steel. In addition the designer would have to consider the fabrication costs which would be substantially the same for A36 and A441, and the lower shipping costs of lighter designs. Smaller foundations might be used for the lighter designs, and total design loads could be reduced when lighter members are used.

Similar designs in the newer A572 steel would be easier because of a single yield point per steel grade and designs probably more economical than A441 steel because of a slightly lower price. Further advances in steel technology can be expected to change relative economies from time to time.

EXAMPLE 21.5

PROBLEM:
Design girders for the 8th floor below the roof of the frame of Example 21.1.

Step No.	Operation	Units	G1	G2	G3	
①	Span L	ft	20	12	28	
②	Live Load w_L	kip/ft	2.4	2.4	2.4	
③	Dead Load w_D	kip/ft	2.88	2.88	2.88	
④	Live Load Reaction $w_L \, L/2$	kip	24.0	14.4	33.6	
⑤	Dead Load Reaction $w_D \, L/2$	kip	28.8	17.28	40.3	
⑥	% Live Load Used (1.0-Red.)	%	0.616	0.770	0.491	
⑦	% LL x $w_L \, L^2$	kip-ft	592	266	925	
⑧	$w_D \, L^2$	kip-ft	1152	415	2260	
⑨	⑦ + ⑧ = $\Sigma \, wL^2$	kip-ft	1744	681	3185	
	Gravity Moments					
⑩	Simple Framing (+) 0.125 x ⑨	kip-ft	218	85.1	398	Fig.21.10a
⑪	Rigid Framing (+) 0.08 x ⑨	kip-ft	140	54.5	255	Fig.21.10d
⑫	" (−) 0.045 x ⑨	kip-ft	78.5	30.6	143	
	The previous steps are constant for all floors					
	Wind+Gravity Moments					
⑬	Wind Moment M_w (portal)	kip-ft	72.0	43.1	100.9	
⑭	0.09375 x ⑨ = 3/4 x1/8 wL^2	kip-ft	164	64.0	299	⎫
⑮	(2.67 x ⑬)^2) ÷ ⑨*	kip-ft	7.9	7.3	8.5	
⑯	⑭ + ⑮ Des. M. Simple Fr. (+)	kip-ft	172	71.3	308	⎬Eq.21.4
⑰	−0.75 x ⑫	kip-ft	−59	−23.0	−107	
⑱	⑯ + ⑰ Des. M. Rigid Fr. (+)	kip-ft	113	48.3	201	⎭
⑲	0.75 x ⑫ = 3/4 Mg. (neg.)	kip-ft	59	23.0	107	
⑳	⑬ + ⑲ Des. M. Rigid Fr. (−)	kip-ft	131	66.1	208	
	Design for Simple Framing					
㉑	Design M (+) ⑩ or ⑯	kip-ft	218	85.1	398	
㉒	Design M (−) ⑬	kip-ft	72.0	43.1	100.9	
㉓	S req (A36) =12.0 x(㉑ or ㉒) ÷ 24	in.³	109	42.6	199	⎫
㉔	Section (Handbook) (A36)	W	W21x55	W16x31	W27x84	
㉕	S req (A36 Comp.) = ㉑ x 12.0 ÷ (1.28 x24)	in.³	85	33.3	156	⎬ A36
㉖	Section for A36 Composite	W	W18x50	W14x26	W24x76	⎭

EXAMPLE 21.5 (Continued)

Step No.	Operation	Units	G1	G2	G3	
	Design for Simple Framing (con't.)					
㉗	S req (A441) = 12.0 x (㉑ or ㉒) ÷ 33	in.^{3}	79.4	31.0	145	A441
㉘	Section for A441	W	W16×50	W14×26	W24×68	
㉙	S req (A441 Comp) = (㉑ x 12.0) ÷ (1.28×33)	in.^{3}	62	24.2	113	
㉚	Section for A441 Composite	W	W16×40	W12×22	W21×62	
㉛	Thickness of A441 less than 3/4 in.?		OK	OK	OK	
	For design of simple framing only, steps ⑪ , ⑫ , ⑰ , ⑱ , ⑲ and ⑳ could be omitted.					
	Design for Rigid Framing					
	For design of rigid framing only, steps ⑩ , ㉑ to ㉛ could be omitted.					
㉜	Design M (+) ⑪ or ⑱	kip-ft	140	54.5	255	
㉝	Design M (−) ⑬ or ⑳	kip-ft	131	66.1	208	
㉞	Design M (+) Composite ㉜ ÷ 1.28	kip-ft	109	42.6	199	
㉟	S req (A36) = (12÷24) (㉜ or ㉝)	in.^{3}	70	33.0	128	A36
㊱	Section for A36	W	W16×45	W14×26	W21×68	
㊲	S req (A36 comp) = (12÷24) (㉝ or ㉞)	in.^{3}	65.5	33.0	104	
㊳	Section for A36 composite	W	W16×45	W14×26	W21×55	
㊴	S req (A441) = (12÷33) (㉜ or ㉝)	in.^{3}	51.0	24.1	92.6	A441
㊵	Section for A441	W	W16×36	W12×22	W18×55	
㊶	S req (A441 comp) = (12÷33) (㉝ or ㉞)	in.^{3}	47.6	24.1	75.6	
㊷	Section for A441 composite	W	W14×34	W12×22	W18×45	
㊸	Thickness of A441 less than 3/4 in.?		OK	OK	OK	

* M_W calculated for step ⑬ is 3/4 of wind moment which would be calculated for specified 20 psf wind load. Coefficient of second term in Eq. 21.4 becomes $3/4 \times 2 \times (4/3)^2 = 2.67$ when wind moments in step ⑬ are used.

EXAMPLE 21.5 (Continued)

Summary of Girder Sizes for Different Solutions of Example 21.5

Type of Design		Girder G1 (20ft)		Girder G2 (12ft)		Girder G3 (28ft)	
		No. of Floors	Member	No. of Floors	Member	No. of Floors	Member
A–36	Simple Framing Non–Composite	Roof 13*	W 18 x 45 W 21 x 55	Roof 13	W 14 x 22 W 16 x 31	Roof 13	W 24 x 68 W 27 x 84
A–441	Simple Framing Non–Composite	Roof 13	W 16 x 36 W 16 x 50	Roof 13	W 12 x 19 W 14 x 26	Roof 13	W 21 x 55 W 24 x 68
A–36	Simple Framing Composite	Roof 13	W 16 x 40 W 18 x 50	Roof 13	W 12 x 19 W 14 x 26	Roof 13	W 21 x 62 W 24 x 76
A–441	Simple Framing Composite	Roof 13	W 16 x 31 W 16 x 40	Roof 13	W 12 x 16.5 W 12 x 22	Roof 13	W 18 x 50 W 21 x 62
A–36	Rigid Framing Non–Composite	Roof 9 2 2	W 14 x 34 W 16 x 45 W 18 x 45 W 18 x 50	Roof 6 2 1 2 2	W 12 x 16.5 W 14 x 22 W 14 x 26 W 16 x 26 W 14 x 30 W 16 x 31	Roof 13	W 18 x 55 W 21 x 68
A–441	Rigid Framing Non–Composite	Roof 10 3	W 14 x 26 W 16 x 36 W 16 x 40	Roof 6 2 2 3	W 12 x 14 W 12 x 19 W 12 x 22 W 14 x 22 W 14 x 26	Roof 13	W 16 x 45 W 18 x 55
A–36	Rigid Framing Composite	Roof 5 2 2 2 2	W 16 x 26 W 16 x 36 W 16 x 40 W 16 x 45 W 18 x 45 W 18 x 50	Roof 3 2 1 2 1 2 2	W 12 x 14 W 12 x 19 W 12 x 22 W 14 x 22 W 14 x 26 W 16 x 26 W 14 x 30 W 16 x 31	Roof 8 3 2	W 18 x 45 W 21 x 55 W 21 x 62 W 21 x 68
A–441	Rigid Framing Composite	Roof 6 1 1 2 3	W 14 x 22 W 14 x 30 W 16 x 31 W 14 x 34 W 16 x 36 W 16 x 40	Roof 4 1 1 2 3	W 10 x 11.5 W 12 x 16.5 W 10 x 19 W 12 x 19 W 12 x 22 W 14 x 22 W 14 x 26	Roof 7 1 1 2 2	W 16 x 36 W 16 x 45 W 18 x 45 W 16 x 50 W 18 x 50 W 18 x 55

*Number of floors refers to the number of floors for which the designated section is minimum size girder. Sections are listed in order from the roof downward.

At this stage in the design the effect of type of framing on the total weight of columns has yet to be determined.

21.7 EFFECTIVE LENGTH OF COLUMNS WITH SEMI-RIGID GIRDER CONNECTIONS

Effective slenderness ratios of axially loaded compression members are given in the AISC Section [1.8] and are described in Chapter 10.   In summary,

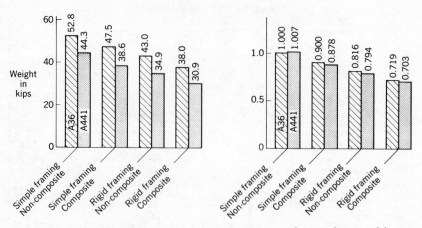

Fig. 21.13 Weight and relative-cost comparison for girders in 14-story building bent.

it is noted that effective slenderness ratios vary between $0.5L/r$ and L/r when sidesway is prevented and may vary from L/r to infinity when only rigid frame action resists sidesway. The coefficient of L/r depends on the relative stiffness of the girders and columns framing together. With fully rigid construction the recommendations of the AISC Specification and Chapter 10 may be followed without change.

1. Theory

To determine the effective length of a column it is necessary to obtain the restraint function:

$$G = \frac{\sum (I_c/L_c)}{\sum (I_g/L_g)} \tag{10.3}$$

The use of the relative stiffness of the girders I_g/L_g in Eq. 10.3 presupposes that the girder-to-column connections are able to maintain right angles between members. When a girder is joined to a column by a flexible connection, less restraint is supplied, and therefore the restraint function will change.

Experimental and theoretical values are available for the moment-rotation behavior of riveted and bolted semi-rigid connections,[4,24] and for welded semi-rigid connections.[19,11,21,18] To use these results in an analysis, a factor γ is defined as the slope of the moment-rotation (M-ϕ) curve of the connection:

$$\gamma = \frac{\phi}{M} \qquad (21.5)$$

The rotation at the end A of a girder will have an amount $M_A\gamma_A$ added to the rotation due to bending. Equations for the end rotations of the girder in

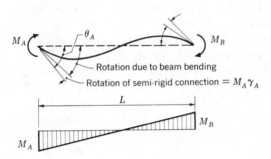

Fig. 21.14 End rotations of a girder with semi-rigid connections.

Fig. 21.14 may be obtained as follows:

$$\theta_A = M_A\gamma_A + \frac{L}{3EI}[M_A - \tfrac{1}{2}M_B] \qquad (21.6)$$

$$\theta_B = M_B\gamma_B + \frac{L}{3EI}[M_B - \tfrac{1}{2}M_A] \qquad (21.7)$$

Assume $\gamma_B = \gamma_A$ from symmetry of the girder connections. The absolute stiffness at end A of the girder is measured by the moment at A required to cause $\theta_A = 1$ and $\theta_B = 0$. This results in an absolute stiffness of

$$K_A = \frac{12(EI/L)}{4(L_A/L) - (L/L_A)} \qquad (21.8)$$

where $L_A = L + 3EI\gamma_A$. The relative stiffness of the girder at A would then be

$$K_{A(\text{rel})} = \frac{3(I_g/L_g)}{4(L_A/L_g) - (L_g/L_A)} \qquad (21.9)$$

This value should be used rather than I_g/L_g.

2. Typical Values

In Table 8A of Ref. 4.24, γ values and allowable moments are given for the standard web-type beam connections of the AISC (Fig. 21.6a). However, substitution of the γ values in Eq. 21.9 would show a girder stiffness of $0.14I_g/L_g$ for a four-rivet standard connection and $0.26I_g/L_g$ for a six-rivet standard connection. These are substantial reductions in girder stiffness. The web connections are not adequate to resist the wind moments for the girders of the sample bent.

To obtain stiffer connections, seat and top angles may be added to the standard AISC connections to give modified connections as shown in Fig. 21.6c. Reference 4.24 has complex equations for evaluating these modified connections. For the frame of Fig. 21.7, modified connections ranging from sizes used for 12-in. girders to sizes used for 27-in. girders are still inadequate to resist the wind moments with simple framing, but they have a moment capacity from 3 to 8 times the comparable web connections. The resulting girder stiffnesses range from $0.79I_g/L_g$ to $0.88I_g/L_g$ with an average of $0.84I_g/L_g$. The inability of these modified connections to resist the wind moments for simple framing shows the difficulty of obtaining simple connections suitable for Type 2 construction. When the wind moments exceed the capacity of the flexible connections it is desirable to consider the use of welded rigid connections or diagonal bracing.

21.8 DESIGN OF COLUMNS

Column loads and moments from the analyses described in Arts. 21.4 and 21.5 may be used in the design of columns. The computation for loads and moments due to gravity load for column C1 of the sample bent of Example 21.1 is illustrated in Example 21.6. Theoretically, effective column lengths need to be used about both axes for the axial load. If the frame is braced with diagonal bracing in the plane perpendicular to the bent, the effective length factor of each column will be less than 1.0 in that plane. If sidesway would be permitted in both directions, two effective lengths would have to be found. Illustrations of such situations are given in Examples 11.5, 11.6, and 11.7.

Examples 21.7 through 21.14 show eight different designs for a typical tier of column C1. A tabular form is used for routine calculations of stresses in members subjected to combined axial compression and bending to meet the requirements of AISC Formulas [1.6-1a] and [1.6-1b]. The significant difference between each of the examples is the effective column length as influenced by factors involving framing and girder stiffness. The effect of framing depends on whether simple, rigid, or semi-rigid framing is used and whether diagonal bracing or composite girders are used. The stiffness of the girders required to carry loads depends on each of these factors as well as on what grade of steel is used. Effective length calculations for each of the examples are contained in steps 10 through 16.

EXAMPLE 21.6

PROBLEM:
   Determine loads and moments due to gravity load for column C I of the frame in Example 21.1 (wind neglected).

SOLUTION:

Tier	Source of Load	C I Gravity Loads		M (k−ft) (caused by floor girders) (All floors)
		P (kips)		
		Dead	Live	
TOP & T−1	Roof Girder	38.2*		Simple Framing
	Roof Spandrel	8.2	0.808	Floor Girder L.L. $0.808 \times 24 \times \frac{d_c}{4} = 4.8\,d_c$ **
	Floor Girder	28.8	x 24.0	
	Floor Spandrel	24.5	= 19.4	Floor Girder D.L. $28.8 \times \frac{d_c}{4}$ = $7.2\,d_c$
	Column (2 stories)	2.8		
	Total Dead	94.3		− Spandrel D.L. $-24.5 \times \frac{d_c}{4}$ = $-6.1\,d_c$
	Total D+L	113.7		Total M $5.9\,d_c$
T−2 & T−3	Total Dead	94.3	0.424	8 in. column 3.9 k−ft.
	2 Floors	106.6	x 3 x 24	10 in. 4.9 k−ft.
	Column (2 stories)	2.8	= 30.5	12 in. 5.9 k−ft.
	Total Dead	203.7		14 in. 6.9 k−ft.
	Total D+L	234.2		
T−4 & T−5	Total Dead	203.7	0.400	
	2 Floors	106.6	x 5 x 24	Rigid Framing
	Column (2 stories)	2.8	= 48.0	$+ \frac{1}{2}$ Girder M (−) + 39.2 k−ft.
	Total Dead	313.1		8 in. column 43.1 k−ft.
	Total D+L	361.1		10 in. 44.1 k−ft.
T−6 & T−7	Total Dead	313.1	0.400	12 in. 45.1 k−ft.
	2 Floors	106.6	x 7 x 24	14 in. 46.1 k−ft.
	Column (2 stories)	2.8	= 67.2	
	Total Dead	422.5		
	Total D+L	489.7		
T−8 & T−9	Total Dead	422.5	0.400	
	2 Floors	106.6	x 9 x 24	
	Column (2 stories)	2.8	= 86.4	
	Total Dead	531.9		
	Total D+L	618.3		
T−10 & T−11	Total Dead	531.9	0.400	
	2 Floors	106.6	x 11 x 24	
	Column (2 stories)	2.8	= 105.6	
	Total Dead	640.3		
	Total D+L	745.9		

* Roof dead load 22.8 kips plus 7.2 kips live load included in this item to facilitate handling of live load reduction.

** d_c = column depth. Moment divided equally above and below joint.

EXAMPLES 21.7-21.14

PROBLEM

For the two-story tier including the 10th and 11th stories below the roof, make eight trial designs in A441 steel for column C1 of the frame in Example 21.1.

(21.7) For a simple or semi-rigid frame with diagonal bracing.
(21.8) For a rigid frame with diagonal bracing.
(21.9) To support A441 rigid, non-composite girders — — unbraced frame.
(21.10) To support A36 rigid, non-composite girders — — unbraced frame.
(21.11) To support A441 rigid, composite girders — – unbraced frame.
(21.12) To support A36 rigid, composite girders — unbraced frame.
(21.13) To support A441 semi-rigid, non-composite girders. — unbraced frame.
(21.14) To support A36 semi-rigid, non-composite girders — unbraced frame.

SOLUTION:

Step No	Operation	Units	21.7 A441 Col. Simple or Semi-Rigid Any Girders, Braced	21.8 A441 Col. Rigid Any Girders, Braced	21.9 A441 Col. Rigid,Non-Comp. A441 Gird. Unbraced	21.10 A441 Col. Rigid,Non-Comp. A36 Gird. Unbraced	21.11 A441 Col. Rigid, Comp. A441 Gird. Unbraced	21.12 A441 Col. Rigid, Comp. A36 Gird. Unbraced	21.13 A441 Col. Semi-Rigid Non-Comp.,A441 Gird,Unbraced	21.14 A441 Col. Semi-Rigid Non-Comp.,A441 Gird,Unbraced	Notes
①	Design Thrust P	kip	746	746	746	746	746	746	746	746	From Prob 21.6
②	Design Moment M	kip-ft	6.9	46	46	46	46	46	6.9	6.9	
③	Trial Column	—	W14x103	W14x111	W14x158	W14x127	W14x127	W14x119	W14x119	W14x111	Handbook Properties
④	A	in²	30.3	32.7	46.5	37.3	37.3	35.0	35.0	32.7	
⑤	I_x	in⁴	1170	1270	1900	1480	1480	1370	1370	1270	
⑥	S_x	in³	164	176	253	202	202	189	189	176	
⑦	r_x	in.	6.21	6.23	6.40	6.29	6.29	6.26	6.26	6.23	
⑧	r_y	in.	3.72	3.73	4.00	3.76	3.76	3.75	3.75	3.73	
⑨	$I_c/L_c =$ ⑤ $\div$ 12.0 Top Joint	in⁴/ft.	(Use Max Value K = 1.0)	—	158	123	123	114	114	106	$L_c = 12$ ft
⑩	$\Sigma I_c/L_c = 2 \times$ ⑨	in⁴/ft.		—	316	246	246	228	228	212	
⑪	$\Sigma I_g/L_g$ (Girder design:Example 21.5)	in⁴/ft.		—	25.8	35.2	1.68x22.3†	1.62x35.2†	0.8x32.7††	0.86x57††	
⑫	$G_T =$ ⑩ $\div$ ⑪ Bott Joint			—	12.3	7.0	6.6	4.0	8.7	4.3	
⑬	$\Sigma I_c/L_c = 2 \times$ ⑨	in⁴/ft.		—	316	246	246	228	228	212	
⑭	$\Sigma I_g/L_g$ (Girder design:Example 21.5)	in⁴/ft.		—	25.8	35.2	1.67x25.8†	1.62x35.2†	0.8x32.7††	0.86x57††	
⑮	$G_B =$ ⑬ $\div$ ⑭			—	12.3	7.0	5.7	4.0	8.7	4.3	
⑯	K (From Figure 10.7b)	—	1.0	1.0	3.3	2.6	2.4	2.05	2.8	2.1	Effective Column Length Factor
⑰	$f_a =$ ① $\div$ ④	ksi	24.7	22.8	16.0	20.0	20.0	21.3	21.3	22.8	Computed Stresses
⑱	$f_b = 12 \times$ ② $\div$ ⑥	ksi	0.51	3.14	2.18	2.73	2.73	2.91	0.44	0.47	

EXAMPLES 21.7-21.14 (Continued)

Step No.	Operation	Units	A441 Col. Simple or Semi-Rigid Any Girders Braced	A441 Col. Rigid Any Girders Braced	A441 Col. Rigid,Non-Comp A441 Gird. Unbraced	A441 Col. Rigid,Non-Comp A36 Gird. Unbraced	A441 Col. Rigid,Comp. A441 Gird. Unbraced	A441 Col. Rigid,Comp. A36 Gird. Unbraced	A441 Col. Semi-Rigid Non-Comp,A441 Gird,Unbraced	A441 Col. Semi-Rigid Non-Comp,A36 Gird,Unbraced	Notes
(19)	Ld/Af	—	173	162	117	144	144	152	153	163	
(20)	Fb=12,000/(19)	ksi	30*	30*	30.5**	33	33	30*	30*	30*	F_b
(21)	L/ry=144÷(8)	—	38.8	38.6	36.0	38.3	38.3	38.4	38.4	38.6	
(22)	Fb (Formula 4)	ksi				Does not govern					Effective Length
(23)	L/rx=144÷(7)	—	23.2	23.1	22.5	22.9	22.9	23.0	23.0	23.2	
(24)	KL/rx=(16) x (23)	—	23.2	23.1	74.3	58.7	55.5	47.0	65.1	48.4	F_a
(25)	Fa (AISC Table [1]) (21) or (24)	ksi	26.0	26.0	19.0	22.9	23.5	24.8	21.8	24.6	AISC Formula (1.6-1a)
(26)	Fe' (AISC Table [2]) (24)	ksi	278	280	27.0	43.3	48.4	68.0	35.2	63.8	
(27)	fa/Fe' (17)÷(26)	—	0.089	0.082	0.594	0.461	0.413	0.314	0.606	0.358	
(28)	1.0 - (27)	—	0.911	0.918	0.406	0.539	0.587	0.686	0.394	0.642	
(29)	Cmfb/((28) x Fb)=(0.85x(18))÷((28)x(20))	—	0.016	0.097	0.150	0.131	0.120	0.120	0.031	0.021	
(30)	fa/Fa = (17)÷(25)	—	0.948	0.878	0.843	0.871	0.852	0.859	0.977	0.929	
(31)	(29) + (30)	—	0.964	0.975	0.993	1.002	0.972	0.979	1.008	0.950	
(32)	OK or NG (Is (31) < 1.0?)	—	OK	OK	OK	§§ OK	OK	OK	§§ OK	OK	
(33)	fa/0.6 Fy = (17)÷0.6 Fy	—	0.822	0.762	0.582	0.666	0.666	0.711	0.711	0.762	Formula (1.6-1b)
(34)	fb/Fb = (18)÷(20)	—	0.017	0.104	0.072	0.083	0.083	0.097	0.015	0.016	
(35)	(33) + (34)	—	0.839	0.866	0.654	0.749	0.749	0.808	0.726	0.778	
(36)	OK or NG (Is (35) < 1.0?)	—	OK	OK	OK	§§ OK	OK	OK	OK	OK	

* Non-Compact Section

** Group 2 Section – Fy = 46 ksi.

† Girder Stiffness Modified for Composite Action.

†† Girder Stiffness Modified for Semi-Rigid Connections.

EXAMPLES 21.7–21.14 (Continued)

Summary of A 441 column sizes for 11th and 12th stories below roof designed as illustrated in Examples 21.7 to 21.14.

Type of Framing and Girder Design	Column No.			
	C 1	C 2	C 3	C 4
BRACED FRAMES				
Simple Framing A36 or A441 Girders Composite or Non-Composite	W 14 x 103	W 14 x 103	W 14 x 127	W 14 x 127
Rigid Framing A36 or A441 Girders Composite or Non-	W 14 x 111	W 14 x 111	W 14 x 150	W 14 x 150
UNBRACED FRAMES				
Rigid Framing A36 Non-Composite Girders	W 14 x 127	W 14 x 119	W 14 x 167	W 14 x 176
Rigid Framing A441 Non-Composite Girders	W 14 x 158	W 14 x 119	W 14 x 176	W 14 x 211
Rigid Framing A36 Composite Girders	W 14 x 119	W 14 x 111	W 14 x 158	W 14 x 167
Rigid Framing A441 Composite Girders	W 14 x 127	W 14 x 119	W 14 x 167	W 14 x 176
Semi-Rigid Framing A36 Non-Composite Girders	W 14 x 111	W 14 x 111	W 14 x 127	W 14 x 127
Semi-Rigid Framing A441 Non-Composite Girders	W 14 x 119	W 14 x 111	W 14 x 142	W 14 x 142

Columns for braced frames are designed in Examples 21.7 and 21.8. To simplify the calculations, the maximum possible effective length factor K of 1.0 was used. If relative girder stiffnesses from the several different designs had been used, K would have varied from about 0.8 to 0.95. However, the allowable stresses for columns with L/r about 23 as used in these examples would increase by only 1 per cent. Available column sizes vary by 5 to 10 per cent, so this refinement would not make it possible to use lighter columns in most cases. The difference in column size between Example 21.7 and 21.8 was a result of the difference in bending moment caused by simple and rigid framing.

Examples 21.9 and 21.10 show calculations for A441 columns to support A441 and A36 non-composite girders in unbraced rigid frames. Because the A36 girder must be larger than the A441 girder to support the same gravity and wind loads, it is able to provide greater restraint to the column. The resulting shorter effective length and higher allowable compressive stress for the column are typical of all the other comparisons for A36 and A441 girders framing into columns.

Columns to support composite A441 and A36 girders for unbraced rigid frames are designed in Examples 21.11 and 21.12. The stiffnesses of the steel girders are multiplied by a factor from Fig. 13.12 which corrects the increase in stiffness due to composite action for the reduction in composite action in regions of negative moment as discussed in Art. 13.10. For all of the girders listed in Example 21.5, the correction factor ranged from 1.54 and 1.84. The increase in stiffness of the girders permitted savings of 20 and 6 per cent in weight of columns as compared with the non-composite designs of Examples 21.9 and 21.10.

The effect of semi-rigid framing is shown in Examples 21.13 and 21.14. The case is designated "semi-rigid" to emphasize the fact that the "simple" connections must be strong enough to resist the sometimes substantial wind moments imposed on the unbraced frame. The beam stiffnesses are reduced by the factor derived in Art. 21.7 to correct for semi-rigid action. Because the girders were proportioned for simple bending, they were large enough in size so that the effective column length considering semi-rigid action was about the same as the other cases of unbraced frames. Because of the low column moment resulting from the semi-rigid framing, the columns selected were relatively light.

Comparison of the several designs following Example 21.14 shows that the most economical column designs result when the frame has diagonal bracing. This is a consequence of the reduction of K to 1.0 and lower moments in columns.

If diagonal bracing could be included without interfering with the occupancy of the building it would be the most effective means of reducing K. The simplest form of diagonal bracing is a pair of crossed diagonal members designed to take tension only. (See Fig. 21.15a.) Such members are usually angle sections or pipes, although round rods or wire ropes might suit the purpose. To prevent "whipping" or vibration, maximum slenderness ratios are usually specified for bracing members (see Art. 6.5). For the same weight and cross-sectional area, angles and pipes will have much greater flexural stiffness than rods and wire ropes. In addition, wire ropes are subject to greater elongation per unit load by virtue of their lower effective modulus of elasticity.

Architectural considerations usually require openings for doors and windows too large to permit the full use of diagonal bracing. In such cases, modified types of bracing as shown in Figs. 21.15b, c, and d may provide

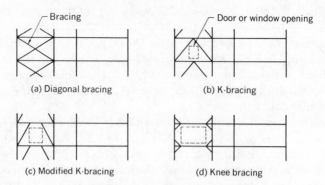

Fig. 21.15 Types of wind bracing.

larger openings while still furnishing adequate stiffness to the structure. These types of bracing must resist compression as well as tension and therefore will be somewhat heavier than the simpler diagonal bracing. However, each of the last three types of bracing can provide intermediate vertical support for the girder, thereby reducing bending moments in both girders and columns.[21.19]

Example 21.15 shows some of the pertinent results of selecting angles for tension diagonals in one panel of a frame. For a 20-ft bay, 5-by-5-by-$\frac{3}{8}$-in. angles weighing 12.3 lb per ft are within the desirable range of slenderness ratio for tension members (see Art. 6.5). The force induced in each diagonal will be proportional to the sway of the frame in a given story. The diagonal being loaded in compression will buckle and carry no more load when both members have reached a stress only one-sixth of the allowable load of the member in tension. Thus it is readily seen why it is worthwhile to consider each diagonal as a "counter" which is only effective in tension. Finally, the example shows that the tension diagonal can supply a shear resistance of 68 kips with a deflection index (or deflection-to-story height ratio) of only 0.0017 even without any assistance from the rigid frame. This story shear would be adequate for the first story of a 16-story frame having the proportions of the sample frame being discussed. At the same time that it reduces the effective length coefficient K to 1.0 or less so that the best advantage can be taken of high-strength steel for columns, diagonal bracing has a very significant effect in reducing the sway of the structure.

In rigid frames without diagonal sway bracing, the relative sizes of the columns and girders at each floor level govern the effective lengths of the columns. If the frame used in the examples were limited to 14 stories in height, its height-to-width ratio would approach 3. For frames of this proportion the strength design of most of the girders is influenced relatively little by wind loads. The girder in each story of a given bay will be very

---

EXAMPLE 21.15

PROBLEM:
  Select diagonal bracing for one story of the bent in Example 21.1. Determine contribution to shear resistance by bracing.

SOLUTION:
  Try angles designed for tension only in 20 ft bay.

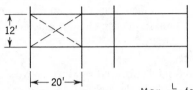

Assume connection at intersection offers no lateral support.

L_d = 23.3 ft = 280 in.

Max $\dfrac{L}{r}$ for tension member = 300

min $r = \dfrac{280}{300}$ in. = 0.93 in.

Try L5 x 5 x $\dfrac{3}{8}$ r = 0.99 in.   A = 3.61 in.²

Allowable Tensile Load (A36 steel) = 22 ksi (3.61 in.²) = 79.5 kip

Compression Buckling Load = $\dfrac{\pi^2 EA}{(L/r)^2} = \dfrac{\pi^2 (29 \times 10^3)(3.61)}{(280/0.99)^2}$ = 12.9 kip.

Change in length of diagonal due to sway of panel.

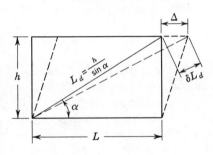

$\delta L_d = \Delta \cos\alpha$

Unit Strain $= \dfrac{\delta L_d}{L_d} = \dfrac{\Delta \cos\alpha}{L_d}$

Force in Diag = AE $\dfrac{\cos\alpha}{L_d} \Delta$

Horiz Shear resisted by diag.

$R_h = R\cos\alpha = AE \dfrac{\cos^2\alpha}{L_d} \Delta$

$R_h = AE \cos^2\alpha \sin\alpha \dfrac{\Delta}{h}$

$\cos\alpha = 0.857$ $\sin\alpha = 0.515$

$R_h = 3.61(29 \times 10^3)(0.857)^2(0.515) \dfrac{\Delta}{h}$

$R_h = 39,600 \dfrac{\Delta}{h}$ (kip)

Δ/h	R_h	Remarks
0	0	
0.001	39.6 kip	
0.0015	59.4	
0.0017	68.0	Allowable
0.0020	79.1	
0.0028	111.2	Yield

---

near to the size of the girder at every other floor since the equal gravity loads on each girder contribute the major portion of the design moment. In contrast, each column resists a thrust approximately equal to the thrust at the top floor times the number of floors above the given column. Therefore, the required column size must increase almost uniformly down the frame.

This means that the sum of the I/L values of girders at all floors is nearly equal, but that the sum of the I/L values of columns increases gradually down the frame. The resulting ratios K, of effective column length to actual unbraced length, range from 1.3 to 3 for interior columns and from 1.4 to 4 for exterior columns.

For typical column sizes having K-values from 3 to 4, such columns can carry only 65 to 80 per cent as much thrust per unit area as columns having K-values of 1.0. Thus, it is obvious that the efficiency of the columns can be improved by reducing the K-values. Such improvements in efficiency are economical if the cost of providing a means of reducing K is less than the reduction in cost of the column member.

In Example 21.9, the heaviest column for an unbraced rigid frame would be needed with A441 non-composite girders while the lightest column may be used with A36 composite girders, as determined in Example 21.12.

The result of a change from A441 non-composite girders to A36 composite girders would be a reduction of 936 pounds in A441 column weight in two stories of the first bay. The girders for two floors would change from 1520 lb of A441 steel to 1800 lb of A36 steel plus some shear connectors. Based on the costs per pound of A36 and A441 steel, the net reduction in cost for the two-story length of column and two 20-ft girders would be about 18 per cent less costs of shear connectors. In addition, there would be a net reduction of 656 lb in weight of the structure plus some reduction in the size of the adjacent interior column and in the lateral deflection of the structure due to wind.

All of these benefits are not obtained without cost, however. The increase in depth of girder which contributes to increased structural efficiency either cuts down on headroom in the building or requires an increase in the height of the building. An increase in the height of the building means increases in column lengths, wall heights and lengths of utility and service lines. Referring to cost data in current issues of *Engineering News-Record* will show that the cost of heating and air conditioning facilities is usually of the same order of magnitude as the cost of the structural frame. Such factors as this make the process of determining the ideal proportions for the structure an exceedingly complex one. Good cooperative planning between the engineer and architect can be valuable in achieving a balance among the various elements in the building.

21.9 DESIGN TABLES

Every engineer should be conscious of the availability of design tables to be used in speeding up the design of large numbers of repetitive elements. Even further, he should be conscious of the possibility of preparing his own design tables for situations which occur frequently in his work.

Examples of design tables readily available are the tables presented in the AISC Manual of Steel Construction.[7.9] Included are tables for the design of beams, columns, connections, plate girders, and composite beams, together with descriptions of the use of each. Typical design tables are prepared by evaluating the capacity of a given rolled shape under a standard load pattern for a wide range of lengths. This is repeated for a complete selection of rolled shapes. While the standard load pattern is generally a very elementary case, the tables can be used for other cases by the application of some simple refinement procedures described with the tables.

Example 21.16 is presented to illustrate the use of design tables for concentrically loaded columns in the design of a beam-column. To use the design tables, a fictitious increment of axial load is formulated from the bending moment and is added to the original axial load. Selection of a column able to carry the increased load as a concentric load should provide a member able to carry the actual combined moment plus axial load. Additional factors presented in the column design table make it possible to allow for the amplification factor used in AISC Equation [1.6-1a] to give a result which is just as accurate as that presented by the interaction formula.

21.10 WIND DRIFT

The wind drift or lateral deflection of the building is another highly indeterminate function of interest to the designer. The results of measurements of wind deflection of the Empire State Building and a model study of that building show that the masonry walls increased the rigidity of the building about 350 per cent above the rigidity of the steel frame.[21.20] With the trend toward lightweight construction and cladding, much of this restraint may be reduced in the future. An ASCE Committee study of wind bracing in tall buildings recommended limiting deflections to 0.002 times the height of a building, although buildings with deflection indexes of upwards of 0.004 or 0.005 have behaved satisfactorily.[3.13]

To compute deflections, it is desirable to use a simplified method which allows computation of the deflection of one story at a time. One such method, using the method of real work, is illustrated in Ref. 3.7 as it was used in the design of Lever House in New York City. Different conditions arising in the bent in Fig. 21.7 make a virtual work method desirable. Using the moments from the portal method of analysis, the deflection of a one-story portion of the bent as shown in Fig. 21.16a may be computed. The set of moments m given in Fig. 21.16b is an arbitrary system of moments in equilibrium with a pair of horizontal unit shears in opposite directions at adjacent floors. The moments M for the same floor due to the real applied loads are given in Fig. 21.16c.

EXAMPLE 21.16

PROBLEM:

Design the beam-column of Example 21.10 using allowable load tables.

(1) Data from Preliminary Design:

$$P = 746 \text{ kip} \qquad M = 46 \text{ kip-ft} \qquad L = 12 \text{ ft}$$

$$\Sigma I_g / L_g = 35.2 \text{ in.}^4/\text{ft for both top and bottom}$$

(2) Try W 14 x 136 of A441 steel.

(3) Find from Column Table II:

A = 40.0 in.^{2}	B_x = 0.186	} Bending
r_x/r_y = 1.67	B_y = 0.520	} Factors
r_y = 3.77 in.	a_x = 237 x 10^6	} Amplification
L_c = 13.3 ft	a_y = 85 x 10^6	} Factors
L_u = 35.5 ft	Also: I_x = 1590 in^4	

(4) Find K values: (Same method as Ex. 21.10 -- steps not repeated here.)

$$K_x = 2.6 \qquad\qquad K_y = 1.0$$

$$K_x L = 2.6(12) = 31.2 \text{ ft} \qquad K_y L = 12.0 \text{ ft}$$

(5) Find Tabular Load, P_T

Equivalent y Axis Length for $K_x L$

$$= \frac{K_x L}{r_x/r_y} = \frac{2.6(12)}{1.67} = 18.6 \text{ ft}$$

P_T Selected from Column Table II versus Larger of 12 ft and 18.6 ft (use 19 ft)

$$P_T = 905 \text{ kip capacity}$$

P_T must equal or exceed combined applied load.

(6) Evaluate Amplification Factor

$$P(K_x L)^2 = 746(31.2 \times 12)^2 = 104.8 \times 10^6$$

$$\frac{a_x}{a_x - P(K_x L)^2} = \frac{237}{237 - 104.8} = 1.79$$

(7) Evaluate AISC Equation $\left[1.6-1a\right]$ modified

$$P + P^1 = P + \left[B_x M_x C_{mx} \left(\frac{F_a}{F_{bx}} \right) \left(\frac{a_x}{a_x - P(K_x L)^2} \right) \right]$$

$$= 746 + \left[0.186(46 \times 12)\, 0.85 \left(\frac{905}{40.0} \right) \left(\frac{1}{33} \right) 1.79 \right]$$

$$P + P^1 = 746 + 107 = 853 < 905 \qquad\qquad\qquad \text{O.K.}$$

NOTE: C_{mx} = 0.85 Sway Case

$$F_a = P_T/A$$

$$F_{bx} = 33 \text{ ksi since } L < L_c$$

(8) Evaluate AISC Equation $\left[1.6-1b\right]$ modified

$$P + P^1 = P \left(\frac{F_a}{0.6 F_y} \right) + \left[B_x M_x \left(\frac{F_a}{F_{bx}} \right) \right]$$

$$= 746 \left(\frac{905}{40.0} \right) \left(\frac{1}{30} \right) + \left[0.186(46 \times 12) \left(\frac{905}{40.0} \right) \left(\frac{1}{33} \right) \right]$$

$$P + P^1 = 564 + 70 = 634 < 905 \qquad\qquad\qquad \text{O.K.}$$

NOTE: If steps (7) or (8) had not been satisfied a new trial column would be selected assuming $P_T > 853$ and $L = 19$ initially.

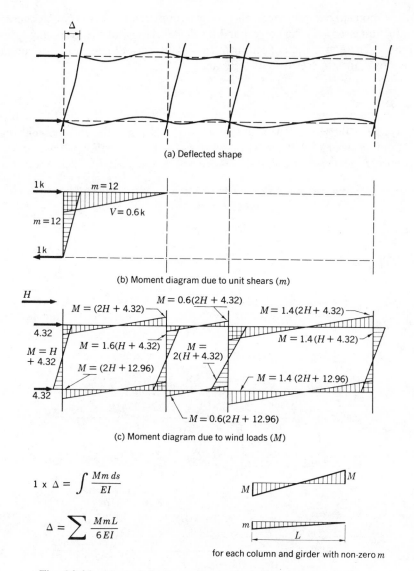

(a) Deflected shape

(b) Moment diagram due to unit shears (m)

(c) Moment diagram due to wind loads (M)

$$1 \times \Delta = \int \frac{Mm\,ds}{EI}$$

$$\Delta = \sum \frac{MmL}{6EI}$$

for each column and girder with non-zero m

Fig. 21.16 Virtual-work method of calculating wind drift.

The external work of the 1-kip loads travelling through the displacement Δ is set equal to the internal work of the m moments acting through the internal bending deformations to give the standard form of the virtual work equation:

$$1 \text{ kip} \times \Delta = \int \frac{Mm\,ds}{EI} \qquad (21.10)$$

All M moment diagram segments are antisymmetrical pairs of triangles acting in the same sense as the corresponding single triangle of the m moment diagram. Therefore, the deflection integral may be evaluated for this special case and put into the form of a summation:

$$E\Delta = \sum \frac{MmL}{6I} \tag{21.11}$$

where M is the end moment due to applied load, m is the end moment due to dummy shear, L is the length of girder or column, and I is the moment of inertia of the girder or column. There will be one term for each girder and

---

EXAMPLE 21.17

PROBLEM:
  Determine the relative deflection due to wind of the 12 th story below the roof of the bent in Example 21.1. Consider flexure only.

SOLUTION:

  (1) Moment Diagram for system in equilibrium with 1 kip shear. (m)

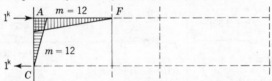

  (2) Moment Diagram due to wind shear. (M)

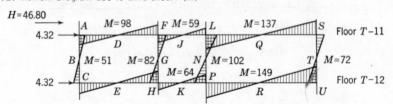

  (3) 1 kip x $\Delta = \Sigma \dfrac{MmL}{6EI}$ $(kip)(in.) = \Sigma \dfrac{(kip-ft)(kip-ft)(ft)(1728\ \frac{in.^3}{ft^3})}{(ksi)\ (in.^4)}$

Member	Section	L(ft)	M (kip–ft)	m(kip–ft)	I (in⁴)	MmL/I
AC	W 14x136	12	51	12	1590.0	4.61
AF	W 18x45	20	98	12	706.5	33.4
					Σ	38.01

  (4) Total Deflection

$$\Delta = \Sigma \frac{MmL}{I} \times \frac{1728}{6E}$$

$$\Delta = \frac{38.01 \times 288}{29 \times 10^3}$$

$$\Delta = 0.378 \text{ in.}, \quad \frac{\Delta}{h} = 0.0026$$

column having an m-moment diagram. This method may be used to determine the deflection index of as many stories as desired. Revision of member sizes may follow if needed. It is possible to revise the method to take account of semi-rigid connections and composite girders.

The sway of one story due to wind is computed in Example 21.17. Comparing the tabulated terms for the beams and girders shows that the deflection caused by bending of the columns is very small compared to that caused by bending of the girders. Thus the importance is emphasized of the effect created by increasing the sizes of the girders. The deflection index of the story is 0.0026 without bracing. Example 21.18 gives an approximate method of calculating the deflection with bracing. Compatibility is established between the elongation of the brace and the sway of the story determined by virtual work. The deflection index of the same story with minimum bracing is 0.00087 or only 33 per cent of the amount without bracing.

By using the axial forces caused by the unit loads and by the real loads, the story deflection contributed by axial strain in the members may also be calculated by the method of virtual work. Unlike the dummy moments which vanish outside the loaded stories, the dummy axial forces in columns continue to the ground. This means that all column axial strains below a given story affect the relative deflection of that story, and the effect becomes ever more important as the height of the building increases.

A similar approach can be used to estimate sway of an unsymmetrical frame due to gravity loads. If approximate girder moments are used as

---

EXAMPLE 21.18

PROBLEM:
  Determine the sway of the story in Example 21.16, if the bracing of Example 21.15 is added to the frame.

SOLUTION:
  Without Bracing: $\Delta = 0.378$ in. when shear H in panel is 51.12 kip.

  With Bracing: $R_h = 39{,}600 \dfrac{\Delta}{h}$ (kips)

       (1) Shear on Rigid Frame with Bracing $= H - R_h$

       (2) Deflection Caused by Reduced Shear on Rigid Frame

$$\Delta = \left(\frac{H - R_h}{H}\right) 0.378 \text{ in.} = \left(1 - \frac{R_h}{H}\right) 0.378$$

       (3) Substitute for H, R_h, & h

$$\Delta = \left(1 - \frac{39{,}600}{51.12 \times 144} \Delta\right) 0.378 = 0.378 - 2.03\Delta$$

$$\Delta = \frac{0.378}{3.03} = 0.125 \text{ in.}$$

       (4) $R_h = 39{,}600 \dfrac{(0.125)}{144} = 34.3$ kip   Shear Force of Brace

       (5) Deflection Index $= \dfrac{\Delta}{h} = \dfrac{0.125}{144} = 0.00087$

         Deflection is only 33% of deflection without bracing.

described in Art. 21.4, the girder moment diagram will be symmetrical. When combined with an anti-symmetrical dummy unit moment diagram, the result is zero contribution to horizontal sway. The column moments also cause counteracting, but unequal, components of deflection so that the net sway is very small.

PROBLEMS

Exercises in the design of multi-story frames are intended to give practice in applying basic enginering principles to new situations which arise in the creation of a useful structure.

Each of the problems in this chapter may be solved for any multi-story bent geometry and loading listed below. The degree of difficulty and magnitude of the problem which is to be attempted may be set by the reader. Different portions of the frame may be designed independently.

Geometry

Complete Problems 21.1 to 21.10 for:
  (a) The frame of Fig. 21.7
  (b) Any frame geometry selected (simple, rigid, semi-rigid)
  (c) The skeleton of a building laid out by an architect and for which the engineer has contracted to do the structural design.

Loading

Work out Problems 21.1 to 21.10 for:
  (d) The loading of Example 21.1
  (e) *American Standard Building Code Requirements for Minimum Design Loads in Buildings and Other Structures*, ASA A58.1
  (f) A local or state building code
  (g) Arbitrary loadings selected by the reader

21.1. Determine girder uniform loads, live load reduction percentages, and concentrated wind loads for a multi-story rigid frame bent.

21.2. Make a wind analysis of the fourteenth (or any other) story below the roof of the building by the portal method.

21.3. Make a wind analysis of the fourteenth (or any other) story below the roof of the building by the cantilever method.

21.4. Design girders for the fourteenth (or any other) story below the roof of the building. Try different steels.

21.5. Determine loads and moments due to gravity load for all stories of a column in the building.

21.6. Determine loads and moments due to wind load for all stories of a column in a building.

21.7. For a two-story tier including the thirteenth and fourteenth story below the roof, select a section for each building column. Try different steels.

21.8. Select diagonal bracing for one story of a building bent.

21.9. Determine the relative deflection of the fourteenth (or any other) story below the roof due to wind by an approximate method.

21.10. Determine the sway of the story in Problem 21.9 if the bracing of 21.8 is added to the frame.

Derivation Problems

21.11. Derive an expression for the effective stiffness of a steel beam made composite with a concrete floor slab by the addition of shear connectors. In negative moment regions, concrete in tension should be considered inactive.

21.12. Derive an expression for the shear resistance of diagonal bracing as shown in Fig. 21.15b (or Fig. 21.15c, or Fig. 21.15d).

21.13. Make an approximate wind analysis of a story of a building using bracing of Problem 21.12.

21.14. Determine the sway of the story in Problem 21.13 by an approximate method.

22

MULTI-STORY FRAMES II

22.1 INTRODUCTION

This chapter is concerned with the behavior of multi-story frames loaded into the inelastic range and with the extent to which the AISC Specification allows the use of plastic design of multi-story frames. Chapter 21 treated the analysis and design of multi-story frames according to allowable-stress methods.

There are essentially five main types of members in a braced multi-story frame.

1. *Beams*, which are members predominantly subjected to bending.
2. *Columns*, which are members predominantly subjected to compression.
3. *Beam-columns*, which are members subjected to both bending and compression.
4. *Bracing*, which may be subjected to either tension or compression.
5. *Connections*, which are subjected to bending, shear, and axial forces.

These members have all been treated individually in previous chapters. An important feature in the analysis of multi-story frames is the interaction of the individual members in resisting loads.

The structural action of a multi-story frame depends on the frame geometry, the type and manner of loading, and the behavior of the component members. One of the objectives of this chapter is the discussion of possible failure modes for multi-story frames.

The concepts involved in the various types of frame action are briefly described, and a plastic design which adheres to the present specification is developed. References 22.1 to 22.6 are the basis of much of this chapter.

A discussion of the status (in 1959) of the research conducted in England on plastically designed multi-story buildings is presented in Ref. 22.7. The reference also gives an example of the design of a four-story building, which was designed plastically. This building, which was erected in 1956, is an extension to the Cambridge University Engineering Laboratory.

Research on plastic design of multi-story frames in the United States led to a summer conference at Lehigh University in 1965.[22.5,22.8,22.9] Lectures were presented on current knowledge and new developments on the subject. Tests to ultimate load of components and frames verified many of the theories presented.[22.10] Even prior to the adoption of plastic design of multi-story frames in specifications and building codes, some buildings designed by the new methods were erected. The first of these was the Stevenson apartments in Bladensburg, Maryland.[22.11,22.12] (See Fig. 22.1.) Further applications of the methods may be anticipated as a result of the publication of a manual on the design of braced frames[22.6] and the adoption of the

Fig. 22.1 A plastically designed braced frame used in the Stevenson Apartments, Bladensburg, Maryland. (Courtesy of Horatio Allison-Robert L. Meyer, Structural Engineers.)

1969 AISC Specification.   The 1969 edition of the ASCE Commentary on Plastic Design in Steel also treats multi-story frames.[22.13]

The extent to which plastic design can be applied to multi-story frames is determined by the current AISC Specification,[1.21] which states that:

1. Bracing must be inserted so that it is unnecessary to rely on frame action to resist horizontal loads and moments due to the lateral deflection of column tops.
2. Bracing must be such that deflections at working loads are not excessive.
3. In the plane of the frame the structure must be designed as a rigid frame.

A design procedure which fulfills these requirements is presented subsequently.   First, however, the structural action of multi-story frames as well as the basic concepts involved will be discussed.

22.2 STRUCTURAL BEHAVIOR OF A MULTI-STORY FRAME

1. Structural Frames

This discussion is concerned with the behavior of rectangular frames subjected to loads applied in the plane of each frame.   The frames are rigidly connected, multi-story, multi-bay structures.   Beams and columns are fabricated of a ductile material such as structural carbon steel which can be strained inelastically without fracture.   The frames can be either symmetrical or unsymmetrical with respect to geometry and cross-sectional properties.   (See Figs. 22.2a and 22.2b.)

Loading can be classified into three basic types.   The frame can be subjected to full dead plus live gravity loads throughout or dead plus live gravity loads on only some of the beams.   It can also be subjected to concurrent gravity and wind loading.   Gravity loads are applied through floor systems to the beams and wind loads are applied through walls to joints at each floor level.   (See Figs. 22.2c, 22.2d, and 22.2e.)

Rigidly connected frames act as a unit, and the type of structural action depends on a combination of geometry and loading.   The structural action of multi-story frames can be classified according to how the frame resists loads.

1. Frame resists gravity loads without sidesway.
2. Frame resists gravity loads with sidesway.
3. Frame resists both gravity and wind loads with sidesway.

Sidesway is the lateral displacement of any panel point on a frame or structural member with respect to the base of the frame or structural member.

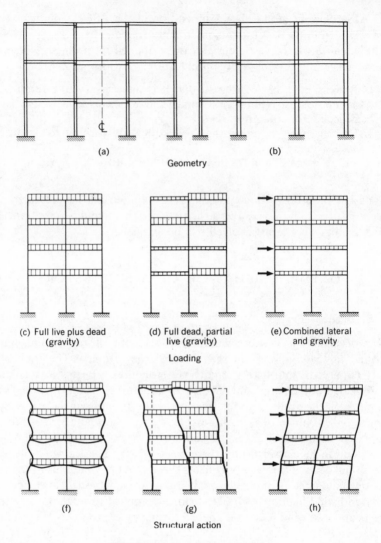

(a)   (b)

Geometry

(c) Full live plus dead   (d) Full dead, partial   (e) Combined lateral
      (gravity)   live (gravity)   and gravity

Loading

(f)   (g)   (h)

Structural action

Fig. 22.2 Multi-story frame.

The first type of structural action (no sidesway) can, in general, occur only when geometrically symmetrical frames are loaded symmetrically (Fig. 22.2f). However, it can also occur when a frame is sufficiently braced so that the effect of lateral deflections is negligible. The structural action which will occur generally is the one which results in sway (types 2 and 3). This structural action can result from non-symmetrical geometry and/or unsymmetrical gravity loading (Fig. 22.2g). The most general types of structural action occur when a frame is subjected to combined gravity and wind loading (Fig. 22h).

A braced frame is one in which lateral motion is resisted by elements other than the rigid framing itself (such as diagonal bracing).

2. Gravity Loading Without Sidesway

If very strong lateral bracing is provided, sidesway due to frame instability or due to lateral wind loads will have a relatively negligible effect on frame strength.   A test result for a three-story, two-bar frame is given in Fig. 22.3.[22.10]   The load vs. vertical beam deflection agrees well with a first order analysis neglecting the overturning effect of vertical loads in the swayed position.   Similar results were obtained from tests of frames with both symmetrical and unsymmetrical vertical loading.[22.14]

The usual failure to be expected in a braced frame at ultimate load is the formation of plastic mechanisms in all beams.   Alternatively, individual columns which have been deforming since the onset of loading may become unstable when the frame as a whole is unable to help them resist additional moments.   These moments are induced by the combination of end moments from beams plus beam shears acting through the eccentricity caused by column curvature.

Full dead plus live loading throughout the structure does not necessarily constitute the worst loading condition for the columns.   If "checkerboard"* live loading is applied in the vicinity of a central column (Fig. 22.4), the bending will often be more critical than would be the case if full loading were applied.   The moments transmitted by the beams to the joints will be larger and the beams will be bent in single curvature—a more severe state than other configurations.   Symmetrical single curvature is the worst possible bending configuration for beam-columns when sidesway is prevented.

3. Gravity Loading With Sidesway

When a frame is not braced, it is possible that under a certain load the frame can resist the load in a swayed position.   The lowest load at which this occurs is known as the buckling or bifurcation load for the frame.   If the structure is slender enough, buckling can occur while the members are still elastic.   If the structure is not slender, buckling will occur when the frame, having deformed into the inelastic range, has undergone a decrease in stiffness which allows it to buckle.   Usually the sway mode of failure is due to shear deformation of the stories, and the relative axial shortening of the

---

* "Checkerboard" loading is loading in which live load is omitted from beams in alternate bays and tiers, causing more severe column bending.

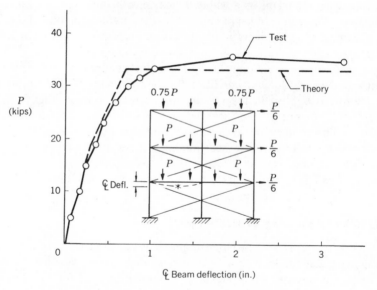

Fig. 22.3   Results of braced frame test.

exterior columns is negligible (Fig. 22.5a). If, however, the building is slender enough, buckling will occur by relative axial shortening of the columns (Fig. 22.5b), even though the stiffnesses of columns and beams are adequate.

When a structure resists gravity loads in a swayed position, the axial forces in the columns become eccentric with respect to the column bases. This

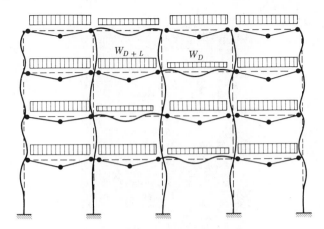

Fig. 22.4 Checkerboard loading.

requires the frame to resist additional bending moments.   Therefore, sway
results in a more serious bending situation for the columns.   Once again
failure takes place when the structure as a whole is no longer able to help the
individual columns resist additional joint moments.

One of a number of tests to study this type of failure is shown in Fig.
22.6.[22.10,10.16]   One plot shows the vertical load versus beam deflection and
a second curve shows the sway deflection caused by vertical loads.   It can

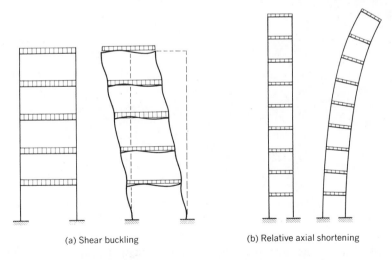

(a) Shear buckling   (b) Relative axial shortening

Fig. 22.5 Shear buckling and axial shortening.

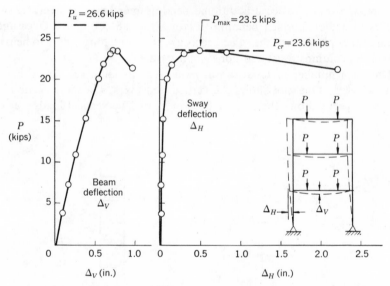

Fig. 22.6 Results of frame-buckling test.

be seen that the maximum load is substantially below the load to cause a mechanism but is in good agreement with a theoretical critical load. (See Chapter 10.) A photograph of the frame tested is also shown. The frame consisted of two identical bents having a ten foot span and a total height of seventeen feet.

4. Gravity Loading and Wind Loading

When horizontal forces are applied to a frame, the frame deforms in a shear-resistant manner. Only the columns can resist horizontal forces. The beams merely hold the columns in the shear-resistant deformed shapes. Failure of unbraced frames occurs due to either of two possibilities: (1) the

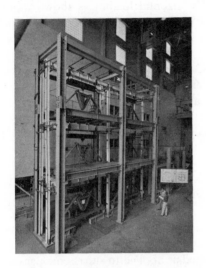

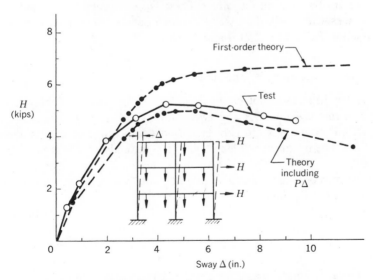

Fig. 22.7 Results of unbraced-frame test.

beams can no longer hold the columns in the necessary bent configuration because hinges form in the beam, or (2) the added moments at the joints produced by the eccentric axial loads are greater than the end moments which the columns can resist. Failure of a braced frame occurs when the bracing can no longer contribute the necessary shear resistance to keep the frame from failing.

The behavior of a two-bay, three-story unbraced frame subjected to combined vertical and horizontal loading is shown in Fig. 22.7.[22.10] The frame had W 6 × 20 columns and had W 12 × 16.5 beams on the two floor levels plus W 10 × 11.5 beams on the roof level. Its overall height and width were both 30 ft. In the graph of horizontal load versus sway deflection, the solid curve of test results falls slightly above the theoretical curve which includes the $P\Delta$ effect. A second theoretical curve which neglects the $P\Delta$ effect falls considerably above the experimental curve. This shows the inadvisability of using first-order theory for the design of unbraced multi-story frames. A photograph of the test setup and specimen accompanies the test curve.

22.3 CONCEPTS OF ANALYSIS

A complete discussion of the structural action and modes of failure of multi-story frames is beyond the scope of this chapter. However, the basic concepts can be illustrated as applied to subassemblages consisting of beams and columns framing into a joint. These basic concepts as applied to subassemblages will now be discussed. Further details of the cases discussed are contained in Refs. 22.1, 22.15, and 22.16.

1. Joint Instability Without Sway

Consider the subassemblage shown in Fig. 22.8. This assemblage consists of two beams OA and OB and two columns OC and OD framing into a joint O. The columns are subjected to a compressive force P and are of length h. The beams are of length L. The far ends of all the members are pinned (at A, B, C, and D). An unbalanced moment exists at joint O. The four members are framed rigidly at joint O so that they are forced to rotate through the same angle.

When an unbalanced moment M_O forces the joint to rotate through an angle θ, each of the four members resists a portion of the moment. The proportion of M_O that can be resisted by each member depends on the end moment vs. rotation relationship of each member. The end moment vs. end rotation relationships of the two beams and two columns are shown in Fig. 22.8. It can be seen from these curves that beam moments M_{OA} and

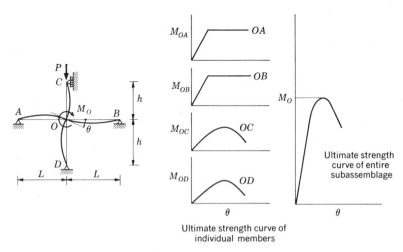

Fig. 22.8   Subassemblage—ultimate strength.

M_{OB} vary linearly with θ until the plastic moment is reached, at which time further rotation is possible without an increase or decrease in end moment. Column moments M_{OC} and M_{OD} vary with θ in the manner shown in the respective curves.   At a certain value of θ, the maximum end moments of the columns are reached, at which time further rotation must be accompanied by a decrease in end moments.   If a column were standing by itself, this maximum moment would correspond to failure.

The carrying capacity of the four members acting as a unit is obtained by adding the values of M_{OA}, M_{OB}, M_{OC}, and M_{OD} for each common value of θ. The resulting plot is also shown (to the right) in Fig. 22.8.

It is important to note that the maximum strength of any one of the four members does not necessarily correspond to the ultimate strength of the subassemblage.   The peak of the M_O vs. θ curve for the subassemblage corresponds to the ultimate strength of the subassemblage.

If the ends of the members OA, OB, OC, and OD were rigidly connected to beams and columns continuous over many supports, the resistance of the subassemblage would be somewhat increased.   If these continuous beams and columns were interconnected by grids of beams and columns, the sub-assemblage would be a portion of a multi-story frame.

In a multi-story frame, the moment at a joint will result from the different loadings and lengths of adjacent beams and eccentric shears.   As is shown in Fig. 22.9, if the frame in the vicinity of column OO' is loaded such that mechanisms form in beams OA and $O'B'$, and beams OB and $O'A'$ remain "continuous," the subassemblage affecting column OO' consists of columns OO', OC, and $O'C'$ and beams OB and $O'A'$.   The effect of the rest of the

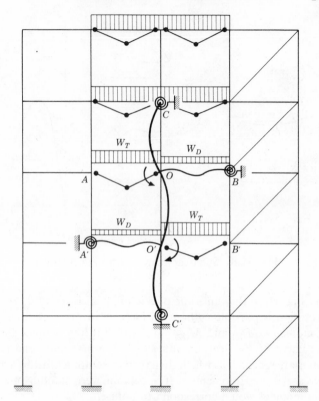

Fig. 22.9 Checkerboard loading column OO'.

frame can be approximated by specifying the end conditions of OC, $O'C'$, OB and $O'A'$.

2. Joint Instability With Sway

If there were no lateral support at the joint of the subassemblage of Fig. 22.8 and no lateral support at the top of column OC, the maximum joint moment M_O would be decreased. As shown in Fig. 22.10a, when the joints are restrained from lateral movement, each column and beam framing into O resists a part of the moment M_O. The moments in the columns (M_{OC}, M_{OD}) result in shears H_{OC} and H_{OD} equal to M_{OC}/h and M_{OD}/h, respectively. These shears are resisted by the support reactions.

On the other hand if there are no supports at the column ends as in Fig. 22.10 then there can be no column shears consistent with equilibrium other than those due to lateral deflection $P\Delta/h$. Further, $P\Delta/h$ adds to M_O. As a result, the columns CO and DO, instead of helping to resist M_O, place a

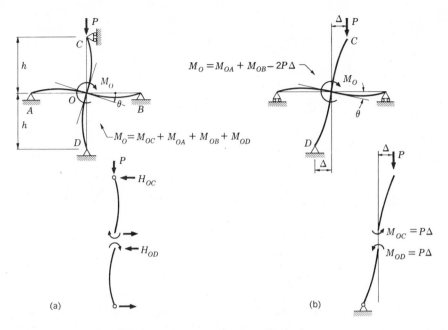

Fig. 22.10 Subassemblage—no lateral supports.

larger burden on the beam ($2P\Delta$ if both columns are of the same cross section).

3. Lateral Instability Due to Vertical Loads

Consider the subassemblage of Fig. 22.11, unsupported laterally and subjected to axial load only. If the axial load is large enough, the subassemblage will fail by lateral sway. This is the problem which is discussed in Art. 10.3. The buckling of such a subassemblage will occur at the largest load for which the structure can be in equilibrium in a vertical or in a swayed position. This implies that there are a set of deformations Δ_U, Δ_L, and θ for which:

1. $M_{OA} + M_{OB} = M_{OC} + M_{OD}$
2. $M_{OC} = P\Delta_U$
3. $M_{OD} = P\Delta_L$

Similarly, for a multi-story frame the buckling loads are the largest loads for which the structure may be in equilibrium either in a vertical or swayed position. This means that there is a swayed position where:

1. All the joints are in equilibrium.
2. There exists shear equilibrium at every story level.

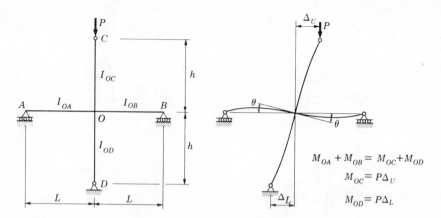

Fig. 22.11　Buckling of a subassemblage.

As an illustration, in Fig. 22.12 the swayed buckled shape of one of the columns of a frame is shown. The frame is a six-story frame with 12-ft-story heights. The span is 12 ft. The beams are all W 12 × 16.5 shapes and the columns are all W 6 × 25. A concentrated load of 455 kips is applied at the top of each column. There are an infinite number of such adjacent deformed positions at the buckling load. However, the ratios of the joint rotations are the same for each, although the moment magnitudes will vary for each possible deformed shape. The deformed position shown is one for a load (455 kips) almost at the buckling load.

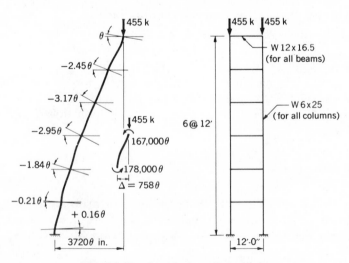

Fig. 22.12　Buckled configuration.

The insertion of adequate bracing at each story level eliminates the possibility of this mode of failure.   The design of such bracing is discussed in Art. 22.5.

4. Lateral Instability Due to Horizontal Forces

Consider the subassemblage shown in Fig. 22.13a.   This subassemblage consists of a beam of length L and of a column of length h subjected to a compressive force P.   A horizontal force H is applied at the joint O.

Only the column can resist the horizontal force H by reacting with a shear equal to M_{OB}/h which is equal to $[H + (P\Delta/h)]$.   The beam merely reacts with a moment M_{OA} to allow the column to deform in a shear-resisting manner. A plot of H/P vs. Δ/h results in a curve with a peak (Fig. 22.13b).   The peak represents the maximum value of H for a given subassemblage subjected to a given column compressive force P.   If the beam is weak and M_{OB} attains the plastic moment M_p of the beam before the peak value of H is reached, the plastic action of the beam will constitute failure.

When bracing is inserted into the frame it will help resist the horizontal force H.   If it is assumed that the axial deformations of members are negligible compared to Δ, both the column top and the bracing will move horizontally a distance Δ, as shown in Figs. 22.13c and d.   In Fig. 22.14, H-Δ curves are shown for the frame and for the bracing.   If the ordinates

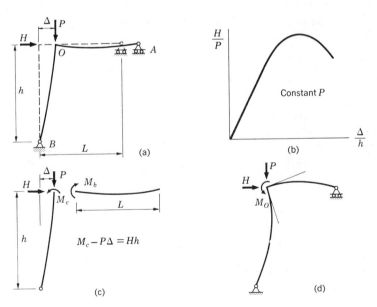

Fig. 22.13 Subassemblage—lateral loads.

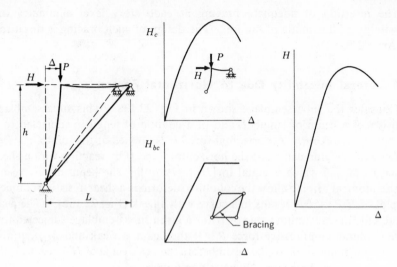

Fig. 22.14 Braced subassemblage.

are added for each common Δ, the resulting curve is the H-Δ curve for the subassemblage as shown to the right.

Figure 22.15 shows H/P_y vs. Δ/h curves for a subassemblage consisting of a W 8 × 31 column, a W 27 × 94 beam, $P = 0.6P_y$, $h = 40r$, and $L = 10d$. The broken curve in Fig. 22.15 shows the behavior if the effect of $P\Delta$ is

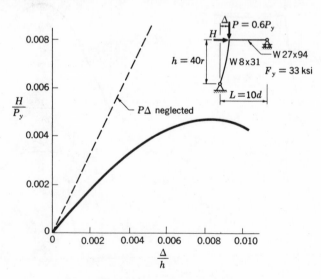

Fig. 22.15 Effect of $P\Delta$.

neglected. The importance of this effect is apparent. (The AISC Specification accounts for it by use of the effective length concept.)

5. Sway Subassemblage with Transverse Loading on Beams

In a real unbraced frame subjected to lateral loads, the beams must carry transverse loads as well as provide assistance to the columns in resisting rotation and sway. The magnitude of transverse loads influences the formation of plastic hinges in the beams. The restraining stiffness of the beams changes with the formation of each plastic hinge. Each column behaves as a restrained column with its restraining stiffness subjected to one, two, or three sudden reductions in value.

The location of an interior sway subassemblage in an unbraced frame is shown in Fig. 22.16. Figure 22.17 shows the manner of isolating the sub-

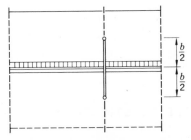

Fig. 22.16 Location of subassemblage in frame.

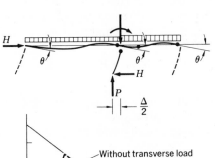

Fig. 22.17 Loading and deformation of subassemblage.

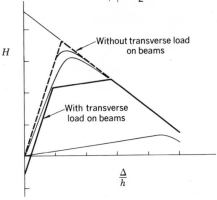

Fig. 22.18 Typical load-deflection curve of interior sway subassemblage.

assemblage in the sway subassemblage method of analysis.[22.17,22.18] All joints in the floor level are assumed to rotate through the same angle θ in the elastic range. Black filled circles indicate the expected locations of some plastic hinges which would cause changes in restraint stiffness as they occur.

A heavy solid curve in Fig. 22.18 gives the horizontal load versus deflection relationship for the subassemblage. A heavy dashed curve gives the load versus deflection relationship of the same subassemblage in the absence of floor loads on the beams. It can be observed that the floor loads cause a substantial reduction in horizontal load capacity. Indicated by fine lines in Fig. 22.18 are theoretical load-deflection curves of the given column for three different values of constant restraint stiffness.[22.8] Each portion of the subassemblage curves is parallel to a corresponding portion of the theoretical restrained column curves. The sharp breaks in the subassemblage curves occur when a plastic hinge forms either in the restraining member or the column. The slope in the subassemblage curve after a break corresponds to the new value of stiffness.

Figure 22.19 shows the load-deflection curves of all three sway subassemblages in a two-bay frame. The sum of the three subassemblage shears for each value of sway results in the story shear versus deflection curve of Fig. 22.19d.

22.4 MODIFICATIONS OF SIMPLE PLASTIC THEORY

Simple plastic theory is a very useful tool which enables a designer to choose member sizes which will give rise to the structural action of his choice. This subject has been treated briefly in Art. 7.8 and has been used to design single-story frames in Chapter 20.

The validity of simple plastic analysis depends on various assumptions. Three critical assumptions are:

1. The deflections are small, so that equilibrium equations can be satisfied in the undeformed state of the structure.
2. Instability does not occur for the frame as a whole or for the individual members.
3. Local and lateral instability of members is prevented.

However, in multi-story frames, the instability of frames can occur due to (1) instability of a joint, or (2) some form of lateral instability. The extent to which these effects limit the validity of simple plastic theory will be examined and the provisions by which the specifications attempt to meet these limitations are indicated.

1. Rotation Capacity

It has been shown in the previous article that compatibility must be

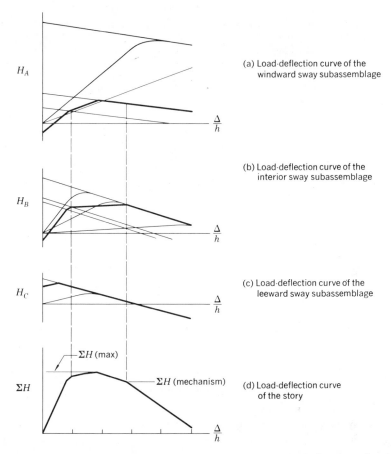

(a) Load-deflection curve of the windward sway subassemblage

(b) Load-deflection curve of the interior sway subassemblage

(c) Load-deflection curve of the leeward sway subassemblage

(d) Load-deflection curve of the story

Fig. 22.19   Determination of the load-deflection curve of a story from the load-deflection curves of the sway subassemblages.

considered in determining the maximum strength of a subassemblage; namely, it must be possible for a stable equilibrium configuration to exist. However, under certain conditions compatibility can be disregarded. For example, if all of the members framing into a joint such as O in Fig. 22.8 have moment-end rotation characteristics as shown for member OA, and as long as it is possible to distribute the joint moment M_O such that the maximum carrying capacity of each member is not exceeded, then the subassemblage can resist safely at least the moment M_O. The lower bound theorem of plastic analysis is satisfied. As can be seen from Fig. 22.20a, the moment-rotation requirement is that once the maximum moment has been reached, further rotation does not involve a decrease in moment. When structural members respond in this manner they are said to have adequate rotation capacity.

Beams designated as "plastic design sections" by AISC can be assumed to have adequate rotation capacity. However, beam-columns which are subjected to relatively high compressive forces generally do not exhibit this property. Beam-columns in general have moment-rotation characteristics as shown in Fig. 22.20b. When this is the case, the lower bound theorem of plastic analysis is not necessarily valid. An example will serve to illustrate this point.

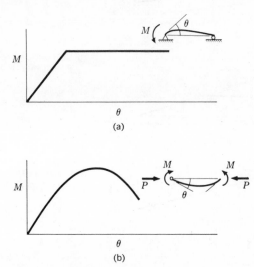

Fig. 22.20 Rotation capacity.

Consider the structure of Fig. 22.21a. According to simple plastic theory, if the moment diagram (shown in Fig. 22.21b) results in equilibrium with the loads, and also nowhere exceeds M_p or $M_{c(max)}$, then the distributed load w is the ultimate load of the structure. The mechanism condition has been satisfied because the beam has been designed so that it will form a mechanism upon application of the distributed load w (see Fig. 22.21a). Thus

$$w = \frac{16M_p}{L^2}$$

Equilibrium has been satisfied as long as

$$M_c + M_b = M_p$$

The maximum moment condition has been satisfied as long as the largest moment on each member has not exceeded the maximum (Fig. 22.21b).

$$M_c \leqslant M_{c(max)}$$

$$M_b \leqslant M_{b(max)} = M_{p(b)}$$

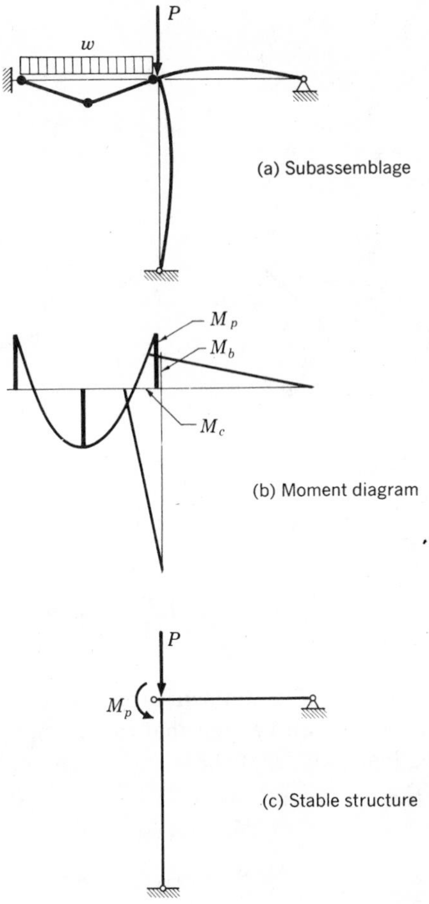

(a) Subassemblage

(b) Moment diagram

(c) Stable structure

Fig. 22.21 Subassemblage with simple plastic theory.

The portion of the structure which still has not failed is shown with its loading in Fig. 22.21c.

In order to test the validity of simple plastic theory (without modification) let the moment-rotation curve for the beam be the inverted curve plotted in Fig. 22.22. Let the inverted curve intersect the ordinate at the value of the unbalanced joint moment H_p. The moment-rotation curve for the column subjected to a compressive force P is shown on the same plot. The rotation at which there would be equilibrium would be a point where the two curves intersect. However, for the particular members whose rotation characteristics are defined by the curves in Fig. 22.22, there is no intersection.

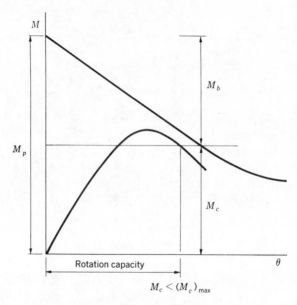

Fig. 22.22   Stability check.

This means that the unbalanced moment which would cause instability is less than M_p. However, it can be seen that the moment diagram results in equilibrium with the loads and that the maximum moment conditions have not been violated:

$$M_b < M_{b(\text{max})}$$

$$M_c < M_{c(\text{max})}$$

Therefore, the fact that the maximum moment for the column is not exceeded was not sufficient to prevent instability. In this instance, there was not sufficient rotation capacity. If there is inadequate rotation capacity, simple plastic theory alone cannot guarantee that a design will not fail by joint instability under the loads. Thus, modification in the approach is needed when sufficient rotation capacity cannot be assured.

Usually for the low slenderness ratios (h/r) which occur in multi-story frames, only columns which are bent in single curvature will not have sufficient rotation capacity to meet the requirements of the simple theory.

2. Design of Columns

The AISC Specification [2.4] requires that columns be designed so that neither of two inequalities are violated. These are Eqs. 11.26 and 11.27, repeated here as

$$\frac{P}{P_0} + \frac{M}{M'} \frac{C_m}{1 - P/P_e} < 1.0 \tag{22.1}$$

$$\frac{P}{P_y} + \frac{M}{1.18M_p} < 1.0 \tag{22.2}$$

As has been shown in Chapter 11, these equations represent approximately the maximum strength of an individual beam-column. The strength is limited in the first case by the possibility of in-plane instability failure or by lateral-torsional buckling and in the second case by complete plastification. The Specification allows the necessary moments, thrusts, and moment gradient on the columns to be determined by a rational analysis.

3. Effect of Lateral Deflection of Panel Points

The effect of lateral deflection of panel points has been discussed for sub-assemblages in Art. 22.3. These effects can be summarized as follows:

1. The ability of a frame to sway makes frame instability possible.
2. Lateral deflections of panel points reduce the ability of columns to help resist unbalanced joint moments.
3. Lateral deflections of panel points reduce the horizontal loads which a frame can resist.

In the design procedure which is subsequently developed, it is assumed that the bracing resists not only the moments P but the horizontal shears as well. This is conservative, but it is required by the present specification for plastically designed multi-story frames. As a result, the plastic design of multi-story frames for combined loading will be equivalent to the design of elastic bracing which connects a system of rigid pinned-end links, as shown in Fig. 22.23. The forces will be horizontal loads H and vertical loads P_0, $P_1, P_2, \ldots P_n$ applied at each panel point, as shown in Fig. 22.23.

22.5 DESIGN PROCEDURE ACCORDING TO PRESENT SPECIFICATION

In this article a design method permitted by the AISC Specification is outlined. The members to be designed are the beams, beam-columns, and the bracing. The design of connections will not be discussed here, but reference is made to Chapters 18 and 19 for their treatment. Connections are assumed to be rigid.

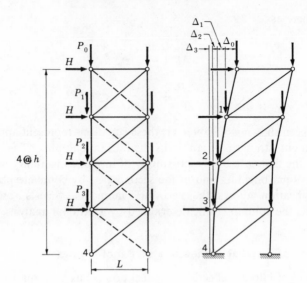

Fig. 22.23 Substitute frame.

1. Beams

The specification allows the plastic design of beams. As a result, all of the beams in a frame can be designed to resist full dead plus live load times a factor of 1.70 as if they were fixed-ended (Fig. 22.24). From equilibrium,

$$M_p = \frac{1.70 w_T L^2}{16} \tag{22.3}$$

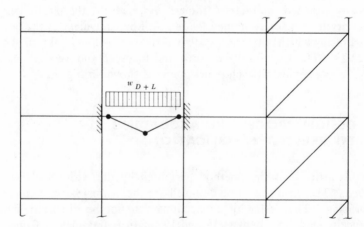

Fig. 22.24 Beam mechanism.

EXAMPLE 22.1

PROBLEM:

Based on the example and computations of Examples 21.1 and 21.2 design all the girders for roof and floor with A-36 steel, using plastic design (Part 2 of AISC specification)

SOLUTION:

Step No	Operation	Units	AB	BC	CD	
1	Span L	ft	20	12	28	
2	Assumed Clear Span $L_g = L.-1.17$	ft	18.83	10.83	26.83	
3	Live Load w_L x 1.70	k/ft	1.22	1.22	1.22	
4	Dead Load w_D x 1.70	k/ft	3.88	3.88	3.88	
5	% Live Load used	%	100	100	100	Root
6	% LL x w_L x 1.70 x L_g^2/16	kip-ft	27.1	9.0	55.1	
7	w_D x 1.70 x L_g^2/16	kip-ft	86.0	28.4	174.6	
8	M_p required	kip-ft	113.1	37.4	229.7	
9	Roof Beam		W 14 x 26	W 8 x 15	W 18 x 40	
10	Weight	lb.	490	162	1073	
11	Clear Span L_g	ft	18.83	10.83	26.83	
12	Live Load w_L x 1.70	k/ft	4.08	4.08	4.08	
13	Dead Load w_D x 1.70	k/ft	4.90	4.90	4.90	
14	% Live Load used	%	61.6	77.0	49.1	Floor
15	% LL x w_L x 1.70 x L_g^2/16	kip-ft	55.7	23.0	90.1	
16	w_D x 1.70 x L_g^2/16	kip-ft	108.6	35.9	220.4	
17	M_p required	kip-ft	164.3	58.9	310.5	
18	Floor Beam		W 18 x 35	W 12 x 16.5	W 21 x 49	
19	Weight	lb.	660	179	1316	

where w_T is the total dead plus live distributed load, L is the length of beam, and M_p is the plastic moment of the cross section. Since $M_p = F_y Z$,

$$Z = \frac{1.70 w_T L^2}{16 F_y} \tag{22.4}$$

The plastic design of beams is based on these assumptions:

1. The effect of axial load on the plastic moment M_p is negligible.
2. The shearing stresses do not affect the yield stress of any fiber.
3. Local and lateral torsional buckling do not occur.

Example 22.1 illustrates the design of girders for the roof and floor of a multi-story frame. Example 22.2 presents the results of computations for gravity axial loads in columns, needed in succeeding examples.

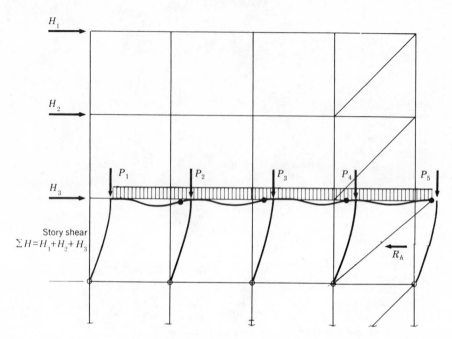

Fig. 22.25   Forces on panel.

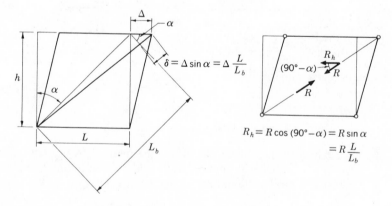

Fig. 22.26   Deformed brace.

EXAMPLE 22.2

PROBLEM:
   Determine the gravity axial loads in all the columns, the summation of all axial loads for each story level, and the wind shear at each story level (in kips).
   The gravity axial loads are determined as in Example 21.6.

Story	P_T C-1	P_T C-2	P_T C-3	P_T C-4	P_T	ΣH
T	40	49	61	52	202	4.8
T-1	114	123	150	142	529	10.6
T-2	178	181	226	215	800	16.3
T-3	234	247	308	290	1080	22.1
T-4	297	313	390	370	1370	27.8
T-5	361	379	473	450	1663	33.6
T-6	425	445	555	529	1955	39.4
T-7	490	511	637	609	2247	45.1
T-8	554	577	719	689	2540	50.9
T-9	618	644	802	768	2832	56.6
T-10	683	710	884	848	3124	62.4
T-11	747	776	966	928	3417	68.2
T-12	811	842	1049	1007	3709	73.9
T-13	875	908	1131	1087	4001	79.7
T-14	940	974	1213	1167	4293	85.4
T-15	1004	1040	1295	1246	4586	91.2
T-16	1068	1106	1378	1326	4878	97.0

The loads P_T and ΣH are working loads.

---

2. Bracing

The forces acting on a braced bent are shown in Fig. 22.25. In the design of bracing it is assumed that the axial shortening of beams and relative axial shortening of columns at each story level is negligible compared to the elongation of the bracing.* For any lateral deflection of a panel Δ, the elongation or shortening of the bracing can then be computed in terms of Δ, L_b, and L. See Fig. 22.26. If δ is the bracing elongation, and ε is the strain in the bracing member,

$$\varepsilon = \frac{\delta}{L_b} = \frac{\Delta L}{L_b^2} \tag{22.5}$$

The reaction force R in the brace is

$$R = \frac{\Delta L}{L_b^2} EA_b \tag{22.6}$$

---

* Shortening of columns and beams in the lower stories as the frame is erected and load on the columns increases could cause bracing members to become slack. To compensate for this, bracing members should be fabricated short, thus causing an initial prestressing of the frame which is relieved when load is added.

The horizontal component of the force R is designated R_h:

$$R_h = \frac{L^2}{L_b^3} EA_b\Delta \tag{22.7}$$

Equation 22.7 determines the shear furnished by a brace of area A_b corresponding to a lateral deflection Δ of the story.

In order to prevent overall frame instability, it is necessary for the bracing to be able to react always with a shear force equal to $\Sigma P_T\Delta$ at each story level. The bracing requirement to prevent frame instability under vertical loading times a load factor F of 1.70 is

$$\frac{L^2}{L_b^3} EA_b\Delta = \frac{\Delta \sum FP_T}{h}$$

$$A_b = \frac{1}{E}\frac{L_b^3}{hL^2}\sum FP_T \tag{22.8}$$

In order to insure that bracing will resist combined wind and vertical load without help from the frame before yielding, the following relationships must hold true

$$\frac{L^2h}{L_b^3} EA_b\Delta = [\Delta \sum FP_w + \sum Hh]$$

$$\frac{L}{L_b^2} E\Delta = F_y \tag{22.9}$$

FP_T is the axial load in a column segment due to gravity loading, when the load factor is 1.70. FP_w is the axial load in a column resulting from gravity plus wind load, when the load factor is 1.30. The first of Eqs. 22.9 insures that the bracing upon deflecting an amount Δ will resist all of the wind force plus the shears due to the axial loads moving through the lateral deflection Δ. The second of Eqs. 22.9 insures that the stress in the bracing will not exceed the yield stress. If the bracing is to resist forces in compression, the second of Eqs. 22.9 must be replaced by

$$\frac{L}{L_b^2} E\Delta = F_{cr}$$

where F_{cr} is the buckling stress in the bracing $[F_{cr} = \pi^2 E/(L_b/r)^2]$.
Combining Eqs. 22.9 yields:

Required area for tension bracing:

$$A_b = \frac{L_b}{L}\frac{\sum H}{F_y} + \frac{1}{E}\frac{L_b^3}{L^2h}\sum FP_w \tag{22.10}$$

Required area for compression bracing:

$$A_b = \frac{L_b}{L}\frac{\sum H}{F_{cr}} + \frac{1}{E}\frac{L_b^3}{L^2h}\sum FP_w \tag{22.11}$$

If a brace is subjected to the full wind load plus shears due to compressive forces which have moved laterally, the deflection at working load is estimated as

$$\frac{\Delta}{h} = \frac{\sum H}{1.30} \frac{1}{\dfrac{L^2 h}{L_b{}^3} EA_b - \sum P_w} \tag{22.12}$$

A value of $\Delta/h = 0.002$ at working load has been recommended in Ref. 3.12. Therefore

$$0.002 > \frac{\sum H}{\dfrac{L^2 h}{L_b{}^3} 1.30 EA_b - \sum 1.30 P_w} \tag{22.13}$$

Bracing then will be designed for the largest value of Eqs. 22.8 and 22.10 (or 22.11) and will be checked by Inequality 22.13.

When the structure is subjected to combined gravity and wind load, the deformed bracing will load the leeward columns of the braced panel. The severity of this additional compressive force on those columns should be checked. However, the load factor for combined loading is 1.30 instead of 1.70, and thus the bending moments as well as the compressive forces due to gravity will be reduced by a factor of approximately 0.75. In the example, columns are designed for gravity loading times a factor of 1.70, and it is assumed that the reduction in design load for combined loading will compensate the cumulative increment of compressive loads on the leeward columns of the braced panels. As braced multi-story frames get taller and taller, columns should be designed for both gravity and combined loading and the more critical design should be used. Examples of additional procedures for bracing design are given elsewhere.[22.5,22.6] These include: consideration of the sway resistance contributed by the frame, consideration of the effect of axial deformation of frame members, design of members in the braced bay, and special procedures for design of a K-type bracing system.

Example 22.3 illustrates the design of tension diagonals for a multi-story frame.

3. Columns

Columns must be designed so that they provide the needed resistance until mechanisms form in the beams. To do this by use of Eqs. 22.1 and 22.2, values for the thrust, maximum moment, and moment gradient in the column need to be determined. It is impossible to determine these values before a frame is designed, so approximations must be used for preliminary member selection. These approximations must seek combinations of loading and boundary conditions which cause the most severe conditions on the member to be designed but which remain within practical extremes possible for the structure.

The worst possible bending configuration for a column subjected to a given compressive force and combination of concurrent end moments and prevented from sway is that of symmetrical single curvature. This is the bending configuration for which the end moments are equal and opposite in sense. A value of $C_m = 1$ is required by the AISC Specification for this case. The bending case in which one end is free of moment is less critical and results in a $C_m = 0.6$. The least critical bending configuration results when the end moments are equal and in the same sense. The Specification limits the value of C_m by specifying that it cannot be smaller than 0.4.

The largest moments at the column ends as well as the worst possible bending condition occur when "checkerboard" live loading is applied in the vicinity of a column. Such a loading is shown in Fig. 22.27 (column AB). In order to simplify computations, only the members framing directly into column AB are considered. The effect of the rest of the frame is simulated by rotational springs at the ends of the restraining members. The rotational springs are taken into account by making assumptions for the end conditions of the restraining members.

Consider the subassemblage of Fig. 22.27. Beams AE and BH have been loaded with full dead plus live load times a factor of 1.70 and have formed plastic mechanisms. This leaves a subassemblage consisting of beams AF, GB, and columns AC, AB, and BD, with unbalanced joint moments $M_{p(AE)}$ and $M_{p(BH)}$.

Studies of probable design ranges for similar subassemblages have shown that the range depends on the combination of beam spans and relative sizes of dead and live loads plus the strength of the beams selected to carry these loads.[22.5] A design aid resulting from such studies is given in Fig. 22.28, which separates interior columns into one group which can be designed for single curvature and another which can be conservatively designed as if they were pin-ended. This chart should be entered with the ratio of the longer to shorter span against the load ratio of dead load for the longer span to total load of the shorter span. The two regions would be consistent with the use of C_m values of 1.0 and 0.6 respectively in Eq. 22.1. Intuition will support this conclusion when it is noted that a short and therefore stiff elastic restraining span in combination with a high dead load on the restraining span will tend to counteract the bending of the column caused by a plastic hinge in the longer beam. An equivalent design chart for exterior columns is given in Fig. 22.29. All exterior columns except possibly the bottom story will be in double curvature. Most will have a large enough end moment ratio to permit the use of the minimum C_m of 0.4. When the total load is very large with respect to dead load, exceeding 2.5 times the dead load, C_m should be used as 0.6.

EXAMPLE 22.3

PROBLEM:
Design the tension diagonals for the 12 ft center bay of the first story (seventeen stories from the top) using the provisions of AISC Section [2.3]

SOLUTION:

$$L_b = 17 \text{ ft}; \quad L = h = 12 \text{ ft}; \quad \frac{L_b}{h} = \frac{17}{12} = 1.417$$

$$\Sigma H = 97.0 \times 1.30 = 126.1^k$$

$$\Sigma FP = 1.70 \times 4878 = 8293^k; \quad 1.30 \times 4878 = 6341^k$$

$$A_b = \frac{1}{E} \frac{L_b^3}{hL^2} \Sigma FP_g = \frac{(17)^3(8293)}{29\times10^3(12)(12)^2} = 0.81 \text{ sq in.} \qquad \text{(frame stability)}$$
$$\text{(Eq. 22.8)}$$

$$A_b = \frac{L_b}{L} \frac{\Sigma H}{F_y} + \frac{1}{E} \frac{L_b^3}{L^2 H} \Sigma FP_w = \frac{17}{12} \frac{(126.1)}{(36)} + \frac{(17)^3(6341)}{29\times10^3(144)(12)}$$

$$= 4.96 + 0.62 = 5.58 \text{ sq in.}$$

Try 1—L5 x 5 x 3/4 with 3/4 bolts

$$A_g = 6.94 \text{ in.}^2; \quad A_n = 6.94 - 0.75(0.875) = 6.28 \text{ in.}^2$$

$$\frac{A_{reqd}}{A_g} = \frac{5.58}{6.94} = 0.81 < 0.85 \qquad \text{AISC Section [1.14.3]} \qquad \underline{O.K.}$$

$$\frac{L_b}{r} = \frac{17 \times 12}{0.97} = 210 < 300 \qquad \text{AISC Section [1.8.5]} \qquad \underline{O.K.}$$

Check Equation 22.13 for deflection

$$0.002 \geq \frac{\Sigma H}{1.30 E \dfrac{L^2 h}{L_b^3} A_b - \Sigma 1.30 P_w}$$

$$\geq \frac{97.0}{\dfrac{1.30(29\times10^3)(12)^2(12)}{(17)^3}(5.58) - 6341} = 0.001434 \qquad \underline{O.K.}$$

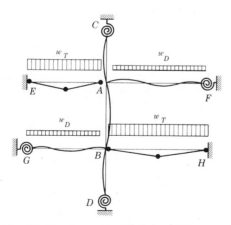

Fig. 22.27 Structural subassemblage.

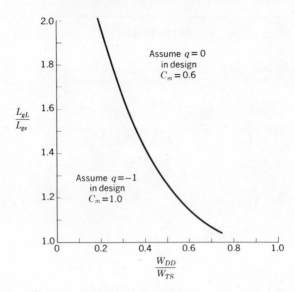

Fig. 22.28 Selection of moment-gradient factor for interior columns.

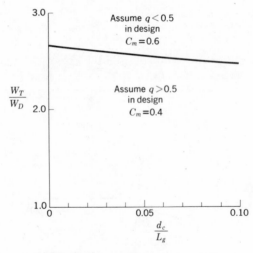

Fig. 22.29 Selection of moment-gradient factor for exterior columns.

Magnitudes of bending moments in columns may be calculated by considering the beam moments and reactions on each side of the column at a joint. Obviously, a symmetrical case with fully loaded equal size beams on each side of an interior column will cause no moment in an interior column. An unbalanced configuration is required to cause severe column bending.

One situation presenting an unbalanced configuration is the case of a long, heavily loaded span opposing a short, lightly loaded span. If both loadings are full factored loads, the difference in plastic moment values and beam reactions will determine the column moment. However a greater unbalance will be caused by checkerboard loading where only factored dead load is present on one of the beams. This configuration is illustrated by Fig. 22.27.

Studies on the effect of various assumed end conditions for the restraining members have come to the conclusion that the fixed end moment should be used for the beam subjected to dead load.[22.5] It should be computed for the clear span between column faces. The plastic moment should be used for the fully loaded beam. A further unbalance in joint moment may be calculated from the difference in beam shears acting at the two faces of the column. The unbalanced moment M_u for the subassemblage of Fig. 22.27 is

$$(M_u)_A = (M_p)_{AE} - \left(\frac{wL^2}{12}\right)_{AF} + \frac{d_c}{2}\left(\frac{wL}{2}\right)_{AE} - \frac{d_c}{2}\left(\frac{wL}{2}\right)_{AF} \qquad (22.14)$$

where d_c is equal to the column depth. In the case that factored dead load could cause the fixed end moment in AF to equal or exceed its plastic hinge value, M_p should be used. When all possible distributions of M_u above and below the joint have been considered, it seems reasonable for design purposes to select the column moment as one-half the unbalanced joint moment.

$$M_c = \tfrac{1}{2}M_u \qquad (22.15)$$

Similar reasoning will lead to an equation replacing Eq. 22.14 for exterior columns. In this case no elastic counterbalancing moment exists unless there is a cantilever beam for a balcony. Frequently a spandrel beam supporting a wall will apply a concentrated load to the outer face of the column. It can be handled in the same manner as a beam reaction.

The selection of trial cross sections can be facilitated by use of a preliminary design formula which converts required moment capacity into an axial load equivalent. One such formula is developed in Ref. 22.6.

$$P_y = P + 2.1\frac{M}{d} \qquad (22.16)$$

but not less than

　　　　$1.12P$　　　for $F_y = 36$ ksi
　　　　$1.18P$　　　for $F_y = 50$ ksi

where P = required axial load capacity, kips
　　M = required major axis end moment capacity, kip ft
　　d = estimated column depth, ft
　　$P_y = AF_y$, kips

Equation 22.16 is derived from Eq. 22.2 and it will result in fairly accurate member selections for h/r_y values of 40 or less.

Trial sections selected should be checked for adequacy by the interaction equations Eq. 22.1 and 22.2.

The procedure for the design of a column is as follows:

1. Choose appropriate subassemblage.
2. Compute M_u (Eq. 22.14).
3. Compute M_c (Eq. 22.15).
4. Compute trial P_y (Eq. 22.16).
5. Choose a trial cross section.
6. Determine C_m.
7. Apply both inequalities 22.1 and 22.2.
8. If either inequality is violated choose a larger cross section.
9. Repeat 5, 7, 8 until the smallest cross section which satisfies both inequalities is found.

In cases where large rotation of plastic hinges is required in beam-columns, Inequalities 22.1 and 22.2 are not adequate to assure sufficient rotation capacity, and additional checks may be needed.

Examples 22.4 and 22.5 are used to illustrate the procedure. Example 22.6 summarizes the designs of Examples 22.1 through 22.5.

22.6 PRELIMINARY DESIGN OF UNBRACED FRAMES

An equilibrium solution permitting preliminary design of members for unbraced frames can be formulated.[22.5]   As in the case of the portal and cantilever methods used in allowable stress design, the frame is reduced to determinacy by a number of statics assumptions and a preliminary analysis can be made without prior knowledge of member properties. This preliminary analysis can be used to select trial member sizes which can then be evaluated by exact methods of analysis.

Data for preliminary design are first obtained by tabulating column loads at each floor for gravity load according to the tributary areas.   Horizontal shears for each story are similarly tabulated according to tributary areas acted upon by wind.   The main remaining steps of the preliminary design are:

1. Sum of column end moments in a story.
2. Sum of girder end moments in a story.
3. Beam moment diagram for transverse load and sway.
4. Moment balancing.
5. Selection of trial members.

Figure 22.30 shows forces used in developing an equilibrium equation for the sum of the column end moments $\sum M_c$ in a story. This equation is formulated by writing an equation for the shear in each column in terms of its end moments M_c, the thrust P, and an assumed sway Δ. All the column

---

EXAMPLE 22.4

PROBLEM :
   Design column C-2, Top and T-1, using the provisions of AISC Specification Section [2.4].
The columns are braced only at the floors in the perpendicular direction :

SOLUTION :
  Joint Moment

$$M = (M_p)_{BA} - \left(\frac{w_D L_g^2}{12}\right)_{BC} + \left[\left(\frac{w_T L_g}{2}\right)_{BA} - \left(\frac{w_D L_g}{2}\right)_{BC}\right]\frac{d_c}{2}$$

$$= 164.3 - \frac{4.90(10.83)^2}{12} + \left[\left(\frac{7.42 \times 18.83}{2}\right) - \left(\frac{4.90 \times 10.83}{2}\right)\right]\frac{7}{12}$$

$$= 141.7 \text{ kip-ft}$$

Moment Ratio in Column

$$\left.\begin{array}{l}\dfrac{L_{g1}}{L_{gr}} = \dfrac{18.83}{10.83} = 1.74 \\[2ex] \dfrac{w_{D1}}{w_{Tr}} = \dfrac{4.90}{8.04} = 0.61\end{array}\right\}\quad\begin{array}{l}\text{From Fig 22.23}\qquad q = 0 \\[1ex] \therefore\ C_m = 0.6\end{array}$$

Subassemblage for all stories except
the first floor

Design Loads

   $P = 123 \times 1.70 = 209$ kip $M = 0.5 \times 141.7 = 70.8$ kip-ft

Try W 14 x 43

   $A = 12.6$ in.^{2} ; $Z_x = 69.7$ in.^{3}; $r_x = 5.82$ in.; $r_y = 1.89$ in.

$$\frac{KL}{r_x} = \frac{144}{5.82} = 25 \ ; \ \frac{KL}{r_y} = \frac{144}{1.89} = 79 \ ; \ F_a = 15.47 \text{ ksi} ; \ F_e' = 238 \text{ ksi}$$

$$P_y = 36. \times 12.6 = 455 \text{ kip} ; \ M_p = \frac{36 \times 69.7}{12} = 209 \text{ kip-ft}$$

$$P_0 = 1.70 \times 15.47 \times 12.6 = 332 \text{ kip} ; \ P_e = 1.92 \times 238 \times 12.6 = 5790 \text{ kip}$$

$$M' = \left[1.07 - \frac{\sqrt{F_y}(L/r_y)}{3160}\right]M_p = \left[1.07 - \frac{\sqrt{36}(79)}{3160}\right]209 = 192 \text{ kip-ft}$$

$$\frac{P}{P_y} = \frac{209}{455} = 0.460 \ ; \ \frac{P}{P_0} = \frac{209}{332} = 0.630 \ ; \ \frac{P}{P_e} = \frac{209}{5790} = 0.0362$$

$$1 - \frac{P}{P_e} = 0.964 \ ; \ \frac{M}{1.18 M_p} = \frac{70.8}{1.18(209)} = 0.287$$

$$\frac{P}{P_y} + \frac{M}{1.18 M_p} = 0.460 + 0.287 = 0.747 < 1.0 \qquad \underline{\text{O.K.}}$$

$$\frac{C_m M}{(1 - \frac{P}{P_e}) M'} = \frac{0.6 \times 70.8}{0.964 \times 192} = 0.230$$

$$\frac{P}{P_0} + \frac{C_m M}{(1 - \frac{P}{P_e}) M'} = 0.630 + 0.230 = 0.860 < 1.0 \qquad \underline{\text{O.K.}}$$

EXAMPLE 22.5

PROBLEM:
  Design exterior column C–I, Top and T–I, per AISC Section [2.4]
SOLUTION:
  Joint Moment

$$M = (M_P)_{AB} + \left[\left(\frac{w_T L_g}{2}\right)_{AB} - (\text{Wall Reaction})\right]\frac{d_c}{2}$$

$$M = 164.3 + \left[\left(\frac{7.41 \times 18.83}{2}\right) - (41.6)\right]\frac{7}{12}$$

$$M = 180.7 \text{ kip-ft}$$

Design Moment $= \dfrac{180.7}{2} = 90.4$ kip-ft

$$P = 114 \times 1.70 = 194 \text{ kip}, \quad C_m = 0.4$$

Try W 14 x 38

$$A = 11.2 \text{ in.}^2; \quad Z_x = 61.6 \text{ in.}^3; \quad r_x = 5.88 \text{ in.}; \quad r_y = 1.54 \text{ in.}$$

$$\frac{KL}{r_x} = \frac{144}{5.88} = 25; \quad \frac{KL}{r_y} = \frac{144}{1.54} = 94; \quad F_a = 13.72 \text{ ksi}; \quad F'_e = 238 \text{ ksi}$$

$$P_y = 36 \times 11.2 = 402 \text{ kip}; \quad M_p = \frac{36 \times 61.6}{12} = 184.5 \text{ kip-ft}$$

$$P_o = 1.70 \times 13.72 \times 11.2 = 261 \text{ kip}; \quad P_e = 1.92 \times 238 \times 11.2 = 5120 \text{ kip}$$

$$M' = \left[1.07 - \frac{\sqrt{F_y}(L/r_y)}{3160}\right]M_p = \left[1.07 - \frac{\sqrt{36}(94)}{3160}\right]184.5 = 164.2 \text{ kip-ft}$$

$$\frac{P}{P_y} = \frac{194}{402} = 0.482; \quad \frac{P}{P_o} = \frac{194}{261} = 0.743; \quad \frac{P}{P_e} = \frac{194}{5120} = 0.038$$

$$1 - \frac{P}{P_e} = 0.962; \quad \frac{M}{1.18 M_p} = \frac{90.4}{1.18(184.5)} = 0.415$$

$$\frac{P}{P_y} + \frac{M}{1.18 M_p} = 0.482 + 0.415 = 0.897 < 1.0 \qquad \text{O.K.}$$

$$\frac{P}{P_o} + \frac{C_m M}{\left(1 - \frac{P}{P_e}\right)M'} = 0.743 + \frac{0.4(90.4)}{0.962(164.2)} = 0.972 < 1.0 \qquad \text{O.K.}$$

Use W 14 x 38

Wall

w_T

Subassemblage for
all stories except
1st floor and top

shears are summed up and then the equation is solved for $\sum M_c$ as follows:

$$\sum M_c = -(\sum H)h - (\sum P)\Delta \qquad (22.17)$$

To use the equation, only the sum of the story shears from all stories above and the sum of the column thrusts need be used. If half the moment is assumed at the top and bottom of each story, the result is the same as assuming inflection points at the center of each column.

The use of Eq. 22.17 in any two successive stories gives a sum of column end moments which must be in equilibrium with the girder moments in the floor level between them. This relationship is expressed in Fig. 22.31. An expression for the sum of girder end moments based on this sketch is given by

EXAMPLE 22.6

Summary of Examples 21.1, 22.2, 22.3, 22.4 and 22.5

Column and Bracing Sizes

	C–1	C–2	C–3	C–4	Bracing
Top	W 14 x 38	W 14 x 43	W 14 x 53	W 14 x 53	I–L 4 x 4 x 3/8
T–1					"
T–2	W 14 x 61	W 14 x 61	W 14 x 78	W 14 x 78	"
T–3					"
T–4	W 14 x 78	W 14 x 78	W 14 x 111	W 14 x 111	"
T–5					"
T–6	W 14 x 111	W 14 x 111	W 14 x 136	W 14 x 127	I–L 4 x 4 x 3/8
T–7					I–L 4 x 4 x 7/16
T–8	W 14 x 119	W 14 x 127	W 14 x 158	W 14 x 158	"
T–9					I–L 4 x 4 x 1/2
T–10	W 14 x 142	W 14 x 142	W 14 x 184	W 14 x 184	I–L 4 x 4 x 5/8
T–11					"
T–12	W 14 x 167	W 14 x 167	W 14 x 219	W 14 x 211	I–L 4 x 4 x 3/4
T–13					"
T–14	W 14 x 184	W 14 x 193	W 14 x 246	W 14 x 237	I–L 5 x 5 x 5/8
T–15					"
T–16	W 14 x 202	W 14 x 202	W 14 x 264	W 14 x 246	I–L 5 x 5 x 3/4

NOTE: Columns are designed by provisions of Part I of the AISC Specification to support girders designed according to Section [2.1].

Bracing is designed plastically

$$\sum M_g = -\tfrac{1}{2}[(\sum M_c)_{n-1} + (\sum M_c)_n] \tag{22.18}$$

where $n - 1$ refers to the story above and n to the story below the girders. Positive end moments act clockwise on the ends of columns and girders.

The typical shape of a girder moment diagram to resist a combination of end moments due to sway and uniformly distributed transverse load is shown in Fig. 22.32. Equations and design charts for all the key values of such moment diagrams may be prepared.[22.5] These make possible the selection of girder members sufficient to provide the necessary moments of resistance. At this stage, the decision of how much of the total $\sum M_g$ to provide in each girder is arbitrary. Experience will, no doubt, result in methods giving the best compromise in economy and structural performance.

After selection of girders, the designer has in hand a set of girder moment diagrams and a sum of column end moments for each story. The moments are all sufficient to resist the external forces on the frame but the moments have not yet been assigned to the individual columns. A first trial would be to distribute moments equally between all columns as shown in Fig. 22.33.

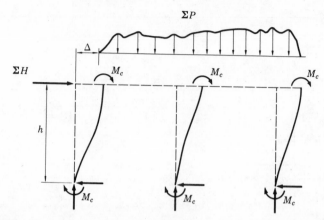

Fig. 22.30 Horizontal shear equilibrium in a story of an unbraced frame.

Here it will be found that the joints are not balanced and therefore some of the column moments have been assigned to the wrong columns. This is remedied by a process called *moment balancing*. This process is simply an orderly method for adjusting and recording changes in the end moments in members to achieve joint equilibrium without destroying the story equilibrium already assured by Eq. 22.17 and 22.18. Comparison of Fig. 22.34 with Fig. 22.33 gives a picture of the changes necessary to obtain joint equilibrium. The left hand joint originally had 11 units of girder moment acting against 22 units of column moment. Approximately half of the unbalance was removed above and below the joint and moved to the center columns where it was needed to increase the column moments sufficiently to balance the girder moments there. Through a coincidence, the right hand joint was initially in equilibrium. Proper use of moment balancing will always result

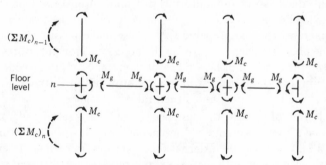

Fig. 22.31 Sum of girder moments in a story of an unbraced frame.

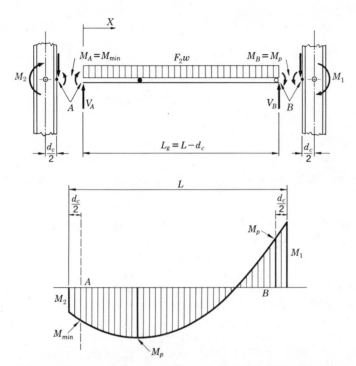

Fig. 22.32 Sway moments on a transversely loaded girder.

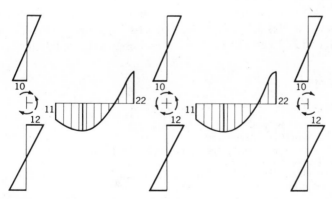

Fig. 22.33 Preliminary moment diagrams prior to moment balance.

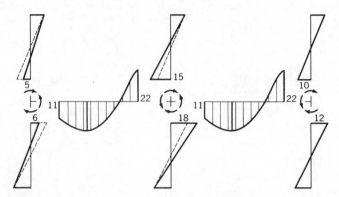

Fig. 22.34 Preliminary moment diagrams resulting from moment balance.

in joint equilibrium at the end of one cycle without disturbing the girder moments or the equilibrium at another floor.

Data available at the end of the moment balancing process are sufficient to allow selection of trial members from the available design aids.[22.5,22.8] To those who doubt the accuracy of some of the assumptions made in making the analysis, it should be pointed out that the equilibrium solution provided a path for all applied loads to travel to the ground without exceeding the plastic limit. The structure itself will probably find a more efficient way of carrying the forces provided no unstable conditions exist. The sway subassemblage method referred to earlier can give an indication of whether an adequate sway Δ was initially assumed. Computer programs for preliminary design can aid in the selection of trial members.[22.19] Computer programs for the analysis of complete frames can be used effectively once trial sections have been selected.

Space permits only this limited treatment of the various problems investigated in the plastic design of steel multi-story frames. References 22.5, 22.6, 22.8, and 22.13 give further examples, design aids, and design shortcuts.

PROBLEMS

22.1. Design a ten-story one-bay frame in which:

(a) The story heights (h) are all equal to 12.5 ft.
(b) The span (L) is 25 ft.
(c) The wind loads are 4.5 kips applied at the level of the roof and 9 kips applied at all other panel points.
(d) The distributed dead plus live load w_T is equal on all stories including the roof and is 1.4 kips per ft.

Use plastic design insofar as the AISC Specification allows its use. Design only beams, columns, and bracing.

22.2. Given a W 8 × 31 column of $h/r_x = 80$ subjected to a compressive force which results in $P/P_y = 0.6$. A W 8 × 31 member whose L/d ratio is equal to 20 frames into the column as shown in Fig. 22.12a. Can this restrained column support its axial force without buckling? How large a horizontal force can it resist? How large a joint moment? Use the column curves given in the figure below. (These curves were generated by the process shown in Example 11.2. They are a family of curves similar to Fig. 11.2. More accurate values may be obtained from Ref. 11.11.) Assume linear M vs. θ relationship for the beam until the moment is equal to M_P.

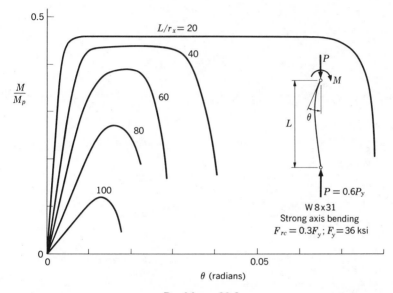

Problem 22.2

REFERENCES

CHAPTER 1. INTRODUCTION

1.1 WRC-ASCE Joint Committee
Commentary on Plastic Design in Steel, ASCE Manual No. 41, 1961. (See also Second Edition, Ref. 22.13.)

1.2 T. C. Kavanagh
New Trends in Design and Construction of Steel Structures, R. P. Davis Lecture Series, West Virginia University, 1968

1.3 T. C. Kavanagh and S. M. Johnson
"Maintenance—The Systems Approach," *Civil Eng.*, July, 1966

1.4 J. A. L. Waddell
Bridge Engineering, Vol. I, Wiley, New York, 1916

1.5 L. S. Beedle
Plastic Design of Steel Frames, Wiley, New York, 1958

1.6 J. Sterling Kinney
Indeterminate Structural Analysis, Addison-Wesley, Reading, Mass., 1957

1.7 E. J. Ruble
Private Communication, April 3, 1963

1.8 E. L. Erickson
Private Communication, May 9, 1963

1.9 J. B. Johnson, C. W. Bryan, and F. E. Tourneare
The Theory and Practice of Modern Framed Structures, 10th ed., Part I, Wiley, New York, 1938

1.10 L. N. Edwards
A Record of History and Evolution of Early American Bridges, University Press, Maine, 1959

1.11 AISC
Iron and Steel Beams, 1873–1952, AISC, New York, 1952

1.12 G. C. Lee and T. V. Galambos
"Post-Buckling Strength of Wide-Flange Beams," *Proc. Am. Soc. Civil Engrs.*, **88** (EM-1), p. 59 (February 1962)

1.13 R. C. Van Kuren and T. V. Galambos
Beam-Column Experiments, Fritz Laboratory Report No. 205A.30, Lehigh University, Bethlehem, Pa. (July 1961)

1.14 G. C. Driscoll, Jr., and L. S. Beedle
"The Plastic Behavior of Structural Members and Frames," *Welding J.*, **36**(6), p. 275-*s* (June 1957)

1.15 G. C. Driscoll, Jr. *et al.*
Plastic Design of Multi-Story Frames, Summer Conference Notes, Fritz Laboratory Report No. 273.30 Lehigh University, Bethlehem, Pa., 1965

1.16 L. S. Beedle, Le-Wu Lu, Lee C. Lim
"Recent Developments in Plastic Design Practice," *ASCE Journal*, **95** (ST-9). p. 1911 (September 1969)

1.17 B. G. Johnston (ed.)
Guide to Design Criteria for Metal Compression Members, 3d ed., Wiley, New York, 1974

1.18 T. R. Higgins
"AISC's Steel Specification Provides New Design Formulas," *Eng. News-Record*, **168**(2), p. 34 (January 1962)

1.19 G. Winter
Commentary on the 1968 Edition of the Specification for the Design of Cold-Formed Steel Structural Members, AISI, New York, 1970

1.20 Research Council on Riveted and Bolted Structural Joints of the Engineering Foundation
"Specifications for Structural Joints Using ASTM A325 Bolts," *Proc. Am. Soc. Civil Engrs.*, **88** (ST-5), p. 11 (October 1962)

1.21 AISC
Specification for the Design, Fabrication and Erection of Structural Steel for Buildings, AISC, New York, 1969 [Part of AISC Manual, Ref. 7.9]

1.22 O. G. Julian
"Synopsis of First Progress Report of Committee on Factors of Safety," *Proc. Am. Soc. Civil Engrs.*, **83** (ST-4), p. 1316 (July 1957)

1.23 AASHO
Standard Specification for Highway Bridges, AASHO, Washington, D.C., 10th ed., 1969

1.24 AISI
Specification for the Design of Cold-Formed Steel Structural Members, New York, 1968

1.25 AREA
Specifications for Steel Railway Bridges, AREA, Chicago, 1969

1.26 AWS
Structural Welding Code, 1st ed., AWS, New York, 1972

1.27 AISE
Specifications for Electric Overhead Traveling Cranes for Steel Mill Service, AISE, Pittsburgh, 1969

1.28 AISE
The Design and Construction of Mill Buildings, AISE Standard No. 13, AISE, Pittsburgh, 1969

1.29 CISC
"Steel Structures for Buildings" (CSA Standard S16-1965), *Structural Steel for Buildings*, Canadian Institute of Steel Construction, Toronto, Ont., 1969

1.30 CSA
General Specification for Welding of Steel Structures (CSA Standard W59.1-1970) Toronto, 1970

1.31 L. S. Beedle
"Ductility as a Basis for Steel Design," in *Engineering Plasticity*, ed. Heyman and Leckie, Cambridge University Press, 1968, p. 41

CHAPTER 2. MATERIALS

2.1 American Society for Testing and Materials
ASTM Standards 1972, Parts 3, 4, and 31, Philadelphia, 1972

2.2 R. L. Brockenbrough and B. G. Johnston
USS Steel Design Manual, United States Steel Corporation, Pittsburgh, Pa. 1968

2.3   S. Timoshenko
      Strength of Materials, Part II, 3d ed., Van Nostrand, Princeton, N.J., March
      1956
2.4   L. S. Beedle and L. Tall
      "Basic Column Strength," *Proc. Am. Soc. Civil Engrs.*, **86** (ST-7), p. 139 (July
      1960)
2.5   D. McLean
      Mechanical Properties of Metals, Wiley, New York, 1962
2.6   G. Winter
      "Properties of Steel and Concrete and the Behavior of Structures," *Proc. Am. Soc.
      Civil Engrs.*, **86** (ST-2), p. 33 (February 1960)
2.7   American Iron and Steel Institute
      Design of Light Gage Cold-Formed Stainless Steel Structural Members, New
      York, 1968
2.8   J. A. Gilligan, G. Haaijer, and R. W. Simon
      New Concepts in Steel Design and Engineering, Design and Engineering Seminar,
      United States Steel Corp., Pittsburgh, Pa., April 1961

CHAPTER 3.   BUILDINGS

3.1   H. Plommer
      Simpson's History of Architecture Development, Vol. I, Longmans Green,
      London, 1958
3.2   L. Grinter
      Theory of modern Steel Structures, Vol. I, Macmillan, New York, 1949
3.3   L. Grinter
      "The Skyscraper," *Fortune*, New York, 1930
3.4   American National Standards Institute
      American Standard Building Code (A58.1-1969), ANSI, New York
3.5   U.S. Weather Bureau
      Snow Load Studies, Supt. of Documents, Government Printing Office, Washing-
      ton, D.C.
3.6   R. W. Abbett
      American Civil Engineering Practice, Vol. III, Wiley, New York, 1957
3.7   E. Gaylord, Jr., and C. Gaylord
      Design of Steel Structures, 2d ed., McGraw-Hill, New York, 1972
3.8   D. W. Taylor
      Soil Mechanics, Wiley, New York, 1948
3.9   B. K. Hough
      Basic Soils Engineering, 2d ed., Ronald Press, New York, 1969
3.10 ASCE Task Committee on Wind Forces
      "Wind Forces on Structures (Final Report)," *Trans. Am. Soc. Civil Engrs.*,
      126 (II), pp. 1124–1198 (1961)
3.11 A. G. Davenport
      Gust Loading Factors, ASCE Environmental Engineering Conference, Preprint
      457, February 6–9, 1967
3.12 ASCE Subcommittee No. 31
      "Wind Bracing in Steel Buildings (Fifth Progress Report)," *Proc. Am. Soc. Civil
      Engrs.*, **62**, pp. 397–412 (March 1936)
3.13 ASCE Subcommittee No. 31
      "Wind Bracing in Steel Buildings (Final Report)," *Trans. Am. Soc. Civil Engrs.*,
      105, pp. 1713–1738 (1940)
3.14 Structural Engineers Association of California
      Recommended Lateral Force Requirements and Commentary, SEAOC, 1966

3.15 R. W. Clough and K. L. Benuska
Nonlinear Earthquake Behavior of Tall Buildings, ASCE Structural Engineering Conference, Reprint 298, January 31–February 4, 1966

3.16 G. V. Berg and J. L. Stratta
Anchorage and the Alaska Earthquake of March 27, 1964, AISI, 1964

3.17 L. Urquhart, C. O'Rourke, and G. Winter
Design of Concrete Structures, pp. 197–200, McGraw-Hill, New York, 1958

3.18 C. Dunham
Planning Industrial Structures, McGraw-Hill, New York, 1948

3.19 Steel Joist Institute
Standard Specifications and Load Tables—Open-Web Steel Joints, SJI, Washington, D.C., 1970

3.20 S. Li
"Metallic Dome Structures," *Proc. Am. Soc. Civil Engrs.*, **88** (ST6), pp. 201–226 (December 1962)

3.21 "Steel Space Frame Folds and Holds Synagogue Roof," *Eng. News-Record*, **168**, pp. 36–37 (April 1961)

3.22 *Cable Construction in Contemporary Architecture*, Bethlehem Steel Corporation Booklet 2264, 1966

3.23 P. Rogers
"Economy in Wide Spans Through Prestressed Steel," *Progressive Architecture*, Reinhold Publishing Co., New York (May 1960)

3.24 "Olympian Valley of Decision," *Steelways*, **16**, AISI, New York (January 1960)

3.25 R. F. Miller
"The Strength of Carbon Steels for Elevated Temperature Applications," *Proc. Am. Soc. Test. Mat.*, **54** (1954)

3.26 AISI
Modern Methods for Fire-Protecting Steel-Framed Buildings, 1965

CHAPTER 4. BRIDGES

4.1 D. Steinman and S. Watson
Bridges and Their Builders, Dover, New York, 1957

4.2 H. Smith
The World's Great Bridges, Harper, New York, 1953

4.3 T. Cooper
"American Railway Bridges," *Trans. Am. Soc. Civil Engrs.*, **21**, p. 2 (1889)

4.4 E. Praeger and T. Kavanagh
"Longest Plate Girder Span in U.S. Completed," *Civil Eng.*, **29** (12), p. 856 (December 1959)

4.5 P. Ferguson
Reinforced Concrete Fundamentals, Wiley, New York, 1958

4.6 I. Lyse and I. Madsen
"Structural Behavior of Battledeck Floor Systems," *Trans. Am. Soc. Civil Engrs.*, **104**, p. 244 (1939)

4.7 W. Scofield, W. O'Brien, and T. Brassell
Modern Timber Engineering, Southern Pine Association, New Orleans, 1954

4.8 C. Norris, R. Hansen, M. Holley, J. Biggs, S. Namyet, and J. Minami
Structural Design for Dynamic Loads, McGraw-Hill, New York, 1959

4.9 AREA, Committee 30
Tests of Steel Girder Spans and a Concrete Pier on the Santa Fe, AREA Bull. No. 58 (1957)

4.10 AREA, Special Committee on Impact
Description and Analysis of Bridge Impact Tests, AREA Bull. No. 46 (450), pp. 189–434 (1945)

4.11 D. Steinman
"Rigidity and Aerodynamic Stability of Suspension Bridges," *Trans. Am. Soc. Civil Engrs.*, **110**, p. 441 (1946)

4.12 F. Farquharson, F. Smith, and G. Vincent
Aerodynamic Stability of Suspension Bridges, University of Washington, Eng. Exp. Sta. Bull. No. 116 (5 parts), 1949–1954

4.13 D. Steinman and C. Gronquist
"Mackinac Bridge, Superstructure Design and Construction," *Civil Eng.*, **29** (1), p. 48 (January 1959)

4.14 E. Durkee
"Designing Economy into Bridge Erection," *Civil Eng.*, **29** (4), p. 60 (April 1959)

4.15 F. Masters and J. Giese
"Findings on the Second Narrows Bridge Collapse at Vancouver," *Civil Eng.*, **29** (2), p. 856 (December 1959)

4.16 B. Farago and W. Chan
"The Analysis of Steel Decks," *Proc. Inst. Civil Engrs.*, **16**, p. 1 (May 1960)

4.17 J. Percy
"A Design Method for Grillages," *Proc. Inst. Civil Engrs.*, **23**, p. 409 (November 1962)

4.18 Lincoln Electric Company
Procedure Handbook of Arc Welding Design and Practice, 11th Ed., Lincoln Electric Co., Cleveland, 1960

4.19 AISC
Structural Shop Drafting, Vol. 3, AISC, 1955

4.20 T. Shedd
Structural Design in Steel, Wiley, New York, 1934

4.21 Portland Cement Association
Handbook of Frame Constants, PCA, Chicago, 1958

4.22 O. Ondra
"Moment-Distribution Constants from Cardboard Analog," *Proc. Am. Soc. Civil Engrs.*, **87** (ST-1), p. 73 (January 1961).

4.23 P. Morice
Linear Structural Analysis, Ronald Press, New York, 1959

4.24 J. Lothar
Advanced Design in Structural Steel, Prentice-Hall, Englewood Cliffs, N.J., 1960

4.25 AISC
Moments, Shears and Reactions—Continuous Highway Tables, AISC, Publication T106, 1966

4.26 M. Salvadori and F. Baron
Numerical Methods in Engineering, Prentice-Hall, Englewood Cliffs, N.J., 1961

4.27 S. Timoshenko and D. Young
Theory of Structures, McGraw-Hill, New York, 1945

4.28 (a) Lubrite Division, Merriman Bros. Inc.
Lubrite, Manuals No. 56, 55-2.
(b) E. I. DuPont
Design of Neoprene Bridge Bearing Pads, A-12505, E. I. DuPont, Wilmington, 1959

4.29 R. Roark
Formulas for Stress and Strain, McGraw-Hill, 4th ed., 1965

4.30 L. Stevens and M. Lay
"Behavior of Indeterminate Triangulated Steel Frameworks," *Trans. Inst. Engrs.*, *Australia*, **CE3** (2), p. 62 (September 1961)

4.31 C. Marsh and J. Sutherland
"Application of Plastic Theory to Latticed Aluminum Frames," *Trans. Eng. Inst. Can.*, **3** (3), (November 1959)

4.32 N. Hoff, B. Boley, S. Nardo, and S. Kaufman
"Buckling of Rigid Jointed Plane Trusses," *Trans. Am. Soc. Civil Engrs.*, **119**, p. 958 (1951)

4.33　J. Ely
　　　"Truss Bridge Project at Northwestern University," *National Engineering Conference Proceedings*, AISC, p. 10, 1960
4.34　R. Barnhoff and W. Mooney
　　　"The Effect of Floor System Participation on Pony Truss Bridges," *Proc. Am. Soc. Civil Engrs.* **88** (ST-4), p. 25 (April 1960)
4.35　AISC
　　　Design Manual for Orthotropic Steel Plate Deck Bridges, AISC, 1963
4.36　J. Chang
　　　"Orthotropic Plate Construction for Short-Span Bridges," *Civil Eng.*, **31** (12), p. 53 (December 1961)

CHAPTER 5.　USE OF MODELS IN DESIGN

5.1　W. J. Eney
　　　"Determining the Deflection of Structures with Models," *Civil Eng.*, **12** (3), p. 150 (March 1942)
5.2　A. H. Finlay
　　　"Deck Participation in Concrete Arch Bridges," *Civil Eng.*, **2** (11), p. 685 (November 1932)
5.3　A. Bull
　　　"Settlement Stresses Found with a Model," *Civil Eng.*, **7** (8), p. 561 (August 1937)
5.4　W. J. Eney
　　　"Fixed-End Moments by Cardboard Models," *Eng. News-Record*, **115** (24), p. 814 (December 1935)
5.5　G. E. Beggs, E. K. Timby, and B. Birdsall
　　　"Suspension Bridge Stresses Determined by Model," *Eng. News-Record*, **108** (23), p. 828 (1932)
5.6　W. J. Eney
　　　"Studies of Continuous Bridge Trusses with Models," *Proc. Soc. Exp. Stress Anal.*, **6** (2), p. 94 (1948)
5.7　L. F. Stephens
　　　"Structural Model Analysis," *Trans. Inst. Civil. Engrs. Ire.*, **88**, p. 57 (1961)
5.8　J. S. Kinney
　　　Indeterminate Structural Analysis, Chap. 14, Addison-Wesley, Reading, Mass., 1957
5.9　R. L. Sanks
　　　Statically Indeterminate Structural Analysis, Part VI, Ronald Press, New York, 1961
5.10　W. J. Eney
　　　"A Large Displacement Deformeter Apparatus for Stress Analysis with Elastic Models," *Proc. Soc. Exp. Stress Anal.*, **6** (2), p. 84 (1948)
5.11　C. Massonnet
　　　"L'Étude des Constructions sur Modèles Réduits sans Emploi de Microscopes: L'Influentiométre du Professeur Eney" (A Study of Models Using the Eney Deformeter), Extrait du Bulletin du Centre d'Études, de Recherches et d'Essais Scientifiques des Constructions du Génie Civil et d'Hydraulique Fluviale, Tome VI, Belgium, 1953
5.12　W. J. Eney
　　　"Model Analysis of Continuous Girders," *Civil Eng.*, **11** (9), p. 521 (1941)
5.13　R. Fleming
　　　"Mechanical Method for Determining Reactions in Continuous Girders," *Eng. News-Record*, **83** (9), p. 428 (1919)

5.14 G. E. Beggs
"An Accurate Mechanical Solution of Statically Indeterminate Structures by Use of Paper Models and Special Gages," *Proc. Am. Concrete Inst.*, **18**, p. 58 (1922)

5.15 G. E. Beggs
"The Use of Models in the Solution of Indeterminate Structures," *J. Franklin Inst.*, **203** (3), p. 375 (1927)

5.16 C. B. McCullough and E. S. Thayer
Elastic Arch Bridges, Chap. VII, Wiley, New York, 1931

5.17 W. J. Eney
"New Deformeter Apparatus," *Eng. News-Record*, **122** (16) p. 221 (February 1939)

5.18 O. Gottschalk
"The Experiment in Statics," *J. Franklin Inst.*, **207** (2), p. 245 (1929)

5.19 O. Gottschalk
"Structural Analysis Based on Principles Pertaining to Unloaded Models," *Trans. Am. Soc. Civil Engrs.*, **103**, p. 1019 (1938)

5.20 A. J. S. Pippard
The Experimental Study of Structures, Edward Arnold, London, 1947

5.21 J. B. Wilbur and C. H. Norris
"Structural Model Analysis," *Handbook of Experimental Stress Analysis*, ed. M. Hetenyi, Wiley, New York, 1950

5.22 M. Hetenyi
Beams on Elastic Foundation, University of Michigan Press, Ann Arbor, 1946

5.23 E. P. Popov
"Successive Approximations for Beams on an Elastic Foundation," *Trans. Am. Soc. Civil Engrs.*, **116**, p. 1083 (1950)

5.24 A. Bull
"Soil Pressures on Subways Analyzed," *Civil Eng.*, **2** (4), p. 263 (April 1932)

5.25 A. J. Durelli and I. M. Daniel
"Structural Model Analysis by Means of Moire Fringes," *Proc. Am. Soc. Civil Engrs.*, **86** (ST-2), p. 93 (1960)

5.26 J. F. Baker, M. R. Horne, and J. Heyman
The Steel Skeleton, Vol. II, *Plastic Behavior and Design*, Chap. 2, Cambridge University Press, Cambridge, England, 1956

5.27 W. A. Litle, D. C. Foster, D. Oakes, P. A. Falcone, and R. B. Reimer
The Structural Behavior of Small-Scale Steel Models, Bulletin No. 10, American Iron and Steel Institute, April 1968

5.28 E. Yarimci and L. W. Lu
Strength and Deformation Modes of Unbraced Multi-Story Frames, Fritz Laboratory Report 273.56, Lehigh University, 1969

CHAPTER 6. TENSION MEMBERS

6.1 G. L. Kulak and J. W. Fisher
"Behavior of Large A514 Steel Bolted Joints," *Proc. Am. Soc. Civil Engrs.*, **95** (ST-9), (1969)

6.2 B. G. Johnston
"Pin-Connected Plate Links," *Trans. Am. Soc. Civil Engrs.*, **104**, p. 2023 (1939)

6.3 W. M. Wilson, W. H. Munse, and M. A. Cayci
A Study of the Practical Efficiency Under Static Loading of Riveted Joints Connecting Plates, University of Illinois Eng. Exp. Sta., Bulletin Series No. 402, 1952

6.4 G. J. Gibson and B. T. Wake
"An Investigation of Welded Connections for Angle Tension Members," *Welding J.*, **21**, p. 44-s (1942)

6.5　V. H. Cochrane
"Calculating Net Section of Riveted Tension Members, and Fixing Rivet Stagger," *Eng. News-Record*, **59**, p. 465 (1908)

6.6　T. A. Smith
"Diagram for Net Section of Riveted Tension Members," *Eng. News-Record*, **73**, p. 892 (1915)

6.7　D. B. Steinman
"Proposes New Specification for Deducting Rivet Holes," *Eng. News-Record*, **78** (11) (June 14, 1917)

6.8　C. R. Young
True Net Sections of Riveted Tension Members, Eng. Res. Bull. No. 2, University of Toronto, Canada, 1921, pp. 233–244.

6.9　V. H. Cochrane
"Rules for Rivet-Hole Deductions in Tension Members," *Eng. News-Record*, **89**, pp. 847–848 (1922)

6.10　F. W. Schutz
"Effective Net Section of Riveted Joints," 2d *Illinois Structural Engineering Conference Proceedings*, p. 54, 1952

6.11　W. M. Wilson
"Discussion of 'Tension Tests of Large Riveted Joints'," *Trans. Am. Soc. Civil Engrs.*, **105**, p. 1268 (1940)

6.12　G. W. Brady and D. C. Drucker
"Investigation and Limit Analysis of Net Area in Tension," *Trans. Am. Soc. Civil Engrs.*, **120**, p. 1133 (1955)

6.13　P. P. Bijlaard
"Discussion of 'Investigation and Limit Analysis of Net Area in Tension'," *Trans. Am. Soc. Civil Engrs.*, **120**, p. 1156 (1955)

CHAPTER 7. BEAMS

7.1　S. P. Timoshenko
History of Strength of Materials, McGraw-Hill, New York, 1953

7.2　F. B. Seely and J. O. Smith
Advanced Mechanics of Materials, Wiley, New York, 1952

7.3　V. Z. Vlasov
Thin-Walled Elastic Beams (translated from Russian), Office of Technical Services, U.S. Department of Commerce, Washington, D.C., 1961

7.4　J. N. Goodier and M. V. Barton
"The Effects of Web Deformation on the Torsion of I-Beams," *J. Appl. Mechanics*, **11**, March 1944

7.5　G. C. Lee
A Survey of Literature on the Lateral Instability of Beams, Bull. No. 63, WRC, New York, 1960

7.6　F. Bleich
Buckling Strength of Metal Structures, McGraw-Hill, New York, 1952

7.7　J. W. Clark and H. N. Hill
"Lateral Buckling of Beams," *Proc. Am. Soc. Civil Engrs.*, **86** (ST-7), p. 175 (July 1960)

7.8　S. P. Timoshenko
Strength of Materials (Part I), Van Nostrand, New York, 1958

7.9　AISC
Steel Construction Manual, 7th ed., AISC, New York, 1970

7.10　S. P. Timoshenko and J. N. Goodier
Theory of Elasticity, McGraw-Hill, New York, 3rd ed., 1969

7.11　C. F. Kollbrunner and K. Basler
Torsion in Structures, Springer-Verlag, New York, 1969

7.12 I. Lyse and B. G. Johnston
      "Structural Beams in Torsion," *Trans. Am. Soc. Civil Engrs.*, **101**, p. 857 (1936)
7.13 G. G. Kubo, B. G. Johnston, and W. J. Eney
      "Non-Uniform Torsion of Plate Girders," *Trans. Am. Soc. Engrs.*, **121**, p. 759 (1956)
7.14 Column Research Council, B. G. Johnston (ed.)
      Guide to Design Criteria for Metal Compression Members, 2nd ed., Wiley, New York, 1966.   (See also Third Edition, Ref. 1.17.)
7.15 T. V. Galambos
      Structural Members and Frames, Prentice-Hall, Englewood Cliffs, N.J., 1968
7.16 K. de Vries
      "Strength of Beams as Determined by Lateral Buckling," *Trans. Am. Soc. Civil Engrs.*, **112**, p. 1245 (1947)
7.17 S. P. Timoshenko and James M. Gere
      Theory of Elastic Stability, McGraw-Hill, New York, 1961
7.18 K. Basler and B. Thürlimann
      "Strength of Plate Girders in Bending," *Proc. Am. Soc. Civil Engrs.*, **87** (ST-6), p. 153 (August 1961)
7.19 G. Winter *et al.*
      "Discussion of 'Strength of Beams as Determined by Lateral Buckling'," *Trans. Am. Soc. Civil Engrs.*, **112**, p. 1245 (1947)
7.20 M. G. Salvadori
      "Lateral Buckling of Eccentrically Loaded I-Columns," *Trans. Am. Soc. Civil Engrs.*, **121**, p. 1163 (1956)
7.21 B. G. Neal
      The Plastic Methods of Structural Analysis, Wiley, New York, 1957
7.22 G. Hodge, Jr.
      Plastic Analysis of Structures, McGraw-Hill, New York, 1959
7.23 J. F. Baker, M. R. Horne, and J. Heyman
      The Steel Skeleton, Vol. II, Cambridge University Press, 1956
7.24 H. J. Greenberg and W. Prager
      "Limit Design of Beams and Frames," *Trans. Am. Soc. Civil Engrs.*, **117**, p. 447 (1952)
7.25 M. G. Lay and T. V. Galambos
      "Inelastic Beams Under Moment Gradient," *Proc. Am. Soc. Civil Engrs.*, **93** (ST1) p. 381 (Feb. 1967)
7.26 Subcommittee on Cover Plates
      "Commentary on Welded Cover-Plated Beams," *Proc. Am. Soc. Civil Engrs.*, **93** (ST4), p. 95 (August 1967)
7.27 AASHO
      Interim Specifications for Highway Bridges, AASHO, Washington, D.C., 1971

CHAPTER 8. PLATE GIRDERS

8.1 K. Basler
      "New Provisions for Plate Girder Design," *National Engineering Conference Proceedings*, AISC, 1961
8.2 K. Basler
      "Strength of Plate Girders in Shear," *Proc. Am. Soc. Civil Engrs.*, **87** (ST-7), p. 151 (October 1961)
8.3 K. Basler
      "Strength of Plate Girders Under Combined Bending and Shear," *Proc. Am. Soc. Civil Engrs.*, **87** (ST-7), p. 181 (October 1961)
8.4 K. Basler, B. T. Yen, J. A. Mueller, and B. Thürlimann
      Web Buckling Tests on Welded Plate Girders, Bull. No. 64, WRC, New York (September 1960)

8.5   C. Dubas
A Contribution of the Study of Buckling of Stiffened Plates, Preliminary Publication, 3d Cong. IABSE, p. 129, 1948
8.6   L. S. Moisseiff and F. Lienhard
"Theory of Elastic Stability Applied to Structural Design," *Trans. Am. Soc. Civil Engrs.*, **106**, p. 1052 (1941)
8.7   E. L. Erickson and N. VanEenam
"Application and Development of AASHO Specifications to Bridge Design," *Proc. Am. Soc. Civil Engrs.*, **83** (ST-4), p. 1320 (July 1957)
8.8   British Standards Institution
British Standard 153: Steel Girder Bridges, Parts 3B and 4, British Standards House, London, 1958
8.9   Deutscher Normenausschuss
DIN 4114, German Buckling Specifications, Blatt 1 and 2, Beuth-Vertrieb GmbH, Berlin and Cologne, July 1952
8.10 K. C. Rockey and D. M. A. Leggett
"The Buckling of a Plate Girder Web Under Pure Bending When Reinforced by a Single Longitudinal Stiffener," *Proc. Inst. Civil Engrs.*, **21**, p. 161 (January 1962)
8.11 Basler and B. Thürlimann
"Plate Girder Research," *National Engineering Conference Proceedings*, AISC, 1959

CHAPTER 9.   COMPRESSION MEMBERS

9.1   E. H. Salmon
Columns, A Treatise on the Strength and Design of Compression Members, Oxford Technical Publications, London, 1921
9.2   J. A. van den Broeck
"English Translation of Euler's *On the Strength of Columns*," *Am. J. Phys.*, **15** (4), p. 309 (July 1947)
9.3   B. G. Johnston
"Inelastic Buckling Gradient," *Proc. Am. Soc. Civil Engrs*, **90** (EM-6), December 1964
9.4   F. R. Shanley
"Inelastic Column Theory," *J. Aeronaut Sci.*, **14** (5) (May 1947)
9.5   L. Tall
"Recent Developments in the Study of Column Behavior," *J. Inst. Engrs.*, Australia, **36**, No. 12, December 1964
9.6   L. Tall
Welded Built-Up Columns, Fritz Laboratory Report No. 249.29, April, 1966
9.7   G. A. Alpsten and L. Tall
"Prediction of Behavior of Steel Columns Under Load," *Proc., Final Report*, "Symposium on Concepts of Safety of Structures and Methods of Design," IABSE, London, September 1969
9.8   ASCE Special Committee
"Steel Column Research," Final Report of the Special Committee (1923–1933), *Trans. Am. Soc. Civil Engrs.*, **98**, p. 1376 (1933)
9.9   D. Sfintesco
"European Steel Column Research" Preprint 502, *ASCE Structural Engrs. Conf.*, Seattle, May 1967
9.10 J. E. Duberg and T. W. Wilder
Inelastic Column Behavior, NACA Technical Note 2267, Washington D.C., January 1951
9.11 L. Tall
The Strength of Welded Built-Up Columns, Ph.D. dissertation, Lehigh University, May 1961; University Microfilms, Inc., Ann Arbor

9.12 B. G. Johnston
"Buckling Behavior Above Tangent Modulus Load," *Proc. Am. Soc. Civil Engrs.*, **87** (EM-6), p. 192 (December 1961)

9.13 L. Tall and F. R. Estuar
"Discussion to Ref. 9.12," *Proc. Am. Soc. Civil Engrs.*, **88** (EM-5), p. 171 (October 1962)

9.14 Column Research Council
The Basic Column Formula, CRC Technical Memorandum No. 1, May 1952

9.15 A. W. Huber and L. S. Beedle
"Residual Stress and the Compressive Strength of Steel," *Welding J.*, **33**, p. 589-s (December 1954)

9.16 N. Tebedge, G. A. Alpsten, and L. Tall
"Measurement of Residual Stresses—A Comparative Study of Methods," *Proc., JBCSA Conference on the Recording and Interpretation of Engineering Measurements*, London, 1972

9.17 A. W. Huber
The Influence of Residual Stress on the Instability of Columns, Ph.D. Dissertation, Lehigh University, May 1956; University Microfilms, Inc., Ann Arbor

9.18 G. A. Alpsten
Thermal Residual Stresses in Hot-Rolled Steel Members, Fritz Laboratory Report No. 337.3, December 1968

9.19 G. A. Alpsten and L. Tall
"Residual Stresses in Heavy Welded Shapes," *Welding J.*, **49**, April 1970

9.20 D. Feder and G. C. Lee
Residual Stress and the Strength of Members of High Strength Steel, Fritz Laboratory Report 269.2, Lehigh University (March 1959)

9.21 L. S. Beedle, T. V. Galambos, and L. Tall
Column Strength of Constructional Steels, Steel Design and Engineering Seminar, U.S. Steel Corp., Pittsburgh, May 1961

9.22 C. H. Yang, L. S. Beedle, and B. G. Johnston
"Residual Stress and the Yield Strength of Steel Beams," *Welding J.*, **31**, p. 205-s (April 1952).

9.23 W. R. Osgood
"The Effect of Residual Stress on Column Strength," *Proceedings First National Congress Applied Mechanics*, June 1951

9.24 Y. Fujita
Built-Up Column Strength, Ph.D. dissertation, Lehigh University, August 1956; University Microfilms, Inc., Ann Arbor

9.25 F. Campus and C. Massonnet
Recherches sur le Flambement de Colonnes en Acier A37, un Profil en Double Tee, Sollicitées Obliquement (The Buckling of Eccentrically Loaded H-Columns of A37 Steel) IRSIA, No. 17, Belgium, April 1956

9.26 C. O'Connor
"Residual Stresses and Their Influence on Structural Design," *J. Inst. Engrs., Australia*, 27(12) (December 1955)

9.27 L. Tall, A. W. Huber, and L. S. Beedle
Residual Stress and the Instability of Axially Loaded Columns, Fritz Laboratory Report 220A.35, Lehigh University, February 1960. Published as Commission X Document, Colloquium, International Institute of Welding, Liège, Belgium, June 1960

9.28 Commission X, International Institute of Welding
Un Inventaire sur le Sujet: Tensions Residuelles et Instabilité (Residual Stress and Instability: A Summary), prepared by Working Group, H. Louis, M. Marincek, and L. Tall, Oslo, July 1962

9.29 L. Tall and D. Feder
"Längsschweissspannungen in Platten und ihr Einfluss auf die Grenzlast von geschweissten Stahlstützen, *Schweissen und Schneiden*, March 1965

9.30   L. Tall
       Stub-Column Test Procedure, Fritz Laboratory Report 220A.36, Lehigh University, February 1961.   Revised as International Institute of Welding Document X-282-61, prepared by Working Group (H. Louis, M. Marincek, and L. Tall), Annual Conference, Oslo, July, 1962

9.31   F. R. Estuar and L. Tall
       "Experimental Investigation of Welded Built-Up Columns," *Welding J.*, **42,** p. 164-*s* (April 1963)

9.32   R. McFalls and L. Tall
       "A Study of Welded Columns Manufactured from Flame-Cut Plates," *Welding J.*, **48,** April 1969

9.33   Y. Kishima, G. A. Alpsten, and L. Tall
       The Strength of ASTM A572(50) Steel Welded Flame-Cut Columns.   Fritz Lab. Report 321.4, July 1970

9.34   F. R. Estuar
       Welding Residual Stresses and the Strength of Heavy Column Shapes, Ph.D. dissertation, Lehigh University, August 1965.   University Microfilms, Inc., Ann Arbor, Michigan

9.35   Y. Fujita
       "Ultimate Strength of Columns with Residual Stresses," *J. Soc. Naval Architects Japan*, January 1960

9.36   N. R. Nagaraja Rao and L. Tall
       "Columns Reinforced Under Load," *Welding J.*, **42,** April 1963

9.37   A. Nitta
       Ultimate Strength of High Strength Steel Circular Columns, Ph.D. dissertation, Lehigh University, June 1960; University Microfilms, Inc., Ann Arbor

9.38   L. Tall
       Residual Stresses in Welded Plates—A Theoretical Study, *Welding J.*, **43,** January 1964

9.39   C. K. Yu and L. Tall
       Welded and Rolled A514 Steel Columns—A Summary Report, Fritz Laboratory Report No. 290.16, June 1970

9.40   E. Odar, F. Nishino, and L. Tall
       Residual Stresses in Rolled Heat-Treated *T-1* Shapes, WRC Bull. no. 121, April 1967

9.41   C. K. Yu and L. Tall
       "A Pilot Study on the Strength of 5 Ni-Cr-Mo-V Steel Columns," *Experimental Mechanics*, **8,** No. 1, January 1968

9.42   R. Bjørhovde and L. Tall
       "Development of Multiple Column Curves," *Proc. CRC/ECCS/JCRC/IABSE Joint Colloquium on Column Strength*, CTICM, Paris, November 1972

9.43   M. W. White and B. Thürlimann
       Study of Columns with Perforated Cover Plates, AREA Bull. No. 531 (September–October 1956)

9.44   G. B. Godfrey
       "The Allowable Stresses in Axially Loaded Steel Struts," *Structural Eng.*, **40** (3), p. 97 (March 1962)

CHAPTER 10.   COMPRESSION MEMBERS IN FRAMES AND TRUSSES

10.1   F. Bleich
       Buckling Strength of Metal Structures, Chaps. VI and VII, McGraw-Hill, New York, 1952

10.2   L. W. Lu
"Stability of Frames Under Primary Bending Moments," *Proc. Am. Soc. Civil Engrs.*, **89** (ST-3), p. 35 (June 1963)

10.3   S. P. Timoshenko and J. M. Gere
Theory of Elastic Stability, Chap. 2, McGraw-Hill, New York, 1961

10.4   T. V. Galambos
"Influence of Partial Base Fixity on Frame Stability," *Trans. Am. Soc. Civil Engrs.*, **126** (II), p. 929 (1961)

10.5   Column Research Council
Guide to Design Criteria for Metal Compression Members, 2d ed., John Wiley, New York, Chaps. 2 and 6, 1966

10.6   L. W. Lu
A Survey of Literature on the Stability of Frames, Bull. No. 81, WRC, New York, September 1962

10.7   G. Winter, P. T. Hsu, B. Koo, and M. H. Loh
Buckling of Trusses and Rigid Frames, Cornell University, Eng. Exp. Sta. Bull. No. 36, 1948

10.8   E. F. Masur
"On the Lateral Stability of Multi-Story Bents," *Proc. Am. Soc. Civil Engrs.*, **81,** Separate No. 672 (April 1955)

10.9   H. L. Su
"On Stability of Two-Dimensional Structural Frameworks," *Proc. Inst. Civil Engrs.*, **16,** p. 143 (June 1960)

10.10   D. E. Johnson
"Lateral Stability of Frames by Energy Method," *Trans. Am. Soc. Civil Engrs.*, **126** (I), p. 176 (1961)

10.11   S. J. McMinn
"The Determination of the Critical Loads of Plane Frames," *Structural Eng.*, **39** (7), p. 221 (July 1961)

10.12   R. H. Wood
"The Stability of Tall Buildings," *Proc. Inst. Civil Engrs.*, **11,** p. 69 (September 1958)

10.13   W. Merchant, C. A. Rashid, A. Bolton, and A. H. Salem
The Behavior of Unclad Frames, Fiftieth Anniversary Conference, The Institution of Structural Engineers, 1958

10.14   L. W. Lu
Inelastic Buckling of Steel Frames, *Proc. Am. Soc. Civil. Engrs.*, **91** (ST-6), p. 185 (Dec. 1965)

10.15   Y. C. Yen, L. W. Lu, and G. C. Driscoll, Jr.
Tests on the Stability of Welded Steel Frames, Bulletin No. 81, WRC, New York (September 1962)

10.16   B. M. McNamee
The General Behavior and Strength of Unbraced Multi-Story Frames Under Gravity Loading, Ph.D. Dissertation, Lehigh University, June 1967, University Microfilms, Inc., Ann Arbor, Michigan

10.17   J. I. Parcel and E. B. Murer
"Effect of Secondary Stresses upon Ultimate Strength," *Trans. Am. Soc. Civil Engrs.*, **101,** p. 289 (1936)

10.18   T. C. Kavanagh
"Effective Length of Framed Columns," *Proc. Am. Soc. Civil Engrs.*, **86** (ST-2), p. 1 (February 1960)

10.19   N. J. Hoff
"Elastically Encastred Struts," *J. Roy. Aeronaut. Soc.*, **40** (309), p. 663 (September 1936)

10.20   N. M. Newmark
"A Simple Approximate Formula for Effective End-Fixity of Columns," *J. Aeronaut Sci.*, **16** (2), p. 116 (February 1949)

10.21 H. E. Wessman and T. C. Kavanagh
"End Restraints on Truss Members," *Trans. Am. Soc. Civil Engrs.*, **115**, p. 1135 (1950)

CHAPTER 11. COMBINED BENDING AND COMPRESSION

11.1 AASHO
Standard Specifications for Highway Bridges, Appendix C, "Formulas for Steel Columns," AASHO, Washington, 1965
11.2 AREA
Specifications for Steel Railway Bridges, Appendix A, "Formulas for Compression Members," AREA, Chicago, 1963
11.3 *German Buckling Specifications*, DIN 4114, Vol. 1, 1952 (J. Jones and T. V. Galambos, trans.), CRC, July 1957
11.4 AISC
Specifications for the Design, Fabrication and Erection of Structural Steel for Buildings, Section 1.6.1, April 17, 1963 and January 1, 1969
11.5 C. S. Gray, L. E. Kent, W. A. Mitchell, and G. B. Godfrey
Steel Designers' Manual, Crosby Lockwood and Son Ltd., London, 1962
11.6 AISI
Light Gage Cold-Formed Steel Design Manual, Art. 3.7, AISI, New York, 1969
11.7 ASCE, Task Committee on Lightweight Alloys
"Suggested Specifications for Structures of Aluminum Alloys 6061-T6, 6062-T6, 6063-T5, 6063-T6," *Proc. Am. Soc. Civil Engrs.*, **88** (ST-6), p. 1 (December 1962)
11.8 R. L. Ketter, E. L. Kaminsky, and L. S. Beedle
"Plastic Deformation of Wide-Flange Beam-Columns," *Trans. Am. Soc. Civil Engrs.*, **120**, p. 1028 (1955)
11.9 F. Bleich
Buckling Strength of Metal Structures, Chap. 1, McGraw-Hill, New York, 1952
11.10 R. L. Ketter
"Stability of Beam-Columns Above the Elastic Limit," *Proc. Am. Soc. Civil Engrs.*, **81**, Separate No. 692 (May 1955)
11.11 M. Ojalvo and Y. Fukumoto
Nomographs for the Solution of Beam-Column Problems, Bull. No. 78, WRC, New York, June 1962
11.12 R. L. Ketter and T. V. Galambos
"Columns Under Combined Bending and Thrust," *Trans. Am. Soc. Civil Engrs.*, **126** (I), p. 1 (1961)
11.13 R. L. Ketter
"Further Studies on the Strength of Beam-Columns," *Proc. Am. Soc. Civil Engrs.*, **87** (ST-6), p. 135 (August 1961)
11.14 M. Ojalvo
"Restrained Columns," *Proc. Am. Soc. Civil Engrs.*, **86** (EM-5), p. 1 (October 1960)
11.15 T. V. Galambos and Y. Fukumoto
"Inelastic Lateral-Torsional Buckling of Beam-Columns," *Proc. Am. Soc. Civil Engrs.*, **92** (ST-2) (April 1966)
11.16 C. K. Yu and L. Tall
A514 Steel Beam-Columns, IABSE Publications (31-II), 1971
11.17 W. J. Austin
"Strength and Design of Metal Beam-Columns," *Proc. Am. Soc. Civil Engrs.*, **87** (ST-4), p. 1 (April 1961)
11.18 ASCE-WRC Committee
Commentary on Plastic Design in Steel, Chap. 7, "Compression Members," ASCE Manual No. 41, 1969

11.19 S. P. Timoshenko and J. M. Gere
       Theory of Elastic Stability, Chap. 1, McGraw-Hill, New York, 1961
11.20 T. V. Galambos
       "Discussion to Ref. 11.17," *Proc. Am. Soc. Civil Engrs.*, **87** (ST-4), p. 1 (April 1961)
11.21 C. Massonnet
       "Stability Considerations in the Design of Steel Columns," *Proc. Am. Soc. Civil Engrs.*, **85** (ST-7), p. 75 (September 1959)
11.22 R. E. Mason, G. P. Fisher, and G. Winter
       "Eccentrically Loaded Hinged Steel Columns," *Proc. Am. Soc. Civil Engrs.*, **84** (EM-4), p. 1792 (October 1958)
11.23 B. G. Johnston (ed.)
       The Column Research Council Guide to Design Criteria for Metal Compression Members, 2nd ed., Chap. 6, Wiley, New York, 1966
11.24 T. V. Galambos and J. Prasad
       Ultimate Strength Tables for Beam-Columns, Bull. No. 78, WRC, New York, July 1962
11.25 T. V. Galambos
       Structural Members and Frames, Chap. 5, Prentice-Hall, Englewood Cliffs, N.J., 1968
11.26 E. H. Gaylord and C. N. Gaylord
       Structural Engineering Handbook, McGraw-Hill, New York, 1968
11.27 R. C. VanKuren and T. V. Galambos
       "Beam-Column Experiments," *Proc. Am. Soc. Civil Engrs.*, **90** (ST-2) (April 1964)
11.28 T. J. Dwyer and T. V. Galambos
       "Plastic Behavior of Tubular Beam-Columns," *Proc. Am. Soc. Civil Engrs.*, **91** (ST-4) (August 1965)
11.29 M. G. Lay and T. V. Galambos
       Tests on Beam and Column Subassemblages, Welding Research Council Bulletin No. 110, Nov. 1965
11.30 J. S. Ellis
       "Plastic Behavior of Compression Members," *J. Mechanics and Physics of Solids*, **6**, 1968
11.31 Lehigh University Staff
       Plastic Design of Multi-Story Frames, Fritz Engineering Laboratory Report No. 273.20 and 273.24 (Lecture notes and design aids), Summer, 1965

CHAPTER 12. COLD-FORMED MEMBERS

12.1 G. Winter
      "Cold-Formed, Light-Gage Steel Construction," *Proc. Am. Soc. Civil Engrs.*, **85** (ST-9), p. 151 November 1959
12.2 E. Griffin
      "Discussion to Ref. 12.6," *Proc. Inst. Civil Engrs.*, **23**, October 1962
12.3 A. Chajes, S. J. Britvec, and G. Winter
      "Effect of Cold-Straining on Structural Sheet Steels," *Proc. Am. Soc. Civil Engrs.*, **89** (ST-2), p. 1 April 1963
12.4 K. W. Karren
      "Corner Properties of Cold-Formed Steel Shapes," *Proc. Am. Soc. Civil Engrs.*, **93** (ST-1), p. 401 February 1967
12.5 K. W. Karren and G. Winter
      "Effects of Cold-Forming on Light-Gage Steel Members," *Proc. Am. Soc. Civil Engrs.*, **93** (ST-1), p. 433 February 1967
12.6 A. H. Chilver
      "Structural Problems in the Use of Cold-Formed Steel Sections," *Proc. Inst. Civil Engrs.*, **20** October 1961

12.7   G. Winter and R. H. J. Pian
       Crushing Strength of Thin Steel Webs, Cornell University, Eng. Exp. Sta. Bull.
       No. 35, Part 1, 1956

12.8   H. N. Hill
       "Lateral Buckling of Channels and Z-Beams," *Trans. Am. Soc. Civil Engrs.*, **119**,
       p. 829 (1954)

12.9   G. Winter, W. Lansing, and R. B. McCalley, Jr.
       Performance of Laterally Loaded Channel Beams, Res. Eng. Structures Suppl.
       (Colston Papers, Vol. II), London, 1949 (Cornell Univ., Eng. Exp. Sta. Reprint
       33)

12.10 L. Zetlin and G. Winter
       "Unsymmetrical Bending of Beams With and Without Lateral Bracing," *Proc.
       Am. Soc. Civil Engrs.*, **81**, Paper 774 (1955)

12.11 R. T. Douty
       A Design Approach to the Strength of Laterally Unbraced Compression Flanges,
       Cornell University, Eng. Exp. Sta. Bull. No. 37, April 1962

12.12 K. Girkmann
       Flächentragwerke (Plate and Shell Structures), 4th ed., Springer-Verlag, Vienna,
       1956

12.13 B. Bresler and T. Y. Lin
       Design of Steel Structures, Wiley, New York, 1960

12.14 G. Winter
       *Stress Distribution in and Equivalent Width of Flanges of Wide, Thin-Walled
       Steel Beams*, NACA Tech. Note 784, 1940

12.15 G. Winter
       Performance of Thin Steel Compression Flanges, IABSE, Third Congress, Liège,
       Prel. Publ. 1948 (Cornell Univ., Eng. Exp. Sta. Reprint 33)

12.16 A. Chajes and G. Winter
       "Torsional-Flexural Buckling of Thin-Walled Members," *Proc. Am. Soc. Civil
       Engrs.* **91** (ST-4), Part 1, p. 103 (August 1965)

12.17 T. B. Pekoz and G. Winter
       "Torsional-Flexural Buckling of Thin-walled Sections Under Eccentric Load,"
       Proc. Am. Soc. Civil Engrs., **95** (ST-5), p. 941 (May 1969)

12.18 T. B. Pekoz and N. Celebi
       Torsional-Flexural Buckling of Thin-Walled Sections Under Eccentric Load,
       Cornell Engineering Research Bulletin, No. 69-1

12.19 V. Hlavacek
       "The Torsional-Flexural Buckling of Thin-walled Open Sections," *Stavebnicky
       Casopis*, **15**, No. 7, p. 385 (1967)

12.20 K. Klöppel and H. Okur
       "Die Vergleichsschlankheiten von mittig gedrückten ⌐─┐ - und ⊏──⊐ -
       Profilen," *Der Stahlbau*, Feb. 1970, p. 61

12.21 G. G. Green, G. Winter, and T. R. Cuykendall
       Light Gage Steel Columns in Wall-Braced Panels, Cornell Univ., Eng. Exp. Sta.
       Bull. No. 35, Part 2, October 1947

12.22 G. Winter
       "Tests on Bolted Connections in Light Gage Steel," *Proc. Am. Soc. Civil Engrs.*,
       82 (ST-2), p. 920, March 1956

12.23 S. P. Timoshenko
       "Theory of Bending, Torsion and Buckling of Thin-Walled Members of Open
       Cross Section," *J. Franklin Inst.*, **239** (3, 4, 5) (1945)

CHAPTER 13. COMPOSITE STEEL-CONCRETE MEMBERS

13.1   I. M. Viest
       "Review of Research on Composite Steel-Concrete Beams," Proc. Am. Soc.
       Civil Engrs., **86** (ST-6) p. 1 (June 1960)

13.2   M. Ros
"Les Constructions Acier-Béton, Système Alpha" (Reinforced Concrete Structures, Alpha System) *L'Ossature Metall.*, **3** (4), pp. 195–208, Brussels, (1934)

13.3   AASHO
Standard Specifications for Highway Bridges, Washington, D.C., 1944

13.4   AISC
Specification for the Design Fabrication and Erection of Structural Steel for Buildings, New York, 1952

13.5   A. O. Adekola
"Effective Widths of Composite Beams of Steel and Concrete," *Struct. Engineer*, **46**, No. 9, September 1968

13.6   I. M. Viest, R. S. Fountain, and R. C. Singleton
Composite Construction in Steel and Concrete, McGraw-Hill, New York, 1958

13.7   Bethlehem Steel Corporation
Properties of Composite Sections for Bridges and Buildings, AIA File No. 13, Catalog 1899, Bethlehem, Pennsylvania

13.8   Bethlehem Steel Corporation
Properties of Composite Sections for Buildings, Handbook No. 2346, Bethlehem, Pennsylvania

13.9   *Alpha Composite Construction Engineering Handbook*,
Porete Manufacturing Company, North Arlington, N.J., 1949

13.10   R. G. Slutter and J. W. Fisher
Fatigue Strength of Shear Connectors, Bulletin No. 5, American Iron and Steel Institute, October 1967

13.11   C. P. Siess, I. M. Viest, and N. M. Newmark
Studies of Slab and Beam Highway Bridges, Part III, "Small Scale Tests of Shear Connectors and Composite T-Beams," University of Illinois, Eng. Exp. Sta. Bull. No. 396, 1952

13.12   R. G. Slutter and G. C. Driscoll, Jr.
"Ultimate Strength of Composite Members," *Conference on Composite Design in Steel and Concrete for Bridges and Buildings Proceedings*, Pittsburgh, Pa., ASCE, 1962

13.13   P. P. Page
"Composite Design for Buildings," *Conference on Composite Design in Steel and Concrete for Bridges and Buildings Proceedings*, Pittsburgh, Pa., ASCE, 1962

13.14   Hugh Robinson
"Tests on Composite Beams with Cellular Deck," *Proc. Am. Soc. Civil Engrs.*, **93** (ST-4) (August 1967)

13.15   J. W. Fisher
"Design of Composite Beams with Formed Metal Deck," *Eng. J., Amer. Inst. of Steel Const.*, **7** (3) (July 1970)

CHAPTER 14. THE WELDING PROCESS

14.1   R. N. Hart
Welding, McGraw-Hill, New York, 1914

14.2   AWS
Welding Handbook, Sec. 1, "Fundamentals of Welding," 6th ed., AWS, New York, 1968

14.3   AWS
Welding Handbook, Sec. 5, "Applications of Welding," 6th ed., AWS, New York, 1973

14.4   Air Reduction Company
Electrode Pocket Guide, Catalog 1318, Air Reduction Co., New York

14.5 AWS
Welding Handbook, Sec. 3, "Special Welding Processes and Cutting," 6th ed., AWS, New York, 1970

14.6 J. E. Norcross
"Electroslag/Electrogas Welding," *Welding J.*, **44**, March 1965

14.7 R. D. Stout and W. D. Doty
Weldability of Steels, 2d ed., Welding Research Council, New York, 1973

14.8 AWS-A5.1
Mild Steel Covered Arc Welding Electrodes, AWS, New York, 1969

14.9 L. Tall
"Residual Stresses in Welded Plates—A Theoretical Study," *Welding J.*, **43** (1) (January 1964)

14.10 L. Tall
Heat Input, Thermal and Residual Stresses in Welded Structural Plates, Fritz Laboratory Report 249.12, Lehigh University, August 1962; Fundamental Research Symposium on Transient Heat Flow, AWS-WRC Meeting, Dallas, September 1961

14.11 AWS
Welding Handbook, Sec. 2, "Welding Processes: Gas, Arc, and Resistance," 6th ed., AWS, New York, 1969

14.12 N. O. Okerblom
The Calculations of Deformations of Welded Metal Structures, Mashgiz, Moscow, 1955 (trans. from Russian by DSIR, London, 1958)

14.13 Y. Ueda
Elastic, Elastic-Plastic, and Plastic Buckling of Plates with Residual Stresses, Ph.D. dissertation, Lehigh University, August 1962; University Microfilms, Inc., Ann Arbor

CHAPTER 15. BRITTLE FRACTURE

15.1 M. E. Shank
A Critical Survey of Brittle Failure in Carbon Plate Steel Structures Other than Ships, Bulletin No. 17, Welding Research Council, New York, January 1954

15.2 H. G. Acker
Review of Welded Ship Failures, Bulletin No. 19, Welding Research Council, New York, November 1954

15.3 W. S. Pellini and P. P. Puzak
Fracture Analysis Diagram Procedures for the Fracture-Safe Engineering Design of Steel Structures, NRL Report 5920, Naval Research Laboratory, Washington, D.C., March 1963

15.4 *Report of Royal Commission into the Failure of Kings Bridge*, Government Printer, Melbourne, Australia, 1963

15.5 W. D. Biggs
The Brittle Fracture of Steel, MacDonald and Evans, Ltd., London, 1960

15.6 F. Forscher
"Crack Propagation," *Welding J.*, **19**, November 1954, pp. 579-s–584-s.

15.7 M. E. Shank, ed.
Control of Steel Construction to Avoid Brittle Fracture, Welding Research Council, New York, 1957

15.8 E. R. Parker
Brittle Behavior of Engineering Structures, Wiley, New York, 1957

15.9 C. F. Tipper
The Brittle Fracture Story, Cambridge University Press, London, 1962

15.10 M. Szczepański
The Brittleness of Steel, Wiley, New York, 1963

15.11 D. C. Drucker and J. J. Gilman, eds.
Fracture of Solids, Interscience Publishers, New York, 1963

15.12 H. D. Greenberg (ed.)
Application of Fracture Toughness Parameters to Structural Metals, Metallurgical Society Conferences, **31**, Gordon and Breach, New York, 1966

15.13 W. J. Hall, H. Kihara, W. Soete, and A. A. Wells
Brittle Fracture of Welded Plate, Prentice-Hall, Englewood Cliffs, N.J., 1967

15.14 W. S. Pellini and P. P. Puzak
Practical Considerations in Applying Laboratory Fracture Test Criteria to the Fracture-Safe Design of Pressure Vessels, NRL Report 6030, Naval Research Laboratory, Washington, D.C., November 1963

15.15 *Fracture Toughness Testing*, Special Technical Publication 381, American Society for Testing and Materials, Philadelphia, 1965

15.16 *Plain Strain Crack Toughness Testing of High Strength Metallic Materials*, Special Technical Publication 410, American Society for Testing and Materials, Philadelphia, 1966

15.17 E. T. Wessel, W. G. Clark, and W. K. Wilson
Engineering Methods for the Design and Selection of Materials Against Fracture, Westinghouse Research Laboratories, Pittsburgh, Pa., June 1966

15.18 G. R. Irwin, J. M. Kraft, P. C. Paris, and A. A. Wells
Basic Aspects of Crack Growth and Fracture, NRL Report 6598, Naval Research Laboratory, Washington, D.C., November 1967

15.19 G. R. Irwin
"Linear Fracture Mechanics, Fracture Transition and Fracture Control," *Engineering Fracture Mechanics*, **1**, 1968, pp. 241–257

15.20 W. S. Pellini
Advances in Fracture Toughness Characterization Procedures and in Quantitative Interpretations to Fracture-Safe Design for Structural Steels, Bulletin No. 130, Welding Research Council, New York, May 1968

15.21 J. M. Barsom and S. T. Rolfe
"Correlations Between K_{Ic} and Charpy V-Notch Test Results in the Transition Temperature Range," *Impact Testing of Metals*, Special Technical Publication 466, American Society for Testing and Materials, Philadelphia, June 1969, pp. 281–302

15.22 H. T. Corten and R. H. Sailors
"Relationship Between Material Fracture Toughness Using Fracture Mechanics and Transition Temperature Tests," *T. & A. M. Report No. 346*, University of Illinois, Urbana, August 1971

15.23 A. A. Wells
"The Status of COD in Fracture Mechanics," *Proc. Canadian Congress on Applied Mechanics*, Calgary, Canada, May 1971.

15.24 P. N. Randall and J. G. Merkle
"Gross Strain Crack Tolerance of Steels," *Nuclear Engineering and Design*, **17** (August 1971), pp. 46–63.

15.25 F. J. Witt and T. R. Mager
"A Procedure for Determining Bounding Values on Fracture Toughness K_{Ic} at Any Temperature," *ORNL-TM-3894*, Oak Ridge National Laboratory, Oak Ridge, Tennessee, October 1972.

CHAPTER 16. FATIGUE

16.1 ASTM
Manual on Fatigue Testing, ASTM STP No. 91, Philadelphia, 1949

16.2 H. S. Reemsnyder

"Procurement and Analysis of Structural Fatigue Data," *Proc. Am. Soc. Civil Engrs.*, **95** (ST-7) (July 1969)

16.3 G. Sines and J. L. Waisman (eds.)
Metal Fatigue, McGraw-Hill, New York, 1959

16.4 H. S. Reemsnyder and J. W. Fisher
"Service Histories and Laboratory Testing," *Proc. Am. Soc. Civil Engrs.*, **94** (ST-12), pp. 2699–2712 (December 1968)

16.5 T. H. Topper, B. I. Sandor, JoDean Morrow
"Cumulative Fatigue Damage Under Cyclic Strain Control," *Journal of Materials*, **4** (1), pp. 189–199 (March 1969)

16.6 T. H. Topper, R. M. Wetzel, JoDean Morrow
"Neuber's Rule Applied to Fatigue of Notched Specimens," *ibid.*, pp. 200–209

16.7 JoDean Morrow, R. M. Wetzel, T. H. Topper
Laboratory Simulation of Structural Fatigue Behavior, A.S.T.M. STP 462, pp. 74–92, 1970

16.8 J. F. Martin, T. H. Topper, G. M. Sinclair
"Computer-Based Simulation of Cyclic Stress-Strain Behavior with Applications to Fatigue," *Materials Research and Standards*, **11** (2), pp. 23–28, 50, 51 (February 1971)

16.9 R. B. Heywood
Designing Against Fatigue of Metals, Reinhold, New York, 1962

16.10 A. F. Madayag (ed.)
Metal Fatigue: Theory and Design, Wiley, New York, 1969

16.11 C. C. Osgood
Fatigue Design, Wiley-Interscience, New York, 1970

16.12 G. M. Rassweiler and W. L. Grube
Internal Stresses and Fatigue in Metals, Elsevier, New York, 1959

16.13 H. J. Grover
Fatigue of Aircraft Structures, NAVAIR 01-1A-13, Government Printing Office, Washington, D. C., 1966

16.14 J. H. Crews, Jr., and H. F. Hardrath
"A Study of Cyclic Plastic Stresses at a Notch Root," *Experimental Mechanics*, **6** (6), pp. 313–320 (June 1966)

16.15 S. S. Manson and M. H. Hirschberg
"Crack Initiation and Propagation in Notched Fatigue Specimens," *Proc. First International Conference of Fracture*, Sendai, Japan, pp. 479–498, 1966

16.16 G. H. Rowe
"Correlation of High-Cycle Fatigue Strength with True Stress–True Strain Behavior," *Journal of Materials*, **1** (3), pp. 689–715 (September 1966)

16.17 H. S. Reemsnyder
"Correlation of Fatigue Limit with True Stress–True Strain Behavior," *Materials Research and Standards*, **7** (9), pp. 390–392 (September 1967)

16.18 T. R. Gurney
Fatigue of Welded Structures, Cambridge University Press, London, New York, 1968

16.19 A. R. Jack and A. T. Price
"Effects of Thickness on Fatigue Crack Initiation and Growth in Notched Mild Steel Specimens," *Acta Metallurgica*, **20**, pp. 857–866, July 1972

16.20 S. J. Klima, D. J. Lesco, J. C. Freche
"Application of Ultrasonics to Detection of Fatigue Cracks," *Experimental Mechanics*, **6** (3), pp. 154–161 (March 1966)

16.21 M. B. P. Allery, G. Birkbeck
"Effect of Notch Root Radius on the Initiation and Propagation of Fatigue Cracks," *Engineering Fracture Mechanics*, **4**, pp. 325–331 (1972)

16.22 R. G. Foreman
Study of Fatigue Crack Initiation from Flaws Using Fracture Mechanics Theory, AFFDL, Wright-Patterson Air Force Base, Ohio, n.d.

16.23   W. G. Clark, Jr.
        *An Evaluation of the Fatigue Crack Initiation Properties of Type 403 Stainless
        Steel in Air and Steam Environments*, 72-1E7-FAILT-P1, Westinghouse Research
        Laboratories, August 25, 1972
16.24   A. R. Jack, A. T. Price
        "The Initiation of Fatigue Cracks from Notches in Mild Steel Plates," *Int. J.
        Fracture Mechanics*, **6** (4), pp. 401–409 (December 1970)
16.25   H. E. Williams, H. Ottsen. F. V. Lawrence, W. H. Munse
        The Effects of Weld Geometry on the Fatigue Behavior of Welded Connections,
        S.R.S. No. 366, University of Illinois, August 1970
16.26   J. M. Barsom
        "Fatigue-Crack Propagation in Steels of Various Yield Strengths," *Trans. ASME*,
        Series B, **93** (4), pp. 1190–1196 (November 1971)
16.27   T. R. Gurney
        "The Effect of Mean Stress and Material Yield Strength on Fatigue Crack
        Propagation in Steels," *Metal Construction and British Welding J.*, pp. 91–96
        (February 1969)
16.28   N. E. Frost, L. P. Pook, K. Denton
        "A Fracture Mechanics Analysis of Fatigue Crack Growth Data for Various
        Materials," *Eng. Fracture Mechanics*, **3**, pp. 109–126 (1971)
16.29   M. Parry, H. Nordberg, R. W. Hertzberg
        "Fatigue Crack Propagation in A514 Base Plate and Welded Joints," Welding
        J., **51** (10), pp. 485s–490s (October 1972)
16.30   T. W. Crooker
        "Effects of Tension-Compression Cycling on Fatigue Crack Growth in High-
        Strength Alloys," *Trans. ASME*, **93**, pp. 893–896 (November 1971)
16.31   S. J. Maddox
        "Fatigue Crack Propagation in Weld Metal and HAZ,"*Metal Construction and
        British Welding J.*, pp. 285–289 (July 1970)
16.32   J. R. Griffiths, J. L. Mogford, C. E. Richards
        "Influence of Mean Stress on Fatigue-Crack Propagation in a Ferritic Weld
        Metal," *Metal Science J.*, **5**, pp. 150–154, 1971
16.33   S. S. Manson
        Thermal Stress and Low-Cycle Fatigue, McGraw-Hill, New York, 1966
16.34   W. H. Munse and L. Grover
        Fatigue of Welded Steel Structures, Welding Research Council, New York, 1964
16.35   H. S. Reemsnyder
        Fatigue Data for Plain and Welded Structural Steels, Bethlehem Steel Corp.,
        Bethlehem, Pa., 1968
16.36   H. S. Reemsnyder
        "Some Significant Parameters in the Fatigue Properties of Welded Joints,"
        Welding J., **48**, pp. 213s–220s (May 1969)
16.37   H. S. Reemsnyder
        "A New Specimen for Fatigue Testing Longitudinal Fillet Weldments," *Proc.
        ASTM*, **65**, pp. 729–735 (1965)
16.38   H. S. Reemsnyder
        "Fatigue Strength of Longitudinal Weldments in Constructional Alloy Steel,"
        Welding J., **44**, pp. 458s–465s (October 1965)
16.39   H. S. Reemsnyder
        Fatigue of Riveted and Bolted Joints—A Literature Survey, Bethlehem Steel Corp.,
        Bethlehem, Pa., 1968
16.40   D. H. Hall and I. M. Viest
        "Design of Steel Structures for Fatigue," *Proc. Am. Soc. Civil Engrs.*, **94** (ST-12),
        pp. 2769–2797 ((December 1968)
16.41   H. Mindlin
        "Influence of Details on Fatigue Behavior of Structures," *Proc. Am. Soc. Civil
        Engrs.*, **94** (ST-12), pp. 2679–2698 (December 1968)

CHAPTER 17. LOCAL BUCKLING

17.1   C. F. Kollbrunner and M. Meister
       Ausbeulen (Buckling of Plates), Springer, 1958

17.2   AISC
       Specification for the Design, Fabrication and Erection of Structural Steel for Buildings, AISC, New York, 1936

17.3   G. Haaijer
       "Plate Buckling in the Strain-Hardening Range," *Trans. Am. Soc. Civil Engrs.*, **124**, p. 117 (1959)

17.4   G. Haaijer and B. Thürlimann
       "Inelastic Buckling in Steel," *Trans. Am. Soc. Civil Engrs.*, **125** (I), p. 308 (1960)

17.5   M. G. Lay
       The Static Load-Deformation Behavior of Planar Steel Structures, Ph.D. dissertation, Lehigh University, 1964; University Microfilms, Ann Arbor, Mich.

17.6   Column Research Council
       Guide to Design Criteria for Metal Compression Members, 2nd ed., Chap. 3, Wiley, 1966

17.7   T. von Kármán, E. E. Sechler, and L. H. Donnell
       "The Strength of Thin Plates in Compression," *Trans. Am. Soc. Mech. Engrs.*, **54** (1932)

17.8   J. R. Jombock and J. W. Clark
       "Postbuckling Behavior of Flat Plates," *Trans. Am. Soc. Civil Engrs.*, **127** (II), p. 227 (1962)

17.9   G. Gerard
       Handbook of Structural Stability, Part IV, NACA Technical Note 3784, Washington, D.C., 1957

17.10 J. M. Frankland
       The Strength of Ship Plating Under Edge Compression, David Taylor Model Basin Report 469, Washington, D.C., 1940

17.11 E. Z. Stowell, G. J. Heimerl, C. Libove, and E. E. Lundquist
       "Buckling Stresses for Flat Plates and Sections," *Trans. Am. Soc. Civil Engrs.*, **117**, p. 545 (1952)

17.12 J. B. Dwight and A. T. Ratcliffe
       "The Strength of Thin Plates in Compression," *Proceedings of Symposium at Univ. College of Swansea on Thin-Walled Steel Structures* (September 1967)

17.13 I. Madsen
       "Report on Plate Girder Tests," *Iron Steel Engr.*, **18** (11) (November 1941)

17.14 J. B. Dwight and K. E. Moxham
       "Welded Steel Plates in Compression," *The Structural Engineer*, **47** (2), p. 49 (February 1969)

17.15 M. Yoshiki, Y. Fujita, and T. Kawai
       "Influence of Residual Stresses on the Buckling of Plates," *J. Soc. Naval Architects Japan*, 107 (1960)

17.16 Committee ofthe Structural Division on Design in Light weightS tructural Alloys
       "ASCE Specifications for Structures of Aluminum Alloy 6016-T6, and Alloy 2014-T6," *Proc. Am. Soc. Civil Engrs.*, **82** (ST3) (May 1956); also in *Aluminum Construction Manual*, The Aluminum Association, New York, 1959

17.17 Aluminum Company of America
       ALCOA Structural Handbook, ALCOA, Pittsburgh, 1958

17.18 N. S. Streletsky *et al.*
       Metal Structures (in Russian), Moscow, 1961

17.19 British Standards Institution
       B.S. 449: The Use of Structural Steel in Buildings, BSI, London, 1959

17.20 B. G. Johnston and G. G. Kubo
       Web Crippling at Seat Angle Supports, Fritz Laboratory Report 192A2, Lehigh
       University (1941)
17.21 I. Lyse and H. J. Godfrey
       "Investigation of Web Buckling in Steel Beams," *Trans. Am. Soc. Civil Engrs.*,
       Paper 1907 (1935)
17.22 T. W. Bossert and A. Ostapenko
       Buckling and Ultimate Loads for Plate Girder Web Plates Under Edge Loading,
       Fritz Engineering Laboratory Report No. 319.1, Lehigh University (June 1967)
17.23 A. Ostapenko, B. T. Yen, and L. S. Beedle
       Research on Plate Girders at Lehigh University, Final Report of the 8th Congress
       of IABSE held in New York, September 1968, IABSE, Zurich (1969)
17.24 J. D. Graham, A. N. Sherbourne, R. N. Khabbaz, and C. D. Jensen
       Welded Interior Beam-to-Column Connections, AISC, New York 1959; also WRC
       Bull. No. 63, August 1960
17.25 C. G. Salmon, D. R. Buettner, and T. C. O'Sheridan
       "Laboratory Investigation of Unstiffened Triangular Bracket Plates," *Proc. Am.
       Soc. Civil Engrs.*, **90** (ST2), p. 257 (April 1964)
17.26 C. G. Salmon
       "Analysis of Triangular Bracket-Type Plates," *Proc. Am. Soc. Civil Engrs.*, **88**
       (EM6), p. 41 (December 1962)
17.27 C. D. Jensen
       "Welded Structural Brackets," *J. Am. Welding Soc.* (Supplement), **15** (10)
       (October 1936)

CHAPTER 18. RIVETED AND BOLTED CONNECTIONS

18.1   American Society of Civil Engineers
       "Bibliography on Bolted and Riveted Joints," *ASCE Manual of Engineering
       Practice*, No. 48, 1967
18.2   Research Council on Riveted and Bolted Structural Joints
       Specifications for Structural Joints Using ASTM A325 or A490 Bolts, March, 1970
18.3   L. Shenker, C. G. Salmon, and B. G. Johnston
       Structural Steel Connections, University of Michigan, June 1954
18.4   W. H. Munse and H. L. Cox
       The Static Strength of Rivets Subjected to Combined Tension and Shear, University
       of Illinois, Eng. Exp. Sta. Bull. No. 437, 1956
18.5   F. E. Graves
       "High Strength Bolting Moves into Fabrication Shops," *Fasteners*, **15** (2, 3)
       Industrial Fasteners Institute, Cleveland, Ohio, 1960
18.6   J. W. Fisher
       High Strength Bolting for Structural Joints, Bethelehem Steel Corp., Booklet
       2534, March, 1969
18.7   W. M. Wilson and W. A. Oliver
       Tension Tests of Rivets, University of Illinois, Eng. Exp. Sta. Bull. No. 210, 1930
18.8   Industrial Fasteners Institute
       Fasteners Standards, 4th ed., Industrial Fasteners Institute, Cleveland, Ohio,
       1965
18.9   E. F. Ball and J. J. Higgins
       "Installation and Tightening of High Strength Bolts," *Trans. Am. Soc. Civil
       Engrs.*, **126** (II), p. 797 (1961)
18.10 C. Batho
       "The Partition of Load in Riveted Joints," *J. Franklin Inst.*, **182** (1916)
18.11 A. Hrennikoff
       "The Work of Rivets in Riveted Joints," *Trans. Am. Soc. Civil Engrs.*, **90** (1934)

18.12   J. W. Fisher and J. L. Rumpf
        "Analysis of Bolted Butt Joints," *Proc. Am. Soc. Civil Engrs.*, **91** (ST-5), October,
        1965
18.13   K. Dornen
        *Die Untersuchung der Schubsteifigkeit von Verbindungen mit hochfest Vorge-
        spannten (HV-) Schrauben im Stahlbau und die daraus sich ergebenden Konstruktiven
        Massnahmen* (The Shear Rigidity of Connections with High Strength Bolts),
        Stahlbau-Verlag, Cologne, 1961
18.14   R. E. Davis, G. B. Woodruff, and H. E. Davis
        "Tension Tests of Large Riveted Joints," *Trans. Am. Soc. Civil Engrs.*, **105**,
        p. 1193 (1940)
18.15   R. A. Bendigo, R. M. Hansen, and J. L. Rumpf
        "Long Bolted Joints," *Proc. Am. Soc. Civil Engrs.*, **89** (ST-6) (1963)
18.16   G. L. Kulak and J. W. Fisher
        "A514 Steel Joints Fastened by A490 Bolts," *Proc. Am. Soc. Civil Engrs.*, **94**
        (ST-10), October 1968
18.17   J. W. Fisher and N. Yoshida
        "Large Bolted and Riveted Shingle Splices," *Proc. Am. Soc. Civil Engrs.*, **96**
        (ST-9), Sept 1970
18.18   E. Chesson and W. H. Munse
        "Riveted and Bolted Joints: Truss-Type Tensile Connections," *Proc. Am. Soc.
        Civil Engrs.*, **89** (ST-1), p. 67 (February 1963)
18.19   T. R. Higgins
        "New Formula for Fasteners Loaded Off Center," *Engineering News-Record*,
        May 21, 1964, pp. 102
18.20   S. F. Crawford and G. L. Kulak
        "Behavior of Eccentrically Loaded Bolted Connections," *Studies in Structural
        Engineering No. 4*, Department of Civil Engineering, Nova Scotia Technical
        College, 1968
18.21   W. H. Munse, E. Chesson, and K. S. Petersen
        "Strength of Rivets and Bolts in Tension," *Proc. Am. Soc. Civil Engrs.*, **85** (ST-3),
        p. 8 (March 1959)
18.22   R. T. Douty and W. McGuire
        "High Strength Bolted Moment Connections," *Proc. Am. Soc. Civil Engrs.*, **91**
        (ST-2), April, 1965
18.23   R. S. Nair, P. Birkemoe, and W. H. Munse
        Behavior of Bolts in Tee-Connections Subjected to Prying Action, Structural
        Research Series No. 353, University of Illinois, September 1969
18.24   Steel Structures Research Committee
        Second Report, Steel Structures Research Committee, Department of Scientific
        and Industrial Research of Great Britain, H.M. Stationery Office, London, 1934
18.25   C. W. Lewitt, E. Chesson, Jr., and W. H. Munse
        *Restraint Characteristics of Flexible Riveted and Bolted Beam-to-Column Connec-
        tions*, Bulletin 500, University of Illinois Engineering Experiment Station, 1969
18.26   R. T. Foreman and J. L. Rumpf
        "Static Tension Tests of Compact Bolted Joints," *Trans. Am. Soc. Civil Engrs.*,
        126 (II), p. 228 (1961)
18.27   J. W. Fisher, P. O. Ramseier, and L. S. Beedle
        "Static Tension Tests of A440 Steel Joints Connected with A325 Bolts,"
        Publications, IABSE, **23**, 1963
18.28   R. Kormanik and J. W. Fisher
        "Bearing-type Bolted Hybrid Joints," *Proc. Am. Soc. Civil Engrs.*, **93** (ST-5),
        June 1966
18.29   G. H. Sterling and J. W. Fisher
        "A440 Steel Joints Connected by A490 Bolts," *Proc. Am. Soc. Civil Engrs.*, **92**
        (ST-3), June 1966

18.30 D. D. Vasarhelyi *et al.*
"Effects of Fabrication Techniques on Bolted Joints," *Proc. Am. Soc. Civil Engrs.*, **85** (ST-3), p. 71 (March 1959)

18.31 W. H. Munse
"Structural Behavior of Hot Galvanized Bolted Connections," *8th International Conference on Hot-Dip Galvanizing*, London, June, 1967

18.32 J. Jones
"Bearing-Ratio Effect on Strength of Riveted Joints," *Trans. Am. Soc. Civil Engrs.*, **123**, p. 964 (1958)

CHAPTER 19. WELDED CONNECTIONS

19.1 F. R. Freeman
"The Strength of Arc-Welded Joints," *Proc. Inst. Civil Engrs.*, **231**, pp. 322-325 (London, 1930)

19.2 American Welding Bureau
Report of Structural Steel Welding Committee, American Welding Bureau, 1931

19.3 L. F. Denara
Survey of Existing Published Information, Appendix D, Report of Weld Panel of the Steel Structures Research Committee, Dept. of Science and Industrial Research. London, 1938

19.4 T. R. Higgins, and F. R. Preece
"Proposed Working Stresses for Fillet Welds in Building Construction," *Welding J.*, **47** (10), p. 429-s (October 1968)

19.5 H. W. Troelsch
"Shear in Welded Connections," *Trans. Am. Soc. Civil Engrs.*, **99**, p. 409 (1934)

19.6 W. H. Weiskopf and M. Male
"Stress Distribution in Side Welded Joints," *Welding J.*, **9** (9), p. 23 (1930)

19.7 F. E. Archer, H. K. Fischer, and E. M. Kitchem
"Fillet Welds Subjected to Bending and Shear," *Civil Eng. Public Works Rev.*, **54** (634) (London, April 1959)

19.8 N. G. Schreiner
"The Behavior of Fillet Welds When Subjected to Bending Stresses," *Welding J.*, **14** (9) (1935)

19.9 J. W. Fisher, G. C. Driscoll, Jr., and L. S. Beedle
Plastic Analysis and Design of Square Rigid Frame Knees, Bull. No. 39, WRC, New York, 1958

19.10 B. G. Johnston and L. F. Green
"Flexible Welded Angle Connections," *Welding J.*, **19** (10) (1940)

19.11 B. G. Johnston and E. H. Mount
"Analysis of Building Frames with Semi-Rigid Connections," *Trans. Am. Soc. Civil Engrs.*, **107**, p. 993 (1942)

19.12 G. J. Gibson and B. T. Wake
"An Investigation of Welded Connections for Angle Tension Members," *Welding J.*, **21** (1), p. 44-s (1942)

19.13 O. W. Blodgett
Design of Welded Structures, J. F. Lincoln Arc Welding Foundation, Cleveland, Ohio, 1966

19.14 I. Lyse and N. G. Schreiner
"An Investigation of Welded Seat Angle Connections," *Welding J.*, **14** (2) (1935)

19.15 A. Stang, M. Greenspan, and W. R. Osgood
"Strength of a Riveted Steel Rigid Frame Having Straight Flanges," R.P. 1130, *Journal of Research of the National Bureau of Standards*, Washington, D.C., 1938

19.16 F. Bleich
Design of Rigid Frame Knees, AISC, July 1943 (reprinted March 1956)

19.17 J. D. Griffiths
Single Span Rigid Frames in Steel, AISC, 1948 (reprinted March 1956)

19.18   A. A. Toprac, B. G. Johnston, and L. S. Beedle
"Connections for Welded Continuous Portal Frames," *Welding J.*, **30** (7), **30** (8), **31** (11) (1951 and 1952)

19.19   J. W. Fisher, G. C. Driscoll, Jr. and F. W. Schutz Jr
"Behavior of Welded Corner Connections," *Welding J.*, **37** (5), p. 216-*s* (May 1958)

19.20   J. W. Fisher and G. C. Driscoll, Jr.
"Corner Connections Loaded in Tension," *Welding J.*, **38** (11), p. 425-*s* (November 1959)

19.21   W. R. Osgood
"A Theory of Flexure for Beams with Nonparallel Extreme Fibers," *J. Appl. Mechanics*, **6** (3) (1939)

19.22   H. C. Olander
"A Method for Calculating Stresses in Rigid Frame Corners," *Trans. Am. Soc. Civil Engrs.*, 119, p. 797 (1954)

19.23   J. W. Fisher, G. C. Lee, J. A. Yura, and G. C. Driscoll, Jr.
Plastic Analysis and Tests of Haunched Corner Connections, Bull. No. 91, WRC, New York, October 1963

19.24   J. W. Peters and G. C. Driscoll
A Study of the Behavior of Beam-to-Column Connections, Fritz Lab. Report 332.2, Lehigh University, June 1968

CHAPTER 20.   SINGLE-STORY RIGID FRAMES

20.1   M. P. Korn
Steel Rigid Frames Manual, J. W. Edwards, Ann Arbor, 1953

20.2   American Institute of Steel Construction
Plastic Design in Steel, AISC, New York, 1959

20.3   G. Winter
"Lateral Bracing of Columns and Beams," *Proc. Am. Soc. Civil Engrs.*, **84** (ST2) p. 1561 (March 1958); see also closing discussion **85** (ST2) p. 179 (February 1959)

20.4   F. E. Burkett
"Effects of Tornados on Building," *Civil Eng.*, **33** (1), p. 51 (January 1963)

20.5   E. F. Masur, I. C. Chang, L. H. Donnell
"Stability of Frames in the Presence of Primary Bending Moments," *Trans. Am. Soc. Civil Engrs.*, **127** (I) p. 736 (1962)

20.6   E. R. Bryan
The Stiffening Effect of Sheeting in Buildings, Preliminary Publication, IABSE, Zurich, Switzerland, 1964

CHAPTER 21.   MULTI-STORY FRAMES I

21.1   C. W. Condit
American Building Art: The Nineteenth Century, Oxford University Press, 1960

21.2   C. W. Condit
American Building Art: The Twentieth Century, Oxford University Press, 1961

21.3   F. R. Khan
"The John Hancock Center," *Civil Eng.*, **37**, No. 10, p. 38 (October 1967)

21.4   F. R. Khan, S. H. Iyengar, J. P. Colaco
"Computer Design of 100-story John Hancock Center," Proc. Am. Soc. Civil Engrs., **92** (ST6), p. 55 (Dec. 1966)

21.5   "Twin Towers to Go 110 Stories," *Eng. News-Record*, **172**, No. 4, Jan. 23, 1964, p. 33–34

21.6   "A 1,450-ft Tube Forms World's Highest Building," *Eng. News-Record*, Vol. 185, No. 6, Aug. 6, 1970, p. 37

21.7   "Plastic Design Cuts Cost of Prototype High Rise," *Eng. News-Record* (July 20 1967)

21.8   B. Sourochnikoff
        "Wind Stresses in Semi-Rigid Connections of Steel Framework," *Trans. Am. Soc. Civil Engrs.*, **115**, p. 382 (1950)

21.9   British Constructional Steelwork Association
        Steel Frames for Multi-Story Buildings, BCSA Publication 16, 1961

21.10   C. H. Norris and J. B. Wilbur
        Elementary Structure Analysis, McGraw-Hill, New York, 1960

21.11   H. Sutherland and H. L. Bowman
        Structural Theory, Wiley, New York, 1950

21.12   H. D. Hauf and H. A. Pfisterer
        Design of Steel Buildings, Wiley, New York, 1949

21.13   O. Gottschalk
        "Simplified Wind-Stress Analysis of Tall Buildings," *Trans. Am. Soc. Civil Engrs.*, **105**, p. 1019 (1940)

21.14   J. I. Parcel and R. B. B. Moorman
        Analysis of Statically Indeterminate Structures, Wiley, New York, 1955

21.15   L. E Grinter
        Theory of Modern Steel Structures, Vol. II, Macmillan, New York, 1949

21.16   A. W. Anderson *et al.*
        "Lateral Forces of Earthquake and Wind," *Trans. Am. Soc. Civil Engrs.*, **117**, p. 716 (1952).

21.17   National Board of Fire Underwriters
        National Building Code, New York, 1955

21.18   O. W. Blodgett
        Design of Welded Structures, James F. Lincoln Arc Welding Foundation, Cleveland, 1966

21.19   H. V. Spurr
        Wind Bracing, McGraw-Hill, New York, 1930

21.20   J. C. Rathbun
        "Wind Forces on a Tall Building," *Trans. Am. Soc. Civil Engrs.*, **105**, p. 1 (1940)

CHAPTER 22.   MULTI-STORY FRAMES II

22.1   M. Ojalvo and V. Levi
        "Columns in Planar Continuous Structures," *Proc. Am. Soc. Civil Engrs.*, **89** (ST1), p. 1 (February 1963)

22.2   V. Levi
        Plastic Design of Braced Multi-Story Frames, Ph.D. dissertation, Lehigh University, 1962; University Microfilms, Inc., Ann Arbor

22.3   T. V. Galambos and M. G. Lay
        "Studies of the Ductility of Steel Members," *Proc. Am. Soc. Civil Engrs.*, **91** (ST4), pp. 125–151 (August 1965)

22.4   R. L. Ketter
        "Plastic Design in Low Buildings," *Trans. Eng. Inst. Can.*, **4** (2), p. 52 (1960)

22.5   G. C. Driscoll *et al.*
        Plastic Design of Multi-Story Frames—Lecture Notes, Fritz Engineering Laboratory, Lehigh University, 1965

22.6   AISI Committee
        Plastic Design of Braced Multi-Story Steel Frames, American Iron & Steel Institute, 1968

22.7   J. F. Baker
       "The Plastic Method of Designing Steel Structures," *Proc. Am. Soc. Civil Engrs.*,
       85 (ST4), p. 57, (April 1959)
22.8   B. P. Parikh, J. H. Daniels, and L. W. Lu
       Plastic Design of Multi-Story Frames—Design Aids, Fritz Engineering Laboratory,
       Lehigh University, 1965
22.9   L. Tall (ed.)
       Plastic Design of Multi-Story Frames—Guest Lectures, Fritz Engineering
       Laboratory, Lehigh University, 1966
22.10 G. C. Driscoll, Jr.
       "Lehigh Conference on Plastic Design of Multi-Story Frames—A Summary,"
       Eng. J., AISC, **3**, No. 2, April 1966
22.11 Anonymous
       "Plastic Design Cuts Cost of Prototype Highrise," *Eng. News-Record*, July 20,
       1967
22.12 H. Allison
       "Plastic Design: A Case History," *Civil Eng.*, **39** (4), p. 61-64, April 1969
22.13 ASCE-WRC Committee
       Plastic Design in Steel (*A Guide and Commentary*), ASCE Manual No. 41,
       Second Edition, 1969
22.14 J. A. Yura and L. W. Lu
       "Ultimate Load Tests on Braced Multistory Frames," *Proc. Am. Soc. Civil Engrs.*,
       95 (ST10), p. 2243 (October 1969)
22.15 V. Levi, G. C. Driscoll, Jr, and L. W. Lu
       "Structural Subassemblages Prevented from Sway," *Proc. Am. Soc. Civil Engrs.*,
       91 (ST5) (October 1965)
22.16 V. Levi, G. C. Driscoll, Jr., and L. W. Lu
       "Analysis of Restrained Columns Permitted to Sway," *Proc. Am. Soc. Civil Engrs.*,
       93 (ST1) (February 1967)
22.17 J. H. Daniels
       Combined Load Analysis of Unbraced Frames, Ph.D. dissertation, Lehigh
       University, 1967; University Microfilms, Inc., Ann Arbor
22.18 J. H. Daniels
       "A Plastic Method for Unbraced Frame Design," *Engineering J.*, AISC, **3**, No.
       4, October 1966
22.19 G. C. Driscoll, Jr., J. O. Armacost, III, and W. C. Hansell
       "Plastic Design of Multistory Frames by Computer," *Proc. Am. Soc. Civil Engrs.*,
       96 (ST1), pp. 17–33 (January 1970)

NOMENCLATURE

SYMBOLS

A — Area of cross section (subscript N denotes net, G denotes gross)

A_b — Nominal cross-sectional area of a single bolt (or rivet)
Area of bracing member

A_c — Area of effective concrete flange in composite member design

A_e — Elastic area

A_{eff} — Effective area

A_f — Area of one flange

A_i — Actual instantaneous area of cross section

A_n — Net area

A_o — Original area of cross section

A_p — Contact area of material gripped by bolt

A_s — Area of steel beam in composite member design

A_{st} — Area of stiffener

A_w — Area of web
Area of weld

a — Spacing of transverse stiffeners or battens (subscript o refers to actual stiffener spacing)
Depth of equivalent compressive stress block in concrete design
Acceleration of the ground
Distance from neutral axis to top concrete fiber in composite member design
Length of plate

B — Coefficient for beam-column interaction equation

b — Width of flange, plate, or stiffener (subscript st denotes stiffener)
Distance between the gage lines of the chords in a laced or battened column
Effective width of concrete slab
Width of rectangular section
Distance between the longitudinal rows of fasteners

b_c — Effective width

b_c' — Modified effective width

b_s — Width of stiffener

C — Compressive force
Ratio of ground acceleration to gravitational acceleration
Ratio between the bending stiffness of a member and its rotational restraint, $C = (EI/L)(1/R)$
Eave thrust coefficient

C_b — Bending coefficient, which is a function of moment gradient

C_c — Slenderness ratio (L/r) of elastic column corresponding to a stress of $F_y/2$; $C_c = \sqrt{2\pi^2 E/F_y}$

C_c, C_s — Distances from the center of the concrete slab and steel section to the shear plane, respectively, in composite member design

C_d	Drag coefficient		light gage cold-formed members
C_m	Equivalent moment coefficient		Lever arm between resultant compression and resultant tension in composite member design
C_v	Ratio of web-buckling stress to shear yield point of web material		
c	Distance from neutral axis to extreme fiber or to top of concrete slab	F	Load factor (subscript c denotes column)
D	Factor which reflects the efficiency of stiffeners furnished in pairs as opposed to one-sided stiffeners	$*F_a$	Allowable compressive axial stress
		F_{as}	Allowable stress for secondary members
	Nominal size of weld	F_b	Allowable bending stress (subscripts x and y refer to bending about x and y axes, respectively)
	Diameter of circular head of eyebar		
	Dead load per unit area		
	Dead weight	F_b'	Reduced bending stress
d	Depth of section (subscripts b and c denote beam and column depths, respectively)	F_{cr}	Buckling (critical) stress
		F_{cw}	Allowable compressive stress in a web under bending
	Diameter of hole		
	Diameter of fastener	F_d	Drag force
d_s	Diameter of stud	F_e	Elastic (or Euler) buckling stress
E	Modulus of elasticity (subscripts c and s refer to concrete and steel moduli; p and m refer to prototype and model moduli, respectively)		
		F_e'	Elastic buckling stress divided by factor of safety
		F_i	Inertial force
E_r	Reduced modulus	F_p	Allowable bearing stress
E_{st}	Strain-hardening modulus		Proportional limit
E_t	Tangent modulus	F_r	Residual stress (additional subscripts c and t refer to compressive and tensile stress, respectively)
$\bar{E}$	Modified tangent modulus which accounts for the effect of shear on column buckling		
			Stress at reduced modulus load
		F_{sr}	Allowable stress range
e	Initial mid-height deflection of a column	F_t	Allowable tensile stress
			Stress at tangent modulus load
	Eccentricity (subscripts 1 and 2 refer to eccentricities with respect to the principal axes; x and y refer to x and y axes)	F_u	Ultimate stress
		F_{ult}	Stress corresponding to maximum load
	Bolt elongation	F_v	Allowable shear stress
	Plate elongation within a pitch p	F_{vu}	Ultimate shear stress
	Limiting edge distance for	F_z, F_y, F_z	Normal stresses in x, y, and z directions

* The letter F with subscript, when used as stress, signifies an allowable stress or other limiting characteristics.

$F_{xy}, F_{yz},$ F_{zx}	Shear stresses on planes perpendicular to the x, y, and z directions	f_{vw}	Shear stress due to warping torsion
F_y	Yield point Yield stress level	f_x	Compressive stress at edge of plate
F_{ys}	Yield stress shear	G	Modulus of elasticity in shear Coefficient for beam-column interaction equation Ratio of stiffness of columns to stiffness of girders
F_{yu}	Upper yield point		
F.S.	Factor of safety (subscripts y and u refer to yielding and ultimate strengths, respectively)		
		G_t	Tangent modulus in shear
F_1, F_2, F_3	Principal stresses	g	Transverse spacing between fastener gage lines (gage distance) Width of tension field strip Acceleration due to gravity Distance from the center of the span to the point of maximum moment
*f	Stress Shape factor		
f_a	Axial stress		
f_b	Bending stress		
f_{bw}	Bending stress due to warping torsion		
f_c	Stress in the top fiber of the concrete slab in composite member design	H	Horizontal reaction Horizontal shear force Coefficient for beam-column interaction equation
f_c'	Compressive strength of concrete at 28 days	h	Clear distance between flanges of a plate girder Story height
f_f	Stress in flange		
$f_{max},$ f_{min}	Maximum and minimum stresses in a single-cycle fatigue test; f_{max} is maximum stress in wide-plate elements having non-uniform stress distribution due to buckling or shear lag	I	Moment of inertia (subscripts c and g refer to column and girder, 1 and 2 to the principal axes, x and y to the x and y axes [strong and weak axes], p and m to prototype and model, respectively) Moment of inertia of the portion of the concrete slab above the interface between steel and concrete (subscripts c and cp mean with and without cover plate, respectively)
f_u	Normal stress on web due to curvature		
f_p	Nominal bearing stress		
f_t	Tensile stress Tension field stress Stress at the top of the steel member for composite member design		
		I_e	Moment of inertia of elastic portion of cross section
f_v	Shear stress	I_f	Moment of inertia of one flange
f_{vs}	Shear between girder web and transverse stiffeners Shear stress due to St. Venant torsion	I_s	Required moment of inertia of a stiffener

* Refer to preceding footnote.

I_{tr} Moment of inertia of transformed composite section

I_{xy} Product of inertia with respect to x and y axes

I_w Warping constant

J Coefficient for beam-column interaction
Polar moment of inertia

K Effective length factor (subscripts x and y refer to column buckling on the strong and weak axes, respectively); KL = effective column length

K Stress intensity factor, or fracture toughness (subscript c denotes critical, and subscript Ic denotes a stable crack growth)

K' Empirical, modified effective length factor which accounts for shear action on a column

K_e Plate buckling coefficient for edge loading

K_f, K_t Stress concentration factor (subscripts f and t refer to experimental and theoretical factors, respectively)

K_{rel} Relative stiffness

K_s Coefficient of slip

K_T Torsional constant

K_1, K_2 Coefficient which is a function of end condition and loading

KL Effective length of member (subscript c refers to column)

k Plate buckling coefficient
Distance from outer face of flange to web toe of fillet
Ratio of the moment of inertia of the prototype to that of the model
Axial load factor for beam-columns ($k^2 = P/EI$)
Ratio of the end moment to the maximum simple-span moment

k_1 Modified effective length factor which accounts for the effect of shear on column buckling
Empirical factor

L Span length (subscripts c and g refer to column and girder spans; x and y refer to strong and weak axes, respectively)
Length of weld
Design live load per unit length
Column length

L_b Length of bearing
Length of bracing member
Unbraced length

L_{cr} Critical buckling length of haunched connection compression flange

L_i Instantaneous gage length

L_m, L_p Unbraced length of model and prototype, respectively

M Bending moment. (Subscripts c and g refer to column and girder, x and y refer to x and y axes, 1 and 2 refer to principal axes, g and w refer to gravity and wind, respectively. Subscript ab refers to moment at a due to load at b. For multistory frames, Mg [with a positive or negative sign] refers to the positive or negative gravity moment in the girder. Subscript des denotes design, subscript T denotes torsional, subscript I denotes impact, and subscript ws denotes wearing surface.)

ΔM Redistributed moment

M_{cr} Bending moment which causes lateral buckling of a beam

M_D Moment due to dead load on the steel member for composite member design

M_d Moment due to dead load on the composite beam

M_F — Fixed-end moment

M_f — Plastic moment capacity of both flanges

M_h — Moment at intersection of the layout lines of a corner connection

M_L — Moment due to live load

M_0 — End moment
External moment applied to a subassemblage

M_p — Plastic moment

M_{px} — Plastic moment at point x

M_u — Ultimate moment

M_w — Wind moment
End moment at working load
Warping bending moment

M_y — Moment at first yield

M_z — Total twisting moment (superscripts w and T refer to warping twisting moment and St. Venant twisting moment, respectively)

m — Number of slip planes
One-half length of rafter in single-story frame
End moment due to dummy shear
Distance from the shear center to the mid-thickness of the channel web
Ratio of difference in flange thickness between the haunched connector and the beam to the beam flange thickness
Length of one gable rafter

N — Coefficient of proportionality
Length of bearing of load
Number of pitches
Number of stress cycles
Number of shear connectors

N.A. — Neutral axis

n — Number of fasteners
Modular ratio (E_s/E_c)
Scale factor in model analysis

n' — Plastic modular ratio

P — Concentrated load (subscripts x and y refer to components in x and y directions, respectively; subscript all denotes allowable)
Applied load
Axial load

P_a — Allowable load
Vertical live load

P_{cr} — Buckling (critical) load

P_e — Elastic buckling load (Euler load) (subscripts x and y refer to buckling about the x and y axis, respectively)

P_g — Gravity load

P_L — A limit load in general

$\mathbb{P}$ — Symbol denotes plate

P_{max}, P_{min} — Maximum tensile load and maximum compressive load, respectively

P_o — Axial load which can be supported when $M = 0$

P_r — Reduced modulus load

P_s — Force in stiffener

P_T — Torsional buckling load about z axis

P_t — Tangent modulus load

P_u — Ultimate load

P_w — Working (allowable) load
Gravity plus wind load

P_y — Axial force corresponding to yield stress level, $= F_y A$
Load at which yielding occurs

p — Fastener pitch
Average static pressure
Bearing pressure on foundation
Impact ratio

Q — First moment of area about the neutral axis (subscripts 1 and 2 refer to principal axes, respectively)
Stress reduction factor, or form factor, for light gage cold-formed columns

Q_c, Q_{cp} — Static moment of the portion of the concrete slab above the interface between steel and concrete in composite member design (subscripts c and cp refer to with and without cover plate, respectively)

Q_w　Warping first moment of area

q　Fatigue notch sensitivity index
Load per stud
Dynamic pressure in unit area
Load per connector
Resultant shear per inch of weld (subscripts x, y, and z refer to x, y, and z components, respectively)

q_u　Ultimate strength of a shear connector

q_v　Shear per inch of weld

R　Concentrated force or reaction
Resultant
Radius of curved flange of a haunched connection
Stress ratio, minimum stress divided by the maximum stress
Radius of curvature
Radius of curve
Reduction in percentage of live loading
End rotational restraint
Total resisting force of fastener

R　Stress range (subscript a refers to allowable)

r　Radius of gyration (subscripts x and y refer to strong and weak axis radii, 1 and 2 refer to principal axes, respectively; subscript T refers to computation including both compression flange and one third of compression web area)
Inside radius of the bend

S　Section modulus (superscript $'$ and subscripts c, cp, t, tp, b, and bp refer to Table 13–1; subscripts x

and y refer to x and y axes, respectively)
Allowable shear per inch of weld
Stress sufficient to cause fatigue failure for certain number of stress cycles
Average stringer spacing
Bolt shear force

S_{max},　Maximum and minimum
S_{min}　stresses cause fatigue failure at N cycles

$S_{max}e$　Equivalent maximum uniaxial stress

S_e　Fatigue limit

S_r　Range of stress, $S_{max} - S_{min}$

S_s　Section modulus of steel beam used in composite member design, referred to tension flange

S_t　Section modulus of transformed composite beam, referred to top of slab

S_{tr}　Section modulus of transformed composite beam, referred to tension flange

S_w　Design capacity of one connector

s　Longitudinal distance (or stagger) between fasteners
Spacing of connectors
Width of tension field strip
Speed of moving loads
Distance along the cross section from the free end

$\bar{s}$　Standard deviation

s.c.　Shear center

T　Temperature
Tensile force
Bolt tension
Theoretical throat dimension of a fillet weld
Applied torque
Concentrated wind force
Horizontal forces for the computation forces in braces
Tensile force in web due to tension field action (subscript w denotes web)

T_b Proof load of a bolt

T_f Tensile force in flange due to tension field action

T_i Initial bolt tension

T_u Eave thrust

t Thickness of flange (subscripts b and c refer to beam and column flange thicknesses, respectively; subscript f denotes flange)
Plate thickness

t' Flange thickness of haunched corner connection

t_s Thickness of stiffener

U Unbuttoning factor

u, v, w Displacements in x, y, and z directions, respectively (subscript ab refers to displacement at a due to load at b)
Column deflection

V Shear force (subscripts 1 and 2 refer to principal axes, x and y refer to x and y axes, respectively. Subscript des denotes design, L denotes live load, and D denotes dead load)
Vertical reaction
Wind velocity
Total lateral force

V_b Shear resisted by beam action

V_f Flange shear
Force due to warping

V_h Total horizontal shear to be resisted by shear connectors

V_p Shear force which causes complete plastification of the web

V_t Shear resisted by tension field action

V_u Ultimate shear force

v Horizontal shear force per unit length
Wind velocity
Column deflection

$\bar{v}$ Deflection due to the transverse forces alone

W Total uniformly distributed load
Moment due to wind load
Weight due to the structure (subscript u denotes ultimate or maximum and subscript w denotes working load)

w Distributed load per unit length (subscripts D, L, and T refer to dead load, live, and total loads, p and y refer to plastic and yield level, respectively; subscript o refers to initial deflection)
Web thickness
Width of plate
Width deducted from a transverse section
Warping coordinate
Width of plate (subscripts g and n refer to gross and net widths, respectively)
Lateral plate deflection

w_T Total dead plus live distributed load per unit length

w_u, w_w, w_y Distributed load per unit length—subscripts u, w, and y refer to ultimate, working, and yielding loads, respectively

X Unknown force

x, y, z Coordinate axes
Coordinates of the point with respect to x, y, and z axes

x', y' Principal axes
Coordinates of the point with respect to the principal axes

x_1, y_1 Principal axes

Z Plastic modulus

α Angle of rotation

β Ratio of end moments
Degree of restraint for plate buckling
Factor which depends on the shape of cross section

	Angle between the inner and the outer flanges of a tapered connection	ϵ_p	Strain rate in plastic region
β_0	Abscissa of the buckling curve for a web plate subjected to bending	ϵ_{st}	Strain at start of strain-hardening
γ	Slope of the moment-rotation curve of the connection	ϵ_y	Strain at yield point
		θ	End slope
			Rotation
γ_s	Ratio of rigidity of stiffener to rigidity of web plate	λ	Column parameter—a function of L/r
Δ	Lateral deflection of a story in a multistory frame (wind drift)		Torsional damping constant
			Length interval along axis of member
	Relative deflections of joints in truss		Plate parameter—a function of b/t
Δ_x	Difference between the simple-span deflection due to the uniform load and the mid-span deflection due to end moments	μ	Poisson's ratio
		ρ	Mass density of air
			Distance from center of gravity to weld
			Distance from centroid of fastener group to a fastener
δ	Initial deflection	ϕ	Curvature
	Elongation of bracing, member, column, etc. (subscripts m and p refer to model and prototype, respectively)		End rotation
			Orientation of angle of the tension field
			Angle of twisting
	Deflection of the flanges toward the neutral axis (curling)		Angle between the point of tangency and the stiffener of the knee
	Maximum deflection of light gage cold-formed cross section	ϕ_0	Simple beam end slope
		ϕ_{st}	Curvature at which strain-hardening commences
ϵ	Strain	ϕ_y	Curvature corresponding to moment at first yield
ϵ_f	Strain in flange		

ABBREVIATIONS

AASHO	American Association of State Highway Officials	AREA	American Railway Engineering Association
AISC	American Institute of Steel Construction	ASCE	American Society of Civil Engineers
AISE	Association of Iron and Steel Engineers	ASME	American Society of Mechanical Engineers
AISI	American Iron and Steel Institute	ASTM	American Society for Testing and Materials
ANSI	American National Standards Institute	AWS	American Welding Society

CISC	Canadian Institute of Steel Construction	ksi	kips per square inch
CSA	Canadian Standards Association	L	Angle
		LL	Live load
CRC	Column Research Council	N.A.	Neutral axis
DL	Dead load	NDT	Nil-ductility temperature
DTT	Ductibility transition temperature	psf	pounds per square foot
		psi	pounds per square inch
DPN	Diamond Penetration Number	s.c.	Shear center
FATT	Fracture appearance transition temperature	S.S.	Single shear
		UM	Universal-mill
F.C.	Flame-cut (or oxygen-cut)	WRC	Welding Research Council
F.S.	factor of safety		
k, kip	kilopound (1000 pounds)	#	pound

GLOSSARY

Allowable Stress Design. A method of proportioning structures which is based on "allowable" or working loads, such that stresses do not exceed prescribed values. The allowable stresses incorporate a factor of safety against one of the limits of structural usefulness.

Beam-Action Shear. The shear force carried by a plate girder web panel at the theoretical shear buckling stress. The shear stress distribution is that predicted by simple beam theory.

Bifurcation. The phenomenon whereby a perfectly straight member may either assume a deflected position or else may remain undeflected; buckling.

Brittle Fracture. Fracture without prior detection of damage or plastic deformation.

Buckling Load. The load at which a perfectly straight member assumes a deflected position.

Compact Shape. A beam which will not experience premature local and lateral-torsional buckling in the plastic range.

Composite Beam. A member consisting of a reinforced concrete slab supported by a structural steel shape, interconnected in such a way that the two act together to resist bending.

Counter. A slender truss diagonal which resists only tension; usually one of a pair which act alternately as the shear on a panel changes direction.

Drift. Lateral sway of a building, due to wind.

Effective Length. The equivalent length (KL_c) used in the Euler formula for computing the strength of a framed column.

Effective Width. The reduced width of the plate or slab which, having a uniform stress distribution, is assumed to produce the same effect on the behavior of a structural member as the actual plate width with its non-uniform stress distribution.

Factor of Safety. As used in allowable stress design, a factor by which limiting stress which represents a designated limit of structural usefulness is divided to determine an allowable stress.

Fail-Safe Design. A design method which accepts a specified amount of cracking but minimizes the probability of crack growth to catastrophic proportions by redundancies, inspection, and repair.

Fatigue. The initiation and growth of a crack due to repetitive stress conditions.

Faying Surface. The contact area of adjacent parts of a connection.

Heat-Affected Zone. That portion of the base metal that has not been melted but whose mechanical properties and microstructure have been changed by the heat of welding.

High-Rise Building. A building with height considerably greater than its floor plan dimensions; skyscraper.

Lateral (or Lateral-Torsional) Buckling. Buckling of a member involving lateral deflection and twist.

Light Gage Cold-Formed Members. Structural members fabricated from light gage sheet steel (thickness, 0.012 to 0.224 in.) without the application of heat.

Limit Design. A design based on any chosen limit of usefulness.

Load Factor. As used in plastic design, a factor by which the working load is multiplied to determine the ultimate load.

Local Buckling. The buckling of a plate component which may precipitate the failure of the whole member.

Maximum Load Design. A design method in which members are selected on the basis of their maximum strength at ultimate load (determined by multiplying the expected loads by a load factor). In steel structures it is utilized in plastic design.

Mechanical Shear Connectors. As used in composite construction, a metal device, attached to the top flange of a steel beam, which is capable of transmitting shear forces between the concrete slab and the steel beam.

Mechanism. A system of members containing a sufficient number of plastic (or real) hinges to be able to deform without a finite increase in load.

Notch Toughness. An arbitrary criterion of resistance to crack growth or fracture.

Plastic Design. A design method for continuous steel beams and frames which defines the limit of structural usefulness as the "ultimate load" associated with the formation of a mechanism. The term "plastic" comes from the fact that the ultimate load is computed from a knowledge of the strength of steel in the plastic range.

Plastic Hinge. A yielded zone found in a structural member when the plastic moment is applied. The beam rotates as if hinged, except for the constant restraining moment M_p.

Plastic Modulus. The resisting modulus of a completely yielded cross section. It is the combined statical moment about the neutral axis of the cross-sectional area above and below that axis.

Plastic Moment. The maximum moment of resistance of a fully yielded cross section.

Plastification. The gradual penetration of yield stress from the outer fiber towards the centroid of a section under increase of moment. Plastification is complete when the plastic moment M_p is attained.

Post-Buckling Strength. The additional load or stress which can be carried by a plate or structural member after buckling.

Redistribution of Moment. A process which results in the successive formation of plastic hinges until the ultimate load is reached. Through transfer of moment which results from the formation of the plastic hinge, the less highly stressed portion of a structure also may reach the M_p value.

Residual Stress. The stresses that are left in a member after it has been formed into a finished product.

Rotation Capacity. The angular rotation which a given cross-sectional shape can accept at the plastic moment value without prior local failure.

Safe-Life Design. A design method which minimizes the probability of crack or fracture initiation but does not necessarily guard against catastrophic failure caused by unexpected damage.

Shape Factor. The ratio M_p/M_y, or Z/S, for a cross section.

Stub Column. A short compression-test specimen, sufficiently long for use in measuring the stress-strain relationship for the complete cross section, but short enough to avoid buckling as a column in the elastic and plastic ranges.

Subassemblage. A connected group of beams and columns which form part of a multi-story frame.

Tension-Field Action. The resistance of a plate girder to externally applied shear forces, in which diagonal tensile stresses develop in the web and compressive forces develop in the transverse stiffeners in a manner analogous to a Pratt truss.

Transition Temperature. The temperature, based on some arbitrary criterion derived from a relatively simple laboratory test, at which the notch toughness changes from a high to a low value.

Ultimate Load. The largest load a structure will support. (In plastic design, it is the load attained when a sufficient number of yield zones have formed to permit the structure to deform plastically without further increase in load.)

Unbraced Length. The distance between braced points of a member.

Unbuttoning. The premature, sequential failure of fasteners, progressing from the ends of the joint inward.

Unsymmetrical Bending. Bending in both principal directions (biaxial bending) and simultaneous torsion.

Web Crippling. The local elastic-plastic failure of the web plate in the immediate vicinity of a concentrated load or reaction.

Weldability. The ease of making a sound and serviceable joint by welding.

Width-Thickness Ratio. The ratio of the width of the outstanding portion of an element to its thickness, an indication of local instability.

Work Hardening. The increase in yield stress due to prior inelastic behavior.

Yield Moment. In a member subjected to bending, the moment required to initiate yielding.

Yield Point. The first point in a material, less than the maximum attainable stress, at which an increase in strain occurs without an increase in stress.

Yield Stress. The stress at which a material exhibits a specified limiting deviation from the proportionality of stress to strain.

Yield Stress Level. The average stress during yielding in the plastic range. It is the stress determined in a tension test corresponding to a strain of 0.005 in./in.

AUTHOR INDEX

Numbers in italic identify entries in the References.

SUBJECT INDEX